W9-ATS-251

FIFTH EDITION

Warm Air Heating for Climate Control

William B. Cooper

Raymond E. Lee

Raymond A. Quinlan

Martin W. Sirowatka

Upper Saddle River, New Jersey
Columbus, Ohio

Library of Congress Cataloging-in-Publication Data

Warm air heating for climate control.—5th ed. / William B. Cooper
[et al.]
p. cm
Includes index.
ISBN 0-13-048390-7
1. Hot-air heating. 2. Heat pumps. 3. Dwellings—Heating and ventilation. William B. Cooper

Editor in Chief: Stephen Helba
Editor: Ed Francis
Production Editor: Christine M. Buckendahl
Production Coordination: Lisa Garboski, *bookworks*
Design Coordinator: Diane Ernsberger
Cover Designer: Jason Moore
Production Manager: Brian Fox
Marketing Manager: Mark Marsden

This book was set in Times by The Clarinda Company, and was printed and bound by R. R. Donnelley & Sons Company. The cover was printed by Phoenix Color Corp.

Pearson Education Ltd.
Pearson Education Australia Pty. Limited
Pearson Education Singapore Pte. Ltd.
Pearson Education North Asia Ltd.
Pearson Education Canada, Ltd.
Pearson Educación de Mexico, S.A. de C.V.
Pearson Education—Japan
Pearson Education Malaysia Pte. Ltd.
Pearson Education, *Upper Saddle River, New Jersey*

15 14 13 12 11

ISBN: 0-13-048390-7

PREFACE

The fifth edition of *Warm Air Heating for Climate Control* has been extensively updated, using the latest information from the heating and cooling industry. The chapter sequence remains the same and gives the technician an understanding of the installation and service of forced-air heating equipment, hydronic systems, heat pumps, and add-on air conditioning. Former Chapters 1 and 2 were combined to allow for the addition of a new Chapter 2 on safety.

The following is a brief description of the major changes in the chapters.

New Chapter on Safety. Chapter 2 was added to provide safety information on handling of tools, using ladders, and working with toxic materials.

Updated Information on Parts Common to All Furnaces. In Chapter 4 we describe the application of the variable air volume Fan Handler. We also updated the humidifier section with new equipment and wiring diagrams.

Additional Information on Integrated Computer Furnace Control Systems. In Chapter 12 we expanded the sequences of operation and troubleshooting of the Honeywell SmartValve system and the White–Rodgers Total Furnace Control system.

New Energy Conservation Information. In Chapter 20 we included the requirements to meet a typical residential energy code. We also provide a means for the reader to evaluate the energy efficiency of a residential window system.

Completely Revised Chapter on Indoor Air Quality. We revised the text of Chapter 21 for easier reading and included new information on HEPA filters, heat recovery ventilation, and germicidal lamps.

Updated Information on Zoning. We included the Honeywell Mastertrol system and the popular California Economizer zone system in Chapter 22. Both systems have detailed application, installation, and adjustment data.

Expanded Hydronics. In Chapter 23 we show how to design a multizone series-loop hot-water system. We also added a hydronic/air heating/air cooling integrated system.

Completely Revised Chapter on Add-On Cooling and Heat Pumps. We start Chapter 24 with a new explanation of the refrigeration cycle and then explain the use of each cycle component. We added some air conditioning service techniques. We also included the various applications of GeoExchange (geothermal) heat pump systems and describe how each is unique in its operation.

Each chapter includes a list of objectives. These are useful to the teacher for class preparation and for curriculum development. The objectives and the study questions also help the student prepare for a successful career in the climate control industry.

CONTENTS

5 Components of Gas-Burning Furnaces 79

6 Basic Electricity and Electrical Symbols 99

7 Schematic Wiring Diagrams 123

8 Using Electrical Test Instruments and Equipment 134

9 Electrical Service Wiring 171

14 Oil Furnace Controls 305

15 Electric Heating 329

16 Estimating the Heating Loads 343

CHAPTER 1

Climate Control

CHAPTER OBJECTIVES

After studying this chapter, the student will be able to:

- Describe the fundamental concepts that form the basis of this book
- Describe the conditions produced by the warm air heating system that are necessary for human comfort

Indoor Climate Control

One of the most challenging professions available to students today is indoor climate control, not because it is a new occupation but because it is an old business that has changed radically. It now requires a high degree of knowledge and skill, and intensive study, and carries a large responsibility. At the same time the profession makes a valuable contribution to the health and welfare of the community.

In the past we considered a satisfactory installation to be one that primarily controlled temperature, and humidity to some degree, and did some filtering of the air. In the new technology we are better equipped to provide comfortable conditions, superior indoor air quality, and equipment and systems that conserve the valuable supply of energy. As an introduction to our study, let us review some of the important fundamental scientific principles that form the basis of indoor climate control:

1. Temperature
2. Pressure
3. Heat
4. Humidity
5. Air
6. Water
7. Laws of thermodynamics

Temperature

Temperature can be defined as the relative hotness or coldness of something. Temperature can be measured accurately using a thermometer.

Thermometers

A *thermometer* is a device for measuring temperature. There are many shapes and sizes of thermometers that use a variety of scales. In general, two types of thermometers are most frequently used. One type is used in the British or English system and is known as the *Fahrenheit* thermometer. The other type is used in the metric system and is called the *Celsius* thermometer. Both these thermometers measure temperature, but using different scales. A comparison of the two thermometer scales is shown in Figure 1.1.

The scale selected for the Celsius thermometer is the easier of the two to understand. Zero degrees (0°) on the Celsius scale is assigned to the freezing point of water, and 100° is assigned to the boiling point of water. There are 100 equal divisions between the 0° point and the 100° mark on the scale. On the Fahrenheit scale, the freezing point of water is 32°, and the boiling point of water is 212°. The Fahrenheit thermometer has 180 equal divisions between these two reference points.

It is important to understand both of these thermometer scales, as some temperature information is given in Fahrenheit degrees and some in Celsius degrees. To pre-

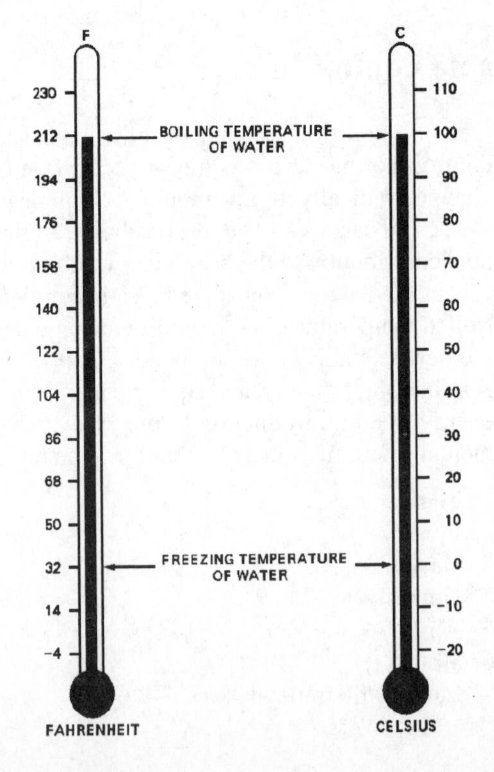

Figure 1.1 Comparison of Fahrenheit and Celsius thermometer scales.

vent confusion about which scale is involved, a capital C is placed after the number of Celsius degrees, and a capital F is placed after the number of Fahrenheit degrees. For example: 40°C means 40 degrees Celsius; 40°F means 40 degrees Fahrenheit.

When working continuously with one scale or the other, there is no problem in recording and using the information; however, sometimes it is necessary to change from one scale to the other. Such *conversions* can be done by using the following formulas:

$$\text{degrees Celsius} = \frac{5}{9} \times (\text{degrees Fahrenheit} - 32°)$$

$$\text{degrees Fahrenheit} = \left(\frac{9}{5} \times \text{degrees Celsius}\right) + 32°$$

Example 1–1 70°F is converted to degrees Celsius as follows:

$$
\begin{aligned}
°C &= \frac{5}{9} \times (°F - 32°) \\
&= \frac{5}{9} \times (70° - 32°) \\
&= \frac{5}{9} \times (38) \\
&= \frac{190}{9} \\
&= 21°C
\end{aligned}
$$

Example 1–2 27°C is converted to degrees Fahrenheit as follows:

$$
\begin{aligned}
°F &= \left(\frac{9}{5} \times °C\right) + 32° \\
&= \left(\frac{9}{5} \times 27°\right) + 32° \\
&= \left(\frac{243}{5}\right) + 32° \\
&= 49° + 32° \\
&= 81°F
\end{aligned}
$$

■

Pressures

Pressure is the weight or force exerted by a substance per unit of area. For example, we speak of city water having a pressure of 75 psi (pounds per square inch). Using correct terminology, we should indicate the pressure in terms of psig (the *g* standing for gauge pressure).

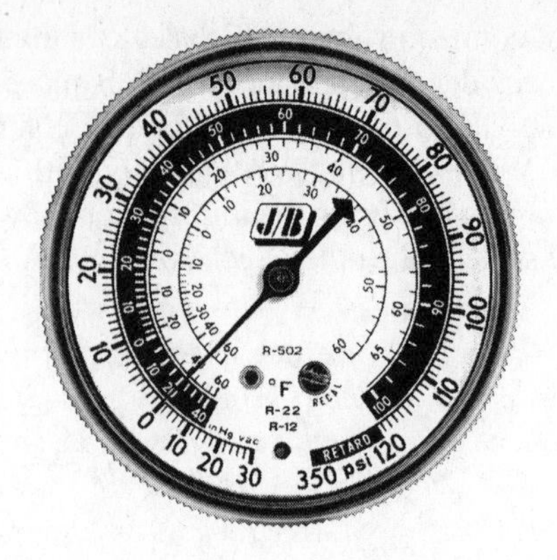

Figure 1.2 Compound gauge with scale 30 in. Hg vacuum and pressures 0–350 psi, including equivalent boiling temperatures. (Courtesy of J/B Industries, Inc.)

There are two general types of pressure: atmospheric and gauge. At sea level we consider the gauge pressure to be 0 psig, and all gauge pressure readings relate to this base. The normal atmospheric pressure at sea level is 14.7 psi, which is usually rounded off to 15 psi. Zero atmospheric pressure is a theoretical place in space where no pressure exists.

In climate control measurements we often use compound gauges (Figure 1.2). On these gauges when we measure pressures below 0 psi, the gauge indicates a vacuum. Instead of measuring a vacuum as a negative pressure, the gauge reads inches of mercury (in. Hg). The "in. Hg" refers to pressure measurements made using a mercury manometer. Atmospheric pressure read on a mercury manometer is 29.92 in. Hg (usually rounded off to 30 in. Hg). Figure 1.2 shows a compound gauge. For a perfect vacuum the gauge would read 30 in. Hg.

There is another scale commonly used to express pressure measurements. Since atmospheric pressure will support a water column 34 ft high (equivalent to 14.7 psi), pressures can be expressed in inches or feet of water. For example, a water pump may have a capacity of 6 gpm (gallons per minute) at a pressure of 10 ft. A fan may have a capacity to deliver 1000 cfm (cubic feet per minute) at a pressure of 1 in.

Heat

Heat is the energy transfer that causes an increase in the temperature of a body when heat is added or a decrease in the temperature of a body when heat is removed, provided there is no change in state in the process. The stipulation "no change in state" requires some explanation. A substance changes its state when it changes its physical form. For example, when a solid changes to a liquid, that is a change in state. When a liquid changes to a vapor, that is also a change in state. *Sublimation* is the change in state from a solid to a vapor without passing through the liquid state. A good example of this is the evaporation of mothballs or the evaporation of dry ice (solid carbon dioxide).

The unit of measurement of heat in the British system is the *British thermal unit* (Btu). One Btu is the amount of heat required to raise the temperature of 1 pound (lb) of water 1°F, as shown in Figure 1.3. Formerly, the metric system unit of heat was the *kilocalorie* (kcal). This is the amount of heat required to raise the temperature of 1 *kilogram* (kg), or 1000 *grams* (g), of water 1°C. This unit of heat is still in use; however, the currently used metric unit or International System (SI) unit for heat, comparable to a Btu, is the joule (J). The joule is defined in terms of work and is equivalent to the movement of a force of one newton (N) a distance of 1 meter (m).

It is important in most cases to note the *rate* of heat change: How fast or how slowly does a piece of heating equipment produce heat? Therefore, a more precise term is *Btu per hour* (Btuh). When values become quite large M is used to represent 1000, so the abbreviation *MBh* represents *thousands of Btu per hour.* The metric unit is the watt (W), which is equivalent to joules per second. For example, a furnace may have an output of 100,000 Btu per hour, or 100 MBh.

Types of Heat

There are two types of heat. One type changes the temperature of a substance, and the other changes its state. The type that changes the temperature is called *sensible heat.* The type that changes the state is called *latent heat.* The formula used to determine the sensible heat added to or removed from a substance is

$$Q = W \times SH \times TD$$

where Q = quantity of heat, Btu
W = weight, lb
SH = specific heat, Btu/lb/°F
TD = temperature difference or change, °F

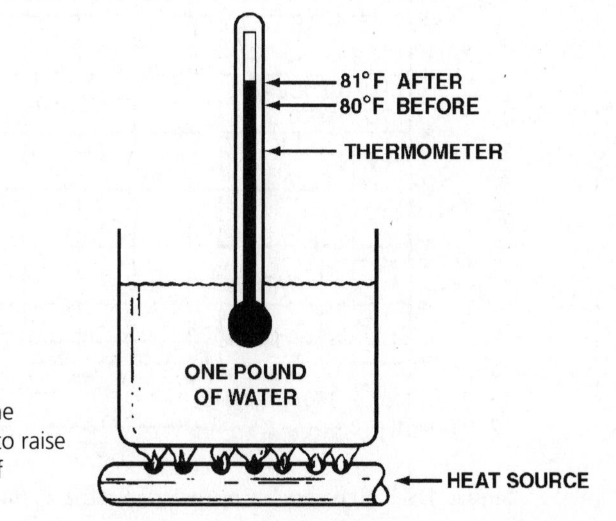

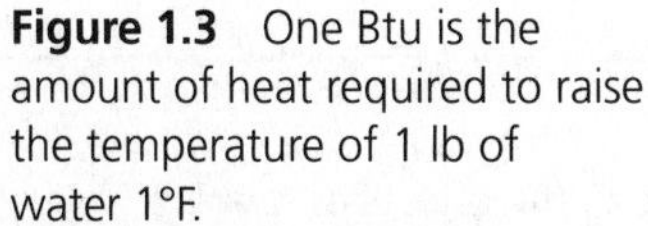

Figure 1.3 One Btu is the amount of heat required to raise the temperature of 1 lb of water 1°F.

The formula used to determine the change in latent heat is

$$Q = W \times L$$

where L is the change in latent heat per pound.

When heat is added to a solid to change it to a liquid, the latent heat is called *heat of fusion*. When the liquid is changed back to a solid, its latent heat of fusion is released to the surroundings. When a liquid is changed to a vapor, the latent heat is called *heat of vaporization.* When the vapor is changed back to a liquid, its latent heat of vaporization is released to the surroundings. Latent heat is always related to a change in state.

A good example for showing the effects of heat is its reaction on water, as shown in Figure 1.4. In this figure, ice is taken at 0°F and heated to steam. Note that there are five parts to the diagram, and the following describe what takes place.

Part 1 Ice at 0°F is heated to 32°F. The amount of heat required is found by substituting values in the sensible heat formula:

$$\begin{aligned} Q &= 1 \text{ lb} \times 0.5 \text{ Btu/lb/°F} \times (32\text{°F} - 0\text{°F}) \\ &= 1 \times 0.5 \times 32 \\ &= 16 \text{ Btu} \end{aligned}$$

Therefore, 16 Btu of sensible heat is added.

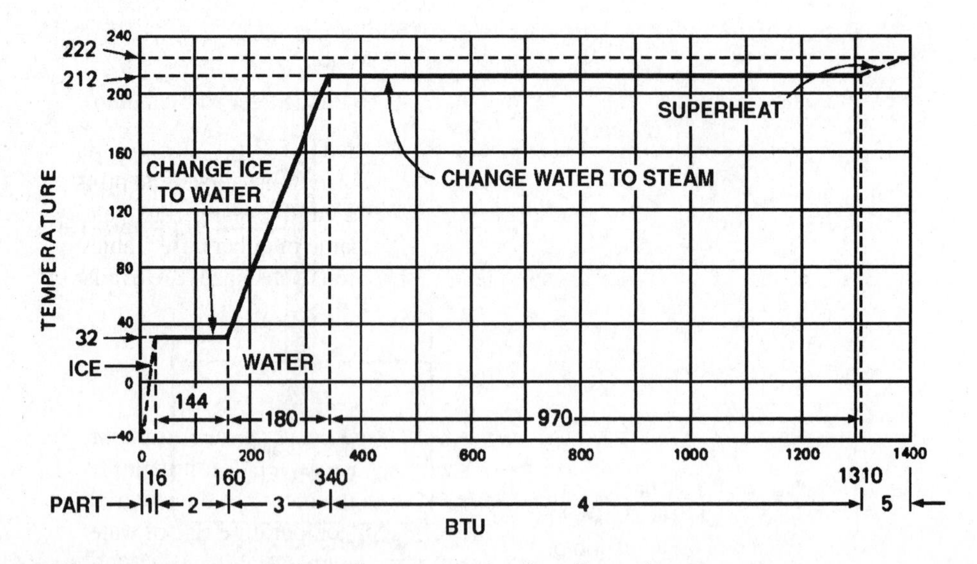

Figure 1.4 Temperature–heat diagram showing the effects of heat on water.

Part 2 Ice at 32°F is changed to water at 32°F. By the latent heat formula, since the heat of fusion of water is 144 Btu/lb,

$$\begin{aligned} Q &= 1 \text{ lb} \times 144 \text{ Btu} \\ &= 144 \text{ Btu} \end{aligned}$$

Therefore, 144 Btu of latent heat is added.

Part 3 Water at 32°F is heated to 212°F. Using the formula and substituting in the values, we obtain

$$\begin{aligned} Q &= 1 \text{ lb} \times 1.0 \text{ Btu/lb/°F} \times (212° - 32°\text{F}) \\ &= 1 \times 1 \times 180 \\ &= 180 \text{ Btu} \end{aligned}$$

Therefore, 180 Btu of sensible heat is added.

Part 4 Water at 212°F is changed to steam. Since the heat of vaporization of water is 970 Btu/lb, using this formula we obtain

$$\begin{aligned} Q &= 1 \text{ lb} \times 970 \text{ Btu} \\ &= 970 \text{ Btu} \end{aligned}$$

Thus, 970 Btu of latent heat is added.

Part 5 The steam is superheated 10°F. *Superheating* means heating steam (vapor) above the boiling point. Since the specific heat of steam is 0.48 Btu/lb/°F, the amount of heat added is determined by the formula

$$\begin{aligned} Q &= 1 \text{ lb} \times 0.48 \text{ Btu/lb/°F} \times (222°\text{F} - 212°\text{F}) \\ &= 1 \times 0.48 \times 10 \\ &= 5 \text{ Btu (rounded to the nearest whole number)} \end{aligned}$$

Therefore, the amount of sensible heat added is 5 Btu.

Both heat of fusion and heat of vaporization are reversible processes. The change in state can occur in either direction. Water can be changed to ice or ice to water. The amount of heat added or subtracted is the same in either case. Tables giving the heat of fusion and the heat of vaporization for various substances are available.

Specific Heat

Except in cases in which a change in state occurs, when most substances are heated they absorb heat, and their temperature rises. The temperature of the substance increases rapidly or slowly, depending on the nature of the material. Because the standard heating unit is defined in terms of the temperature rise of water, it is fitting to use water as a basis for comparison for the heat-absorbing quality of other substances.

WATER	1.00
ICE	0.50
AIR (DRY)	0.24
STEAM	0.48
ALUMINUM	0.22
BRICK	0.20
CONCRETE	0.16
COPPER	0.09
GLASS (PYREX)	0.20
IRON	0.10
PAPER	0.32
STEEL (MILD)	0.12
WOOD (HARD)	0.45
WOOD (PINE)	0.67

THESE VALUES MAY BE USED FOR COMPUTATIONS WHICH INVOLVE NO CHANGE OF STATE.

Figure 1.5 Specific heat values for some common substances.

The definition of specific heat is much like the definition of Btu. The specific heat of a substance is the amount of heat required to raise the temperature of 1 lb of a substance 1°F. Figure 1.5 lists specific heat values for various substances.

These specific heat values make it easy to calculate the amount of heat added to or removed from a substance when the temperature rise or drop is known. The amount of heat required is equal to the weight of the substance, times the specific heat, times the rise in temperature in degrees Fahrenheit.

Example 1–3 How much heat is required to heat 10 lb of aluminum from 50°F to 60°F?

Solution: Btu = pounds of aluminum × specific heat × °F temperature rise

$$= 10 \text{ lb} \times 0.22 \text{ Btu/lb/°F} \times (60°\text{F} - 50°\text{F})$$
$$= 10 \times 0.22 \times 10$$
$$= 22 \text{ Btu}$$

■

Transfer of Heat

Three different ways to transfer heat are by convection, conduction, and radiation, as shown in Figure 1.6. Modern heating plants make use of all three ways to transfer heat.

Convection is the circulatory motion in air caused by the rising of the warmer portions and the sinking of the denser, cooler portions. For convection to take place, there

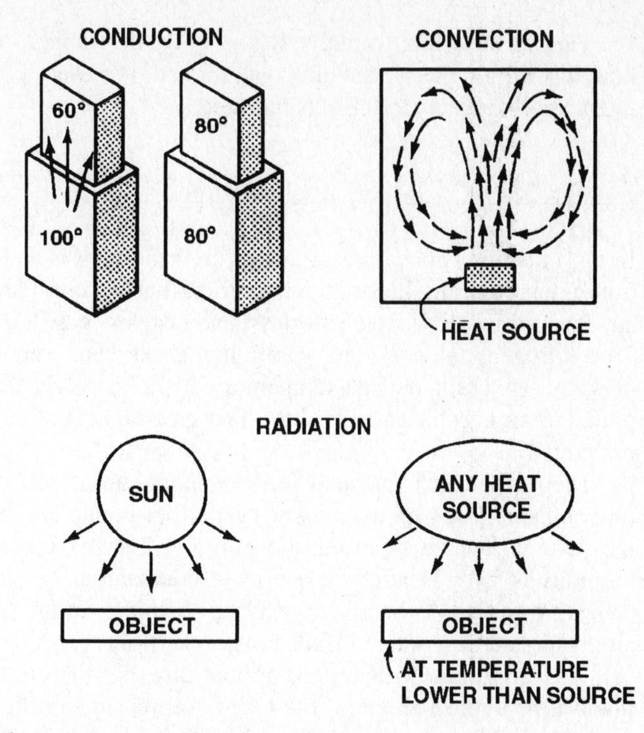

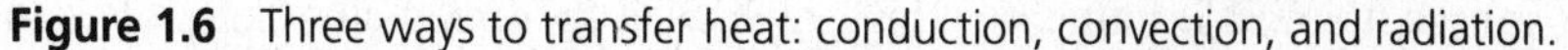

Figure 1.6 Three ways to transfer heat: conduction, convection, and radiation.

must be a difference in temperature between the source of heat and the surrounding air. The greater the difference in temperature, the greater the movement of air by convection. The greater the movement of air, the greater the transfer of heat.

Conduction is the flow of heat from one part of a material to another part in direct contact with it. The rate at which a material transmits heat is known as its *conductivity.* The amount of heat transmitted by conduction through a material is determined by the surface area of the material, the thickness, the temperature between two surfaces, and the conductivity.

Radiation is the transfer of heat through space by wave motion. Heat passes from one object to another without warming the space in between. The amount of heat transferred by radiation depends on the area of the radiating body, the temperature difference, and the distance between the source of heat and the object being heated.

Humidity

Humidity is the amount of water vapor within a given space. There are two types of humidity: absolute and relative. *Absolute humidity* is the weight of water vapor per unit of volume. *Relative humidity* is the ratio of the weight of water vapor in 1 lb (or in 1 kg) of dry air compared with the maximum amount of water vapor 1 lb (or 1 kg) of air can hold at a given temperature, expressed as a percent.

During the winter months, it is usually desirable to increase the percentage of humidity within the space being conditioned. For every pound of water that is evaporated, approximately 970 Btu is required.

Air

In most heating applications, transferring heat to air is an important part of the conditioning process. It is desirable, therefore, that we consider some of the properties of air. Pure air is an invisible, odorless, and tasteless gas. It contains about 21% oxygen, 78% nitrogen, and 1% other gases, including water vapor. Standard air at sea-level pressure (14.7 psia) and at a temperature of 72°F, weighs 0.0725 lb per cubic foot. One pound of air occupies about 14 ft^3. The specific heat of air is approximately 0.24 Btu per pound.

Moisture in air is in vapor form. Standard air at 72°F that is saturated with moisture contains 118.4 grains of water vapor per pound (7000 grains = 1 pound). When the same air contains half the moisture (59.2 gr/lb), it can be stated that the relative humidity is 50%. Heating air causes it to expand and weigh less per cubic foot, thus causing it to rise. As heat is transferred to objects in the space the temperature of the air drops, and the cooler air falls toward the floor.

It is important in heating applications that the circulation of air in the space involve the entire volume of the room, to prevent stratification and uneven temperatures. In most installations this is achieved by forced circulation. Because air is invisible, it can be unnoticeably contaminated. An extreme case of air contamination is described in Chapter 21. A number of people died and many became ill at a convention of legionnaires at a large hotel in Philadelphia in 1976. The cause of the illness was airborne bacteria that circulated through the air-conditioning system. Good air-quality measures could have prevented this tragedy.

Water

Water is another substance that can be used to transfer heat from the source to the space being conditioned, as shown in Chapter 23, Hydronic Heat. It is therefore important to consider some of the properties of water.

Pure water is transparent to some degree, depending on its thickness and surrounding light conditions. The appearance of water can be misleading, as it can contain foreign substances that greatly affect its possible uses. Most communities have access to testing facilities. Water weighs 62.5 lb per cubic foot, and one cubic foot is equivalent to 8.33 gallons. Water may be found in any one of its three states: solid, liquid, or gas. The specific heat of water is 1.0. The specific heat of ice is 0.5.

There are two conditions that cause the HVAC technician special concern:

1. When water is heated, as in a hot water heating system, it expands. There must be expansion space in a closed system with an air cushion to allow for this increase in volume. In addition it is required by code to have an automatic relief valve to

release excess pressure. Both of these protective devices must be in good operating condition at all times.
2. When water freezes, it expands. The expansion can be so great that it will break almost any container, including piping.

There are numerous methods of preventing damage due to freeze-up, depending on the application. One way is to drain the system before it is subjected to freezing conditions. Another way is to add an antifreeze solution to the water being circulated in a hot-water heating or cooling system. The amount of antifreeze added will be determined by the percent by volume necessary to prevent freezing in the region where the system is installed.

Laws of Thermodynamics

Thermodynamics is the science that deals with the relationships between heat and mechanical energy. *Energy* is the ability to do work. If a force is applied to an object that moves it a given distance, work has been performed. Heat is a form of energy. Other forms of energy include light, chemical, mechanical, and electrical.

Since heat is a form of energy, it follows the natural laws that relate to energy. These laws are useful in the study of heating. From the laws of thermodynamics, two helpful facts are derived:

1. Energy can be neither created nor destroyed, but it can be converted from one form of energy to another.
2. Heat flows from hot to cold.

An example of the conversion of energy is the changing of electrical energy to heat in an electric heating unit. An example of the flow of heat from hot to cold is illustrated in Figure 1.7.

Conditions That Affect Comfort

Comfort is the absence of disturbing or distressing conditions—it is a feeling of contentment with the environment. The study of human comfort concerns itself with two aspects:

1. How the body functions with respect to heat
2. How the area around a person affects the feeling of comfort

Human Requirements for Comfort

The body can be compared to a heat engine. A heat engine has three characteristics:

1. It consumes fuel.
2. It performs work.
3. It dissipates heat.

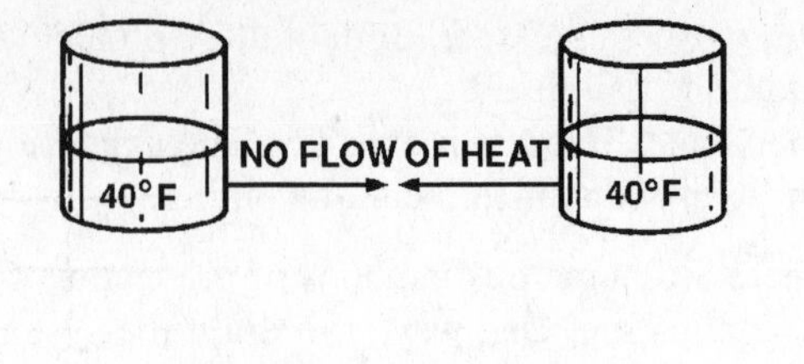

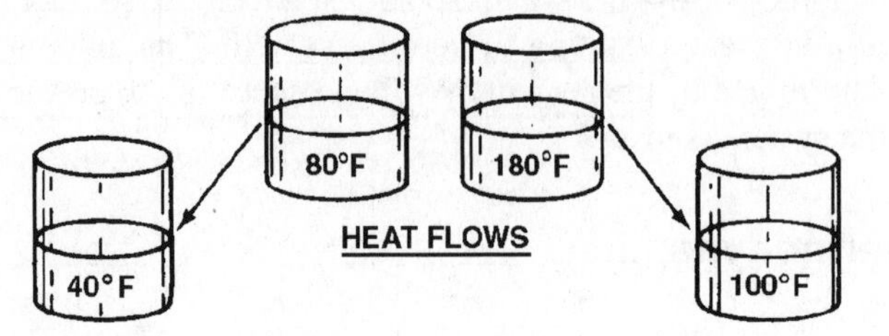

Figure 1.7 Heat flows from hot to cold.

The body consumes fuel in the form of food, which produces energy to perform work. The body also dissipates heat to the surrounding atmosphere. The body has a unique characteristic in that it maintains a closely regulated internal body temperature of 97 to 100°F (36 to 38°C). Any excess energy that is not used to produce work or to perform essential body functions is expelled to the atmosphere. It is therefore important that any loss of body heat be at the proper rate to maintain body temperature. If the body loses heat too fast, a person has the sensation of being cold. If the body loses heat too slowly, a person has the sensation of being too warm.

There are two additional factors that must be taken into consideration:

1. *Amount of activity of a person.* The greater the activity, the more heat produced by the body that must be dissipated to the atmosphere (Figure 1.8).
2. *Amount of clothing worn by a person.* Clothing is a form of insulation, and insulation slows down the transfer of heat. Therefore, the warmer the clothing, the greater the insulation value and the less heat dissipated. Thus, heat that would normally be lost to the atmosphere is used to warm the surface of the body.

Heat Transfer

Excess heat must be transferred from the body to the area around it. This transfer is accomplished in three ways (Figure 1.9):

1. Convection
2. Radiation
3. Evaporation

ACTIVITY	TOTAL HEAT ADJUSTED* BTUH	SENSIBLE HEAT BTUH	LATENT HEAT BTUH
SCHOOL	420	230	190
OFFICE	510	255	255
LIGHT WORK	640	315	325
DANCING	1280	405	875
HEAVY WORK	1600	565	1035

*ADJUSTED TOTAL GAIN IS BASED ON NORMAL PERCENTAGE OF MEN, WOMEN AND CHILDREN FOR THE APPLICATION LISTED.

Figure 1.8 Heat from people.

Convection For convection to take place, the area around the body must be at a lower temperature. The greater the difference in temperature, the greater the movement of convection currents around the body and the greater the heat transfer. If the temperature difference is too great, a person will feel the sensation of being cold. If the temperature difference is too small, a person will feel the sensation of being hot.

Radiation The body radiates heat to cooler surfaces just as the sun radiates heat to the earth. The greater the temperature difference between the body and the exposed surfaces, the greater the rate of heat transfer. It has been found that the most comfortable conditions are maintained if the inside surface of the outside walls and windows

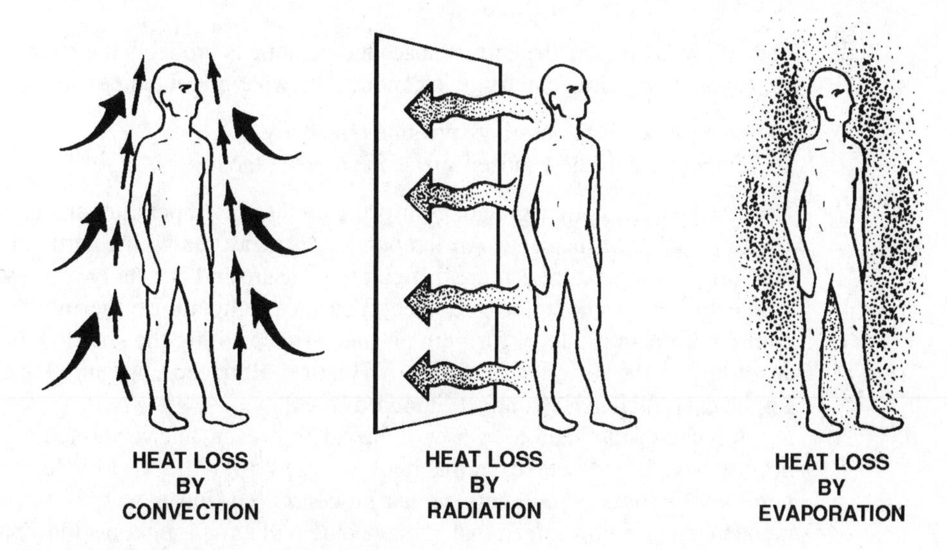

Figure 1.9 Three ways the body loses heat.

are heated to the room temperature. If these surface temperatures are too low, the body radiates heat to them too fast, and this produces a feeling of discomfort.

Evaporation Water (or moisture) on the surface of the skin enters the surrounding air by means of evaporation. Evaporation is the process of changing water to vapor by the addition of heat. As the water absorbs heat from the body it evaporates, thus transferring heat to the air. If the rate of evaporation is too great, the skin has a dry, uncomfortable feeling. The membranes of the nose and throat require adequate humidity to maintain their flexible condition. If the rate of evaporation is too slow, the skin has a clammy, sticky feeling. Evaporation of moisture requires approximately 970 Btu of heat per pound of water evaporated. Heat required for evaporation is lost from the body, thereby cooling it.

Space

Space is defined as the enclosed area in which one lives or works. Conditions for comfort are maintained within the space by mechanical heating and cooling. The requirement for comfort is that the space supply a means for dissipating the heat from the body at the proper rate to maintain proper body temperature.

The conditions for comfort in the space may be divided into two groups: thermal and environmental. The thermal conditions include temperature, relative humidity, and air motion. The environmental conditions include clean air, freedom from disturbing noise, and freedom from disagreeable odors. Because of the importance of the thermal factors, these will be discussed first.

Temperature

The ability to produce the correct space temperature is probably the most important single factor in providing comfortable conditions. Two types of temperature are considered:

1. Temperature of the air that surrounds the body
2. Temperature of the exposed surface enclosing the room

For the maximum degree of comfort, both the air temperature and the surface temperature should be the same. Air temperature is measured with an ordinary thermometer. Surface temperature is more difficult to measure and is usually read by using either an electronic thermometer or a special surface-temperature thermometer. The best comfort temperature level for both air and surface within the space is 76°F (24.5°C), according to the American Society of Heating, Refrigerating, and Air Conditioning Engineers (ASHRAE) Comfort Standard 55-80.

In view of the need to conserve energy, however, interior temperature levels for heating have been lowered. In this book we use 70°F (21.1°C) as a design inside temperature. In cases when energy must be conserved, lower temperatures are recommended. It is acknowledged that some people will have to put on additional clothing to remain comfortable at this temperature.

Relative Humidity

As defined earlier, *relative humidity* indicates the percent water vapor present in the air compared with the maximum amount that the air can hold at the same temperature. Two types of psychrometers are used to determine the relative humidity, the *sling psychrometer* and the *power psychrometer,* which are shown in Figure 1.10. The difference between the instruments is that the sling psychrometer is manual, and the power psychrometer utilizes a fan powered by a battery.

To take readings using a sling psychrometer, dip the wick on the wet bulb thermometer in water (distilled, if possible). (Use only one dipping per determination of relative humidity, but never dip between readings.) The evaporation of the moisture on the wick is the determining factor of the wet bulb reading. Whirl the sling psychrometer for 30 seconds. Quickly take the reading on the wet bulb thermometer first, then read the dry bulb and record the readings. Continue to whirl the psychrometer, taking readings at 30-second intervals for five successive readings, recording the temperatures each time until the lowest readings for wet bulb and dry bulb have been obtained.

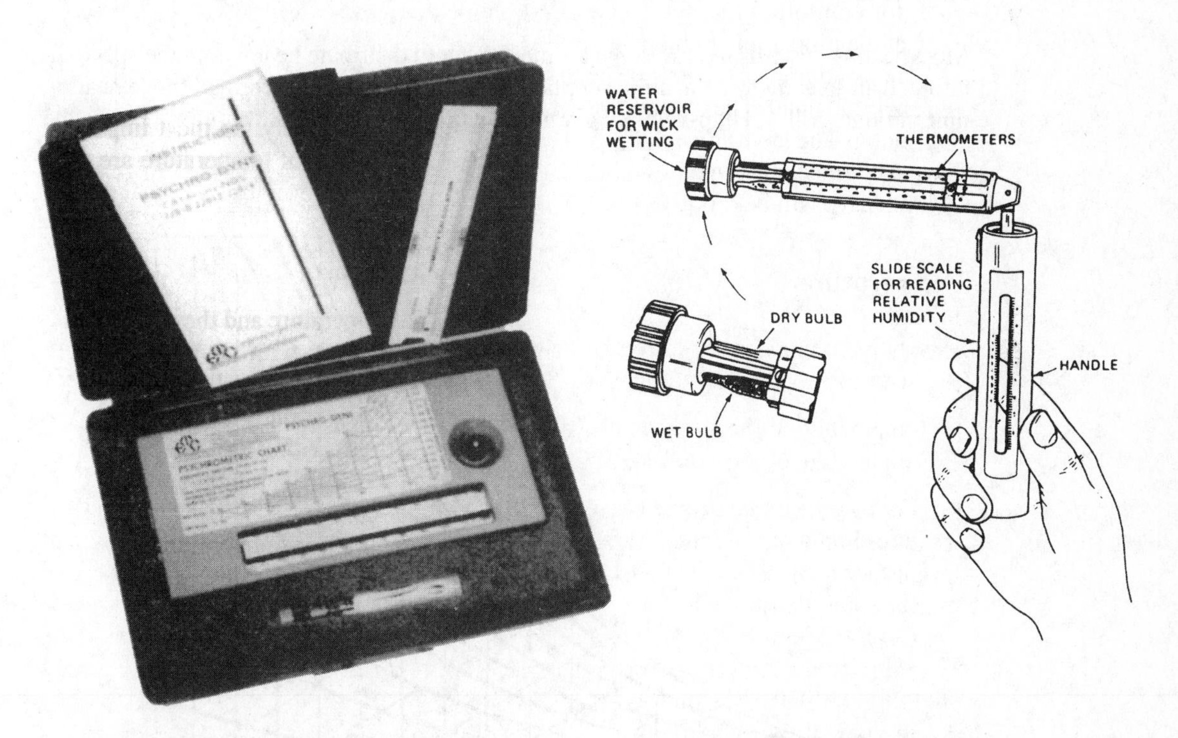

PSYCHRO-DYNE POWER PSYCHROMETER.
(COURTESY, ENVIRONMENTAL TECTONICS CO.)

SLING PSYCHROMETER.
(COURTESY, BACHARACH, INC.)

Figure 1.10 Two types of psychrometers. (a) Courtesy of Environmental Tectonics Co. b) Courtesy of Bacharach, Inc.)

The power psychrometer is designed to provide a 15-ft/s airflow over the thermometers. When the fan operates, the wet bulb temperature will be rapidly lowered. A reading should not be taken until the temperature drops to its lowest point. This may take from 1 to 2 minutes, depending on the dryness of the air. Take two readings. Use a psychrometric chart or table to obtain the relative humidity (rh).

Figure 1.11 is a simplified psychrometric chart that can be used to determine the relative humidity of a sample of air. Note that the wet bulb lines slope downward to the right, the dry bulb lines are vertical, and the relative humidity lines curve upward to the right. Any point on the chart, therefore, represents some wet bulb, dry bulb, and relative humidity condition.

Example 1–4

What is the relative humidity for the condition of 75°F db (dry bulb) and 60°F wb (wet bulb)?

Solution: Locate 75°F db on the baseline and trace vertically upward until it meets the 60°F wb line. The relative humidity at this point is 40%. ■

Air Motion

Most heating systems require some air movement to distribute heat within the space. In a forced air heating system, air is supplied through registers and returned to the heating unit through grilles. High-velocity air, measured in feet per minute (ft/min.), enters the

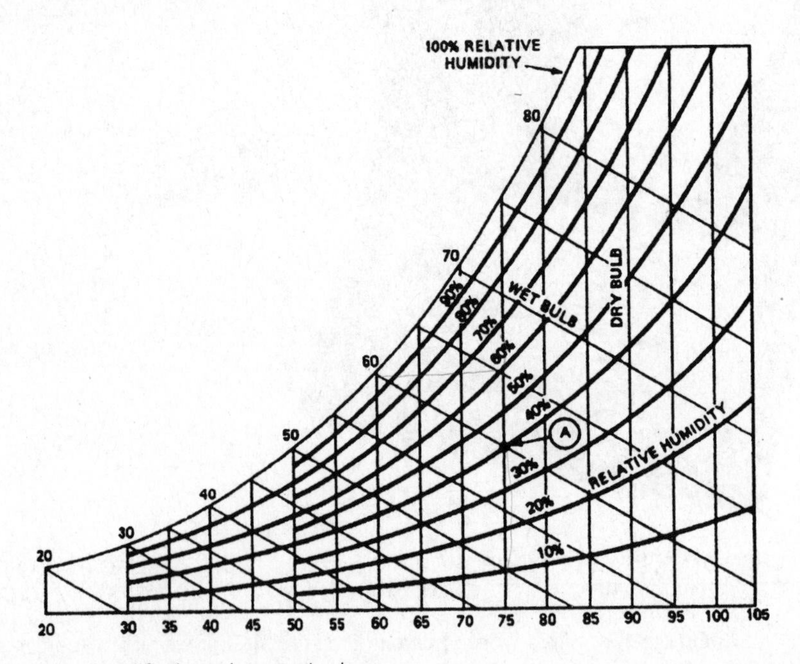

Figure 1.11 Simplified psychrometric chart.

room through the supply registers. Air at these velocities may be required to heat cold exterior walls properly. The velocity is greatly reduced by natural means before the air reaches the occupants of the room. According to the ASHRAE Comfort Standard, provided that other recommended conditions are maintained (76°F db, 40% rh), the air that comes in contact with people should not exceed a velocity of 45 ft/min or 0.23 m/s.

Air Quality

Filtration is very important today because residential and commercial buildings are more energy efficient. Consequently, there is less infiltration. To compensate, more outdoor ventilation air must be brought in; however, if the indoor air can be cleaned through filtration, less outdoor air needs to be brought in. Because the outdoor air must be heated in the winter and cooled in the summer, less outdoor air means lower operational costs. ASHRAE has established the standards for ventilation for acceptable indoor air quality. The standard establishes a recommended outdoor air minimum of 15 cfm per person. For offices the minimum is 20 cfm per person. The outdoor air may enter because of infiltration, ventilation, or a combination of both. Filters must be selected with care if the quality of the indoor air is to be maintained.

Freedom from Disturbing Noise

Noise or objectionable sound can be airborne or travel through the structure of a building. Airborne noise can be caused by mechanical equipment that is located too near a building's occupants or by the transmission of mechanical noise through connecting air ducts. Noise that travels through the structure of a building is caused by the vibration of equipment mounted in direct contact with the building.

To remove noise problems, equipment should be properly selected from a noise-level standpoint. Where noise or vibration does exist, equipment should be properly isolated (separated from sound-transmitting materials) (Figure 1.12). Sound can be measured by acoustical instruments. The unit of sound-level measurement is the decibel (dB).

Freedom from Disagreeable Odors

Any closed space occupied by people will develop odors. Ventilation is the most effective means of reducing odors. Some outside air leaks into a building through the cracks around windows and through the building construction. In modern construction, occasionally this is not enough to provide adequate ventilation.

A provision can be made to provide an additional quantity of outside air. Ventilation air should not be confused with combustion air, since combustion air must be provided to burn the furnace fuel properly. This must not in any way subtract from the air (oxygen) available for the occupants of the structure.

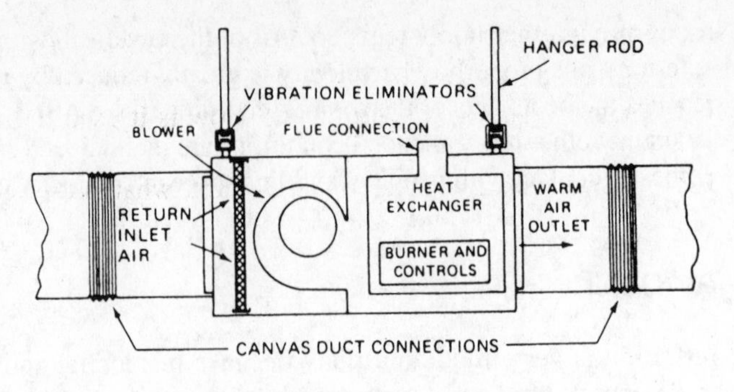

Figure 1.12 Noise and vibration eliminators. (By permission of *ASHRAE Handbook*)

STUDY QUESTIONS

Answers to the study questions may be found in the sections noted in brackets.

1-1 What are the three important requirements in the construction of indoor climate systems? *[Indoor Climate Control]*
1-2 Define *temperature. [Temperature]*
1-3 Describe the two types of scales for measuring temperature. *[Thermometers]*
1-4 Give the formula for converting degrees Fahrenheit to degrees Celsius, and degrees Celsius to degrees Fahrenheit. *[Thermometers]*
1-5 Give two definitions for the British thermal unit (Btu), one for Fahrenheit temperatures, and one for Celsius temperatures. *[Heat]*
1-6 Give the formulas for calculating sensible and latent heat. *[Types of Heat]*
1-7 Define *specific heat* in terms of British units. *[Specific Heat]*
1-8 Name the three ways to transfer heat. *[Transfer of Heat]*
1-9 What is the difference between absolute and relative humidity? *[Humidity]*
1-10 Define *energy. [Laws of Thermodynamics]*
1-11 What is meant by a *change of state? [Types of Heat]*
1-12 Give the specific heats of ice, water, and steam. *[Specific Heat]*
1-13 Define *comfort. [Conditions That Affect Comfort]*
1-14 How can the body be compared to a heat engine? *[Human Requirements for Comfort]*
1-15 What is normal body temperature? *[Human Requirements for Comfort]*
1-16 What are the conditions for body comfort from the standpoint of heat loss? *[Human Requirements for Comfort]*
1-17 How does the body transfer heat? *[Human Requirements for Comfort]*
1-18 How does the temperature of exposed room surfaces affect comfort? *[Temperature]*
1-19 How much heat is required to evaporate 1 lb of water? *[Evaporation]*

1-20 How does the activity of a person affect the amount of heat dissipated? *[Human Requirements for Comfort]*

1-21 What is a sling psychrometer? How is it used to measure relative humidity? *[Relative Humidity]*

1-22 Referring to the psychrometric chart, what is the relative humidity for 75° db and 60° wb? *[Relative Humidity]*

CHAPTER 2

Safety Analysis and Identification

OBJECTIVES

After studying this chapter, the student will be able to:

- Identify environmental and hazardous situations
- Assist in creating a safe workplace

Introduction

The safety of the service technicians and customers is of great concern to contractors as well as to product manufacturers. Being aware of safety, both on and off the job site, is everybody's responsibility. Standards set forth by the Occupational Safety and Health Administration (OSHA), by federal legislation, and by local codes are designed to ensure safety as well as compliance in the workplace. The employer or employee must know that making the workplace safe is both the correct and most cost-effective thing to do over the long term. Many other issues are as important—selling goods or services, employing or working with the best people, keeping good customer relations, and selecting the best materials for the project. The most important objective in any safety program is to be aware of the work surroundings.

Standards can be designed to be cost effective in that they employ the least expensive protective measures capable of reducing or eliminating a significant risk, but they must be implemented. One of the best methods of conducting a job analysis/hazard assessment survey is to simply start by looking over the area before starting an installation or service project and asking the following questions:

1. Is the area safe to work in and clear of debris?
2. Is there adequate lighting suitable for work to be done?

3. Is there an electrical hazard present?
4. Is there a suitable fire extinguisher available?

A safety program cannot be effective unless one accepts the responsibility of implementing it. Safety suggestions are not as difficult to put in place as one might think—they are mostly common sense. The success of safety in the workplace depends on the sincere, constant, and cooperative effort of all people and their active participation and support.

Health and Safety

Notification of Accident or Injury

When possible, an employee should inform the employer before visiting a medical facility for treatment of a work-related injury or illness and report back to the employer as soon as possible after receiving treatment. The employer will meet with the employee, investigate the incident, complete a report of the incident, and place a copy of the report in both the company's and employee's file.

Clothing

All eye protection needs to meet American National Standards Institute (ANSI) Z87.1-1989 standards. The style of the safety glasses chosen should relate directly to the work being performed. All head protection needs to meet ANSI standards, coordinate with the eye protection, and relate directly to the work being performed (Figure 2.1).

For maximum safety the clothing worn should comply with the following:

1. Pants that are tight fitting and not too long
2. Outerwear that protects against rain and/or cold
3. No neckties or scarves
4. No jewelry such as finger rings, or neck or wrist chains
5. Gloves of proper size and material to meet safety standards (Figure 2.2).

Confined Space

A confined space is one that has a very limited entrance or exit. This space can be a room, a piece of mechanical equipment, or a storage tank, for example. A person should never enter a confined space alone. The buddy system must be employed so constant communication can be maintained at all times. Proper safety equipment should be available to remove the person in case of loss of consciousness or injury. Before entry, the worker must ensure that all process piping, mechanical equipment, electrical equipment, or any other source of energy that could create a potentially hazardous condition is rendered inoperable. If there is any question of safe entrance, the proper authorities must be contacted.

Figure 2.1 Head and eye protection. (Courtesy of Uvex.)

Ventilation

Mechanical ventilation is required whenever the workspace indicates the presence of a hazardous atmosphere, which can be created by the use of chemicals, welding materials, toxic gases, refrigerants, and natural gas. The forced ventilation must be directed so that it will clear the immediate area and will stay on until the work is completed. The type of work to be performed will determine the requirements for the use of eye, face, and skin protection. In the handling of asbestos or other types of hazardous chemicals, one must "suit up" properly and wear respiratory protection equipment. The respirator must be a self-contained breathing apparatus approved by the National Institute of Occupational Safety and Health (NIOSH).

Caution: Compressed air should never be used for ventilation or cleaning debris from one's clothing.

Brazing materials contain toxic metals that may be emitted in vapor form. Cadmium is used in some brazing rods, and it is highly toxic. Breathing of these vapors must be avoided. Brazing in tight spaces should be performed only with an exhaust fan pulling fumes from the area.

Containers of solvent cements and cleaners used to join plastic pipe and tubing list stringent safety warnings (Figure 2.3). These warnings must be read and heeded. Vapors of these products may create immediate and long-term health problems. At times the vapors may also be explosive.

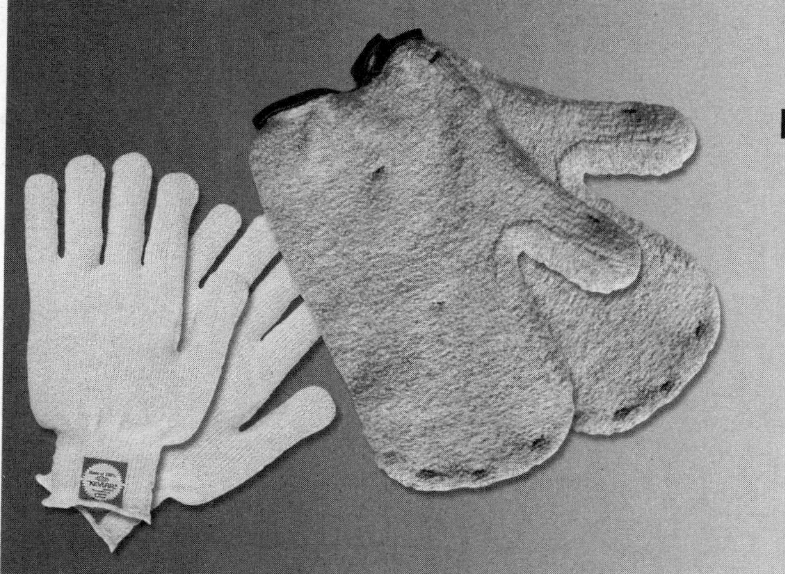

Figure 2.2 Heat protective gloves. (Courtesy of Kevlar)

FOR CPVC or PVC or ABS • CLEAR

CLEANER

LIMPIADOR PARA CPVC o PVC o ABS • TRANSPARENTE

DANGER:/PELIGRO:

EXTREMELY FLAMMABLE • HARMFUL OR FATAL IF SWALLOWED • VAPOR HARMFUL • MAY IRRITATE EYES AND SKIN • MAY BE ABSORBED THROUGH SKIN • VAPORS MAY CAUSE FLASH FIRES

EXTREMADAMENTE FLAMABLE • NOCIVO O FATAL SI SE INGIERE • VAPORES NOCIVOS • PUEDE IRRITAR LOS OJOS Y LA PIEL • PUEDE SER ABSORBIDO A TRAVES DE LA PIEL • LOS VAPORES PUEDEN CAUSAR INCENDIOS

MADE IN **USA** 8 fl. oz (1/2 Pt) 237 ml HECHO EN E U

HAZARDOUS INGREDIENTS: Methyl Ethyl Ketone 78-93-3 and Acetone 67-64-1 Store in a cool, dry, well ventilated place. Close container after use. Vapors may ignite explosively. Use only with adequate ventilation. If forced air ventilation is used, be sure it does not cause a fire hazard from the solvent vapors. If adequate ventilation can not be provided, wear a NIOSH-approved respirator for organic solvents. Wear safety glasses with side shields. Avoid skin contact; wear rubber gloves. If swallowed, drink water. DO NOT INDUCE VOMITING, call a physician or poison control center immediately. If inhaled and headache, dizziness, respiratory discomfort, or intoxication are experienced, get freh air and obtain medical attention if ill feelings persist. Wash thoroughly after use andbefore eating. In case of eye or skin contact, immediately flush with water for 15 minutes and seek medical attention if irritation persists. Long term, repeated overexposure to solvents may cause damage to the brain and nervous system, reproductive and respiratory systems, mucous membranes, liver and kidneys. For more information, obtain a copy of MSDS from distributor.

KEEP OUT OF REACH OF CHILDREN.

Figure 2.3 Plastic pipe cleaner usage warnings.

Air-conditioning refrigerants in small quantities are generally not toxic, except for R-123. Refrigerants do displace oxygen and are heavier than air. Inhalation may cause asphyxiation, because the refrigerant fills up and stays in the bottom of the lungs. When exposed to flame or excessive heat, such as during brazing or in a compressor burnout, refrigerants decompose to form hydrochloric and hydrofluoric acids. The acid vapors damage the lining of the lungs, which may be fatal. During brazing, if the flame of the torch turns a fluorescent green, shut off the torch immediately and evacuate and ventilate the work area. Do not smoke in a refrigerant environment.

Excessive exposure to air-conditioning refrigerant can also create a cardiac sensitivity, which may result in a heart attack. Emergency treatment under these conditions should avoid the normally administered adrenalin therapy. Emergency medical personnel must be advised that the heart attack may have been precipitated by exposure to refrigerant.

Natural gas has an additive that smells like a rotten egg. The odor is added to the natural gas by the utility companies to aid in determining gas leaks. If the smell of gas is detected, all sources of heat and ignition should be shut off and the area ventilated. Checking for the leak should be done with soap or an electronic detector, **never** with a flame. Natural gas is lighter than air and tends to collect high in a room. It is not toxic, but it displaces air and thus may cause suffocation if not ventilated. Natural gas is explosive and has been known to completely flatten entire residential buildings.

Liquid petroleum (LP) gas differs from natural gas in that it is heavier than air and therefore collects along the ground. Should a leak occur in LP piping or equipment, immediate action should be taken to completely ventilate the area. A spark or standing pilot could ignite the LP gas, causing an explosion or fire.

Ladders

Ladders are available for three work classifications. Type I industrial ladders have a load limit of 250 lb for both the worker and the tools being used. Type II commercial ladders have a load limit of 225 lb. Type III household ladders should not be used for work applications. Ladders are constructed of wood, aluminum, or fiberglass. Weight and expected service length are considered when deciding on a material. The ladder must have approved antiskid feet (Figure 2.4). When a ladder is being used in work on the roof of a building, the ladder must extend a minimum of 3 ft above the roof's edge and be tightly secured. The feet of the ladder should be placed out from the wall a distance that is one-fourth the height to the roof's edge (Figure 2.5). In many cases, an L shape is drawn on the side of the ladder, which indicates the correct angle when the lines are vertical and horizontal (Figure 2.6). Working on roofs when there are strong winds or slippery conditions should be avoided. If, for any reason, the work cannot be postponed, special precautions should be taken.

Scaffolds

OSHA requires that scaffolds less than 45 in. wide have rails around the working platform whenever the work level is over 4 ft. Wider scaffolds require rails whenever the work level is above 10 ft. Wheels must always be locked while a worker is on the scaffold.

Lifting

When lifting it is important to use the knees and not the back. The back should be kept as straight as possible at all times to avoid injury. Lifting improperly can cause serious back injuries. There are proven methods for picking up heavy objects, which shift the pressure from the back to the legs. If a piece of equipment, like a furnace, is being

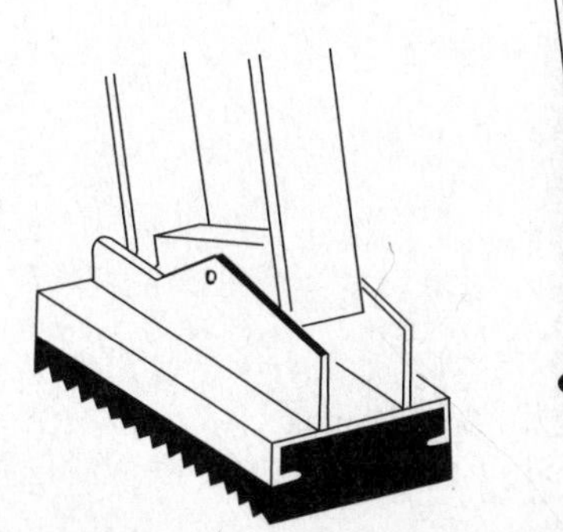

Figure 2.4 Antiskid safety feet on bottom of ladder.

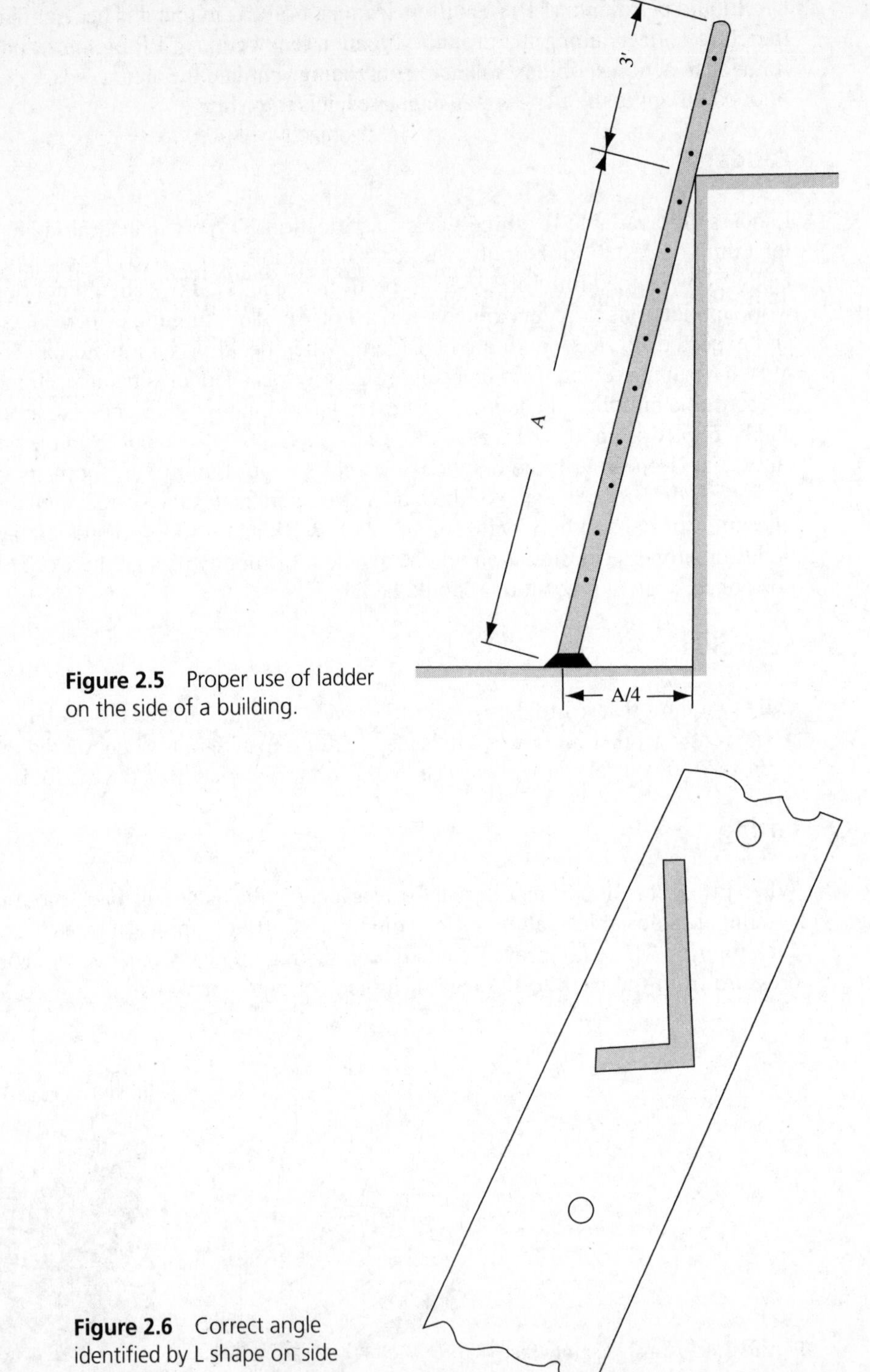

Figure 2.5 Proper use of ladder on the side of a building.

Figure 2.6 Correct angle identified by L shape on side of ladder.

moved down steps into a basement, it is important that the path to the final location be checked before the unit is moved. If a load is too heavy for a worker to handle alone, assistance must be requested before proceeding.

When a two-wheeled hand truck is used, it is important to secure the load with straps or bungee cords. In some cases, very heavy objects will need to be moved on rollers.

Tools

Hand and portable power tools must be properly maintained. If found defective, they must be repaired or replaced before any work is started. Tools are always designed for a specific purpose and should be used only for the work for which they were designed. Technicians must review and follow all safe operating instructions provided by the manufacturer. Before using any tool, the technician should check for obvious signs of damage or wear that could make the tool hazardous to use.

Hand Tool Types

Hammers Hammers are percussion tools. Striking an object can propel parts of a hammer, causing serious injury. Safety rules to follow are:

Inspect handles and replace them if they are broken, cracked, or splintered.

Inspect the head, which can be secured with wedges.

Screwdrivers The object being worked on should be secured, **never** held in the hand, on the lap, or under the arm. The screwdriver should not be used if the handle is cracked or the shank is bent.

Knives The blade must always be kept sharpened. Dull blades require force to cut, reducing safe control of the knife. A fixed blade knife must be carried in a sheath or other protective cover. A folding knife that cannot be locked in place must be used in a manner that keeps the blade from folding on the fingers.

Chisels Drift pins, punches, star drills, or wedges must never be used if the head is mushroomed. The striking end must be ground with a crowned radius and beveled edge.

Portable Power Tools

Safety precautions when operating portable power tools require certified eye protection and also hearing protection if the tool is excessively loud. If a cord or hose must temporarily lie on the floor, it must be protected, or the work area must be cordoned off.

Electrical Drills, grinders, routers, sanders, and saws must be grounded or double insulated. Never use tools on which the grounding plug has been removed or altered

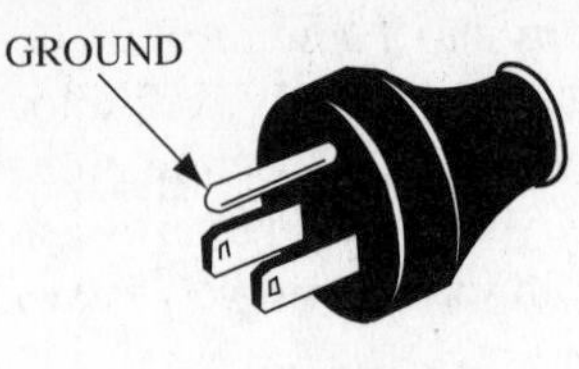

Figure 2.7 Grounded plug.

(Figure 2.7). **Do not** use the electrical cord to lift or to lower tools. Power actuated tools such as nailers or drivers should be tested before use to ensure that the safety devices are working properly. Fasteners should not be driven into brittle or very hard materials such as cast iron, rock, hollow tiles, or glass block.

Fire Hazards

Special precautions must be taken in the use of brazing, soldering, and cutting torches. Welding processes involving portable equipment using electric arc, open flames, or any other work producing a high temperature can all provide an ignition source for a building fire. The technician must examine the area in which the work is to be performed and select the specific equipment to prevent a possible fire. Shields must be used to protect other people in nearby areas from arc welding flashes. Ample portable extinguishing equipment should be readily available when performing a task where a possible fire hazard exists. Since most portable extinguishers discharge completely in as little as 8 to 10 seconds they need to be serviced and recharged after each use. Unused extinguishers need to be inspected for damage and full charge at least once a month. Disposable units, once discharged, cannot be recharged.

Fire consists of four basic elements—fuel, oxygen, heat, and chain reaction. The elimination of any one of these components will extinguish the fire. All technicians should be totally familiar with the manufacturer's instructions on any extinguisher that they may be required to use. When operating a fire extinguisher an easy key word to remember is PASS:

P Pull (safety pin)
A Aim (at base of flame)
S Squeeze (lever)
S Sweep (side to side)

Fire Extinguisher Guidelines

Guidelines are set forth by federal and national organizations, such as the National Fire Protection Association (NFPA). Fire extinguishers have been divided into five classes, each designated by a class letter and a picture symbol that identifies the types of fires on which they can be used. Technicians should be familiar with the classes and picture symbols before the use of a fire extinguisher is required (Figure 2.8).

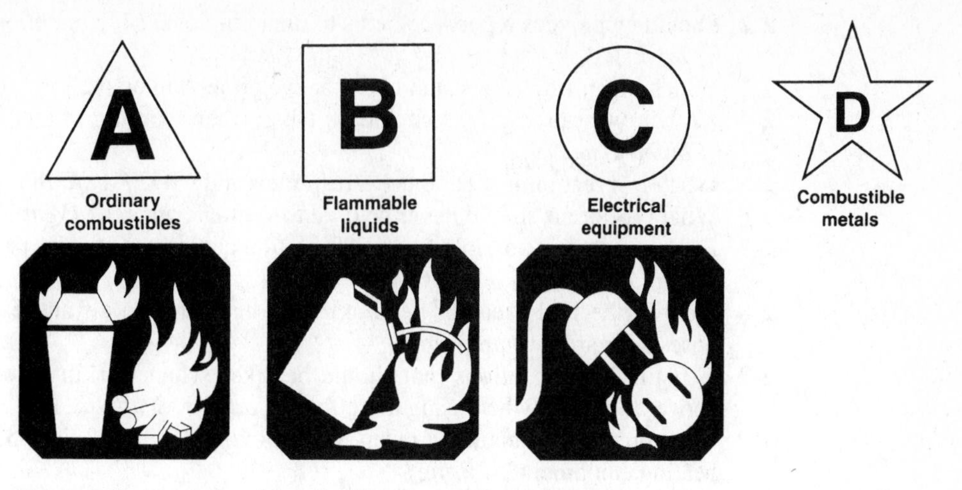

Figure 2.8 Fire extinguisher classifications and symbols.

Types of Fires and Classes

CLASS A	Ordinary combustibles: wood, paper, rubber, fabrics, and many plastics.
CLASS B	Flammable liquids and gases: gasoline, oils, paint, lacquer, and tar.
CLASS C	Fires involving energized electrical equipment: power tools, wiring, appliances, electric motors, etc.
CLASS D	Combustible metals or combustible metal alloys
CLASS K	Fires in cooking appliances that involve combustible cooking media, such as, but not limited to, vegetable or animal oils and fats.

Note: ABC Multipurpose dry chemical is electrically nonconductive; thus, it is also suitable for attacking fires of electrical origin.

A diagonal red slash through the picture symbol indicates a potential danger if the extinguisher is used on that particular type of fire. A missing symbol means that the extinguisher has not been tested for that class of fire.

All fires of any nature should be reported to the local fire department. Once a fire has started, the first step is to clear all the people from the area and second to call the fire department from a safe location. Most technicians are not trained for and do not have the equipment for fighting a major fire.

STUDY QUESTIONS

Answers to the study questions may be found in the sections noted in brackets.

2-1 Which federal legislative act affects the heating and air-conditioning trade? *[Introduction]*

2-2 Should employees report accidents to their employer? *[Notification of Accident or Injury]*
2-3 Which organization sets standards for eye protection on the job? *[Clothing]*
2-4 Explain what precautions should be taken when working in a confined space. *[Confined Space]*
2-5 When is it recommended to use a respirator and why? *[Ventilation]*
2-6 What type of air should never be used to ventilate an area? *[Ventilation]*
2-7 Explain how it is possible for a refrigeration compressor to give off toxic gases. *[Ventilation]*
2-8 Identify the fuel used for heating that is heavier than air and explain why it poses a hazard. *[Ventilation]*
2-9 Explain the precautions that should be taken when working more than 10 ft above floor level when using a scaffold. *[Ladders, Scaffolds, and Lifting]*
2-10 What precaution should be taken when moving a piece of air-conditioning and heating equipment? *[Lifting]*
2-11 When can a tool be hazardous? *[Tools]*
2-12 Why should electrical equipment be properly grounded? *[Tools]*
2-13 Where should fastener equipment not be used? *[Tools]*
2-14 Why can't all fire extinguishers be used to fight all fires? *[Fire Hazards]*
2-15 What elements constitute a fire? *[Fire Hazards]*

CHAPTER 3

Combustion and Fuels

OBJECTIVES

After studying this chapter, the student will be able to:

- Compare the heating qualities of various fuels
- Determine the conditions necessary for efficient utilization of fuels
- Use combustion test instruments
- Evaluate the results of combustion testing
- Determine the changes that must be made to reach maximum combustion efficiency

Combustion

Combustion is the chemical process in which oxygen is combined rapidly with a fuel to release the stored energy in the form of heat. Three conditions are necessary for combustion to take place:

1. *Fuel:* consisting of a combination of carbon and hydrogen
2. *Heat:* sufficient to raise the temperature of the fuel to ignition (burning) point
3. *Oxygen:* from the air, combined with the elements in the fuel

Figure 3.1 illustrates the conditions necessary for combustion. The fuel can be gas (such as natural gas), liquid (such as fuel oil), or solid (such as coal). Two elements all fuels have in common are hydrogen and carbon. Fuel must be heated to burn. For example, a pilot burner (small flame) can be used to ignite gas burners; electric ignition (an electric spark) is used to ignite oil; and, usually, a wood-burning fire is used to ignite coal. An example of a pilot burner and spark igniter is shown in Figure 3.2. Air containing oxygen must be present for burning of fuel to take place. As an example, a burning candle can be extinguished by placing a glass jar around it to enclose it (Figure 3.3). The candle goes out when it no longer has oxygen to support combustion.

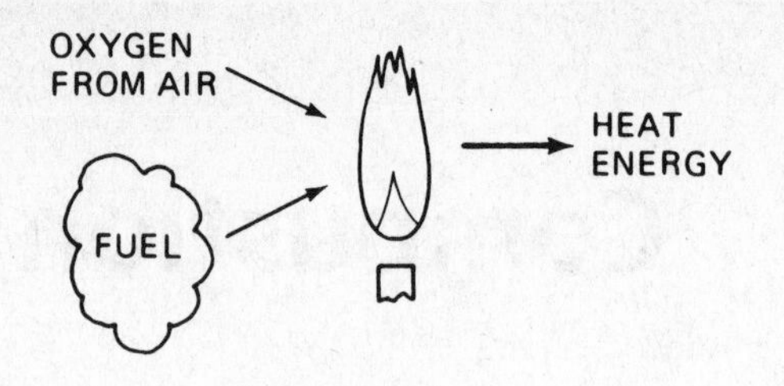

Figure 3.1 Conditions necessary for combustion.

Types of Combustion

There are two types of combustion: complete combustion and incomplete combustion. Complete combustion must be obtained in all fuel-burning devices. Incomplete combustion is dangerous. *Complete combustion* results when carbon combines with oxygen to form carbon dioxide (CO_2), which is nontoxic and can be readily exhausted to the atmosphere. The hydrogen combines with oxygen to form water vapor (H_2O), which also can be exhausted harmlessly to the atmosphere.

$$CH_4 + O_2 \rightarrow CO_2 + 2H_2O$$

Incomplete combustion results when a lack of sufficient oxygen causes the formation of undesirable products, including:

- Carbon monoxide (CO)
- Pure carbon or soot (C)
- Aldehyde, a colorless volatile liquid with a strong unpleasant odor (CH_3CHO)

Both carbon monoxide and aldehyde are toxic and poisonous. Soot causes coating of the heating surface of the furnace and reduces heat transfer (useful heat). Thus, the heating service technician must adjust the fuel-burning device to produce complete combustion of the fuel.

Caution: To prevent the dangers of incomplete combustion sufficient air must be provided for proper combustion to take place.

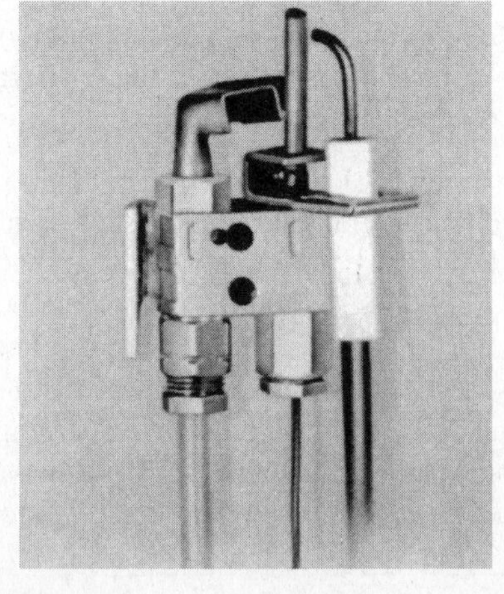

Figure 3.2 Example of pilot burner and spark igniter. (Courtesy of White-Rodgers, Division of Emerson Electric Company)

Figure 3.3 Candle flame extinguished by lack of oxygen.

Air consists of about 21% oxygen and 79% nitrogen by volume (Figure 3.4). The nitrogen in air dilutes the oxygen, which would otherwise be too concentrated to breathe in its pure form. Nitrogen is an *inert* element, which means that it remains in a pure state without combining with other elements under ordinary conditions. For example, in a furnace, when nitrogen is heated to temperatures higher than 2000°F, it does not react with the elements of the fuel. It enters the furnace with the combustion air and leaves through the chimney as pure nitrogen.

Flue Gases

The flue gases of a furnace operating to produce complete combustion contain:

- Carbon dioxide
- Water vapor
- Nitrogen
- Excess air

Carbon dioxide and water vapor are the products of complete combustion. Nitrogen remains after oxygen in the combustion air is consumed by the fuel. Excess air is supplied to the fuel-burning device to guard against the possibility of producing incomplete combustion. Normally, furnaces are adjusted to use 5 to 50% excess air. The effect of the amount of excess air on the CO_2 in the flue gases is shown in Figure 3.5.

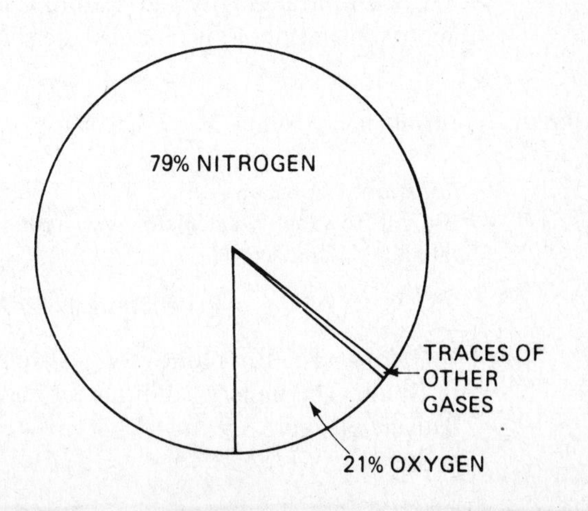

Figure 3.4 Chemical composition of air.

APPROXIMATE MAXIMUM CO_2 VALUES FOR VARIOUS FUELS WITH DIFFERENT PERCENTAGES OF EXCESS AIR

TYPE OF FUEL	MAXIMUM THEORETICAL CO_2 PERCENT	PERCENT CO_2 AT GIVEN EXCESS AIR VALUES		
		20%	40%	60%
GASEOUS FUELS				
NATURAL GAS	12.1	9.9	8.4	7.3
PROPANE GAS (COMMERCIAL)	13.9	11.4	9.6	8.4
BUTANE GAS (COMMERCIAL)	14.1	11.6	9.8	8.5
MIXED GAS (NATURAL AND CARBURETED WATER GAS)	11.2	12.5	10.5	9.1
CARBURETED WATER GAS	17.2	14.2	12.1	10.6
COKE OVEN GAS	11.2	9.2	7.8	6.8
LIQUID FUELS				
NO. 1 AND 2 FUEL OIL	15.0	12.3	10.5	9.1
NO. 6 FUEL OIL	16.5	13.6	11.6	10.1
SOLID FUELS				
BITUMINOUS COAL	18.2	15.1	12.9	11.3
ANTHRACITE	20.2	16.8	14.4	12.6
COKE	21.0	17.5	15.0	13.0

Figure 3.5 Effect of excess air on the CO_2 in the flue gases. (By permission from the *ASHRAE Handbook*)

Heating Values

Each fuel, when burned, is capable of producing a given amount of heat, depending on the constituents of the fuel. This information is useful in determining the heating capacity of a furnace. If the heating value of a unit of fuel produces a given amount of Btu, and the number of units of fuel burned per hour is known, the input rating of the furnace can be calculated. The Btu ratings for units of natural gas are shown in Figure 3.6. The typical gravity and heating value of fuel oil are shown in Figure 3.7, and the approximate Btu ratings for coal are shown in Figure 3.8.

Example 3–1 An oil furnace burning No. 2 fuel oil uses 0.75 gal/h. What is the input rating of the furnace?

Solution The heating value of No. 2 oil from Figure 3.7 is between 141,800 and 137,000 Btu/gal. To simplify calculations, we round the rating to 140,000 Btu/gal. Therefore, the input rating of the furnace is

$$140{,}000 \text{ Btu/gal} \times 0.75 \text{ gal/h} = 105{,}000 \text{ Btuh input}$$ ■

Note: The Btu ratings for gas are given in Btu/ft^3, for oil in Btu/gal, and for coal in Btu/lb. The units used differ for each fuel because of the form in which they are delivered (Figure 3.9).

NO.	CITY	HEAT VALUE, BTU/CU FT	SPECIFIC GRAVITY
1	ABILENE, TEX.	1121	0.710
2	AKRON, OHIO	1037	0.600
3	ALBUQUERQUE, N.M.	1120	0.646
4	ATLANTA, GA.	1031	0.604
5	BALTIMORE, MD.	1051	0.590
6	BIRMINGHAM, ALA.	1024	0.599
7	BOSTON, MASS.	1057	0.604
8	BROOKLYN, N.Y.	1049	0.595
9	BUTTE, MONT.	1000	0.610
10	CANTON, OHIO	1037	0.600
11	CHEYENNE, WYO.	1060	0.610
12	CINCINNATI, OHIO	1031	0.591
13	CLEVELAND, OHIO	1037	0.600
14	COLUMBUS, OHIO	1028	0.597
15	DALLAS, TEX.	1093	0.641
16	DENVER, COLO.	1011	0.659
17	DES MOINES, IOWA	1012	0.669
18	DETROIT, MICH.	1016	0.616
19	EL PASO, TEX.	1082	0.630
20	FT. WORTH, TEX.	1115	0.649
21	HOUSTON, TEX.	1031	0.623
22	KANSAS CITY, MO.	945	0.695
23	LITTLE ROCK, ARK.	1035	0.590
24	LOS ANGELES, CALIF.	1084	0.638
25	LOUISVILLE, KY.	1034	0.506
26	MEMPHIS, TENN.	1044	0.608
27	MILWAUKEE, WIS.	1051	0.627
28	NEW ORLEANS, LA.	1072	0.612
29	NEW YORK CITY	1049	0.595
30	OKLAHOMA CITY, OKLA.	1080	0.615
31	OMAHA, NEB.	1020	0.669
32	PARKERSBURG, W. VA.	1049	0.592
33	PHOENIX, ARIZ.	1071	0.633
34	PITTSBURGH, PA.	1051	0.595
35	PROVIDENCE, R.I.	1057	0.601
36	PROVO, UTAH	1032	0.605
37	PUEBLO, COLO.	980	0.706
38	RAPID CITY, S.D.	1077	0.607
39	ST. LOUIS, MO.	–	–
40	SALT LAKE CITY, UTAH	1082	0.614
41	SAN DIEGO, CALIF.	1079	0.643
42	SAN FRANCISCO, CALIF.	1086	0.624
43	TOLEDO, OHIO	1028	0.597
44	TULSA, OKLA.	1086	0.630
45	WACO, TEX.	1042	0.607
46	WASHINGTON, D.C.	1042	0.586
47	WICHITA, KAN.	1051	0.690
48	YOUNGSTOWN, OHIO	1037	0.600

***Average analyses obtained from the operating utility company supplying the city; the supply may vary considerably from these data – especially where more than one pipeline supplies the city. Also, as new supplies may be received from other sources, the analyses may change.**

Figure 3.6 Btu ratings for units of natural gas. (By permission from the *ASHRAE Handbook*)

GRADE NO.	GRAVITY, API	WEIGHT, LB PER GALLON	HEATING VALUE, BTU PER GALLON
1	38-45	6.95 -6.675	137,000-132,900
2	30-38	7.296-6.960	141,800-137,000
4	20-28	7.787-7.396	148,100-143,100
5L	17-22	7.94 -7.686	150,000-146,800
5H	14-18	8.08 -7.89	152,000-149,400
6	8-15	8.448-8.053	155,900-151,300

(API AMERICAN PETROLEUM INSTITUTE RATING)

Figure 3.7 Typical gravity and heating values for standard grades of fuel oil. (By permission from the *ASHRAE Handbook*)

Losses and Efficiencies

When fuel is burned in a furnace a certain amount of heat is lost in the hot gases that rise through the chimney. Although this function is necessary to dispose of the products of combustion, the loss should be minimized to allow the furnace to operate at its highest efficiency. Air entering the furnace at room temperature, or lower, is heated to fuel-gas temperatures. These temperatures range from 350 to 600°F, depending on the design of the furnace and its adjustment by a service technician.

If the amount of heat lost is 20%, the efficiency of the furnace is 80%. The calculation of the combustion efficiency is based on knowing the temperature and the carbon dioxide content of the flue gases. Knowing the efficiency of the furnace makes it possible for a heating service technician to calculate the output of the furnace. The following formula is used:

$$\text{Btuh input} \times \%\ \text{efficiency} = \text{Btuh output}$$

Example 3–2 The input of an oil furnace is 105,000 Btuh. Its efficiency is 80%. What is its output?

Solution

$$105{,}000\ \text{Btuh} \times 0.80\ \text{efficiency} = 84{,}000\ \text{Btuh output}$$

■

RANK	BTU PER LB AS RECEIVED
ANTHRACITE	12,700
SEMIANTHRACITE	13,600
LOW-VOLATILE BITUMINOUS	14,350
MEDIUM-VOLATILE BITUMINOUS	14,000
HIGH-VOLATILE BITUMINOUS A	13,800
HIGH-VOLATILE BITUMINOUS B	12,500
HIGH-VOLATILE BITUMINOUS C	11,000
SUBBITUMINOUS B	9,000
SUBBITUMINOUS C	8,500
LIGNITE	6,900

Figure 3.8 Approximate Btu ratings for various types of coal. (By permission from the *ASHRAE Handbook*)

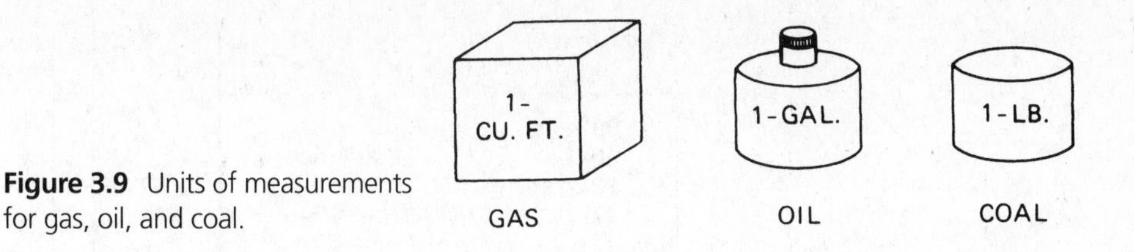

Figure 3.9 Units of measurements for gas, oil, and coal.

Types of Flames

Basically, there are two types of flames: yellow and blue (Figure 3.10). Pressure-type oil burners burn with a yellow flame. Modern Bunsen-type gas burners burn with a blue flame. The difference is due mainly to the manner in which air is mixed with the fuel. A yellow flame is produced when gas is burned by igniting fuel gushing from an open end of a gas pipe, such as may be seen in fixtures used for ornamental decoration. A blue flame is produced when approximately 50% of the air requirement is mixed with the gas prior to ignition. This is called *primary air.* A Bunsen burner uses this arrangement. The balance of air, called *secondary air,* is supplied during combustion to the exterior of the flame. Air adjustments are discussed in Chapter 5.

Improper gas flames are the result of inefficient or incomplete combustion and can be caused by:

- Excess supply of primary air
- Lack of secondary air
- Impingement of the flame on a cool surface (Figure 3.11)

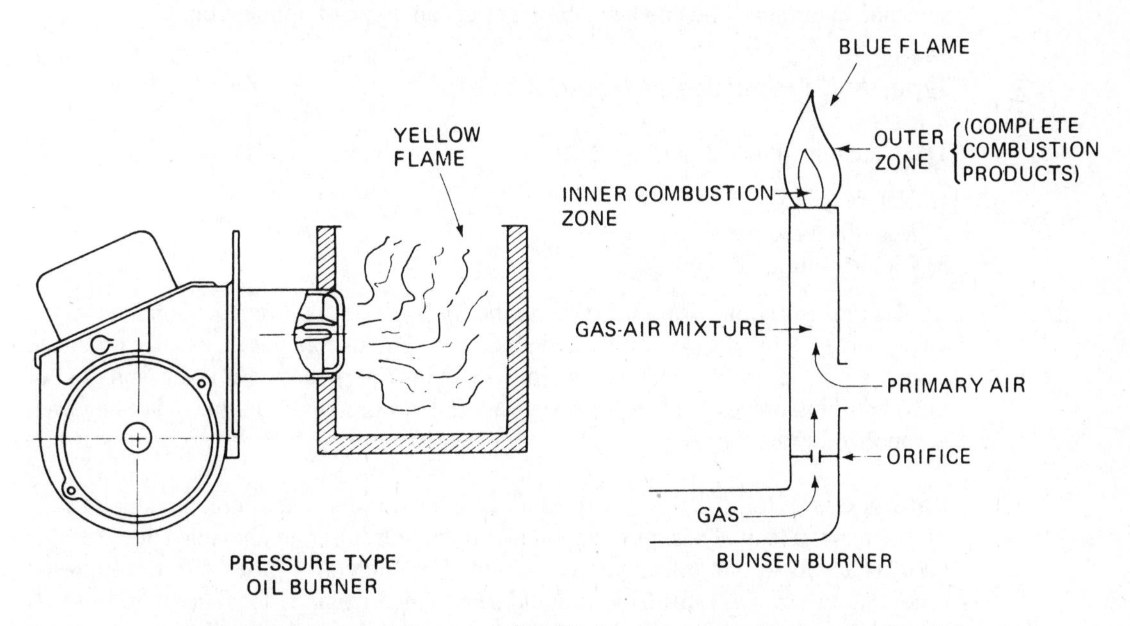

Figure 3.10 Types of flames produced by pressure-type oil burner and Bunsen burner. [Courtesy of Robertshaw Control Company (Bunsen Burner)]

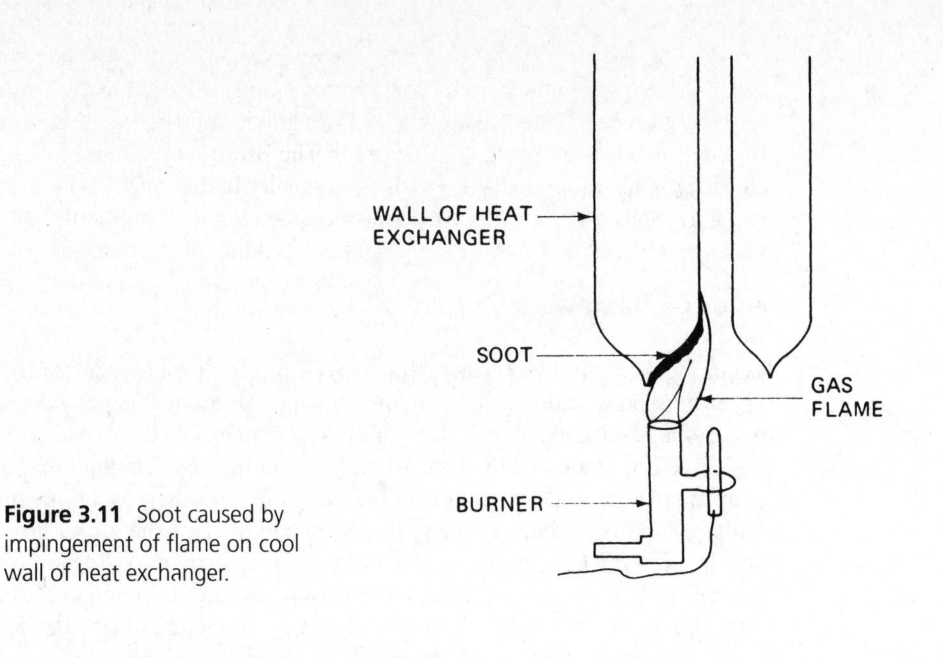

Figure 3.11 Soot caused by impingement of flame on cool wall of heat exchanger.

Fuels

Fuels are available in three forms: gases, liquids, and solids. Gases include natural gas and liquid petroleum; fuel oils are rated as grades 1, 2, 4, 5, and 6; and coals are of various types, mainly anthracite and bituminous. Each fuel has its own individual Btu heat content per unit and its own desirable or undesirable characteristics. The fuel selection is usually based on availability, price, and type of application.

Types and Properties of Gaseous Fuels

There are three types of gaseous fuels:

1. Natural gas
2. Manufactured gas
3. Liquid petroleum (LP)

Natural gas comes from the earth in the form of gas and often accumulates in the upper part of oil wells. Manufactured gases are combustible gases, usually produced from solid or liquid fuel and used mainly for industrial process. LP is a by-product of the oil refining process. It is so named because it is stored in liquid form. LP, however, is vaporized when burned.

Natural Gas Natural gas is nearly odorless and colorless. Therefore, an odorant such as a mercaptan (any of various compounds containing sulfur and having a disagreeable odor) is added so that a leak can be sensed. The content of gases differs somewhat according to locality. Information should be obtained from the local gas company relative to the specific gravity and the Btu/ft^3 content of the gas available.

The specific gravity affects piping sizes. *Specific gravity* is the ratio of the weight of a given volume of substance to an equal volume of air or water at a given temperature and pressure. Thus gas with a specific gravity of 0.60 weighs six-tenths, or three-fifths, as much as an equal volume of air. The Btu/ft^3 content of gas varies from 900 to 1200, depending on the locality, but it is usually in the range 1000 to 1050 Btu/ft^3.

The chief constituent of natural gas is methane. Commonly called *marsh gas,* methane is a gaseous hydrocarbon that is a product of decomposition of organic matter in marshes or mines or of the carbonization of coal. Natural gas comprises 55 to 95% methane and the remainder, other hydrocarbon gas.

Manufactured Gas Manufactured gas is produced from coal, oil, and other hydrocarbons. It is comparatively low in Btu/ft^3, usually in the range of 500 to 600. It is not considered an economical space-heating fuel.

Liquid Petroleum There are two types of liquid petroleum: propane and butane. Propane is more useful as a space-heating fuel, since it boils at −40°F and therefore can readily be vaporized for heating in a northern climate. Butane boils at about 32°F.

Propane has a heating value of 21,560 Btu/lb or about 2500 Btu/ft^3. Butane has a heating value of 21,180 Btu/lb or about 3200 Btu/ft^3. When LP gas is used as a heating fuel, the equipment must be designed to use this type of gas. When ordering equipment, the purchaser must indicate which type of gaseous fuel is being used. Conversion kits are available for converting natural gas furnaces to LP gas when necessary. It is important to follow the manufacturer's instructions with great care.

Note: Propane and butane vapors are generally considered more dangerous than those of natural gas, since they have higher specific gravities (propane 1.52, butane 2.01). Because the vapor is heavier than air it tends to accumulate near the floor, thereby increasing the danger of an explosion upon ignition.

Types and Properties of Fuel Oils

Fuel oils are rated according to their Btu/gal content and API gravity (shown in Figure 3.7). The API gravity is an index selected by the American Petroleum Institute. There are six grades of oil: Nos. 1, 2, 4, 5 (light), 5 (heavy), and 6. Note that the lighter-weight oils have a higher API gravity.

Grade No. 1: light-grade distillate prepared for vaporizing-type oil burners.

Grade No. 2: heavier distillate than grade No. 1. It is manufactured for domestic pressure-type oil burners.

Grade No. 4: light residue or heavy distillate. It is produced for commercial oil burners using a higher pressure than domestic burners.

Grade No. 5 (light): residual-type fuel of medium weight. It is used for commercial burners that are specially designed for its use.

Grade No. 5 (heavy): residual-type fuel for commercial oil burners. It usually requires preheating.

Grade No. 6: also called *Bunker C;* a heavy residue used for commercial burners. It requires preheating in the tank to permit pumping and additional preheating at the burner to permit atomization (breaking up into fine particles).

Types and Properties of Coals

There are four different types of coal: anthracite, bituminous, subbituminous, and lignite. Coal is constituted principally of carbon, with the better grades having as much as 80% carbon. The types vary not only in Btu/lb but also in their burning and handling qualities.

Anthracite Anthracite is a clean, hard coal. It burns with an almost smokeless short flame. It is difficult to ignite but burns freely when started. It is noncaking, leaving a fine ash that does not clog grates and ash removal equipment.

Bituminous Bituminous coals include a wide range of coals, varying from high grade in the East to low grade in the West. It is more brittle than anthracite coal and readily breaks up into small pieces for grading and screening. The length of the flame is long, but varies for different grades. Unless burning is carefully controlled, much smoke and soot can result.

Subbituminous This coal has a high moisture content and tends to break up when dry. It can ignite spontaneously when stored. It ignites easily and burns with a medium flame. It is desirable for its noncaking characteristic and for the fact that it forms little soot and smoke.

Lignite Lignite coal has a woody consistency, is high in moisture content and low in heating value, and is clean to handle. Lignite has a greater tendency to break up when dry than subbituminous coal. Because of its high moisture content, it is difficult to ignite. It is noncaking and forms little smoke and soot.

Efficiency of Operation

Combustion is a chemical reaction resulting in the production of a flame. The fuels used for heating consist chiefly of carbon. In combustion, carbon and oxygen are combined to produce heat. Instruments are used to measure how well the equipment performs the process of combustion. The heating service technician makes whatever adjustments are necessary to produce the most efficient operation. The products of combustion that leave the fuel-burning unit through the flue are:

- Carbon dioxide (CO_2)
- Carbon monoxide (CO)
- Oxygen (O_2)

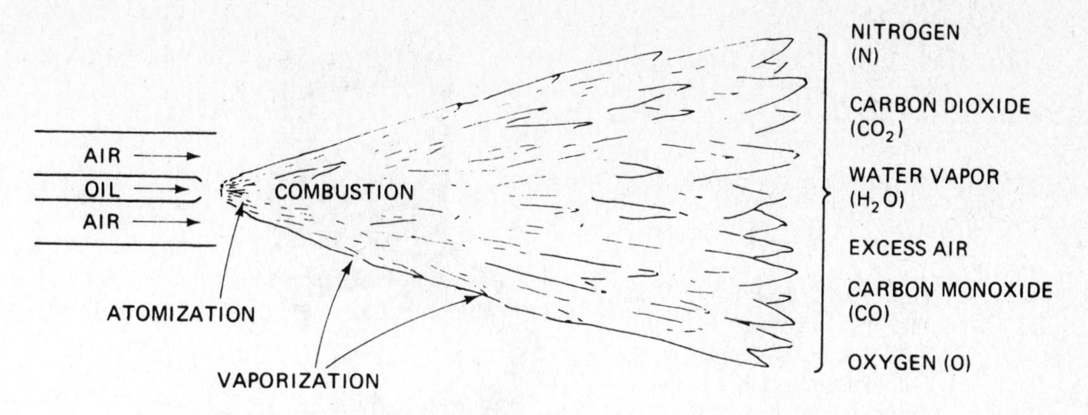

Figure 3.12 Products of combustion.

- Nitrogen (N)
- Water vapor (H_2O)

An outline of the products of combustion is shown in Figure 3.12.

The amount of CO_2 measured in the flue gases is an indication of the amount of air used by the fuel-burning equipment. It is desirable to have a high CO_2 measurement, as this indicates a hot fire. The maximum CO_2 measurement for oil with no excess air is 15.6%; for natural gas, 11.8%. However, it is not practical, with field-installed equipment, to reach these high CO_2 readings. In practice, oil furnaces should have a CO_2 reading of between 10 and 12%; natural gas furnaces should read between 8¼ and 9½%. The CO_2 in the flue gases is the indicator used (along with stack temperature) to measure the efficiency of combustion.

Instruments Used in Testing

Instruments used for combustion testing are available from a number of manufacturers. Although the illustrations for the instruments described here have been provided by specific companies, instruments that perform similar functions can be obtained from various other manufacturers. The instruments and their functions are:

- *Draft gauge:* used for measuring draft (Figure 3.13)
- *Smoke tester:* used for determining smoke scale (Figure 3.14)
- *Flue gas analyzer:* used for testing CO_2 content (Figure 3.15)
- *Stack thermometer:* used for determining stack temperature (Figure 3.16)

Complete combustion kits (Figure 3.17) are available with many of the preceding items. The location of the various parts of the heating plant equipment, and sample holes for testing, are shown in Figure 3.18.

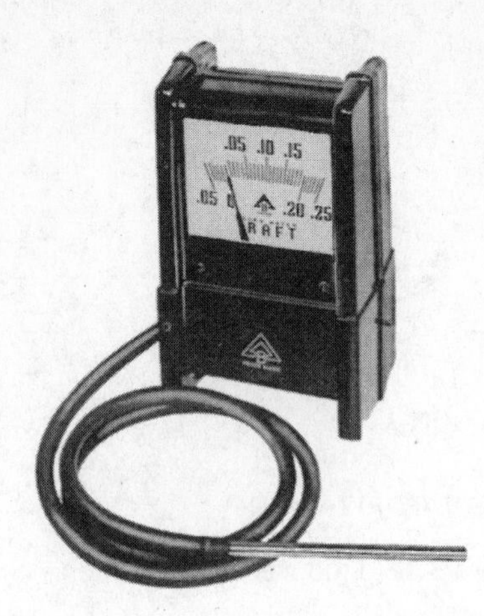

Figure 3.13 Draft gauge model MZF, dry type, range +0.05 to −0.25 in. water, supplied with 5-in. draft tube and 9 ft. of rubber tubing. (Courtesy of Bacharach, Inc.)

Testing Procedures

Tests should be made in the following order:

1. Draft measurement
2. Smoke sample
3. CO_2 content
4. Stack temperature

The equipment should be run for a minimum of 5 to 10 minutes before testing to stabilize operating conditions. A record should be kept both before adjustments are made and after the tests.

Draft Test

The draft gauge reads in inches of water column (W.C.). For example, a reading of 0.02 in. W.C. means two-hundredths of an inch of water column. Always calibrate the instrument before using it. Place the instrument on a level surface and adjust the indicator needle to read zero.

Drill a ¼-in. hole in the firebox door (Figure 3.18) and two ¼-in. holes in the flue at the point just after it leaves the heat exchanger. The holes in the flue must be placed between the heat exchanger and the draft regulator on an oil furnace and between the heat exchanger and the draft diverter on a gas furnace. One of the two holes in the flue is used for the stack thermometer, the other is used for the CO_2 and smoke tests. Occasionally, an additional ¼-in. hole is drilled in the flue pipe near the chimney to measure chimney draft. This is normally used only when it is necessary to troubleshoot draft problems.

The draft should always be negative when measured with the draft gauge, since the flue gases are moving away from the furnace. The draft over the fire, measured at the firebox door, should be −0.01 to −0.02 in. W.C. (Figure 3.19) The difference between the draft reading in the flue at the furnace outlet and the draft-over-the-fire reading indicates the heat-exchanger leakage.

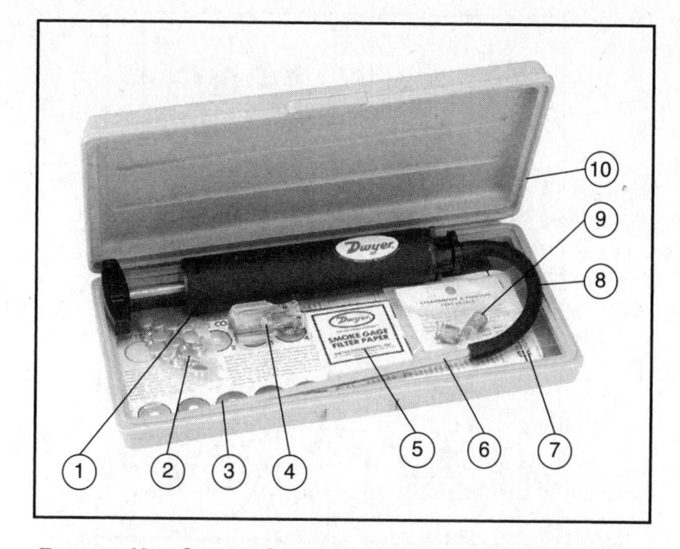

Easy-to-Use Smoke Gage Kit enables you to quickly balance maximum CO_2 with minimum smoke for clean, efficient combustion. Simply take an actual sample of the smoke being tested and compare it to the standards on the Smoke Chart included. Just 10 full strokes of the pump produces an accurate test sample.

Popular with professional heating engineers, the No. 920 Smoke Gage Kit includes these quality components:

(1). Dwyer Smoke Gage Pump – Fast working; the first known unit to conform with ASTM D-2156-XX standards for testing smoke density in flue gases from burning distillate fuels. Filter paper clamps instantly into pump inlet.
(2). Hole Plugs – Packet of 20 to fill awl holes.
(3). Smoke Chart – For easy comparison with samples taken. Laminated plastic, wipes clean.
(4). Awl-Pierces smoke pipe for thermometer and sampling tubes. Large, comfortable plastic handle.
(5). Smoke Gage Filter Paper – An exclusive, time-saving Dwyer development. Roll of filter paper is contained in dispenser box, kept clean and convenient.
(6). Metal Terminal Tube – 8½" long, heavy gauge brass.
(7). How-to-Use Instructions – Simple with step-by-step illustrations.
(8). Rubber Tubing – Flexible, long-lived.
(9). Spring Holders – A superior design for holding thermometer and sampling tubes.
(10). Kit Case – Tough, durable, one piece high density polyethylene with living hinge and clasp. Foam liner protects contents.

Figure 3.14 Smoke gauge kit. (Courtesy of Dwyer Instruments, Inc.)

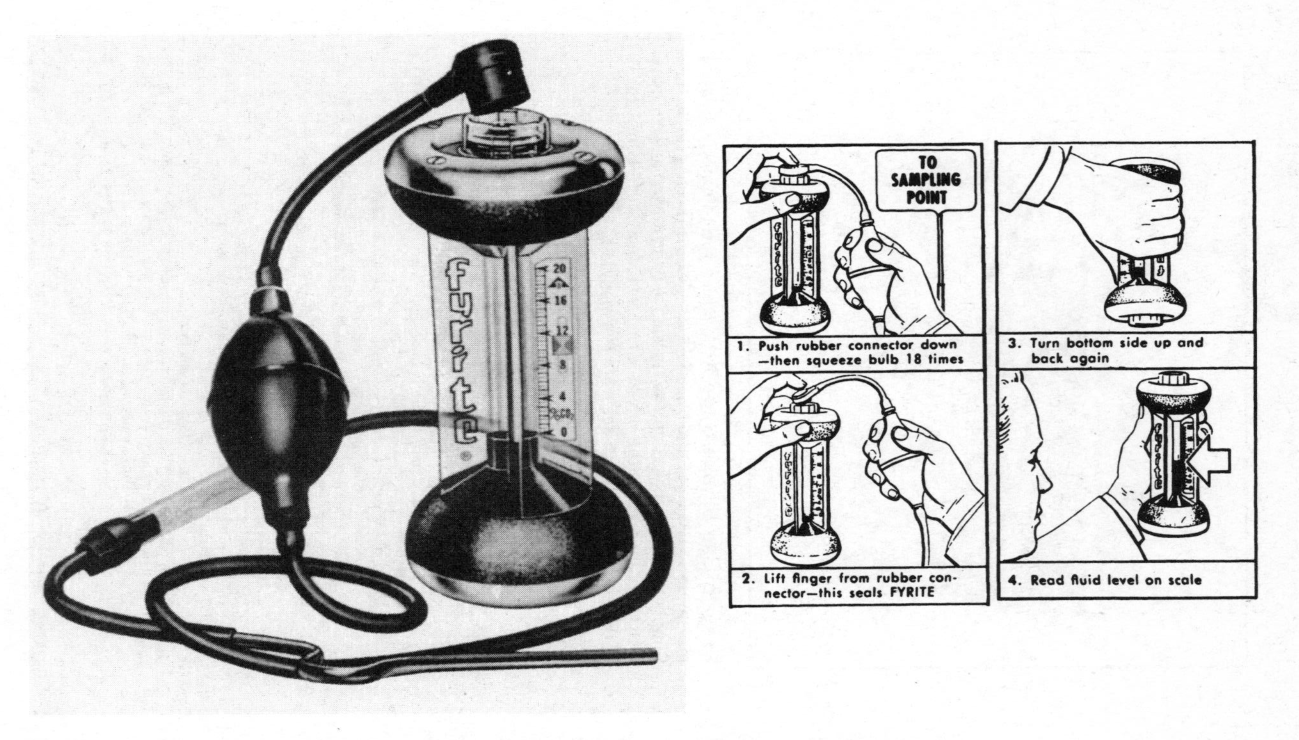

Figure 3.15 Fyrite gas analyzers, available for measuring carbon dioxide or oxygen, can be exposed to temperatures from −30 to 150°F, and gases up to 850°F may be tested. Kit 10-5001 CO_2 (range 0 to 20%) and Kit 10-5012 O_2 (range 0 to 21%). Case is included. (Courtesy of Bacharach, Inc.)

Figure 3.16 Tempoint dial thermometer, bimetal type, accurate from −40 to +1000°F. Designed to read stack temperatures; recalibration possible. (Courtesy of Bacharach, Inc.)

Quick and Easy-to-Use Dwyer Combustion Test Kit.

Measure all four factors that govern overall efficiency: CO_2 content, draft, stack temperature and smoke. Accuracy to ±½%CO_2 fully meets the requirements of the professional Heating Engineer. Yet an apprentice can follow the simple, illustrated step-by-step instructions.

Add a small pitot tube, which will fit into the kit case, for air velocity measurements.

STOCKED MODEL

No. 1200-A Combustion Test Kit....

(1). New easy-to-carry, impact resistant polyethylene Carrying Case with handle.
(2). No. 1101 CO_2 Indicator – Clear acrylic plastic, virtually unbreakable.
(3). Aspirator Bulb, Filter and Tubing Assembly.
(4). Filter Wool – Extra supply in bag for about 1000 tests.
(5). No. 171 Draft Gage – Continuously indicating, permanently accurate. 0 to .25" W.C. range in .01 divisions.
(6). Dwyer Smoke Gage Pump – Fast-working, the first known unit to conform with ASTM D-2156-XX standards for testing smoke density in flue gases.
(7). Smoke Filter Paper – A new, exclusive Dwyer development.
(8). Magnehelic® Gage – 0-.25" W.C. and 300 to 2000 FPM. Included in Model 1200-B kit below.
(9). Awl – Pierces smoke pipe for thermometer and sampling tubes.
(10). CO_2 Absorbent Solution – Extra supply for many tests.
(11). No. A-503 Stack Thermometer – Dial type, stainless steel 200° to 1000° F".
(12). Gage Oil – Long-lasting extra supply.
(13). Spring Holders – For thermometer and terminal tubes.
(14). Hole Plugs – Packet to fill awl holes.
(15). Smoke Chart – Laminated plastic, for comparison with samples taken.
(16). Dwyer Combustion Efficiency Slide Rule Computer – Simple to operate. (17). Operating Instructions – Simple, illustrated. Show step by-step use.
(18). Combustion Data Cards and Parts List – Plastic. weight only 6 lbs. 8 oz.

STOCKED MODEL

No. 1200-B Combustion Test Kit....

Includes portable Dwyer Magnehelic® Draft Gage with large 4" dial instead of No. 171 Inclined Draft Gage, 0 to .25" W.C. range in .01" divisions. Also has air velocity scale with 300 to 2000 F.P.M, range. No leveling necessary; no gage oil required, Kit is otherwise identical to No. 1200-A, Weight only 7 lbs. 5 oz.

OPTIONAL: On special order, the portable Dwyer No. 460 Air Meter may be substituted for the No. 171 Inclined Draft Gage.

Figure 3.17 Combustion test kit. (Courtesy of Dwyer Instruments, Inc.)

Testing Procedures

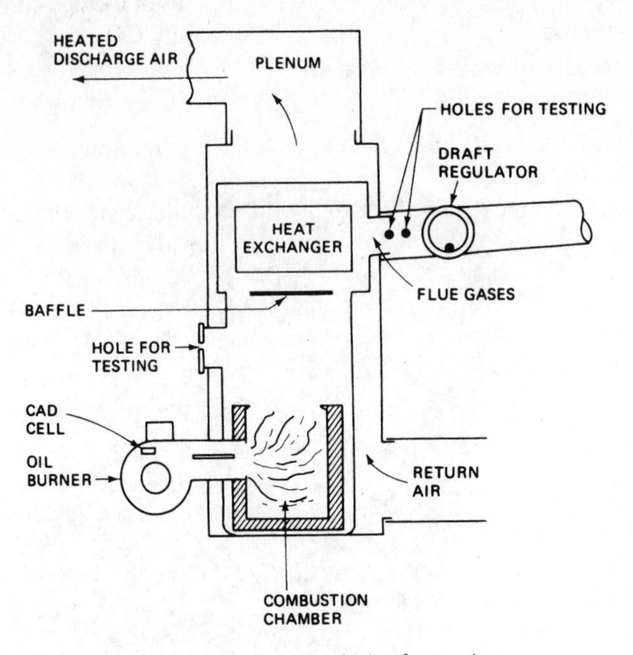

Figure 3.18 Heating plant equipment, showing holes for testing.

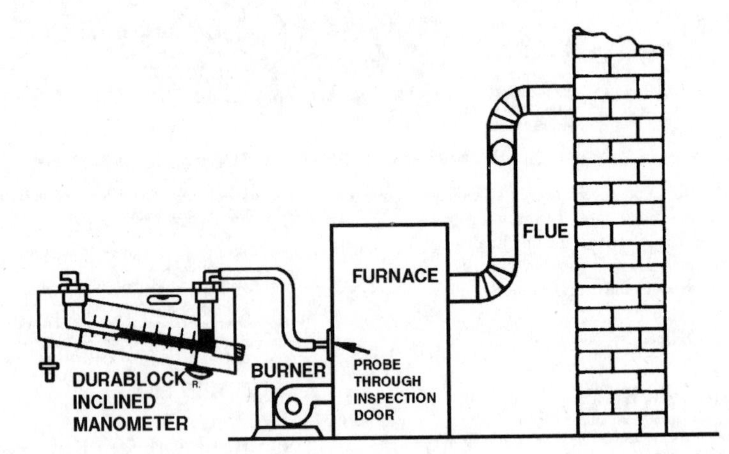

Figure 3.19 Using a draft gauge to adjust for efficient combustion. (Courtesy of Dwyer Instruments, Inc.)

Carbon Dioxide Test

An analyzer for CO_2 uses potassium hydroxide (KOH), a chemical that has the property of being able to absorb large quantities of CO_2. A known volume of flue gas is run into the tester. Since KOH will absorb only CO_2, the reduction in volume of the flue gas is an indication of the amount of CO_2 absorbed by the KOH solution. Instructions for using the tester shown in Figure 3.15 are as follows:

1. Set the instrument to zero by adjusting the sliding scale to the level of the fluid in the tube.
2. Insert the sample tube in the flue opening. Place the rubber connector on top of the instrument and depress to open the valve. Collapse the bulb 18 times to fill the instrument with flue gas and then release the valve.
3. Tip the instrument over and back twice and hold at a 45° angle for 5 seconds.
4. Hold the instrument upright and read the CO_2 content on the scale. Release the pressure by opening the valve when the test is complete.

Stack Temperature Test

A stack thermometer is used to determine stack temperature. Insert the thermometer in the flue hole as shown in Figure 3.18. Operate the burner unit until the temperature rise, as read on the thermometer, is no more than 3°F/min (indicating stability); then, read the final stack temperature.

Smoke Test

The burner flame should not produce excessive smoke. Smoke causes soot (carbon) to collect on the surfaces of the heat exchanger, reducing its heat-transfer rate. A soot deposit of ⅛ in. can cause a reduction of 10% in the rate of heat absorption. Smoke is measured with a smoke tester. A sample is collected by pumping 2200 cm^3 (10 full strokes with the sampling pump) through a 0.38-cm^2 filter paper. The color of the sample filter paper spot is compared with a standard graduated smoke scale (Figure 3.14). Spot No. 0 is white and indicates no smoke. Spot No. 9 is darkest and represents the most extreme smoke condition. Between 0 and 9 the scale has 10 different grades. It is most desirable to produce smoke at a scale of No. 1 or 2 but not 0. Older-style or conversion units could read at high as No. 4. No. 0 indicates excess air supply to the burner and low efficiency. Readings higher than No. 2 indicate the production of excessive carbon and soot. Normally, this test is performed only on oil-burning equipment.

Combustion Efficiency

Combustion efficiency, expressed in percentage, is a measure of the useful heat produced in the fuel compared with the amount of fuel available. A furnace that operates at 80% efficiency is therefore one that loses 20% of the fuel value in the heat that

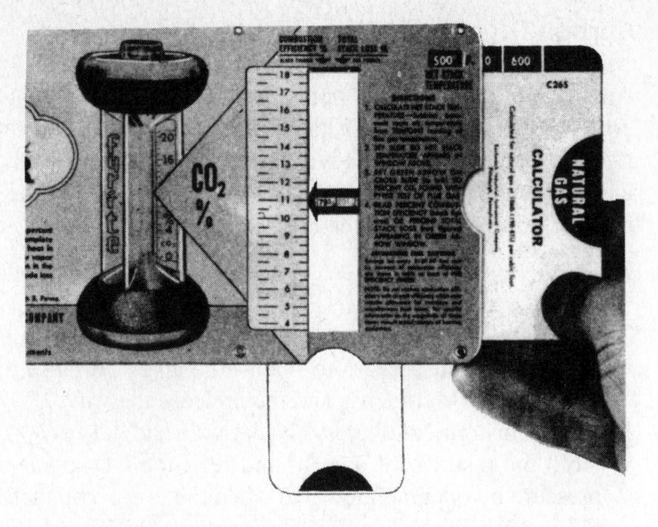

Figure 3.20 Combustion efficiency slide rule. (Courtesy of Bacharach, Inc.)

goes out through the chimney. High-efficiency furnaces have an operating efficiency of up to 96%.

A combustion efficiency slide rule is a useful tool (Figure 3.20). After determining the stack (flue) temperature and CO_2 content, and by using the appropriate slide on the rule, a technician can determine the combustion efficiency. Note that combustion efficiency is based on net stack temperature. Net stack temperature is the reading taken on the stack thermometer minus the room temperature (the temperature of the air entering the furnace). For example, assume that the net stack temperature for an oil burner is found to be 500°F, and the CO_2 content is determined as 9%. Using the slide rule, move the large slide to the right so that 500 appears in the small window at the upper right marked "net stack temperature." Then, move the small vertical slide until the arrow points to reading 9 on the CO_2 scale. Through the window in the arrow, read the figures 80 (black) and 20 (red). This means that the combustion efficiency is 80%, and the stack loss is 20%. Figure 3.21 shows, for example, how increasing the combustion efficiency from 80% to 85% would save $5.90 of every $100 of fuel cost.

Analysis And Adjustments

Usually, the tests are made after routine maintenance has been performed. The combustion tests indicate the condition of the equipment and point to further service or adjustments that may be necessary.

Oil-Burning Equipment

On oil-burning equipment, adjustments are made in the air supply to the burner. Air is reduced to maintain a smoke test within the No. 1 to No. 2 limits while providing the highest possible CO_2 content in the flue gases. If an efficiency rating of 75% or better

From An Original Efficiency of	To An Increased Combustion Efficiency of:							
	55%	60%	65%	70%	75%	80%	(85%)	90%
50%	9.10	16.70	23.10	28.60	33.30	37.50	41.20	44.40
55%	–	8.30	15.40	21.50	26.70	31.20	35.30	38.90
60%	–	–	7.70	14.30	20.00	25.00	29.40	33.30
65%	–	–	–	7.10	13.80	18.80	23.50	27.80
70%	–	–	–	–	6.70	12.50	17.60	22.20
75%	–	–	–	–	–	6.30	11.80	16.70
(80%)	–	–	–	–	–	–	(5.90)	11.10
85%	–	–	–	–	–	–	–	5.60

Figure 3.21 Fuel saved by increasing efficiency. (Courtesy of R. W. Beckett Corporation.)

cannot be obtained as a result of adjustments in the air supply, further service on the burner is required. Listed next are some common causes of low CO_2 content and smoky fires.

1. Incorrect air supply
2. Combustion chamber air leaks
3. Improper operation of draft regulator
4. Insufficient draft
5. Burner "on" periods too short
6. Oil does not conform to burner requirements
7. Air-handling parts defective or incorrectly adjusted
8. Firebox cracked
9. Spray angle of nozzle unsuitable
10. Nozzle worn, clogged, or incorrect type
11. Gallons/hour rate too high for size of combustion chamber
12. Ignition delay due to defective stack control
13. Nozzle spray or capacity unsuited to type of burner
14. Oil pressure to nozzle improperly adjusted
15. Nozzle loose or not centered
16. Electrodes dirty, loose, or incorrectly set
17. Leaking cutoff valve
18. Rotary burner motor running under speed

Some conditions that produce poor efficiency, such as those indicated next, can be determined by visual inspection.

1. *Check burner shutdown.* A flame should last no longer than 2 seconds after the burner shuts down.

2. *Check the flame.* It should be symmetrically shaped and centered in the combustion chamber. It should not strike the walls or floor of the combustion chamber.
3. *Check for air leaks.* There should be no leakage around the burner tube where it enters the combustion chamber.
4. *Check the burner operating period.* Burning periods of less than 5 minutes' duration usually do not produce efficient operation.

Gas-Burning Equipment

On gas-burning equipment, the primary air is reduced by adjusting the air shutter to provide a blue flame. The secondary air is a fixed quantity on most furnaces. Inspection should be made, however, to determine if the secondary air restrictor has become loose or perhaps been removed. A correction should be made. If the flame does not adjust properly to produce a blue flame, and an efficiency rating of 75% or better cannot be obtained, further service is required.

Some causes of poor efficiency are:

1. *Unclean burner ports.* To clean the burner ports, remove the lint and dirt.
2. *Insufficient gas pressure.* The draft diverter should be checked with the burner in operation. Air should be moving into the opening, thus indicating a negative pressure at the inlet. This condition can be verified by holding a lighted match at the opening.
3. *Improper gas pressure.* The gas pressure should be 3½ in. W.C. at the manifold for natural gas. Check the gas meter for input into the furnace. It should read the same as the input indicated on the furnace nameplate. If the gas pressure reads between 3 and 4 in. W.C. and still does not produce the rated input, the orifice should be cleaned or replaced with one of proper size. For LP the gas pressure should be 11 in. W.C. at the manifold on the furnace.
4. *Improper "on" and "off" periods of the furnace.* "On" periods of less than 5 minutes are too brief to produce high efficiency.

STUDY QUESTIONS

Answers to the study questions may be found in the sections noted in brackets.

3-1 What type of process is combustion? Describe the action that takes place. *[Combustion]*

3-2 What are the three conditions necessary for combustion? *[Combustion]*

3-3 What action takes place during complete combustion? What elements are combined and what products are produced? *[Types of Combustion]*

3-4 What is the cause of incomplete combustion? What are the products formed? *[Types of Combustion]*

3-5 What is the percentage of oxygen in the air? *[Types of Combustion]*

3-6 What is the range of heating values for natural gas, fuel oil, and coal? *[Heating Values]*

3-7 What is the maximum efficiency for natural gas and oil fuels? *[Losses and Efficiencies]*

3-8 What is the proper color for an oil flame? a gas flame? *[Types of Flames]*

3-9 Should the draft over the fire for an oil burner be positive, negative, or neutral? *[Draft Test]*

3-10 What is the proper draft measurement over the fire for an oil furnace with a draft regulator? *[Draft Test]*

3-11 In testing for CO_2 in the flue gas with a potassium hydroxide analyzer, how many times should the bulb be collapsed? *[Carbon Dioxide Test]*

3-12 What should be the combustion efficiency on a standard oil-burning furnace? *[Combustion Efficiency]*

3-13 What dollar savings for every $100 of fuel cost would be realized in increasing the efficiency from 60% to 80%? *[Combustion Efficiency]*

3-14 Name 10 conditions that can cause an oil burner to produce low CO_2 and a smoky fire. *[Oil-Burning Equipment]*

3-15 Name three conditions that can cause a gas burner to operate at low efficiency. *[Gas-Burning Equipment]*

CHAPTER 4

Parts Common to All Furnaces

OBJECTIVES

After studying this chapter, the student will be able to:

- Identify the various components of forced warm air heating furnaces
- Adjust the speed of a furnace fan
- Provide proper service and maintenance for common parts of a heating system

How a Furnace Operates

A warm-air furnace is a device for providing space heating. Fuel, a form of energy, is converted into heat and distributed to various parts of a structure. Fuel-burning furnaces provide heat by combustion of the fuel within a heat exchanger. Air is passed over the outside surface of the heat exchanger, which transfers the heat from the fuel to the air. In a fuel-burning furnace, the products of combustion are exhausted to the atmosphere through the flue passages connected to the heat exchanger.

Because no products of combustion are formed in an electric furnace, air passes directly over the electrically heated elements without the use of a heat exchanger. A forced-air furnace uses a fan to propel the air over the heat exchanger and to circulate the air through the distribution system. A furnace is considered a residential type when its input is less than 250,000 Btuh.

Basic Components

A forced warm air furnace has six basic components:

1. Heat exchanger
2. Fuel-burning device
3. Cabinet or enclosure

4. Fan and motor
5. Air filters
6. Controls

The fuel-burning device is discussed in Chapters 5 and 13. The controls are discussed in Chapters 10, 11, and 14.

Arrangement of Components

Four different designs of forced air furnaces are in common use. Each requires a different arrangement of the basic components:

1. Horizontal
2. Upflow or highboy
3. Downflow or counterflow
4. Multiposition Furnace

The *horizontal* furnace is used in attic spaces or other locations where the height of the furnace must be kept as low as possible (Figures 4.1 and 4.2). Air enters at one end of the unit through the fan compartment, is forced horizontally over the heat exchanger, and then exits at the opposite end.

The *upflow* or *highboy* design is used in the basement or in a first-floor equipment room where floor space is at a premium (Figures 4.3 and 4.4). The fan is located below the heat exchanger. Air enters at the bottom or lower sides of the unit and leaves at the top through a warm-air plenum.

The *downflow* or *counterflow* design is used in houses having an under-the-floor type of distribution system (Figure 4.7). The fan is located above the heat exchanger. The return-air plenum is connected to the top of the unit. The supply-air plenum is connected to the bottom of the unit.

The *multiposition* furnace is a new design that allows the contractor to select one furnace to fit most installations. The cabinet is such that the combustion chamber and component can be changed from a horizontal right or left to an upflow right or left and also a downflow right or left, as shown in Figures 4.5, 4.6, and 4.7.

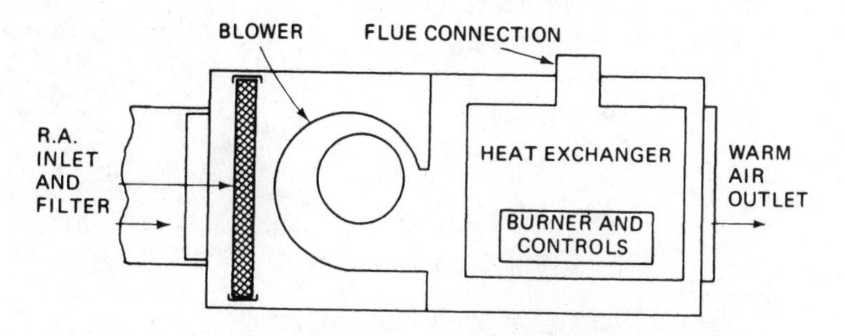

Figure 4.1 Horizontal forced warm air furnace. (By permission from the *ASHRAE Handbook*)

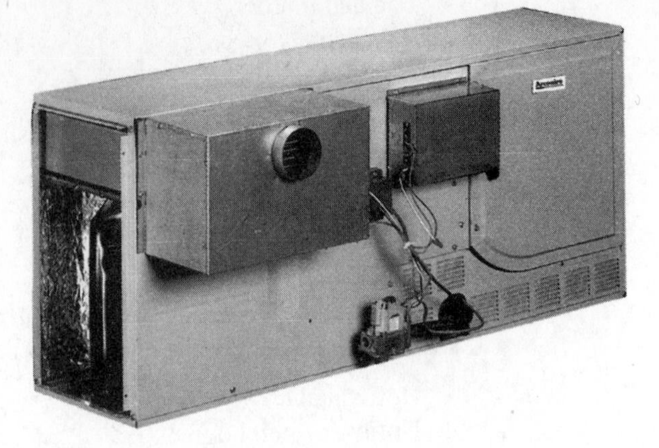

Figure 4.2 Horizontal forced warm air furnace, gas-fired. (Courtesy of ARCOAIRE Air Conditioning & Heating)

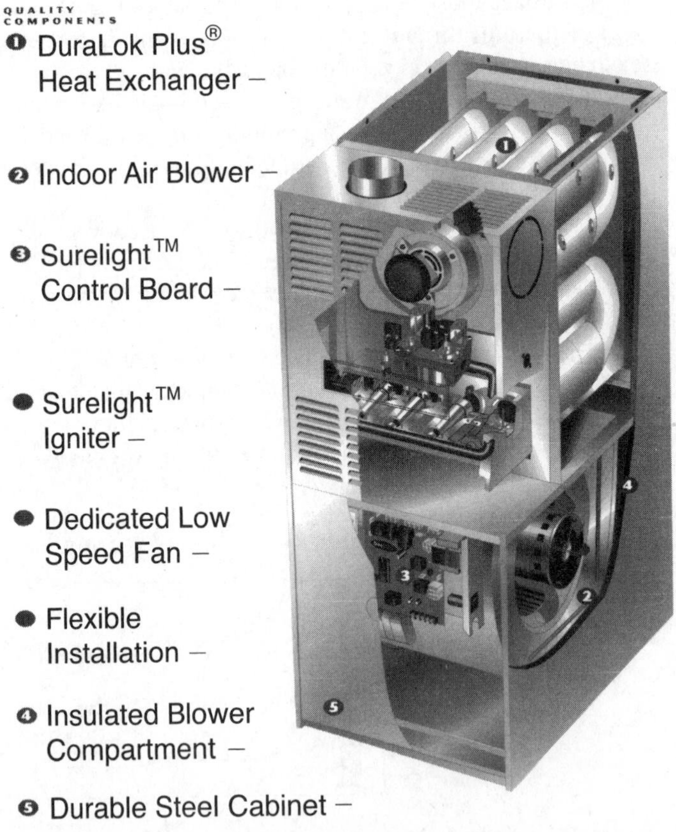

QUALITY COMPONENTS

❶ DuraLok Plus® Heat Exchanger –

❷ Indoor Air Blower –

❸ Surelight™ Control Board –

● Surelight™ Igniter –

● Dedicated Low Speed Fan –

● Flexible Installation –

❹ Insulated Blower Compartment –

❺ Durable Steel Cabinet –

Figure 4.3 Upflow (highboy) forced warm air furnace, gas-fired. (Courtesy of Lennox Industries Inc.)

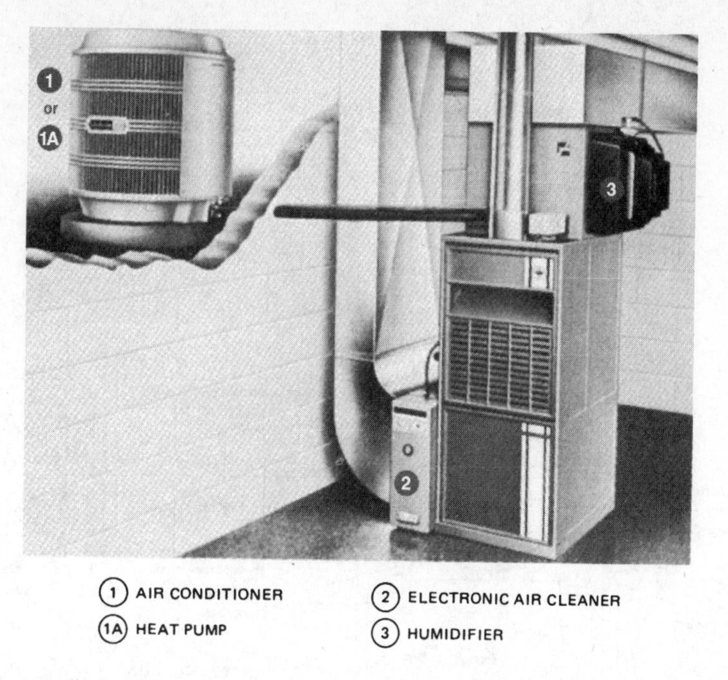

Figure 4.4 Installed upflow warm air furnace, gas-fired, including air conditioner, air cleaner, and humidifier. (Courtesy of Carrier Corporation)

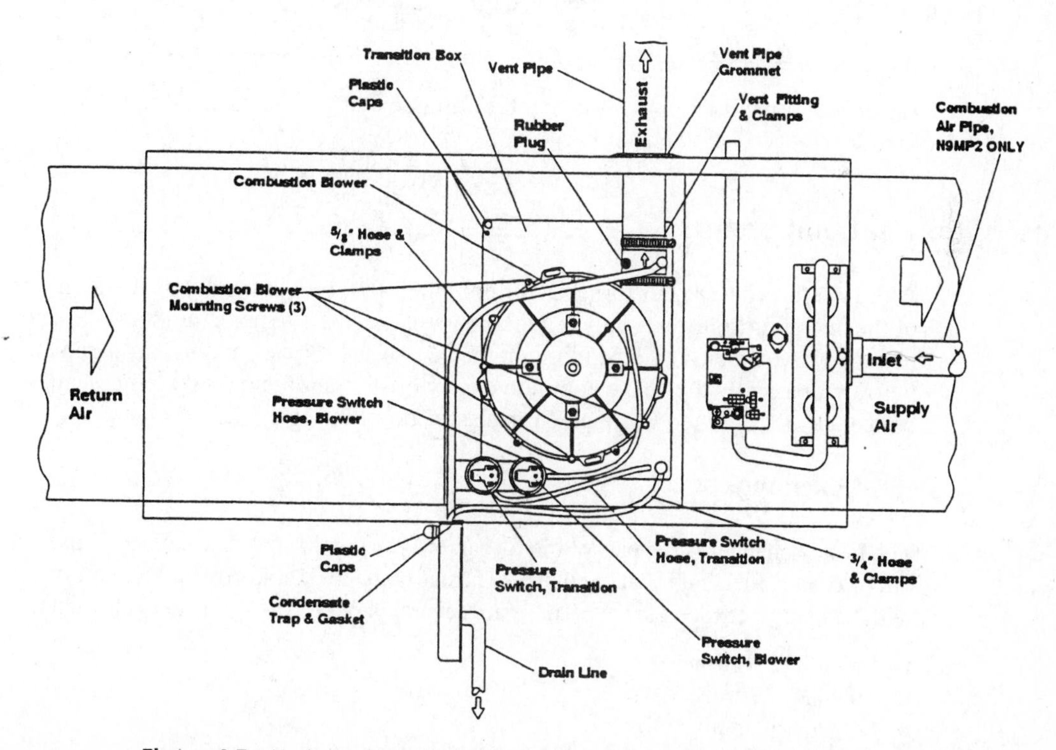

Figure 4.5 Horizontal installation for right- or left-side venting. (Courtesy of International Comfort Products Corporation, USA)

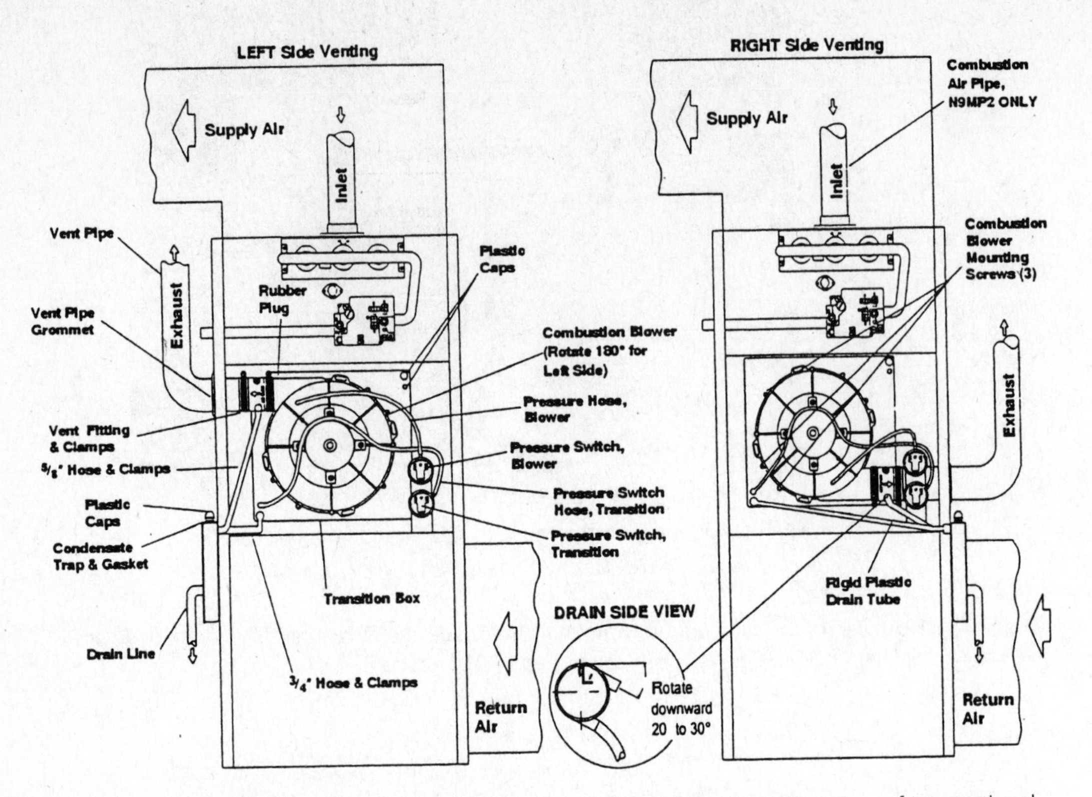

Figure 4.6 Upflow installations for right- or left-side venting. (Courtesy of International Comfort Products Corporation, USA)

Description of Components

Although forced warm air furnaces differ in their fuel-burning equipment and in many of the operating controls, many other components are the same or similar. These components vary somewhat, depending on the size of the furnace and the arrangement of parts, but basically they have common characteristics and functions. Components of a typical gas-fired forced warm air furnace are shown in Figure 4.8.

Heat Exchanger

The heat exchanger is the part of the furnace where combustion takes place. It uses gas, oil, or coal as fuel. The heat exchanger is usually made of cold-rolled, low-carbon steel with welded or crimped seams. There are several general types of heat exchangers:

- Individual section
- Tubular
- Serpentine (clamshell)
- Cylindrical

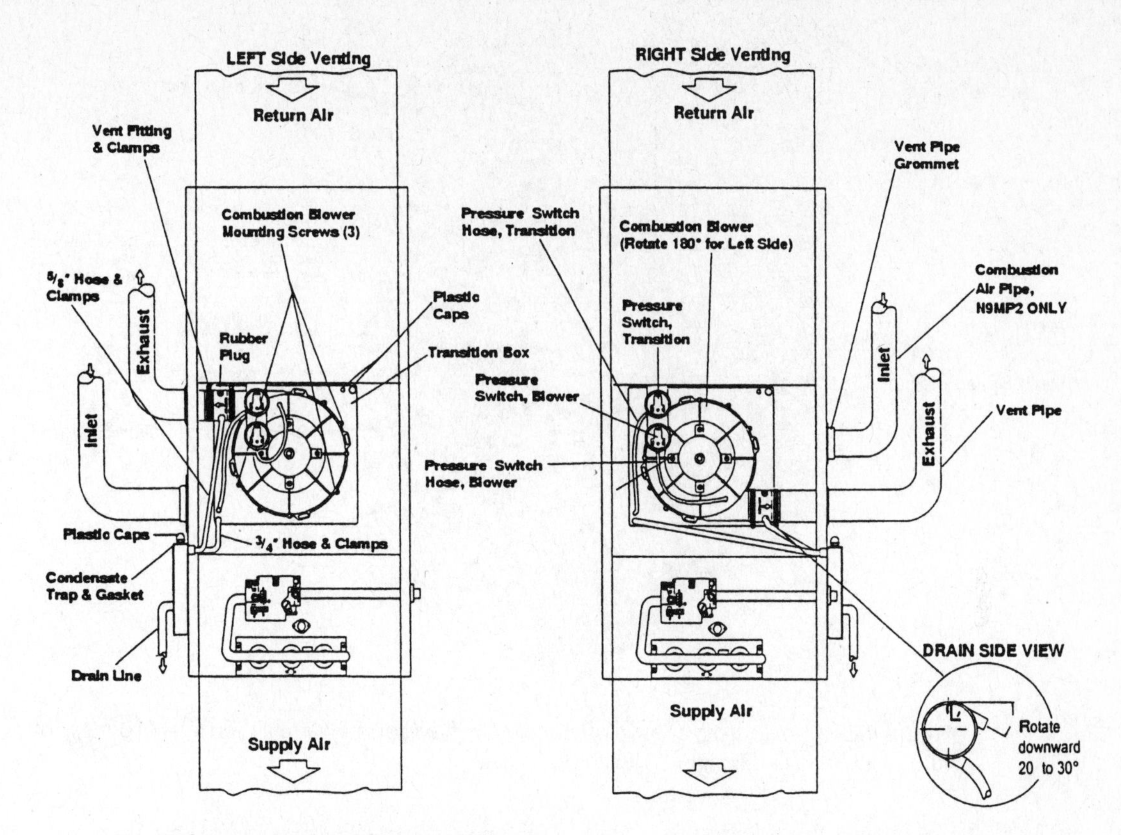

Figure 4.7 Downflow installations for right- or left-side venting. (Courtesy of International Comfort Products Corporation, USA)

Individual-Section Heat Exchanger This type of heat exchanger has a number of separate heat exchangers (Figure 4.9). Each section has individual burners (fuel-burning devices). These sections are joined together at the bottom so that a common pilot can light all burners. The sections are joined together at the top so that flue gases are directed to a common flue. The individual-section type of heat exchanger is used only on gas-burning equipment. The tubular and serpentine heat exchangers are variations of the individual-section type (Figures 4.10 and 4.11).

Cylindrical Heat Exchanger This type of heat exchanger has a single combustion chamber and uses a single-port fuel-burning device (Figure 4.12). The cylindrical type of heat exchanger is used on gas, coal, and oil units. Many heat exchangers for coal, gas, and oil have two types of surfaces:

1. Primary
2. Secondary

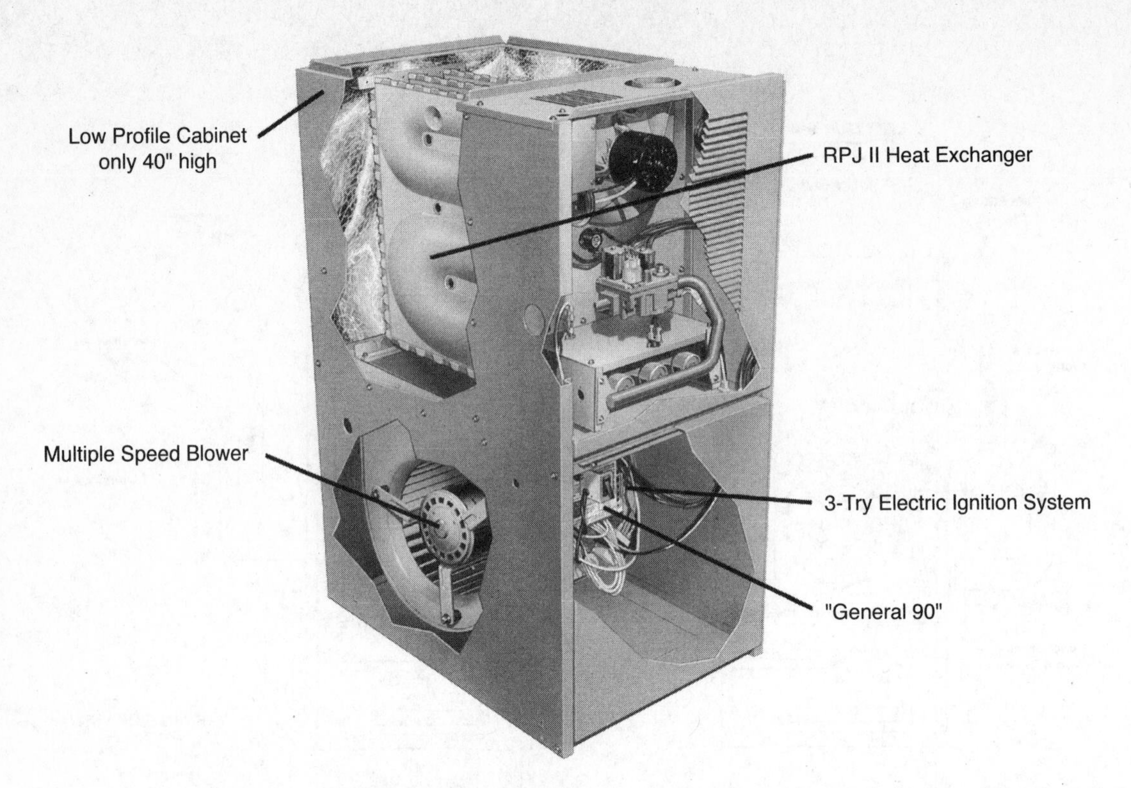

Figure 4.8 Components of a typical upflow gas-fired forced warm air furnace. (Courtesy of ARCOAIRE Air Conditioning & Heating)

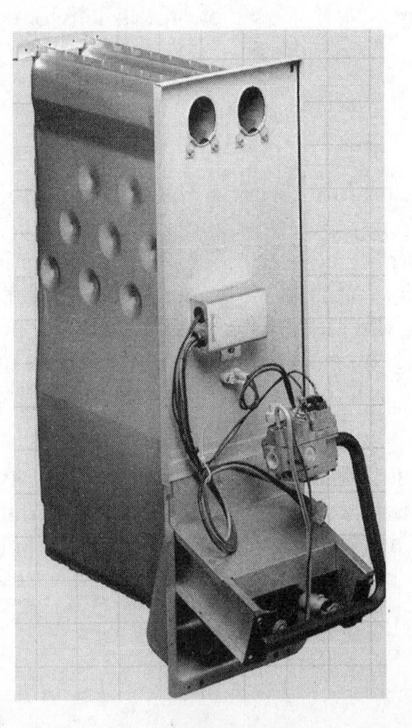

Figure 4.9 Individual-section heat exchanger. (Courtesy of ARCOAIRE Air Conditioning & Heating)

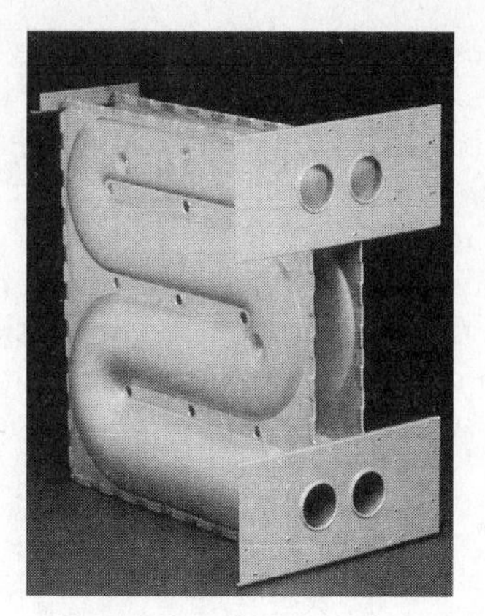

Figure 4.10 Serpentine (clamshell). (Courtesy of ARCOAIRE Air Conditioning & Heating)

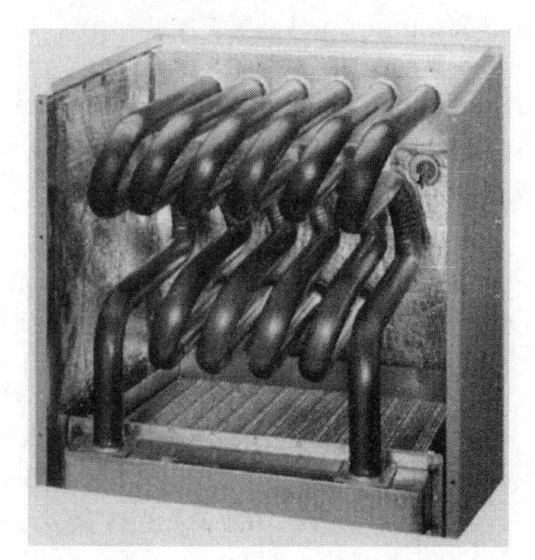

Figure 4.11 Tubular heat exchanger. (Courtesy of Amana Refrigeration, Inc.)

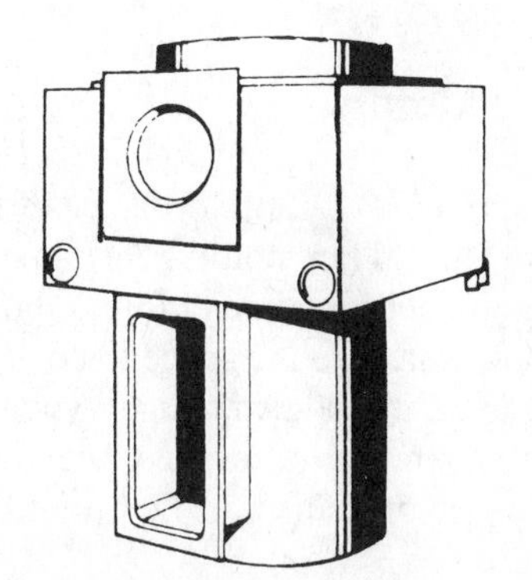

Figure 4.12 Cylindrical heat exchanger.

The *primary surface* is in contact with or in direct sight of the flame and is located where the greatest heat occurs. The *secondary surface* follows the primary surface in the path between the burner and the flue. This surface is used to extract as much heat as possible from the flue gases before the products of combustion exit through the flue to the chimney.

The fuel used by the furnace has a definite relation to the amount of chimney draft required. Coal, because of the high resistance of its fuel bed, requires a relatively high draft. Oil requires a relatively low draft. Gas requires a balanced draft, which is effective only when the furnace provides heat.

Cabinet

Cabinets for forced warm air furnaces serve many functions. They provide:

- An attractive exterior
- Insulation in the area around the heat exchanger
- Mounting facilities for controls, fan and motor, filters, and other items that they enclose
- Access panels for parts requiring service
- Connections for the supply and return-air ducts
- An airtight enclosure for air to travel over the heat exchanger

Cabinets for gas and oil furnaces are usually rectangular in shape and have a baked enamel finish. Cabinets are insulated on the interior with aluminum-based mineral wool, fiberglass, or a metal liner so that the surface will not be too hot to the touch. Insulation also makes it possible to place flammable material in contact with the cabinet without danger of fire.

Complete units are furnished so that a minimum amount of assembly time is required for installation. Usually, only duct and service connections need be made by the installer. Knockouts (readily accessible openings) are provided for gas and electric connections. The bolts used for holding the unit in place during shipment are also used for leveling the unit during installation. On highboy units, return-air openings in the fan compartment can be cut in the cabinet sides or bottom to fit installation requirements.

Fans and Motors

Much of the successful performance of a heating system depends on the proper operation of the fan and fan motors. The fan moves air through the system, obtaining it from return air and outside air, and forces it over the heat exchanger and through the supply distribution system to the space to be heated.

Fans used are the centrifugal type with forward-curved blades (Figure 4.13). Air enters through both ends of the wheel (double inlet) and is pumped or compressed (the buildup of air pressure) at the outlet. Fans and motors must be the proper size to

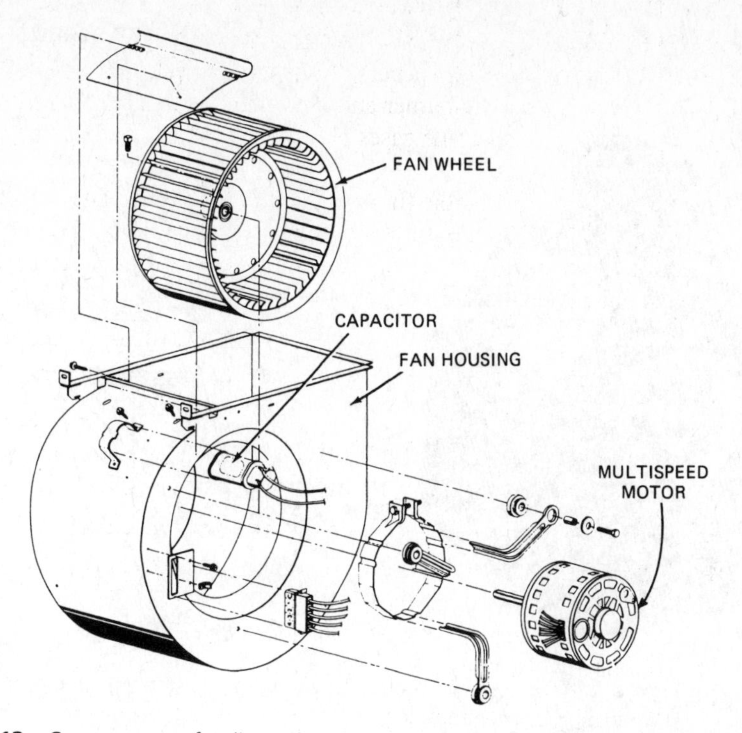

Figure 4.13 Components of a direct-drive centrifugal fan. (Courtesy of DNP Company)

deliver the required amount of air against the total resistance produced by the system. Depending on the quantity of air moved and the resistance of the system, a motor of the proper size is selected.

Because the resistance of the system cannot be fully determined beforehand, it is usually necessary to make some adjustment of the air quantities at the time of installation. Generally, the higher the speed of the fan, the greater the cubic-feet-per-minute output, thereby requiring more power and often a higher-horsepower motor.

Types of Fans Two types of fan motor arrangements are used:

1. Belt-drive
2. Direct-drive

The *belt-drive* arrangement uses a fixed pulley on the fan shaft and a variable-pitch pulley on the motor shaft (Figure 4.14). The speed of the fan is directly proportional to the ratio of the pulley diameters. The following equation is used:

$$\text{fan speed (rev/min)} = \frac{\text{diameter of motor pulley (in.)}}{\text{diameter of fan pulley (in.)}} \times \text{motor speed (rev/min)}$$

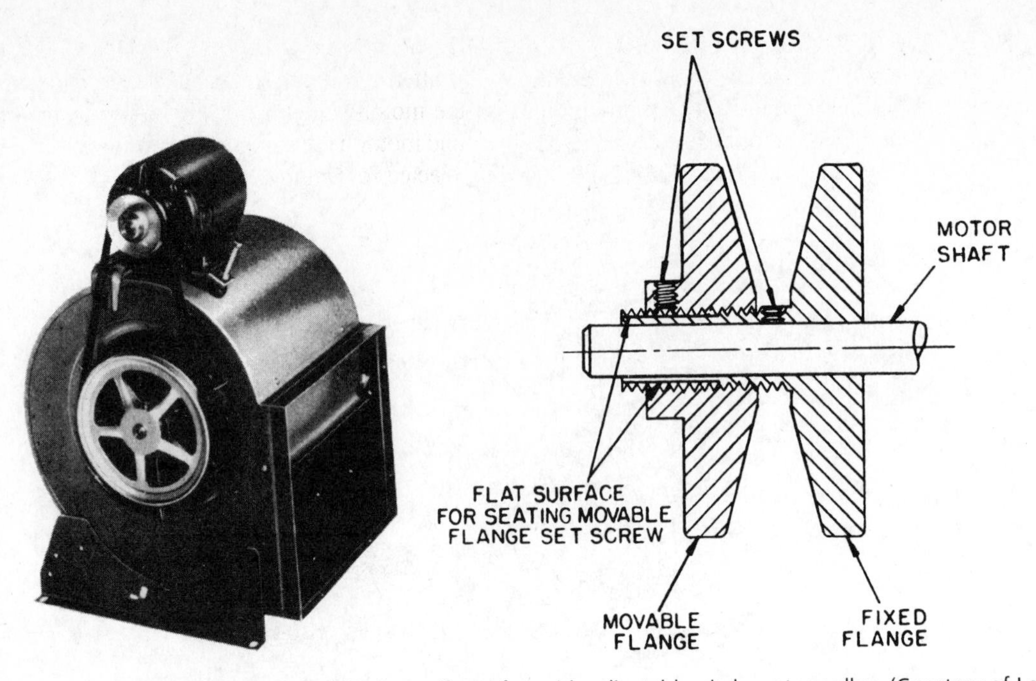

Figure 4.14 Belt-driven centrifugal fan with adjustable pitch motor pulley. (Courtesy of Lau Division, Phillips Industries, Inc.)

Example 4–1 What is the fan speed for a 3-in. motor pulley, a 9-in. fan pulley, and an 1800-rev/min motor?

Solution:

$$\text{fan speed} = \frac{3\ \text{in.}}{9\ \text{in.}} \times 1800 = 600\ \text{rev/min}$$

■

The diameter of a variable-pitch pulley can be changed by adjusting the position of the outer flange of the pulley. This arrangement permits adjustment of the speed up to 30%. For changes greater than this, a different pulley must be used.

On the belt-drive arrangement, provision is made for regulating the belt tension and isolating the motor from the metal housing. The belt tension should be sufficient to prevent slippage. The motor mounts (usually rubber) provide isolation to prevent sound transmission.

The amperage draw of the motor is an indication of the amount of loading of the motor. The maximum loading is indicated on the motor nameplate as full-load amperes (FLA). If the FLA value is exceeded, a larger motor must be used. Additional information on measuring amperage draw is given in Chapter 8. If the motor is not drawing its full capacity in FLA, it is an indication that the motor is underloaded. This may be caused by the additional resistance of dirty filters.

In the *direct-drive* arrangement the fan is mounted on an extension of the motor shaft. Fan speeds can be changed only by altering the motor speed, which is accomplished by the use of extra windings on the motor, sometimes called a *tap-wound,* or *multispeed,* motor (Figure 4.15). A tap-wound motor has a series of connections that provide different speeds. Often, one speed is selected for heating and another, for cooling.

Fan Handler

Whole-house temperature balance can be achieved when the central fan is run on a continuous basis; however, during the heating-off cycle, cold drafts and noise make this

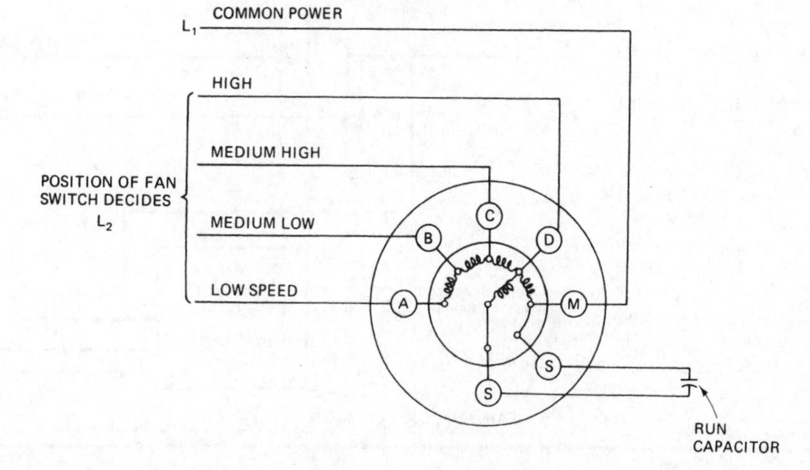

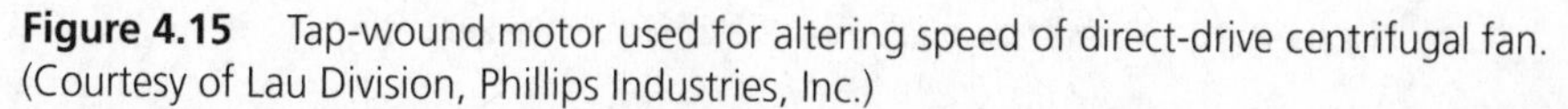

Figure 4.15 Tap-wound motor used for altering speed of direct-drive centrifugal fan. (Courtesy of Lau Division, Phillips Industries, Inc.)

constant operation of a full-speed fan less desirable. Fan handler controls provide a range of variable air flow as needed, from super slow to full speed. Figure 4.16 illustrates the change in fan speed with the change in conditioning temperature. The field-installed controls are designed to operate with both shaded pole and permanent split capacitor motors. They are mounted to the side of the blower housing and wired in series with the existing fan motor, as shown in Figure 4.16. Additional benefits are obtained with the constant operation of accessories such as filters, electronic air cleaners, humidifiers, and germicidal lamps.

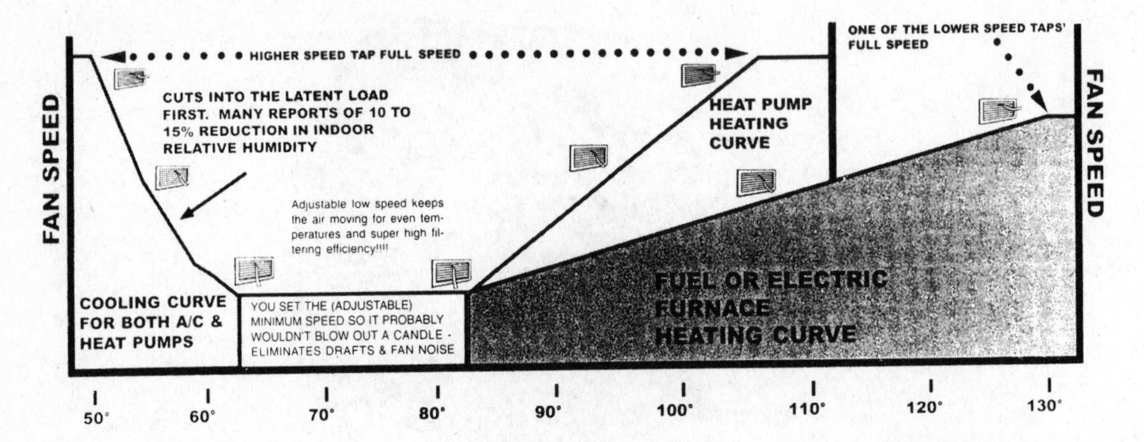

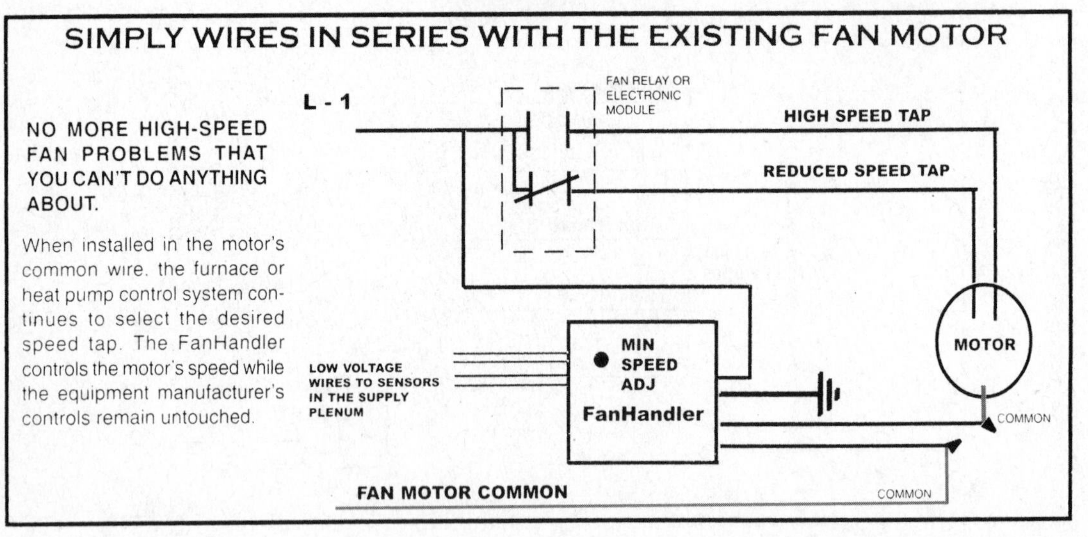

Figure 4.16 Operation and wiring of the Fan Handler system. (Courtesy of Fan Handler Inc.)

Filters

One of the desirable qualities of a forced warm air heating system is its ability, through air filters, to remove certain portions of the dust and dirt that are carried by the air. There are three types of residential air filters: disposable (throwaway), permanent (cleanable), and electronic, as shown in Figure 4.17.

Disposable filters are made of oil-impregnated fibers. When dirty, they are non-cleanable and are replaced with new ones. Permanent filters are made of metal or specially constructed fibrous material. The dirt is collected on the surface of the filtering material as the air passes through. When properly cleaned, these filters can be reused. Electronic filters remove dust particles by first placing an electrical charge on the dust particle and attracting the dust to the collector plates having the opposite charge. They are usually used with cleanable-type prefilters because the electronic filter removes only extremely small particles.

Figure 4.18 shows the size of various particles. The size is indicated in microns (μm). A micron is 1/25,400 in., or 0.001 millimeter (mm). An ordinary throwaway or cleanable home air filter will remove particles down to 10 μm. An electronic air filter will remove particles from 0.1 to 10 μm.

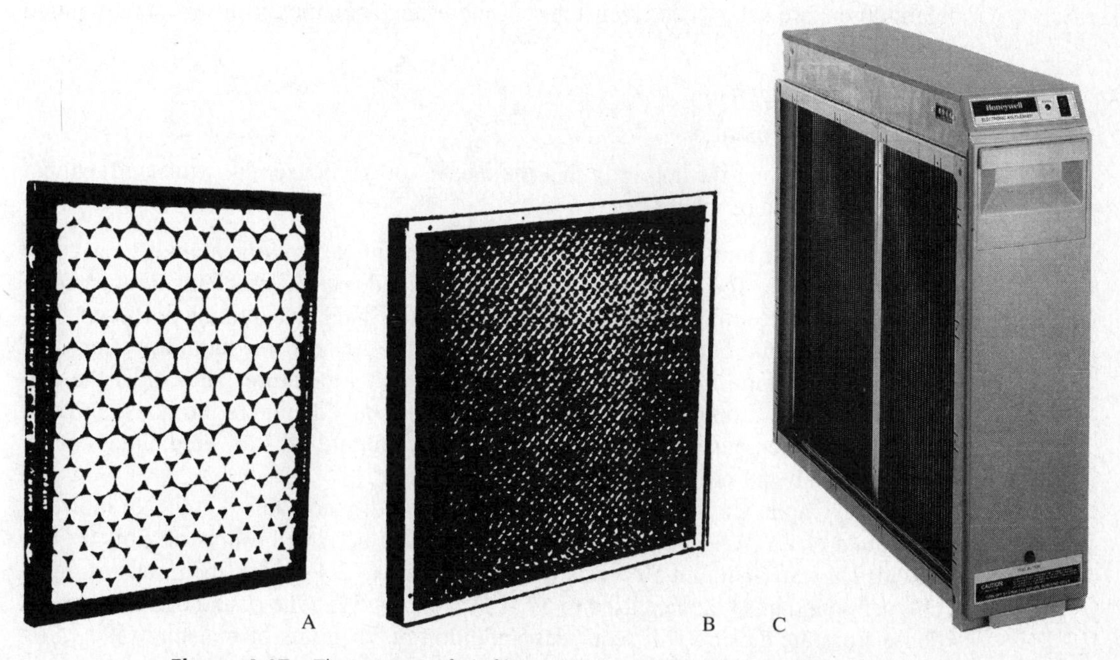

Figure 4.17 Three types of air filters: (a) disposable; (b) permanent; (c) electronic

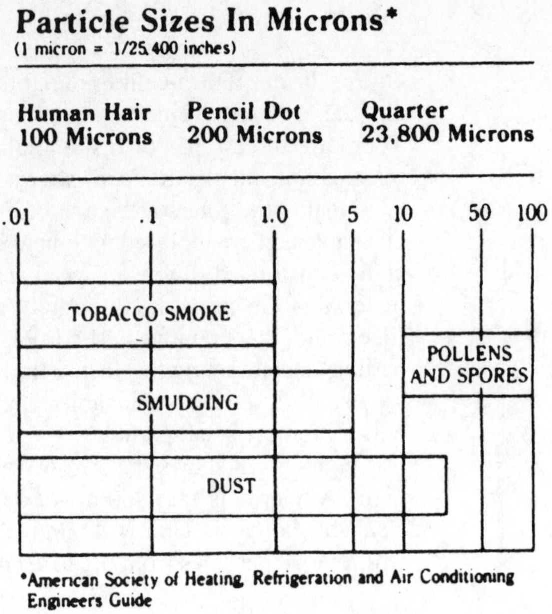

Figure 4.18 Particle size in microns. (Courtesy of Research Products Corporation)

Humidifiers and Humidistats

Humidifiers are used to add moisture to indoor air. The amount of moisture required depends on:

Outside temperatures

House construction

Amount of relative humidity that the interior of the house will withstand without condensation problems

It is desirable to maintain 30 to 50% relative humidity. Too little humidity can cause furniture to crack. Too much humidity can cause condensation problems. From a comfort and health standpoint, a range as wide as 20 to 60% relative humidity is acceptable.

The colder the outside temperature, the greater the need for humidification. The amount of moisture the air will hold depends on its temperature. The colder the outside air, the less moisture it contains. Air from the outside enters the house through infiltration. When outside air is warmed its relative humidity is lowered, unless moisture is added by humidification (Figure 4.19).

For example, air at 20°F and 60% relative humidity contains 8 grains of moisture per pound of air. When air is heated to 72°F, it can hold 118 grains of moisture per pound. Thus, to maintain 50% relative humidity in a 72°F house, the grains of moisture per pound must be increased to 59 ($118 \times 0.5 = 59$). One pound of air entering a house from the outside will require the addition of 51 grains of moisture ($59 - 8 = 51$). The amount of infiltration depends on the tightness of the windows, doors, and other parts of the building's construction.

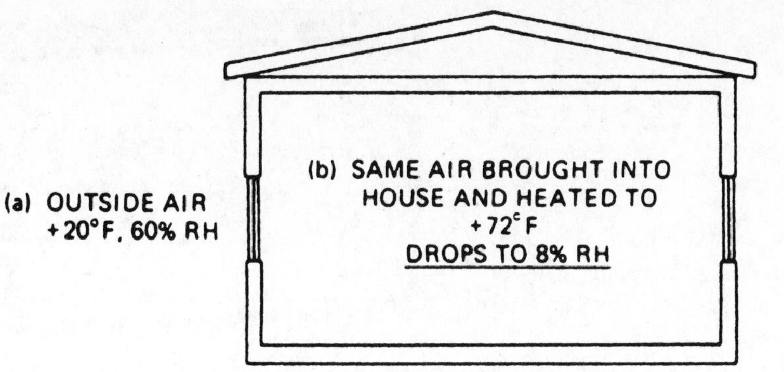

INDOOR RELATIVE HUMIDITY TABLE

OUTDOOR RELATIVE HUMIDITY	0°	10°	20°	30°	40°	50°
100%	6	9	14	21	31	46
80%	5	7	11	17	25	37
60%	3	5	(8%)	13	19	28
40%	2	4	6	8	12	18
20%	1	2	3	4	6	10

OUTDOOR TEMPERATURE

THE ABOVE CHART SHOWS WHAT HAPPENS TO THE HUMIDITY IN YOUR HOME WHEN YOU HEAT COLD WINTER AIR TO 72° ROOM TEMPERATURE. EXAMPLE: WHEN OUTDOOR TEMPERATURE IS 20° AND OUTDOOR RELATIVE HUMIDITY IS 60%, INDOOR RELATIVE HUMIDITY DROPS TO ONLY 8%. YOU NEED 30% TO 45% FOR COMFORT.

Figure 4.19 Effect of temperature on relative humidity. (Courtesy of Humid-Aire Corporation)

It is impractical in most buildings to maintain high relative humidity when the outside temperature is low. Condensation forms on the inside of the window when the surface temperature reaches the dew point temperature of the air. The *dew point temperature* is the temperature at which moisture begins to condense when the air temperature is lowered.

Humidistats It is desirable to control the amount of humidity in the house by the use of a humidistat. A humidistat is a device that regulates the "on" and "off" periods of humidification (Figure 4.20). The setting of a humidistat can be changed to comply with changing outside temperatures.

Figure 4.20 Humidity control. (Courtesy of Invensys Climate Controls)

Figure 4.21 defines the terms *tight house, average house,* and *loose house.* Based on the type of construction and the house size this chart gives the gallons per day (GPD) required to maintain an acceptable level of humidity.

Types of Humidifiers The rotating-drum evaporative humidifier (Figure 4.22) has a slowly revolving drum covered with a polyurethane pad partially submerged in water. As the drum rotates it absorbs water. The water level in the pan is maintained by a float valve. The humidifier is mounted on the side of the return-air plenum. Air from the supply plenum is ducted into the side of the humidifier. The air passes over the wetted

Humidity requirements in GPD based on house size and type of construction (from ARI Guideline F)

TYPE OF CONSTRUCTION	SIZE OF HOUSE (sq. ft.)*					
	500	1000	1500	2000	2500	3000
TIGHT	2.1	4.2	6.4	8.5	10.6	12.7
AVERAGE	3.3	6.5	9.8	13.1	16.3	19.6
LOOSE	4.6	9.2	13.8	18.4	23.0	27.6

*Based on 8 ft. ceiling height

Tight House

With insulated walls and ceilings, vapor barriers, weather stripping on doors and windows, and snug doors, windows and fireplace damper (1/2 air change per hour)

Average House

Insulated walls and ceilings with vapor barriers, but loose storm doors, windows and fireplace damper (1 air change per hour)

Loose House

Without insulation, storm doors, storm windows, weather stripping or vapor barriers (2 air changes per hour)

Figure 4.21 Humidity requirements in GPD. (Courtesy of General Filters, Inc.)

MODEL 81

Drum-type humidifier with capacity for most homes. Installs on vertical warm air supply or cold air return (recommended) plenum.
12" wide x 11-1/2" high x 11" deep
8" wide x 7-1/2" high
6" diameter (Bypass Pipe Assembly Supplied)
Temp °F / GPD 80° 7.0 100° 11.3 120° 16.0 140° 18.0 160° 21.7
Forced warm air heating systems create a pressure differential that diverts some heated air through a bypass duct to the humidifier. There, the warm air absorbs moisture from a water-soaked, drum-mounted evaporator sleeve and then rejoins the primary warm air stream for distribution throughout the living area. The drum-mounted evaporator sleeve picks up moisture as it slowly rotates in a water reservoir. A float valve controls the level of water in the reservoir.
81-15 Evaporator sleeve. Split drum allows easy separation for removal of polyurethane foam sleeve for cleaning.
Unique, time-proven GENERALAire float valve assembly meters water into reservoir through precision machined orifice. Saddle valve and 1/4" plastic tubing included.

Figure 4.22 Rotating-drum evaporative humidifier. (Courtesy of General Filters, Inc.)

surface, absorbs moisture, then moves into the return-air plenum. The evaporative capacity of this unit is 16 GPD with 120°F air.

The fan-powered evaporative humidifier (Figure 4.23) is mounted on the supply-air plenum. Air is drawn in by the fan, forced over the wetted core, and delivered back into the supply-air plenum. The water flow over the core is controlled by a water valve. A humidistat is used to turn the humidifier on and off, controlling both the fan and the water valve. The control system is set up so that the humidifier can operate only when the furnace fan is running. Installation of this model is quick and easy. It fits on a 14-in. duct and requires only 1/2 in. of overhead clearance. The evaporative capacity is 19 GPD with 120°F air.

The flow-through or evaporative humidifier (Figure 4.24) is mounted on either the vertical warm-air supply or the cold-air return plenum. Forced-air heating systems create a pressure differential that diverts some heated air to the humidifier through a bypass duct. There the warm air absorbs moisture from a water-soaked evaporative pad and then rejoins the warm-air stream for distribution throughout the

Cover removal is simple. No need to disconnect water or drain lines. And only 1/2" of overhead clearance is required.

The exclusive Trion Evaporator Pad lifts out easily for replacement.

Quick & Easy Installation

Five screws secure the mounting base to the duct. The attractive cover snaps over the base and is secured on the bottom by just two screws. Easily fits 14" duct and requires only 1/2" overhead clearance.

Humidistat; Transformer; Saddle Valve; Mounting Template; and Mounting Hardware are provided. (Note: Installer supplies 1/4" O.D. copper water supply tubing; 1/2" drain hose to length; and hose clamp.)

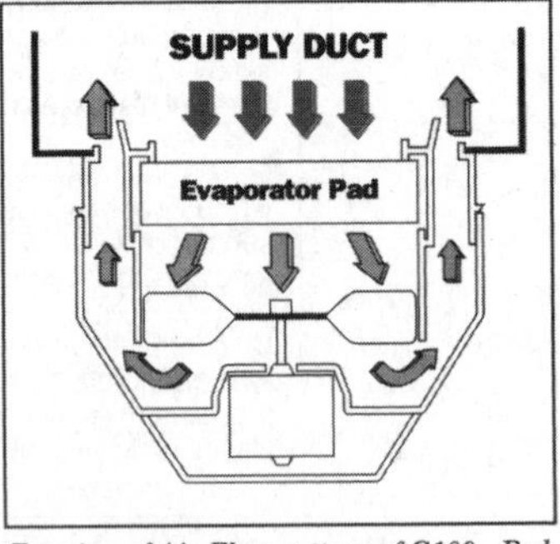

Top view of Air Flow pattern of G100. Red arrows show warm, dry air; blue arrows show humidified air.

Specifications

EVAPORATIVE CAPACITY
19 gallons per day (.79 gallons per hour) @ 120°F duct temperature (ARI Std 610). Humidifies up to 3000 sq. ft. of residential space.

DIMENSIONS
13" W X 18-3/4" H X 11-1/2" D (Includes bottom-mounted solenoid valve.)

MOUNTING
Duct Opening 12-1/4" W X 15-1/2" H - fits on a 14" metal or fiberglass supply duct.

ELECTRICAL SUPPLY REQUIREMENTS
Unit Power: 2.0 A @ 115 VAC, 50/60 Hz, 1/20 HP Motor
Control Power: 0.5 A @ 24 VAC, 50/60 Hz

HUMIDISTAT
Adjustable from 20-80% R.H.

WATER SUPPLY
Brass solenoid valve rated for 150 psig. Cleanable/replaceable internal filter protects solenoid. (For water pressure greater than 100 psig., a pressure-reducing valve is required.)

HOUSING MATERIALS
Non-corrosive, high-temperature engineering-grade plastic.

TWO-YEAR WARRANTY
(excludes evaporator pad)

Figure 4.23 Flow-through fan powered humidifier. (Courtesy of TRION Inc.)

Model 1042
Power Humidifier

APPLICATION	Flow-through humidifier that's ideal for most homes. Installs on vertical warm air supply or cold air return plenum.
UNIT SIZE	15" wide x 11-1/2" high x 9" deep
PLENUM OPENING	10-1/4" wide x 9-1/2" high
BYPASS DUCT	6" diameter (collar supplied)
EVAPORATIVE OUTPUT CAPACITY @ .20" PRESSURE DIFFERENTIAL (except 747 rated at 800 FPM duct velocity) ABOVE AT STANDARD INDUSTRY RATING CONDITIONS	Temp °F / GPD: 80° 8.0; 100° 12.2; 120° 16.4; 140° 19.2; 160° 23.2
EVAPORATIVE PRINCIPLE GENERALAire Humidifiers are 100% evaporative, adding moisture in the form of water vapor only. They are clean operating. The nuisance of airborne water droplets or mineral dust is avoided.	Forced warm air heating systems create a pressure differential that diverts some heated air to the humidifier through a bypass duct. There, the warm air absorbs moisture from a water-soaked evaporator pad and then rejoins the primary warm air stream for distribution throughout the living area. A solenoid valve and patented distribution tray meter and disperse water uniformly across the evaporator pad. Unevaporated water drains from the unit.
EVAPORATIVE MEDIA	990-13 Evaporator Pad. Designed for high output with minimal hard water scale accumulation. UL listed.
WATER SUPPLY	Dependable stainless steel solenoid valve with monel wire mesh filter that protects the precision machined Teflon flow-control orifice. Saddle valve supplied. Use 1/4" copper tubing.
DRAIN CONNECTION	High capacity 5/8" drain and hose prevents clogging. Includes 15' vinyl hose.

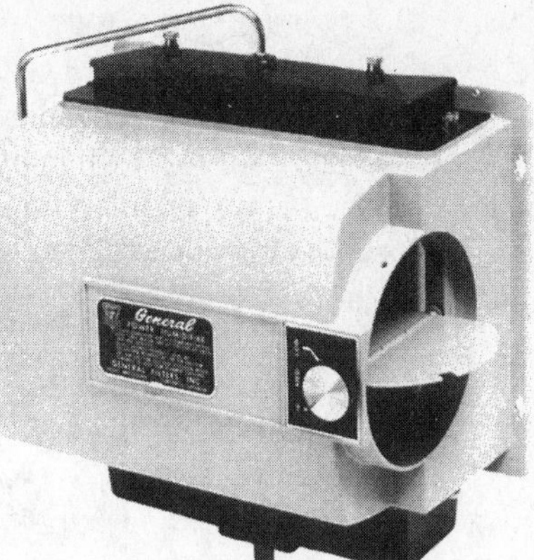

Unique patented water distribution trough for complete wetting of the evaporator pad.

CONSTRUCTION, MATERIALS AND STYLING	Attractive cabinet made from durable, non-corrosive, high temperature materials for long service life.
VERSATILITY	Right or left hand bypass installation. May be installed on up-flow, counter-flow or horizontal furnaces or air handlers.
SERVICEABILITY	Thumbscrews allow easy access to distribution trough, drain pan and evaporative pad for quick, easy cleaning and service.
EASY INSTALLATION – GENERALAire humidifiers are specially designed for fast, easy installation	Most important components are included. Outboard and keyholed screw holes and simple thumbscrew assembly speeds installation. Patented water distribution tray reduces need for critical leveling.

Figure 4.24 Flow-through bypass humidifier. (Courtesy of General Filters, Inc.)

home. Installation can be right or left hand, on any type of furnace—upflow, counterflow, or horizontal flow.

Figure 4.25 is another example of a flow-through bypass humidifier. This type is popular because it is durable, is essentially trouble free, and does not accumulate contaminants, because the unevaporated water drains from the unit at the end of each cycle. This unit is quick and easy to install and easily fits on a 10-in. duct and requires only 1/2 in. of overhead clearance. The mounting can be either left or right hand. A cleanable/replaceable internal filter protects the solenoid valve.

The high-capacity steam power humidifier (Figure 4.26) uses an electrical heating element immersed in a water reservoir to evaporate moisture into the furnace supply-air plenum. A thermal fan interlock control allows the unit to humidify air without heat from the furnace. This humidifier is an excellent choice for heat pumps and high-efficiency heating systems. Two models are shown: the 60-1 is 120 V, with an output of 13 GPD, and the 60-2 is 240 V, with an output of 17 GPD.

The G200's compact design allows it to fit on ducts as narrow as 10 inches...

...and the removable cover makes evaporator pad replacement easy.

Specifications

EVAPORATIVE CAPACITY
12 gallons per day @ 120°F (supply duct temperature).
Humidifies up to 1500 sq. ft. of residential space.

DIMENSIONS
11-1/2" W X 16" H X 9-1/2" D
(Includes bottom-mounted solenoid valve.)

BYPASS
6" dia., flexible or hard duct

MOUNTING
Duct Opening 8-5/8" W X 10-3/4" H; Fits on a 10" duct; Left- or right-hand mounting on sheet metal or fiberglass duct, Supply or Return.

ELECTRICAL
Control Power 0.33 A @ 24 VAC, 50/60 Hz.

HUMIDISTAT
Adjustable from 20-80% R.H.

WATER SUPPLY
Cleanable/replaceable internal filter protects brass solenoid rated for 150 psig. (A pressure reducing valve is required for water pressure greater than 100 psig.)

HOUSING MATERIALS
Non-corrosive, high-temperature engineering-grade plastic

TWO-YEAR WARRANTY
On parts (excludes evaporator pad)

Quick & Easy Installation
Four screws secure the mounting base to the duct, then the attractive cover snaps easily into place and is secured with a single fastener. Configures for Left- or Right-Hand Mounting. Easily fits on a 10" duct and requires only 1/2" overhead clearance.

Humidistat; Transformer; Saddle Valve; Mounting Template; and Mounting Hardware are provided. (Note: Installer supplies 1/4" O.D. copper water supply tubing; 1/2" drain hose to length; hose clamp; 6" flexible or hard duct; and duct hardware.)

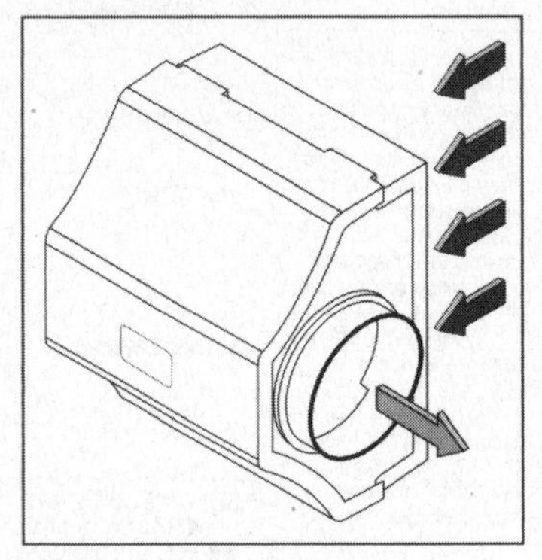

Air Flow pattern of G200 mounted on Supply Duct. Red arrows show warm, dry air; blue arrow shows humidified air.

Figure 4.25 Flow-through bypass humidifier. (Courtesy of TRION Inc.)

For those humidifiers with water reservoirs already installed, an automatic flushing timer (Figure 4.27) reduces maintenance. Whenever a reservoir is part of the system there is a tendency to build up minerals, dissolved solids, and other contaminants. This automatic flushing timer drains and refills the reservoir with clean, fresh water every 12 hours. (See the two installation diagrams.) The kit includes parts to adapt to most humidifiers on the market today.

The air pressure switch, Figure 4.28, is designed to control humidifier operation by sensing changes in air pressure within the furnace plenums, when the furnace is in operation. The switch contains normally open, single-throw contacts to supply 120 Vac to the humidifier when the furnace blower runs. Check the important note below the figure.

A current-sensing relay, shown in Figure 4.29, is used to detect the amperage of the blower's speed that is used for heating. Once current is sensed, a set of normally open contacts close, completing the path for the humidifier circuit.

High-Capacity Steam Power Humidifiers

Excellent choice for heat pumps and high-efficiency heating systems

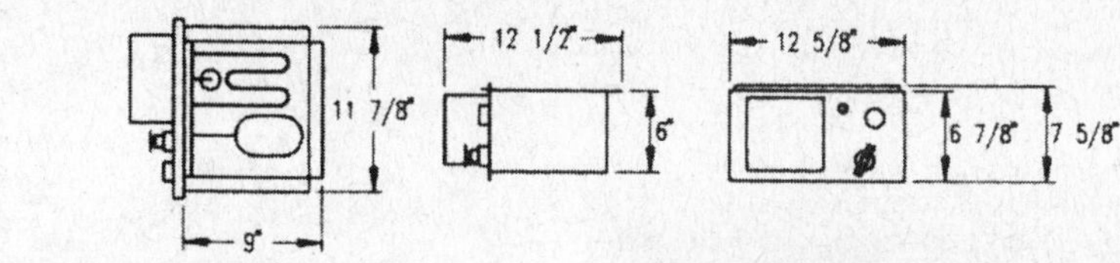

Specifications Model 60-1

Moisture Output: 13 gallons per day (G.P.D.)
(4.5 lbs. of H2O evaporated per hour)
1.5 kw 120V. 12.5 amps incoloy sheathed element

Specifications Model 60-2

Moisture Output: 17 gallons per day
(5.9 lbs. of H2O evaporated per hour)
2.0 kw 240V. 8.3 amps incoloy sheathed element

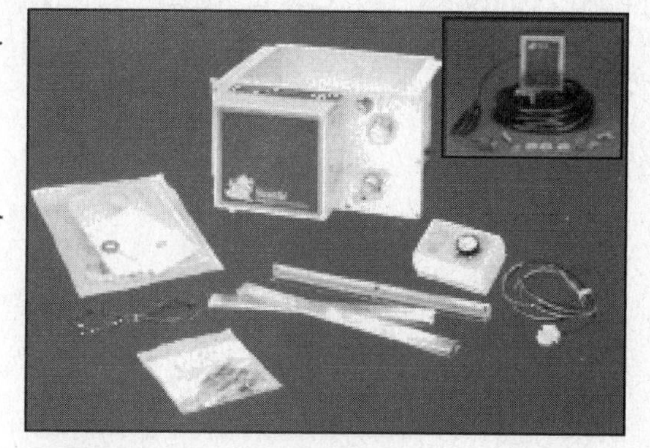

Installation Package includes all adapters and mounting components for typical installation. Relay sometimes required for independent fan operation is not supplied. Comes complete with patented Skuttle* Universal humidistat which can either be wall or duct mounted.

- *Thermal fan interlock control allows unit to humidify air without heat from furnace.*
- *Compact size ideal where space is at a premium.*
- *Cleaning is made easy with a one-piece service drain petcock.*
- *Corrosion resistant low water safety cut-off switch and built-in overflow protection.*
- *Uses minimum water.*
- *Does not contaminate heated air.*

Typical Installation

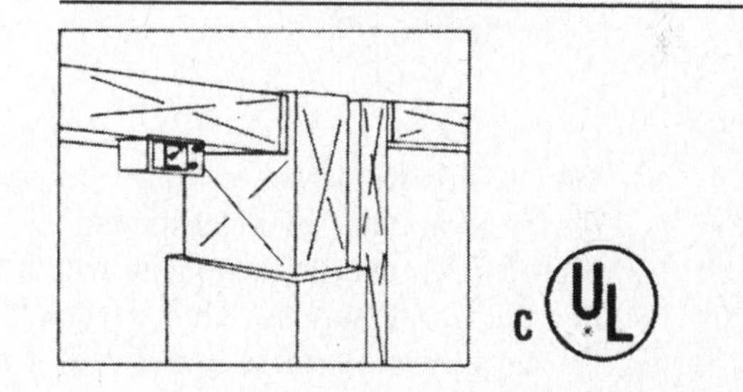

c UL

The frequency of cleaning your humidifier will depend primarily on the "hardness" (Mineral Content) of the water supplied to the unit. Because of the vast difference in water hardness throughout the country, it is nearly impossible to predict the amount of time that will lapse before the humidifier requires cleaning. This type of humidifier can require considerable maintenance. Proper maintenance will help the humidifier operate at a higher efficiency, reducing operational and maintenance repair cost. To minimize humidifier maintenance, an optional Skuttle* Humidifier Flushing Timer may be added to your installation.

Figure 4.26 High-capacity steam power humidifier. (Courtesy of Skuttle Indoor Air Quality Products)

Automatic Flushing Timer OOS-HAFT-000

Flushes reservoir-type humidifiers twice a day.

Skuttle's Automatic Flushing Timer uses a fraction of the water that a Flow-Thru type humidifier uses during normal operation. Once every 12 hours, the flushing timer drains and refills the reservoir with clean, fresh water. This method helps maintain evaporative efficiency by reducing build-up of minerals and dissolved solids in the reservoir. Each Flushing Timer kit includes parts to adapt to most humidifier brands on the market today.

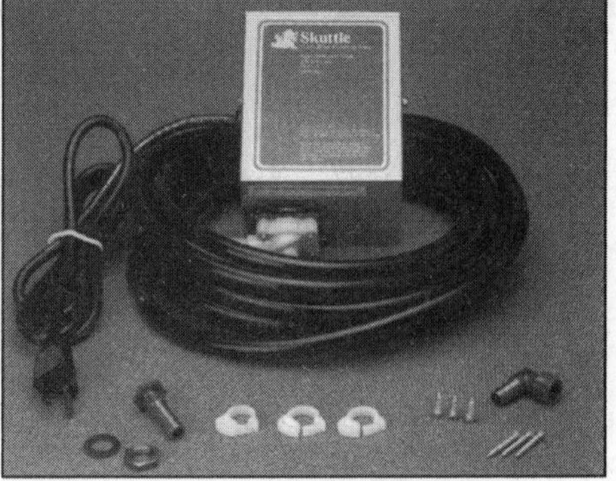

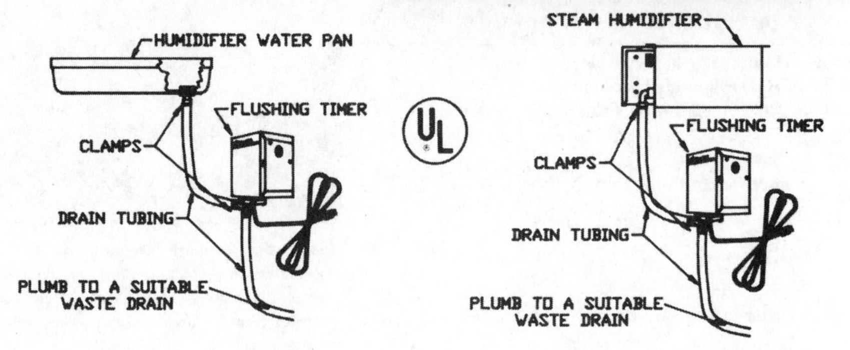

Figure 4.27 Automatic flushing timer. (Courtesy of Skuttle Indoor Air Quality Products)

Wiring diagrams are shown in Figures 4.30, 4.31, and 4.32. Figure 4.30 is for systems with a single-speed blower motor. Figure 4.31 is for systems with a two-speed blower motor; and includes the model 12500 air pressure switch shown in Figure 4.28. Figure 4.32 covers situations in which wiring within the furnace is not practical. Note that the air pressure switch is needed here, too.

Problems can arise in the use of humidifiers when the water hardness (mineral content) is too great. Water treatment tablets can be added to humidifiers when the water hardness is more than 10 grains of dissolved mineral particles per gallon. A kit is available for testing the water.

In Figure 4.26 a residential-size steam power humidifier is shown. Figure 4.33 shows an electronic steam humidifier that has application in large homes, computer rooms, offices, and medical clinics. This unit combines the purity of steam humidification with the precision of electronic monitoring and control. The circuitry continuously monitors the mineral content of the water. Model 6000-1 is 115 Vac, and 6000-2 is 208–230 Vac.

MODEL 12500

AIR PRESSURE SWITCH

FOR INSTALLATION ON ANY FORCED AIR HEATING AND COOLING SYSTEM

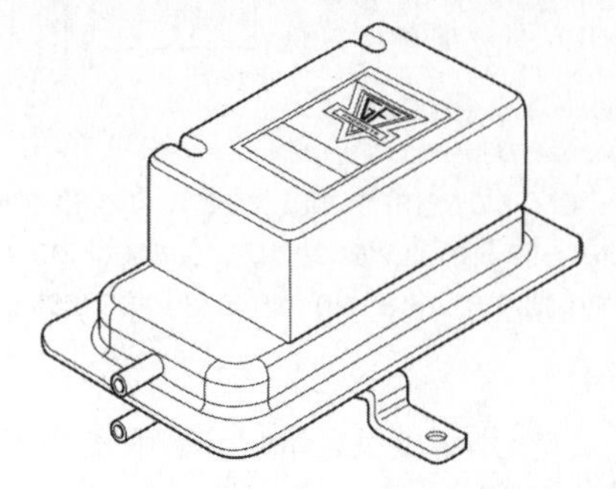

This Air Pressure Switch is designed to control Humidifier operation by sensing changes in air pressure within the furnace plenums when the furnace is in operation. The Air Pressure Switch can be actuated by a positive pressure, negative pressure or a differential pressure of .05" W.C. or more. The Air Pressure Switch contains normally open, single throw contacts to supply 120V A.C. to the humidifier or humidifier transformer when the furnace blower runs.

IMPORTANT: SINGLE POLE, SINGLE THROW ON/OFF SWITCH MUST BE PLACED BETWEEN PRESSURE SWITCH AND POWER SUPPLY TO TURN HUMIDIFIER OFF DURING COOLING SEASON.

Figure 4.28 Air-pressure switch. (Courtesy of General Filters, Inc.)

Current Sensing Relay A50

For Interfacing HVAC Equipment and Humidifiers.

The A50 is easily installed on any 120 or 240 VAC multi-speed blower, by attaching it to the common lead wire, or to the proper colored wire for the speed of the blower motor. The relay is then wired in series with the 24VAC humidity control circuit to activate the humidifier during blower operation. Attachment is made easy by the removal of one screw and placing the selected wire under the steel bracket. The wire is protected from the contact with the bracket by a rubber sponge to prevent damage.

Figure 4.29 Current-sensing relay. (Courtesy of Skuttle Indoor Air Quality Products)

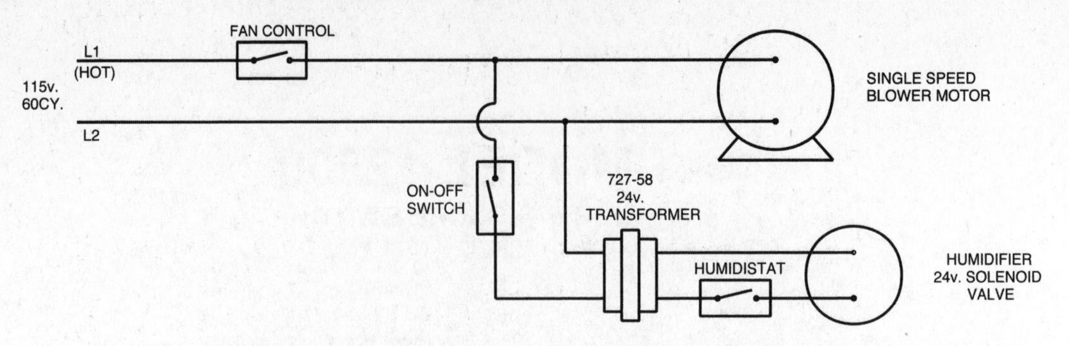

On furnaces with single speed blowers, mount the 24v. transformer on a junction box with 115v. primary leads connected in parallel with the blower circuit. Connect the leads from the 24v. solenoid valve to the terminal screws on the transformer. If a humidistat is to be installed, connect it in series with the 24v. circuit, as shown above.

Figure 4.30 Wiring diagram for furnace with single-speed blower. (Courtesy of General Filters, Inc.)

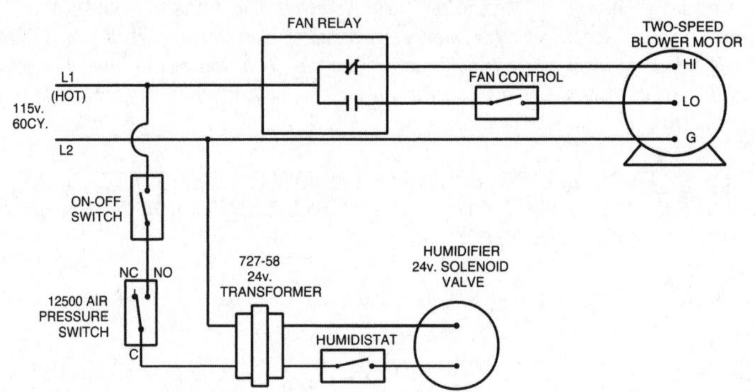

On furnaces with a two speed blower, wired as shown above, it is necessary to eliminate voltage feedback from the low speed motor windings during the cooling season. Install a 12500 Air Pressure Switch in series with the transformer primary and on/off switch as shown.

Figure 4.31 Wiring diagram for furnace with two-speed blower. (Courtesy of General Filters, Inc.)

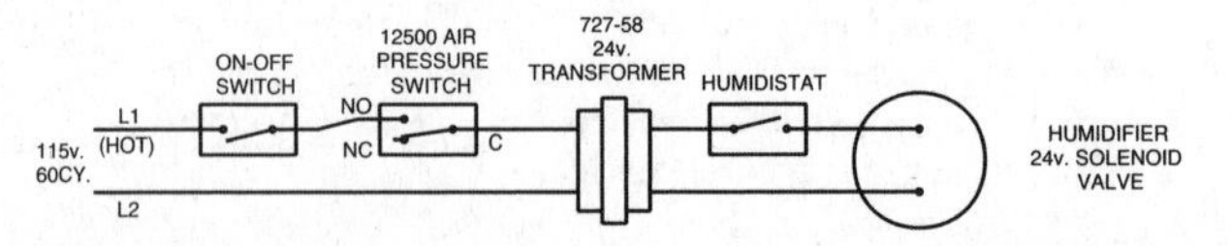

When wiring within the furnace is not practical, the humidifier and a 12500 Air Pressure Switch may be wired from a continuous 115 volt power source. Install the on/off switch and Air Pressure Switch in series with the transformer primary on the hot or black wire. The Air Pressure Switch will detect furnace operation and supply power to the humidifier accordingly.

Figure 4.32 Wiring diagram when wiring within furnace is not practical. (Courtesy of General Filters, Inc.)

6000 Series Electronic Steam Humidifier

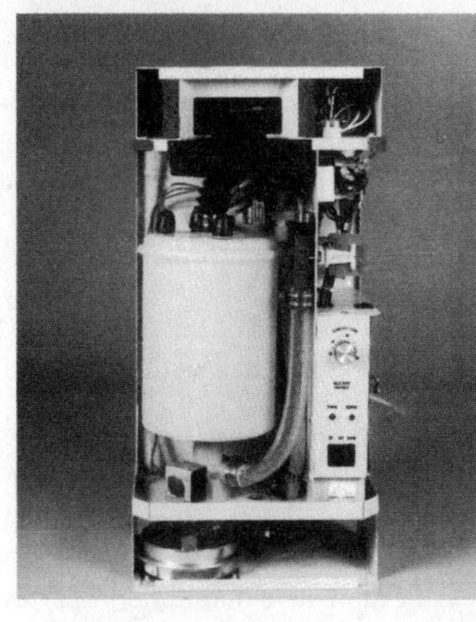

Herrmidifier's 6000 Series Electronic Steam Humidifiers combine the purity of steam humidification with the precision of electronic monitoring and control. Exclusive circuitry continuously monitors the mineral content of water to provide optimum performance and trouble-free operation.

Applications

- Large Homes
- Computer Rooms
- Laboratories
- Clean Rooms
- Offices
- Paper Storage & Handling
- Clinics/Medical Offices
- Resort Homes & Condos

MODEL NUMBER	6000-1	6000-2
Max. Humidification Capacity	4 lbs/hr (11.5 gpd)	8 lbs/hr (23 gpd)
Distribution Method	Built-In Blower	Built-In Blower
Humidistat	Built-In	Built-In
Differential Pressure Switch	Built-In	Built-In
Steam Plume Length	3'*	5'*
Voltage	115 VAC	208-230 VAC
Phase	Single	Single
Cycle	50/60 Hz	50/60 Hz
Max. Operating Amps	11.5	11.5
Power Consumption	1.33 Kw	2.66 Kw
Required Circuit	15 Amps	15 Amps
Shipping Weight	28 lbs.	28 lbs.
Cabinet Dimensions	22" H x 11" W x 7-1/4" D	

Figure 4.33 Electronic steam humidifier. (Courtesy of TRION Inc.)

STUDY QUESTIONS

Answers to the study questions may be found in the sections noted in brackets.

4-1 What are the four different configurations of parts available on forced warm air furnaces? *[Arrangement of Components]*

4-2 Where is the fan located in respect to the heat exchanger on each type of furnace? *[Arrangement of Components]*

4-3 Describe the two types of heat exchangers on standard furnaces. *[Heat Exchanger]*

4-4 What material is used to construct the standard heat exchanger? *[Heat Exchanger]*

4-5 What are the two types of fan-motor arrangements? *[Fans and Motors]*

4-6 If a motor has a 3-in. pulley, and the blower a 12-in. pulley, using a motor operating at 1200 rev/min, what is the speed of the fan? *[Types of Fans]*

4-7 How much can the fan speed be adjusted with a variable-pitch motor pulley? *[Types of Fans]*

4-8 Name and describe the three types of air filters. *[Filters]*

4-9 What is the range of particles removed by the electronic air filter? *[Filters]*

4-10 How is the inside humidity requirement affected by the outside temperature? *[Humidifiers and Humidistats]*

4-11 What are the three types of evaporative humidifiers? *[Types of Humidifiers]*

4-12 Where does the steam power humidifier get its steam? *[Types of Humidifiers]*

4-13 What is the purpose of the flushing device? *[Types of Humidifiers]*

CHAPTER 5

Components of Gas-Burning Furnaces

OBJECTIVES

After studying this chapter, the student will be able to:

- Identify the components of a gas-burning assembly
- Adjust a gas burner for best performance
- Provide service and maintenance for a gas-burning assembly

Gas-Burning Assembly

The function of a gas-burning assembly is to produce a proper fire at the base of the heat exchanger. To do this, the assembly must:

- Control and regulate the flow of gas
- Assure the proper mixture of gas with air
- Ignite the gas under safe conditions

To accomplish these actions, a gas-burning assembly consists of four major parts or sections. These parts are:

1. Gas valve
2. Flame proving circuit
3. Manifold and orifice
4. Gas burners and adjustment

An illustration of a complete gas burner manifold assembly is shown in Figure 5.1.

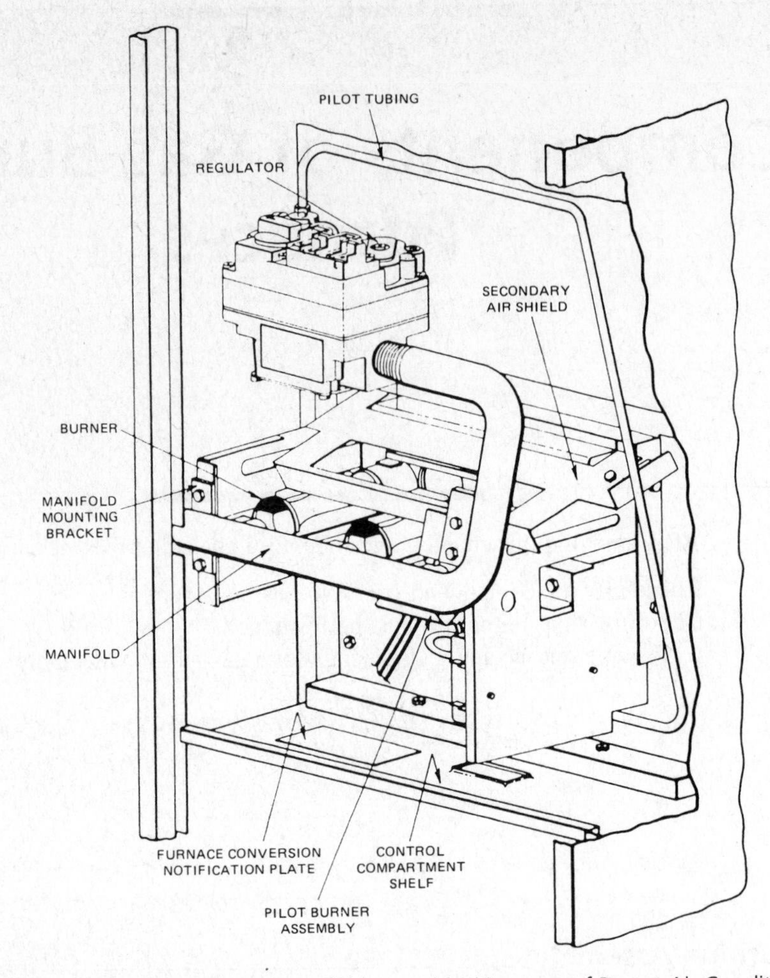

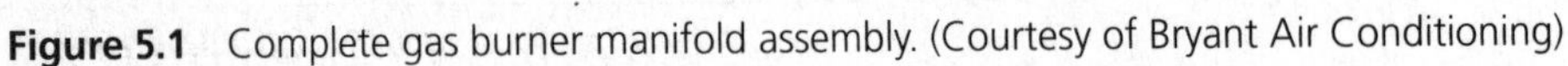
Figure 5.1 Complete gas burner manifold assembly. (Courtesy of Bryant Air Conditioning)

Gas Valve

The gas valve section consists of a number of parts, with each performing a different function. On older units these operations were entirely separate. On modern units these parts are all contained in a combination gas valve (CGV) and include:

- Hand shutoff valve
- Pressure-reducing valve
- Safety shutoff equipment
- Operator-controlled automatic gas valve

Figure 5.2 shows the separate parts of the gas valve section installed along the manifold, which is sometimes called a *gas regulator train.* Figure 5.3 shows the CGV that includes these separate parts, each of which is replaceable if trouble should develop.

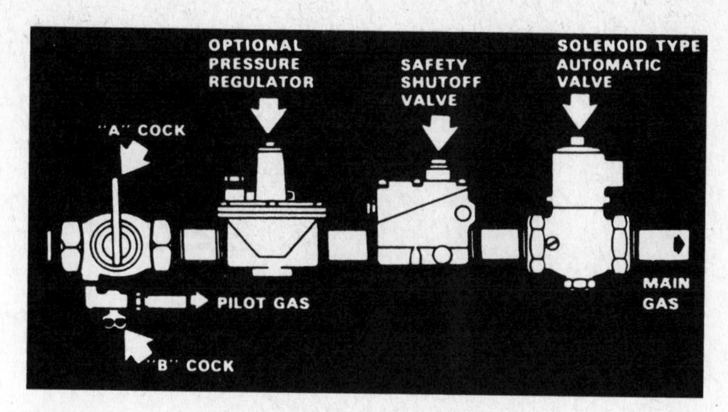

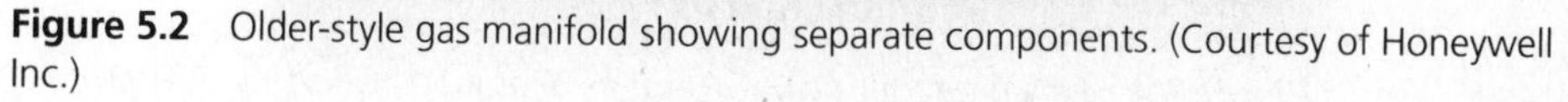

Figure 5.2 Older-style gas manifold showing separate components. (Courtesy of Honeywell Inc.)

Hand Shutoff Valves Referring to Figure 5.2, note that the main shutoff valve or gas cock is composed of two parts, an A cock and a B cock. The A cock is used to turn the main gas supply on and off manually. The B cock is used to turn the gas supply to the pilot on and off manually.

These gate- or ball-type valves are open when the handle is parallel to the length of the pipe and closed when the handle is perpendicular to the length of the pipe. They should be either on or off and not placed in an intermediate position. If it is desirable to regulate the supply of gas, it should be done by adjusting the pressure regulator.

Pressure-Reducing Valve The pressure-reducing valve, or pressure regulator, decreases the gas pressure supplied from the utility meter at approximately 7 in. W.C. to 3½ in. W.C. The regulator maintains a constant gas pressure at the furnace to provide a constant input of gas to the furnace. On an LP system a regulator is supplied at the tank and at the gas valve on the furnace.

As shown in Figure 5.4, the spring exerts downward pressure on the diaphragm that is connected to the gas valve. The spring pressure tends to open the valve. When the gas starts to flow through the valve, the downstream pressure of the gas creates an upward pressure on the diaphragm, tending to close it. Thus the spring pressure and the downstream gas pressure oppose each other and must be equal for the pressure to remain in equilibrium. To adjust the gas pressure the cap is removed and the spring

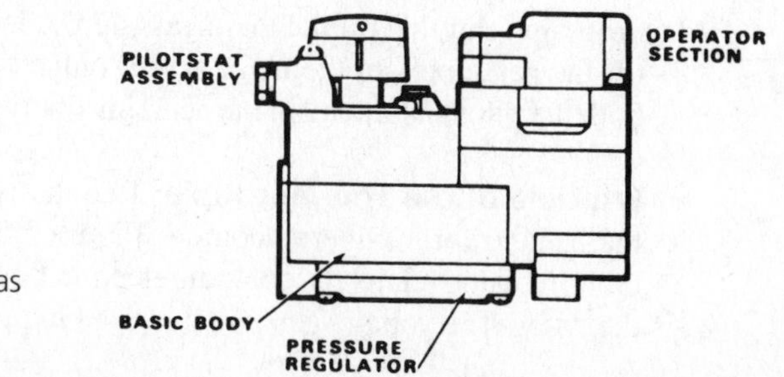

Figure 5.3 Combination gas valve (CGV). (Courtesy of Honeywell Inc.)

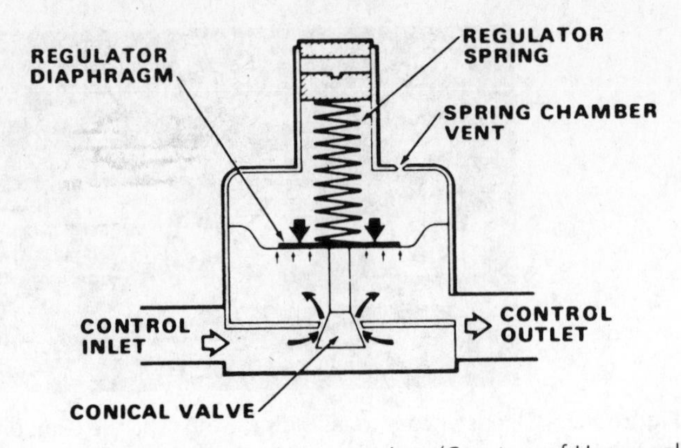

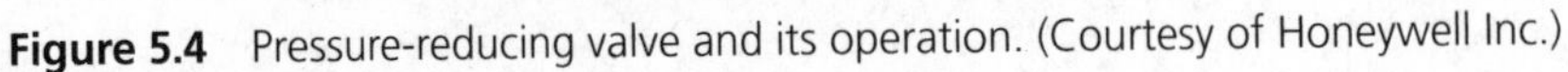
Figure 5.4 Pressure-reducing valve and its operation. (Courtesy of Honeywell Inc.)

tension is adjusted. Also note that the vent maintains atmospheric pressure on the top of the diaphragm.

Safety Shutoff Equipment The safety shutoff equipment consists of:

- Pilot burner
- Thermocouple or thermopile
- Pilotstat power unit

The pilot burner and the thermocouple or thermopile are assembled into one unit, as shown in Figures 5.12 to 5.14. This unit is placed near the main gas burner. The thermocouple is connected to the pilotstat power unit. The pilotstat power unit may be located in the CGV or installed as a separate unit in the gas regulation train, as shown in Figure 5.2.

The pilot burner has two functions:

1. It directs the pilot flame for proper ignition of the main burner flame.
2. It holds the thermocouple or thermopile in correct position with relation to the pilot flame.

The thermocouple or thermopile extends into the pilot flame and generates sufficient voltage to hold in the pilotstat. The thermopile generates additional voltage, which also operates the main gas valve.

The pilotstat power unit is held in by the voltage generated by the thermocouple. If the voltage is insufficient, indicating a poor or nonexistent pilot light, the power unit drops out, shutting off the main gas supply. These units are of two types: one that shuts off the gas supply to the main burner only, and one that shuts off the supply of gas to both the pilot and the main burner. Units burning LP gas always require 100% shutoff.

Principle of the Thermocouple Two dissimilar metal wires are welded together at the ends to form a thermocouple (Figure 5.5). When one junction is heated and the other remains relatively cool, an electrical current is generated, which flows through the wires. The voltage generated is used to operate the pilotstat power unit.

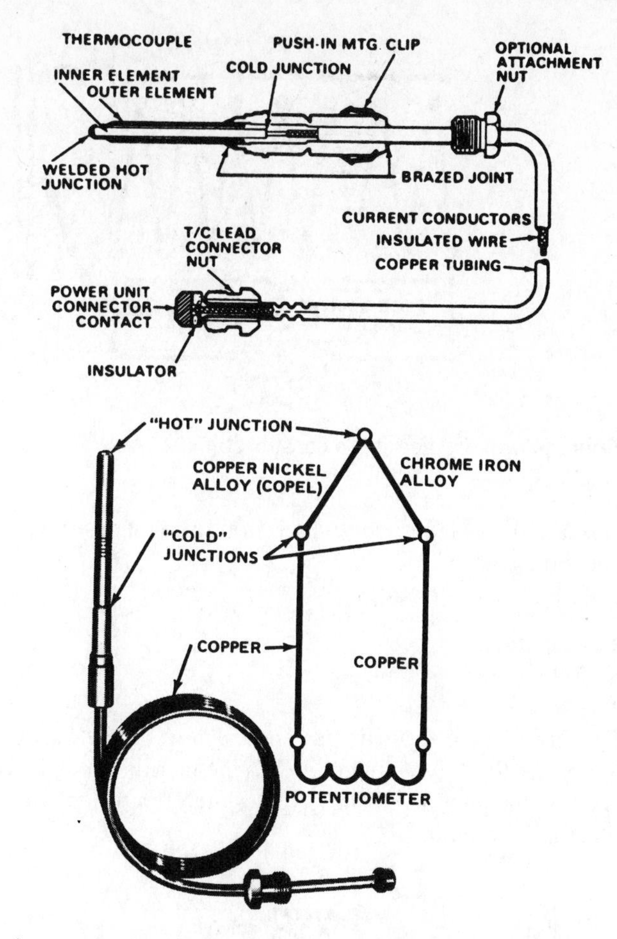

Figure 5.5 Basic construction. (Courtesy of Honeywell Inc.)

Thermopile The thermopile is a number of thermocouples connected in series (Figure 5.6). One thermocouple may generate 25 to 30 millivolts (mV). A thermopile may generate as high as 750 or 800 mV (1000 mV equals 1 V).

Power Unit The power unit is energized by the voltage generated by the thermocouple or thermopile. This voltage operates through an electromagnet. The voltage generated is sufficient to hold in the plunger against the pressure of the spring. The position of the plunger against the electromagnet must be manually set. If the thermocouple voltage drops due to a poor or nonexistent pilot, the plunger is released by the spring. This action either causes the plunger itself to block the flow of gas or operates an electrical switch that shuts off the supply of gas. The plunger must be manually reset when the proper pilot is restored. Figure 5.7 shows the pilotstat mechanism installed in a CGV.

Operator-Controlled Automatic Gas Valve The main function of the operator is to control the gas flow. Some operators control the gas flow directly; some regulate the

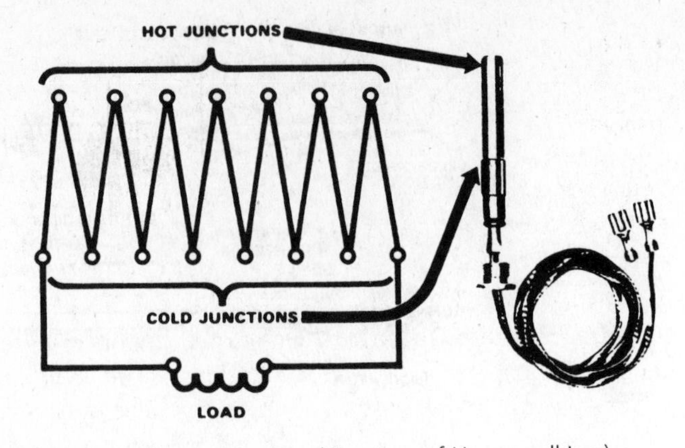

Figure 5.6 Basic thermopile construction. (Courtesy of Honeywell Inc.)

pressure on a diaphragm, which in turn regulates the gas flow. The various types of operators are:

- Solenoid
- Diaphragm
- Bulb

Solenoid Valve Operators These units (Figure 5.8) employ electromagnetic force to operate the valve plunger. When the thermostat calls for heat, the plunger is raised, opening the gas valve. These valves are often filled with oil to eliminate noise and to serve as lubrication.

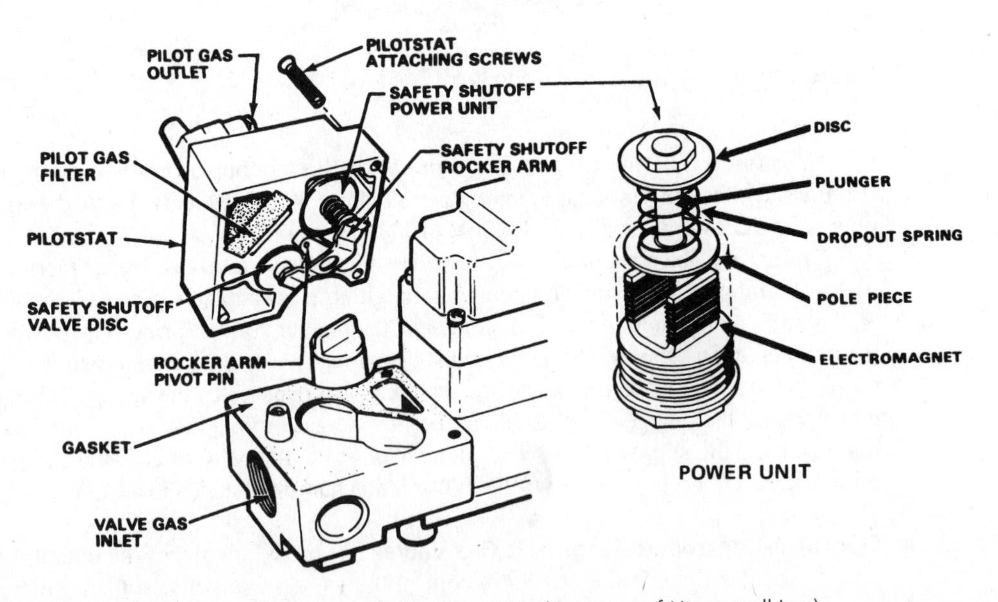

Figure 5.7 Pilotstat mechanism in a CGV. (Courtesy of Honeywell Inc.)

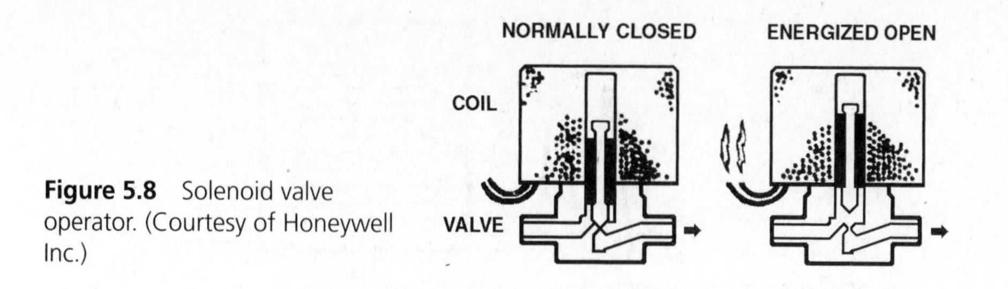

Figure 5.8 Solenoid valve operator. (Courtesy of Honeywell Inc.)

Diaphragm Valve Operators On these operators (Figure 5.9) gas is used to control the pressure above and below the diaphragm. The diaphragm is attached to the valve. To close the valve, gas pressure is released above the diaphragm. With equal pressure above and below the diaphragm, the weight of the valve closes it. To open the valve, the supply of gas is cut off above the diaphragm, and gas pressure is vented to atmosphere. The pressure of the gas below the diaphragm opens the valve. These valves also can be filled with oil.

Bulb Valve Operators Bulb operators (Figure 5.10) use the expansion of a liquid-filled bulb to provide the operating force. A snap-action disk is also incorporated to speed up the opening and closing action.

Redundant Gas Valve The redundant gas valve, shown in Figure 5.11, is used on furnaces with electronic ignition and circuit boards. This valve has two or more internal valves operating in series. All of the main burner gas must flow through both valves before reaching the gas manifold. Either valve, operating independently of the other, can interrupt all main burner gas flow if the need arises. Usually, the first valve

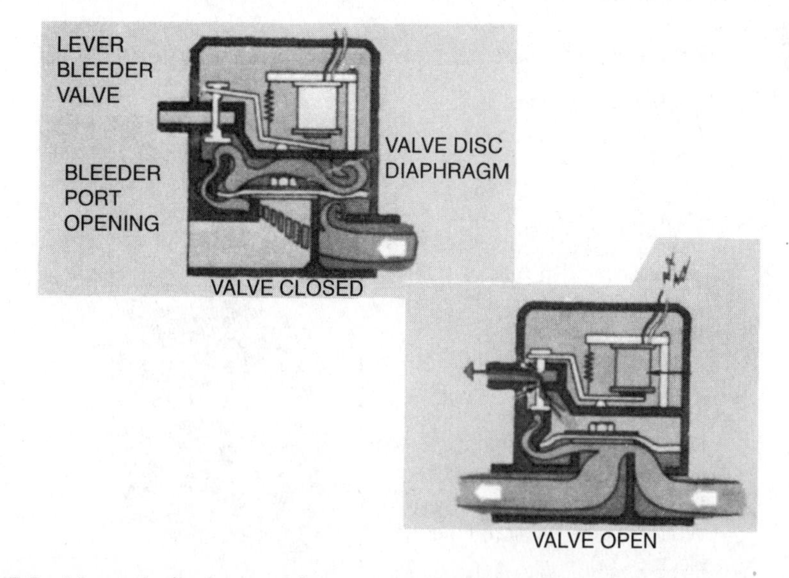

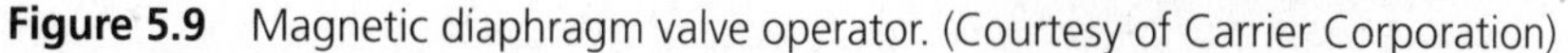

Figure 5.9 Magnetic diaphragm valve operator. (Courtesy of Carrier Corporation)

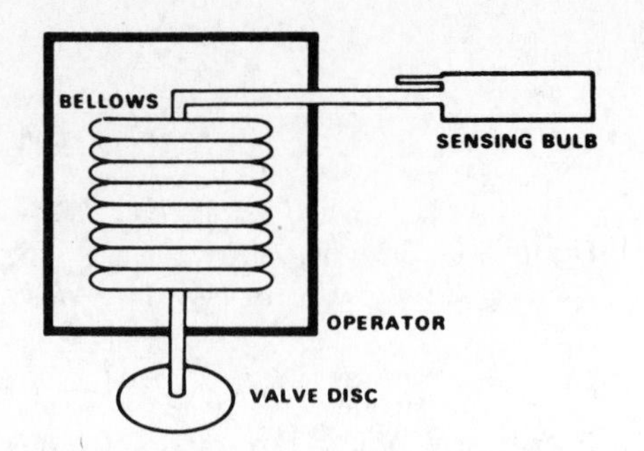

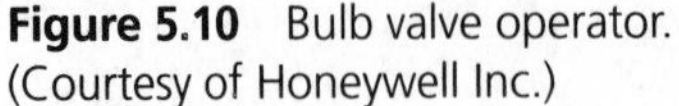
Figure 5.10 Bulb valve operator. (Courtesy of Honeywell Inc.)

is operated by a solenoid, and the second valve is opened like a diaphragm valve or with a bimetal heat motor. A knob is provided for manual shutoff and must be kept in the "on" position at all times during normal operation.

When the furnace is equipped with a spark ignition pilot, the pilot tap is normally located between the two valves. The circuit board energizes the first valve on a call for heat, prior to pilot ignition. The second valve is energized when the pilot has been verified.

On furnaces where an electric heating element, called a hot surface ignitor (HSI), lights the main burner directly, the internal valves are often energized simultaneously. Therefore, the redundancy is for safety only.

Safety Pilot

The function of the safety pilot is to provide an ignition flame for the main burner and to heat the thermocouple to provide safe operation.

Pilot Burner There are two types of pilot burners:

1. Primary-aerated
2. Non-primary-aerated

Figure 5.11 Redundant gas valve for intermittent and direct spark application. (Courtesy of White–Rogers Division, Emerson Electric Co.)

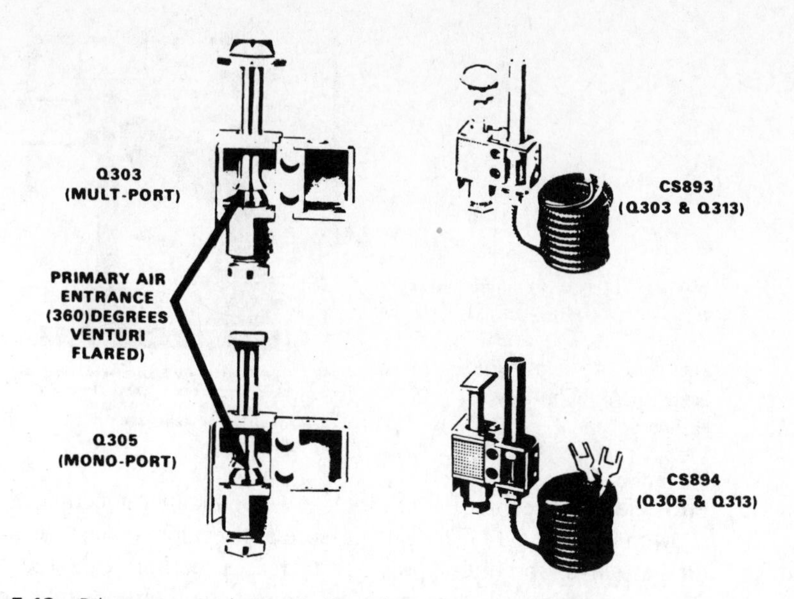

Figure 5.12 Primary-aerated pilot designs. (Courtesy of Honeywell Inc.)

On the *primary-aerated pilot* (Figure 5.12), the air is mixed with the gas before it enters the pilot burner. The disadvantage of this system is that dirt and lint tend to clog the screened air opening. The air opening must be cleaned periodically.

On the *non-primary-aerated pilot* (Figure 5.13), the gas is supplied directly to the pilot, without the addition of primary air. All the necessary air is supplied as secondary air. This eliminates the need for cleaning the air passages.

Pilot Burner Orifice The orifice is the part that controls the supply of gas to the burner. The drilled opening permits a small stream of gas to enter the burner. The amount that enters is dependent on the size of the drilled hole and the manifold pressure. Some cleaning of this opening may be required. This can be done by blowing air through the orifice or by using a suitable non-oily solvent.

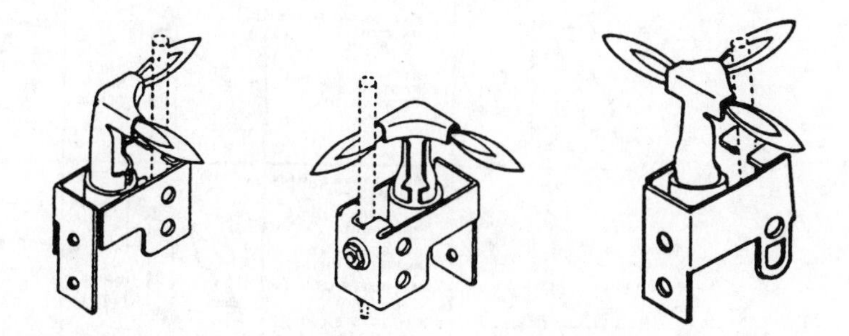

Figure 5.13 Non-primary-aerated pilot designs. (Courtesy of White–Rodgers Division, Emerson Electric Co.)

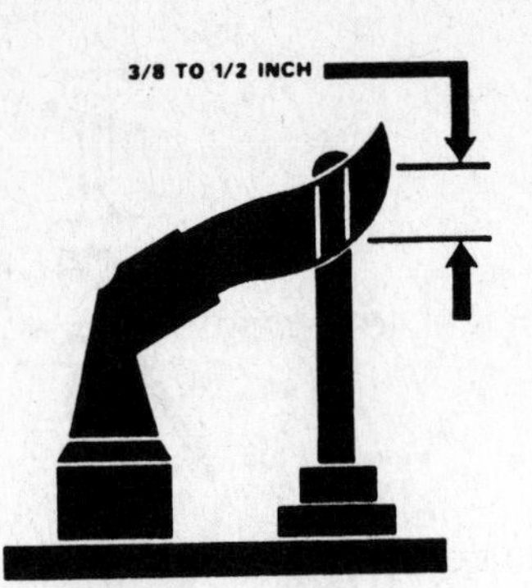

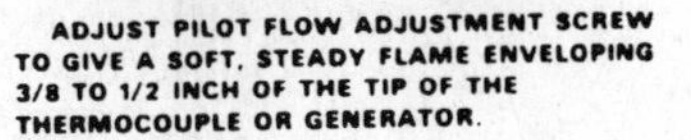

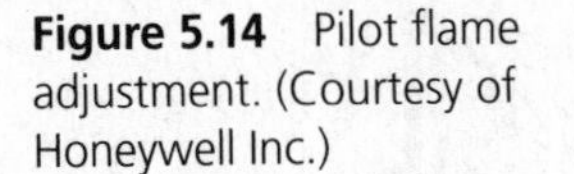
Figure 5.14 Pilot flame adjustment. (Courtesy of Honeywell Inc.)

Pilot Flame The pilot flame should envelop the thermocouple ⅜ to ½ in. at its top, as shown in Figure 5.14. The gas pressure at the pilot is the same as the inlet pressure to the gas valve. Too high a pressure decreases the life of the thermocouple. Too low a pressure provides unsatisfactory heating of the thermocouple tip. Poor flame conditions are shown in Figure 5.15.

Mercury Flame Sensor

A mercury flame sensor (Figure 5.16) is a mechanical device that proves the existence of a pilot flame. It converts the heat of the pilot flame to motion that is used to open and close a set of electrical contacts. The sensor consists of a bulb, capillary,

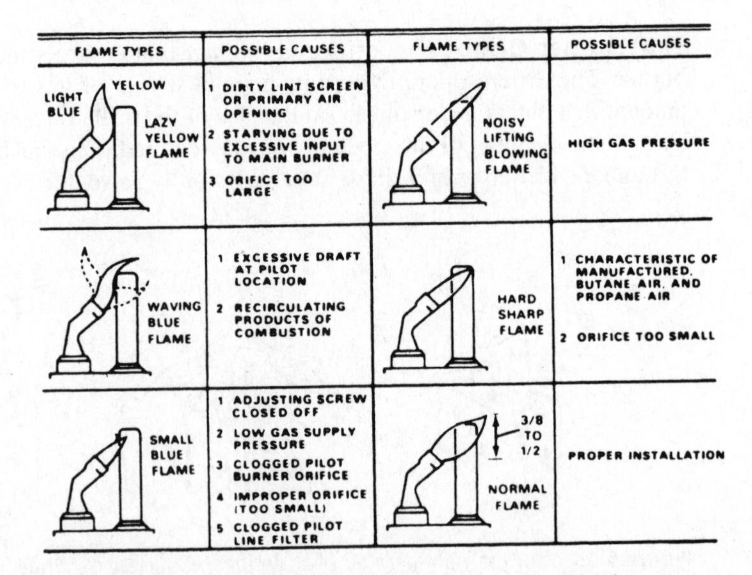

FLAME TYPES	POSSIBLE CAUSES	FLAME TYPES	POSSIBLE CAUSES
YELLOW, LIGHT BLUE — LAZY YELLOW FLAME	1 DIRTY LINT SCREEN OR PRIMARY AIR OPENING 2 STARVING DUE TO EXCESSIVE INPUT TO MAIN BURNER 3 ORIFICE TOO LARGE	NOISY LIFTING BLOWING FLAME	HIGH GAS PRESSURE
WAVING BLUE FLAME	1 EXCESSIVE DRAFT AT PILOT LOCATION 2 RECIRCULATING PRODUCTS OF COMBUSTION	HARD SHARP FLAME	1 CHARACTERISTIC OF MANUFACTURED, BUTANE AIR, AND PROPANE AIR 2 ORIFICE TOO SMALL
SMALL BLUE FLAME	1 ADJUSTING SCREW CLOSED OFF 2 LOW GAS SUPPLY PRESSURE 3 CLOGGED PILOT BURNER ORIFICE 4 IMPROPER ORIFICE (TOO SMALL) 5 CLOGGED PILOT LINE FILTER	3/8 TO 1/2 — NORMAL FLAME	PROPER INSTALLATION

Figure 5.15 Poor pilot flame conditions. (Courtesy of Honeywell Inc.)

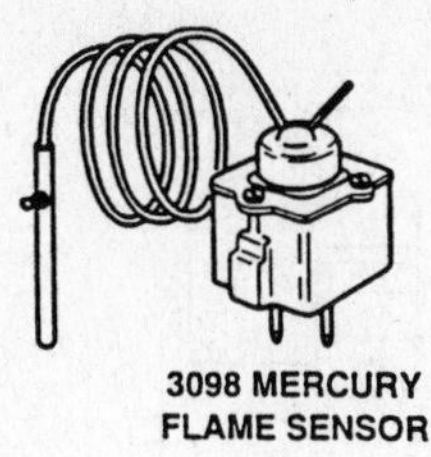

Figure 5.16 A mercury flame sensor. (Courtesy of White–Rogers Division, Emerson Electric Co.)

diaphragm, snap-switch mechanism, and mercury. When the bulb is heated by the pilot flame, the mercury vaporizes, causing pressure in the capillary and the diaphragm. Movement of the diaphragm causes the snap switch to open to one set of contacts and close to another set. These contacts control the pilot valve and the main valve.

Bimetal Flame Sensor

A bimetal flame sensor (Figure 5.17) is a mechanical device that proves the existence of a pilot flame. It converts the heat of the pilot flame to motion that is used to open and close a set of electrical contacts. The sensor consists of a set of contacts and a bimetal snap-switch mechanism. When the sensor is heated by the pilot flame, the bimetal warps in one direction. Movement of the bimetal causes the snap switch to open to one set of contacts and close to another set. These contacts control the pilot valve and the main valve.

Electronic Flame Sensor (Flame Rectification System)

A flame rectification system (Figure 5.18) is the most common flame proving system being used today. It is used universally by many heating control companies. These systems are seen on furnaces of the indirect ignition (utilizing a pilot) and direct ignition (no pilot) types.

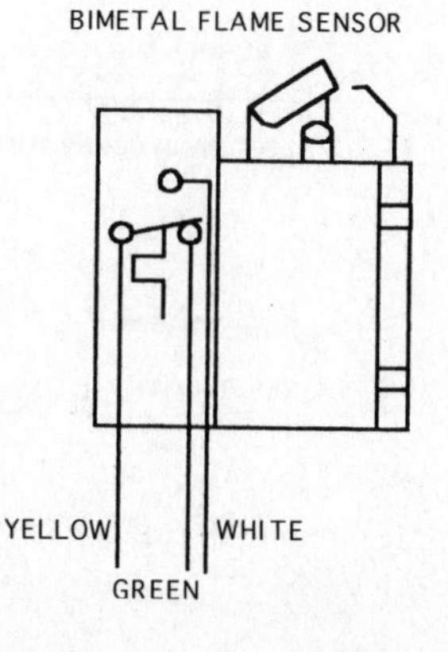

Figure 5.17 A bimetal flame sensor.

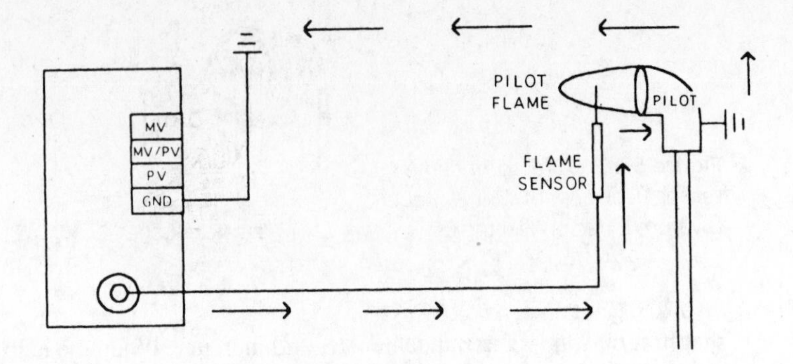

Figure 5.18 A flame rectification system.

Flame rectification systems use an electronic control module that sends an ac voltage to the flame sensor. Through the flame, alternating current is converted to a direct current that flows through the pilot or burner flame and completes the circuit to the grounded burner. From the burner, current flows back to the grounded control module, thus completing a circuit. The flow of current through the flame completes and proves burner operation, and the system will stay lit for the complete cycle of the furnace. The current that flows in this circuit is measured in microamps (μA), one-millionth of an amp. Many systems range from 0.2 up to 15 μA.

Manifold and Orifice

The manifold delivers gas equally to all the burners. It connects the supply of gas from the gas valve to the burners and is usually made of ½- to 1-in. pipe. Connections can be made either to the right or left, depending on the design, as shown in Figure 5.19.

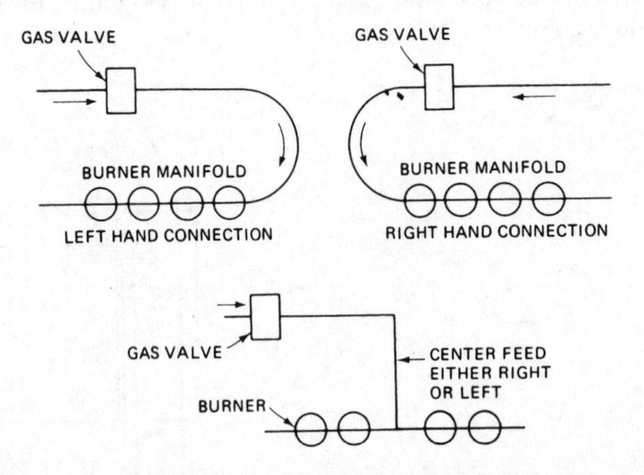

Figure 5.19 Gas manifold piping arrangements. (Courtesy of White–Rogers Division of Emerson Electric Co.)

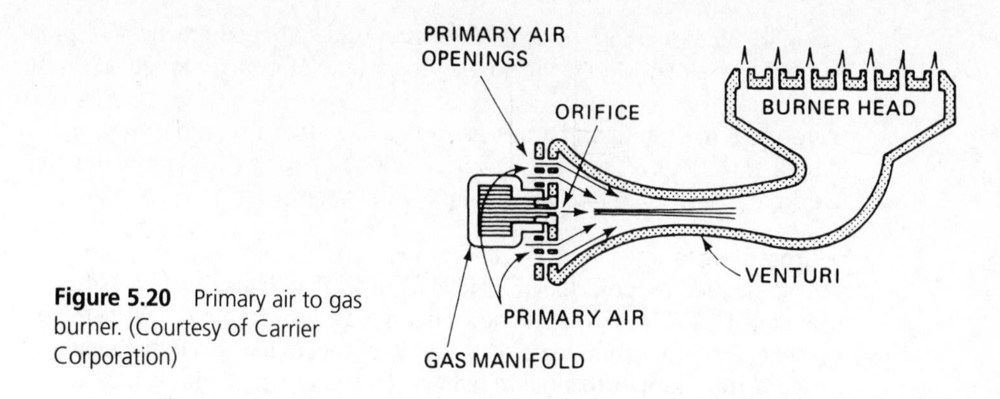

Figure 5.20 Primary air to gas burner. (Courtesy of Carrier Corporation)

The orifice used for the main burners is similar to the pilot orifice, but larger. The drilled opening permits a fast stream of raw gas to enter the burner. The orifice is sized to permit the proper flow of gas. Items to be considered in sizing the orifice are:

- Type of gas
- Pressure in manifold
- Input of gas required for each burner

Gas Burners and Adjustment

Primary and Secondary Air Most gas burners require that some air be mixed with the gas before combustion. This air is called *primary air* (Figure 5.20). Primary air constitutes approximately half of the total air required for combustion. Too much primary air causes the flame to lift off the burner surface. Too little primary air causes a yellow flame.

Air that is supplied to the burner at the time of combustion is called *secondary air* (Figure 5.21). Too little secondary air causes the formation of carbon monoxide. To be certain that enough secondary air is provided, most units operate on an excess of sec-

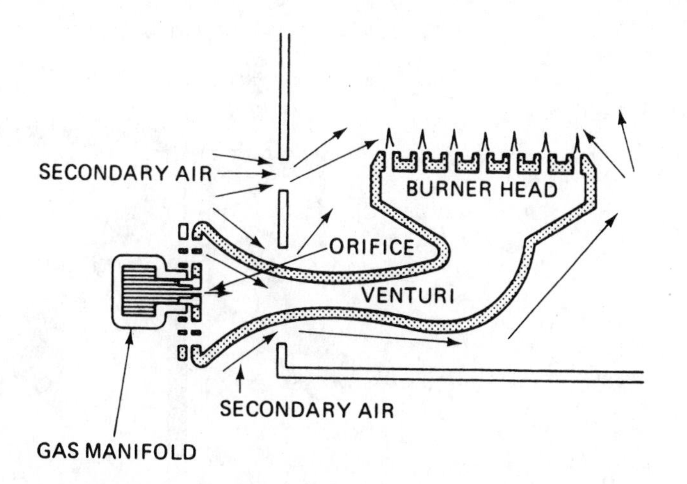

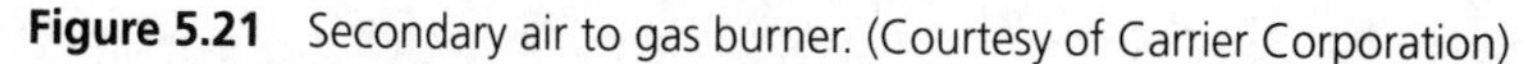
Figure 5.21 Secondary air to gas burner. (Courtesy of Carrier Corporation)

ondary air. An excess of about 50% is considered good practice. To produce a good flame it is essential to maintain the proper ratio of primary to secondary air.

Venturi In a multiport burner, gas is delivered into a venturi (mixing tube), creating a high gas velocity. The increase in gas velocity creates a sucking action that causes the primary air to enter the tube and mix with the raw gas.

Crossover Igniters The crossover or carryover igniter is a projection on each burner near the first few ports that permits the burner that lights first to pass flame to the other burners. There is a baffle, as shown in Figure 5.24, that directs gas to the first few ports of the burner to assure the proper supply of fuel to the area of the igniter.

Note: Burners that fail to light properly can cause rollout, a dangerous condition in which flames emerge in the area containing the gas valve.

Draft Diverter The draft diverter (Figure 5.22) is designed to provide a balanced draft (slightly negative) over the flame in a gas-fired furnace. The bottom of the diverter is open to allow air from the furnace room to blend with the products of combustion. In case of blockage or downdraft in the chimney, the flue gases vent to the area around the furnace.

Note: The American Gas Association (AGA) requires a rigid test to be certain that under these conditions complete combustion occurs and there is no danger of suffocation of the flame.

Types of Burners There are four types of burners, classified according to the type of burner head:

1. Single-port or inshot
2. Drilled-port
3. Slotted-port
4. Ribbon

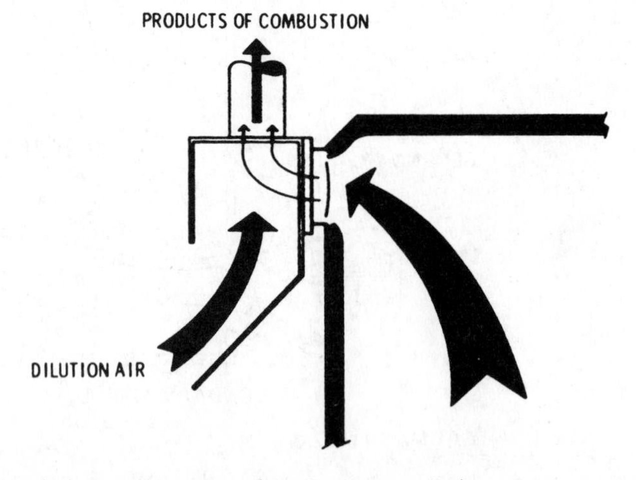

Figure 5.22 Draft diverter. (Courtesy of Carrier Corporation)

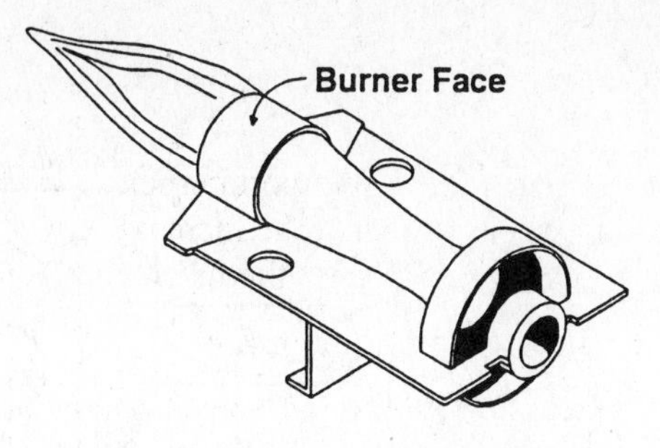

Figure 5.23 Single-port or inshot burner. (Courtesy of International Comfort Products Corporation, USA)

Single-Port or Inshot Burner The single-port is the simplest type (Figure 5.23). Fuel and air are mixed inside the burner tube and ignited at the outlet of the burner. After ignition, the flame is directed into the heat exchanger.

Drilled-Port Burner The drilled-port burner is usually made of cast iron with a series of small drilled holes. The size of the hole varies for different types of gas.

Slotted-Port Burner The slotted-port burner (Figure 5.24) is similar to the drilled-port burner, but in place of holes, it uses elongated slots. It has the advantage of being able to burn different types of gas without change of slot size. It is also less susceptible to clogging with lint.

Ribbon Burner The ribbon burner has a continuous opening down each side, and when it is lit, the flame has the appearance of a ribbon. The burner itself is made either of cast iron or fabricated metal. The ribbon insert is usually corrugated stainless steel.

Adjusting Gas Pressure

It is extremely critical to have the properly designed gas pressure to the furnace to ensure the correct Btu rating output for the furnace. If a furnace has too high a gas

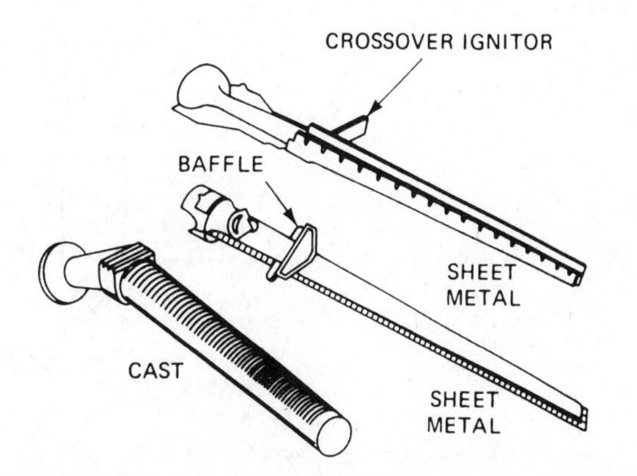

Figure 5.24 Slotted-port burners. (Courtesy of Carrier Corporation)

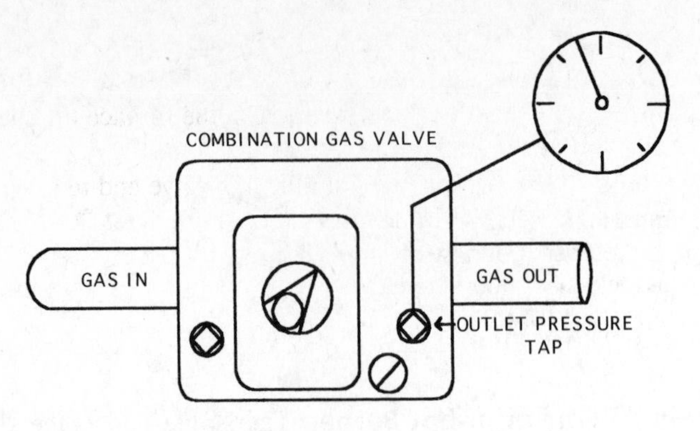

Figure 5.25 Measuring gas manifold pressure.

pressure, it will overheat and can crack the heat exchanger. If a furnace has too low a gas pressure, it can cause condensation and corrosion to form in the heat exchanger.

To test and adjust gas pressure on an LP or natural gas–fired furnace, use the following procedures:

1. Check the furnace name plate for required operating gas pressure. (Commonly 3.5 in. W.C. for natural gas and 10 to 11 in. W.C. for LP)
2. Turn the main gas valve to the "off" position.
3. Locate the setscrew for the outlet gas pressure on the combination gas valve, shown in Figure 5.25. (The reading can also be taken at the manifold if a pressure tap is present.)
4. Remove the setscrew and install the manometer.
5. Turn the main gas valve to the "on" (in-line) position and start the furnace.
6. When the furnace ignites, a pressure reading will be present on the gauge.
7. If the gas pressure is not the recommended gas pressure of the furnace, locate the gas pressure regulator adjustment screw on the gas valve (Figure 5.26). Remove the

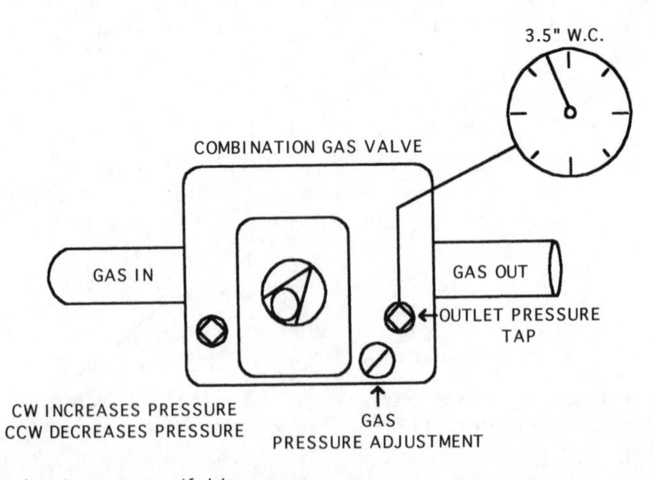

Figure 5.26 Adjusting gas manifold pressure.

cap and adjust the gas pressure. (Clockwise increases pressure and counterclockwise decreases pressure.)

8. After adjusting the gas pressure turn the furnace off and turn the main gas valve to the "off" position.
9. Remove the manometer from the gas valve and reinsert the plug to the outlet pressure tap.

Maintenance and Service

The main items requiring service on gas-burning equipment are the:

- Gas burner assembly
- Pilot assembly
- Automatic gas valve

After the equipment has been in use for a period of time, each of these items may require service, depending on the conditions of the installation.

Gas Burner Assembly

Problems that can arise include:

- Flashback
- Carbon on the burners
- Dirty air mixture
- Noise of ignition
- Gas flames lifting from ports
- Appearance of yellow tips in flames

When the velocity of the gas–air mixture is reduced below a certain speed, the flame will flash back through the ports. This is undesirable and must be corrected. The solution is to close the primary air shutter as much as possible and still maintain a clear blue flame (Figure 5.27). Burners that become clogged with carbon due to flashback conditions must be cleared in order to prevent interference with the proper flow of gas.

If the burner flame becomes soft and yellow-tipped, it is an indication that cleaning is required. Items requiring cleaning are the shutter opening, venturi, and the burner head itself. In some cases these can be cleaned with the suction of a vacuum cleaner. In other cases, a thin, long-handled brush may be used to reach completely inside the mixing tube.

Noise of ignition is usually caused by delayed or faulty ignition. Conditions that must be checked and corrected, if called for, are location of the pilot, poor flame travel, or poor distribution of the flame on the burner itself. Lifting flames are usually caused by gas pressure that is too high or by improper primary air adjustment. The gas pressure for natural gas at the manifold should be 3.5 in. W.C. Primary air should be adjusted to provide a blue flame.

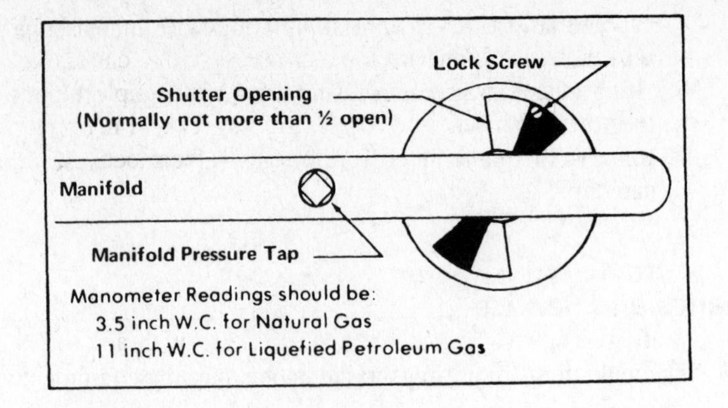

PRIMARY AIR ADJUSTMENT

To adjust primary air supply, turn main gas supply **ON** and operate unit for 15 to 20 minutes. Then loosen set screw and rotate shutter on burner bell. Open shutter until the yellow tips just disappear. Lock shutter in position by tightening set screw. Turn main gas supply **OFF**, let cool completely, and re-check operation and flame characteristics from a cold start.

Figure 5.27 Primary air adjustment.

The appearance of yellow tips on the flame indicate incomplete combustion, which can cause sooting of the heat exchanger. The following items should be checked and corrected if necessary:

1. *Gas pressure.* Pressure should be 3.5 in. W.C. for natural gas.
2. *Clogged orifices.* Inspect and clean if required.
3. *Air adjustment.* Use minimum primary air required for blue flame.
4. *Alignment.* Gas stream should move down the center of the venturi tube.
5. *Flues.* Air passages must not be clogged.
6. *Air leaks.* Air from fan or strong outside air current can cause a yellow flame.

Pilot Assembly

Pilot problems can be caused by:

- Clogged air openings
- Dirty pilot filters
- Clogged orifice
- Defective thermocouples

Dirt and lint in the air opening are the most common problems. Cleaning solves these problems. Pilots for manufactured gases use pilot filters. The cartridge must be replaced periodically. Occasionally, it is necessary to clean the pilot orifice, which may become clogged from an accumulation of dirt, lint, carbon, or condensation in the lines.

If the pilot flame does not heat the thermocouple adequately, not enough dc voltage will be generated, and the electromagnet will not energize sufficiently to permit the opening of the automatic gas valve. The thermocouple can be checked, as shown in Figure 8.3.

Automatic Gas Valve

Many gas valves are of the diaphragm type with an electromagnetic or heat motor–operated controller. The two principal problems that can occur with gas valves are:

1. A valve that will not open
2. Gas leakage through the valve

First, the pilot must be checked to determine that the proper power is being produced by the thermocouple. If the gas valve will not open after the lighting procedure shown on the nameplate is used, proceed as follows: Check to determine if power is available at the valve. If power is available, connect and disconnect the power and listen for a muffled click, which indicates that the lever arm is being actuated. Check to be sure that the vent above the diaphragm is open. If the valve still does not operate, replace the entire operator or valve top assembly. If gas continues to leak through the valve outlet or the vent opening after the valve is deenergized, replace the entire valve head assembly.

STUDY QUESTIONS

Answers to the study questions may be found in the sections noted in brackets.

5-1 What are the functions of the gas-burning assembly, and what are its component parts? *[Gas-Burning Assembly]*

5-2 On the modern gas valve, there are two shutoff cocks. What are they used for? *[Hand Shutoff Valves]*

5-3 What is the correct natural gas pressure leaving the gas valve? *[Pressure-Reducing Valve]*

5-4 What are the component parts of the safety shutoff equipment? *[Safety Shutoff Equipment]*

5-5 What are the functions of the pilot burner? *[Safety Shutoff Equipment]*

5-6 How much voltage does the thermocouple generate? *[Principle of the Thermocouple]*

5-7 What does the acronym CGV represent? *[Gas Valve]*

5-8 What type of valve operator has a delayed action feature? *[Operator-Controlled Automatic Gas Valve]*

5-9 Describe the operation of the two types of pilot burners. *[Pilot Burner]*

5-10 What type of valve requires 100% shutoff? *[Safety Shutoff Equipment]*
5-11 What is the function of the power unit? *[Safety Shutoff Equipment]*
5-12 Describe a proper pilot flame. *[Pilot Flame]*
5-13 What factors are considered in sizing the main burner orifice? *[Manifold and Orifice]*
5-14 Describe what is meant by primary and secondary main burner air. *[Primary and Secondary Air]*
5-15 What is the purpose of the venturi? *[Venturi]*
5-16 What is the function of the draft diverter? *[Draft Diverter]*
5-17 Name and describe the various types of main gas burners. *[Types of Burners]*
5-18 Describe the service required for gas-burning equipment. *[Maintenance and Service]*
5-19 Describe the various problems that can arise with the gas burner assembly, pilot assembly, and automatic gas valve. *[Maintenance and Service]*

CHAPTER 6

Basic Electricity and Electrical Symbols

OBJECTIVES

After studying this chapter, the student will be able to:

- Identify the basic electrical symbols used in wiring diagrams for electrical load devices
- Determine the types of circuits used to connect electrical loads

Electrical Terms

A knowledge of certain electrical terms is necessary in working with electrical power. Most power used for warm air heating systems is alternating current (ac). Some of the common terms used in describing alternating current are:

Volts: measure of electromotive force (EMF) or pressure being supplied to cause the electrical current to flow

Amperes: measure of the flow of current (electrons) through a conductor

Ohms (Ω): measure of the resistance to current flow through a conductor

Watts: measure of power consumed by an electrical load

Power factor: fraction obtained by dividing watts by the product of volts times amperes

Ohm's Law

Ohm's law is used to predict the behavior of electrical current in an electrical circuit. Simply stated,

$$E = IR$$

where E = electromotive force EMF, volts
I = intensity of current, amperes
R = resistance, ohms

Ohm's law can be stated in three ways (Figure 6.1). An example of the use of Ohm's law is shown in Figure 6.2. Given $E = 120$ V and $R = 10\ \Omega$ ($I = E/R$), the current flow in the circuit is 12 A.

A voltmeter can be used to measure the voltage between L_1 and L_2. An ohmmeter can be used to measure the resistance of the load but should be used *only when the power is turned off.* An ammeter can be used to measure current.

Series Circuits

A series circuit is one in which resistances are wired end to end like a string of boxcars on a train. In a series circuit the amperage stays the same throughout the circuit. The voltage is divided among the various loads. The total resistance of the circuit is the sum of the various resistances in the circuit. Thus, in Figure 6.3,

$$E_T = E_1 + E_2$$
$$I_T = I_1 = I_2$$
$$R_T = R_1 + R_2$$

Given $E_T = 120$, $R_1 = 10$, $R_2 = 10$, then

1. $R_T = R_1 + R_2 = 10 + 10 = 20\ \Omega$
2. $I_T = \dfrac{E_T}{R_T}$ (Ohm's law) $= \dfrac{120}{20} = 6$ A
3. $I_T = I_1 = I_2 = 6$ A
4. $E_1 = I_1 \times R_1 = 10 \times 6 = 60$ V
5. $E_2 = I_2 \times R_2 = 10 \times 6 = 60$ V

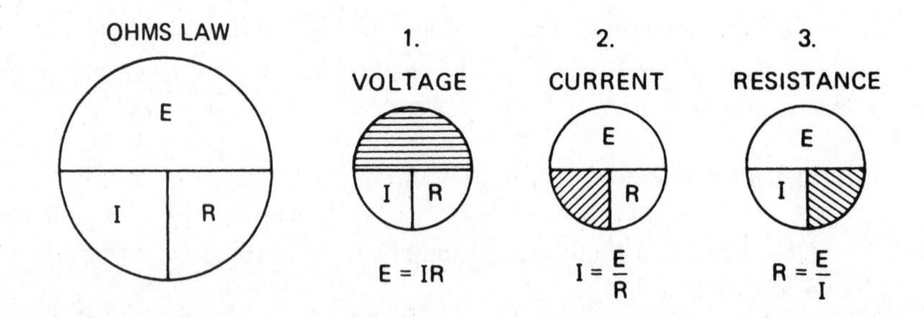

Figure 6.1 Ohm's law.

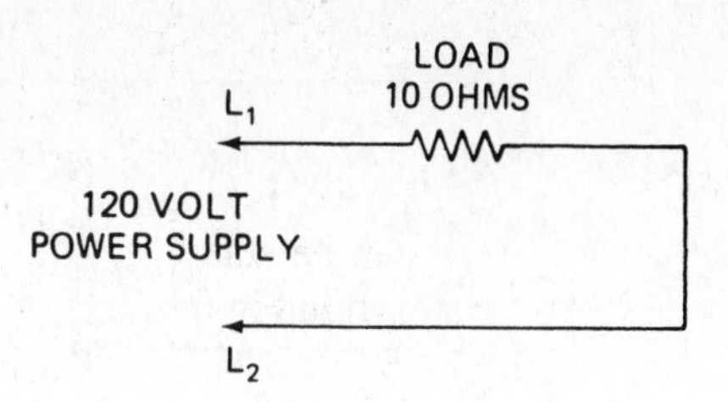

Figure 6.2 Simple electrical circuit.

Parallel Circuits

In a parallel circuit each load provides a separate path for electricity, and each path may have different current flowing through it. The amount is determined by the resistance of the load. The voltages for each load are the same. The total current is the sum of the individual load currents. The reciprocal of the total resistance is equal to the sum of the reciprocals of the individual resistances. Thus, in Figure 6.4, we have

$$E_T = E_1 = E_2$$

$$I_T = I_1 + I_2$$

$$\frac{1}{R_T} = \frac{1}{R_1} + \frac{1}{R_2}$$

or for two resistances,

$$R_T = \frac{R_1 \times R_2}{R_1 + R_2}$$

Given $E_1 = 120$, $R_1 = 10$, $R_2 = 10$, then

1. $E_T = E_1 = E_2 = 120\text{ V}$
2. $I_1 = \frac{E_1}{R_1} \text{(Ohm's law)} = \frac{120}{10} = 12\text{ A}$
3. $I_2 = \frac{E_2}{R_2} = \frac{120}{10} = 12\text{ A}$

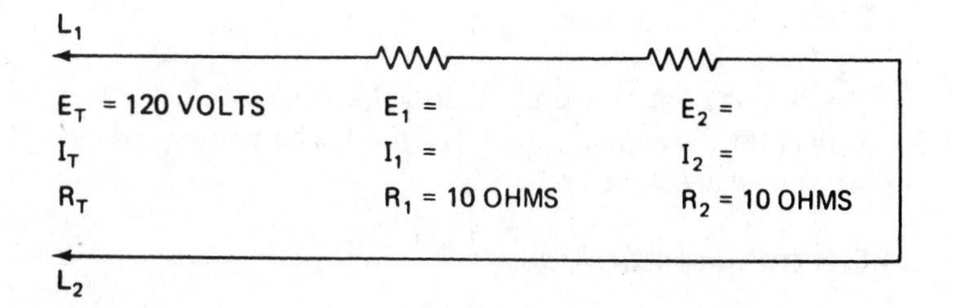

Figure 6.3 Series circuit.

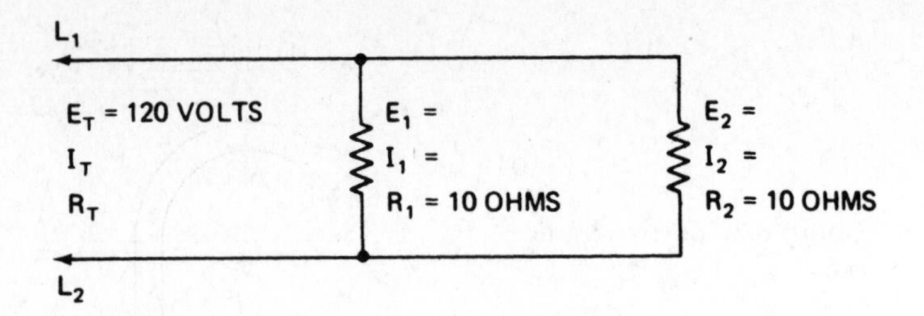

Figure 6.4 Parallel circuit.

4. $I_T = I_1 + I_2 = 12 + 12 = 24$ A
5. $R_T = \dfrac{E_T}{I_T} = \dfrac{120}{24} = 5\ \Omega$

or

$$R_T = \frac{R_1 \times R_2}{R_1 + R_2} = \frac{10 \times 10}{10 + 10} = \frac{100}{20} = 5\ \Omega$$

Electrical Potential

Electrical potential is the force that produces the flow of electricity. It is similar to the action of a pump in a water system that causes the flow of water. The battery in a flashlight is a source of electrical potential. The unit of electrical potential is the *volt.* Voltage is measured with a voltmeter.

Resistance

Resistance is the pressure exerted by the conductor in restricting the flow of current. In a water system, the flow is limited by the size of the pipe. In an electrical system, the flow of current is limited by the size of the conducting wire or by the electrical devices in the circuit. Resistance in an electrical circuit can be a means of converting electrical energy to other forms of energy, such as heat, light, or mechanical work. The unit of resistance is the ohm (Ω). Resistance is measured with an ohmmeter.

Power

Power is the use of energy to do work. An electrical power company charges its customers for the amount of energy used. The unit of power is the *watt.* Power is measured with a wattmeter.

Electromagnetic Action

Magnetic action is the force exerted by a magnet (Figure 6.5). A magnet has two poles: north and south. Like poles repel, and unlike poles attract. The force of a mag-

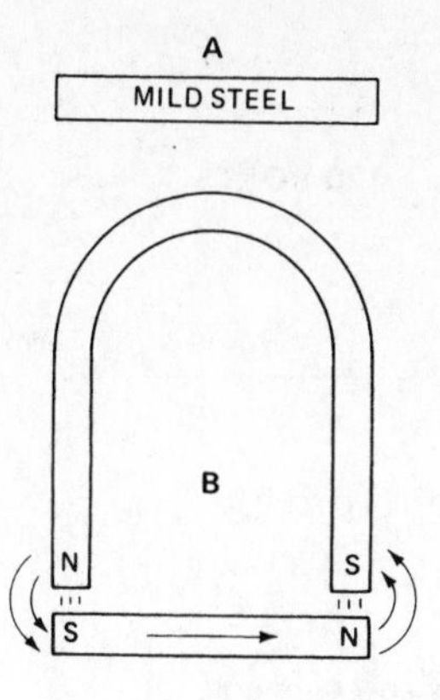

Figure 6.5 Magnetic action.

net is dependent on its strength and the distance between the magnet and the metal it is affecting.

Current

Current is a term used in electricity to describe the rate of electrical flow. Electricity is believed to be the movement of *electrons,* small electrically charged subatomic particles, through a *conductor.* A conductor is a type of metal through which electricity will flow under certain conditions. These conditions are described later in the chapter. Current flow is measured in amperes with an electrical instrument called an ammeter.

There are two types of electric current: *direct current* and *alternating current.* In direct current (dc), the electrons move through the conductor in only one direction. This is the type of current produced by a battery. In alternating current (ac), the electrons move first in one direction and then in the other, alternating their movement, usually 60 times per second [called 60 cycles or 60 hertz (Hz)]. Alternating current is the type available for residential use and is supplied by an electric power company.

Electromagnetic action is the force exerted by the magnetic field of a magnet (Figure 6.6). Current flowing through a conductor creates a magnetic field around the conductor. This action is extremely useful in construction of electrical devices.

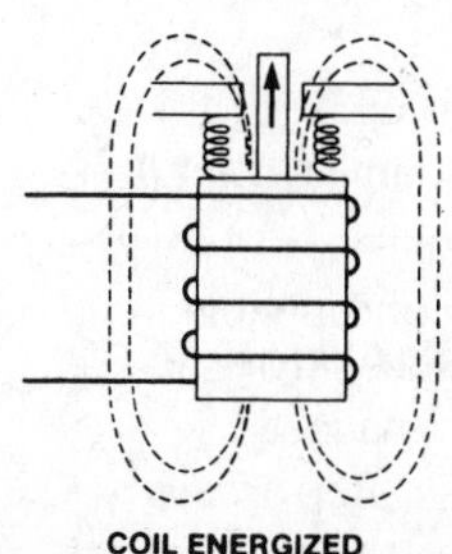

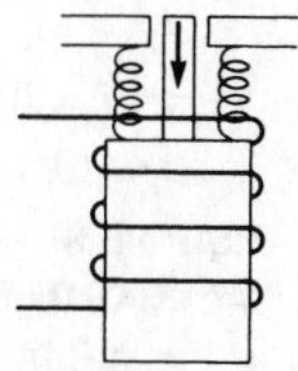

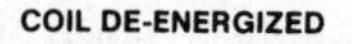

Figure 6.6 Electromagnetic action. (Courtesy of I-T-E Electrical Products)

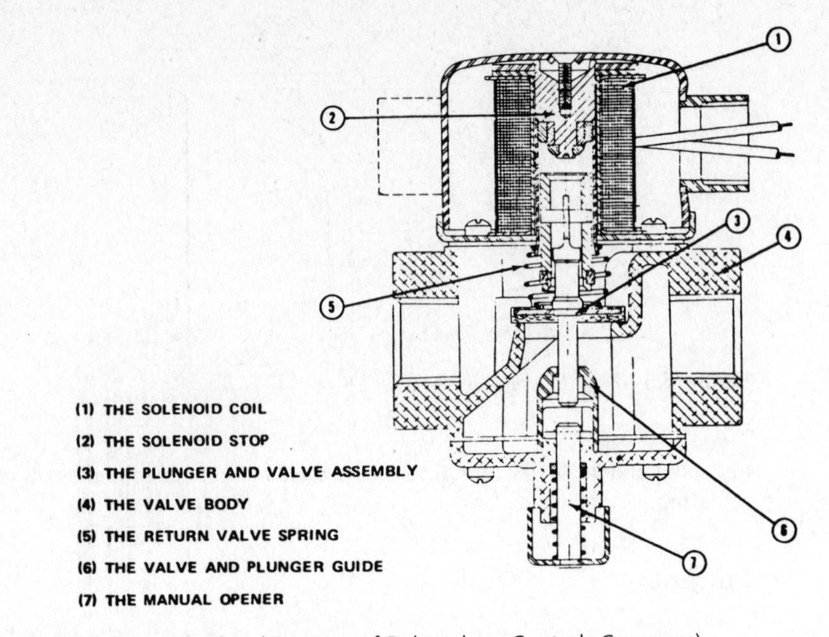

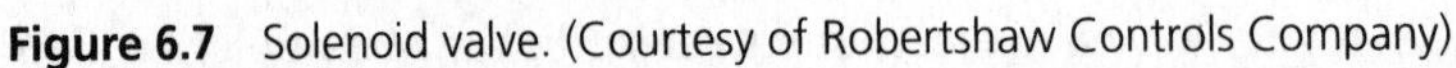

Figure 6.7 Solenoid valve. (Courtesy of Robertshaw Controls Company)

If an insulated conductor is coiled, and alternating current is passed through it, two important effects are produced:

1. A magnetic effect with directional force is produced in the field inside the coil. This force can move a separate metal plunger. The resultant action is called the *solenoid effect.* A solenoid is shown in Figure 6.7.
2. A magnetic field effect can transfer current to a nearby circuit, causing current to flow. This is called *electromagnetic induction* (Figure 6.8).

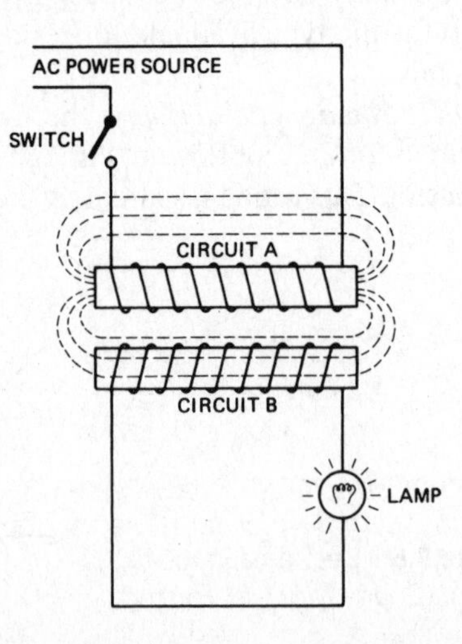

Figure 6.8 A flow of current in circuit A induces a flow of current in circuit B by electromagnetic induction.

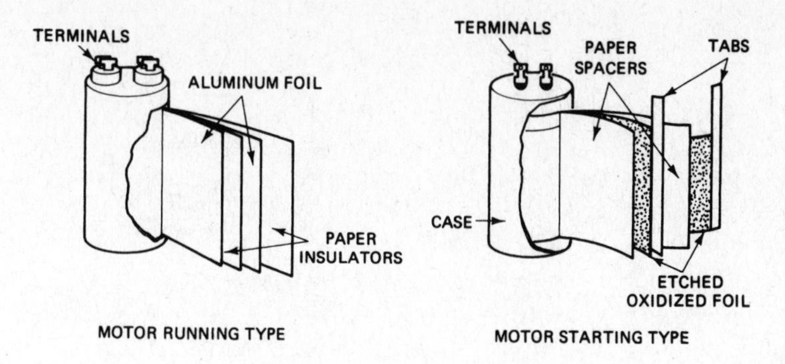

Figure 6.9 Capacitors. (Courtesy of Sprague Electric Co.)

The solenoid effect is used to operate switches and valves. The electromagnetic induction effect is used to operate motors and transformers.

Capacitance

Capacitance is the charging and discharging ability of an electrical device, called a capacitor, to store electricity. Capacitors are made up of a series of conductor surfaces separated by insulation (Figure 6.9). The unit of capacitance is the *farad* (F). Most capacitors for furnace motors are used to increase the starting power of the motor. These capacitors are rated in microfarads (μF).

Circuits

Electrical devices are used in a circuit to perform various functions. A circuit is an electrical system that provides the following:

- Source of power
- Path for power to follow
- Place for power to be used (load)

An electrical circuit is shown in Figure 6.10.

The source of power can be a battery that stores direct current or a connection to a power company's generator that supplies alternating current. The path is a continuous electrical conductor (such as copper wire) that connects the power supply to the load and from the load back to the power supply. The load is a type of electrical resistance that converts power to other forms of energy, such as heat, light, or mechanical work.

A switch can be inserted into the circuit to connect (make) or disconnect (break) the flow of current as desired, to control the operation of the load. Thus, a light switch is used to turn a light on and off by making or breaking its connection to the power supply.

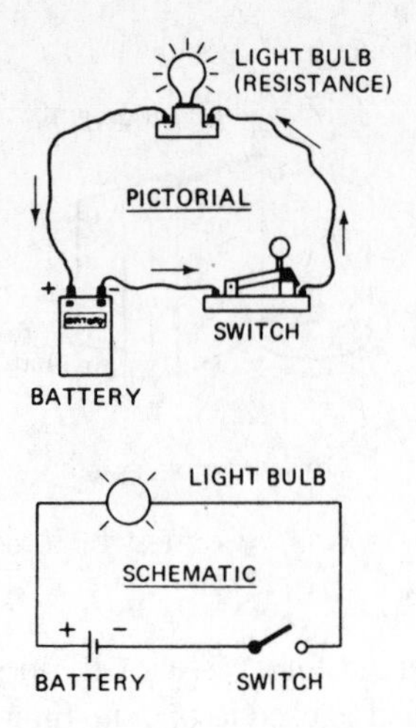

Figure 6.10 Electrical circuit.

A single-phase power supply, such as that used for heating equipment, always consists of two wires. For 120-V ac power, one wire is called *hot* (usually black), and the other wire is called *neutral* (usually white). The hot wire can be found by using a test light attached to two wire leads (Figure 6.11). Touch one lead to the hot wire and the other to the ground (a metal pipe inserted in the ground), and the bulb will light. Touch one lead to the neutral wire and the other to the ground, and the bulb will not light.

In a circuit using ac power, the assumption is made that power moves from the hot wire through the load to the neutral wire (although the current does alternate). This assumption is made only to simplify tracing circuits and locating switches.

Caution: Switches are always placed between the hot wire and the load, not between the load and the neutral wire. Thus, when the switch is open (disconnecting the circuit) it is safe to work on the load. If the switch were on the neutral side of the

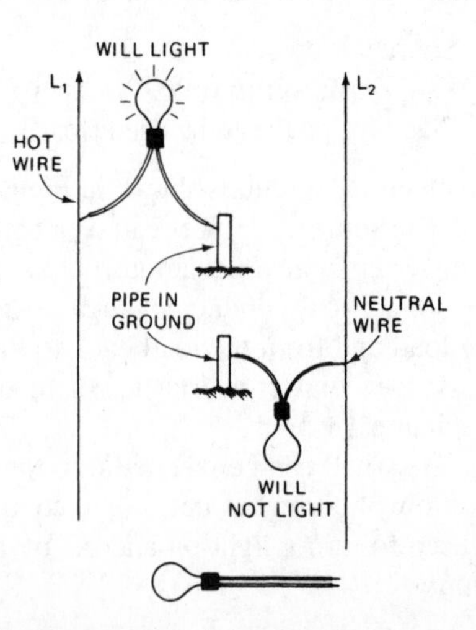

Figure 6.11 Use of test light.

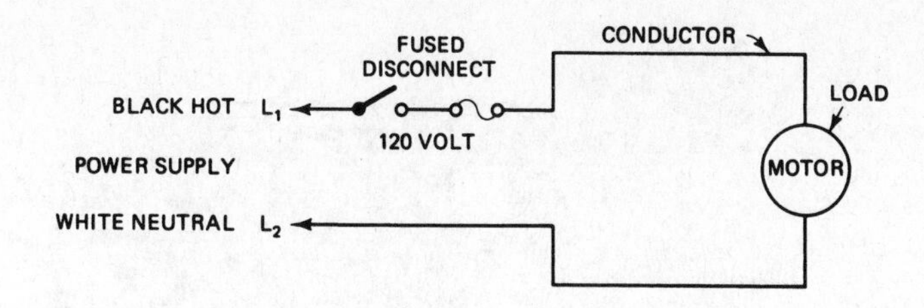

Figure 6.12 Typical electrical circuit, including a source of power, a path for the current to flow in, and a load.

load, touching the load could complete the circuit to ground. This dangerous condition could cause current to flow through a person's body with serious or fatal results.

A circuit consists of a source of electrical current, a path for it to follow, and a place for it to go (electrical load). Switches are inserted in the path of the current to control its flow either manually or automatically. Electrical devices are connected in circuits to produce a specific action, such as operating a fan motor or a firing device on a furnace.

There are two basic types of circuits: series and parallel. Electrical devices can be connected in either of these ways or in a combination of both. In Figure 6.12, with 120 V ac as a power source, the two wires are termed L_1 (hot) and L_2 (neutral). The hot wire is usually black and the neutral wire is usually white. Most 120-V installations include a third wire (color coded green) for the earth ground. These systems require a three-prong plug and outlet.

The fuse inserted in the hot side of the line is usually incorporated in the switching device known as a *fused disconnect.* When the load is in some type of residence, a thermal fuse is used. In case of a shorted circuit or overload, the fuse melts and disconnects the power automatically. When the type of load is a motor, a time-delay fuse is used.

Electrical Devices

Electrical devices are chiefly of two types (Figure 6.13):

1. Loads
2. Switches

Loads have resistance and consume power. Loads usually transform power into some other form of energy. Examples of loads are motors, resistance heaters, and lights. *Switches* are used to connect loads to the power supply or to disconnect them when they are not required.

Loads and Their Symbols

The common loads used on heating equipment are described next, along with the symbol used to represent them in the schematic wiring diagram.

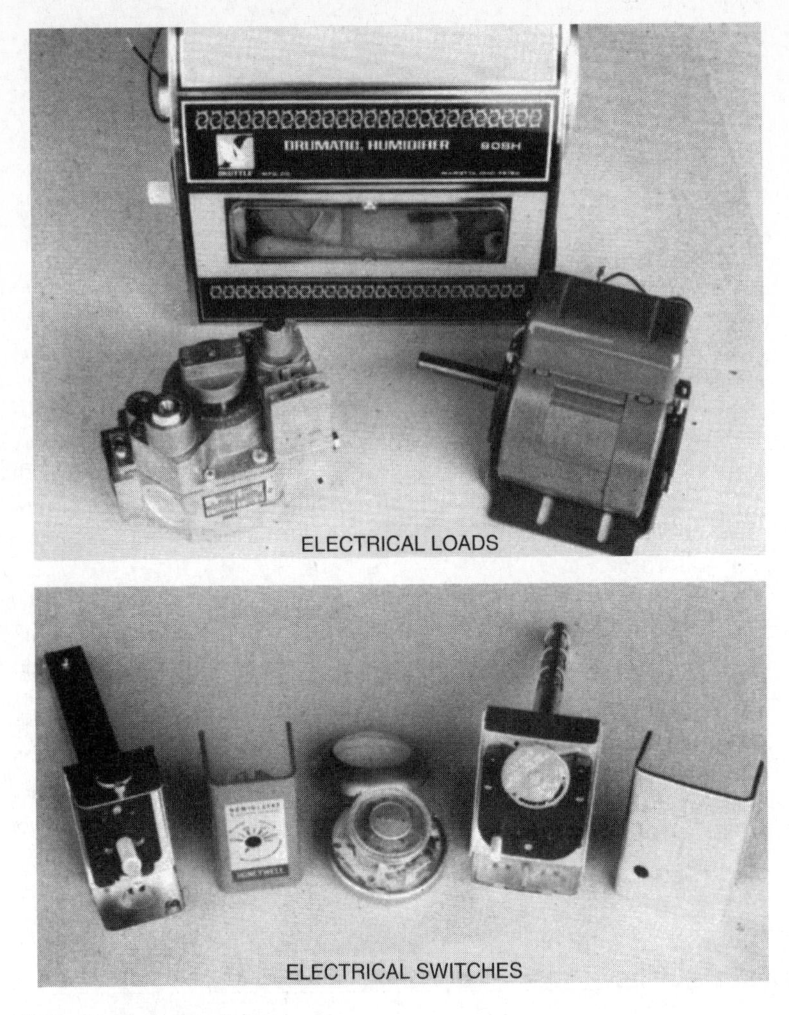

Figure 6.13 Loads and switches.

Motors The electrical motor is usually considered the most important load device in the electrical system. It can be represented in two ways: by a large circle or by the internal wiring (Figure 6.14). To better understand the electrical motor application, refer to the legend on the wiring diagram. The symbol may represent a fan motor, an oil burner motor, or an automatic humidifier motor. When current flows through the motor, the motor should run.

Solenoids The second most important device is the solenoid (Figure 6.15). When the current flows through the solenoid (coil or wire), magnetism is created. The solenoid is a device designed to harness and use magnetism. It is most frequently used to open and close switches. It is also used to open and close valves. It is common practice to

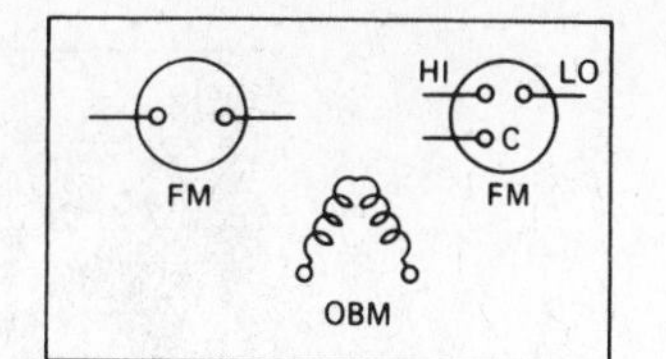

LEGEND

FM = FAN MOTOR
FM = FAN MOTOR, MULTIPLE SPEED
OBM = OIL BURNER MOTOR

Figure 6.14 Electric motor and symbols. (Courtesy of Essex Group, Controls Division, Steveco Products, Inc.)

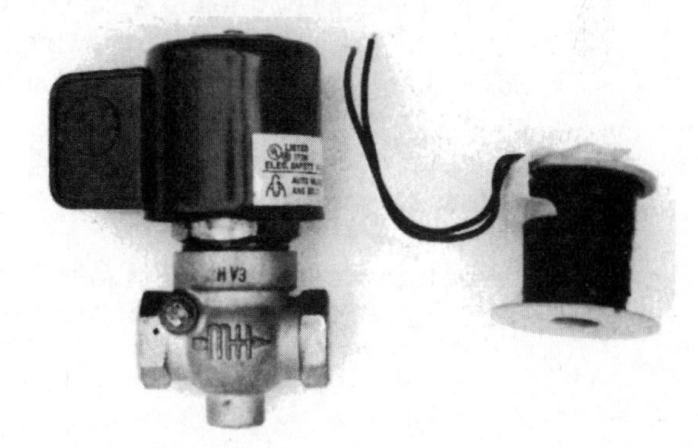

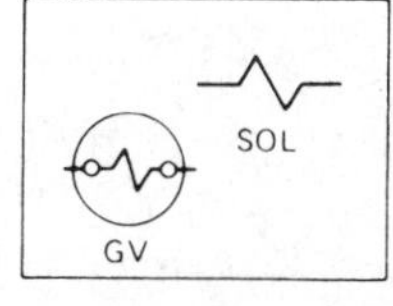

LEGEND

SOL = SOLENOID (RELAY COIL)
GV = SOLENOID (GAS VALVE)

Figure 6.15 Solenoid and symbols. (Courtesy of White–Rodgers Division, Emerson Electric Co.)

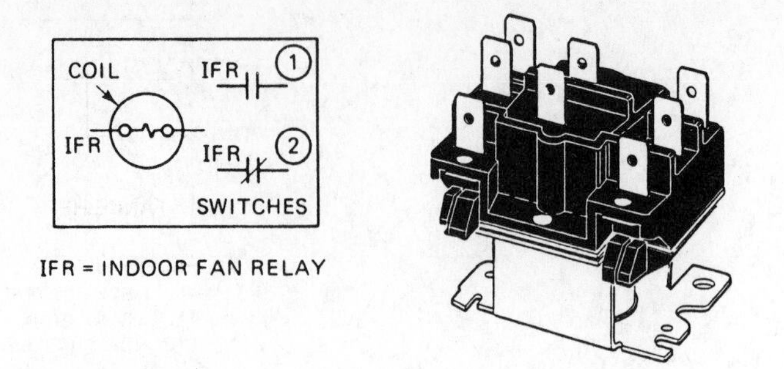

Figure 6.16 Indoor fan relay switch and symbols; switch 1 open and switch 2 closed. (Courtesy of White–Rodgers Division, Emerson Electric Co.)

use letters under the symbol to abbreviate the name of the device and to provide a reference to the legend. For example, GV under the symbol would be shown in the legend to mean gas valve.

Relays A relay is a useful application of a solenoid. By flowing current through the solenoid coil, one or more mechanically operated switches can be opened or closed. The solenoid coil is located in one circuit, and the switches are usually in separate circuits. A relay is identified by the symbols shown in Figures 6.16 and 6.17. There are at least two parts to the relay: the coil and the switch (or switches). The switch may be in one part of the diagram and the coil in another. The two symbols are identified as belonging to the same electrical device by the letters above them.

When the switch has a diagonal line across it, it is a closed switch; without the diagonal line, the switch is open. All wiring diagrams show relay switches in their normal position, the position in which no current is applied to the solenoid coil. Thus, in a diagram, when no current is flowing through the coil, the open switches are called *normally open (NO),* and the closed switches are called *normally closed (NC).* When current is applied to the coil, all the related switches change position (Figure 6.17). There are several types of relays used in heating work. They differ in the number of NO and NC switches.

Resistance Heaters Another form of load device commonly used is the resistance heater. In the resistance heater, electricity is converted to heat. Heat in an electric fur-

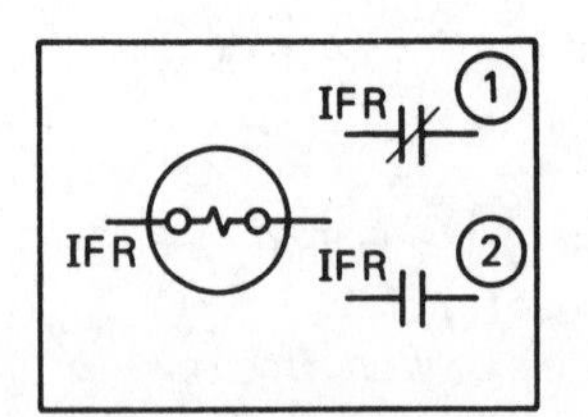

Figure 6.17 Indoor fan relay switch and symbols; switch 2 open and switch 1 closed. (Courtesy of Essex Group, Controls Division, Steveco Products, Inc.)

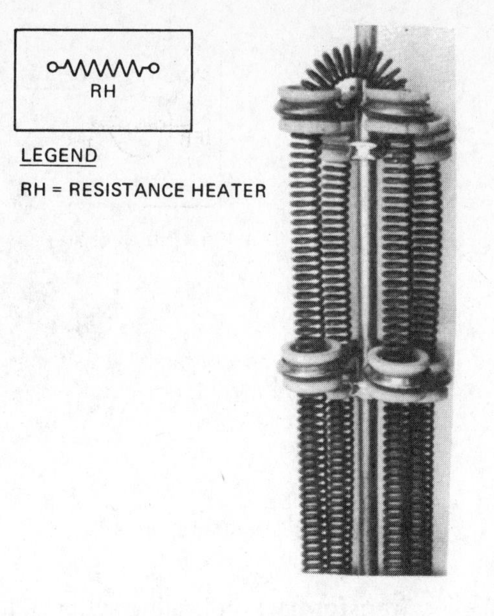

Figure 6.18 Resistance heater and symbols.

nace is produced by electricity. Heat is also used to control switches. The lower the resistance, the greater the amount of heat that is produced. The symbol for a resistance heater is a zigzag line (Figure 6.18). A letter is used under the heater symbol to designate its use, as indicated in the legend.

Heat Relays The symbols and letters for a heat relay are shown in Figure 6.19. In heating circuits, particularly for electric heating, resistance heaters are used to operate switches. The advantage of this type of relay is that it provides a time delay in operating the switch. When current is supplied to the heater, it heats up a bimetal element located in another circuit. Bimetal elements are made by bonding together two pieces of metal that expand at different rates. As the bimetal heats, its shape changes, thereby closing or opening a switch.

Lights Lights are a type of load. They have resistance to current flow. A signal light is often used to indicate an electrical condition that cannot otherwise be readily observed. The color of the light is often indicated by a letter on the symbol (Figure 6.20).

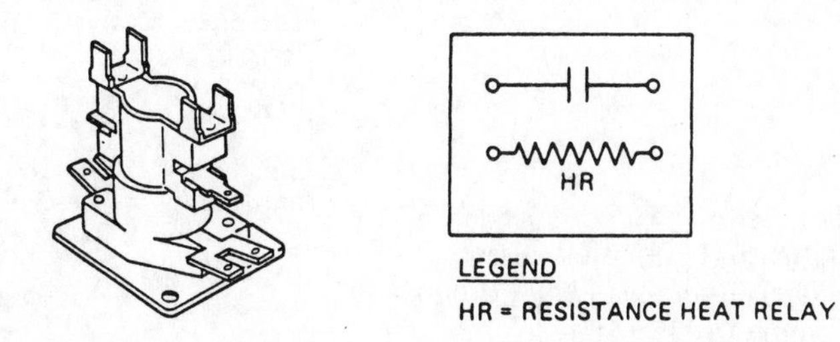

Figure 6.19 Heat relay and symbols. (Courtesy of White–Rodgers Division, Emerson Electric Co.)

Figure 6.20 Light and symbol.

Transformers In heating systems, it is often desirable to use two or more different voltages to operate the system. The fan must run on line voltage (usually 120 V), but the thermostat circuit (control circuit) can often best be run on low voltage (24 V). A transformer is used to change from one voltage to another. The legend and symbols for transformers are shown in Figure 6.21.

Switches and Their Symbols

Loads perform many functions; however, switches perform only one: to start and stop the flow of electricity. Electrical switches are classified according to the force used to operate them: manual, magnetic (solenoid), heat, light, or moisture. All wiring diagrams show the position of the switches when the operating solenoid (relay coil) or mechanism is deenergized.

A thermostat is assumed to be an NO switch. A cooling thermostat makes (closes) an electrical circuit on a rise in temperature. A heating limit switch is considered an

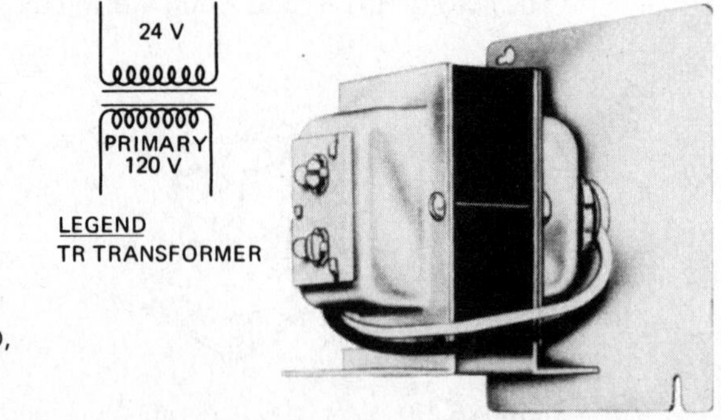

Figure 6.21 Transformer and symbol. (Courtesy of Essex Group, Controls Division, Steveco Products, Inc.)

NC switch, since it is normally closed when the system is in operation and opens only when excessively high temperatures are reached in the furnace.

It is important to identify the type of force that operates a switch. Only then can its normal position be accurately determined. The simplest type of switch is one that makes (closes) or breaks (opens) a single electrical circuit. Other switches make or break several circuits. The switching action is described by:

- Number of poles (number of electrical circuits through the switch)
- The throw (number of places for the electrical current to go)

The following abbreviations are often used to designate the types of switching action.

SPST: single-pole single-throw

SPDT: single-pole double-throw

DPST: double-pole single-throw

DPDT: double-pole double-throw

These designations and their symbols are shown in Figure 6.22.

The common types of switches used for heating equipment controls are described, together with the symbols used to represent them.

Manual Switches Manual disconnect switches are hand operated and are usually used to disconnect the electrical power supplying an air-conditioning unit. Disconnect switches that contain no fuses, or the bladed type, are referred to as *safety switches.* These switches are used to shut off the electrical power to the unit whenever service is required or for seasonal shutdown.

The fusible disconnect shown in Figure 6.23 serves two functions. It is used as a safety switch and also contains fuses that are rated to protect the equipment from exces-

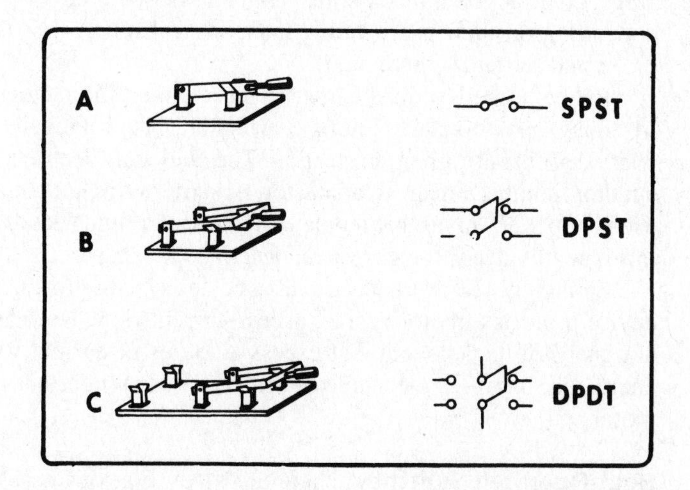

Figure 6.22 Designations and symbols for switching actions. (Courtesy of Carrier Corporation)

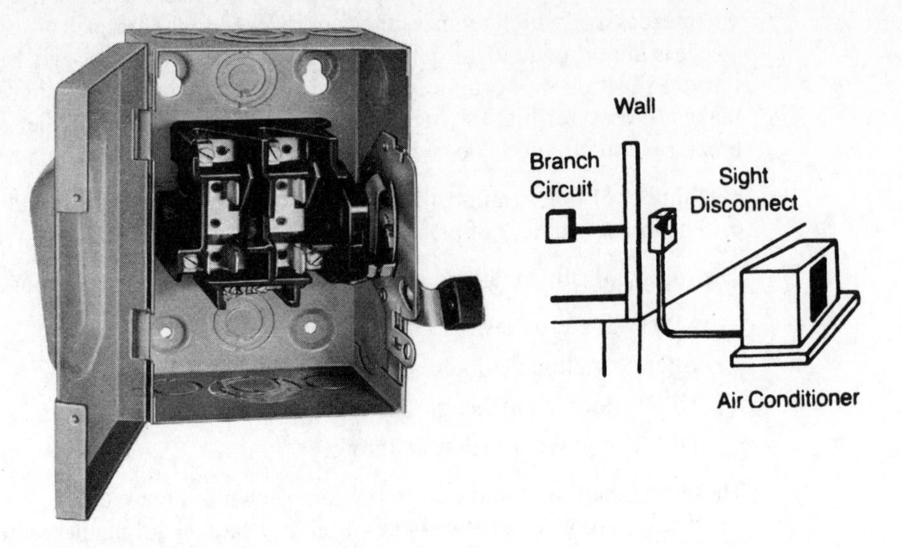

Figure 6.23 General-duty safety switches. (Courtesy of Square D Company)

sive electrical overload. Note that for proper equipment protection, the fuses must be sized for each unit's amperage draw.

Magnetic Switches Magnetic or solenoid switches are electrically operated switches using the force of the magnetic effect to operate the switch. To produce the required amount of power to operate the switch, the wire is coiled, creating a strong magnetic effect on a metal core. Magnetic or solenoid switches have various names, depending on their use. Among these are relays (Figure 6.24), contactors (Figure 6.25), and starters (Figure 6.26).

Relays were described earlier in this chapter. Contactors are relatively large electric relays used to start motors. Starters are also relatively large relays that include overload (excess current) protection. The National Electrical Code® specifies whether a motor requires a manual, contactor, or starter switch. Solenoid switches may also be used to operate valves that regulate the flow of a fluid (liquid or vapor). An example of this type of valve is the gas valve, described in Chapter 5.

Some types of overloads can also be described as magnetic switches. An overload device protects a motor against excess current flow. Normal amounts of current will not energize the solenoid, but excess amounts of current will. When the solenoid is energized, it trips a mechanically interlocked switch that cuts off the power to the motor.

Heat-Operated Switches There are many types of heat-operated switches. In all cases, heat is the force that operates the switch. These switches include thermostats, fan and limit controls, heat relays, fuses, overloads, and circuit breakers.

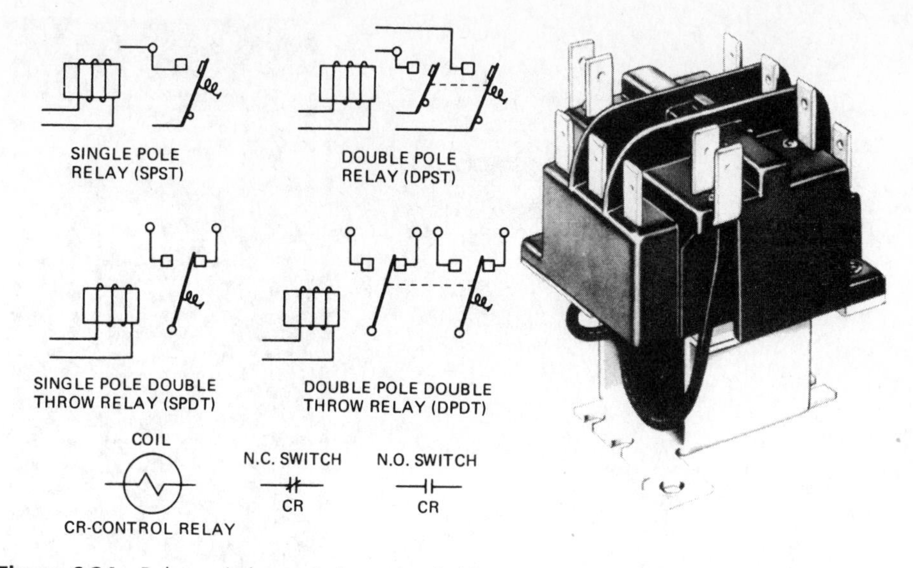

Figure 6.24 Relay switch, symbols, and switching action. (Courtesy of Essex Group, Controls Division, Steveco Products, Inc.)

Thermostats A heating thermostat (Figure 6.27) makes (closes) on a drop in room temperature. The intensity of heat affects the bimetal, changing its shape, which in turn actuates a switch. The thermostat is usually shown as an NO switch. Thermostats are discussed in Chapter 10.

Fan and Limit Controls Fan and limit controls (Figure 6.28) have a bimetal sensing element that protrudes into the warm air passage of a furnace. The fan control makes

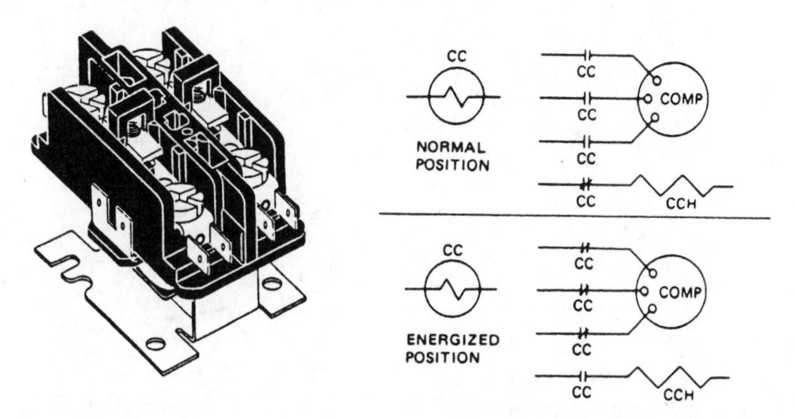

Figure 6.25 Contactor switch and symbol. (Courtesy of White–Rodgers Division, Emerson Electric Co.)

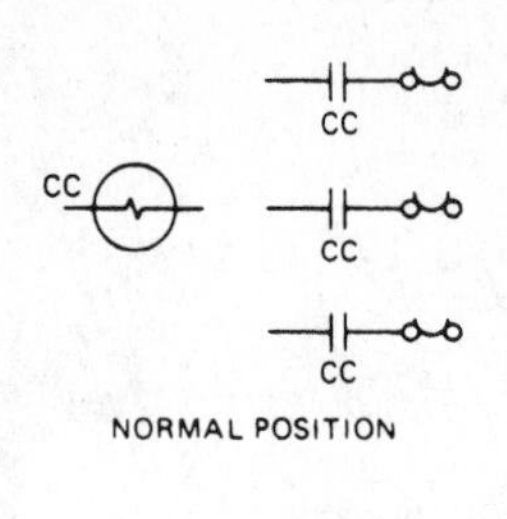

Figure 6.26 Starter switch and symbol.

on a rise in temperature. It is considered to be an NO switch because with the furnace shut down, the fan control will have an open switch. The limit control protects the furnace against excessively high air temperatures, turning off the firing devices when a predetermined temperature is reached. The limit control is diagrammed as an NC switch.

Heat Relays Heat relays (Figure 6.29) are similar to magnetic or solenoid relays in function; however, the mechanical action takes place as a result of heat produced in an electrical resistance. When the resistance coil is energized, heat is produced. This heat is applied to one or more bimetal elements. As the bimetal elements expand they either break or make an electrical circuit.

The heat relay is a delayed-action switch. This means that the switch requires some amount of time to change position. Depending on the construction, the time delay may be 15, 30, 45 seconds, or more. This feature permits staging or sequencing

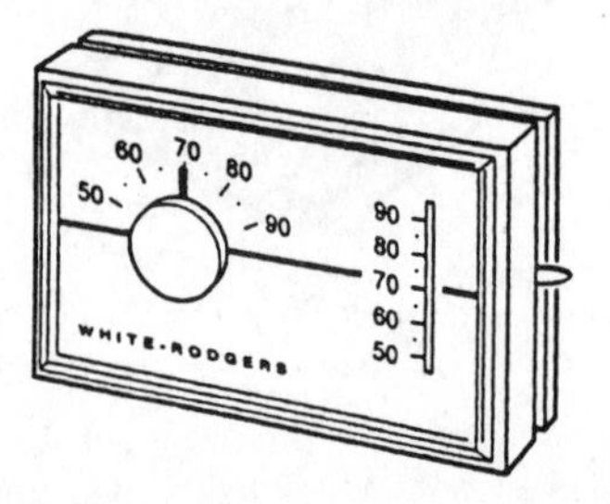

Figure 6.27 Thermostat switch and symbol. (Courtesy of White–Rodgers Division, Emerson Electric Co.)

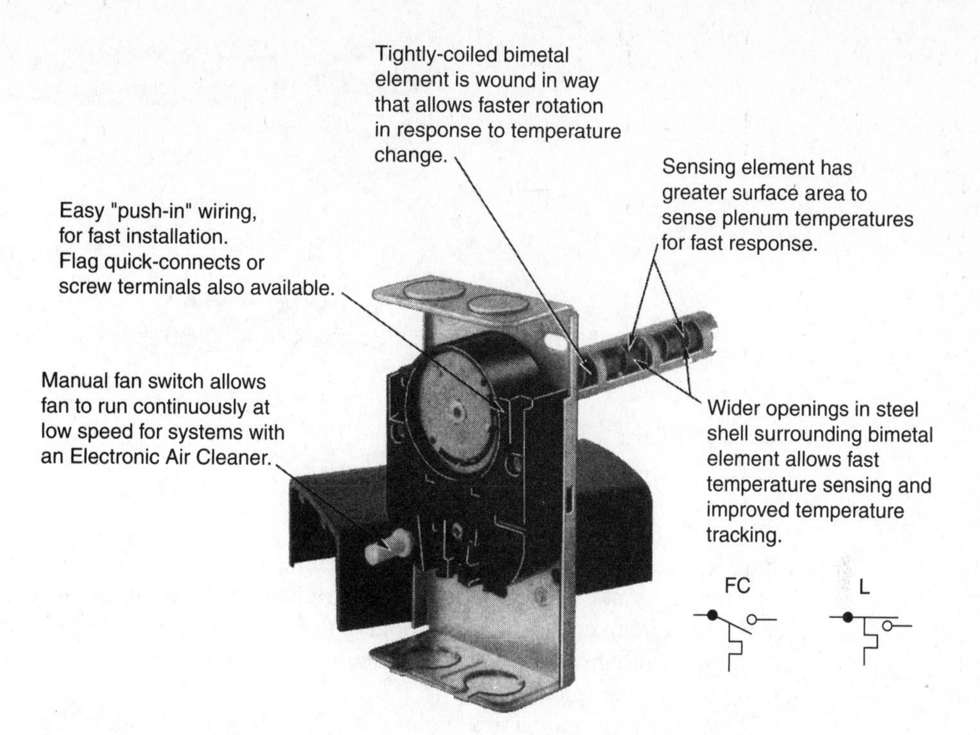

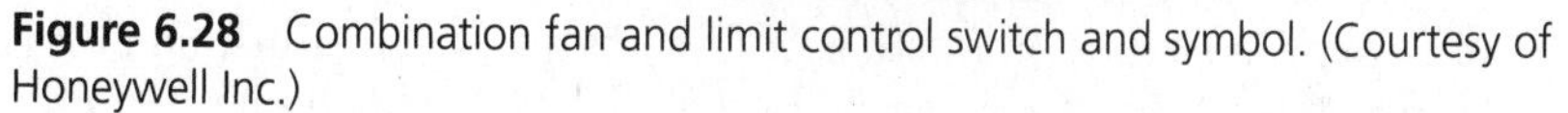

Figure 6.28 Combination fan and limit control switch and symbol. (Courtesy of Honeywell Inc.)

loads (turning loads on automatically at different times). Because the inrush current to a load, when first turned on, is usually many times greater than its running current, sequencing is important. Sequencing permits using a smaller power service to the appliance. A sequencer is used on an electric furnace to turn on the electric heating elements in steps.

Fuses Fuses (Figure 6.30) are placed in an electrical circuit to cut off the flow of current when there is an overload or a short. An *overload* is current in excess of the circuit

Figure 6.29 Heat relay switch and symbol. (Courtesy of Honeywell Inc.)

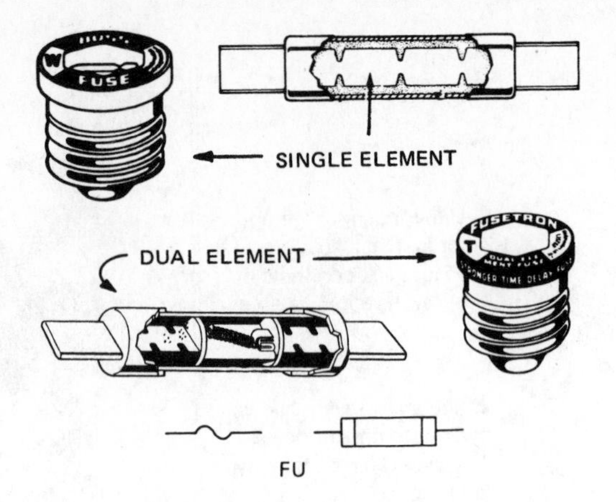

Figure 6.30 Fuses.

design. A *short* is a direct connection between the two wires of a power supply so that the current does not pass through a load. In either case, the fuse will heat and melt, breaking the circuit continuity and stopping the flow of current. When the fault is over-current, the melting of the fuse takes place slowly. When the fault is a short, the melting of the fuse takes place quickly.

Some circuits require a time-delay fuse. An example is a motor circuit. The inrush of current in starting (a fraction of a second) is so great that an ordinary fuse would "blow" before the motor reached running speed. This initial motor current is called *locked-rotor amperes* (LRA). The running amperes or *full-load amperes* (FLA) are much less. Because of the special design of delayed-action fuses, they may be selected on the basis of FLA. According to the National Electrical Code®, fuses and wiring can be sized on the basis of 125% of FLA.

Overloads Overloads (Figure 6.31) can be constructed in many ways. All overloads are designed to stop the flow of current when safe limits are exceeded. Overloads differ from fuses in that they do not have one-time use as does a fuse. When a fuse melts, it must be replaced to return the circuit to operation. When an overload senses excess current, a switch is opened, breaking the flow of current. Some of these switches are reset automatically when the current returns to normal; others must be manually reset. An overload is considered an NC switch, since it is closed when no current is flowing through the circuit and during normal operation of the equipment.

Circuit Breakers The main power circuit to a heating furnace must be protected against excess current flow by a fuse or a circuit breaker (Figure 6.32). Either one of these switches will disconnect the power supply if the equipment draws excess current. The circuit breaker is a type of overload device placed in the power supply that will "trip" (open the circuit) in the event of excessive current flow. When the fault is corrected, the circuit breaker can be reset manually to restore the circuit to its original condition.

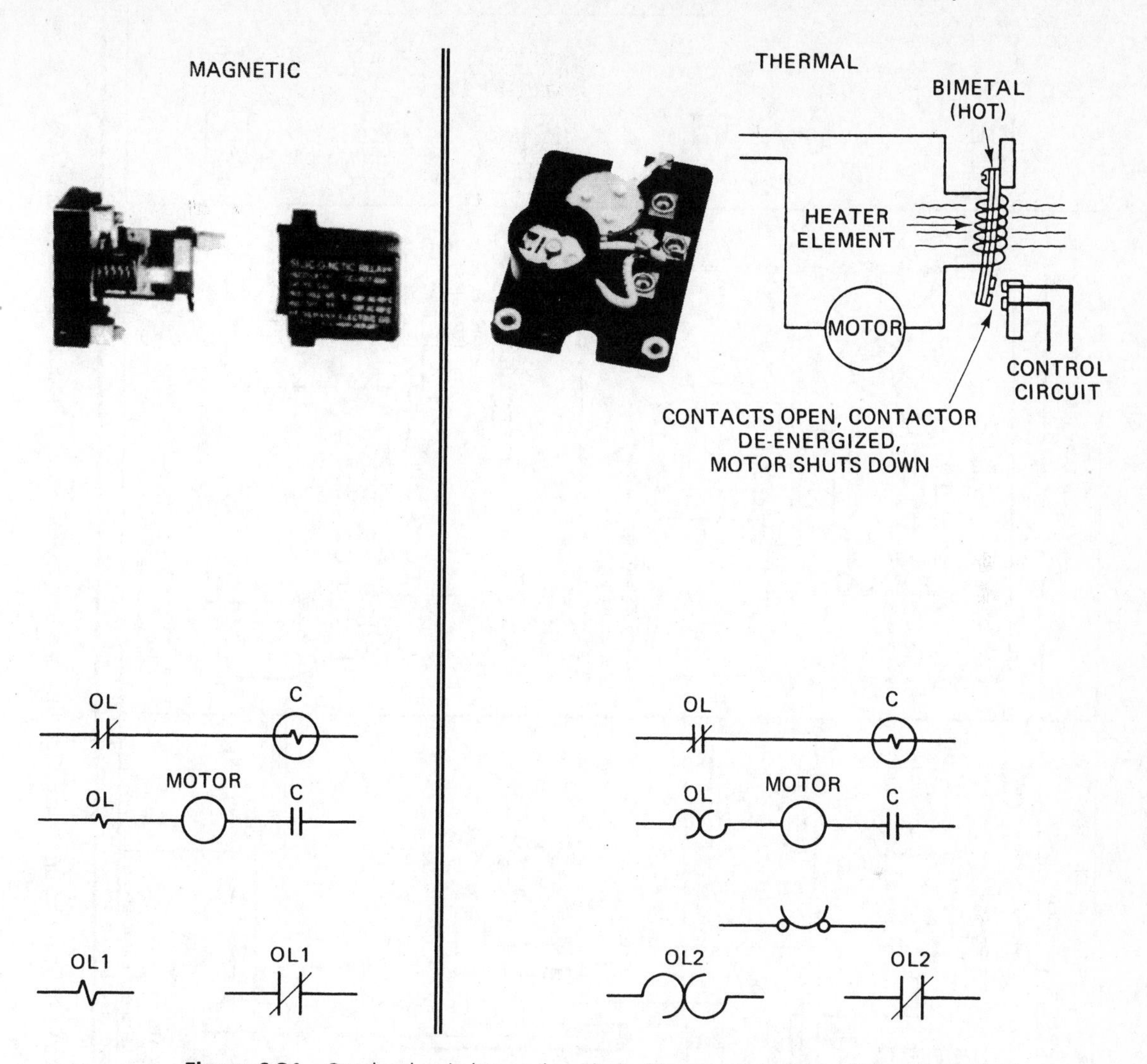

Figure 6.31 Overload switches and symbols. (Courtesy of Motors & Armatures, Inc.)

Figure 6.32 Circuit breaker switch. (Courtesy of Sears)

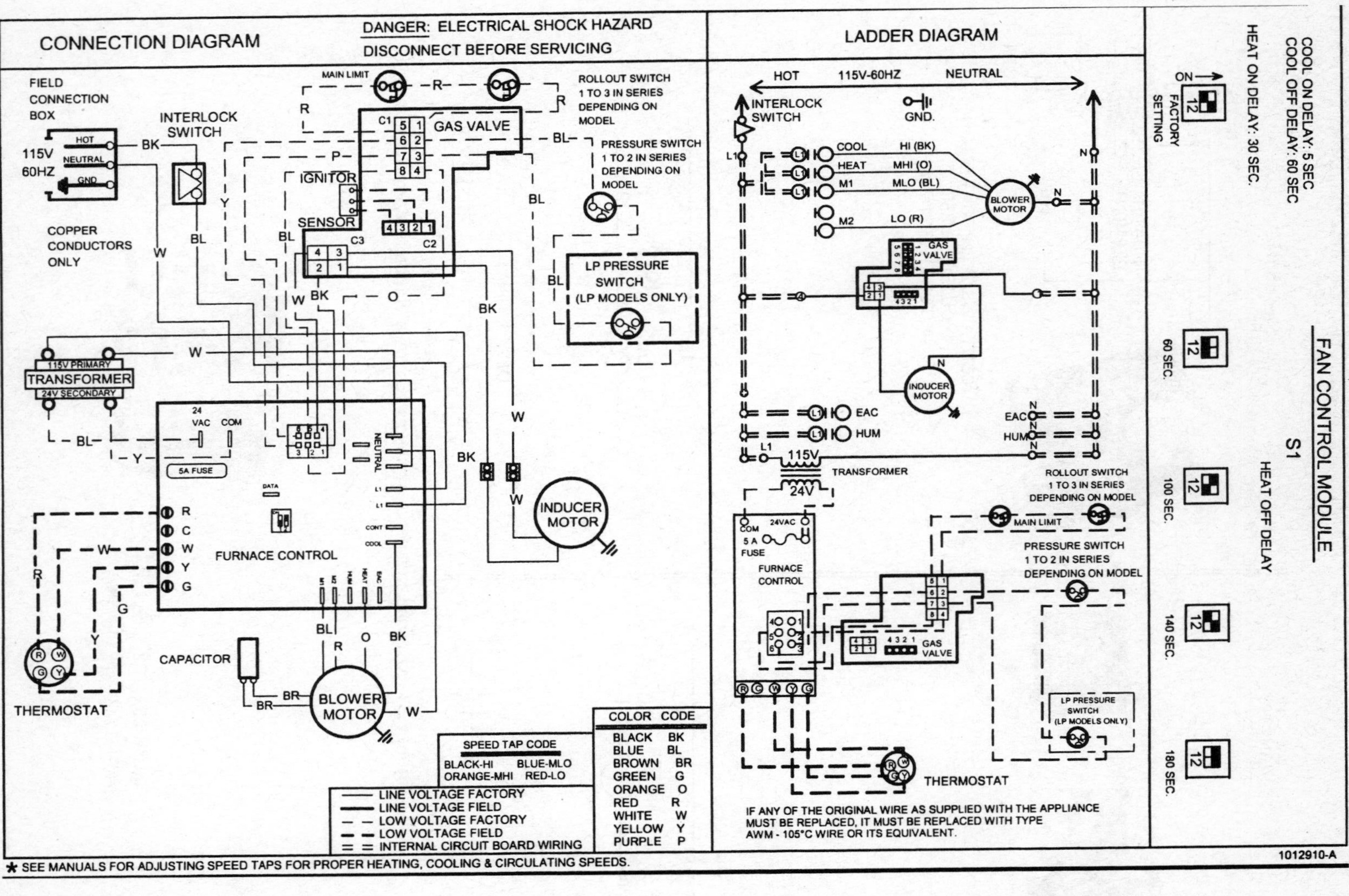

Figure 6.33 Connection and schematic wiring diagrams. (Courtesy of International Comfort Products Corporation, USA)

Light-Operated Switches Some switches are activated by light. An example is the cad cell used on the primary control of an oil burner. A *cad cell* is a cadmium sulfate sensing element. In the presence of light, the electrical resistance of a cad cell is about 1000 Ω; in darkness, its resistance is about 100,000 Ω. The cad cell is located in the draft tube of an oil burner and senses the presence (or proof) of a flame. If the cad cell is in darkness, there is no flame; if the cad cell is lighted, the flame is proof that the burner has been started by the ignition system. It therefore acts as a safety device.

In the electrical circuit the cad cell is placed in series with a relay coil. If the cad cell does not sense adequate light from the flame, the relay will not be energized. If the cad cell senses the flame, its resistance is reduced and the relay is energized, which allows the burner to continue running.

Moisture-Operated Switches The presence of moisture in the air can be used to operate a switch. Certain materials like nylon, expand when moist and contract when dry. Thus, this material can be used to sense the relative humidity in the air. The change in the length of a strand of nylon can be used to operate a suitable switch. The switch will turn a humidifier on or off to maintain the desired relative humidity in a space.

Electrical Diagrams

The greatest advances in the design and use of forced warm air heating equipment came after the invention of automatic control systems. It was then no longer necessary to operate the equipment manually. Automatic controls regulate the furnace to maintain the desired temperature conditions. Forced air heating owes its very beginning to the use of electricity to drive the fan motor.

A wiring diagram describes an electrical system. There are generally two types of diagrams for warm air heating units. A *connection diagram* shows the electrical devices in much the same way as they are positioned on the equipment (Figure 6.33). Lines are connected to the electrical terminals to show the paths the electric current will follow. A *schematic diagram* (Figure 6.33) uses symbols to represent the electrical devices. The arrangement of the symbols in the diagram, using ladder-type connecting lines, indicates how the system works. It shows the sequence of operation. This diagram is essential to the service technician in troubleshooting problems in heating systems.

STUDY QUESTIONS

Answers to the study questions may be found in the sections noted in brackets.

6-1 Define volt, ampere, ohm, and watt. *[Electrical Terms]*

6-2 State Ohm's law and give an example of its use. *[Ohm's Law]*

6-3 How does a series circuit differ from a parallel circuit? *[Series Circuits; Parallel Circuits]*

6-4 State the formulas for series and parallel circuits. *[Series Circuits; Parallel Circuits]*

6-5 Describe how a solenoid coil uses electromagnetic action. *[Electromagnetic Action]*

6-6 How are capacitors rated? *[Capacitance]*

6-7 What is an electrical circuit? *[Circuits]*

6-8 How do loads differ from switches? *[Electrical Devices]*

6-9 Give the electrical symbols for motors, heaters, solenoids, relays, lights, and transformers. *[Loads and Their Symbols]*

6-10 How does a relay operate? *[Relays]*

6-11 What do the acronyms SPST, SPDT, DPST, and DPDT mean? *[Electrical Switches and Their Symbols]*

6-12 What is the difference between a connection and a schematic wiring diagram? *[Electrical Diagrams]*

CHAPTER 7

Schematic Wiring Diagrams

OBJECTIVES

After studying this chapter, the student will be able to:

- Read and construct a schematic electrical wiring diagram

Purpose of a Schematic

A schematic wiring diagram consists of a group of lines and electrical symbols arranged in ladder form to represent the individual circuits controlling or operating a unit. The lines represent the connecting wires. The electrical symbols represent loads or switches. The rungs of the ladder represent individual electrical circuits. The unit can be an electrical–mechanical device such as a furnace or air conditioner.

Understanding a Schematic

To understand a schematic diagram, a student must know:

- The purpose of each electrical component
- Exactly how the unit operates, both mechanically and electrically
- The sequence of operation of each electrical component

Sequence refers to the condition in which the operation of one electrical component follows another to produce a final result. It represents the order of events as they occur in a system of electrical controls. All electrical systems must be wired by making connections to power and to each electrical device. A schematic diagram is important in assisting a service technician in wiring the equipment, in locating connections for testing circuits, and in analyzing the operation of the control system. Occasionally, however, a schematic diagram may be unavailable, so it is essential that a service technician be able to construct one.

Following is the legend to the schematic diagrams used in this chapter:

CR	Control relay
FC	Fan control
FD	Fused disconnect
FM	Fan motor
FR	Fan relay
FS	Fan switch
GV	Gas valve
IFR	Indoor fan relay
L	Limit
LA	Limit auxiliary
L_1, L_2	Power supply
NC	Normally closed
NO	Normally open
T	Thermostat
TR	Transformer
Y, R, G, W	Terminals of thermostat

Schematic Construction

Connection and schematic wiring diagrams are shown in (Figures 7.1 and 7.2). To draw a schematic and to separate the individual circuits on the unit itself, a service technician must be able to trace (or follow) the wiring of each individual circuit. The method of tracing a circuit is as follows:

1. Starting at one side of the power supply (L_1), the technician goes through the resistance (load), and returns to the other side of the power supply (L_2). This is a *complete circuit.*

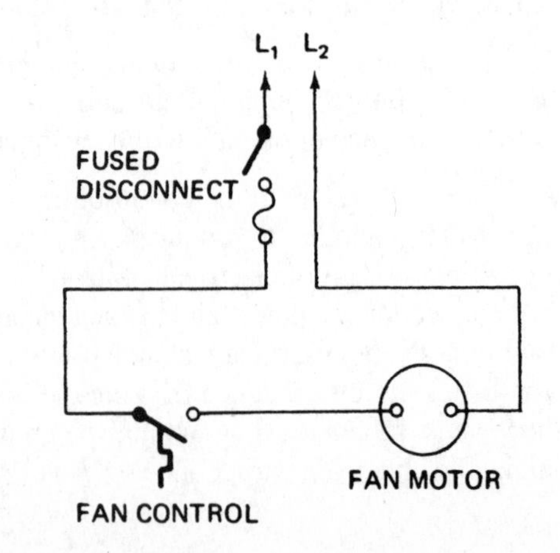

Figure 7.1 Connection wiring diagram.

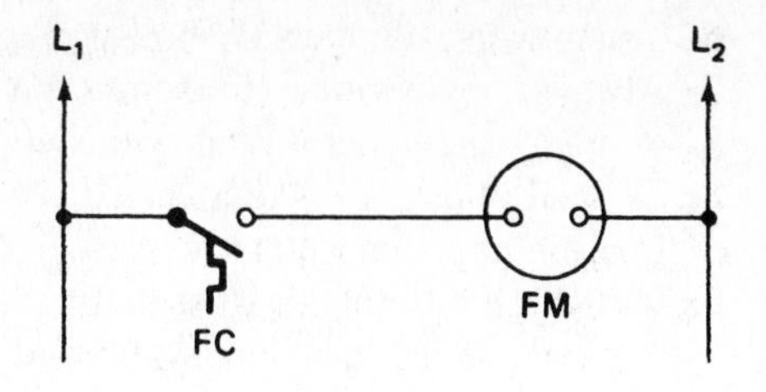

Figure 7.2 Schematic wiring diagram.

2. If a technician starts at L_1 and goes through the resistance and cannot reach L_2 or returns back to L_1 this is an *open circuit.*
3. If a technician starts at L_1 and reaches L_2 without passing through a resistance (load), this is a *short circuit.*

Rules for Drawing: Vertical Style

1. Use letters on each symbol to represent the name of the component.
2. List in the legend the names of all components represented by letters.
3. When using a 120-V ac power supply*, show the hot line (L_1) on the left side and the neutral line (L_2) on the right side of the diagram.
4. When using a 120-V power supply, you must place the switches on the hot side (L_1) of the load.
5. Mark relay coils and their switches (Figure 7.3) with the same (matched) symbol letters.

*In 208-, 230-, 240-, or 440-V single-phase circuits, there are two hot wires and no neutral wire. In these circuits, switches can be placed on either side of the load.

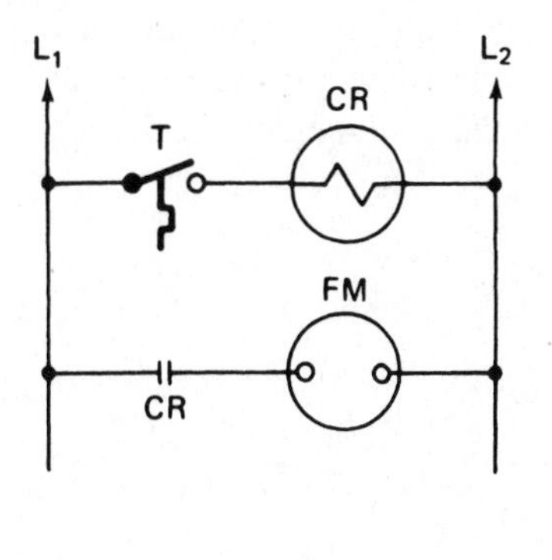

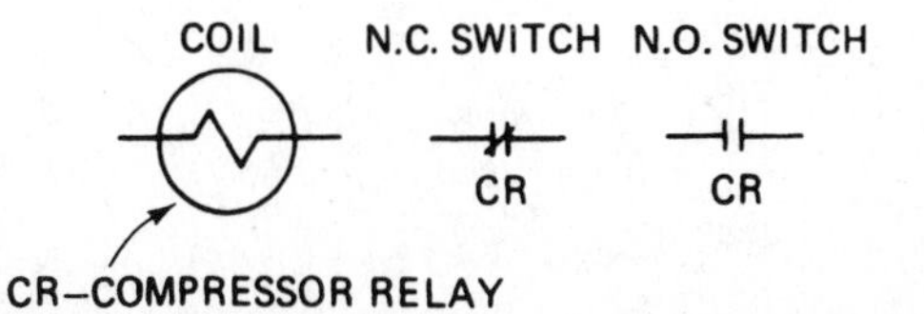

Figure 7.3 Relay coils and switches.

6. You may use numbers to show wiring connections to controls or terminals.
7. Always show switches in their normal (de-energized) position.
8. You may show thermostats with switching subbases, primary controls for oil burners, and other more complicated controls by terminals only in the main diagram. Use subdiagrams when necessary to show the internal control circuits.
9. It is common practice to start the diagram showing line-voltage circuits first and low-voltage (control) circuits second (Figure 7.4).

Rules for Drawing: Horizontal Style

Assuming that either a completely wired unit or a connection wiring diagram is available, the following is a step-by-step procedure for drawing a schematic:

Referring to the Connecting Wiring	*Drawing the Schematic Diagram*
Locate the source of power. Trace it back to the disconnect switch.	Draw two horizontal lines* representing L_1 and L_2, with sufficient space between them for the diagram. Draw in the disconnect switch.
Trace each circuit on the diagram, starting with L_1 and returning to L_2.	Draw each circuit on the diagram using a vertical line to connect L_1 through the switches, through the load to L_2.
Determine the names of each switch and load.	Make a legend by listing the names of the electrical components, together with the letters representing each. The letters are used on the diagram to identify the parts.

*Vertical lines could also be used to represent L_1 and L_2 in which case horizontal lines would then be used to represent the individual circuits.

A connection diagram usually shows the relative position of the various controls on the unit. A schematic diagram makes it possible to indicate the sequence of operation of the electrical devices. A verbal description of the electrical system may also reveal some useful facts about the operation of the unit. Thus, the connection diagram, the schematic diagram, and the verbal description are all useful in understanding how the unit operates electrically.

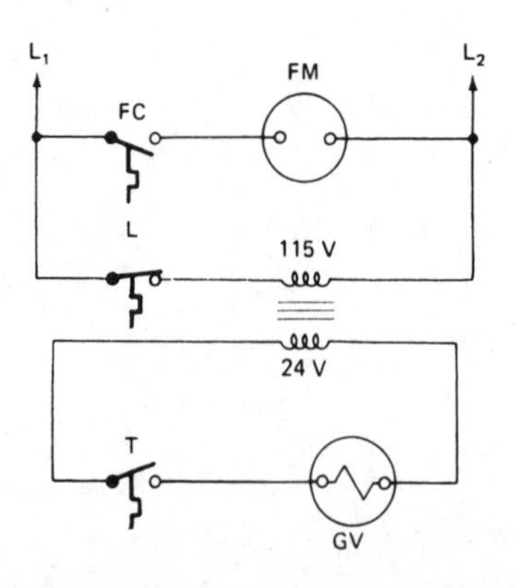

Figure 7.4 Line-voltage and low-voltage circuits schematic diagram.

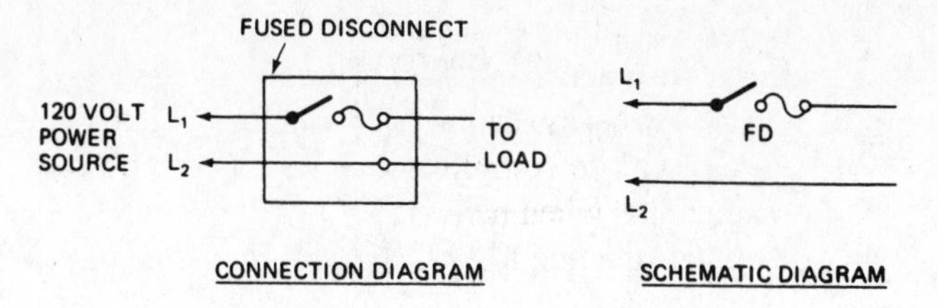

Figure 7.5 Power supply to furnace.

To show how these three are related, a connection diagram and a schematic diagram will be constructed from the description of each circuit.

Power Supply *Description:* The power supply is 120-V, 60-Hz, single-phase (Figure 7.5). A fused disconnect is placed in the hot line to the furnace to disconnect the power when the furnace is not being used. The power supply consists of a hot wire and neutral wire. The circuit has a 20-A fuse.

Circuits *Description:* A fan motor operating on 120 V is placed in circuit 1 of Figure 7.6. It is controlled by a fan control. This fan control is located in the plenum or in the furnace cabinet near the heat exchanger. It is adjustable, but to conserve energy it is usually set to turn the fan on at 110°F (43°C) and off at 90°F (32°C). This permits the furnace to heat up before the fan turns on to deliver heated air to the building.

Note that the fan control is part of a combination fan and limit control: Both use the same bimetal heat element. The limit control settings are higher than the fan control settings.

Description: the primary of a transformer is placed in a separate 120-V circuit, circuit 2 (Figure 7.7). The transformer will be used to supply 24-V (secondary) power to the control circuits.

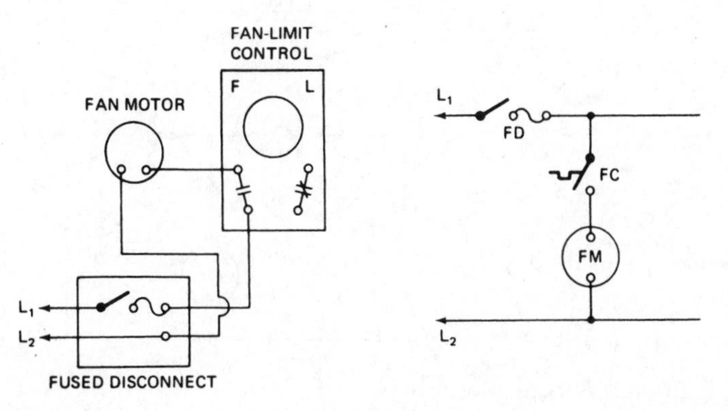

Figure 7.6 Fan circuit.

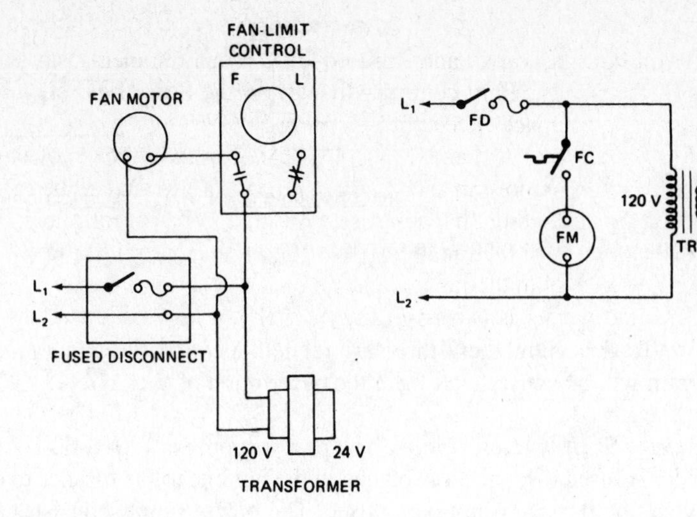

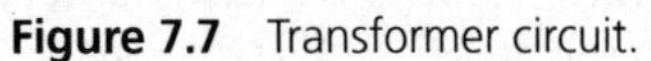

Figure 7.7 Transformer circuit.

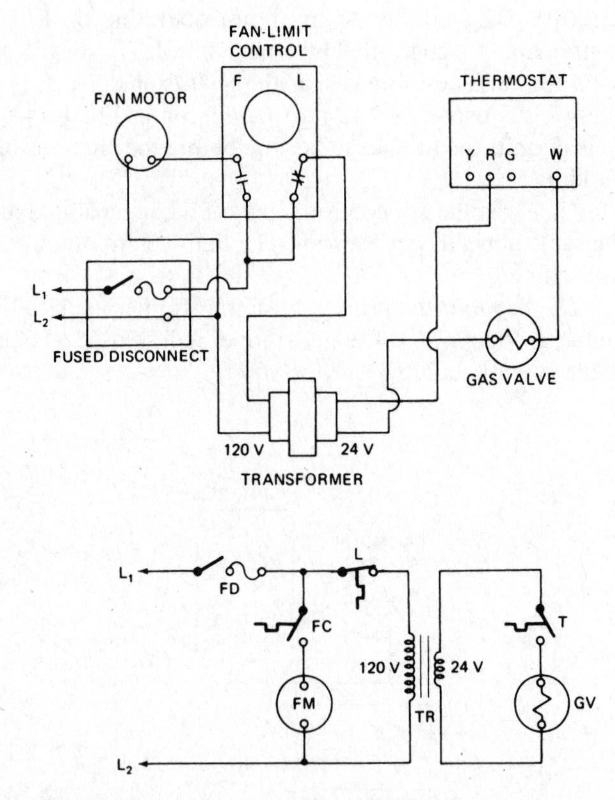

Figure 7.8 Gas valve circuit.

Description: This is a 24-V gas valve circuit (Figure 7.8). Circuit 3 includes the thermostat, gas valve, and limit control. When the thermostat calls for heat, the gas valve opens. The limit control will shut off the flow of gas should the air temperature leaving the furnace exceed 200°F (93°C).

Description: A manual switch, located in the subbase of the thermostat, is connected to an indoor fan relay, circuit 4, so that the fan can be manually turned on for ventilation even though the gas is off (Figure 7.9). The relay switch is in the 120-V circuit parallel with the fan control, and the solenoid coil of the relay is in the 24-V circuit in series with the fan switch (FS).

Field wiring is shown in Figure 7.10 for high-efficiency heating only and for a combination heating and cooling unit.

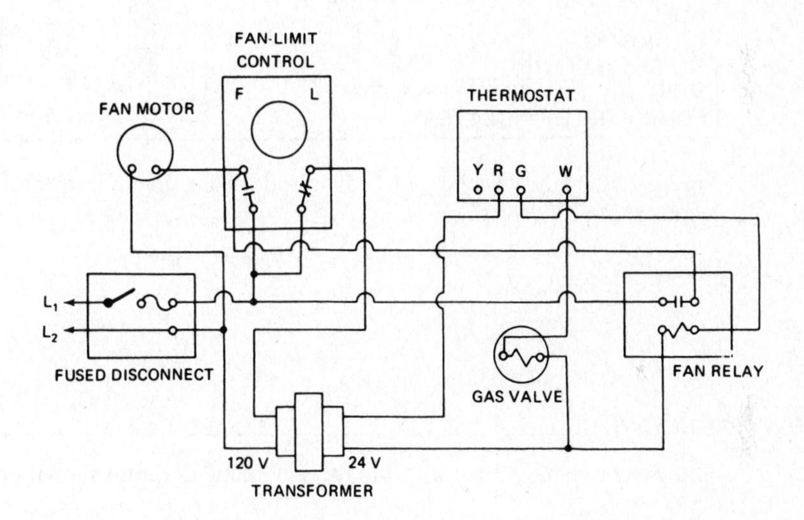

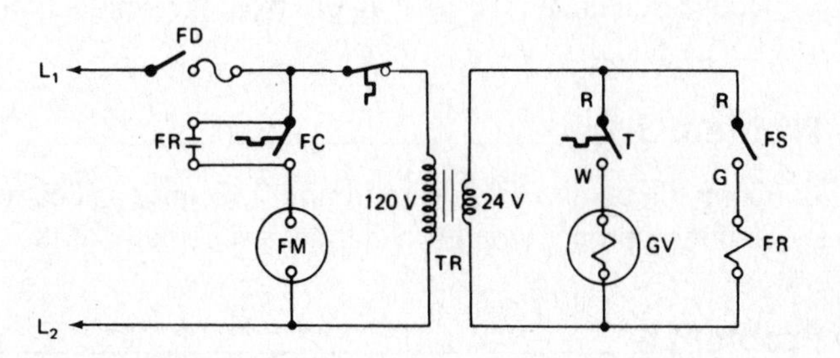

Figure 7.9 Manual fan circuit.

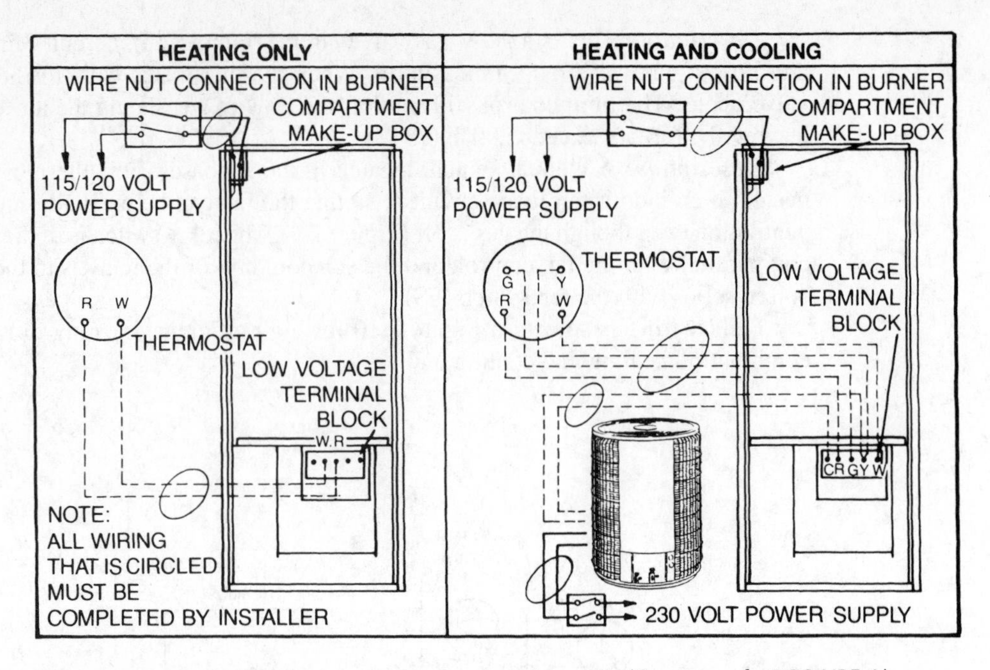

Figure 7.10 Typical wiring for a heating and cooling unit. (Courtesy of ARCOAIRE Air Conditioning & Heating)

Review Problem 1

Draw a connection diagram from the schematic diagram shown in Figure 7.11.

Review Problem 2

Draw a schematic wiring diagram from the connection diagram shown in Figure 7.12.

Review Problem 3

Copy the components from Figure 7.13 onto a separate sheet of paper, connect them into their proper circuits, and draw a schematic of the system.

Review Problem 4

Copy the components from Figure 7.14 onto a separate sheet of paper, connect them into their proper circuits, and draw a schematic of the system.

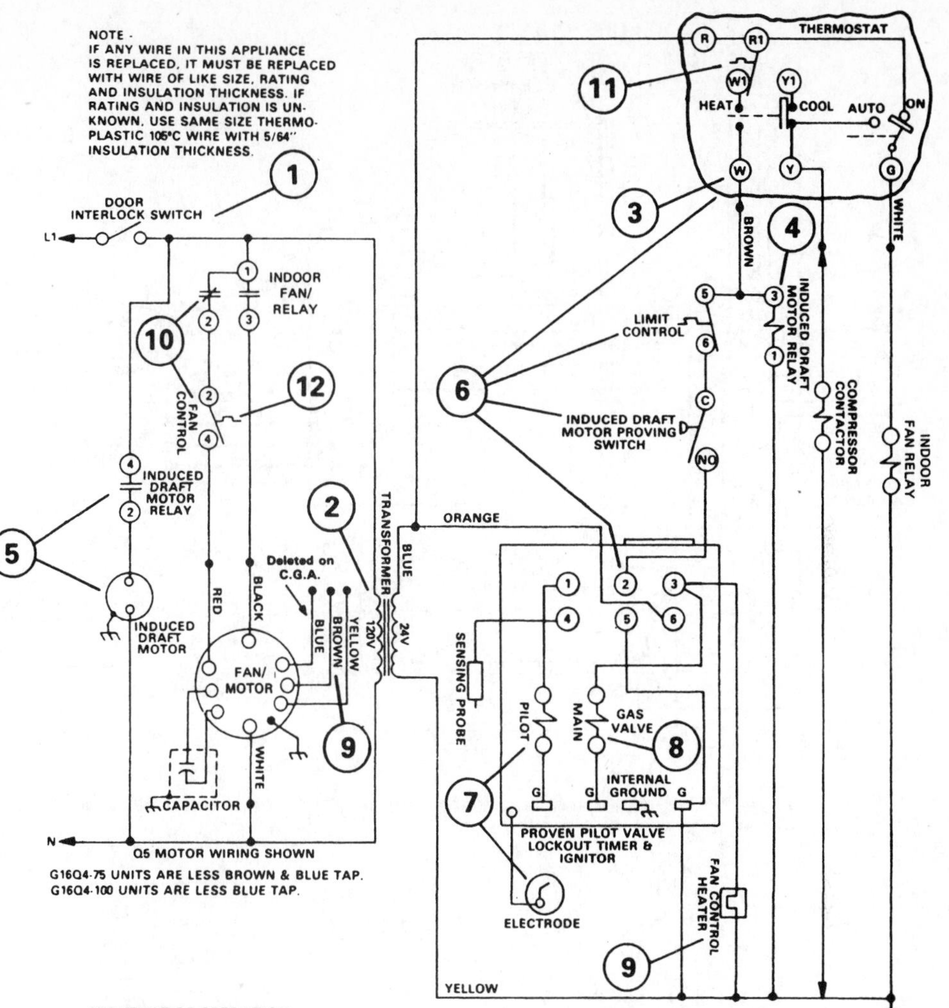

SEQUENCE OF OPERATION

1. LINE POTENTIAL FEEDS THROUGH DOOR INTERLOCK. ACCESS PANEL MUST BE IN PLACE TO ENERGIZE UNIT.
2. TRANSFORMER PROVIDES 24 VOLTS TO POWER CONTROL CIRCUIT.
3. ON A HEATING DEMAND, THERMOSTAT BULB MAKES PROVIDING 24 V AT "W" LEG.
4. INDUCED DRAFT MOTOR RELAY IS ENERGIZED FROM "W" LEG.
5. INDUCED DRAFT MOTOR RELAY N.O. CONTACTS CLOSE AND ENERGIZE THE INDUCED DRAFT MOTOR.
6. WHEN THE INDUCED DRAFT MOTOR COMES UP TO SPEED, THE INDUCED DRAFT MOTOR PROVING SWITCH CLOSES COMPLETING THE CIRCUIT FROM "W" LEG OF THERMOSTAT THROUGH LIMIT CONTROL TO IGNITION CONTROL TERMINAL 2.
7. PILOT GAS VALVE AND PILOT IGNITION SPARK ARE ENERGIZED.
8. AFTER PILOT FLAME HAS BEEN PROVEN BY IGNITION CONTROL, MAIN GAS VALVE IS ENERGIZED AND SPARK IS DE-ENERGIZED. (MAIN GAS VALVE WILL OPEN ONLY ON PROOF OF PILOT FLAME.)
9. AS THE MAIN GAS VALVE IS ENERGIZED, THE FAN CONTROL HEATER IS ACTIVATED.
10. IN APPROXIMATELY 30–80 SECONDS N.O. FAN CONTROL CONTACTS CLOSE ENERGIZING BLOWER MOTOR FROM N.C. BLOWER MOTOR RELAY CONTACTS TO THE HEATING SPEED TAP.
11. AS HEATING DEMAND IS SATISFIED, THERMOSTAT HEAT BULB BREAKS DE-ENERGIZING IGNITION CONTROL, GAS VALVE, AND FAN CONTROL HEATER.
12. BLOWER MOTOR CONTINUES RUNNING UNTIL FURNACE TEMPERATURE DROPS BELOW FAN CONTROL SET POINT.

Figure 7.11 Gas-fired upflow furnace schematic diagram.

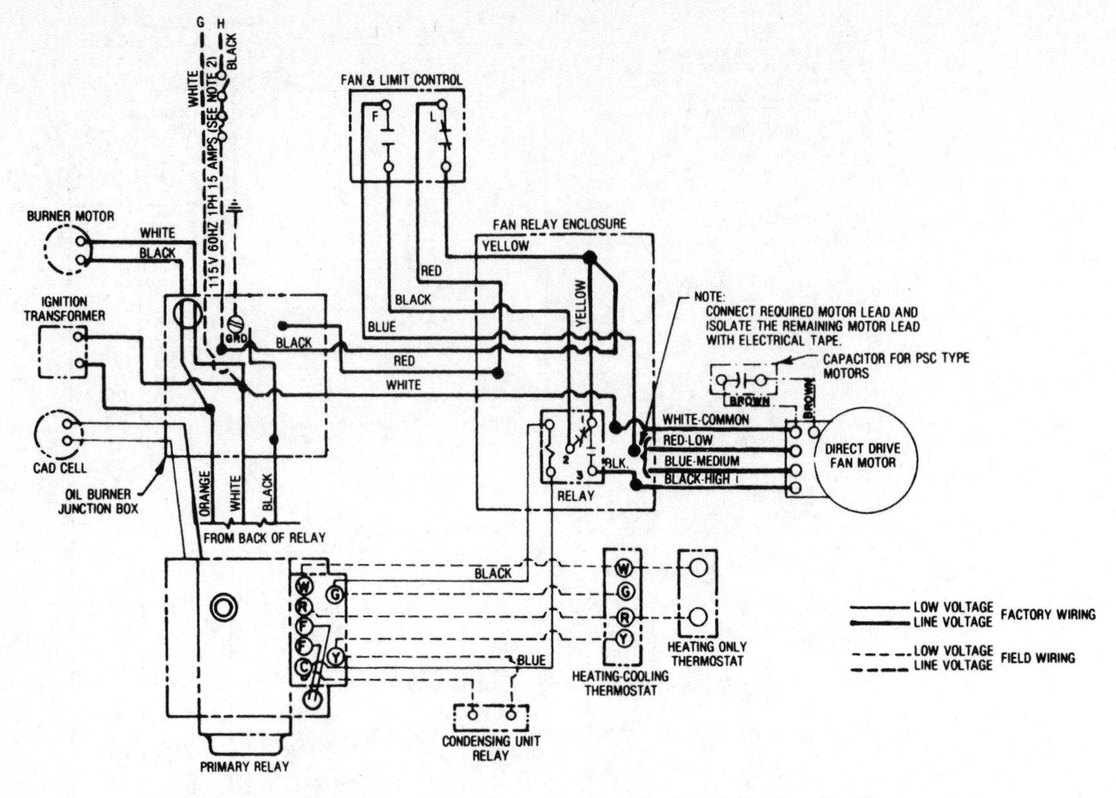

Figure 7.12 Heating–cooling connection diagram for oil-fired upflow furnace.

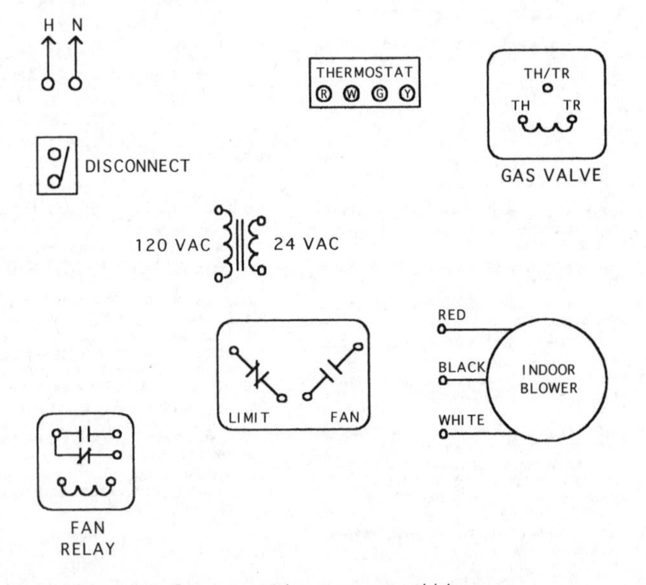

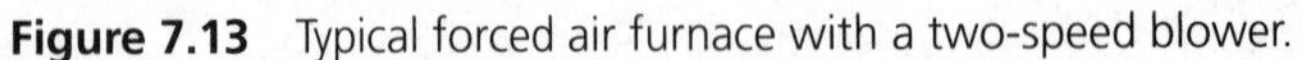
Figure 7.13 Typical forced air furnace with a two-speed blower.

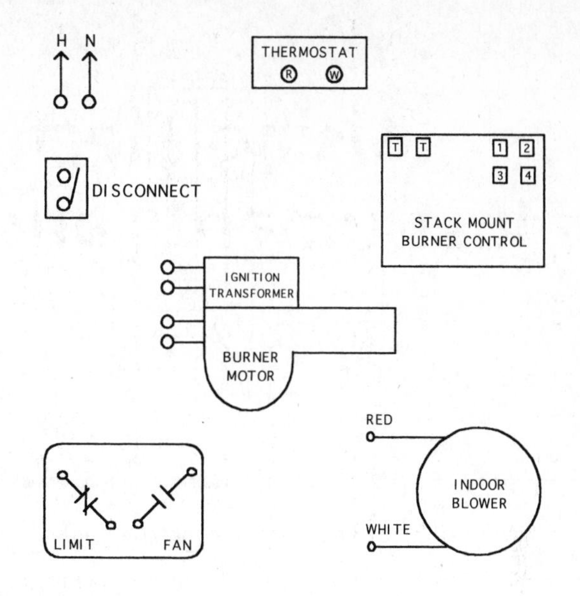

Figure 7.14 Oil-fired furnace with burner mount stack control.

STUDY QUESTIONS

Answers to the study questions may be found in the sections noted in brackets.

7-1 What is the purpose of a schematic wiring diagram? *[Understanding a Schematic]*

7-2 If the power lines are vertical, what does each horizontal line represent? *[Rules for Drawing]*

7-3 What essential electrical devices must be shown in each circuit? *[Rules for Drawing]*

7-4 With two vertical lines representing the power supply, which line is usually shown "hot"? *[Rules for Drawing]*

7-5 With L_1 hot and L_2 neutral, where should the fused disconnect be placed? *[Rules for Drawing]*

7-6 How is the position of two switches shown if both must be on to have the load operate? *[Rules for Drawing]*

7-7 If one switch operates two loads, how is that drawn? *[Rules for Drawing]*

CHAPTER 8

Using Electrical Test Instruments and Equipment

OBJECTIVES

After studying this chapter, the student will be able to:

- Use common electrical test instruments for service or troubleshooting

Electrical Test Instruments

To test the performance of a heating unit, or to troubleshoot service problems, requires the use of instruments. Because a heating unit has many electrical components, a heating service technician should:

- Know the unit electrically
- Be able to read and use schematic wiring diagrams
- Be able to use electrical test instruments

The reading and construction of schematic wiring diagrams was discussed in Chapter 7. The use of schematic diagrams is discussed in this chapter.

Component Functions

To know the unit electrically means to understand exactly how each component functions. Figure 8.1 shows a simple electrical heating wiring arrangement for a forced-air gas system. Two views are shown:

1. Schematic diagram
2. Connection diagram

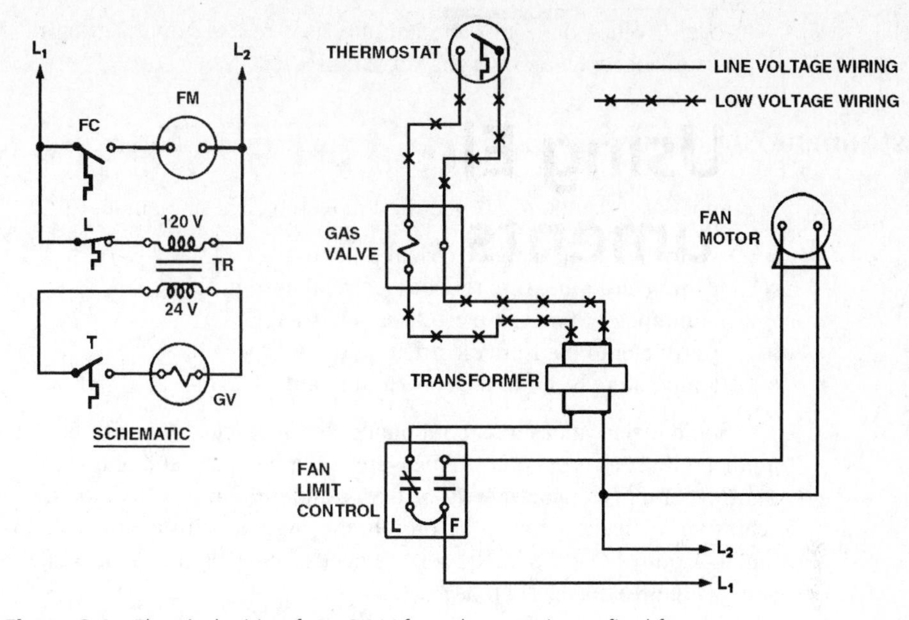

Figure 8.1 Electrical wiring for a 24-V forced warm air gas-fired furnace.

Three load devices are shown:

1. Fan motor
2. Transformer
3. Gas valve

The fan motor turns the fan when it is supplied with power, in this case 120-V, single-phase, 60-Hz current. The transformer, which has a primary voltage of 120 V and a secondary voltage of 24 V, is energized whenever power is supplied. A solenoid gas valve opens when it is energized and operates on 24-V power.

There is at least one automatic switch in each circuit. The fan has one, the transformer has one, and the gas valve has two. The following is a description of the types of switches used:

- The fan control turns on the fan when the bonnet temperature rises to the cut-in setting. It turns off the fan when the bonnet temperature drops to the cutout setting.
- The limit control is a safety device that turns off the power to the transformer when the bonnet temperature rises to the cutout setting. It turns on the power to the transformer automatically when the bonnet temperature drops to the cut-in setting.
- A combination gas valve has a built-in pilotstat. The pilotstat is a safety device operated by a thermocouple (not shown). When the pilot is burning properly, the NO contacts are closed, permitting the gas valve to open in response to the thermostat.
- The thermostat is a low-voltage (24-V) switch operated by a bimetal sensing element. When the room temperature drops, and the thermostat reaches its cut-in setting, the contacts close, and the gas valve opens (provided that the pilot flame is

proven). When the room temperature rises to the cutout setting of the thermostat, the contacts open and the gas valve closes.

Instrument Functions

A number of instruments are required in testing. These include:

- A voltmeter to measure electrical potential
- An ammeter to measure rate of electrical current flow
- An ohmmeter to measure electrical resistance
- A wattmeter to measure electrical power
- A temperature tester to measure temperatures

Some meters measure a combination of characteristics. For example, a clamp-on ammeter that has provisions to measure amperes, volts, and ohms is available. A VOM multimeter that can measure volts, ohms, and milliamperes is available. The following scales are typical; however, before purchasing an instrument it is important to verify that its features match specific service needs. For example, do its scales allow the measured values to be clearly read?

1. *DC millivolt scales:* 0 to 50, 0 to 500, 0 to 1500. These scales are used to read voltages on thermocouples and thermopiles. A millivolt is one-thousandth (1/1000) of a volt.
2. *AC voltage scales:* 0 to 30, 0 to 500. These scales are used to read low-voltage control circuits and line-voltage circuits.
3. *AC amperage scales:* 0 to 15, 0 to 75. These scales are used for measuring amperage drawn by various low-voltage and line-voltage loads.
4. *Ohm scales:* 0 to 400, 0 to 3000. These scales are used for resistance readings on all types of circuits and for continuity checks on all systems.
5. *Wattage scales:* 0 to 300, 0 to 600, 0 to 1500, 0 to 3000. These scales are used to measure power input to a circuit or system.
6. *Temperature scale:* 0° to 1200°F. This scale is used to check return-air temperatures, discharge-air temperatures, and stack temperatures.

Using Meters

When using meters, the technician should consider the following points:

1. Always use the highest scale first, then work down until midscale readings are obtained. This prevents damage to the meter.
2. Always check the calibration of a meter before using it. For example, on an ohmmeter short together the two test leads and turn the zero adjust knob until the needle reads zero resistance.
3. When using a clamp-on ammeter, increase the sensitivity of the instrument by wrapping the conductor wire around the jaws. The sensitivity will be multiplied by the number of turns taken. For example: 10 loops of single-strand insulated thermostat wire can be wrapped around the jaws of an ammeter, as shown in Figure 8.2. If the

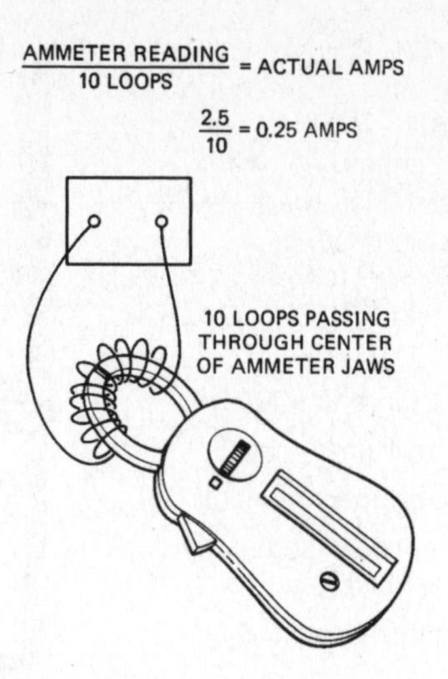

Figure 8.2 Method used to read low-amperage values accurately using a clamp-on ammeter.

meter is set to read on the 1- to 5-A scale, a reading of 2.5 A would be divided by 10 to arrive at the true reading, 0.25 A.

4. When using a clamp-on ammeter, be sure that the clamp is around only one wire. If it encloses two wires, the current may be flowing in opposite directions, which can cause a zero reading even though current is actually flowing.
5. Always have a supply of the required meter fuses on hand. Certain special fuses may be difficult to obtain on short notice.
6. When dc millivolts being delivered by a thermocouple in a circuit are being measured, an adapter is required to access the internal connection. See Figure 8.3 for application.

Figure 8.3 Adapter being used for thermocouple measurements.

Figure 8.4 The M1K is a highly compact and economical multimeter, ideal for quick checks and troubleshooting.

15 Ranges
500 ac and dc
Resistance to 500 kΩ
DC milliamps
Decibel measurements
Color-coded and mirrored scale plate
Test leads included

(Courtesy of UEi)

7. The multimeter can be used to check the output of pilot generators and thermocouples. See Figure 8.4.
8. The digital multimeter shown in Figure 8.5 is used when testing burner equipment. It checks the flame signal current on systems using rectifying flame rod, photocell, infrared, or ultraviolet flame detectors.

Figure 8.5 Digital volt–ohm–ampere multitester. The kit includes the LT17 meter with rubber holster, leads, thermocouple, current clamp, and case. (Courtesy of Fieldpiece Instruments, Inc.)

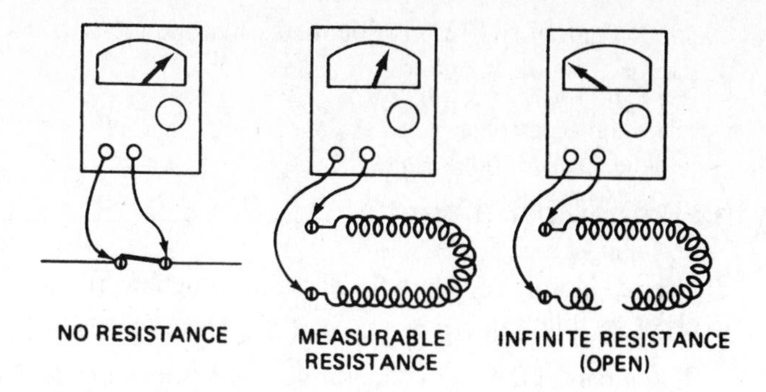

Figure 8.6 Three types of readings taken with an ohmmeter.

Three types of readings that can be taken with an ohmmeter are shown in Figure 8.6:

1. *No resistance.* Closed switch gives full-scale deflection, indicating 0 Ω.
2. *Measurable resistance.* Measurable resistance produces smaller deflection or a specific ohm reading on the meter.
3. *Infinite resistance (open).* There is no complete path for the ohmmeter current to flow through, so there is no deflection of the meter needle. The resistance is so high that it cannot be measured. The meter shows infinity (∞), representing an open circuit.

Note: It is important to remember when using an ohmmeter that it has its own source of power and must never be used to test a circuit that is "hot" (connected to power). All power must be off or the ohmmeter can be seriously damaged.

One good example of the use of an ohmmeter is in testing fuses (Figure 8.7). The fuse is removed from the power circuit and the ohmmeter is used to make a continuity

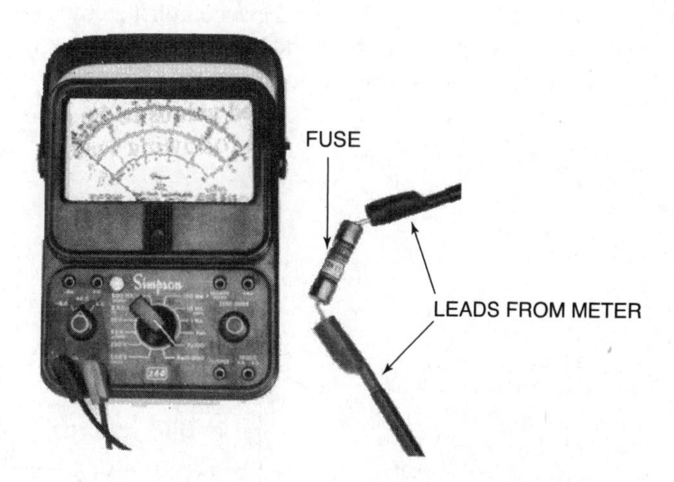

Figure 8.7 Ohm scale of multimeter being used for testing fuses. (Courtesy of Simpson Electric Company)

check. If continuity (0 Ω) is obtained when the meter leads are touching the two ends of the fuse, the fuse is good.

Selecting the Proper Instrument

The most helpful procedure for deciding which instrument to use in troubleshooting is to check as follows:

1. If any part of the unit is operating, the voltmeter or ammeter is normally used.
2. If the unit is inoperable because of a blown fuse, an ohmmeter may be used to test the circuit for a short or a grounded condition.

When attempting to measure the resistance of any component, the possibility of obtaining an incorrect reading always exists when a part is wired into the system. In a parallel circuit, when testing with an ohmmeter, it is necessary to disconnect one side of the component being tested to avoid the possibility of an incorrect reading of the resistance (Figure 8.8).

A wattmeter measures both amperage and volts simultaneously. This is necessary in an ac circuit because watts take into consideration the power factor. The *power factor* indicates the percentage of the volts times the amperes that a consumer actually pays for. Because the voltage and current peaks do not usually occur at the same time, wattage is normally less than the product of the volts times the amperes.

Power Calculations

It may be necessary to calculate the power, measured in watts, consumed by a furnace or air-conditioning unit. This can be accomplished by taking both voltage and amperage readings at the equipment and then inserting these values into the following formula. The power factor can be obtained from the electric company.

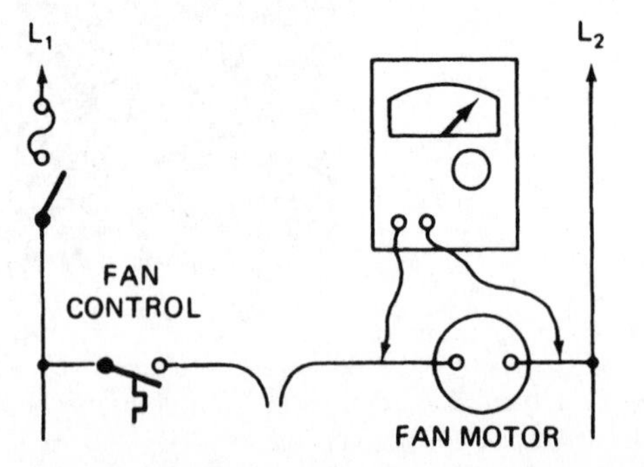

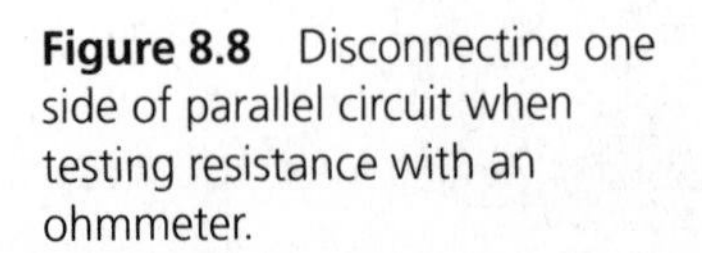

Figure 8.8 Disconnecting one side of parallel circuit when testing resistance with an ohmmeter.

$$P = E \times I \times \text{PF}$$

$$\begin{aligned} \textit{where } P &= \text{power, watts} \\ E &= \text{EMF, volts} \\ I &= \text{current, amperes} \\ \text{PF} &= \text{power factor} \end{aligned}$$

Substitute the sample values $E = 120$ V, $I = 12$ A, and PF = 0.90 into the formula:

$$P = 120 \text{ V} \times 12 \text{ A} \times 0.90 \text{ power factor} = 1296 \text{ W}$$

or approximately 1.3 kilowatts (kW), where k = 1000.

Clamp-on Volt–Ohm–Ammeters

The clamp-on style of volt–ohm–ammeter is popular with service personnel. The instrument is easy to use, and the variable scales are easy to read. See Figures 8.9 to 8.11 for specifications. Figure 8.12 shows their application in troubleshooting motors.

Voltprobe Tester

Figure 8.13 shows an instrument that is designed for measuring ac/dc voltage and resistance only. Some checks that can be made with a Voltprobe tester are shown in Figure 8.14.

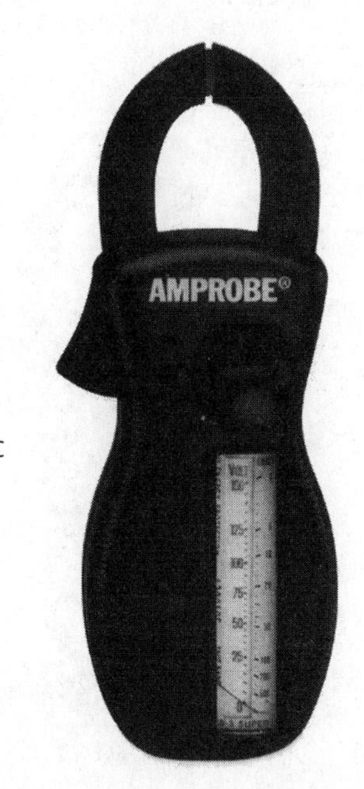

Figure 8.9 RS-3 Super Rotary Scale clamp-on volt–ohm–ammeter.

Ranges: 0–150/300/600 V ac

0–6/15/40/100/300 A ac

Operates on frequencies of 50–400Hz for use in industrial environments.

Drop proof from 5 ft for increased durability.

(Courtesy of Amprobe)

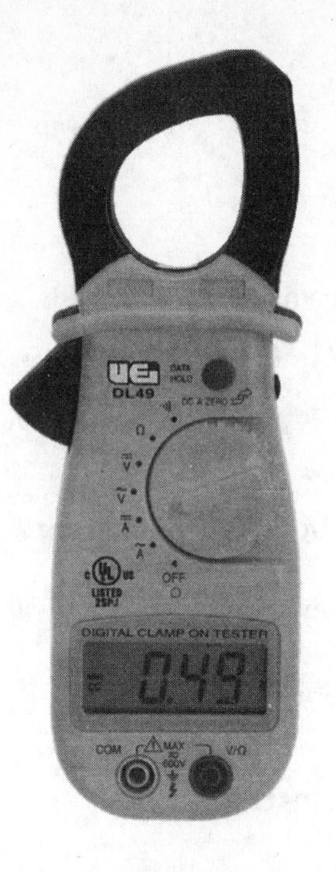

Figure 8.10 Digital snap-around volt–ohm–ammeter. The autoranging DL49 is ideal for HVAC/R technicians, electricians and plant maintenance professionals.

Ranges: 0–600 V ac
0–400 A ac and dc
0–600 V dc
0–40 MΩ

(Courtesy of UEi)

Figure 8.11 Digital clamp-on volt–ohm–ammeter.

Ranges: 0–200/400/600 A ac
0–200/1000 A dc
0–200/600 V ac
0–200/600 V dc
0–200 Ω
Holds display

(Courtesy of Fluke Corporation)

Specifications

Function	Range & Resolution	Fluke 30
AC current	600A 400A 200.0A	N/A ±(1.3%+3) 0-100A: ±(1.8%+8) 100-200A: ±(1.4%+0)
DC current	1000A 1000A 200.0A	N/A
AC volts	200.0V/600V	±(1.2%+3)
DC volts	200.0V/600V	N/A
Frequency range		50-60 Hz
Resistance	200.0Ω	±(1.2%+5)
Crest factor, AC current		N/A
Operating temperature		-10°C to 50°C (14°F to 122°F)
Storage temperature		-20°C to +60°C (-4°F to 140°F)
Battery life		200 hrs. continuous, alkaline
Size		213 mm x 93 mm x 45 mm 8.4i x 3.75i x 1.77i
Weight		0.45 kg (1lb.)
Warranty		one year

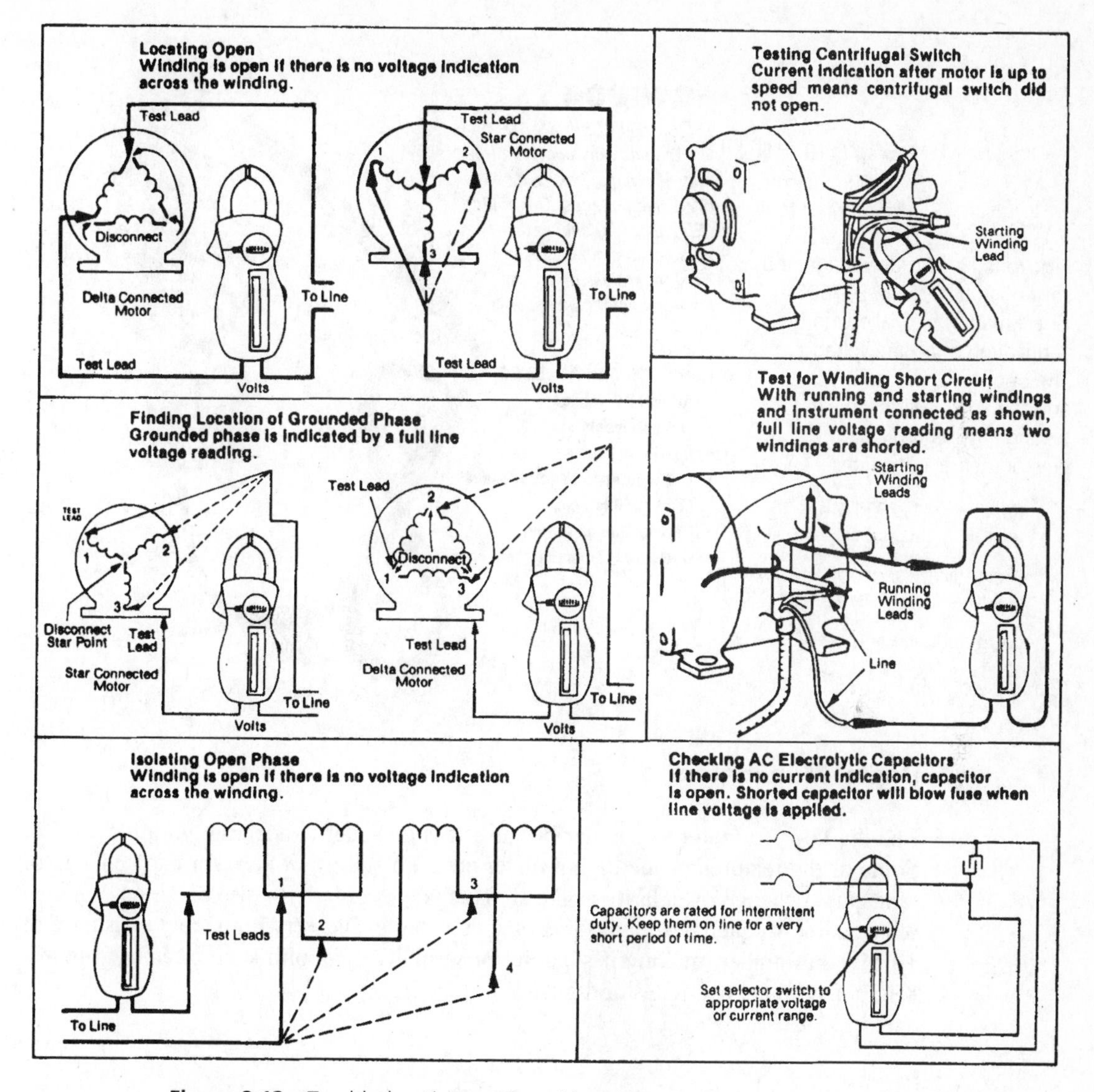

Figure 8.12 Troubleshooting motors with a clamp-on instrument. (Courtesy of Amprobe Instruments)

ECT600A

The ECT600A CHECKMATE™ voltage and continuity tester is fast, allowing quick verification of AC line voltages. The bright LED indicator panel and the audible signal make testing easy regardless of environmental conditions.

Features

- **12 to 600 Volts AC and DC**
- **AC and DC voltage measurements**
- **Continuity tester**
- **LED indication of voltage level**
- **Loud audible tone**
- **Battery-free operation**
- **One year limited warranty**

Figure 8.13 Ac/dc voltage and continuity tester. The ECT600A Checkmate is fast, allowing quick verification of ac line voltages. The bright LED indicator panel and the audible signal make testing easy regardless of environmental conditions.

Ranges: 0–12/600 V ac and dc

LED indication of voltage level

Loud audible tone

(Courtesy of UEi)

Digital Multimeters

The digital-type meter is popular because it is considered accurate within ±0.05%. Some of the features of the digital meter are auto setting of zero for each change of range, auto change of polarity as the signal changes polarity, noise rejection, removal of an ac line-frequency signal riding on a dc voltage, and very high input resistance of 10 MΩ and higher, making it suitable for virtually any solid-state circuitry. Figures 8.15 and 8.16 are examples of digital multimeters.

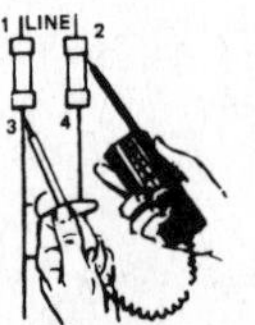

Figure 8.14 Application of Voltprobe tester. (Courtesy of Amprobe Instruments)

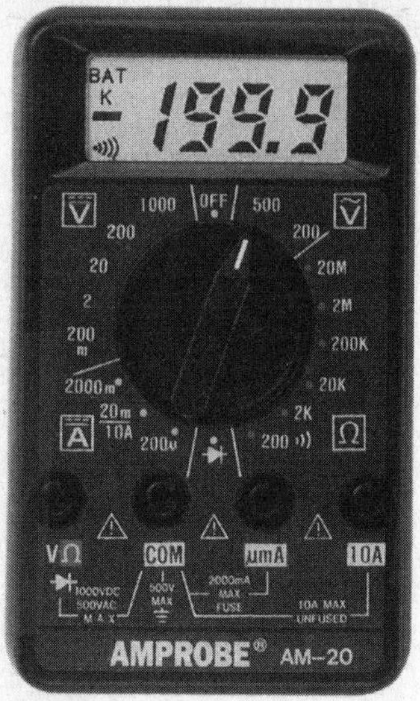

Figure 8.15 Digital volt–ohm–amp multimeter, with temperature ranges.

Ranges: 200 mV/2/20/200/1000 V dc

200/500 V ac

200 μA/20 mA/2000 mA/10 A dc

200 μA/2 mA/20 mA/200 mA/20 A ac

200/2k/20k/200k/2M/20M ohm

Continuity and diode test

(Courtesy of Amprobe)

Volt–Ampere Recorder

When a permanent record of volts and amperes is required, a portable recorder can be used. This is helpful for detecting and interpreting intermittent electrical problems. The recorder shown in Figure 8.17 can record up to three variables on the same paper chart.

Troubleshooting a Furnace Electrically

Testing Part of a Furnace

To troubleshoot a unit electrically, assume, for example, that service is required on a gas-fired forced-air furnace, with wiring shown in Figure 8.1. The complaint is that the furnace does not heat properly. The fan will not run. The gas burner is cycling because the limit control is cycling on and off.

The first step is to review the schematic wiring diagram and eliminate from consideration any circuits that appear to be operating properly. In this case the transformer and the gas valve circuits are eliminated. Now the testing can be confined to the fan motor circuit.

Because part of the furnace does run, a voltmeter should be used for testing. By the process of elimination the source of the trouble is found. First, measure the voltage across the ends of the circuit where the connection is made to L_1 and L_2 (Figure 8.18). The voltmeter reads 120 V, which is satisfactory. Second, with one meter lead on L_2, place the other meter lead between the fan control and the motor (Figure 8.19). The

Figure 8.16 Pocket-size volt–ohm–ampere multimeter offers the functions and accuracy typically found in full-size meters. It can be kept in the shirt pocket for fast and sure work of initial diagnosis.

Ranges: 0–750 V ac and dc

0–5 A ac and dc

Continuity

Resistance to 30 MΩ

Autoranging

Data hold

(Courtesy of UEi)

Keep the full-featured DM5B in your shirt pocket to make fast and sure work of initial diagnostics.

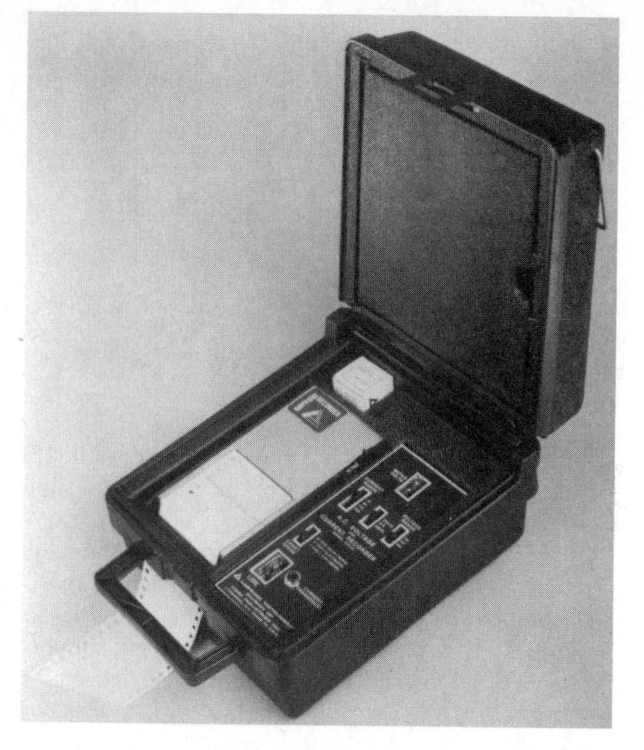

Figure 8.17 Portable stripchart volt/ammeter recorder.

Ranges: 0–15/300/600 V ac

0–15/60/150/300 V ac

Voltage and amperage can be recorded on the same chart. Supplied with power cord, voltage meter cord, clamp on current transducer, roll of chart paper. (Courtesy of Amprobe Instruments)

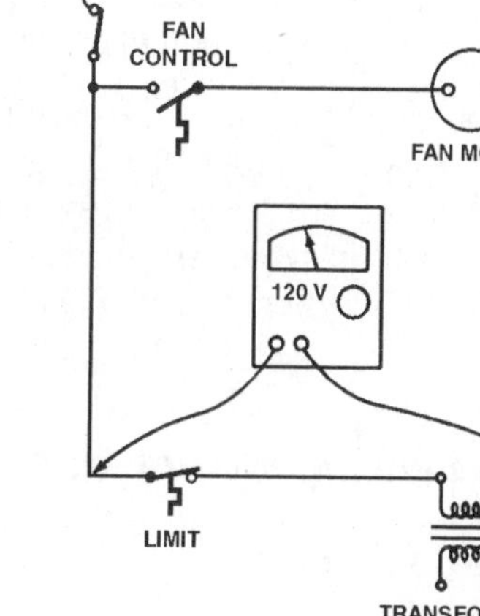

Figure 8.18 Measuring line voltage.

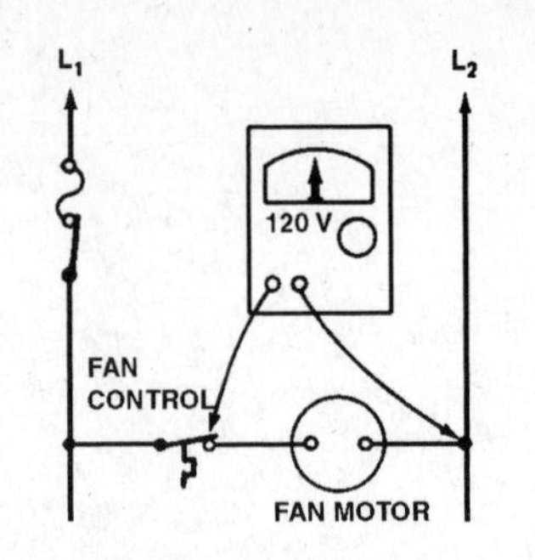

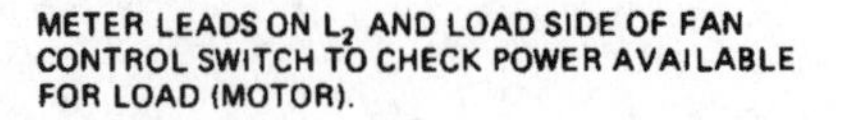

Figure 8.19 Placement of meter leads.

meter reads 120 V. The bonnet temperature is 150°F. The fan control is set to turn the fan on at 120°F. Therefore, the fan control should be made, and this checks out as satisfactory.

If the fan motor still does not run after the proper power has been supplied, it is an indication that the fan motor is defective and must be replaced. The process of elimination can be used on any circuit where troubleshooting is required, regardless of the number of switches. Where power is being supplied to the load and the load does not operate, the load device is defective and needs to be replaced.

Testing a Complete Unit

In troubleshooting a complete unit, the technician can make numerous electrical tests with the instruments. By the process of elimination, it is important to rule out as many of the circuits as possible that are not causing trouble. The methods of testing various heating systems are indicated in the following diagrams:

1. Thermopile, self-generating, gas heating control system (Figure 8.20)
2. 24-V gas heating control system (Figure 8.21)
3. Line-voltage gas heating control system (Figure 8.22)
4. Oil burner control system (Figure 8.23)
5. Electric heating system (Figure 8.24)

Efficiency and Air-Balancing Instruments

Spark Ignition Tester

Figure 8.25 shows an instrument specifically designed to test residential intermittent spark ignition devices and commercial flame safety controls.

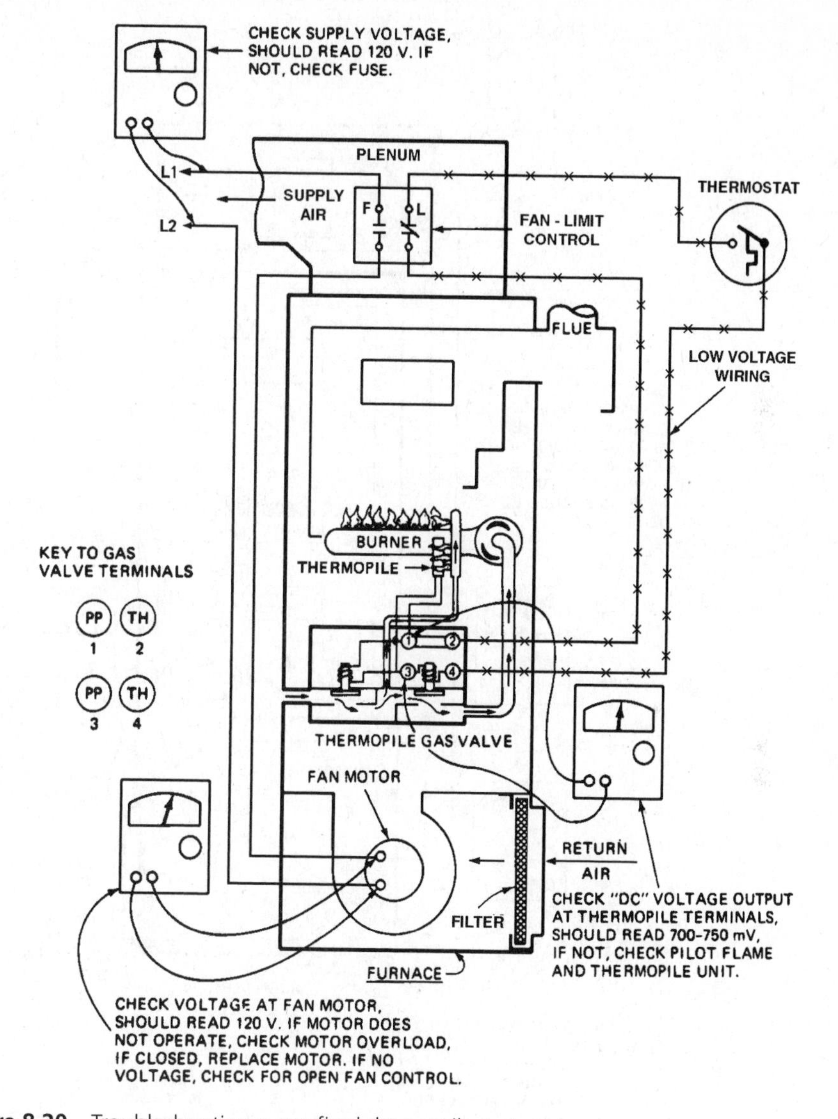

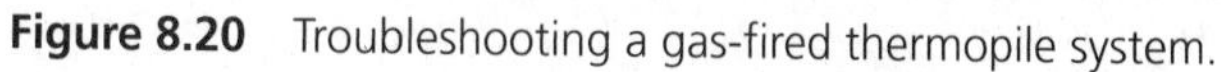

Figure 8.20 Troubleshooting a gas-fired thermopile system.

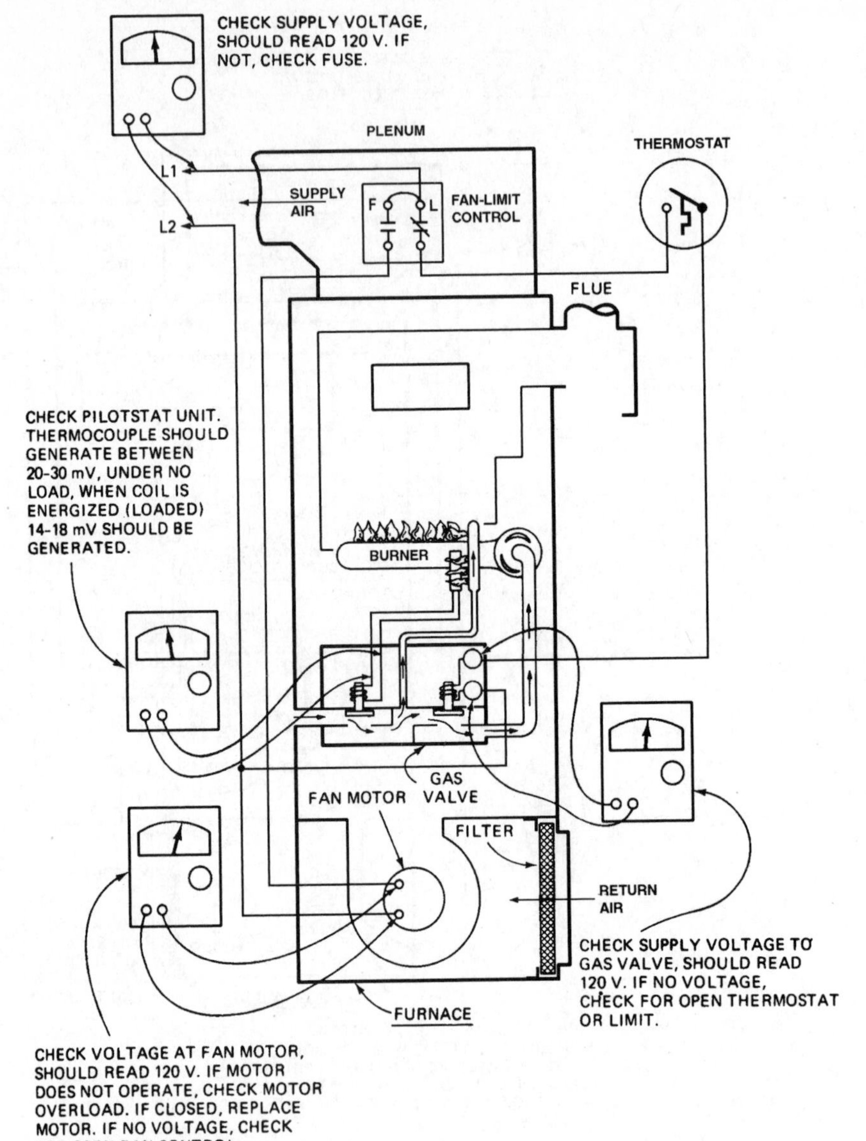

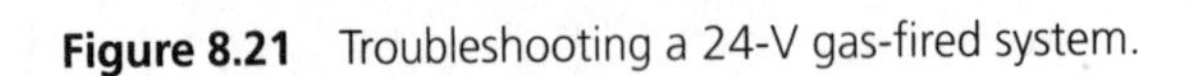

Figure 8.21 Troubleshooting a 24-V gas-fired system.

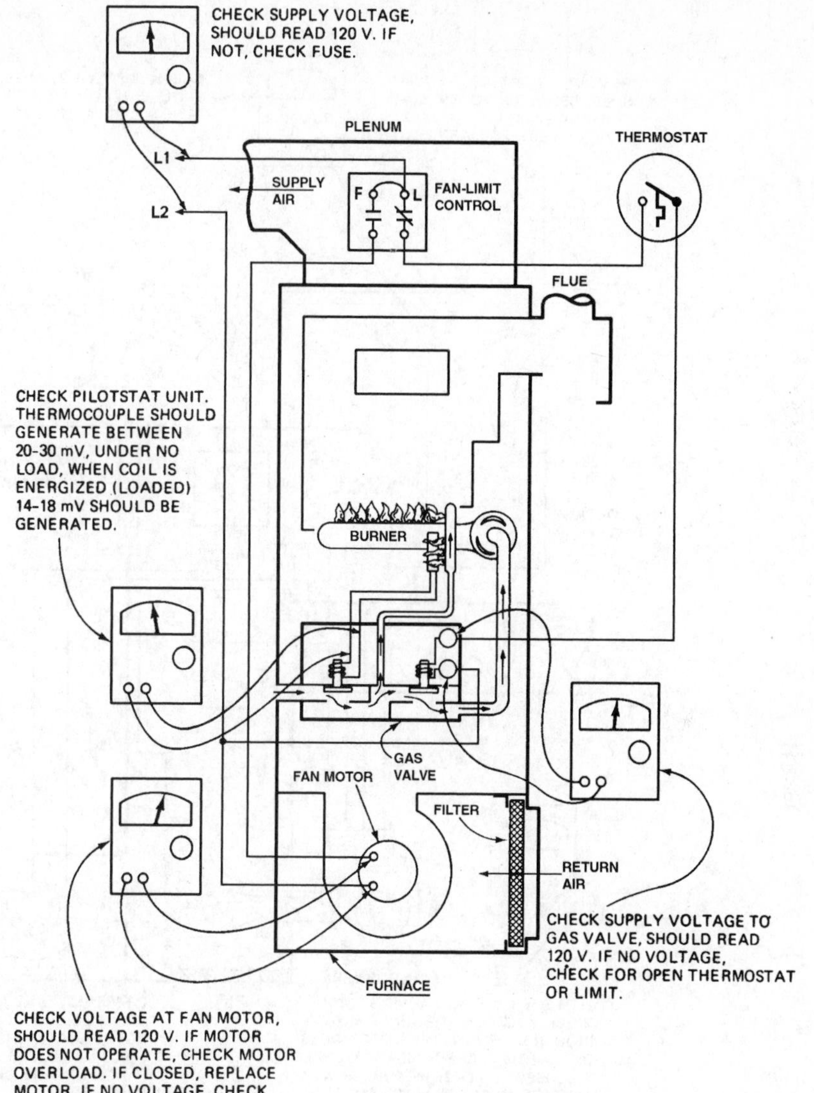

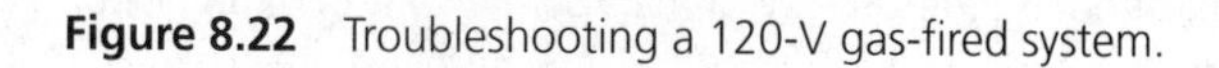
Figure 8.22 Troubleshooting a 120-V gas-fired system.

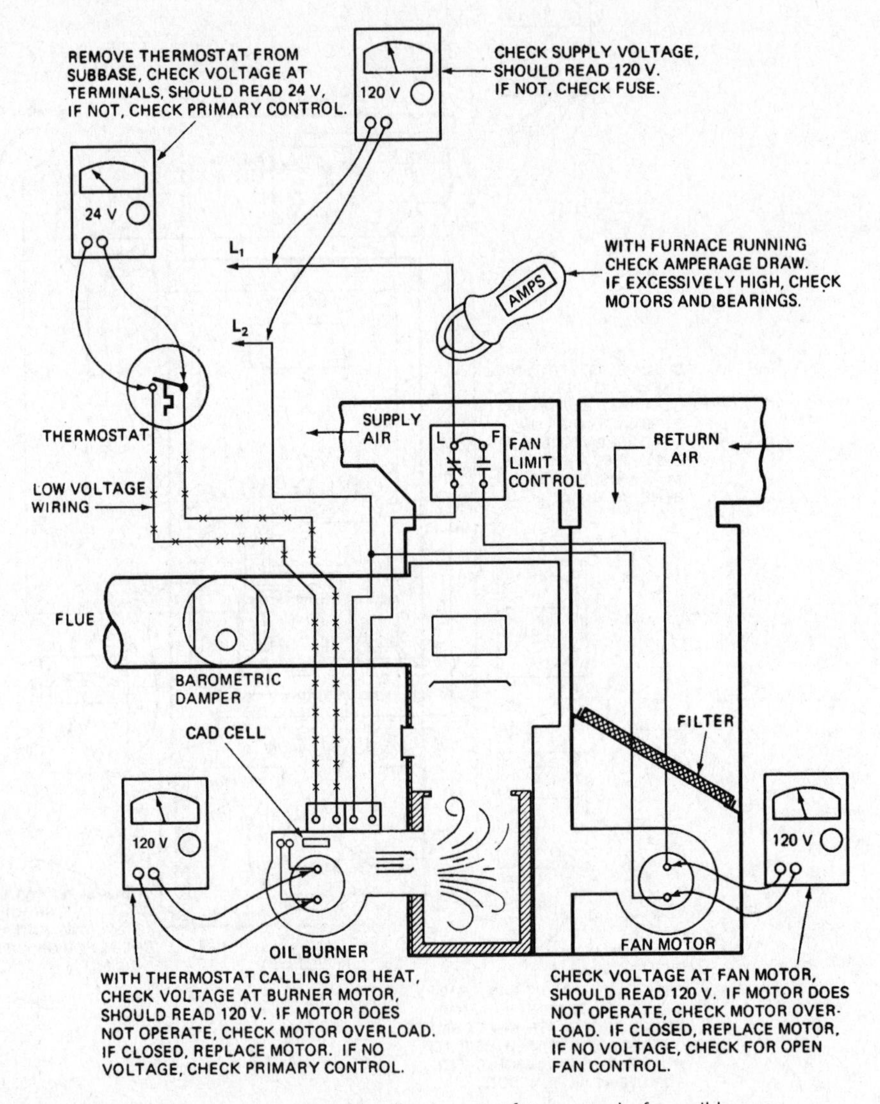

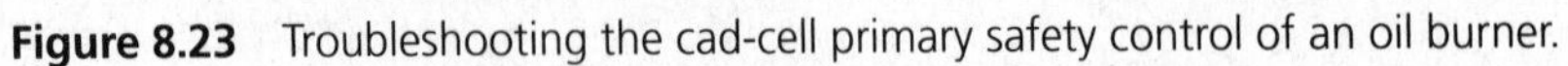

Figure 8.23 Troubleshooting the cad-cell primary safety control of an oil burner.

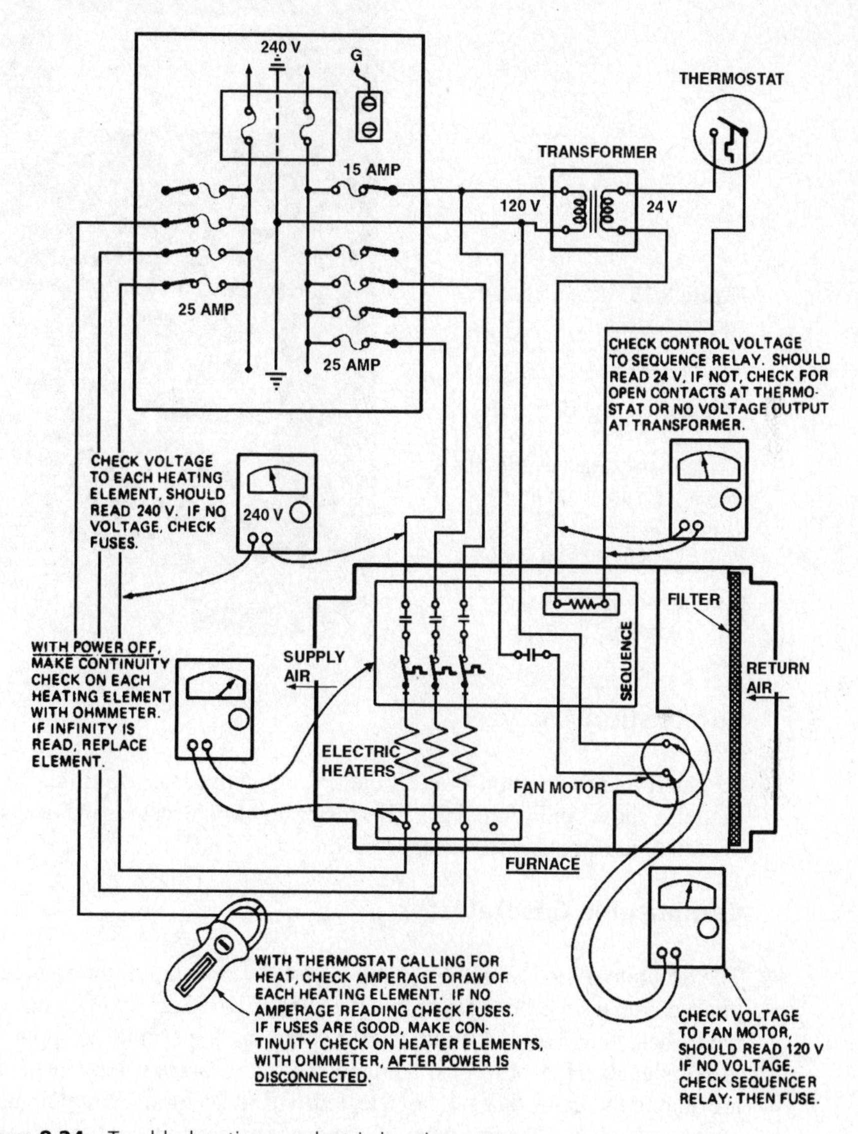

Figure 8.24 Troubleshooting an electric heating system.

Figure 8.25 Spark ignition control tester.

Ranges: 0–2 μA, 0–8 μA
Volts ac and dc

Test lead package includes flame simulator lead, microampere probe, spare resistor for setting flame current on S86 controls.

(Courtesy of JamesKamm Technologies)

Combustion Analyzer

To determine the combustion efficiency, an analyzer is required. A popular handheld model is shown in Figure 8.26. This tester quickly measures the oxygen in the flue gas. With this information the furnace efficiency is calculated.

Combustible Gas Detector

The accurate detection of combustible gases is extremely important and can be accomplished with a solid-state electronic detector. Figure 8.27 shows a handheld model capable of detecting leaks as small as 50 to 1000 parts per million (ppm) with audible and visual signals. Hydrocarbons, halogenated hydrocarbons, alcohols, ethers, and ketones are among the gases that can be detected. This detector is extremely useful as a general-purpose tool in any environment where gasoline, propane, natural gas, or fuel oil is used.

Temperature Testers

Temperature readings are often required when performing service work. Three types of temperature testers are shown in Figures 8.28 to 8.30. The model shown in Figure 8.28

Figure 8.26 Digital electronic combustion tester that measures O_2, stack and ambient temperatures, and calculates CO_2, efficiency and excess air. It has 99 memory locations to store and recall test results. Some of the optional upgrades include a CO sensor and infrared remote printer. (Courtesy of UEi)

has 15-ft remote probes and can read from −50° to 300°F. Both Fahrenheit and Celsius temperatures can be read on the digital model shown in Figure 8.29. The infrared thermometer shown in Figure 8.30 is also digital and reads from −40° to 300°F.

Air Meters

There are many types of testers used to read and measure air velocity, pressure, and volume. Possible readings can include the following supply or return: grille velocity,

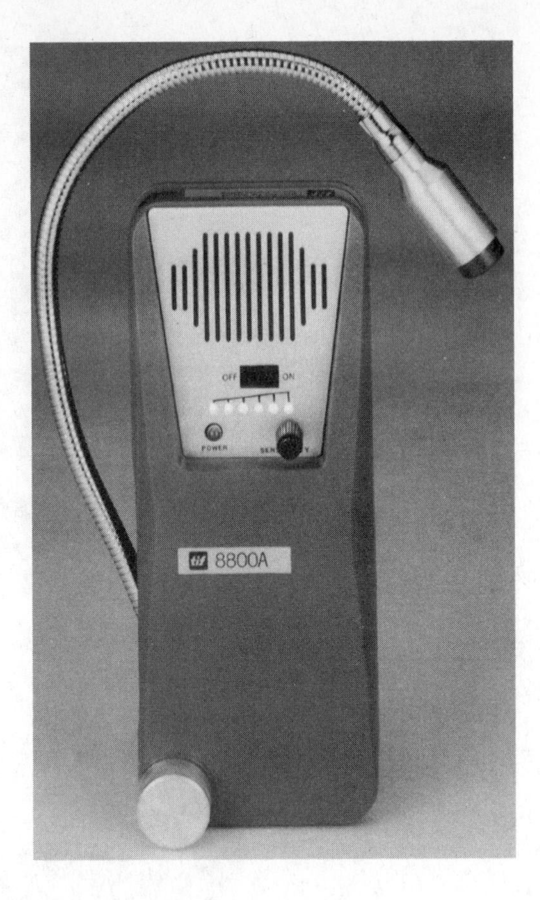

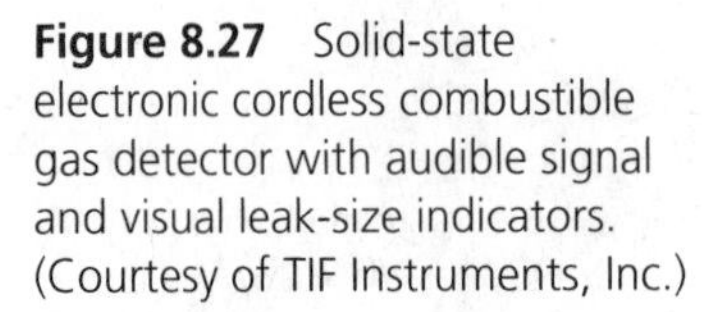

Figure 8.27 Solid-state electronic cordless combustible gas detector with audible signal and visual leak-size indicators. (Courtesy of TIF Instruments, Inc.)

duct air velocity, and pressure drops across cooling coils or air filters. It is also possible to measure furnace drafts and exhaust hood face velocities (see Figures 8.31 to 8.34).

Manometer

To test a gas furnace for proper operating gas pressures, a manometer is used. There are two basic types of manometers, a gauge type and a U-tube type. The gauge type connects to the gas manifold and uses the gas pressure to deflect the needle to obtain a pressure reading (see Figure 8.35).

A U-tube manometer is a plastic U-shaped tube that uses water for measurement. The water is halfway in the tube and set to a zero reading before being connected to a system. When gas pressure is applied, this will push the water column up to readable pressure. The pressure reading is the difference in height, which is the sum of the readings above and below zero. See Figure 8.36, where the pressure reading shown is $1\frac{3}{4}$ in. + $1\frac{3}{4}$ in. = $3\frac{1}{2}$ in.

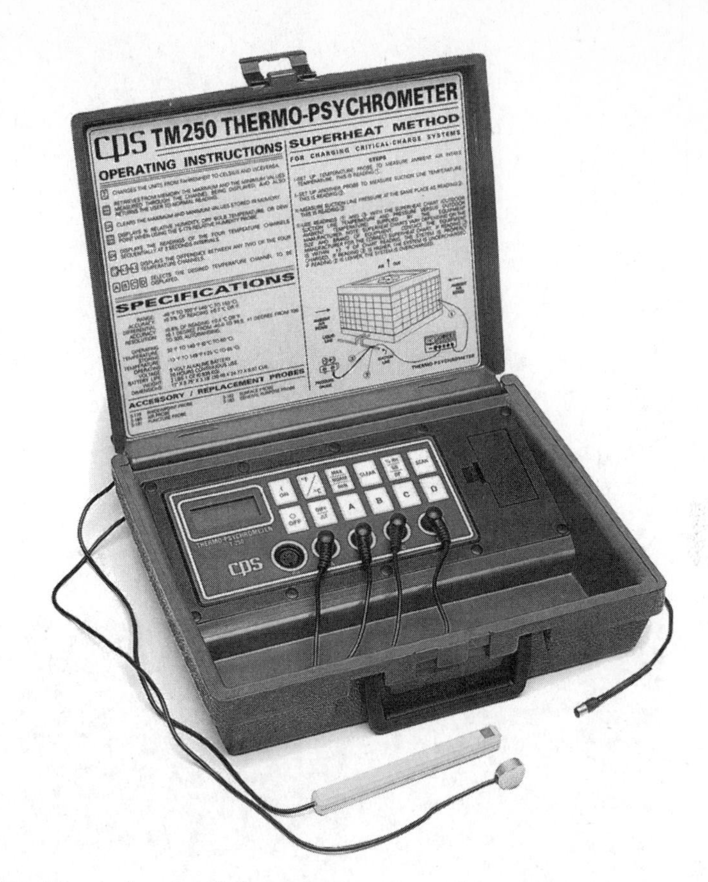

Figure 8.28 Four-station digital thermometer. CPS Model TM250C. (Courtesy of CPS Products, Inc.)

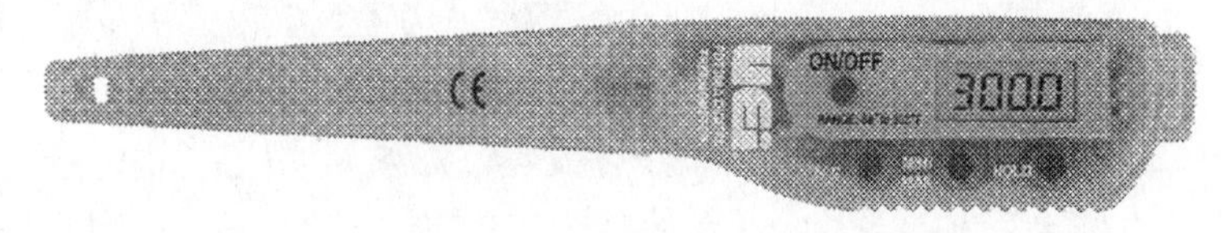

Figure 8.29 Digital thermometer. The PDT300A has a °F and °C selector switch, magnetic mount, and length extension, and is battery powered.

Ranges: −58° to 302°F

−50° to 150°C

(Courtesy of UEi)

Figure 8.30 Model H25-886MTh RAYTEK. Non-contact infrared thermometer. (Courtesy of Raytek Corporation)

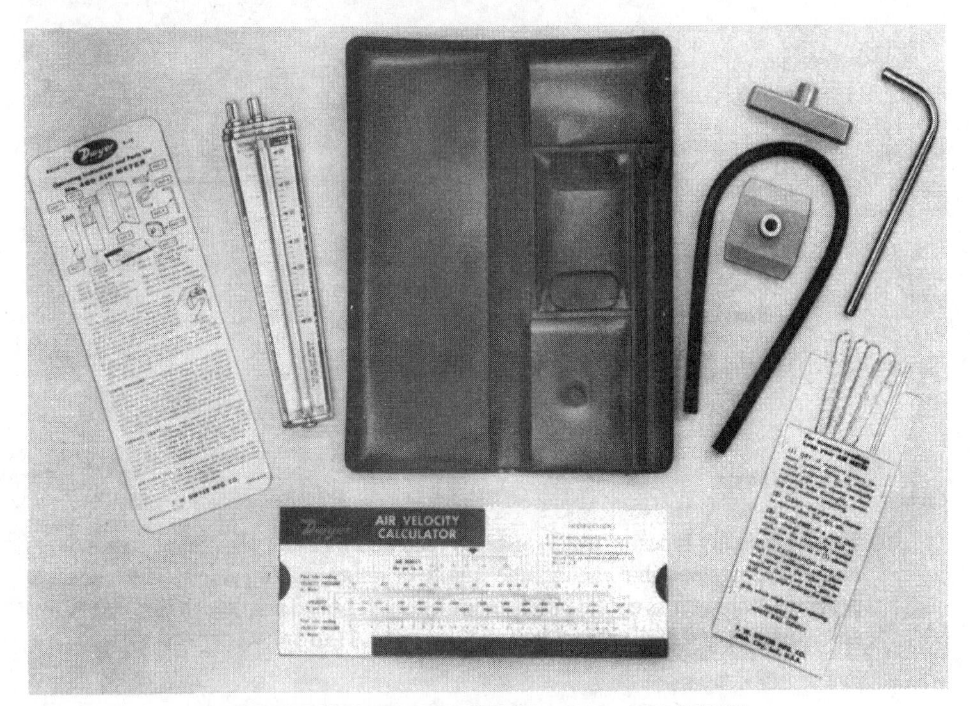

Figure 8.31 Direct-reading air meter kit; includes meter, probes, and air-velocity calculator. (Courtesy of Dwyer Instruments, Inc.)

Figure 8.32 Standard magnehelic gauge. The Model 2000-00AV magnehelic gauge reads air pressure differential in inches of water column on a large 4" diameter dial.

Range: 0–0.50 in. W.C.

(Courtesy of Dwyer Instruments, Inc.)

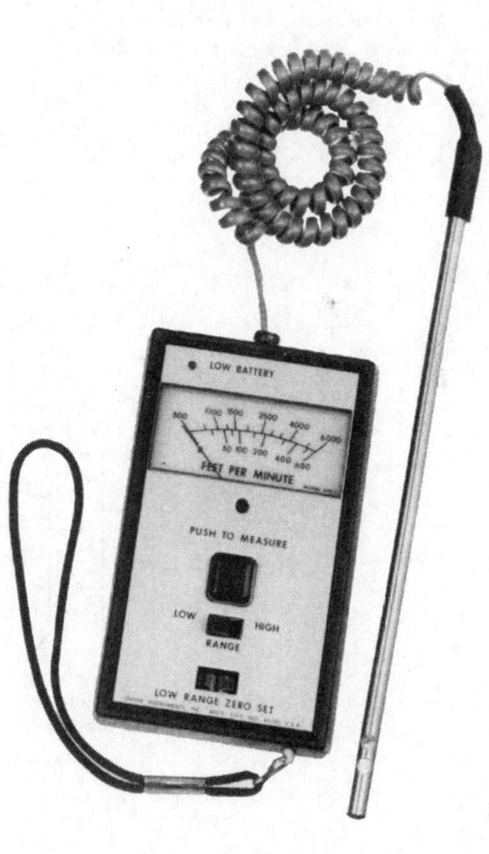

Figure 8.33 Handheld thermal anemometer, battery powered with low-battery indicator. (Courtesy of Dwyer Instruments, Inc.)

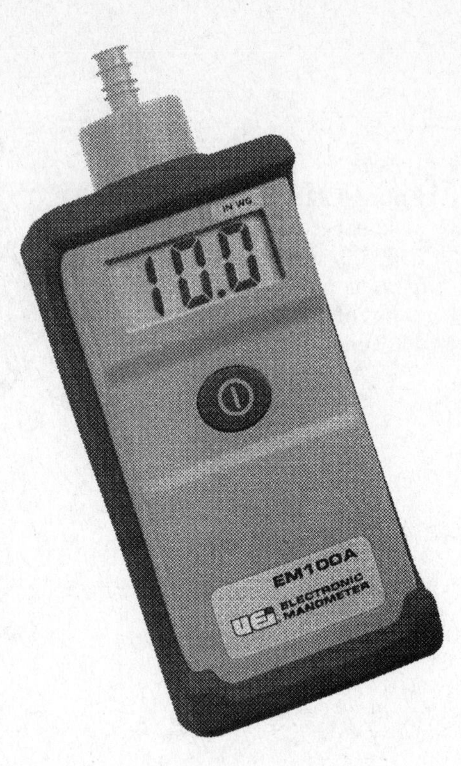

Figure 8.34 Electronic manometer. The EM100A is an electronic manometer that provides a continuous display of the measured positive or negative pressure. It is used for applications such as air balancing and pressure switch testing.

Ratings: −20 to +20 in. W. C.

(Courtesy of UEi)

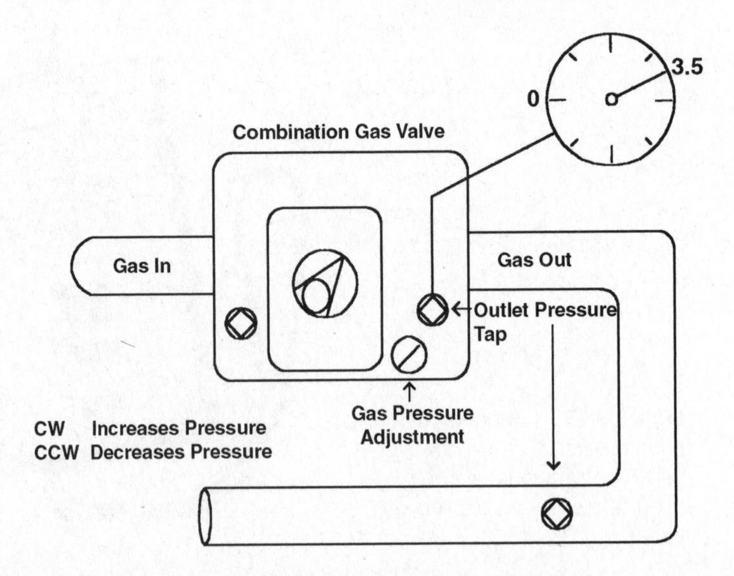

Figure 8.35 Measuring gas pressure using a gauge-type manometer.

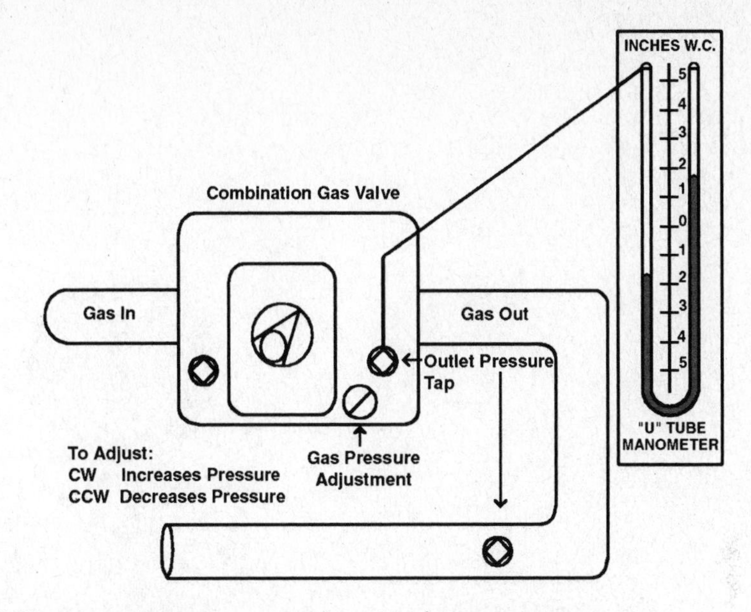

Figure 8.36 Measuring gas pressure using a U-tube manometer.

Heat Pump and Add-on Air-Conditioning Service Instruments and Testers

Many specialized instruments and equipment items are required in the servicing of heat pumps and add-on air conditioners. Included in this group are gauge manifolds, vacuum pumps, recovery equipment, and leak detectors. Figure 8.37 shows a heat pump manifold set comprising one compound gauge, one high-pressure gauge, and a second high-pressure gauge for direct pressure check of the switching assembly. Standard add-on air-conditioning manifolds will not have the third gauge.

Superheat as described in Chapter 24 is one of the best methods for verifying the correct refrigerant charge of an add-on air-conditioning system. A superheat gauge (Figure 8.38) can be connected to the gauge manifold for an easy reading of superheat without calculations.

Vacuum Pumps

A vacuum pump is used to evacuate all air and moisture from the refrigerating system. The moisture is removed as a vapor after it has boiled at the very deep vacuum produced by the pump. This requires a correctly sized pump. As a rule of thumb, the pump's cfm rating squared equals the maximum system tonnage. Thus, 3 cfm pump is rated for 9 tons, and a 5 cfm is rated for 25 tons (Figure 8.39).

To assure a deep vacuum, it must be measured. The most accurate measurement is with a micron gauge. The compound gauge shows markings for 30 in. of vacuum.

Figure 8.37 Heat-pump manifold set. (Courtesy of Robinair Division, SPX Corporation)

Each inch is equal to 25,400 microns. A good micron gauge can read down to 50 microns of vacuum (Figure 8.40). Evacuation is considered complete when a system holds at 500 microns. Although the compound gauge indicates that a vacuum is being produced, it does not provide an accurate measurement for the moisture being boiled off.

Air-Conditioning Tools

When the total required refrigerant charge is known or when recovery work is being done, a digital scale is useful for measuring the amount of refrigerant to charge into the system or into the recovery tank. Most digital scales are accurate to the nearest ½ oz. (Figure 8.41).

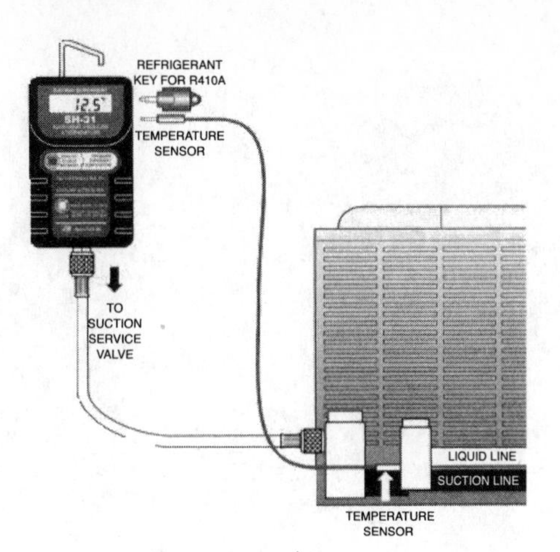

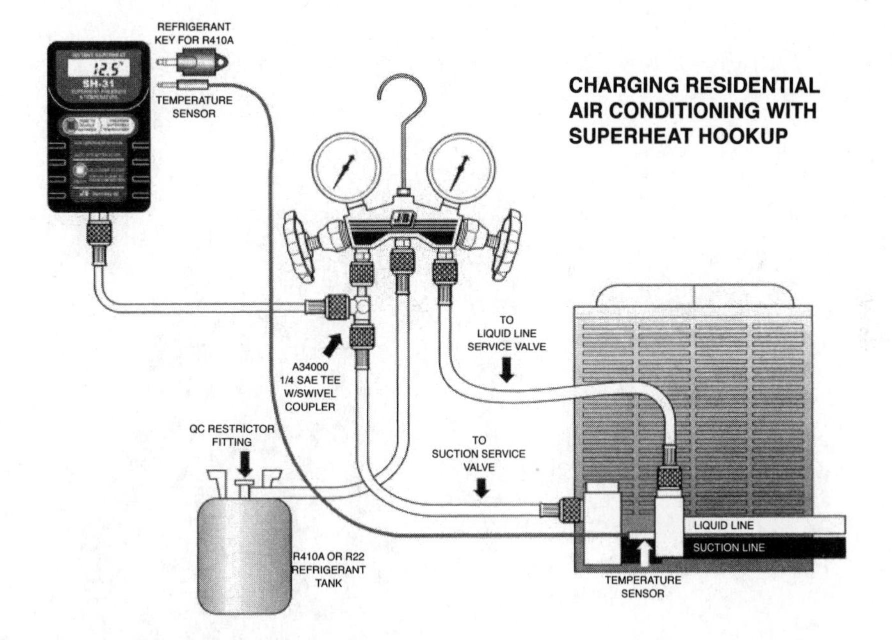

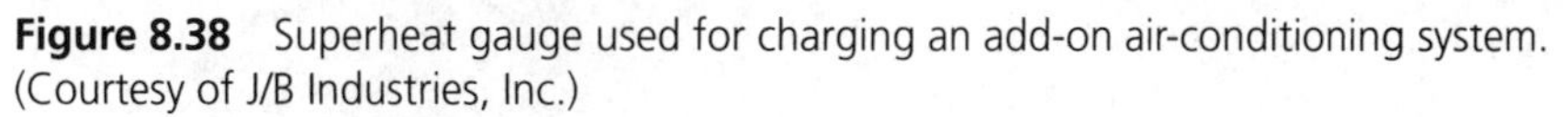
Figure 8.38 Superheat gauge used for charging an add-on air-conditioning system. (Courtesy of J/B Industries, Inc.)

Figure 8.39 Vacuum pumps of 3 and 7 cfm capacity. (Courtesy of J/B Industries, Inc.)

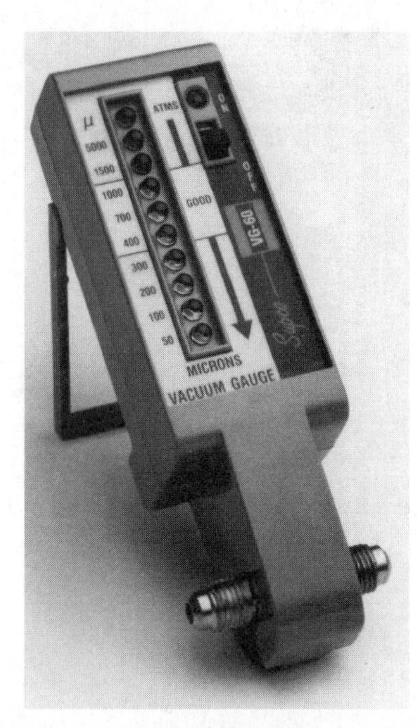

Figure 8.40 Micron gauge. (Courtesy of Supco Sealed Units Parts Co., Inc.)

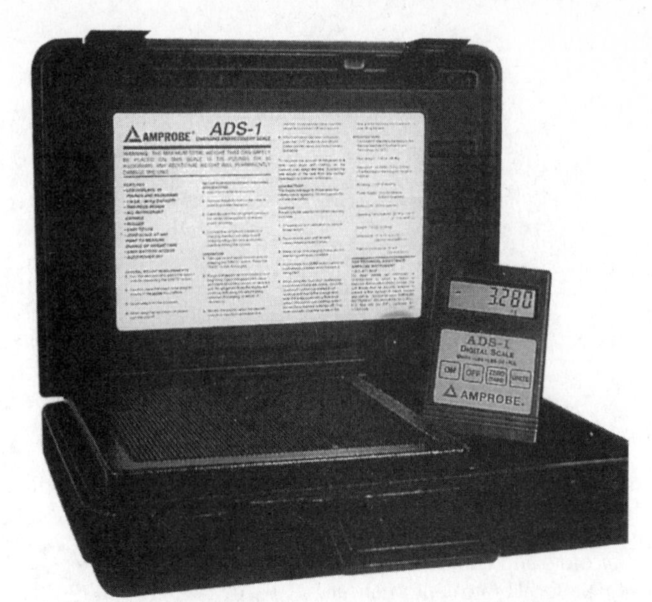

ADS-1 Digital Scale

Features & Benefits

- Recovery & charging made easy
- Weighing capacity up to 176 lbs. (80 kg)
- Accuracy ±0.031 lb. (0.0140 Kg)
- LCD display in Pounds, Pounds & Ounces and Kilograms
- Uses load-cell technology, giving the technician an accurate reading every time no matter where the refrigerant tank is located on the scale
- Handles all refrigerants
- Rugged compact design for field use-comes standard with carrying case for equipment protection
- Two piece design for easier use
- Instructions and 9-volt battery included
- Calibrated to Standards traceable to the National Institute of Standards and Technology (N.I.S.T)
- For operation, scale must be removed from case

Figure 8.41 Digital scale. The ADS-1 digital scale has a capacity of 176 lb. (80 kg) with an accuracy of 0.031 lb (0.0140 kg). The LCD displays the weight in pounds, pounds and ounces, and in kilograms. Uses a 9-V battery. (Courtesy of Amprobe)

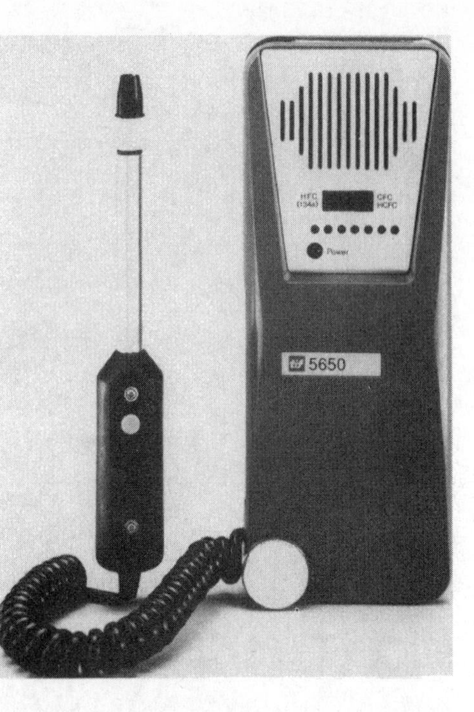

Figure 8.42 Refrigerant leak detector, pump style, HFC/HCFC/CFC, with visual leak-size indicators and audible beeping signal. Recalibrates automatically in contaminated atmospheres. Battery powered. (Courtesy of TIF Instruments, Inc.)

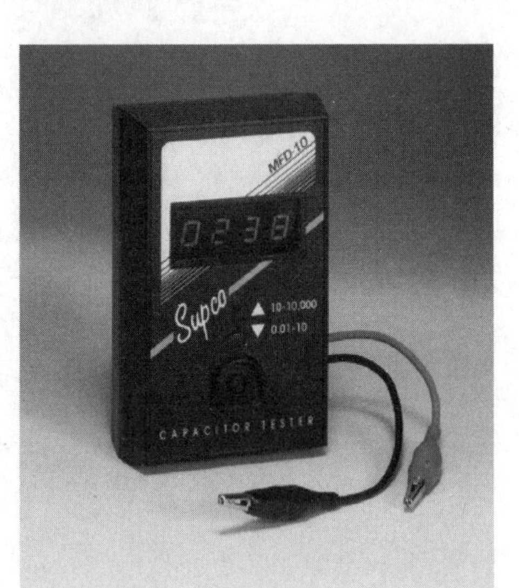

Figure 8.43 Digital capacitor tester. The MFD10 measures the capacitance value from 0.01 to 10,000 μF. Checks for open and shorted capacitors. Can be used as a continuity tester. Checks run, start, and dc electrolytic capacitors. Identifies unmarked capacitors. Uses four AA batteries. (Courtesy of Supco Sealed Unit Parts Co., Inc.)

Refrigerant leaks may be found using several methods. A soapy solution can be placed on the suspected leak area, which will produce bubbles at the exact site of the leak. An older method was to use a halide torch. This torch sensed the presence of chlorine in the refrigerant. Many new refrigerants are chlorine free, and a leak cannot be determined with the torch. An electronic leak detector (Figure 8.42) capable of finding leaks of all refrigerants used in air conditioning, is a very reliable method of leak detection. Most detectors have an audible alarm that increases in beeping frequency as the leak size increases.

Motors used in heat pumps and air conditioning frequently use capacitors. A digital capacitor tester (Figure 8.43) helps to determine if the capacitor is operating within

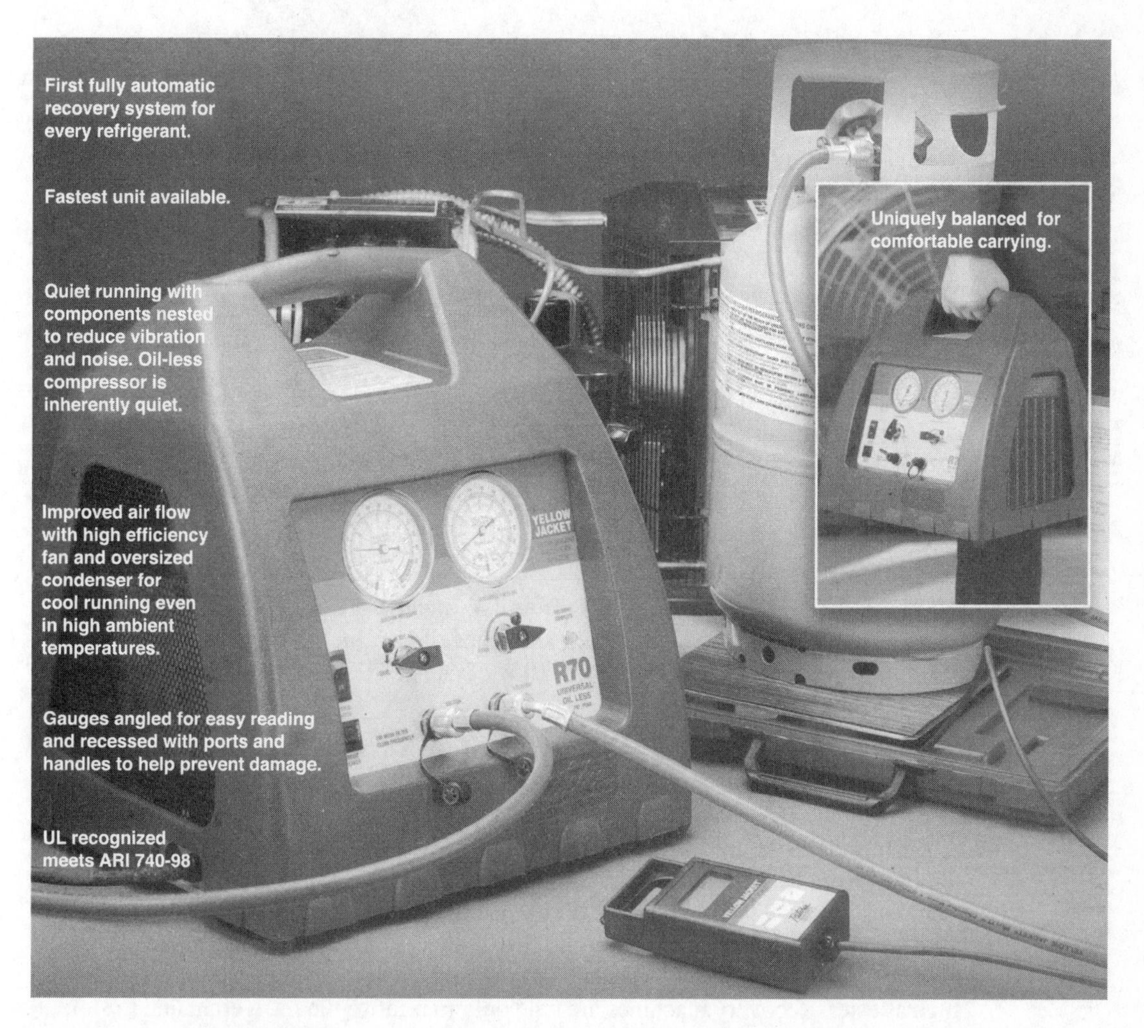

Figure 8.44 Universal refrigerant recovery system. Oil-less compressor works with every refrigerant. (Courtesy of Yellow Jacket, Ritchie Engineering Company, Inc.)

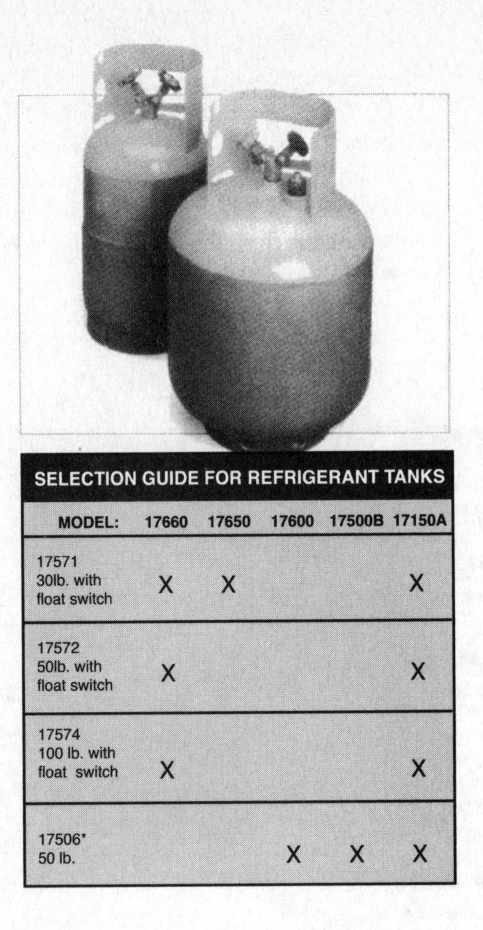

SELECTION GUIDE FOR REFRIGERANT TANKS					
MODEL:	17660	17650	17600	17500B	17150A
17571 30lb. with float switch	X	X			X
17572 50lb. with float switch	X				X
17574 100 lb. with float switch	X				X
17506* 50 lb.			X	X	X

Figure 8.45 Refrigerant recovery tanks. (Courtesy of Robinair Division, SPX Corporation)

the range specified. Attach the clips from the meter leads to the capacitor terminals, and press the button to display the exact capacitor value on the LED readout. Run capacitors must test out within +/−10% of the rated capacitance. Start capacitor should be at the rated capacitance or +20%.

Refrigerant Recovery Equipment

Since July 1, 1992, all refrigerants must be recovered into containers when they are removed from heat-pump and air-conditioning systems. Many recovery and recycling machines are available. The most useful machines are constructed so that they can be used for all refrigerants. This is not the case with some machines that can be used with a refrigerant of only one oil type, such as mineral oil, but cannot used with another oil such as ester. Recovery machines are able only to remove the refrigerant and to store it in a container without processing. The more expensive recycling units are able to clean the refrigerant after recovery. Units should be selected based on refrigerants to be recovered, size of air conditioner to be emptied, and weight (Figure 8.44). Refrigerant

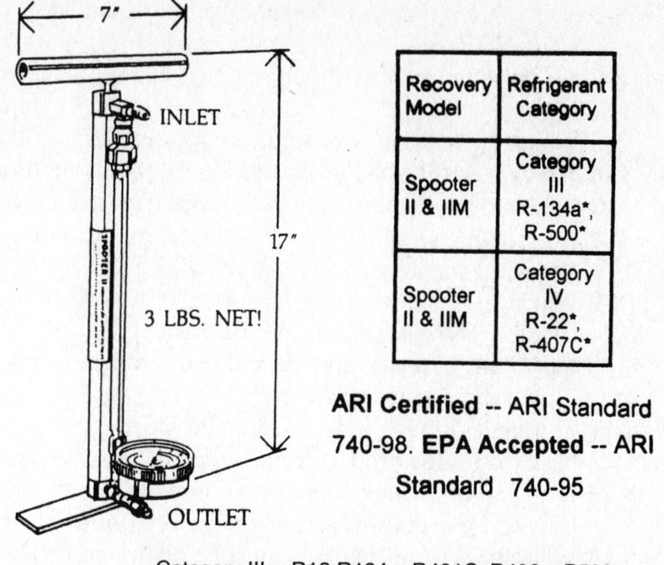

Recovery Model	Refrigerant Category
Spooter II & IIM	Category III R-134a*, R-500*
Spooter II & IIM	Category IV R-22*, R-407C*

Category III -- R12,R134a, R401C, R406a, R500
Category IV -- R401a, R409A, R401B, R412A, R411A, R407D, R22, R411B, R407C, R402B, R408A, R509

- You Can Use Your Spooter As A Control Testing Pump
- You Can Use Your Spooter As An Oil Pump
- You Can Use Your Spooter As A Pressure Test Pump
- You Can Use Your Spooter As A Vacuum Pump
- You Can Use Your Spooter For Transferring Refrigerants From One Cylinder To Another

Figure 8.46 Hand pump–style recovery device. Available for most popular refrigerants. (Courtesy of ICOR International Inc.)

may be placed only into approved recovery cylinders, which normally are color coded with a yellow top on a gray body (Figure 8.45).

An alternative recovery unit for smaller air-conditioning equipment is a hand-pump device as shown in Figure 8.46. It is EPA certified as an active recovery machine and works with most popular refrigerants.

STUDY QUESTIONS

Answers to the study questions may be found in the sections noted in brackets.

8-1 Explain the functions of the voltmeter, ammeter, ohmmeter, and wattmeter. *[Instrument Functions]*

8-2 In purchasing a meter, it is important to consider which scales on each of the commonly used meters? *[Selecting the Proper Instrument]*

8-3 Name three important points to be remembered in using electrical test instruments. *[Selecting the Proper Instrument]*

8-4 Describe the types of readings that can be taken with an ohmmeter. *[Using Meters]*

8-5 Which meter has its own source of power? *[Using Meters]*

8-6 Which meter is good for testing fuses? *[Using Meters]*

8-7 Which meter requires calibration each time it is used? *[Instrument Functions]*

8-8 How many wires should be placed in the jaws of a clamp-on ammeter? *[Selecting the Proper Instrument]*

8-9 Which special device is required to measure dc millivolts delivered by a thermocouple? *[Selecting the Proper Instrument]*

8-10 If a voltmeter reads zero when placed across a switch when the power is on, what is the condition of the switch? *[Selecting the Proper Instrument]*

8-11 In troubleshooting a defective unit, which circuits can the technician eliminate from the testing procedure? *[Testing a Complete Unit]*

8-12 If no part of the unit operates because of a blown fuse, which instrument should be used for testing? *[Selecting the Proper Instrument]*

8-13 What types of refrigerants must be recovered before opening a heat-pump system? *[Refrigerant Recovery Equipment]*

8-14 What instrument should be used to accurately measure the depth of vacuum? *[Vacuum Pumps]*

CHAPTER 9

Electrical Service Wiring

OBJECTIVES

After studying this chapter, the student will be able to:

- Describe the electrical pathway from the generator of electricity to the residence
- Evaluate the external residential electrical service
- Determine the wiring requirements for a heating and air-conditioning system

Electrical Transmission

Electric energy usage throughout the world has quadrupled in the last 20 years, both in industrial as well as residential applications. Automated equipment, robotics, air conditioning required by computer mainframes, and human comfort are examples of the electrical demands on the utility companies. The residential market has to an extent followed the industrial trend by adding electrical appliances such as refrigerators, freezers, televisions, computers, home offices and air conditioning, along with sophisticated lighting systems. All these devices put a tremendous load on the electrical utility providers. The electrical power providers use steam, which is the prime mover of the generator, which produces the electricity to the power lines, as shown in Figure 9.1. The energy source to generate steam can be oil, gas-fired boilers, coal, geothermal, wind, as well as nuclear reactors. The electrical power produced by a power plant is also fed into a grid system, which provides backup power to assist others on the grid in case of power interruption in their area, which can be caused by lightning or other types of severe storms.

Main Power Supply

The usual source of the main power supply to a building is an entrance cable, consisting of three wires. On newer services one of these wires is black (hot), one is red (hot),

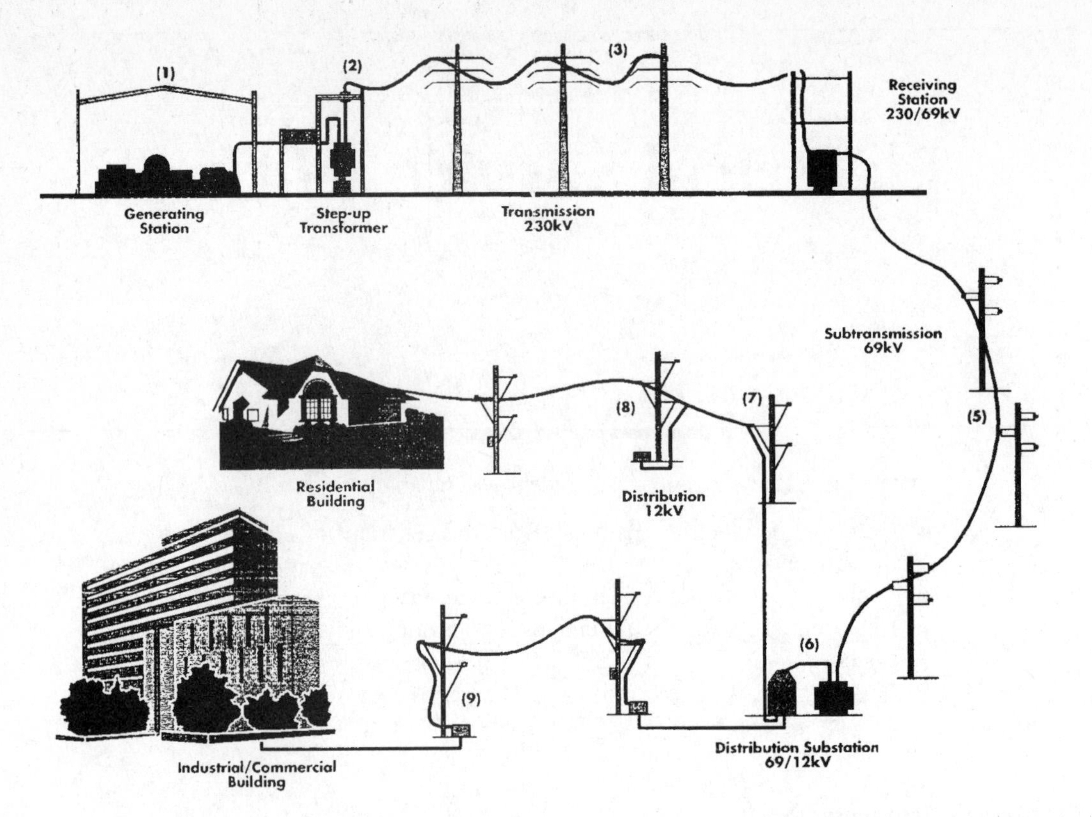

After electricity is produced at the generating station **(1)**, it begins its journey to customers at the station's switchyard. A "step-up" transformer **(2)** converts to a high voltage - 230 kilovolt (kV) to 500 kV - to travel on transmission lines **(3)**.

These lines are linked to receiving stations **(4)**, where the electricity is "stepped down" to lesser voltages, such as 69 kV. From these receiving stations, electricity travels over subtransmission lines **(5)** to distribution substations **(6)**.

The power is again stepped down to 12 kV, and travels on distribution lines **(7)**, either overhead or underground. At regular intervals, the lines carry power through transformers **(8)** where it is again stepped down to the proper voltage for home or business use.

Larger commercial and industrial customers may be linked directly to the subtransmission systems, and have their own substations **(9)**.

Figure 9.1 The distribution of electricity from generating station to residential, commercial, and industrial buildings. (Courtesy of SRP Research and Communications Services)

and the third is white (neutral). The size of these entrance wires determines the maximum size of the service panel that can be installed.

Entrance Cable The wires in an entrance cable connect the incoming power lines with the building wiring. The cable is connected through the meter and socket to the main service panel, as shown in Figure 9.2. Note that the neutral wire at the main panel is connected to an ground.

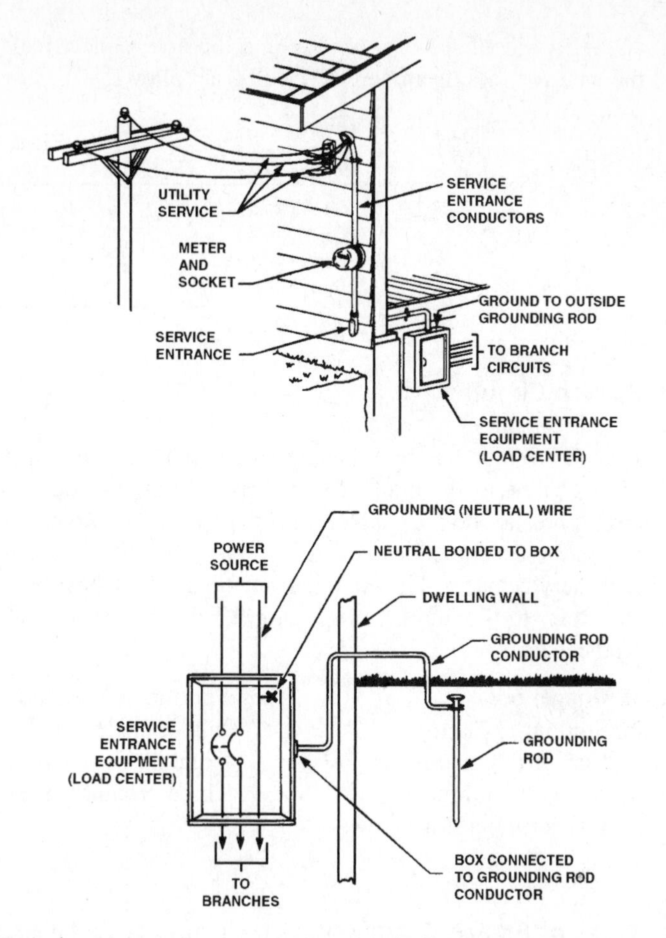

Figure 9.2 Typical service entrance arrangement including one with neutral ground. (Courtesy of ITT Electrical Products)

When it is necessary to increase the size or capacity of the service, an additional fuse panel or subpanel can be added to the main service panel. This is possible, however, only when the entrance wiring is large enough to permit the extra load.

Building's Ampere Service A building's ampere service is the size or capacity of the power supply available from the main panel for electrical service within the building. The service is usually 60, 100, 150, or 200 A. If a home has an electric range, electric dryer, and central air conditioning, 150-A service is needed. For homes having electric heating, 200-A service is required. The minimum service for a modern home is 100 A. Many older homes have only a 60-A service. The service must always be large enough to supply the needs of the connected loads.

All residential services require a three-wire electrical power supply. The size of the wire for various ampere services is as follows:

Copper Wire	*Aluminum and Copper-Clad Wire*
60-A No. 6	No. 6
100-A No. 4	No. 2
150-A No. 1	No. 0
200-A No. 2/0	No. 3/0

Branch Circuits

Branch circuits are the divisions of the main power supply that are fused and connected to the loads in a building. The main power supply is connected to the fuse or circuit breaker box, as shown in Figure 9.3. The two hot lines are fused. The voltage across the two hot lines is usually 208 or 240 V. The neutral line is grounded at the main service panel. The voltage from either of the hot lines to the neutral line is 120 V.

Refer to the numbers in Figure 9.3 for the following components of the electric load center:

1. Copper bus
2. Straight-in mains
3. Convertible mains
4. Main circuit breaker
5. Interior mounting screws
6. Split branch neutral
7. Indoor enclosure design
8. Combination neutral ground bars
9. Three ground bar mounting locations

Types of Branch Circuits Branch circuits can be either 208/240 V (Figure 9.4) or 120 V (Figure 9.5). Branch circuits for 208/240 V have fuses in both hot lines. Branch circuits of 120 V have one fuse in the hot line and none in the neutral. The main fuse box or circuit breaker can have both 208/240-V and 120-V branch circuits. The maximum fuse size on a 120-V branch is 20 A. The 20-A 120-V branch requires No. 12 wire minimum. The 15-A 120-V branch requires No. 14 wire minimum, or as required by local codes.

Grounding

The electrical terms *grounding, grounded,* and *ground* may be confusing. *Grounding* is the process of connecting to the earth a wire or other conductor from a motor frame or metal enclosure to the electrical utility system or other conducting material. Grounding is done chiefly for safety, and all electrical equipment should be *grounded.* Modern three-prong plugs and receptacles for 120-V circuits have a grounding wire (Figure 9.6).

Figure 9.3 Electric load center. (Courtesy of Square D, Schneider Electric)

A *ground* is the common return circuit in electrical equipment whose potential is zero. A ground permits the current to get through or around the insulation to normally exposed metal parts that are hot, or "live." A proper ground protects a user from electrical shock should a short circuit occur.

Low Voltage

All load devices are designed to operate at a specified voltage that is marked on the equipment. The voltage may be 120, 208, or 240 V. Most electrical equipment will tolerate a variation of voltage from 10% above to 10% below rated (specified) voltage. Thus, a motor rated at 120 V will operate with voltages between 108 and 132 V.

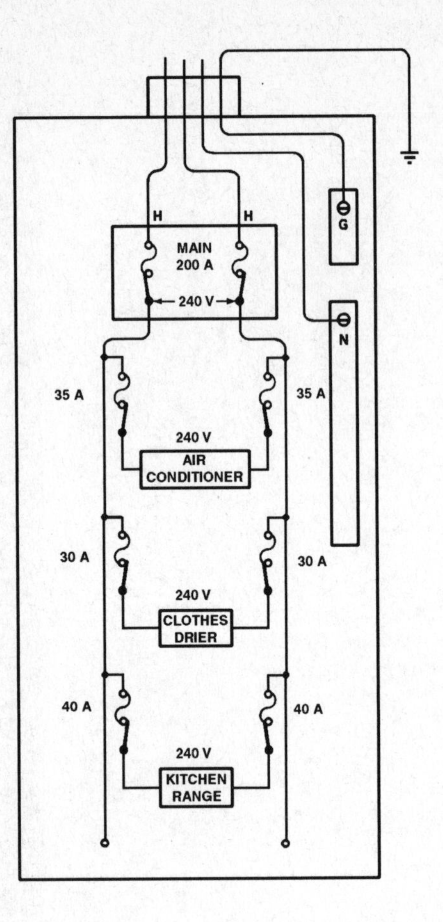

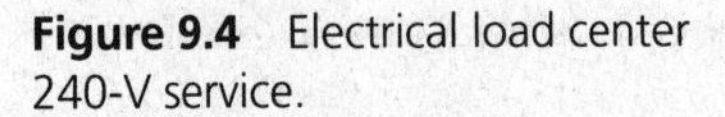
Figure 9.4 Electrical load center 240-V service.

Low or high voltage is usually considered to be any voltage that is not within the tolerated range. Thus, if a 120-V motor is supplied with 100-V power, the voltage will be too low and the motor will probably fail to operate properly. Low voltage is a much more common problem than high voltage, because all current-carrying wire offers a resistance to flow, which results in a voltage drop. For example, a 50-ft length of No. 14 wire would have a $3\frac{1}{2}$-V drop when carrying 15 A. This drop is not excessive, since the resultant voltage is within the 10% allowable limit. If a circuit is overloaded, the current rises above the rated current-carrying capacity of the wire, and the voltage drop can easily exceed the 10% limit established for the load. The low-voltage condition that results from overloaded circuits can cause motors to fail or burn out.

All power supplies should be checked during peak load conditions to determine if proper voltage is being supplied. This should be done at the service entrance, at the furnace disconnect, and at load devices. If there is a problem at the service entrance, the power company should be contacted. If there is a problem of overloaded circuits in the building, the owner should be notified and an electrician's services secured. If

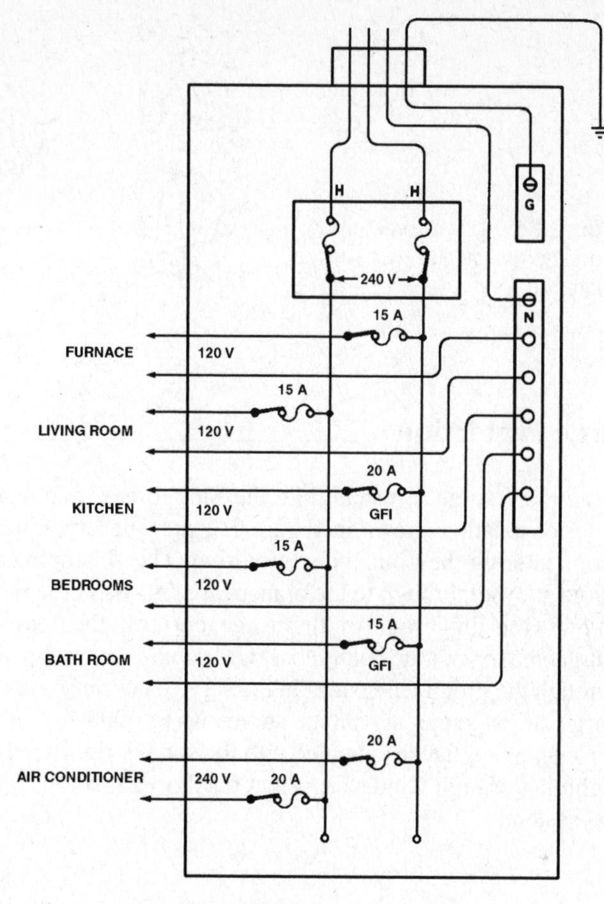

Figure 9.5 General-purpose electrical load center.

there is a problem in the furnace wiring, a service technician should look for defective wiring and/or check for proper sizes of wires and transformers.

Load Center

The existing main electrical panel may be suitable or it may need replacing, depending on power requirements for the equipment being installed. On new installations, the building plans should include adequate main panel service.

There are two types of panels: those with fuses (old style; Figure 9.7) and those with circuit breakers for the individual circuits (Figure 9.8). The circuit-breaker type of panel has the advantage that the breaker can simply be reset rather than having to have the fuses replaced, making it more convenient to service when an overload occurs. Both types of panels are available with delayed-action tripping to permit loads to draw extra current when starting without shutting down the power supply.

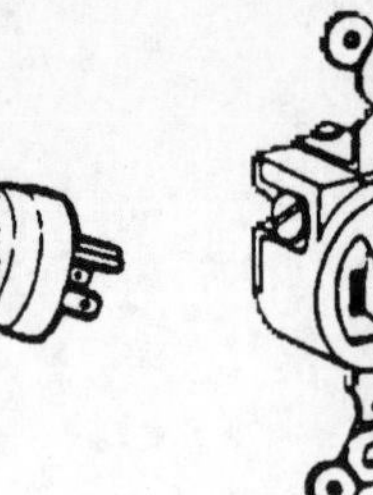

Figure 9.6 120-V three-prong plug and receptacles. (Courtesy of Sears)

Electrical Surge Protection

Secondary surge arresters, like the Surgebreaker shown in Figure 9.8, and also the exterior location shown in Figure 9.9, prevent large electrical surges up to 20,000 V from entering the branch circuit wiring. The "clamping voltage"—the amount of the surge allowed through to the branch circuits—is in the range of 500 to 1350 V depending on the amperage of the surge current, the length of the wire between the Surgebreaker and the neutral bar. The remaining voltage in the surge is taken to ground through the grounded neutral in the service entrance load center. Surge suppressors, or surge strips, suppress voltage surges up to 6000 V so that no more than 330 V gets through to appliances plugged into the surge strip. Electrical line surges are caused by lightning or high wind storms that result in a rush of high voltage through the electrical system.

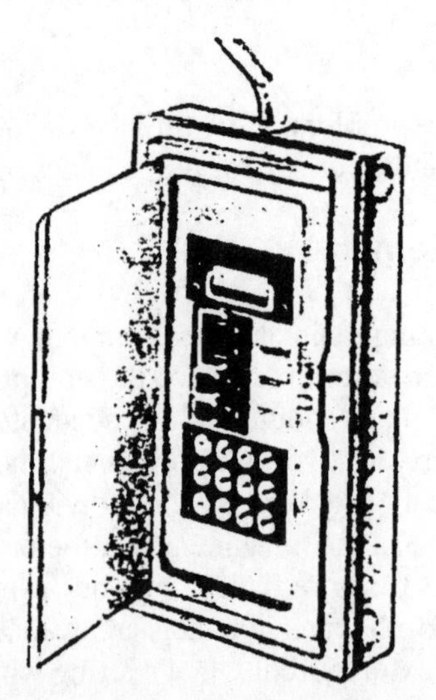

Figure 9.7 Fuse-type load center. (Courtesy of Sears)

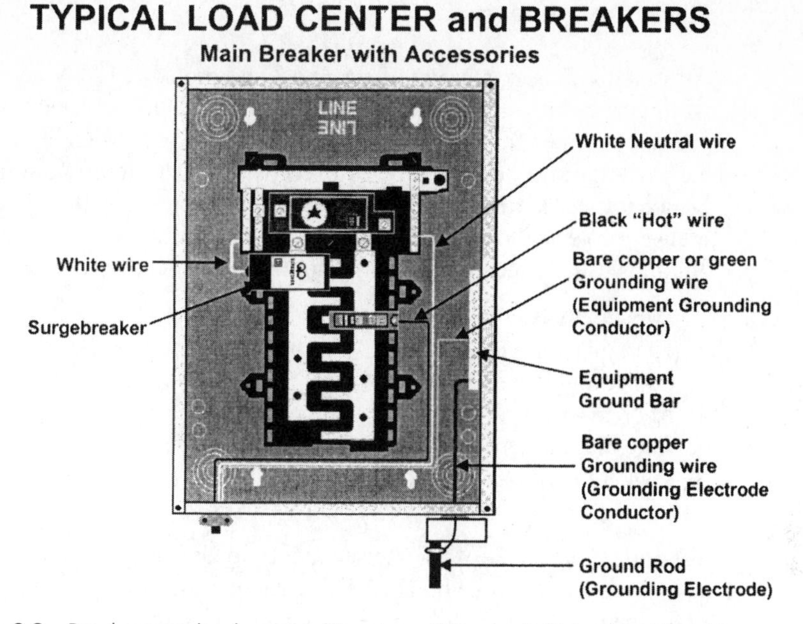

Figure 9.8 Breaker-type load center. (Courtesy of Square D, Schneider Electric)

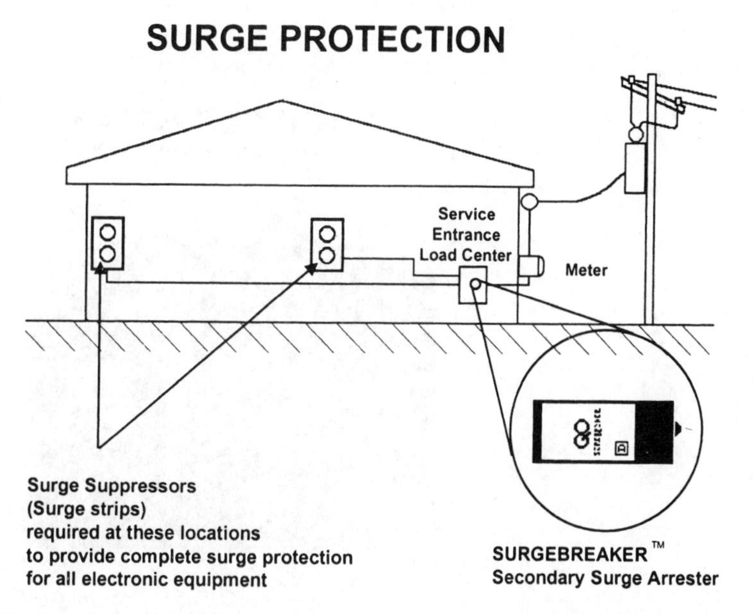

Figure 9.9 Location of secondary surge arrester. (Surgebreaker) (Courtesy of Square D, Schneider Electric)

Fused Disconnect

A fused disconnect is used to disconnect and protect a 120-V ac circuit and is located within the electrical load center. It is sometimes called a *service switch* because it permits a convenient means for disconnecting the power to the branch circuit. In Figure 9.10 overcurrent devices are required by Underwriters Laboratories (UL) and the National Electric Code® (NEC) to protect conductors and their insulations. The circuit breaker protects the electrical line from an overload or a short in the circuit. A short circuit is usually much higher in magnitude than the overload protection.

1. Tripped breaker
2. Thermal overload
3. Center toggle mechanism
4. Arc barrier
5. Wire binding screws
6. Lugs for (2) #14 to 10 Conductors

Secondary Load Center The secondary load center is connected below the service side of the main service breaker. This type has no main circuit breaker and is normally used downstream from the entrance panel. In Figure 9.11 the separate branch circuit breakers are wired in the same manner and serve the same purpose as the branch breakers in the main load center. The illustration identifies the main components necessary to install a subpanel. Wiring is in accordance with NEC requirements for secondary load centers and must also conform to local codes.

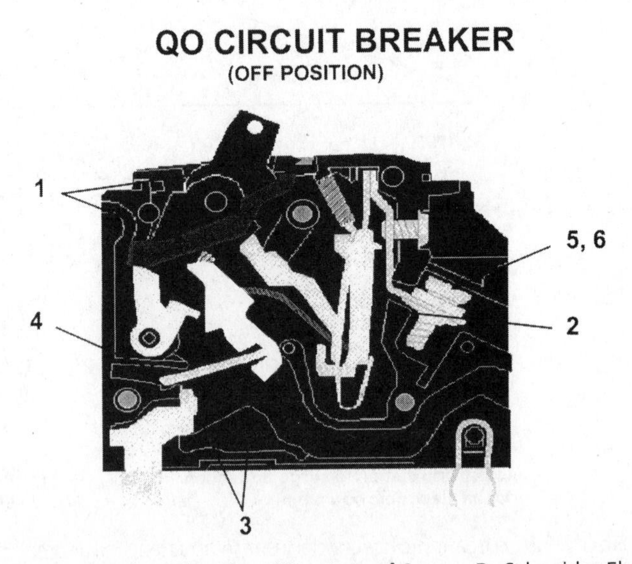

Figure 9.10 Overcurrent circuit breaker. (Courtesy of Square D, Schneider Electric)

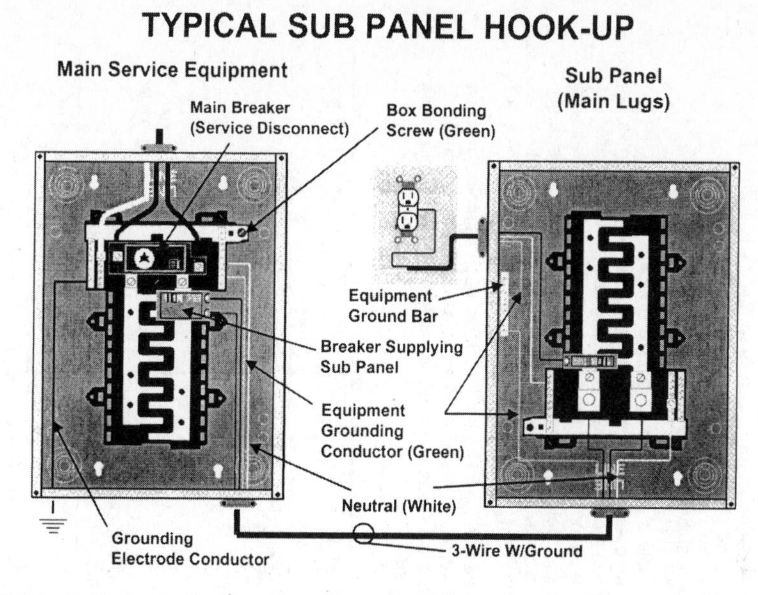

Figure 9.11 Typical subpanel hookup. (Courtesy of Square D, Schneider Electric)

Cable

Cable has a protective shield that encloses electrical wires. In most areas, all line-voltage wiring must be enclosed in approved cable with wire connections made in an approved junction box.

There are four types of cable (Figure 9.12):

1. Indoor-type plastic-sheathed
2. Dual-purpose plastic-sheathed
3. Flexible armored
4. Thin-wall and rigid conduit

The conduit shown in Figure 9.13 is a nonmetallic material, which is used when a single wire in an electrical circuit is exposed to an open area. There are two other types of conduit, metal and plastic. Each application is governed by the NEC or local area codes. Every wire that is run through conduit must be at least large enough to match the ampere rating of the circuit breaker, which directly feeds the appliance.

Adapter The adapter shown in Figure 9.14 is used to convert a two-wire circuit to a three-wire circuit. In order to protect the three-wire service, the receptacle box must be properly connected to the main load center ground bar.

Wire Sizes

The larger the wire diameter, the more current (amperes) it can carry. Sizes and types of wire are established in both the National Electrical Code and in local codes. Wire size is determined by location, loading, and use.

The American Wire Gauge (AWG) system is the standard system for measuring wire size (gauge).
A gauge number is inverse to its size, e.g., number 14 wire is smaller than number 10 wire.

Every wire must be at least large enough to match the ampere rating of the circuit breaker or fuse which directly feeds it, except for the bare ground wires in some multi-wire cables.

Each wire is marked with its size. In addition, multi-wire cables are marked with the number of wires it contains; e.g., a #14 cable with two wires is marked "14-2." Normally, a bare ground wire is also included. The marking for a #14 cable with two wires and a bare ground wire is "14-2 with ground."

Select the wire size you need from table below.

Older style cable is covered with a spiral armor of steel (referred to as Type AC armored cable). In today's homes, wire is more often in a plastic or woven jacketed non-metallic cable called Type NMC. Working with armored cable or conduit wiring requires techniques and tools which may be beyond the skills of the do-it-yourselfer, so this bulletin deals only with non-metallic cable installations.

Larger wires are required for hot locations, such as attics, or where there are more than three wires in a cable or conduit. See the National Electrical Code and local codes for details and for other ampere ratings.

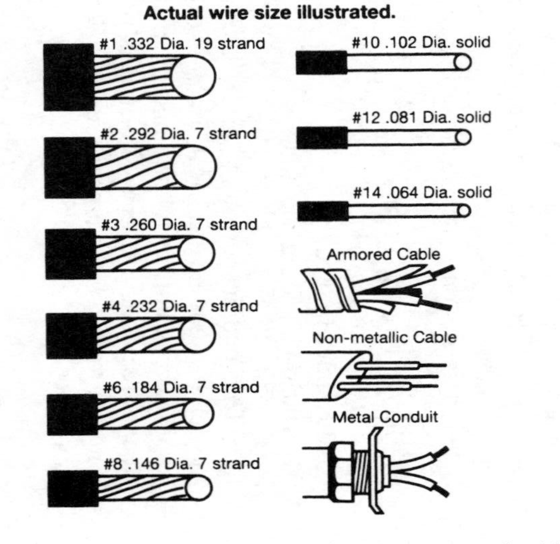

Circuit Breaker or Fuse Ampere Rating	Minimum Copper Wire Size	Minimum Aluminum Wire Size	Circuit Breaker or Fuse Ampere Rating	Minimum Copper Wire Size	Minimum Aluminum Wire Size
FEEDERS AND BRANCH CIRCUITS					
15	14	—	60	4	3
20	12	—	70	4	2
25	10	—	100	1	0
30	10	8	125	1	00
40	8	6	150	0	000
50	6	4	200	000	250 KCMIL
THREE WIRE SERVICE CABLES OR OTHER WIRES WHICH CARRY ALL THE CURRENT FOR THE BUILDING					
100	4	2	150	1	000
125	2	0	200	00	0000

NOTE: Conductors carry a specific temerature rating based on the type of insulation employed on the conductor. Common insulation types can be found in Table 310-13 of the National Electrical Code and corresponding ampacities can be found in Table 310-16.

When a conductor is chosen to carry a specific load, the user/installer or designer must know termination ratings for the equipment involved in the circuit.

Figure 9.12 Types of cable. (Courtesy of Square D, Schneider Electric)

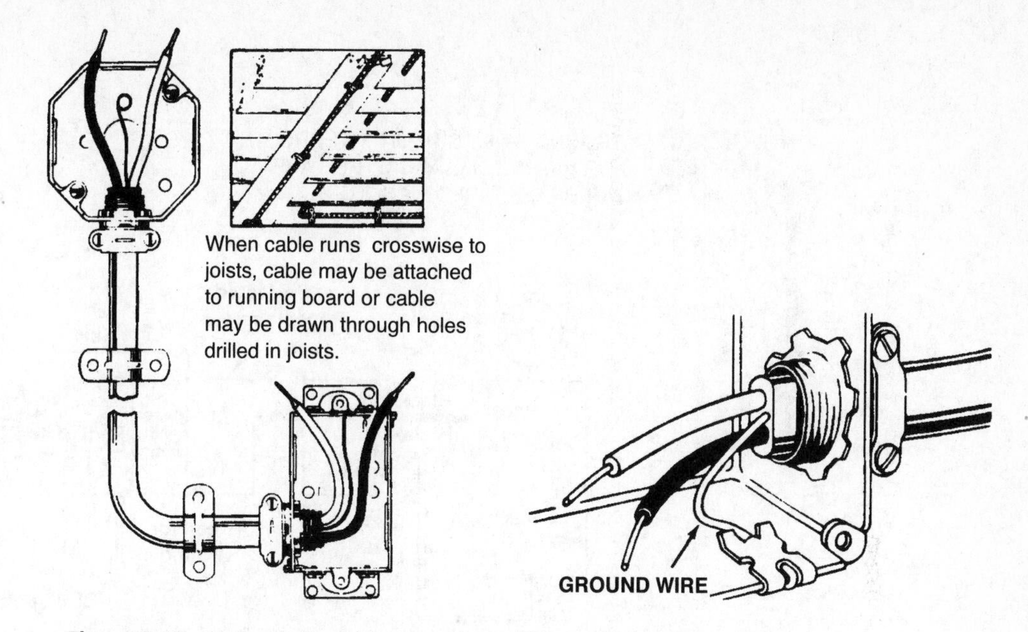

Figure 9.13 Nonmetallic cable connectors. (Courtesy of Sears)

Ground-fault Interrupters

Fuses and circuit breakers protect circuits and wire against overloads and short circuits but not against current leakage. Small amounts of leakage can occur without blowing a fuse or tripping a breaker. Under certain conditions the leakage can be hazardous.

A relatively new product adapted for residential use is called the *ground-fault circuit interrupter* (GFCI) (see Figures 9.15 and 9.16). The GFCI is designed to detect a small current leakage and to interrupt the power supply quickly enough to prevent a serious problem. After the problem is corrected, the GFCI can be reset, and power at that point will be restored.

The NEC recommends the use of GFCIs on all outdoor circuits. Local codes are especially strong on their use in conjunction with swimming pools that have any electrical connections.

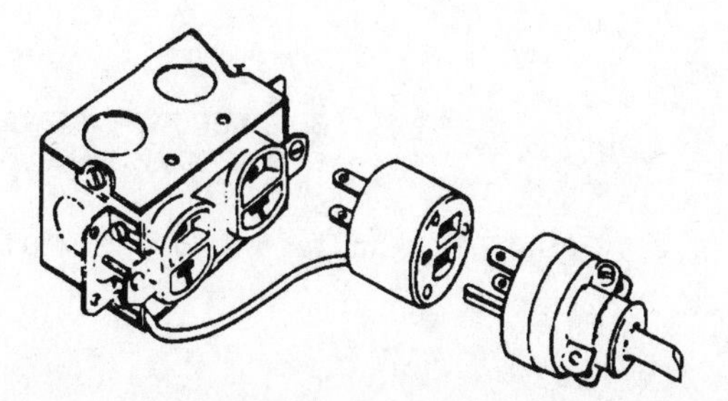

Figure 9.14 Adapter for conversion of two-wire to three-wire system. (Courtesy of Sears)

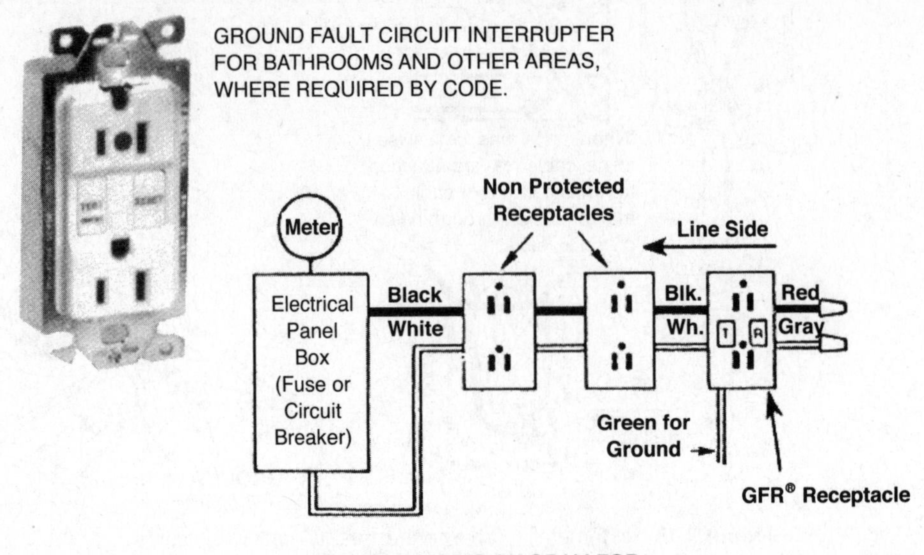

Figure 9.15 Indoor ground-fault circuit interrupters. (Courtesy of Bryant Electric)

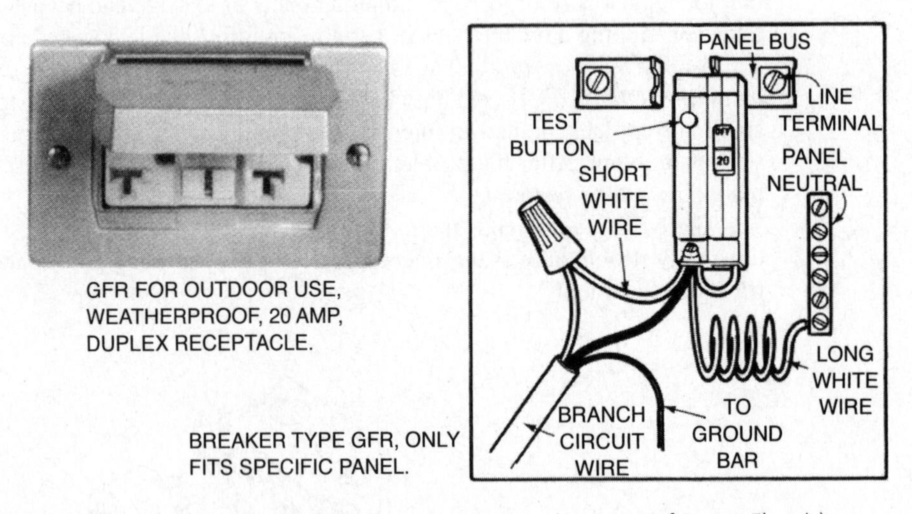

Figure 9.16 Outdoor ground-fault circuit interrupters. (Courtesy of Bryant Electric)

External Wiring of a Gas Furnace and an Air-Conditioning System

Three power supplies are of concern when a furnace and air-conditioning system are installed in a home, namely, 120 V for the furnace circuit, 24 V for the control circuit, and 220 V for the outdoor air-conditioning condensing unit (Figure 9.17).

120-V Circuit

The line voltage (120 V) is brought to the furnace to operate the blower and the primary side of the transformer. A hot leg is fed from a 20-A circuit breaker accompanied by a neutral leg and a ground. At the furnace, the hot leg must be controlled by a disconnect switch. (Because many high-efficiency furnaces use condensate pumps to remove the water created in the combustion process, a switch with an electrical outlet may be used to supply voltage to the pump.)

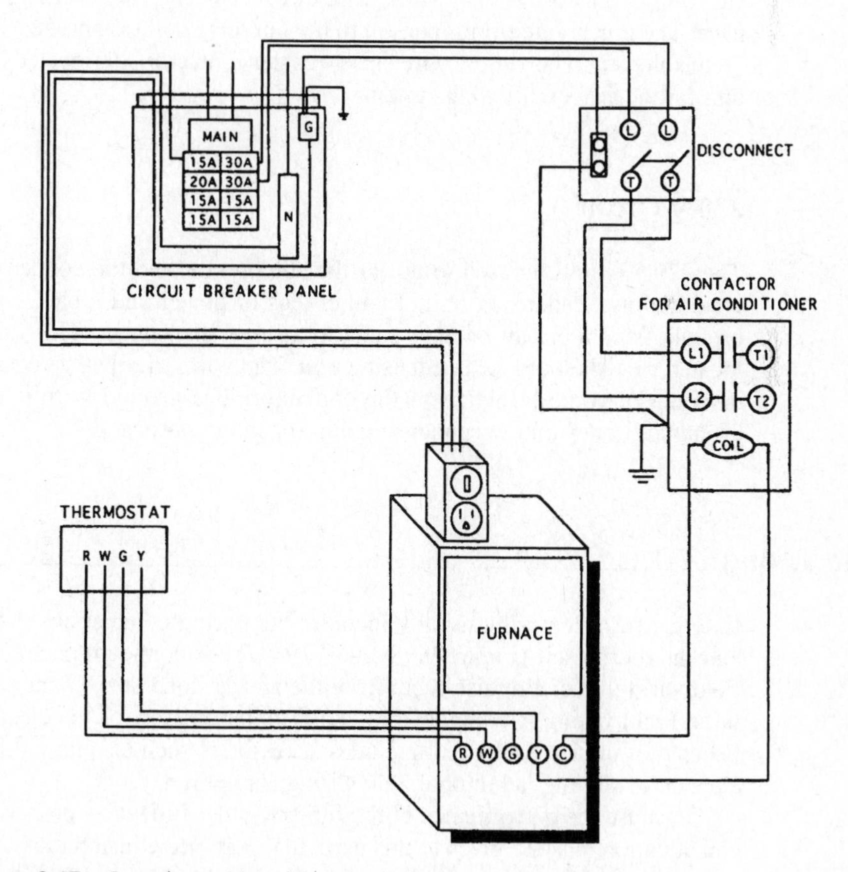

Figure 9.17 Complete wiring diagram of a furnace and air conditioner.

For proper polarity, it is necessary to have the black wire, the white wire, and the bare wire connected to the appropriate connections at the breaker panel. The black wire (hot leg) should be connected to the circuit breaker, the white wire (common or neutral leg) should be connected to the neutral bus at the breaker panel, and a bare or green wire (ground) should be connected to the ground bus.

24-V Circuit

When the primary side of the transformer is energized with 120 V, this induces 24 V on the secondary side of the transformer.

The 24-V circuit is used to control the heating and cooling circuits by energizing a fan relay coil, a contactor coil, and a gas valve. When installing a furnace and air-conditioning system, it is necessary to connect four-conductor thermostat wire from the thermostat to the terminal block on the furnace. The four wires are color coded for proper installation. The red wire supplies 24 V to the thermostat from the secondary side of the transformer. The white wire directs current to the heating circuit at the furnace. The green wire brings current to the fan relay coil (to energize the cooling speed on the blower). The yellow wire energizes the contactor coil located in the condensing unit for the air-conditioning system.

220-V Circuit

The 220-V circuit is used to power the compressor and the condenser fan at the air-conditioning condensing unit. Two hot legs of power and a bare grounding wire are brought from a circuit breaker panel to a fused disconnect. This fused disconnect is located near the outdoor condensing unit. The wires are then directed from the fused disconnect to the line side of the contactor. The ground wire is connected to the grounding screw on the condensing unit for safety purposes.

Field Wiring

External furnace wiring usually includes the wiring connections to the power supply and the thermostat (Figure 9.18 and 9.19). These connections are also described as field wiring, to distinguish them from the wiring done at the factory by the manufacturer. Field wiring is completed by the installation crew or the electrician on the job. When the furnace equipment includes accessories such as a humidifier, electronic air cleaner, or cooling, additional field wiring is required.

Because the performance of the furnace is dependent on proper field wiring, special attention must be given to this particular part of the installation. Most manufacturers supply detailed instructions for connecting the thermostat and accessories to the

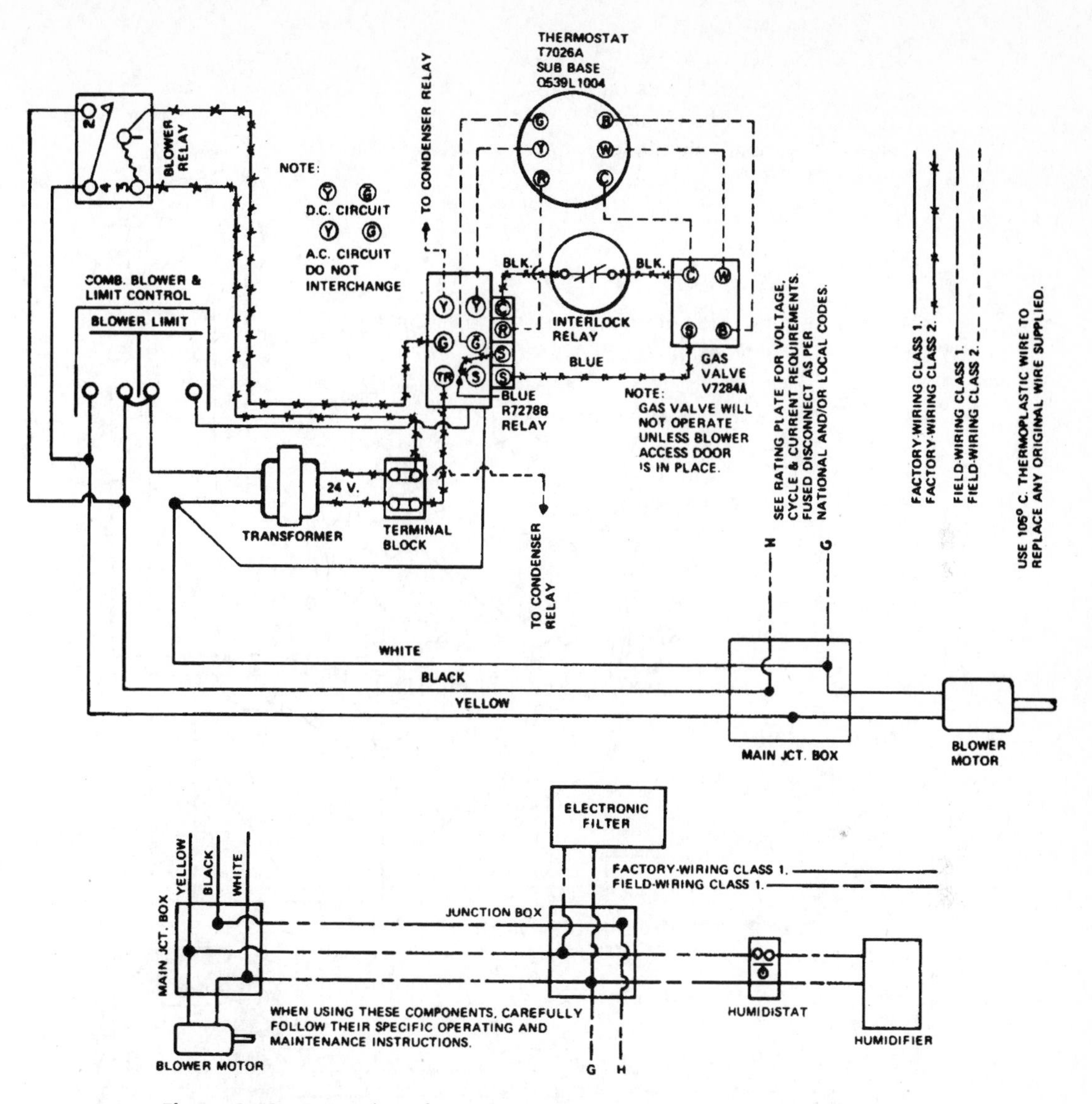

Figure 9.18 External gas furnace wiring. (Courtesy of Luxaire, Inc.)

furnace. Most field-wiring problems occur because of improper or inadequate connections to the building's power supply.

Most gas and oil furnaces operate on a 20-A, 120-V single-phase ac branch circuit. When electric heating or cooling equipment is installed, a 240-V single-phase ac circuit is required. The amperes of service needed depend on the size of the electric furnace or cooling unit.

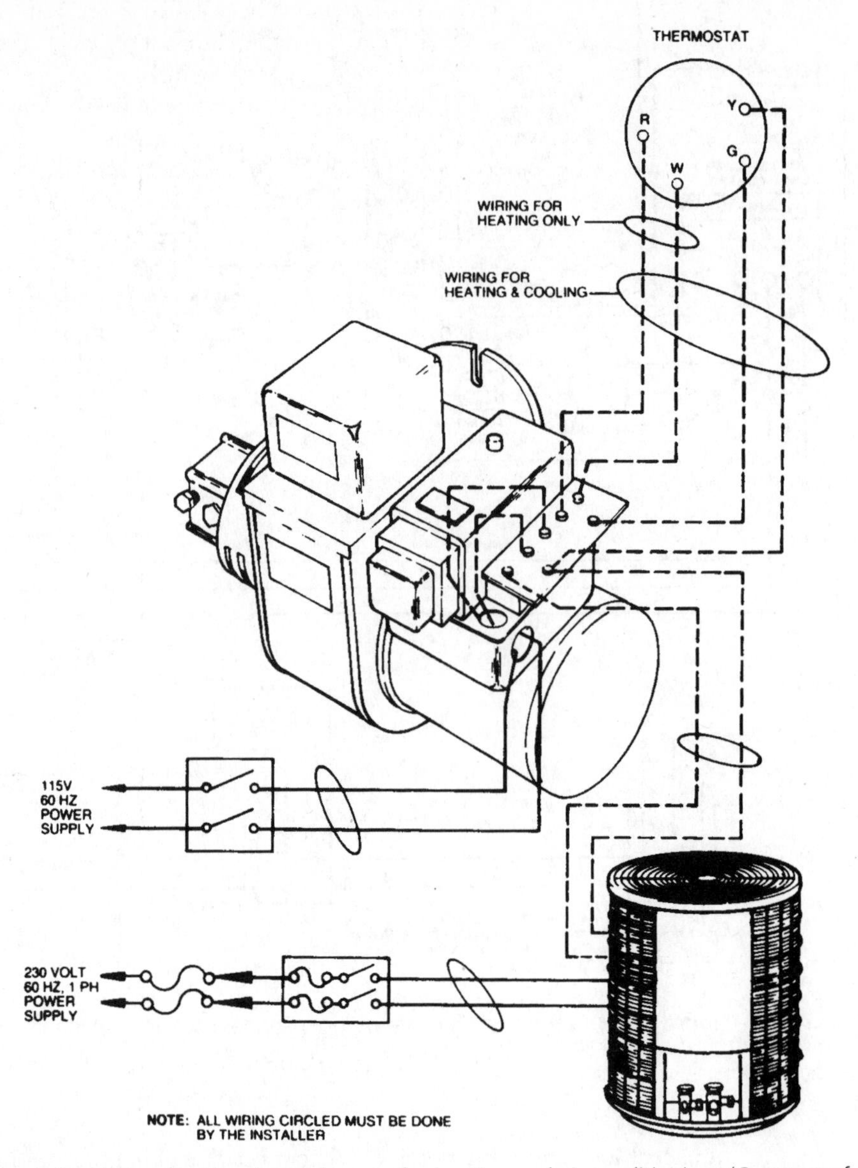

Figure 9.19 External wiring for oil furnace for heating and air conditioning. (Courtesy of ARCOAIRE Air Conditioning & Heating)

Analyzing the Power Supply

A building's electrical system must be analyzed to determine whether the existing power supply is adequate. All wiring must be installed in accordance with the National Electric Code® and any local codes or regulations that apply. The copper electrical wires that are field installed will conform with the temperature limitation for 95°F/35°C rise wire when installed in accordance with instructions. Specific electrical data is given on the furnace rating plate. A power supply separate from all other circuits must be provided, and overcurrent protection and disconnect switch must be installed per local/national electrical codes. The switch should be reasonably close to the unit for convenience in servicing.

NOTE: The Furnace's Control System Depends on Correct Polarity of the Power Supply and a Proper Ground Connection. The power supply must be connected as shown on the unit wiring label on the inside of the blower compartment door and Figures 9.20 and 9.21. The black furnace lead must be connected to the *L*1 (hot) wire from the power supply. The white furnace lead must be connected to neutral. Also, the green equipment ground wire must be connected to the power supply ground.

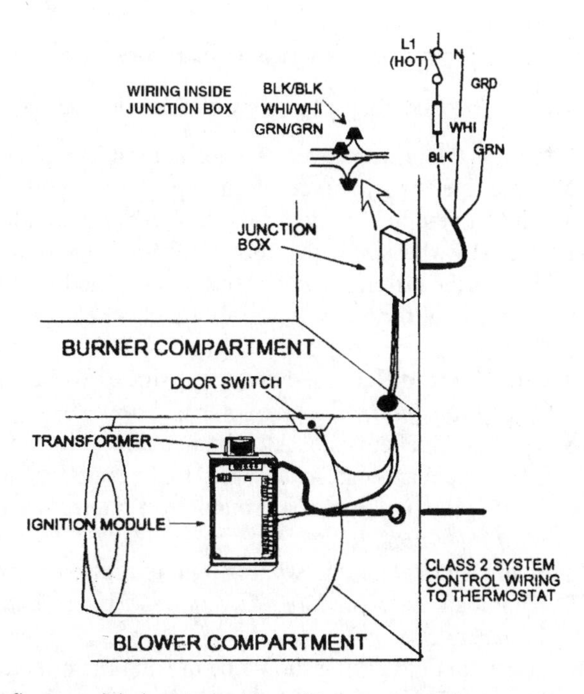

Figure 9.20 Upflow model electrical wiring. (Courtesy of Coleman-Unitary Products Group, York International)

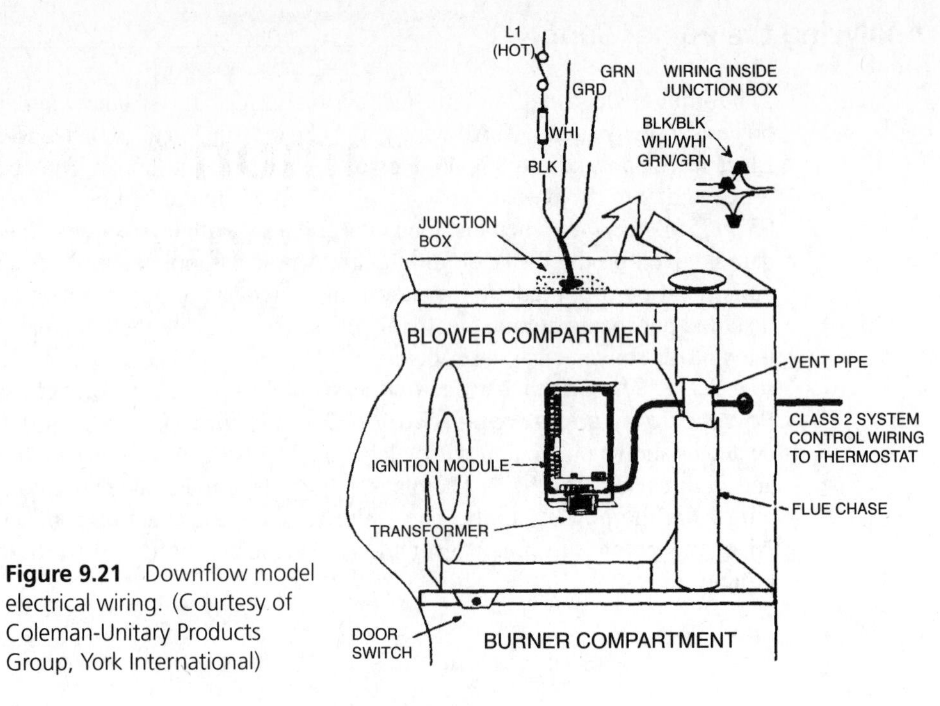

Figure 9.21 Downflow model electrical wiring. (Courtesy of Coleman-Unitary Products Group, York International)

STUDY QUESTIONS

Answers to the study questions may be found in the sections noted in brackets.

9-1 What is the prime mover of the generator? *[Electrical Transmission]*

9-2 Name five energy sources used to produce steam. *[Electrical Transmission]*

9-3 Where in the load center can 220 V, 208/240 be identified? *[Branch Circuits]*

9-4 Generally what size wire is used in a 20-A branch circuit? *[Types of Branch Circuits]*

9-5 What is the difference between *grounded* and *grounding? [Grounding]*

9-6 When should the power supply be checked for low-voltage conditions? *[Low Voltage]*

9-7 Explain the difference between a surge strip and a surgebreaker. *[Electrical Surge Protection]*

9-8 A circuit breaker protects the branch circuit from what? *[Fused Disconnect]*

9-9 Is the secondary load center wired to the main feed line? *[Secondary Load Center]*

9-10 What is the function of a ground-fault interrupter in a branch circuit? *[Ground-Fault Interrupter]*

9-11 In an electrical circuit, what is used to reduce the voltage? *[24-V Circuit]*

9-12 What load center amperage is required for residential electric heating? *[Building Amperage Service]*

9-13 Electrical field wiring should be of what material to conform to temperature limitations of 95°F/35°C? *[Analyzing the Power Supply]*

9-14 Explain the method used to determine the correct polarity of an electrical circuit. *[Analyzing the Power Supply]*

CHAPTER 10

Controls Common to All Forced-Air Furnaces

OBJECTIVES

After studying this chapter, the student will be able to:

- Identify the common types of electrical control devices used on forced warm air furnaces
- Evaluate the performance of common types of controls in the system

Functions of Controls

The function of automatic controls is to operate a system or unit in response to some variable condition. Automatic controls are used to turn the various electrical components (load devices) on and off. These devices operate in response to a controller that senses the room temperature or humidity, thereby activating the equipment to maintain the desired conditions. Load devices that make up a heating unit are:

- Fuel-burning device or heater
- Fan
- Humidifier
- Electrostatic air cleaner
- Cooling equipment

Each of these load devices has some type of switch or switches that automatically or manually cause the device to operate. The automatic switches respond to variable conditions from a sensor. A *sensor* is a device that reacts to a change in conditions and is then capable of transmitting a response to one or more switching devices. For example, a thermostat senses the need for heat and switches on the heating unit. When the thermostat senses that enough heat has been supplied, it switches off the heating unit.

Components of Control Systems

The control system for a forced warm air heating unit comprises five elements:

1. Power supply
2. Controllers
3. Limit controls
4. Primary controls
5. Accessory controls

Power Supply The power supply furnishes the necessary current at the proper voltage to operate the various control devices. The fused disconnect and the transformer are parts of the power supply.

Controllers Controllers sense the condition being regulated and perform the necessary switching action on the proper load device. The controller group includes such devices as the thermostat, the humidistat, and the fan control.

Limit Controls Limit controls shut off the firing device or heater when the maximum safe operating temperature is exceeded.

Primary Controls The gas valve, the flame detector relay for an oil burner, and the sequencer for electric furnaces are types of primary controls. A primary control usually includes some type of safety device. For example, on an oil furnace primary control, the burner will be shut off if a flame is not produced or goes out.

Accessory Controls Used to add special features to the control system. One of these is the fan relay, which permits the fan switch on a 24-V thermostat to operate a 120-V fan.

Thermostats

Thermostats control the source of heat to closely maintain the selected temperature in the space being conditioned. A heating system should provide even (consistent) temperatures for the comfort of a building's occupants. It is said that people can sense a change of approximately 1½°F in temperature. A well-regulated system will provide less than 1°F variation.

A sensing element is usually bimetallic, usually copper and invar bonded together. When heated, copper has a more rapid expansion rate than invar, so when the bimetal is heated, it changes shape. In a heating thermostat, this movement is mechanically connected to a switch that closes on a drop in temperature and opens on a rise in temperature (Figure 10.1). Bimetallic elements are constructed in various shapes (Figure 10.2). A spiral-wound bimetallic element (Figure 10.3) is compact in construction. For this reason it is the element used in many thermostats.

Switching action should take place rapidly to prevent arcing, which damages the switch contacts. A magnet is used to provide rapid action. The most common type of switching action in use is the mercury tube arrangement (Figure 10.4). The electrical

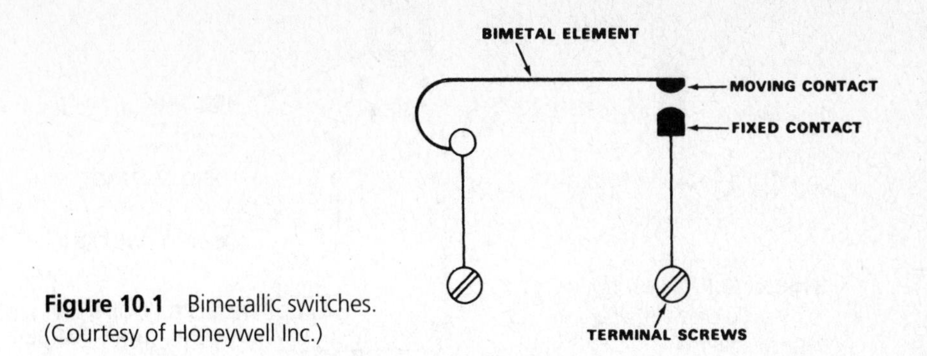

Figure 10.1 Bimetallic switches. (Courtesy of Honeywell Inc.)

contacts of the switch are inside the tube together with a globule of mercury. The electrical contacts are located at one end of the tube. When the tube is tipped in one direction, the mercury makes an electrical connection between the contacts and the switch is closed. When the tube is tipped in the opposite direction, the mercury flows to the other end of the tube and the switch is opened.

Electronic Temperature Sensing

Thermistors use various materials whose resistance decreases when their temperature rises, and increases when their temperature falls. Electronic sensing devices have no moving parts. As the temperature changes, their resistance changes. An electronic device (thermostat) converts these changes in resistance into a control signal that turns burners or compressors on and off. Different styles of thermistors and their symbol are shown in Figure 10.5.

Figure 10.6 shows another type of electronic temperature-sensing device, called a *resistance bulb,* which is a coil of fine wire wound around a bobbin.* The resistance of

*A bobbin is a small round device or cylinder.

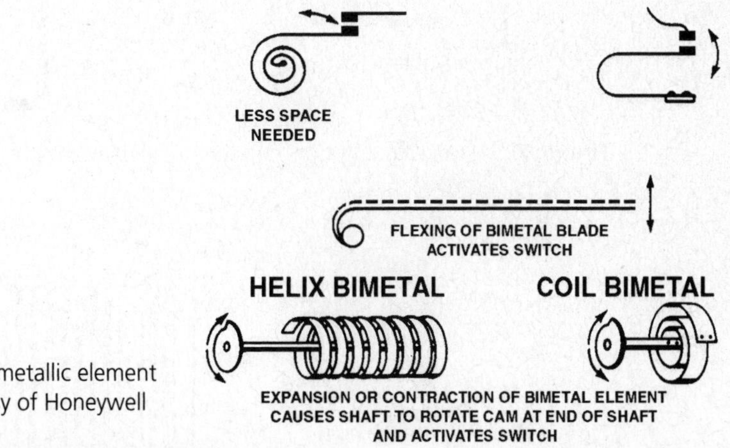

Figure 10.2 Bimetallic element shapes. (Courtesy of Honeywell Inc.)

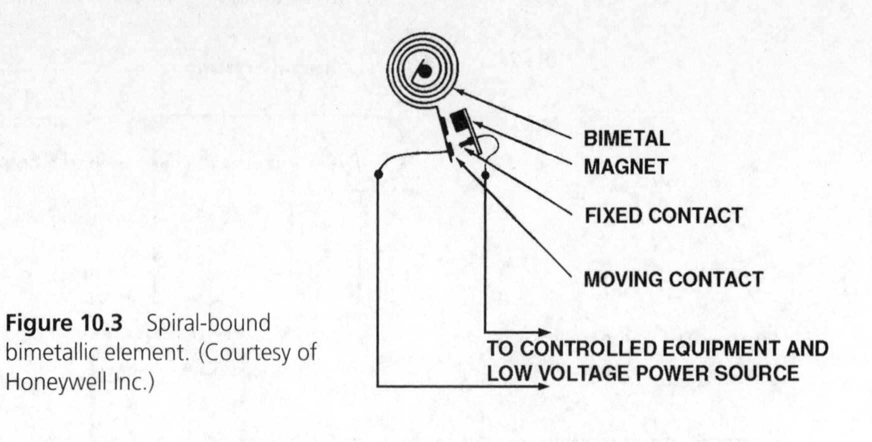

Figure 10.3 Spiral-bound bimetallic element. (Courtesy of Honeywell Inc.)

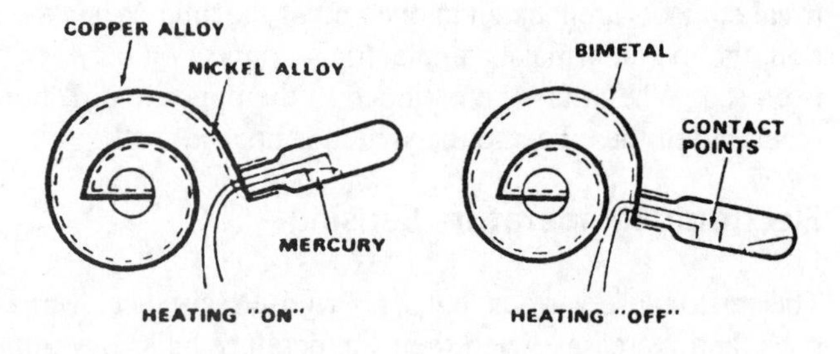

Figure 10.4 Mercury tube switching action. (Courtesy of Honeywell Inc.)

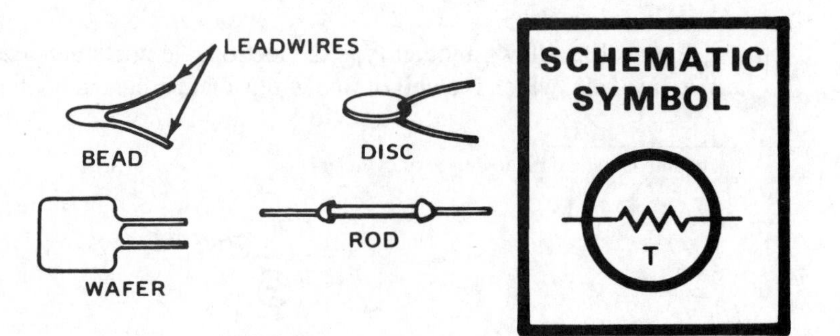

Figure 10.5 Typical thermistors and their schematic symbol. (Courtesy of Honeywell Inc.)

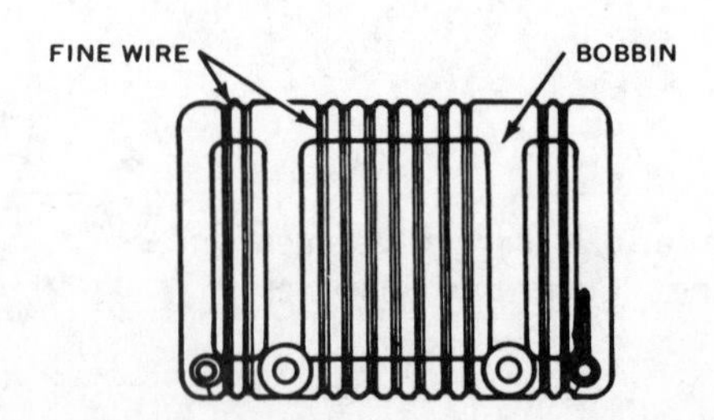

Figure 10.6 Resistance bulb. (Courtesy of Honeywell Inc.)

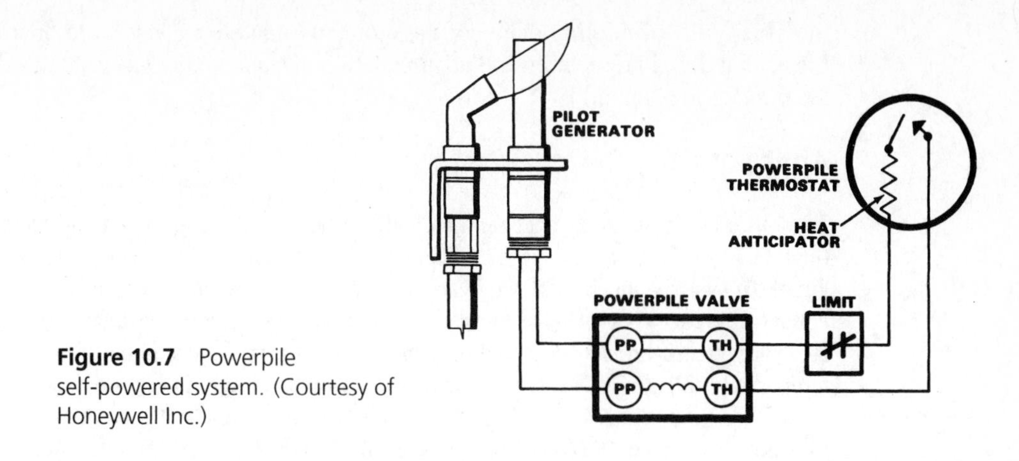

Figure 10.7 Powerpile self-powered system. (Courtesy of Honeywell Inc.)

the wire increases as the temperature rises and decreases as the temperature falls, which is the opposite of a thermistor.

Types of Current

Most residential thermostats are of the low-voltage type (24 V). Low-voltage wires are easier to install between the thermostat and the heating unit. The construction of a low-voltage thermostat provides greater sensitivity to changing temperature conditions than does a line-voltage (120-V) thermostat. The materials can be of lighter construction and easier to move, with less chance of arcing. Some self-generating systems (Figure 10.7) use a millivolt (usually 750 mV) power supply to operate the thermostat circuit. These thermostats are very similar in construction to low-voltage thermostats.

Purpose

Some thermostats are designed for heating only or for cooling only. Other thermostats are designed for a combination of both heating and cooling (Figure 10.8). With the mercury tube design, a set of contacts is located at one end of the tube for heating and

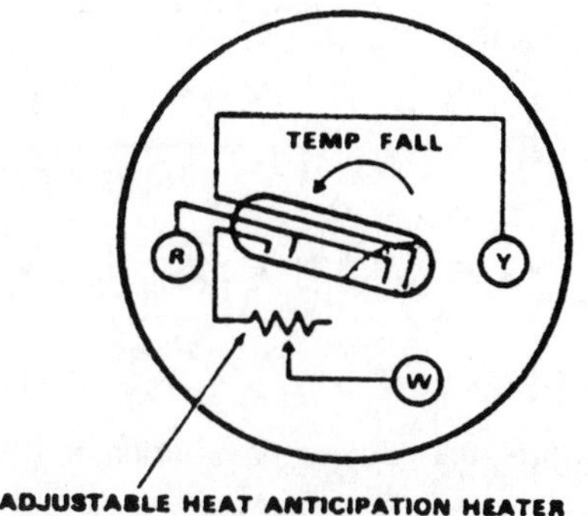

Figure 10.8 Contacts in the mercury bulb. (Courtesy of Honeywell Inc.)

the other end for cooling. The cooling contacts make on a rise in temperature and break on a drop in temperature. The tube with both sets of contacts is attached to a single bimetallic element.

Anticipators

There is a lag (time delay) between the call for heat by the thermostat and the amount of time it takes for heat to reach a specific area. The differential of the thermostat plus the heat lag of the system can cause a wide variation in room temperature. There is always a differential (difference) between the temperature at which the thermostat makes and the temperature at which it breaks (opens). For example, the thermostat may call for heat when the temperature falls to 70°F and shut the furnace off when the temperature rises to 72°F.

To provide closer control of room temperature, a heat anticipator is built into the thermostat (Figure 10.9). A heat anticipator consists of a resistance heater placed in series with the thermostat contacts. On a call for heat, current is supplied to the heater. This action heats the bimetallic element, causing it to respond somewhat ahead of the actual rise in room temperature. By supplying part of the heat with the anticipator, less room heat is required to meet the room thermostat setting. Thus, the furnace shuts off before the actual room temperature reaches the cutout point on the thermostat. Although the furnace continues to supply heat for a short period after the unit shuts off, the actual room temperature does not exceed the setting of the thermostat because the anticipator assists in producing even heating and prevents wide temperature variations.

Anticipators are also supplied for thermostats that control cooling. A cooling anticipator (Figure 10.10) is placed in parallel with the thermostat contacts. Thus heat is supplied to the bimetal on the off cycle (when cooling is off) and serves to decrease the length of the shutdown period.

Thermostat Anticipator Setting Proper control of the indoor air temperature can be achieved only if the thermostat is calibrated to the heating and/or cooling cycle. A vital part of this calibration is related to the thermostat heat anticipator. Anticipators for the cooling operation are generally preset by the thermostat manufacturer and require no adjustment. Anticipators for the heating operation are of two types, preset and adjustable. Those that are preset will not have an adjustment scale and are generally marked accordingly.

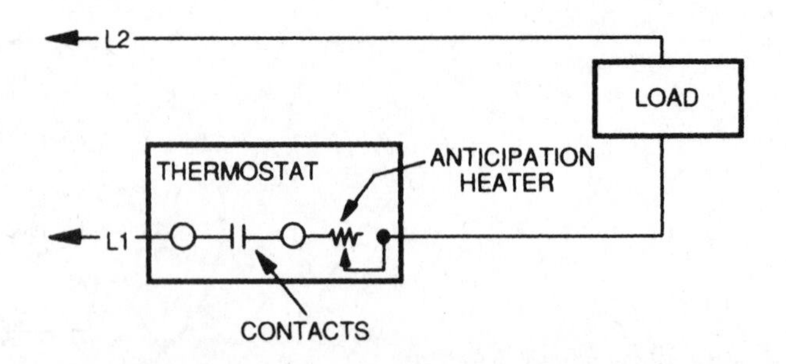

Figure 10.9 Series heat anticipator. (Courtesy of Honeywell Inc.)

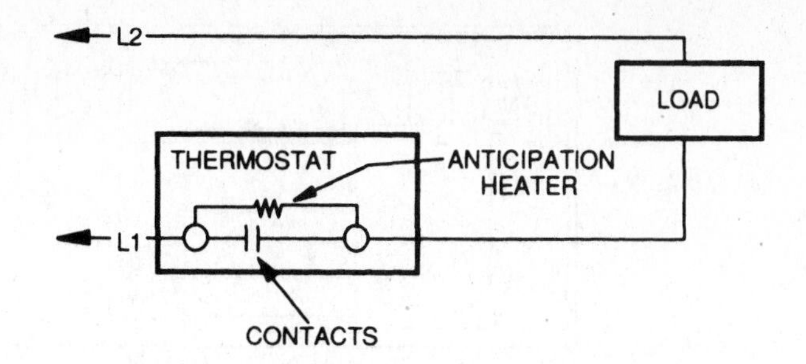

Figure 10.10 Parallel cooling anticipator. (Courtesy of Honeywell Inc.)

Thermostat models having a scale, as shown in Figure 10.11, must be adjusted to each application. The proper thermostat heat anticipator setting is a minimum of 0.8 A for furnace operation only. A lower setting will result in short cycling of the furnace and in extreme cases will cause a complaint of "no heat." To increase the length of the cycle, increase the setting of the heat scale. To decrease the length of the cycle, decrease the setting of the heat scale.

A third type uses a cycle-rate adjustment (see Figure 10.12). Generally, the rate is factory set for forced-air furnaces. Consult the thermostat instruction sheet for details. In many cases this adjustment setting can be found in the thermostat instruction sheet. If this information is not available, or if the correct setting is questioned, the following procedures should be followed:

Preferred Method. Use a low-scale ammeter such as an ampcheck or milliammeter. Connect the meter across terminals R and W on the subbase (RH and W1 on the multistage thermostat subbase).

Step 1. Wrap 10 loops of single-strand insulated thermostat wire around the prongs of an ammeter. Set the scale to the lowest ampere scale.

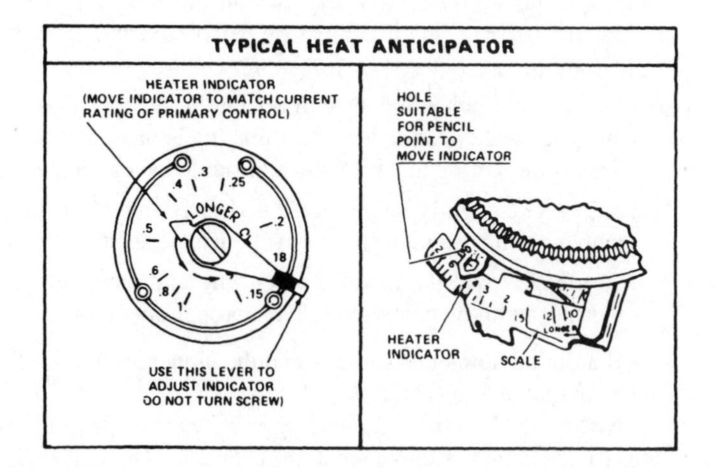

Figure 10.11 Typical heat anticipator. (Courtesy of ARCOAIRE Air Conditioning & Heating)

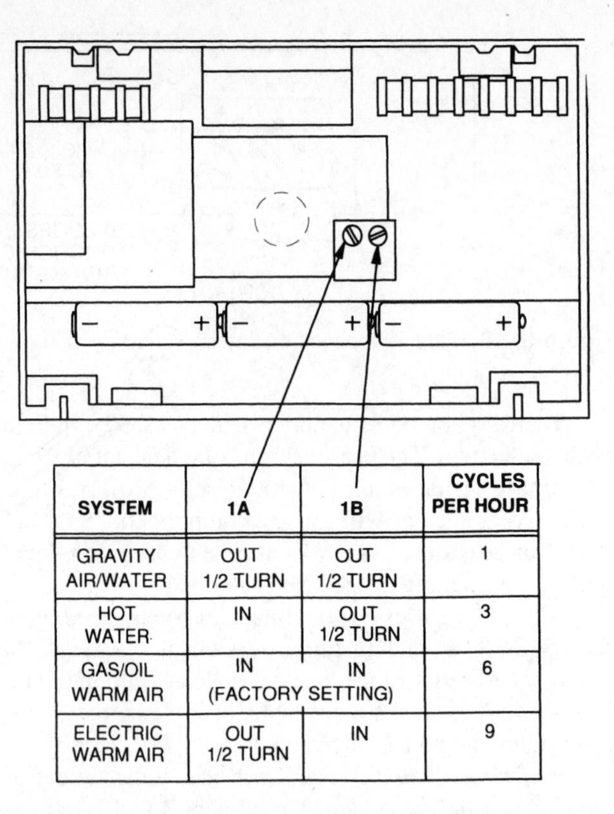

SYSTEM	1A	1B	CYCLES PER HOUR
GRAVITY AIR/WATER	OUT 1/2 TURN	OUT 1/2 TURN	1
HOT WATER	IN	OUT 1/2 TURN	3
GAS/OIL WARM AIR	IN (FACTORY SETTING)	IN	6
ELECTRIC WARM AIR	OUT 1/2 TURN	IN	9

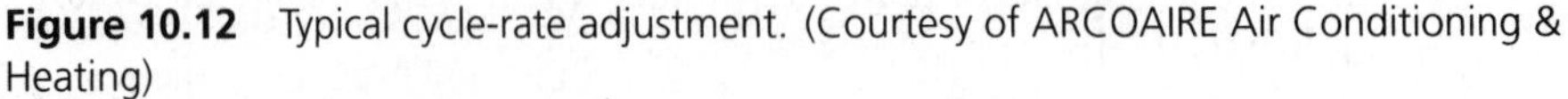

Figure 10.12 Typical cycle-rate adjustment. (Courtesy of ARCOAIRE Air Conditioning & Heating)

Step 2. Connect the uninsulated ends of this wire jumper across terminals R and W on the subbase (RH and W1 on the multistage thermostat subbase) (see Figure 10.13). This test must be performed with the thermostat removed from the subbase.

Step 3. Let the heating system operate in this position for about 1 min. Read the ammeter scale. Divide whatever reading is indicated by 10 (for 10 loops of wire). This is the setting at which the adjustable heat anticipator should be set.

Step 4. If a slightly longer cycle is desired, move the heat anticipator pointer to a higher setting. Slightly shorter cycles can be achieved by moving to a lower setting.

Step 5. Remove the meter jumper wire and reconnect the thermostat. Check the thermostat in the heating mode for proper operation.

If a digital ammeter is used, read the ampere draw direct from the meter. Steps 1 and 3 are then not required.

Note: The length of the heating cycle can also be affected by the fan control settings (of the combination fan and limit control). The fan "on" and "off" settings should be checked at this point.

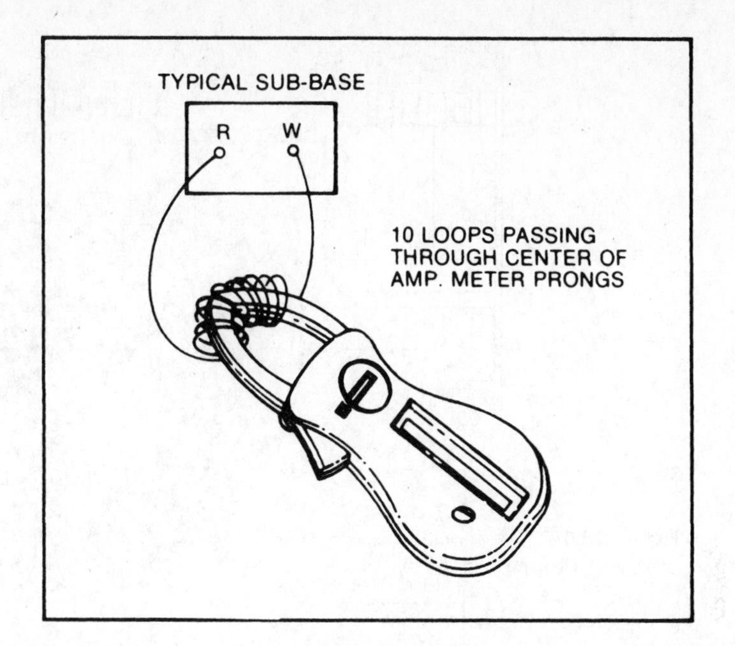

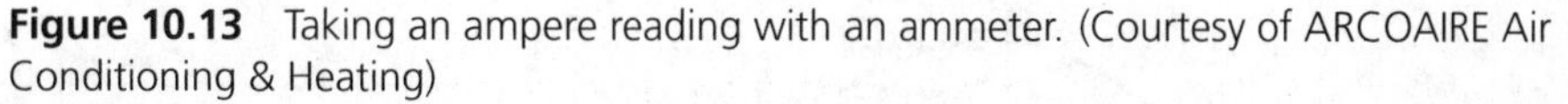

Figure 10.13 Taking an ampere reading with an ammeter. (Courtesy of ARCOAIRE Air Conditioning & Heating)

Subbases

Most thermostats have some type of subbase that serves as a mounting plate (Figure 10.14). A subbase provides a means for leveling and fastening a thermostat to the wall and contains electrical connections and manual switches. Manual switches offer a homeowner the choice of HEAT-OFF-COOL and FAN AUTO-ON. Switches on a subbase increase the number of functions that can be performed by a thermostat (Figure 10.15). Although a thermostat has only a sensing element and one or two automatic switches, the addition of the subbase manual switches provides a choice of many switching actions.

Application and Installation

The following are recommended procedures in the application and installation of thermostats.

- Select the thermostat that best meets customers' needs.
- Select the thermostat designed for the specific type of application: heat only, heat cool, heat pump, and so on.
- Select the location.
- Mount the thermostat approximately 5 ft above the floor.
- Select the anticipator setting that matches the equipment.
- Select the proper location for the outdoor sensor, if used.
- Seal any openings in the wall under the mounting plate to eliminate infiltration.

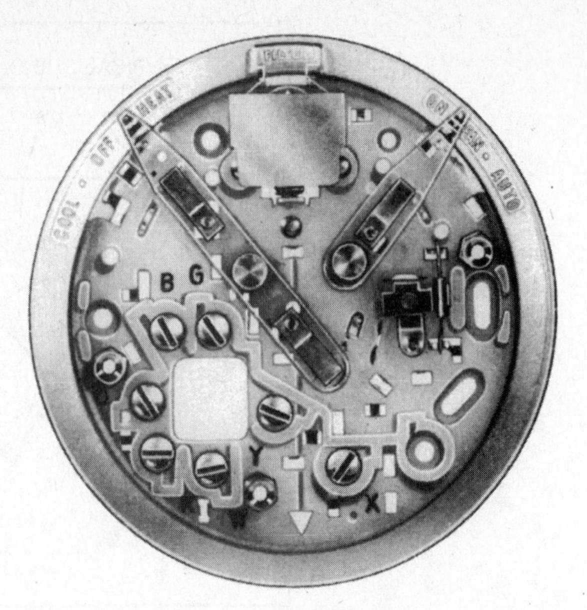

Figure 10.14 Thermostat subbase. (Courtesy of Honeywell Inc.)

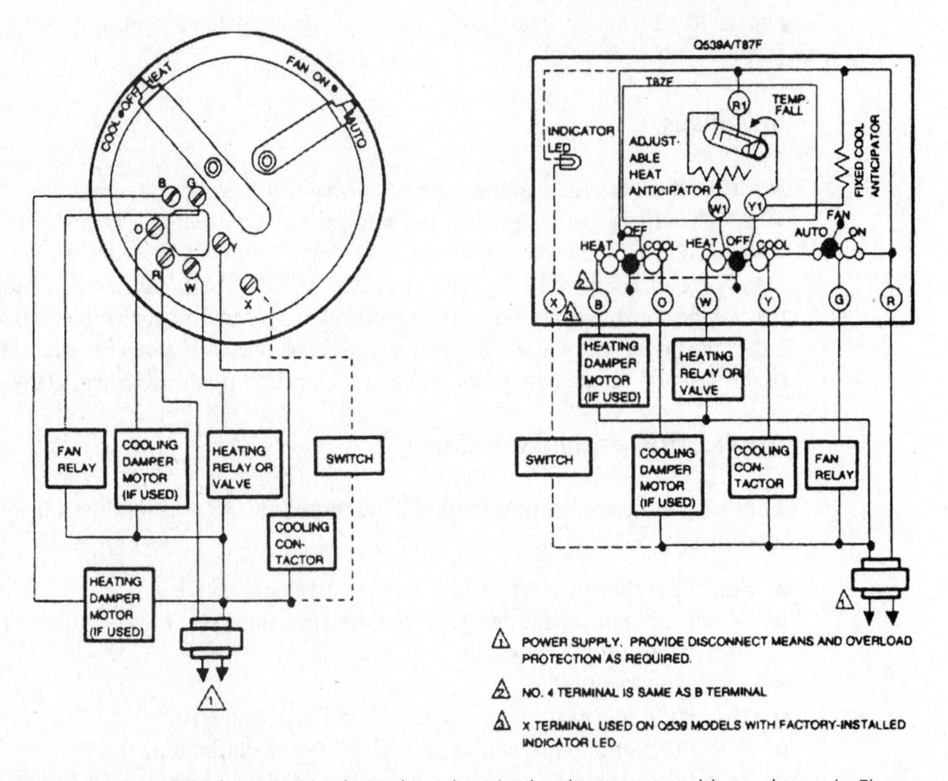

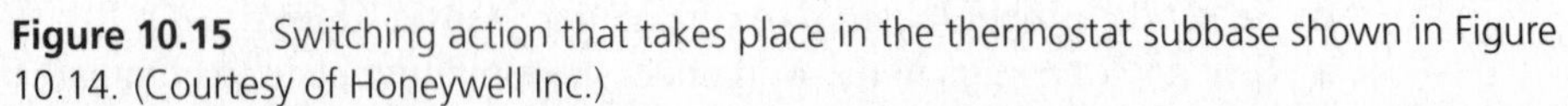

Figure 10.15 Switching action that takes place in the thermostat subbase shown in Figure 10.14. (Courtesy of Honeywell Inc.)

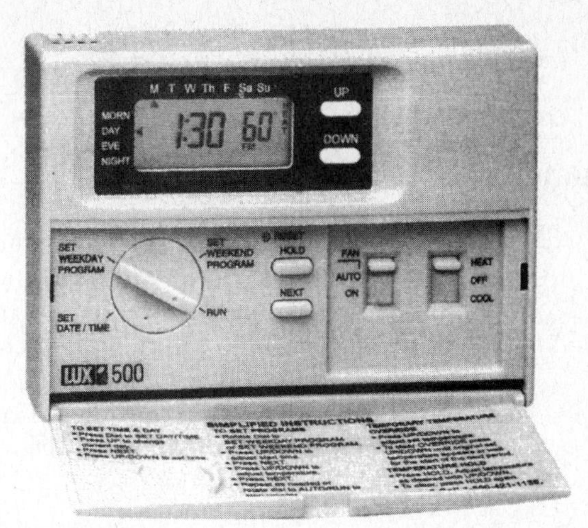

Figure 10.16 Seven-day programmable thermostat with digital display of temperature and time settings. (Courtesy of LUX Products Corporation)

Programmable Thermostats

The programmable thermostat is a solid-state microcomputer control with 7-day programming. This thermostat is designed with a full range of temperature setback–setup–offset capabilities. A battery is used to maintain the program and time during periods of power interruption. In the thermostat shown in Figure 10.16 there is a built-in cooling time delay to protect the compressor from damage due to short cycling.

Temperature setback during heating and setup during cooling maximizes energy savings during unoccupied periods and during the night hours. The thermostat and control module shown in Figure 10.17 contains optional features that perform several functions:

1. *Controls startup times:* automatically varies the daily startup time depending on building load. If the outdoor air temperature is suitable, the building can be precooled.

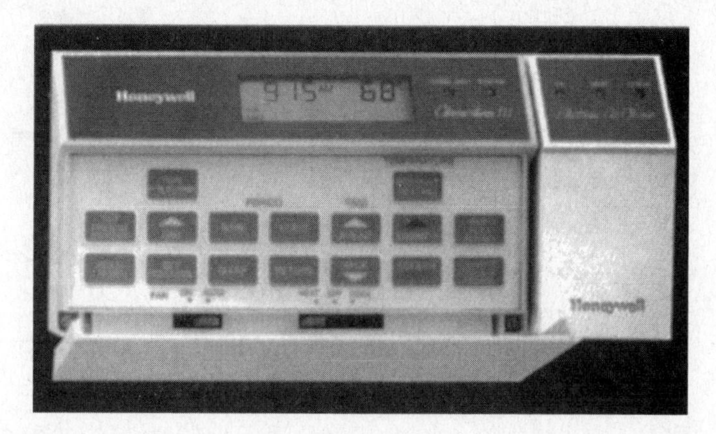

Figure 10.17 Seven-day programmable thermostat for heating, ventilation, and air-conditioning equipment. (Courtesy of Honeywell Inc.)

2. *Controls damper and fan operation:* closes outdoor air dampers and allows fan operation only on a call for heating or cooling during unoccupied periods.

Humidistats

A *humidistat* (Figure 10.18) is a sensing control that measures the amount of humidity in the air and provides switching action for the humidifier. The sensing element is made of nylon ribbon. This material expands when moist and contracts when dry. This movement is used to operate the switching mechanism.

The sensing element for the humidistat can be located either in the return-air duct or in the space being conditioned (Figure 10.19). The humidistat is wired in such a way that humidity is added only when the fan is operating.

Fan and Limit Controls

Fan (blower) and limit controls have separate functions in the heating system. They are discussed together, since both can use the same sensing element and are therefore often combined into a single control.

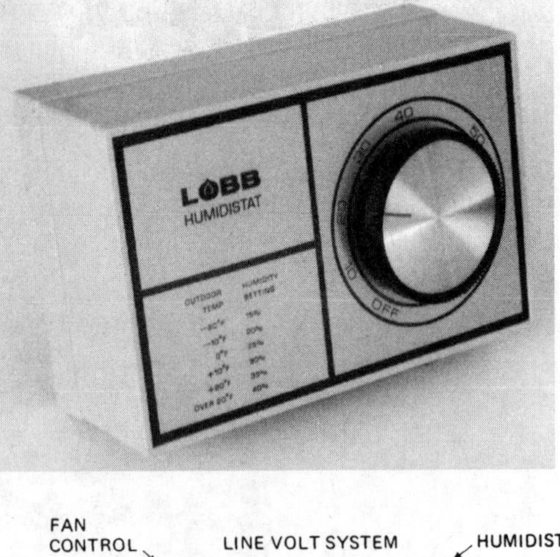

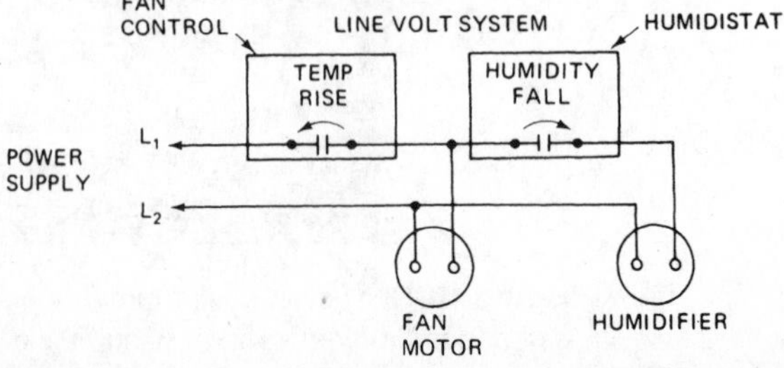

Figure 10.18 Humidistat.

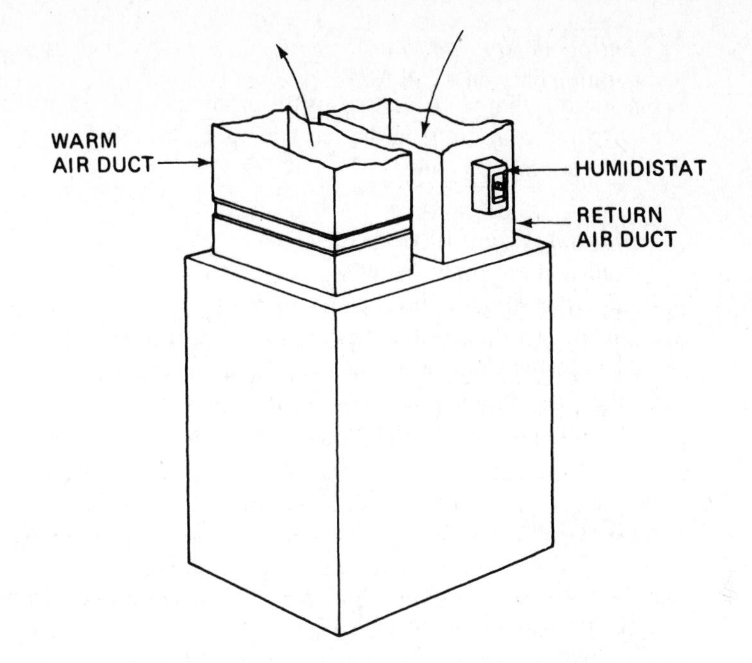

Figure 10.19 Duct installation of humidistat.

The insertion element (sensor) can be made in a number of forms (Figure 10.20). Some are bimetallic and some are hydraulic. The hydraulic element is filled with liquid that expands when heated, moving a diaphragm connected to a switch. The sensing element for these controls is inserted in the warm-air plenum of the furnace. It must be located in the moving airstream where it can quickly sense the warm air temperature rise. Sensing elements are made in different lengths to fit various applications. When equipment is replaced, the original insertion length should always be duplicated.

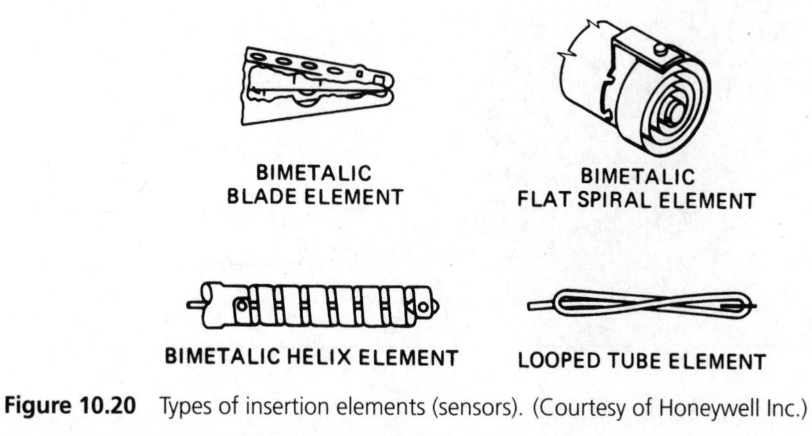

Figure 10.20 Types of insertion elements (sensors). (Courtesy of Honeywell Inc.)

Fan Controls

A fan control (Figure 10.21) senses the air temperature in the furnace plenum. To prevent discomfort, it turns on the fan when the air is sufficiently heated. There are two general types of fan controls:

- Temperature-sensing
- Timed fan start

Temperature-Sensing Fan Control These controls depend on the gravity heating action of the furnace to move air across the sensing element. When the air temperature reaches the fan-on temperature (usually, about 110°F), the fan starts. When the thermostat is no longer calling for heat, the fan continues to run to move the remaining heat out of the furnace. The fan stops at the fan-off temperature setting of the fan control (usually about 90°F). Most fan controls are adjustable so that changes in fan operation can be made to fit individual requirements. The fan control circuit is usually line voltage (120 V). The fan control switch is placed in series with the power to the fan motor. Some fan controls have a fixed differential. The fan-on temperature can be set, but the differential (fan-on minus fan-off temperature) is set at the factory. The fixed differential is normally 20° to 25°F.

Timed Fan Start This control (Figure 10.22) is used in a downflow or horizontal furnace where gravity air movement over the heat exchanger cannot be depended on to warm the sensing element. This control has a low-voltage (24-V) resistance heater,

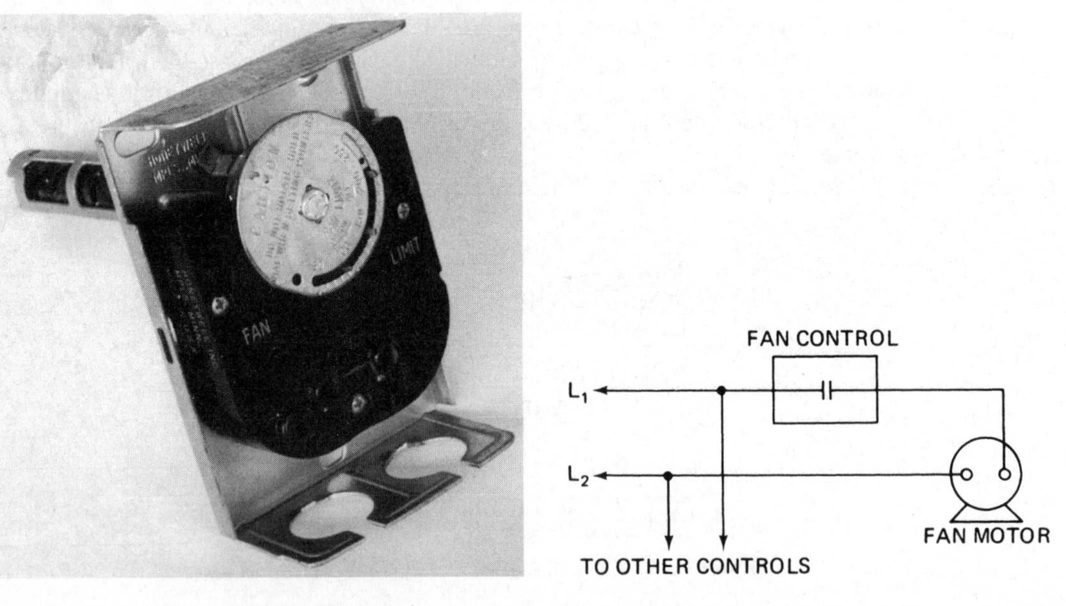

Figure 10.21 Fan control.

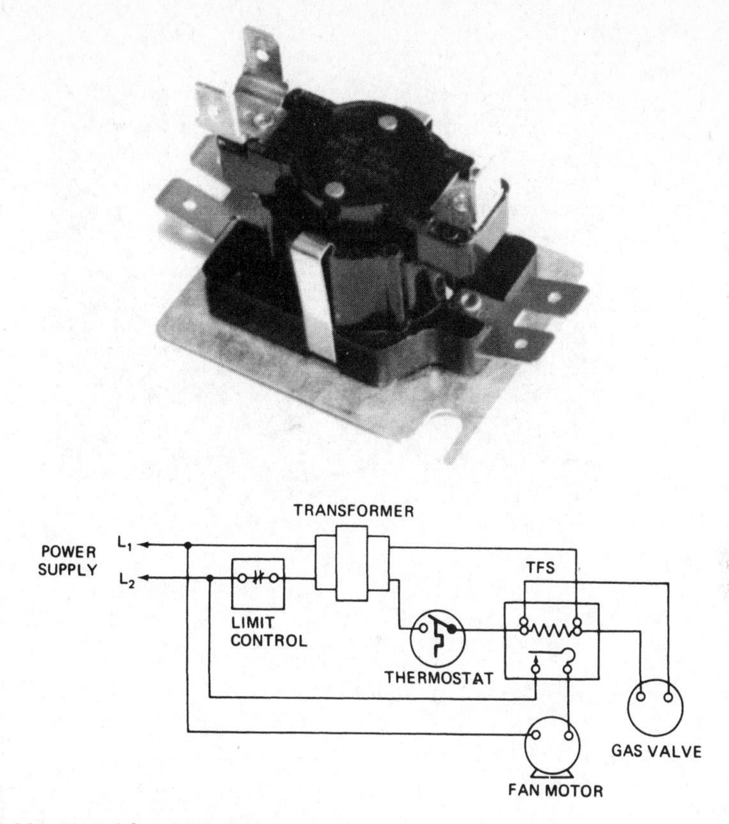

Figure 10.22 Timed fan start control.

which is energized when the thermostat calls for heat. The fan starts approximately 60 seconds after the heat is turned on. Usually, the fan is turned off by a conventional temperature-sensing fan control. It is important that the manufacturer's instructions be followed for setting the heat anticipator in the thermostat.

Limit Controls

The limit control is a safety device that shuts off the source of heat when the maximum safe operating temperature is reached (Figure 10.23). Like the fan control, it senses furnace plenum temperature or air temperature at the outlet. Warm-air furnace limit controls are usually set to cut out at 200°F and cut in automatically at 175°F.

On a gas furnace the line-voltage limit control is placed in series with the transformer. On an oil furnace the line-voltage limit control is placed in series with the primary control. On an electric furnace, the limit controls are placed in series with the heating elements. On a horizontal or downflow furnace, a secondary limit control is required (Figure 10.24). This control is located above the heating element and is usually set to cut out the source of heat when the sensing temperature reaches 145°F.

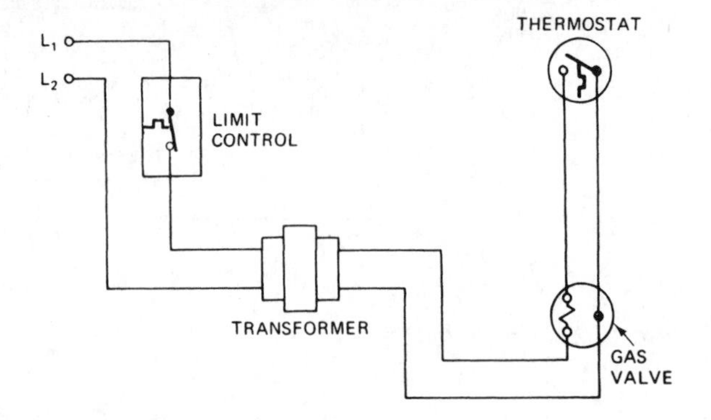

Figure 10.23 Limit control.

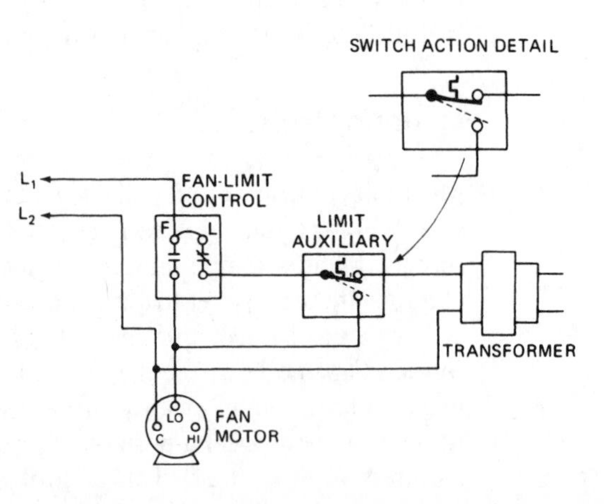

Figure 10.24 Secondary limit control.

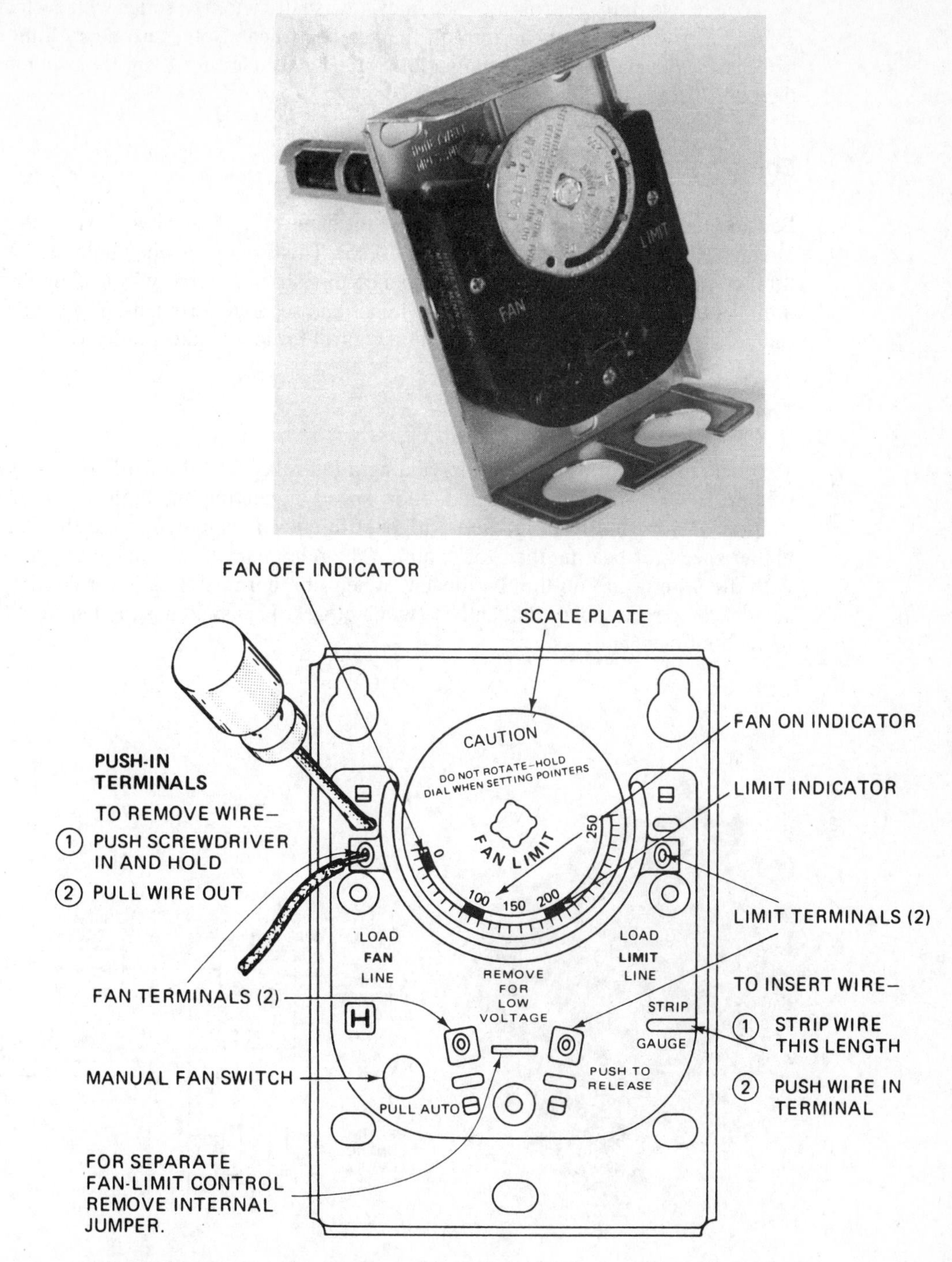

Figure 10.25 Combination fan and limit control. (Courtesy of Honeywell Inc.)

Where two limit controls are used, they are usually wired in series with each other. Thus, either limit control can turn off the source of heat. Some secondary limit controls have a manual reset (not automatic). Some have an arrangement for switching on the fan when the limit contacts are opened.

Combinations

Because the fan and limit controls both use the same type of sensing element, they are often combined into one control (Figure 10.25). This is helpful when both the fan and limit controls are of line-voltage type, since this simplifies the wiring. The fan and limit controls, though, have separate switches and separate terminals. The fan control can be wired for line voltage and the limit control for low voltage, if desired.

Two-Speed Fan Controls

Two fan speeds can be obtained by adding a fan relay to a standard control system (Figure 10.26). The fan is operated at low speed on heating and high speed for ventilation or air conditioning. Some high-efficiency furnaces operate the fan at a higher speed for heating than for cooling. The relay has a low-voltage coil in series with the *G* terminal on the thermostat. When the thermostat calls for cooling, the switch in the relay opens and an NO switch closes. This switches the fan from heating to cooling speed.

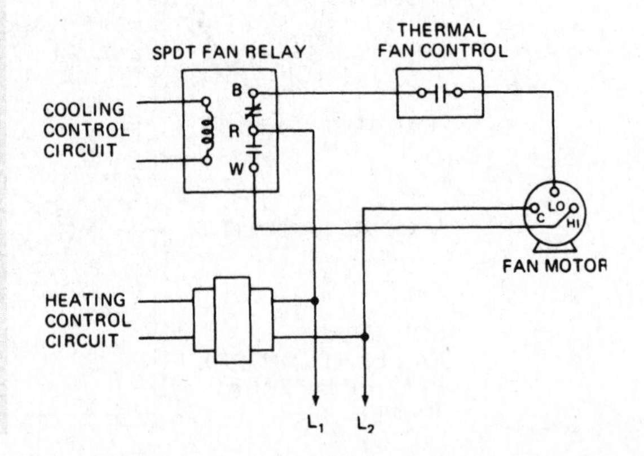

Figure 10.26 Two-speed fan control.

STUDY QUESTIONS

Answers to the study questions may be found in the sections noted in brackets.

10-1 Describe the function of an automatic control system. *[Functions of Controls]*
10-2 Name and describe the five elements of a control system. *[Functions of Controls]*
10-3 What is the smallest temperature difference that most people can sense? *[Thermostats]*
10-4 What metals are used in constructing bimetal sensors? *[Thermostats]*
10-5 What is the rating of the low-voltage circuit used on thermostat circuits? *[Thermostats]*
10-6 What voltage is used in a self-generating system? *[Types of Current]*
10-7 How is the differential of a thermostat determined? *[Anticipators]*
10-8 How are the terminals on a thermostat identified by color code? *[Thermostats]*
10-9 What are the additional features in a programmable thermostat? *[Programmable Thermostats]*
10-10 What type of sensing element does a humidistat have? *[Humidistats]*
10-11 Why are fan and limit controls both placed on the same electrical device? *[Fan and Limit Controls]*
10-12 Where is a timed fan-start switch used? *[Timed Fan Start]*
10-13 How does a two-speed fan control operate? *[Two-Speed Fan Controls]*
10-14 Where is the heat anticipator located, and how is it adjusted? *[Anticipators]*

CHAPTER 11

Gas Furnace Controls

OBJECTIVES

After studying this chapter, the student will be able to:

- Identify the various electrical controls and circuits on a gas-fired furnace
- Determine the sequence of operation of the controls
- Service and troubleshoot the gas heating unit control system

Use of Gas Furnace Controls

Broadly speaking, gas furnace controls are the electrical and mechanical equipment that operates the unit manually or automatically. The controls used depend on:

1. Type of fuel or energy
2. Type of furnace
3. Optional accessories

The controls differ for gas, oil, and electric heat sources. They differ somewhat for the upflow and downflow units. They also become more complex as accessories such as humidifiers, electrostatic filters, and cooling, are added.

Circuits

Control circuits for a gas warm air heating unit are:

1. Power supply circuit
2. Fan circuits
3. Pilot circuit
4. Fuel-burning or heater circuit
5. Accessory circuits

Both line-voltage and low-voltage circuits are required for some components. For example, a relay used to start a fan motor may have a low-voltage coil, but the switch that starts the fan motor operates on line voltage.

Schematic diagrams are used in the study of circuits. Following is the key to the legends used on schematics described in this chapter:

CAP	Capacitor
CC	Compressor contactor
DPDT	Double-pole double-throw switch
DPST	Double-pole single-throw
DSI	Direct spark ignition
EAC	Electrostatic air cleaner
EP	Electric pilot
FC	Fan control
FM	Fan motor
FD	Fused disconnect
FR	Fan relay
G	Fan relay terminal
GV	Gas valve
H	Humidistat
HS	Humidification system
HU	Humidifier
L	Limit
LA	Limit auxiliary
L_1, L_2	Power supply
N	Neutral
NC	Normally closed
NO	Normally open
PPC	Pilot power control
R	24-V power terminal
SPDT	Single-pole double-throw
SPST	Single-pole single-throw
TFS	Timed fan start
THR	Thermocouple
TR	Transformer
W	Heating terminal
Y	Cooling terminal

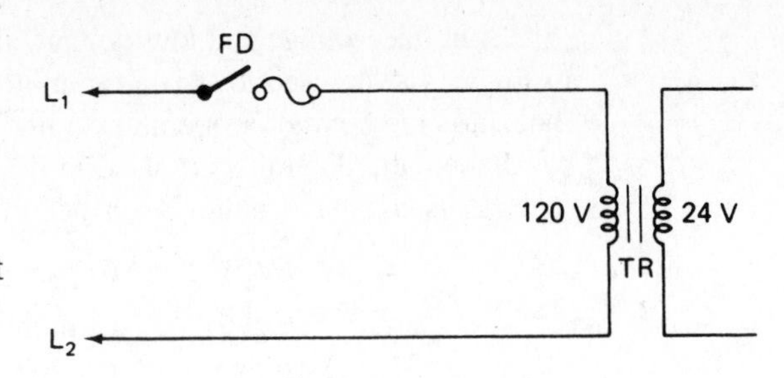

Figure 11.1 Power supply circuit with fused disconnect (FD) and transformer (TR).

Power Supply

The power supply circuit (Figure 11.1) usually consists of a 15-A or 20-A source of 120-V ac power, a fused disconnect (FD) in the hot line, and a transformer (TR) to supply 24-V ac power to the low-voltage controls.

A single source of power is adequate unless a cooling accessory is added, in which case a separate source of 208/240-V ac power is required. Power for cooling requires a DPST disconnect with fuses in each hot line. An additional low-voltage transformer may be required if the one used for heating does not have sufficient capacity.

Fan

The line-voltage circuit for a single-speed fan consists of a fan control (FC in series with the fan motor (FM) connected across the power supply (Figure 11.2). Many heating units have multiple-speed fans. One speed is used for heating, and the other is used for cooling or ventilation. The operation of the fan with a two-speed motor requires two circuits (Figure 11.3). There is a line-voltage circuit to power the fan motor, and a low-voltage circuit to operate the switching relay.

Under normal operation in heating, the fan motor (FM) is controlled by the fan control (FC). When the fan is operating on cooling speed, the fan switch on the sub-base of the thermostat can be moved to the "on" position. This action closes a contact between R and G in the thermostat. Current flows through the coil of the cooling fan

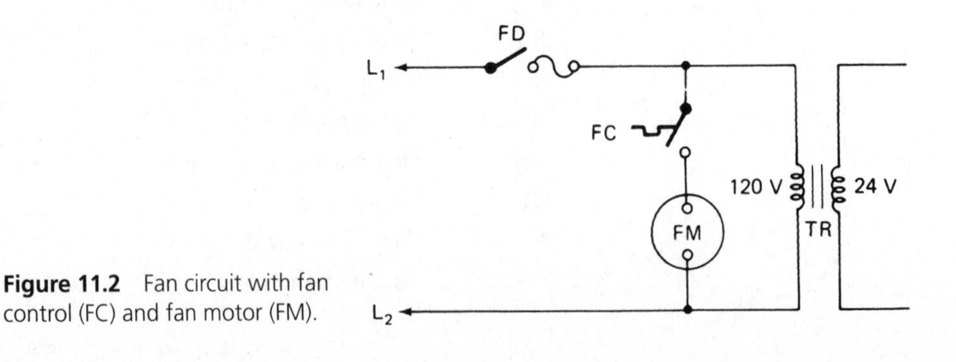

Figure 11.2 Fan circuit with fan control (FC) and fan motor (FM).

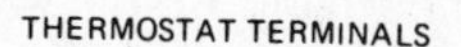

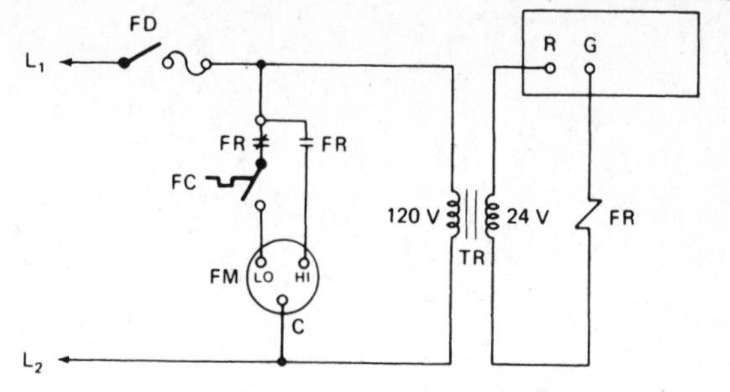

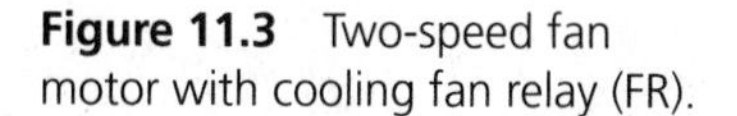
Figure 11.3 Two-speed fan motor with cooling fan relay (FR).

relay (FR), opening the NC switch and closing the NO switch operating the fan motor on cooling speed.

On a downflow or horizontal unit, the fan motor can be operated by a timed fan start relay (TFS) (Figure 11.4). On these units, with the fan not running and the sensing element of the fan control located in the air outlet, the standard fan control cannot be used to start the fan.

On a call for heating, R and W make in the thermostat. The heater of the timed fan start (TFS) is energized. After about 45 seconds, the fan motor is operated by the TFS switch. When the thermostat is satisfied and R and W break, the fan motor continues to operate until the leaving air temperature reaches the cutout temperature setting of the fan control (FC).

Pilot

The gas safety circuit is powered by a thermocouple (THR) (Figure 11.5). When a satisfactory pilot is established the thermocouple generates approximately 30 mV to energize the safety control portion of the combination gas valve. Thus the gas cannot enter the burner until a satisfactory pilot is established. Most wiring diagrams omit the

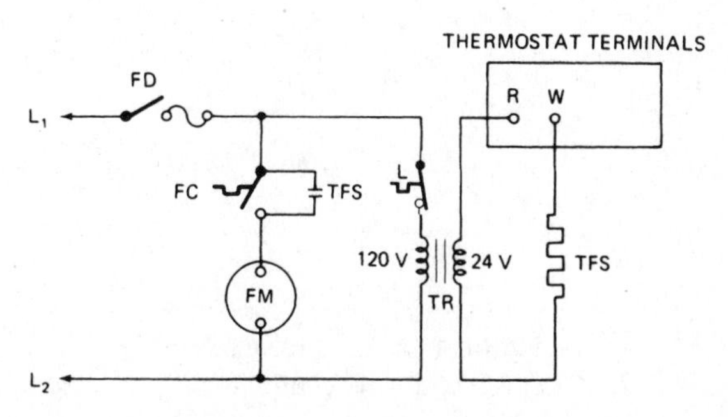

Figure 11.4 Fan control for downflow furnace using timed fan starts (TFS).

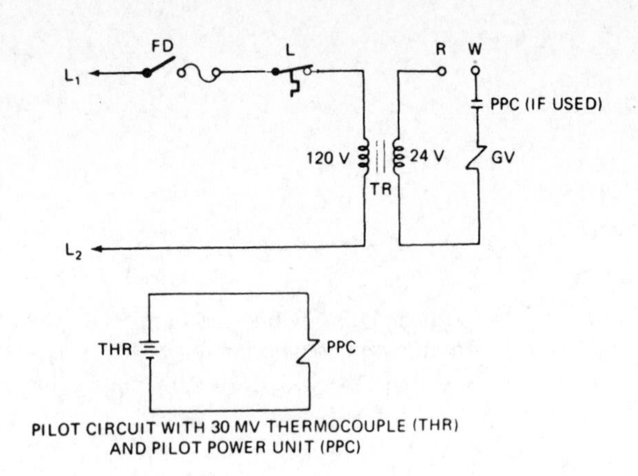

Figure 11.5 Pilot circuit with 30-mV thermocouple (THR) and pilot power control (PPC).

pilot circuit, since it is common to all combination gas systems and need not be repeated in each diagram.

Gas Valve

The gas valve circuit is a low-voltage circuit (Figure 11.6). When the thermostat calls for heat, R and W make, energizing the gas valve (GV) because the limit control (L) is normally closed.

Should excessive temperatures occur (usually 200°F in the plenum), the limit control opens the circuit to de-energize and close the gas valve. The limit control automatically restarts the burner when its cut-in temperature (usually 175°F) is reached. On a downflow or horizontal furnace, a limit auxiliary (LA) control is used (Figure 11.7). The LA control is located in a position to sense gravity heat from the heating elements. If the fan fails to start, the LA control will sense excess heat (usually set to cut out at 145°F).

Limit controls can be line voltage and placed in series with the transformer (Figure 11.8). Some line-voltage limit controls have a separate set of contacts that make when the limit breaks, turning the fan on if, for any reason, it is not running. Limit controls can also be placed in series with the low-voltage heating control cir-

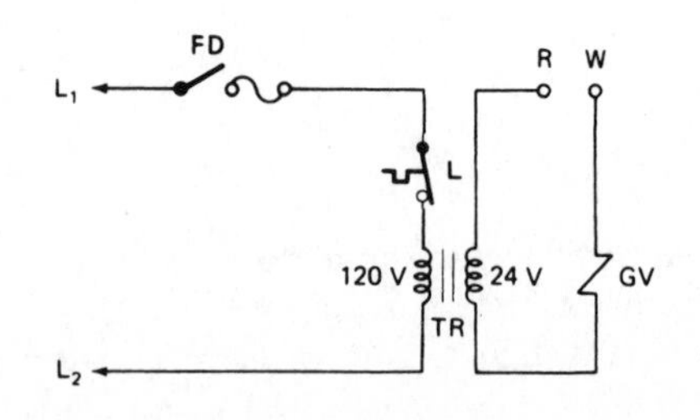

Figure 11.6 Gas valve (GV) circuit with limit control (L).

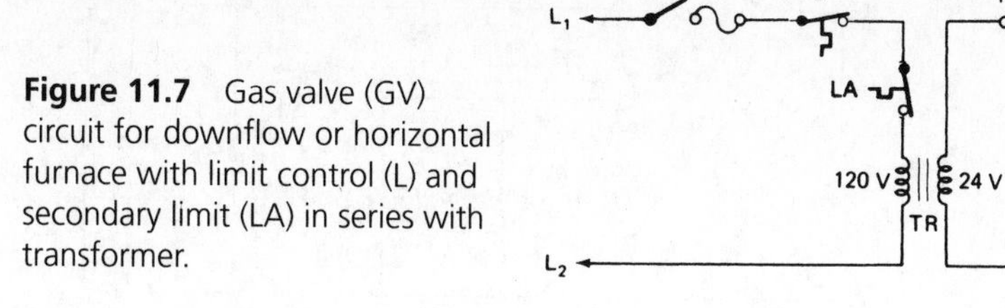

Figure 11.7 Gas valve (GV) circuit for downflow or horizontal furnace with limit control (L) and secondary limit (LA) in series with transformer.

cuit. If the limit control opens, this de-energizes the heating circuit and close the gas valve.

Accessories

Figure 11.9 shows the schematic of accessory circuits. The humidistat (H) and humidifier (HU) are wired in series and connected to power only when the fan is running. The electrostatic air cleaner also operates only when the fan is running. These accessories are supplied with line-voltage power.

The cooling contactor (CC) is in series with the Y connection on the thermostat. When R and Y make, the cooling contactor coil is energized, operating the cooling unit.

Typical Wiring Diagrams

The following schematic wiring diagrams illustrate the control circuits found on many heating furnaces. A schematic aids a service technician in understanding how the control system operates. Many manufacturers, however, supply only connection wiring diagrams for their equipment. It may therefore be necessary for a technician to construct a schematic in order to separate the circuits for testing and for diagnosing service problems.

Unfortunately, few standards exist that require manufacturers to conform in making connection wiring diagrams. If service technicians know the most common controls and their functions, with practice they can interpret almost any diagram. When a diagram is not available, a technician can also prepare one to suit the equipment found on the job.

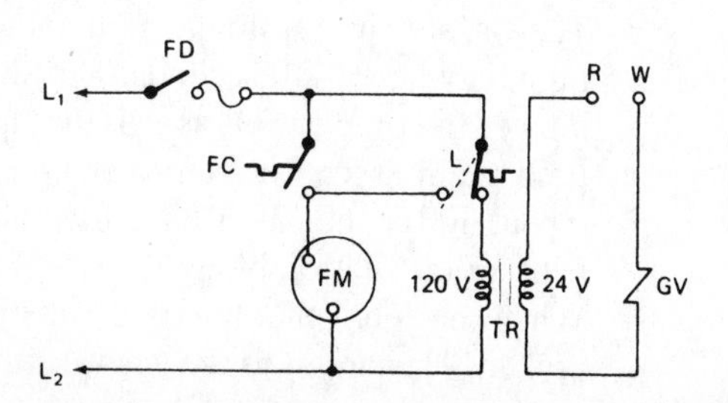

Figure 11.8 Single-pole double-throw (SPDT) limit control with set of contacts to operate fan-on limit action.

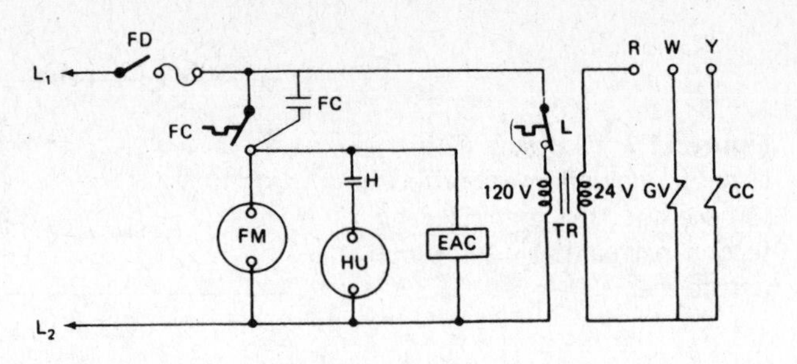

Figure 11.9 Accessory circuits.

Furnace Sequence of Operation

The following is a review of some typical connection diagrams supplied by various manufacturers for their equipment.

Upflow

The schematic diagram is shown in Figure 11.10; the sequence of operations follows.

This furnace is equipped with an electric hot surface burner ignition system.

Caution: This furnace does not have a pilot. It is equipped with an ignition device that automatically lights the burners. **Do not** try to light the burners by hand.

In response to a call for heat by the room thermostat, a hot glowing igniter lights the burners at the beginning of each operation cycle.

1. When the furnace control board is activated, the vent blower is turned on. A circuit is also made through the NO pressure switch contacts.
2. As the vent blower increases in speed the contacts of the pressure switch close and complete the electrical circuit to the igniter.
3. During the next 30 seconds, the vent blower brings fresh air into the heat exchanger and the igniter begins to glow. At the end of this period, the gas valve opens and the burners light.
4. After the burners light, a separate sensor acts as a flame probe to check for the presence of flame. As long as a flame is present, the system will monitor it and hold the gas valve open.
5. The elapsed time from the moment the room thermostat closes to when the burners light may be 30–40 seconds. This delay is caused by the time required for the vent blower to come to full speed, the time required for the igniter to heat up, and the time required for fresh air to be brought into the heat exchanger.
6. About 30 seconds after the burners have lighted, the fan switch closes and the furnace air circulation blower runs.
7. When the room thermostat is satisfied, the circuit to the furnace control board is broken. The circuit to the gas valve is broken and the burners are extinguished. The

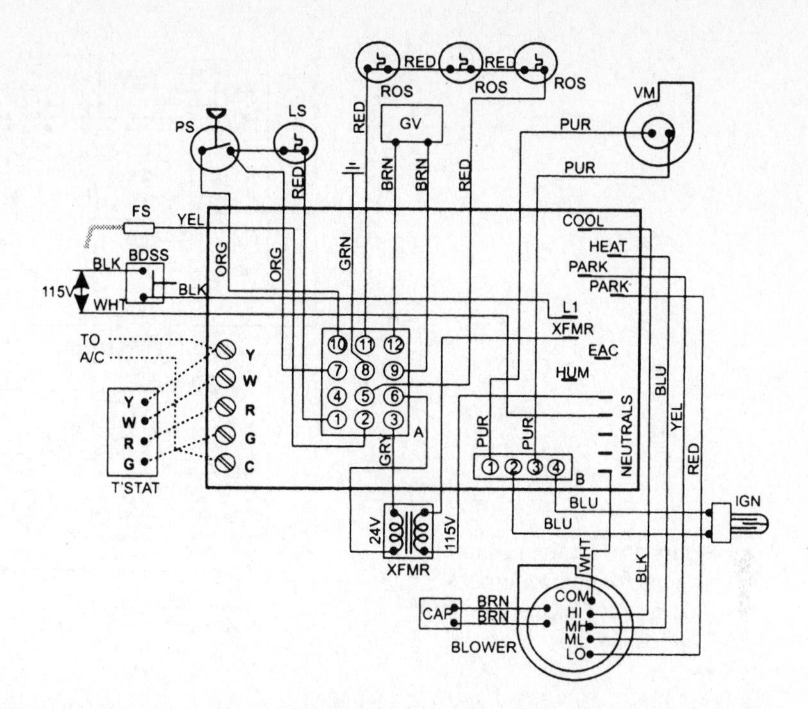

Figure 11.10 Schematic diagram for upflow furnace. (Courtesy of Coleman–Unitary Products Group)

vent blower continues to run for a few seconds. Then, the furnace control board keeps the circulating blower running for a period of time to allow additional heat to be drawn from the heat exchanger.

Figure 11.11 shows the matching connection wiring diagram for Figure 11.10. This type of wiring diagram displays the various components in their approximate location.

- Blower motor
- Vent fan
- Pressure switch and limit switch
- Gas valve

Twinning

When two furnaces are installed using the same duct system, it is very important that the two furnace circulating air blowers operate simultaneously. If one blower starts before the second blower, the duct system will become pressurized with air, and the second blower will be made to turn backward. During heating operation, this will

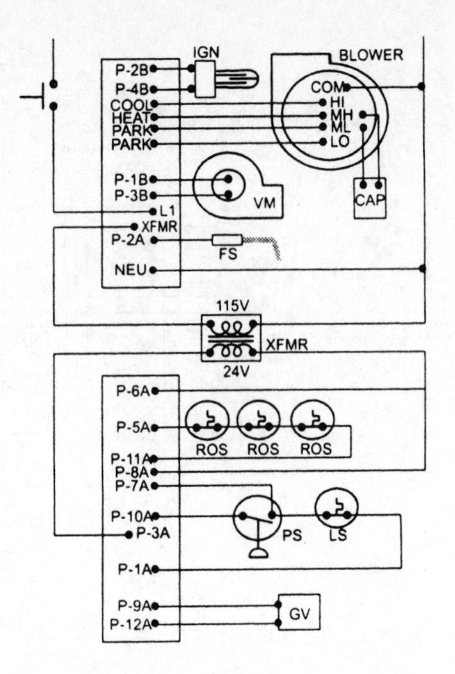

Figure 11.11 Connection diagram for upflow furnace. (Courtesy of Coleman–Unitary Products Group)

cause overheating of the second furnace, possibly causing an unsafe condition and damage to the furnace. The furnace control board has a terminal marked TWIN that can be used to cause two furnace blowers to operate together.

If two furnaces are to be twinned using a single wall thermostat, connect an isolation relay as shown in Figure 11.12.

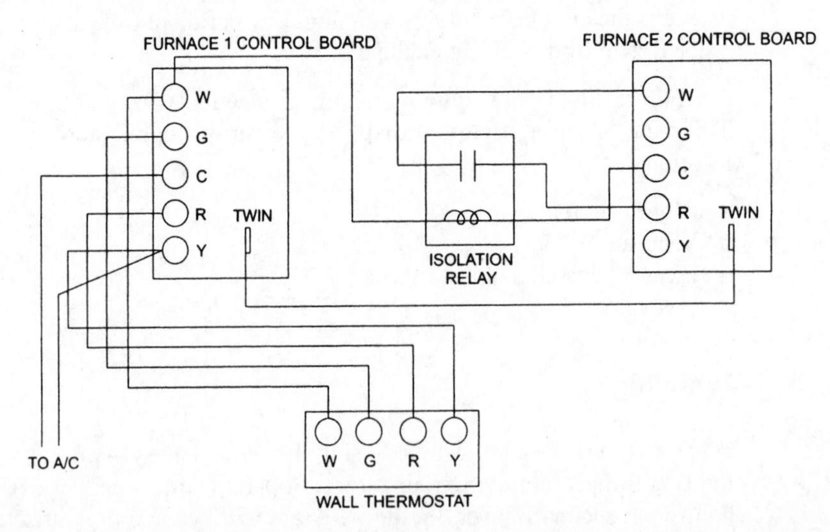

Figure 11.12 Twin connection wiring diagram for two furnace applications. (Courtesy of Coleman–Unitary Products Group)

Downflow

The controls in a downflow gas furnace (Figure 11.13) include:

- Thermostat
- Gas valve
- Combination fan and limit control
- Auxiliary limit
- Fan motor with run capacitor
- Transformer

Note that Figures 11.13 and 11.14 do not show the timed fan start, which is common to many downflow (counterflow) units. On certain units, the fan control is located

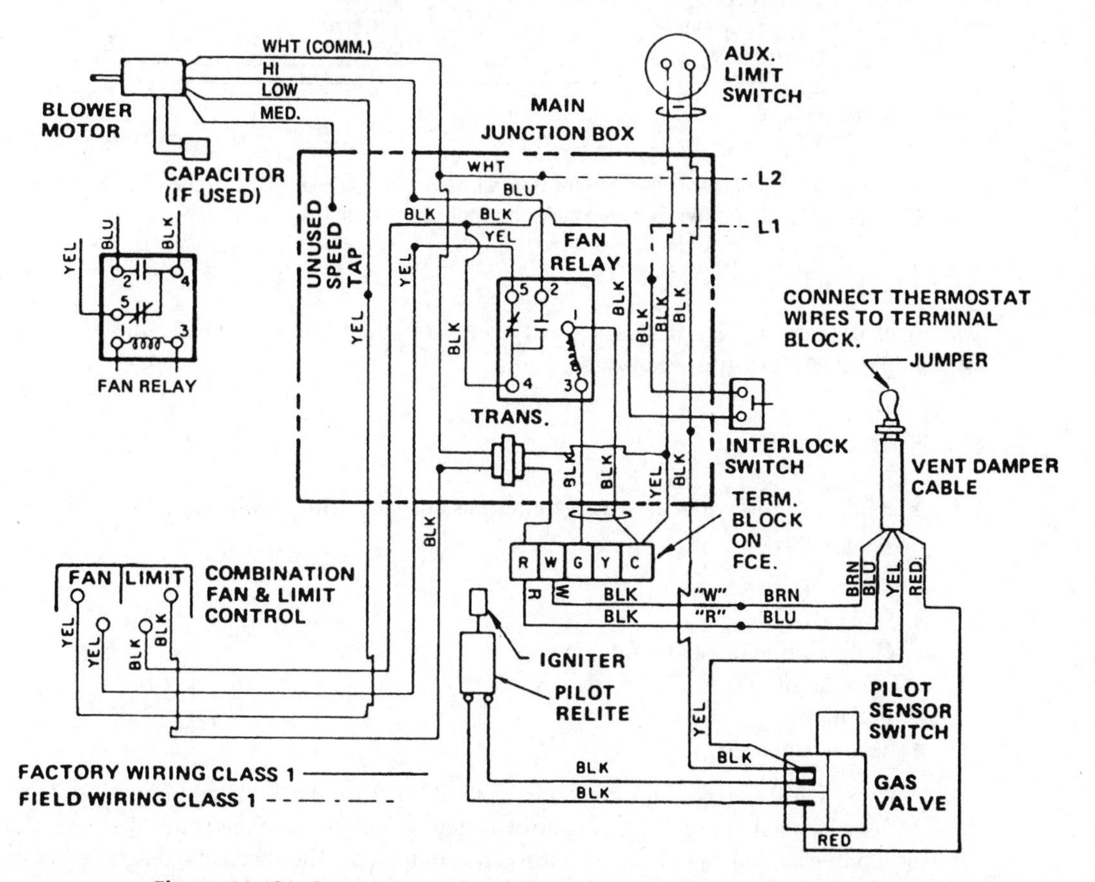

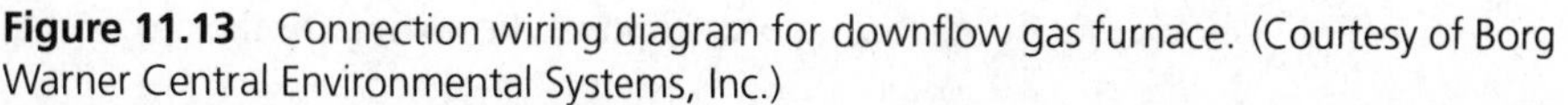

Figure 11.13 Connection wiring diagram for downflow gas furnace. (Courtesy of Borg Warner Central Environmental Systems, Inc.)

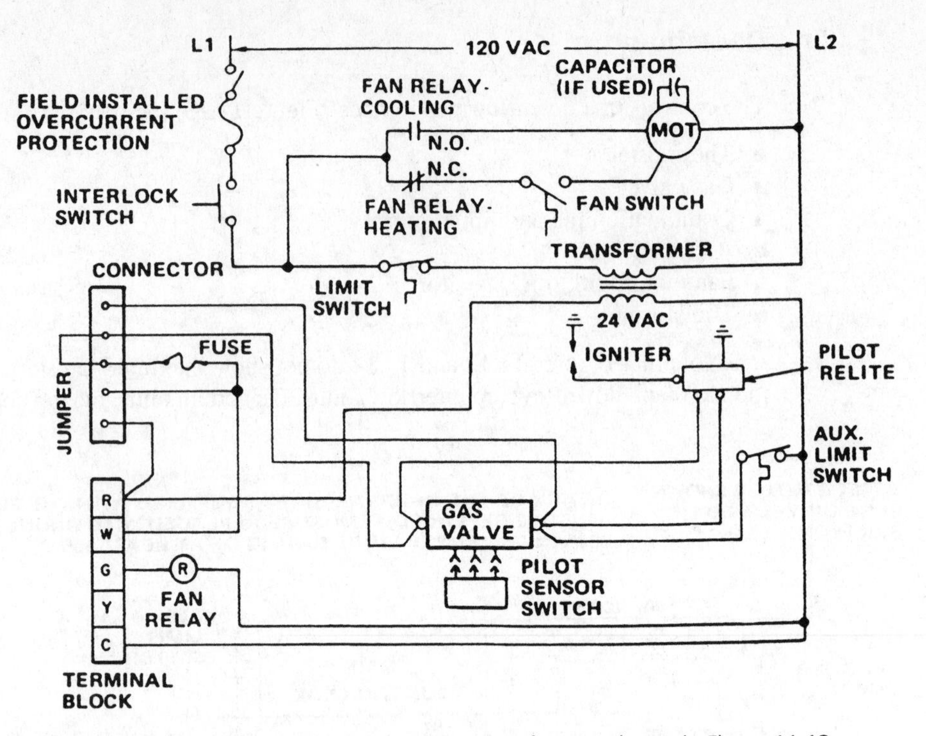

Figure 11.14 Schematic diagram for downflow gas furnace shown in Figure 11.13. (Courtesy of Borg Warner Central Environmental Systems, Inc.)

in such a position that it senses gravity heat from the heating elements, thus eliminating the need for a timed fan start.

Horizontal Flow

A connection drawing for a horizontal gas furnace with cooling accessory (Figure 11.15) includes:

- Thermostat
- Gas valve
- Combination fan and limit control
- Auxiliary limit
- Fan motor
- Transformer

A schematic diagram for this unit can be constructed as shown in Figure 11.16.

One special feature of this control system is that the purchaser has the option of installing electronic air cleaner controls and equipment. The schematic diagram, Figure 11.17, shows the proper locations to connect the wiring for the electronic air cleaner.

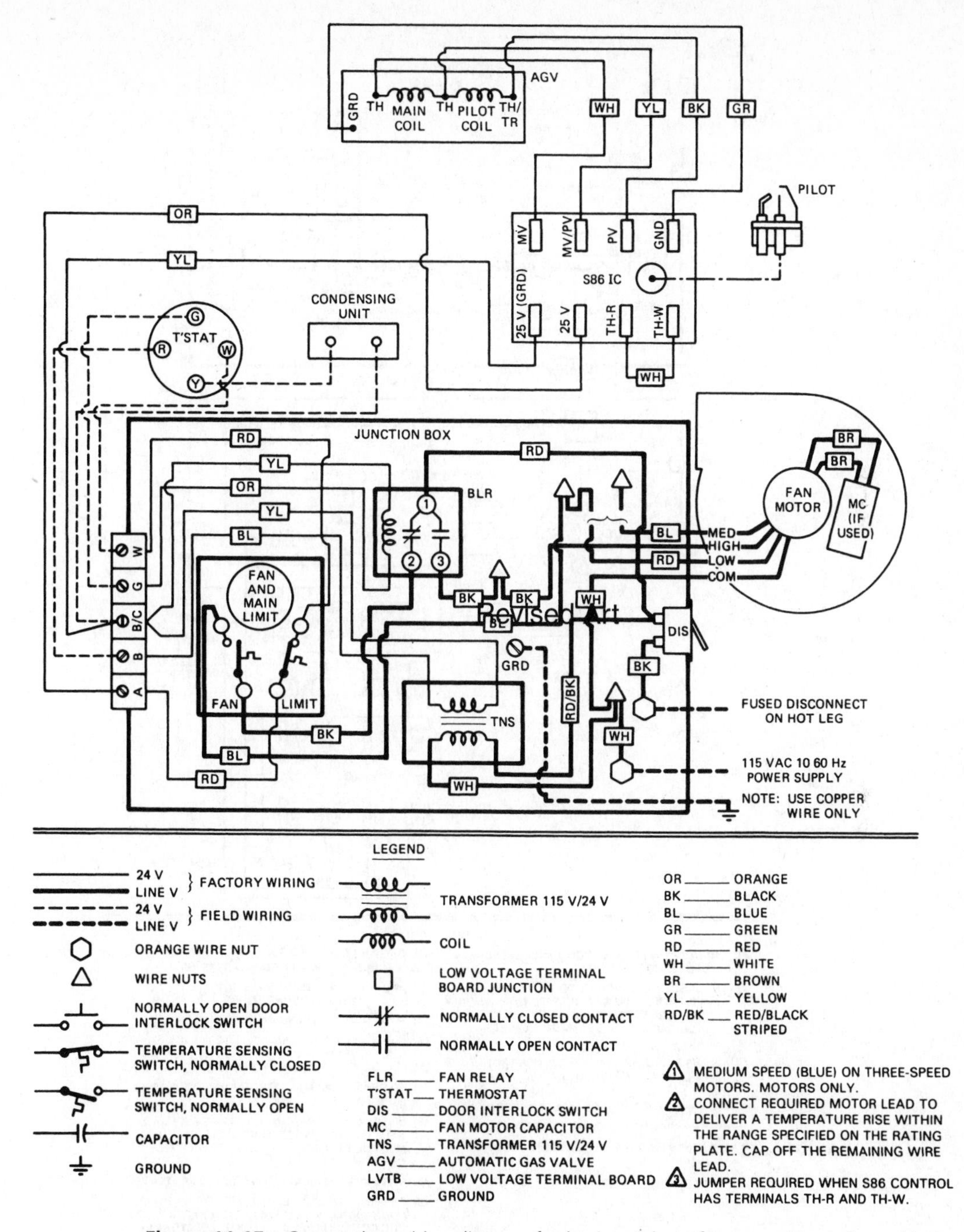

Figure 11.15 Connection wiring diagram for horizontal gas furnace with cooling added. (Courtesy of Borg Warner Central Environmental Systems, Inc.)

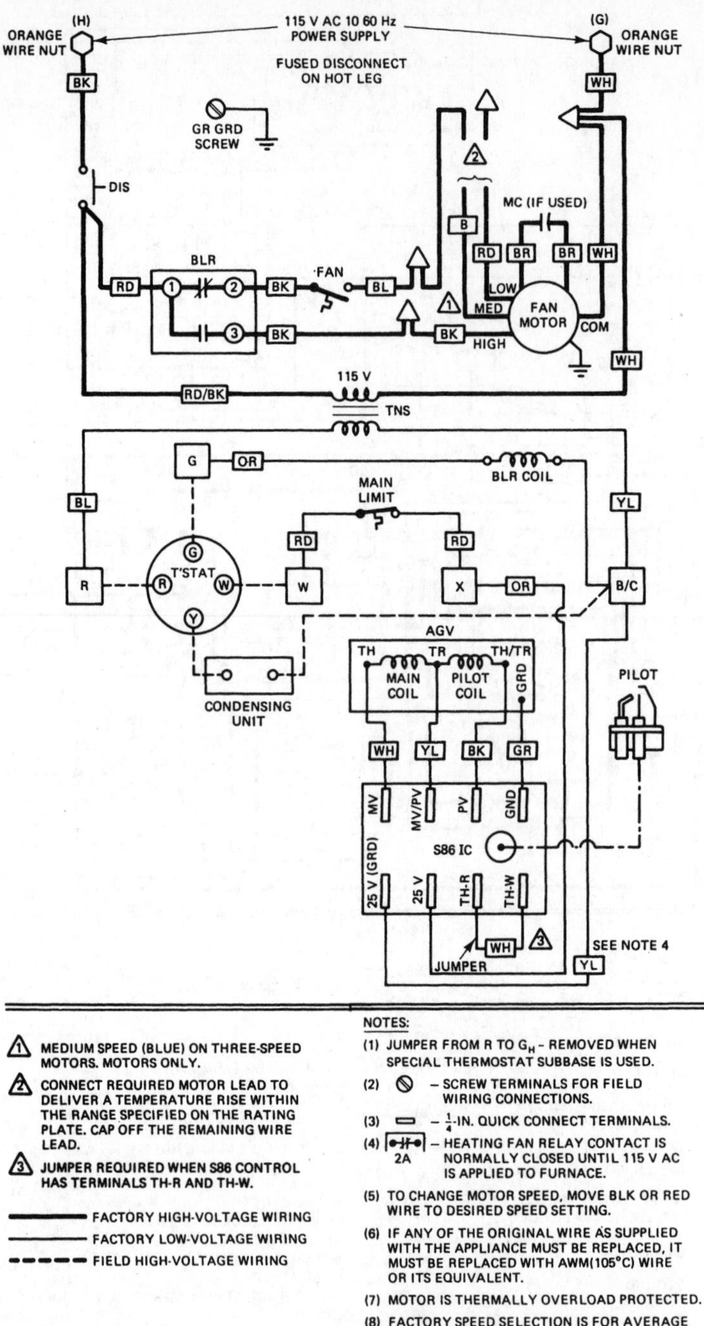

Figure 11.16 Schematic wiring diagram for horizontal gas furnace with cooling added. (Courtesy of Borg Warner Central Environmental Systems, Inc.)

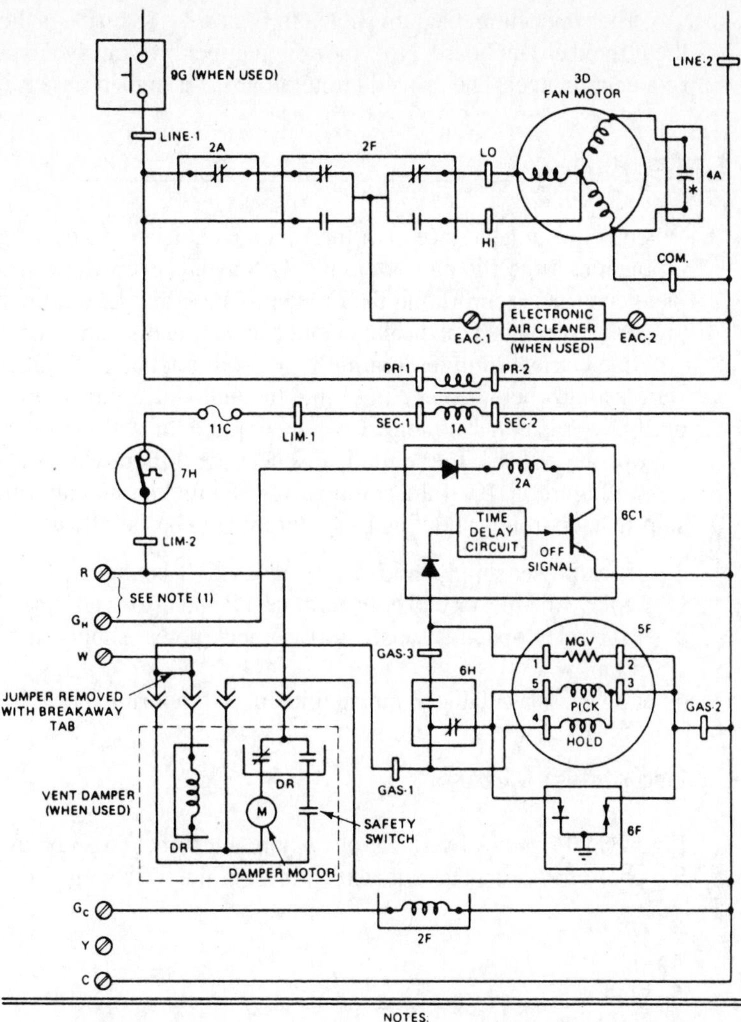

1A – TRANSFORMER 115/24
2A – RELAY - HEAT (SPST-NC)
2F – RELAY - COOL (DPDT)
3D – FAN MOTOR
4A – RUN CAPACITOR
5F – GAS VALVE
6C1 – PRINTED CIRCUIT BOARD
6F – PILOT IGNITER
6H – SAFETY PILOT (FLAME SENSING)
7H – LIMIT SWITCH (SPST-NC)
9G – FAN DOOR SWITCH (SPST-N.O.)
11C – FUSIBLE LINK
11E – GROUND LUG

FACTORY HIGH-VOLTAGE WIRING
FACTORY LOW-VOLTAGE WIRING
FIELD HIGH-VOLTAGE WIRING

NOTES:

(1) JUMPER FROM R TO G_H - REMOVED WHEN SPECIAL THERMOSTAT SUBBASE IS USED.
(2) – SCREW TERMINALS FOR FIELD WIRING CONNECTIONS
(3) – $\frac{1}{4}$ IN QUICK CONNECT TERMINALS
(4) 2A – HEATING FAN RELAY CONTACT IS NORMALLY CLOSED UNTIL 115-V AC IS APPLIED TO FURNACE.
(5) TO CHANGE MOTOR SPEED, MOVE BLK OR RED WIRE TO DESIRED SPEED SETTING.
(6) IF ANY OF THE ORIGINAL WIRE AS SUPPLIED WITH THE APPLIANCE MUST BE REPLACED, IT MUST BE REPLACED WITH AWM(105°C) WIRE OR ITS EQUIVALENT.
(7) MOTOR IS THERMALLY OVERLOAD PROTECTED
(8) FACTORY SPEED SELECTION IS FOR AVERAGE CONDITIONS. SEE INSTALLATION INSTRUCTIONS FOR OPTIMUM SPEED SELECTION. MOTOR MAY BE 3 OR 4 SPEED.
(9) SYMBOLS ARE AN ELECTRICAL REPRESENTATION ONLY.

Figure 11.17 Wiring diagram for upflow gas furnace with printed circuit control center. (Courtesy of Bryant Air Conditioning/Heating)

The connection diagram shown in Figure 11.18 displays the wiring connections to the printed circuit board. Note the vent damper breakaway tab provided for installation of a vent damper. The printed circuit board is designed for ease of service in the field.

Testing and Servicing

When performing service on a furnace, a technician makes every effort to pinpoint the problem to a specific part or circuit. This avoids extra work in checking over the complete control system to find the problem. For example, if all parts of a furnace operate properly except the fan, the fan circuit can be tested separately to locate the malfunction.

In electrical troubleshooting, it is good practice to "start from power," that is, to start testing where power comes into the unit and continue testing in the problem area until power is either no longer being supplied, or with power available, the load does not operate. When either point is reached, the difficulty is located. For example, referring to Figure 11.10, if the complaint is that the fan will not run on cooling speed, the step-by-step troubleshooting procedure would be as follows:

1. Check power supply at load side of fused disconnect.
2. Remove thermostat and jump terminals *R* and *G* to determine if the relay will "pull in."
3. If the relay operates satisfactorily, check power supply at cooling-speed terminals of fan.
4. If power is available at fan, and it still will not run, motor is defective.

Use of Test Meters

Figure 11.19 shows five locations where a test meter can be used on a 24-V gas heating control system. These meters check out the following circuits:

- Power
- Fan
- Pilot
- Gas valve
- Transformer

No accessories are shown, but if these circuits exist on equipment being serviced, meter readings can be taken at these parts in addition to the ones shown.

Figure 11.20 illustrates a furnace installation, gas piping and controls, along with the self-diagnostic electronic control board located in the blower section. The electronic furnace control is equipped with a diagnostic light that flashes when there is a service problem with the furnace. The number of times the light flashes indicates the location of the problem.

LED Lighting Schedule

Steady off—Normal operation

One flash—False flame sense; check for stuck-open gas valve.

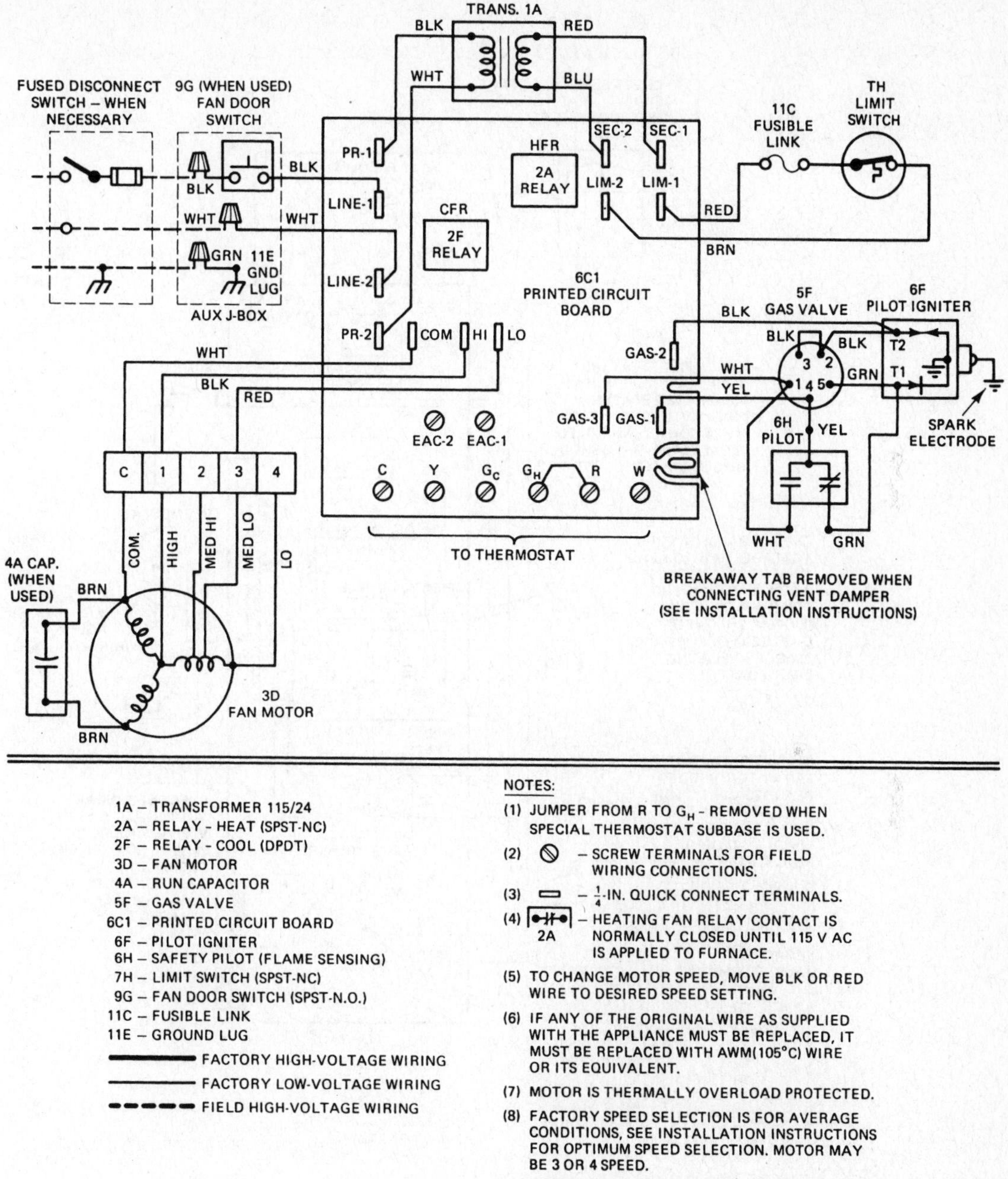

1A – TRANSFORMER 115/24
2A – RELAY - HEAT (SPST-NC)
2F – RELAY - COOL (DPDT)
3D – FAN MOTOR
4A – RUN CAPACITOR
5F – GAS VALVE
6C1 – PRINTED CIRCUIT BOARD
6F – PILOT IGNITER
6H – SAFETY PILOT (FLAME SENSING)
7H – LIMIT SWITCH (SPST-NC)
9G – FAN DOOR SWITCH (SPST-N.O.)
11C – FUSIBLE LINK
11E – GROUND LUG

FACTORY HIGH-VOLTAGE WIRING
FACTORY LOW-VOLTAGE WIRING
FIELD HIGH-VOLTAGE WIRING

NOTES:

(1) JUMPER FROM R TO G_H - REMOVED WHEN SPECIAL THERMOSTAT SUBBASE IS USED.

(2) ⊘ – SCREW TERMINALS FOR FIELD WIRING CONNECTIONS.

(3) ▭ – $\frac{1}{4}$-IN. QUICK CONNECT TERMINALS.

(4) 2A – HEATING FAN RELAY CONTACT IS NORMALLY CLOSED UNTIL 115 V AC IS APPLIED TO FURNACE.

(5) TO CHANGE MOTOR SPEED, MOVE BLK OR RED WIRE TO DESIRED SPEED SETTING.

(6) IF ANY OF THE ORIGINAL WIRE AS SUPPLIED WITH THE APPLIANCE MUST BE REPLACED, IT MUST BE REPLACED WITH AWM(105°C) WIRE OR ITS EQUIVALENT.

(7) MOTOR IS THERMALLY OVERLOAD PROTECTED.

(8) FACTORY SPEED SELECTION IS FOR AVERAGE CONDITIONS, SEE INSTALLATION INSTRUCTIONS FOR OPTIMUM SPEED SELECTION. MOTOR MAY BE 3 OR 4 SPEED.

(9) SYMBOLS ARE AN ELECTRICAL REPRESENTATION ONLY.

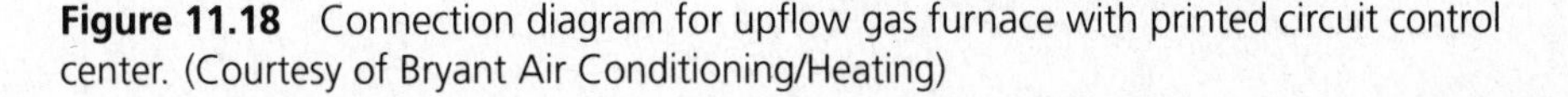

Figure 11.18 Connection diagram for upflow gas furnace with printed circuit control center. (Courtesy of Bryant Air Conditioning/Heating)

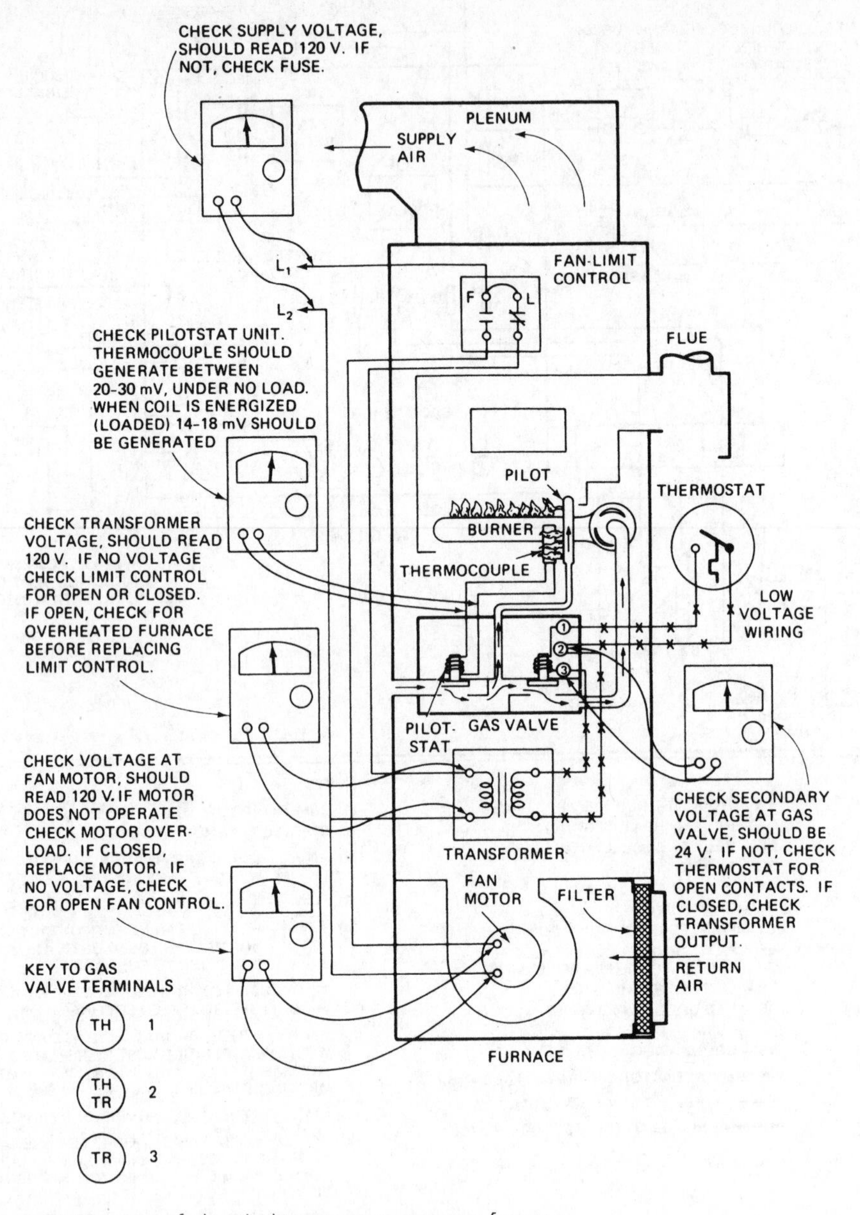

Figure 11.19 Use of electrical test meters on a gas furnace.

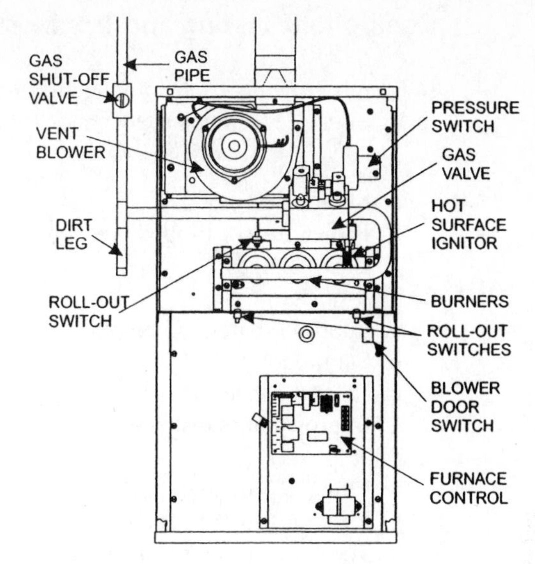

Figure 11.20 Typical installation for an upflow furnace with an electronic solid-state board. (Courtesy of Coleman–Unitary Products Group)

Two flashes—Pressure switch stuck closed; check for shorted wires, bad pressure switch.

Three flashes—Pressure switch failed to close; check vent blower, pressure switch, vent blockage, disconnected pressure hose.

Four flashes—Limit switch open; check for open limit switch, loose connections in limit circuit.

Five flashes—Rollout switch open; check for open rollout switches, loose connections in rollout switch circuit.

Six flashes—One-hour pressure switch lockout—pressure switch has cycled four times in a single call for heat; check for vent blockage, loose connections in a pressure switch circuit.

Seven flashes—One-hour ignition lockout—burner failed to light in three tries; check gas flow, gas pressure, gas valve operation, flame sensor.

Eight flashes—One-hour ignition lockout—five recycles in a single call for heat; check gas flow, gas pressure, gas valve operation, flame sensor.

Nine flashes—Reversed line voltage polarity; check incoming power wiring for proper polarity.

Steady on—Gas valve energized with no call to heat from thermostat, or control failure.

Note: The service technician should check the manufacturer's specifications before making a decision on the problem indicated by the electronic diagnostic system.

Component Testing and Troubleshooting

In troubleshooting, each component (or circuit) has the potential for certain problems that the service technician can check for or test. Some of these potential problems are:

1. Power supply
   a. Switch open
   b. Blown fuse or tripped breaker
   c. Low voltages
2. Thermostat
   a. Subbase switch turned off
   b. Set too low
   c. Loose connection
   d. Improper anticipator setting
3. Pilot
   a. Plugged orifice on pilot
   b. Pilot blowing out due to draft
   c. Too little or too much gas being supplied to pilot
4. Thermocouple
   a. Loose connection
   b. Too close or too far from flame
   c. Defective thermocouple
5. Transformer
   a. Low primary voltage
   b. Defective transformer
   c. Blown fuse in 24-V circuit
6. Limit control
   a. Switch open
   b. Loose connection
   c. Defective control
7. Gas valve
   a. Defective pilot safety power unit
   b. Defective main gas valve
8. Fan control
   a. Switch open
   b. Improper setting
   c. Defective control

Thermocouple Circuit

Millivoltmeter probes can be connected in the thermocouple circuit, as shown in Figure 11.21, using the special adapter. If the meter reading is less than 17 mV, there is insufficient power to operate the pilot safety power unit. This must be corrected either by improving the flame impingement on the thermocouple or by replacing the thermocouple.

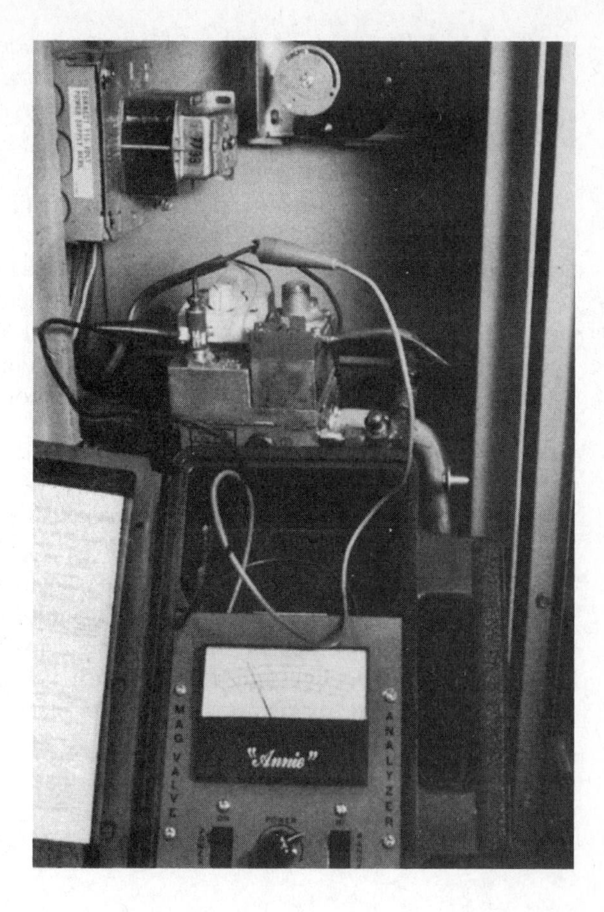

Figure 11.21 Testing a thermocouple circuit.

Burner Ignition

There are three types of gas burner ignition systems:

1. Pilot ignition system (intermittent)
2. Direct-spark ignition system
3. Silicon carbide ignition system

Pilot Ignition

The pilot ignition system (intermittent) is designed for low-voltage application on all types of residential gas-fired heating equipment.

Principle of Operation The thermostat powers the control to open the pilot burner gas valve and to provide the ignition spark simultaneously. As soon as the pilot flame is established the spark ceases, and the main burner gas valve is energized. Should the flame not be established within a predetermined period, the system provides safety shutdown. (Sparking stops, and pilot gas flow is interrupted.)

Electronic flame-sensing circuitry in the igniter detects the presence or absence of the pilot burner flame. If the flame is not established during the trial-for-ignition period, the system closes the pilot gas valve and locks out. If the burner flame is extinguished during the duty cycle, the main gas valve will close and the igniter will retry ignition before going into lockout.

Proper location of the electrode assembly is important for optimum system performance. It is recommended that there be approximately a ⅛-in. gap between the electrodes and the pilot, as shown in Figure 11.22. This figure also shows the typical retrofit wiring for a pilot ignition system. Figure 11.23 shows the ignition system used in a heat–cool application. This type of installation can be used with a two-stage gas valve.

Direct-Spark Ignition

Direct-spark ignition (DSI) does away with the pilot altogether and lights the main burner with an electric spark. Operation of the burner depends on continuous detection of the main burner flame by an electronic flame-sensing system. Unlike the intermittent pilot ignition, DSI is required to light a significantly larger gas flow. Safety demands that ignition and flame detection occur in a few seconds.

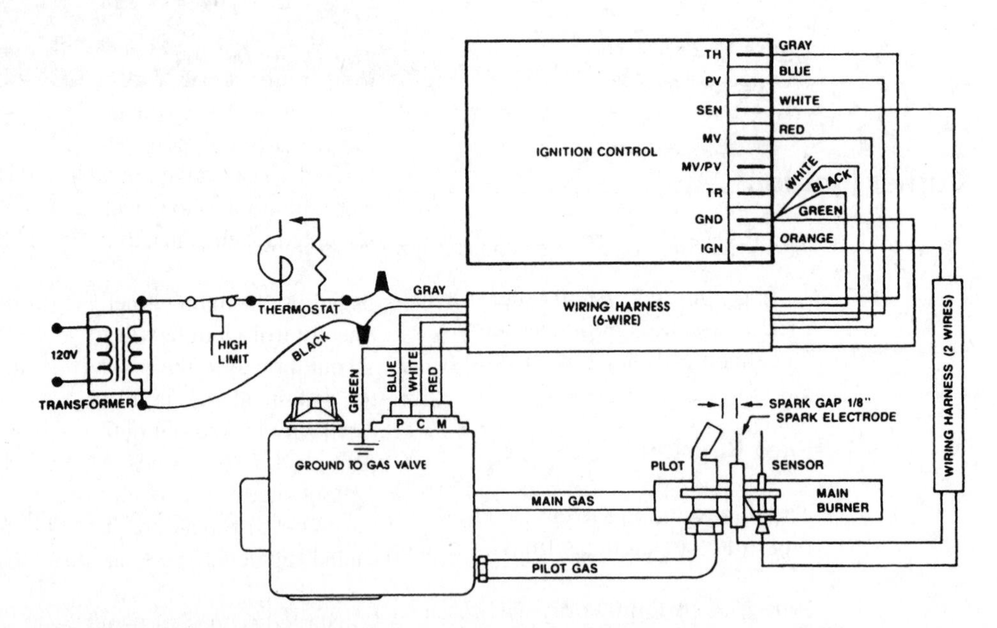

Figure 11.22 Typical retrofit wiring of an electronic ignition control. (Courtesy of Robertshaw Controls Company)

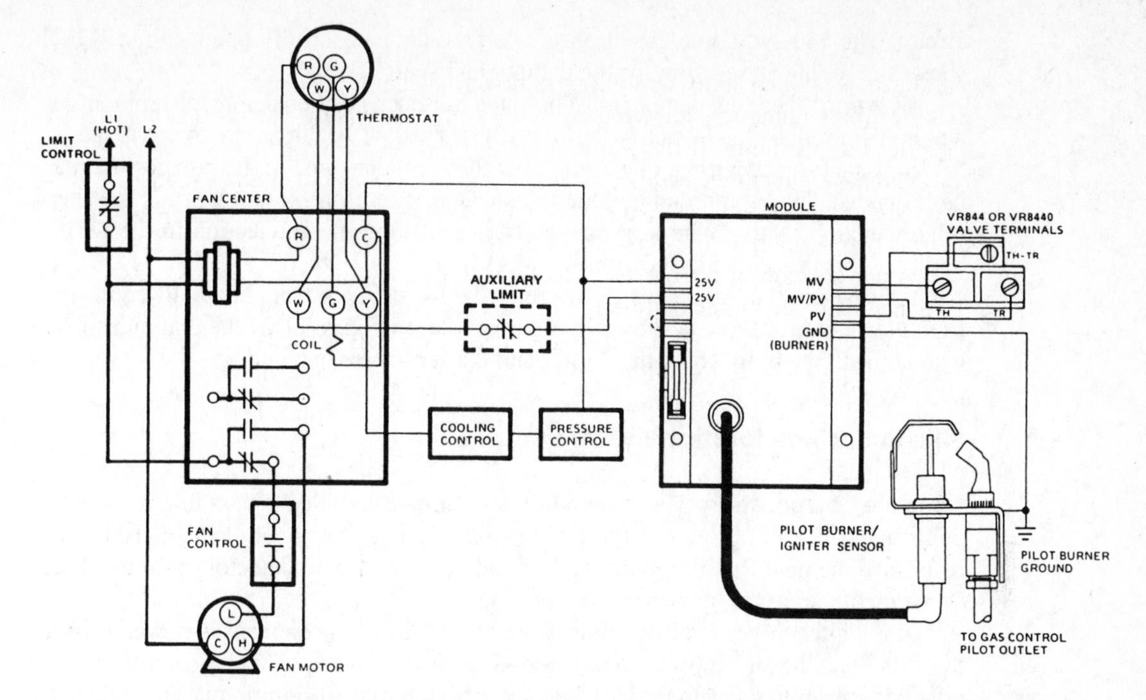

Figure 11.23 Wiring diagram for heat-cool application of a pilot spark ignition system. (Courtesy of Honeywell Inc.)

Principle of Operation If the main burner lights, a circuit is completed from the flame sensor through the flame to the burner head and ground. This current flow proves the presence of the main burner flame and resets the lockout timer and at the same time interrupts the spark ignition circuit. The gas valve remains open as long as there is a call for heat and as long as the module continues to detect the presence of the main flame. Should the current flow be interrupted (flame out), trial for ignition begins again.

If the safety lockout timing period ends before the main burner lights or before the flame sensor establishes enough current, the control module will go into safety lockout. When the module goes into safety lockout, power to the pulse generator is interrupted, the gas control circuit is interrupted, and the alarm circuit relay (if the module is so equipped) is energized. The module will stay locked out until it is reset by turning the thermostat down below room temperature for 30 seconds. Applying a DSI system is a complex engineering task best accomplished by the equipment manufacturer. Field retrofitting the DSI controls is not a practical procedure. There are, however, times when components of DSI systems need replacing. Parts are available for this kind of job.

Ignition control units are sometimes installed with vent dampers. When a vent damper is in the system, the thermostat wire is connected to one side of the *end switch* circuit. The end switch is closed only when the damper is fully open. Figure 11.24 shows the wiring connection for the damper end switch.

Warning: The vent damper must be in the full-open position before the ignition system is energized. Failure to verify this may cause a serious health hazard to occupants.

In some furnace wiring, one terminal of the secondary side of the transformer may be grounded. This could damage the transformer. Therefore, it is important to determine which gas valve wire is grounded before making the connection to the wiring harness. This procedure is shown in Figure 11.25.

Ignition and flame-sensing hardware for DSI systems is available in different configurations. One of these is shown in Figure 11.26. In Figure 11.27 the diagram shows a combined system in which the igniter and the sensor are one unit.

Silicon Carbide Ignition System

Principle of Operation The system utilizes a silicon carbide element that performs a dual function of ignition and flame detection. The igniter is an electrically heated resistance element that thermally ignites the gas. The flame detector circuit utilizes flame rectification for monitoring the gas flame.

The igniter serves two functions. On a call for heat, the element is powered from the 120-V ac line and allowed to heat for 45 seconds (typical). Then, the main valve is powered, permitting gas to flow to the burner for the trial-for-ignition period, which is

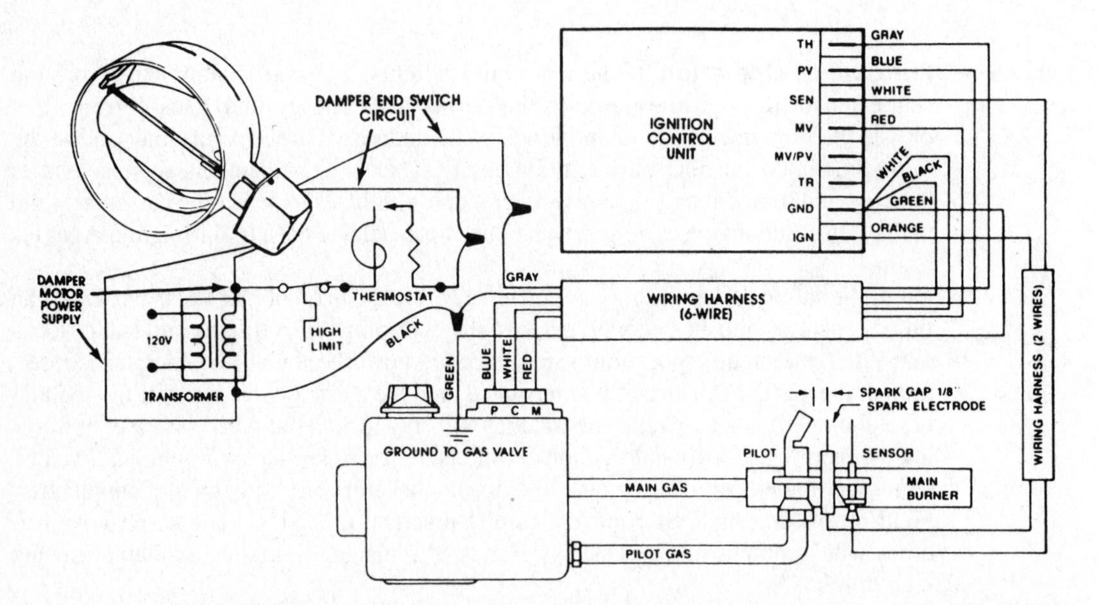

Figure 11.24 Wiring diagram for installation of ignition system with flue damper. (Courtesy of Robertshaw Controls Company)

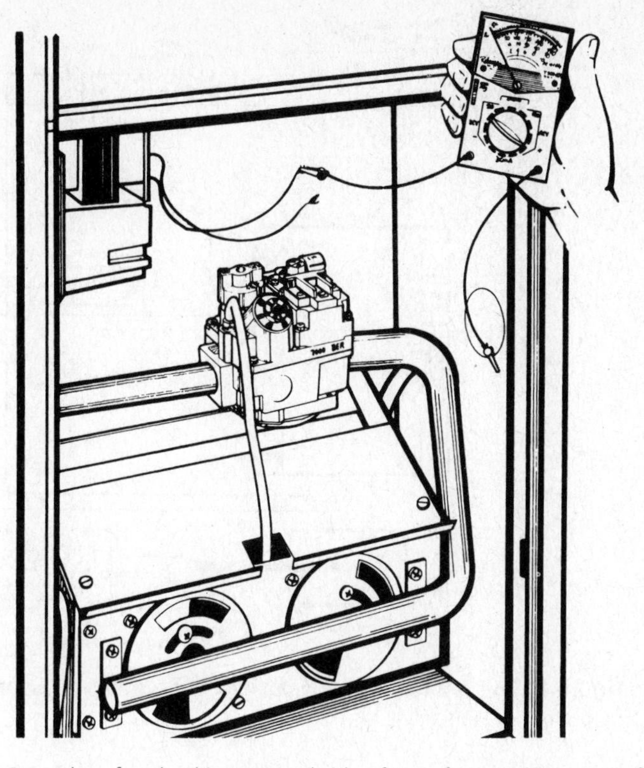

Figure 11.25 Procedure for checking ground side of transformer. (Courtesy of Robertshaw Controls Company)

typically 7 seconds long, although other timings are available. At the end of this period, the igniter is switched from its heating function to that of a flame probe that checks for the presence of flame. If flame is present, the system will monitor it and hold the main valve open. If flame is not established within the trial-for-ignition period, the system will lock out, closing the main valve and shutting off power to the igniter. If a loss of flame occurs during the heating cycle, the system will recycle through the ignition sequence.

Silicon Carbide Ignition Proper location of the silicon carbide igniter is important for optimum system performance. It is recommended that the igniter be temporarily mounted using clamps or other suitable means so that the system can be tested before it is permanently mounted. The igniter should be located so that its tip extends ¼ in. through the center of the flame and about ½ in. above the base of the flame, as shown in Figure 11.28. Wiring diagrams for the silicon carbide systems are shown in Figures 11.29 and 11.30.

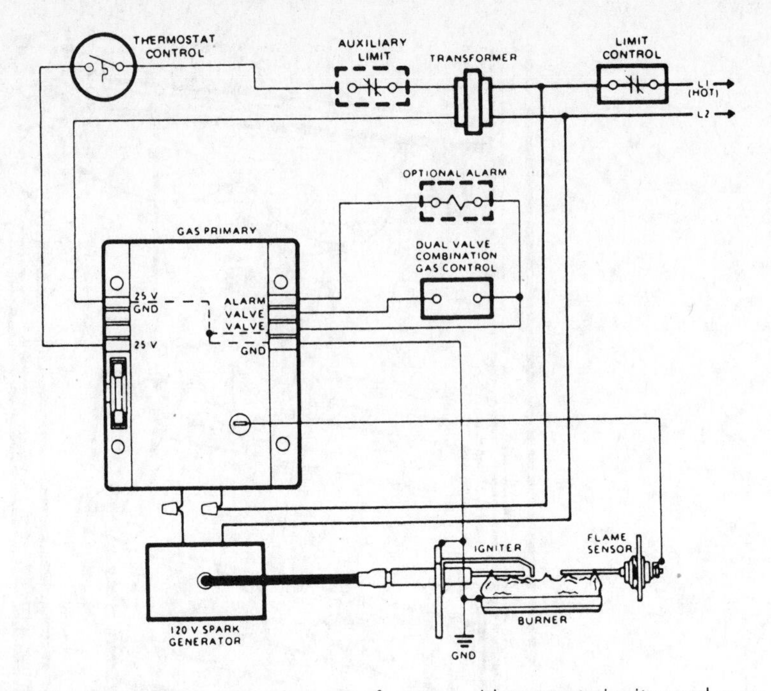

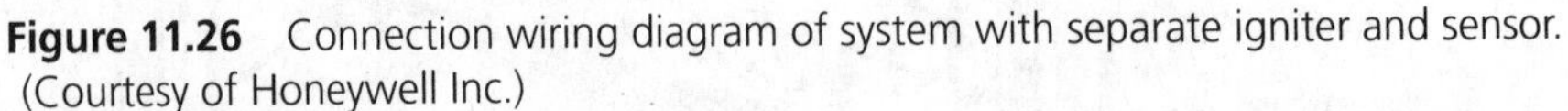

Figure 11.26 Connection wiring diagram of system with separate igniter and sensor. (Courtesy of Honeywell Inc.)

Operation and Troubleshooting of a Honeywell Universal Intermittent Pilot Module S8610U

The Honeywell S8610U module in Figure 11.31 is a general replacement control for many pilot spark ignition modules used on furnaces today. It provides ignition sequence, flame monitoring, and safety shutoff for central heating systems.

Sequence of Operation for the Honeywell S8610U

1. On a call for heat the heating thermostat closes from R to W and brings 24 V to the TH-W terminal on the S8610U control module.
2. 24 V energizes the pilot valve from the PV and MV/PV terminals on the S8610U control module. At the same time 13,000 V is fed to the igniter from the SPARK terminal on the S8610U control module.
3. When the pilot lights, the flame rectification system proves the pilot by sending a micro amp signal (minimum of 1 μA) to the SENSE terminal on the S8610U.
4. After the pilot has been proven, 24 V energizes the main gas valve from terminals MV and MV/PV on the S8610U control module.
5. When the thermostat is satisfied, its contacts open, and the pilot and main gas valve close, ending the burner cycle.

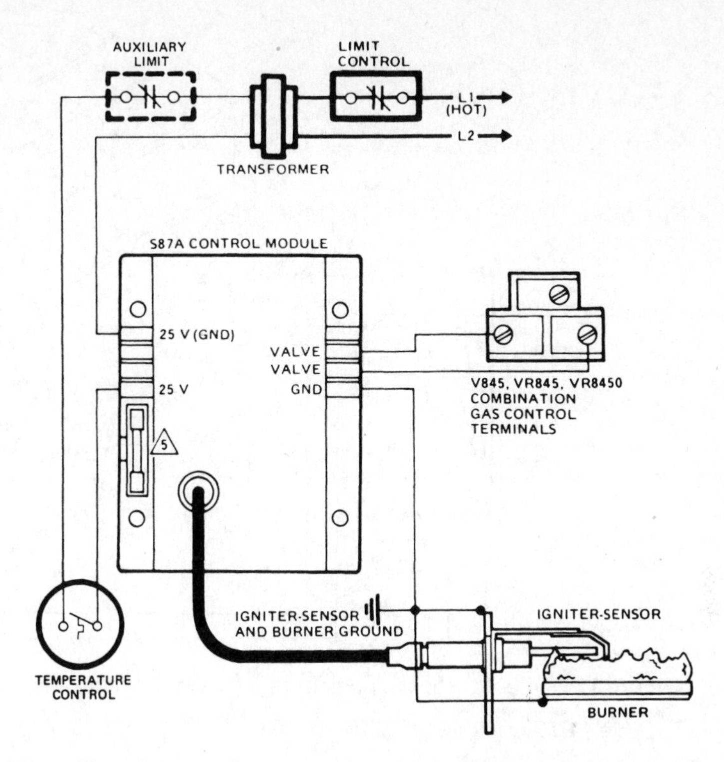

Figure 11.27 Wiring diagram of system with combined igniter–sensor. (Courtesy of Honeywell Inc.)

Operation of a Johnson G779 Universal Intermittent Pilot Ignition Control

The Johnson G779 control (see Figure 11.32) replaces many existing intermittent pilot ignition controls made by various manufacturers. It is a safety control designed for indirect ignition systems that provides precise timing functions, draft-tolerant burner supervision, ignition retries, and a choice of separate or integral spark and sensor combination.

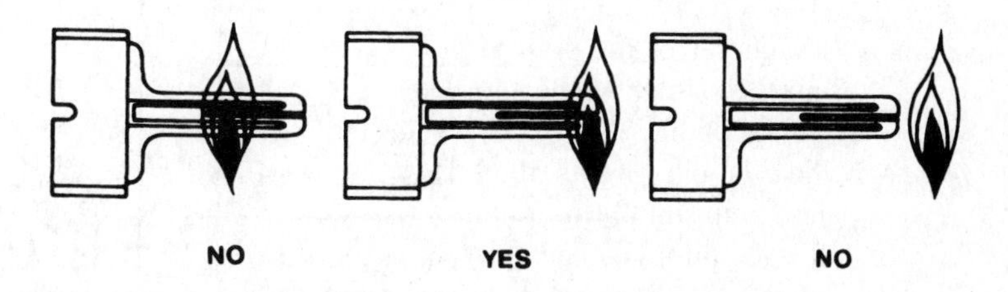

Figure 11.28 Proper location of silicon carbide igniter element in relation to flame. (Courtesy of Fenwal Incorporated)

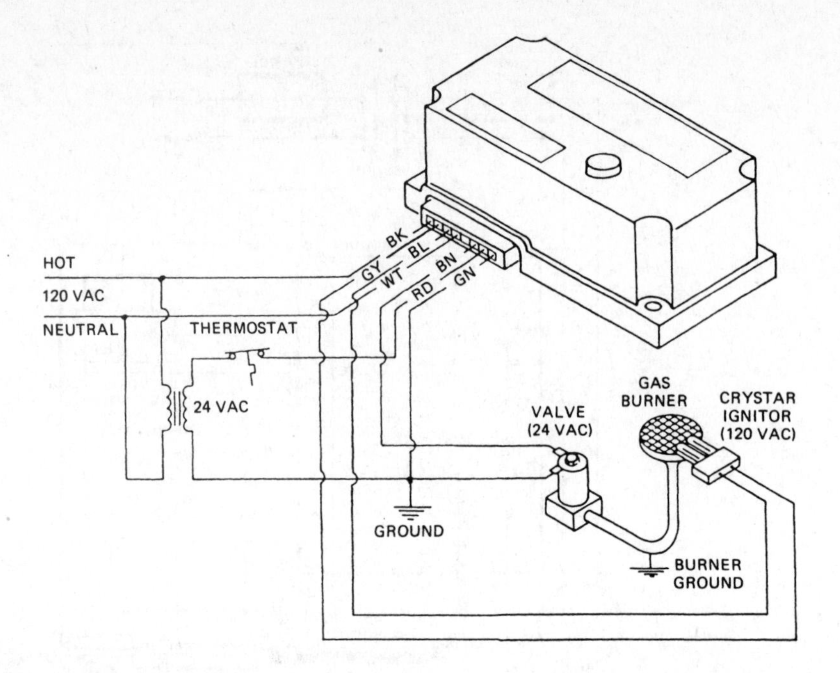

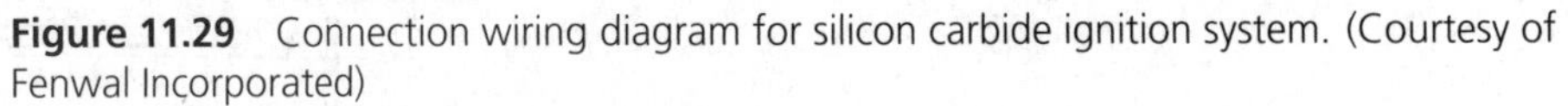
Figure 11.29 Connection wiring diagram for silicon carbide ignition system. (Courtesy of Fenwal Incorporated)

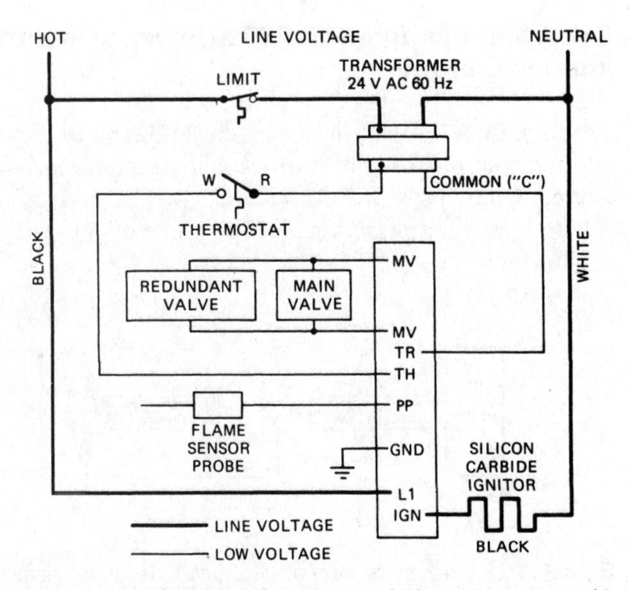

Figure 11.30 Schematic wiring diagram for silicon carbide igniter system. (Courtesy of White–Rodgers Division, Emerson Electric Co.)

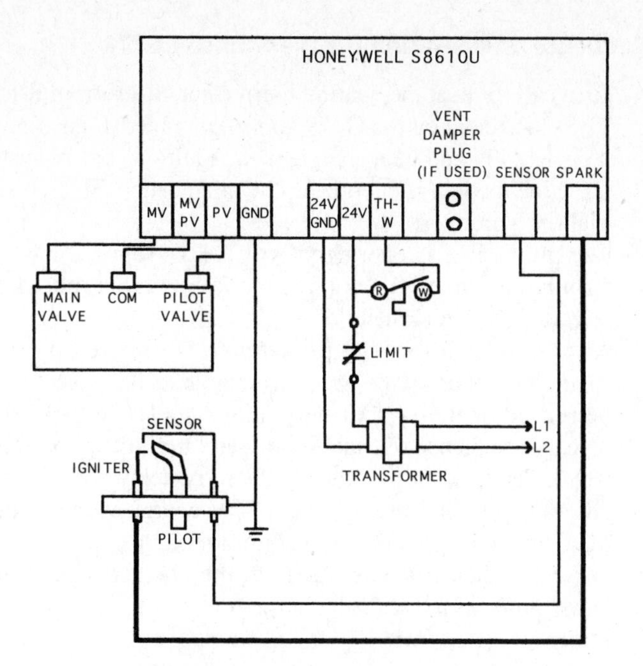

Figure 11.31 Honeywell general replacement module. (Courtesy of Honeywell Inc.)

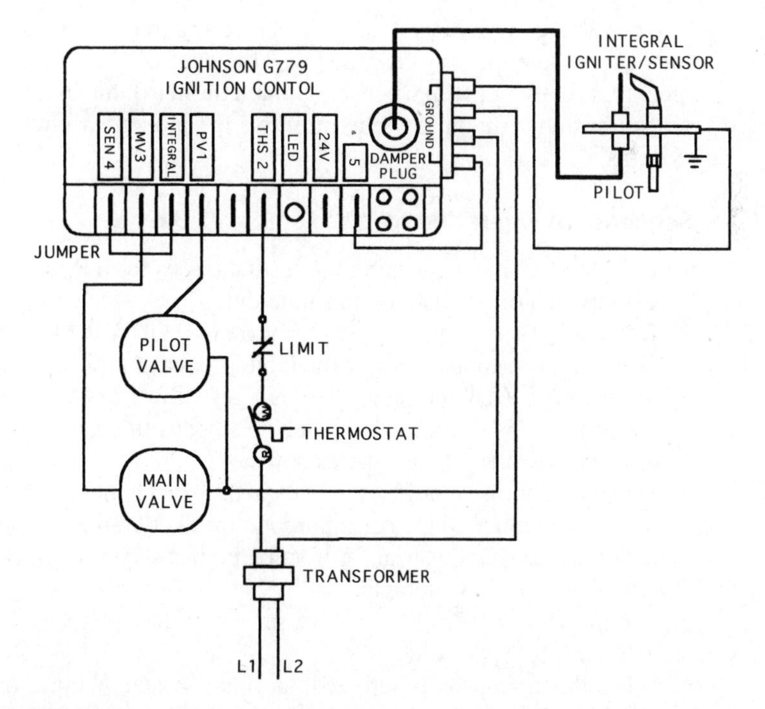

Figure 11.32 Johnson G779 Universal intermittent pilot ignition control system. (Courtesy of Johnson)

Sequence of Operation for the Johnson G779

1. On a call for heat the heating thermostat closes from R to W and brings 24 V to the THS-2 terminal on the G779 control module. (If an automatic vent damper is used, it opens fully and then energizes the ignition control system.)
2. Once the module is energized, the diagnostic LED lights, and the control begins its trial for ignition.
3. The pilot valve is energized with 24 V from the PV1 terminal and the Ground terminal on the G779 control module. At this time, the igniter is energized for 3 seconds to ignite the pilot gas.
4. After pilot ignition, the pilot flame is detected by the flame sensor (separate spark/sensor) or by the igniter (integral spark/sensor). A minimum of 0.3 μA must be sensed or the system will enter into 100% lockout. At this time the system repeats a sequence of trial–delay–retry until the pilot lights or the thermostat is satisfied. The LED flashes during the retry/delay period.
5. After the pilot has been proven, 24 V energizes the main gas valve from terminals MV3 and No. 5 on the G779 control module.
6. When the thermostat is satisfied, the contacts open, and the pilot and main gas valve close, which ends the burner cycle.

Operation of a Fenwal 05-31 Hot Surface Ignition System

The Fenwal 05-31 HSI system control (see Figures 11.33 and 11.34) is widely used for many residential furnaces today. This control uses a patented hot surface element to directly ignite and sense the main burner flame of a gas fired system. It provides ignition sequence, flame monitoring, and safety shutoff for central heating systems.

Sequence of Operation for the Fenwal 05-31

1. On a call for heat the heating thermostat closes from R to W and brings 24 V to the TH terminal on the 05-31 control module.
2. Once the TH terminal has been powered by 24 V, the hot surface igniter is energized for 20 seconds to allow the igniter to reach ignition temperature.
3. Twenty-four V is brought to the main gas valve from terminals MV1 and MV2 on the Fenwal 05-31 control module. This opens the gas valve, the igniter lights the gas, and the main burners are established.
4. After main burner ignition, the Fenwal 05-31 switches the hot surface element from the ignition mode to the sensing mode for the duration of the cycle. If the flame is not sensed during this trial for ignition period, the system goes into a lockout mode after three ignition attempts.
5. A minimum of 0.75 μA must be sensed by the sensor igniter or the system will enter into 100% lockout.
6. When the thermostat is satisfied, terminals R and W open, and the main gas valve closes, shutting down the burner cycle.

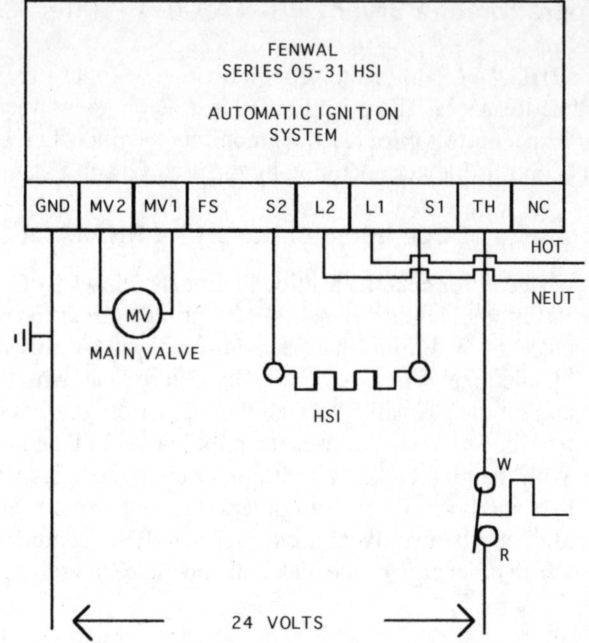

Figure 11.33 Local sensing wiring diagram. (Courtesy of Fenwal Incorporated)

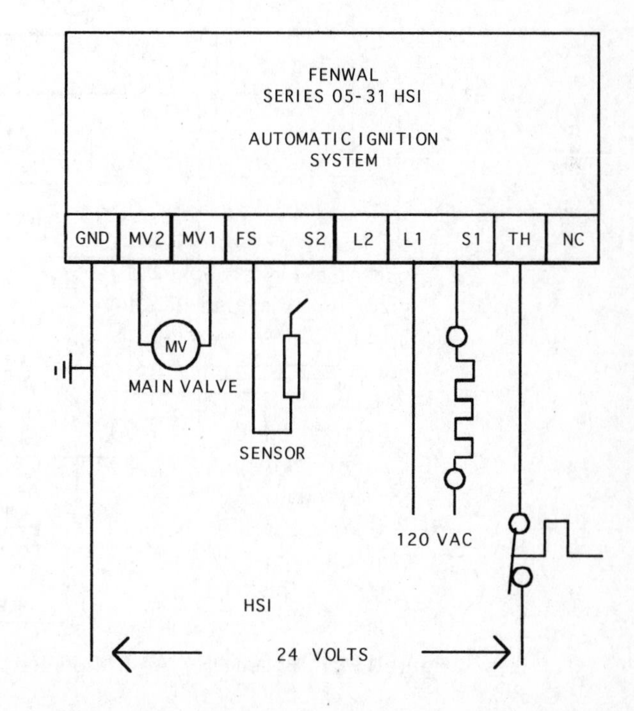

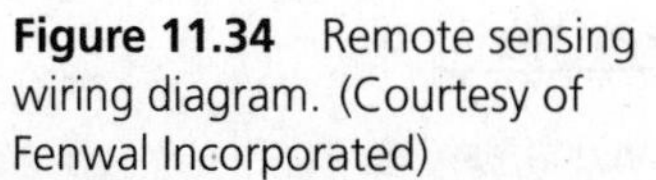
Figure 11.34 Remote sensing wiring diagram. (Courtesy of Fenwal Incorporated)

Operation of a Bryant HHA4AA-011 Printed Circuit Board

The HHA4AA-011 printed circuit board (see Figure 11.35) is used only on Bryant and Carrier furnaces. The circuit board is used on an indirect ignition system that on a call for heat, controls pilot ignition, monitors the pilot flame, energizes the main gas valve, and controls blower operation by the use of a solid-state control circuit.

Sequence of Operation for the Bryant HHA4AA-011 Printed Circuit Board

1. On a call for heat the heating thermostat closes from R to W and brings 24 V to the W terminal on the HHA4AA-011 printed circuit board.
2. Once the W terminal has been powered by 24 V, this starts the sequence of operation by energizing the GAS1 and GAS2 terminals with 24 V which energizes the pick coil on the gas valve through the NC set of contacts on the bimetal pilot sensor. At the same time, it energizes the pilot igniter and the hold coil on the gas valve.
3. With the pick coil and pilot igniter energized, gas flows to the pilot and is ignited by a spark from the pilot igniter. The pilot flame heats the bimetal pilot sensor, and after approximately 45 seconds, a set of NC contacts changes to the open position, which de-energizes the pick coil and the pilot igniter. The hold coil stays energized,

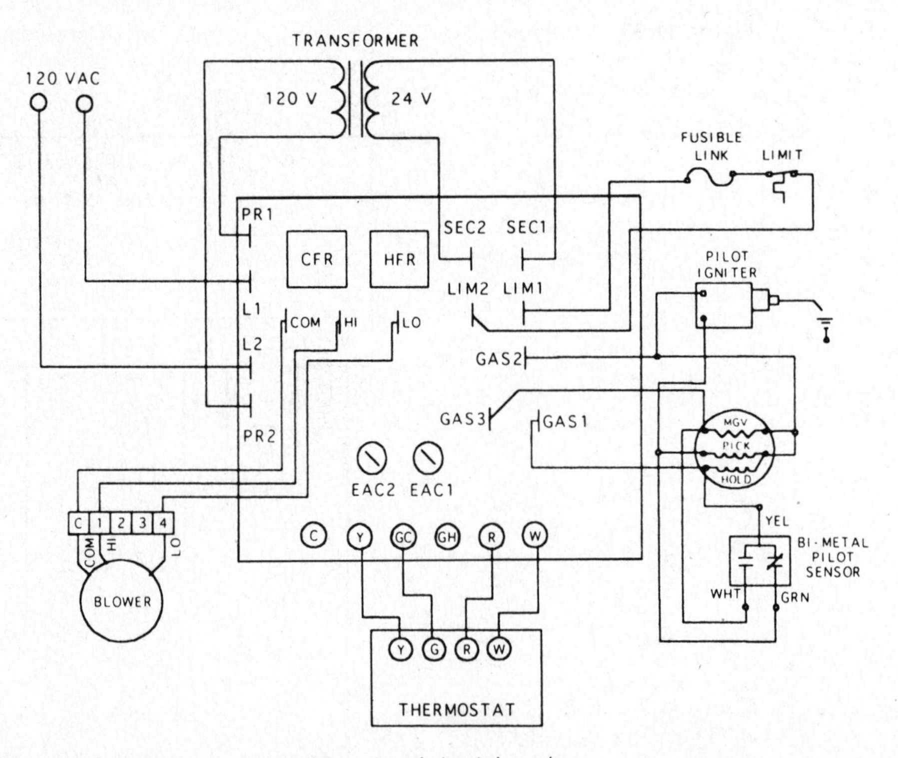

Figure 11.35 Bryant HHA4AA-011 printed circuit board.

since it is directly controlled from the GAS1 terminal and is not controlled by the bimetal flame sensor. The hold coil keeps the pilot valve open, which in turn keeps the pilot lit and ready for the ignition of the main gas burners.

4. The bimetal pilot sensor also has a set of NO contacts that close once the pilot flame is sensed, approximately 45 seconds after pilot ignition. This proves the existence of the pilot and energizes the main gas valve. The main gas valve feeds gas to the main burners, and the gas is ignited by the pilot.
5. When the main gas valve is energized, 24 V is sent to terminal GAS3 from the main gas valve, which starts the solid-state timed delay circuit. After 45 seconds, the fan energizes on the heating speed.
6. When the thermostat is satisfied, terminals R and W open, which de-energizes the (hold) pilot valve and the main gas valve. The blower operates for 45 seconds and then shuts down, completing the cycle.

White–Rodgers Mercury Flame Sensor

The White–Rodgers mercury flame sensor (see Figure 11.36) is used on an indirect ignition system that on a call for heat, controls pilot ignition, monitors the pilot flame, and energizes the main gas valve.

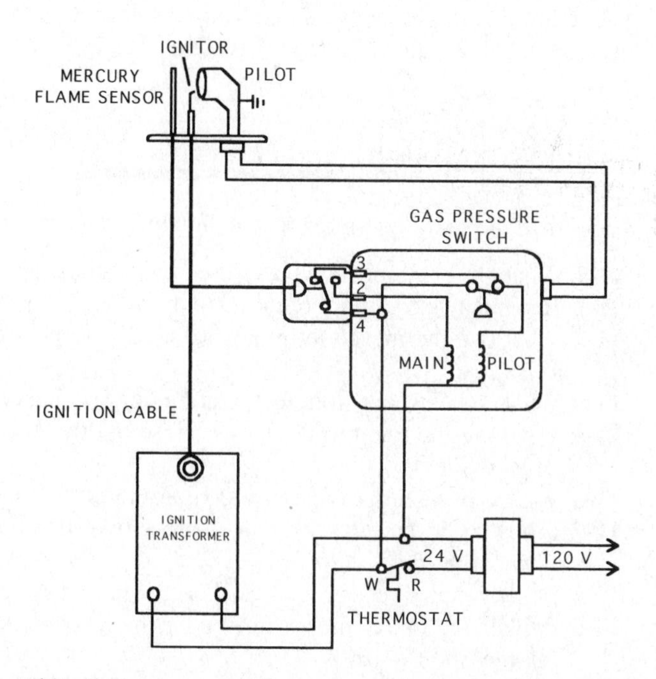

Figure 11.36 White–Rodgers mercury flame sensor. (Courtesy of White–Rodgers Division, Emerson Electric Co.)

Sequence of Operation for the White–Rodgers Mercury Flame Sensor Ignition System

1. On a call for heat the heating thermostat closes from R to W and brings 24 V to the ignition transformer, which creates a spark for the ignition of the pilot gas. At the same time, current flows to the No. 4 terminal on the mercury flame sensor. From terminal No. 4, current flows to terminal No. 3, through a closed set of contacts. From terminal No. 3 current flows to the pilot valve, which opens the valve and allows gas to flow to the pilot.
2. The flow of gas to the pilot causes the gas pressure switch to close and the igniter to light the pilot gas.
3. The flame from the pilot heats the sensor and causes the mercury to vaporize. This applies pressure on the diaphragm of the sensor, causing the normally open (hot) contacts to close and energizing the main gas valve. The NC (cold) contacts open, breaking the circuit to the pilot valve coil. The gas pressure switch is in parallel with the NC (cold) contacts, which keeps the pilot valve energized during burner operation.
4. When the thermostat is satisfied, terminals R and W open, which de-energizes the pilot valve, the main gas valve, and the ignition transformer, ending the burner cycle.

Honeywell Integrated Control System

A description of the Honeywell one- and two-stage SmartValve integrated control systems is found in Chapter 12, High-Efficiency Furnaces.

STUDY QUESTIONS

Answers to the study questions may be found in the sections noted in brackets.

11-1 What power supply should be used on a standard gas furnace? *[Power Supply]*

11-2 How many circuits are required to operate a two-speed fan? *[Fan]*

11-3 What two thermostat terminals must close to provide a call for heating? *[Gas Valve]*

11-4 What is the usual cutout temperature of a secondary limit control? *[Gas Valve]*

11-5 What are the two terminals that close on the thermostat to call for cooling? *[Accessories]*

11-6 What type of furnace uses an auxiliary limit (LA) control? *[Gas Valve]*

11-7 In testing a fan relay, what voltage should be applied to the fan relay coil? *[Typical Wiring Diagrams]*

11-8 Describe the three types of gas burner ignition systems. *[Burner Ignition]*

11-9 Explain the function of a furnace having an electronic diagnostic light system. *[Diagnostic Control]*

11-10 In an electronic diagnostic system, if the LED light flashes three times, what is the problem? *[Diagnostic Control]*

CHAPTER 12

High-Efficiency Furnaces

OBJECTIVES

After studying this chapter, the student will be able to:

- Select a furnace based on its energy requirements
- Read a wiring diagram for a high-efficiency furnace and determine the sequence of operation
- Install and service furnaces with energy-saving components
- Troubleshoot a solid-state control system

78 to 80% AFUE Furnaces

The minimum requirement for an energy-saving furnace is 78% annual fuel utilization efficiency (AFUE). The testing procedure was developed by the U.S. Department of Energy (DOE). All furnace manufacturers make units in the 78 to 80% AFUE category. Furnaces with efficiencies as high as 96 to 97% AFUE, known as *high-efficiency furnaces* (HEFs), also are available. In many areas the increased cost of an HEF can be justified in view of the additional fuel saving.

In this chapter we study both the "78–80s" and the HEF units to see how these efficiencies are achieved. One of the important functions of the heating–cooling technician is to assure that these ratings are maintained not only in the original installation but also throughout the life of the equipment.

Design Characteristics

There is considerable similarity in various manufacturers' models, although there are individual differences that may be important to the customer. A typical unit is shown in Figure 12.1. Some of the principal components are identified:

Induced draft blower (Figure 12.2)

Tubular heat exchanger (Figure 12.3)

1. Integrated Furnace Control
2. Induced Draft Blower
3. Ignition System
4. Gas Valve and Manifold
5. In-shot Burners
6. Heat Exchanger
7. High Efficiency Blower Unit
8. Easy Access Washble Filter

Figure 12.1 Cutaway view of typical 80% AFUE furnace. (Courtesy of Rheem MFG. Co., Air Conditioning Division)

Hot surface igniter (Figure 12.4)

In-shot gas burner (Figure 12.5)

Solid-state control board (Figure 12.6)

Components

A description of the essential parts of these units, as supplied by various manufacturers, follows.

Figure 12.2 Induced draft blower. (Courtesy of Rheem MFG. Co., Air Conditioning Division)

Heat Exchangers The heat exchangers (HXs) are all single pass, with a separate HX for each burner, assembled together into a single part. These units carry an extended warranty ranging from 10 years to the lifetime of the original purchaser. One manufacturer indicates that its HX units have more space between sections than those of their competitors, thus offering less air resistance to the blower.

Gas Burner Lighting Arrangement There are basically five types of gas burner lighting arrangements: standing pilot, pilot spark ignition, pilot hot surface ignition, direct spark ignition, and direct hot surface ignition.

Controls Most units are equipped with solid-state control boards, which when defective are replaced as a single unit. These units can also be diagnostic, to assist in troubleshooting. There are various ways to control the blower. Continuous blower operation is always an option. For heating, a timed fan start is the usual arrangement. Termination is generally by temperature. A limit control is always provided to protect against overheating in case the fan does not start.

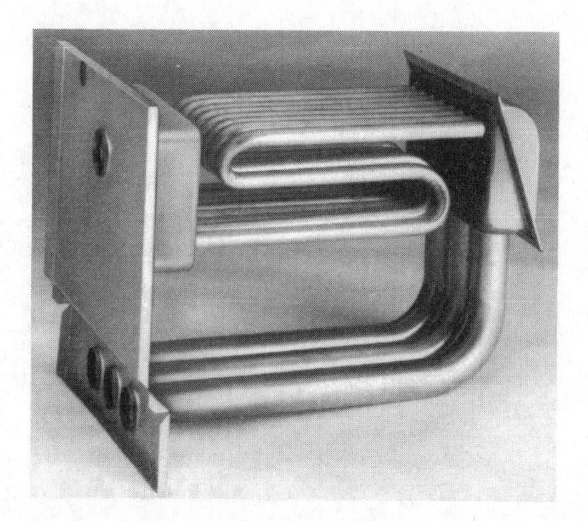

Figure 12.3 Tubular heat exchanger. (Courtesy of Rheem MFG. Co., Air Conditioning Division)

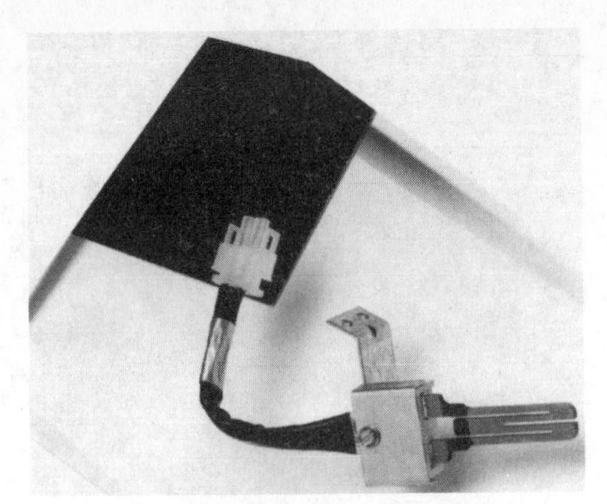

Figure 12.4 Hot-surface igniter. (Courtesy of Rheem MFG. Co., Air Conditioning Division)

Figure 12.5 In-shot gas burner. (Courtesy of Rheem MFG. Co., Air Conditioning Division)

Figure 12.6 Solid-state control board. (Courtesy of Rheem MFG. Co., Air Conditioning Division)

Gas Valves Some units have slow-opening gas valves to prevent noisy startup, whereas others are fast opening.

Draft Arrangements The majority of units are equipped with induced draft fans that require a differential pressure switch to prove that the fan is running. One manufacturer does not use an induced draft fan, but its units are equipped with a draft diverter.

Gas Burners The in-shot burner is the most popular type, although a number of manufacturers use the multiple-port type. A rollout safety switch is provided to close off the gas supply in case the flue is blocked.

Control Sequence

A typical control sequence is shown in Figures 12.7 to 12.13. The wiring diagrams have been simplified for easier understanding. The legend for the wiring diagrams is as follows:

ALS	Auxiliary limit switch
BLWM	Blower motor
CFR	Cooling blower relay
DSS	Draft safeguard switch
F	Fuse
FL	Fusible link
GV	Gas valve
HFR	Heating blower relay
IDM	Induced draft motor
IDR	Induced draft relay
LS	Limit switch
PRS	Pressure switch
TRAN	24-V transformer

The various stages of operation are illustrated in the following figures. The descriptions of the action can be traced on the diagrams.

Ready to Start (Figure 12.7) The unit is in the ready-to-start mode. The door interlock switch is closed, and power is available at three electrical devices:

1. L_1 and L_2 are connected to the line side of the transformer, which is energized.
2. One side of the low-voltage power is connected to the heating blower relay.
3. One side of the low-voltage power is connected to the induced draft relay through the fusible link and limit switches.

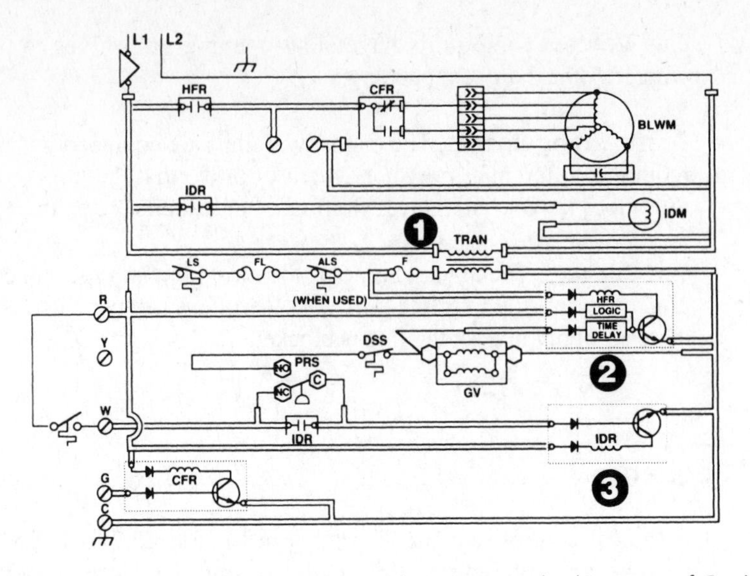

Figure 12.7 Wiring diagram for furnace in ready-to-start mode. (Courtesy of Carrier Corporation)

Call for Heat, IDR Energized (Figure 12.8) On a call for heat (1) the R and W contacts on the thermostat close, (2) connecting the other side of the low-voltage power to the induced draft relay, (3) through the NC pressure switch contacts.

Call for Heat, IDM Energized (Figure 12.9) Note that the induced draft relay has two sets of NO contacts:

1. One set of contacts close to form a holding circuit for the induced draft relay.
2. The other set of contacts close to start the induced draft motor.

Call for Heat, Gas Valve Energized (Figure 12.10) Assuming that the vent system is free from blockage or restriction and that the inducer fan is operating satisfactorily:

1. The pressure switch senses the required difference in pressure and switch positions.
2. The induced draft relay stays energized through its holding circuit.
3. With the pressure switch in its new position, power is supplied to the gas valve through the closed draft safeguard switch. If the pilot is burning satisfactorily, the gas valve opens and the burners light.

Call for Heat, Fan Motor Energized (Figure 12.11)

1. At the same time the power is supplied to the gas valve, power is also applied to the time-delay circuit of the heating blower relay. After 45 seconds the contacts of the heating relay close.
2. This applies power to the selected winding of the blower motor through the NC cooling blower relay contacts, and the blower starts.

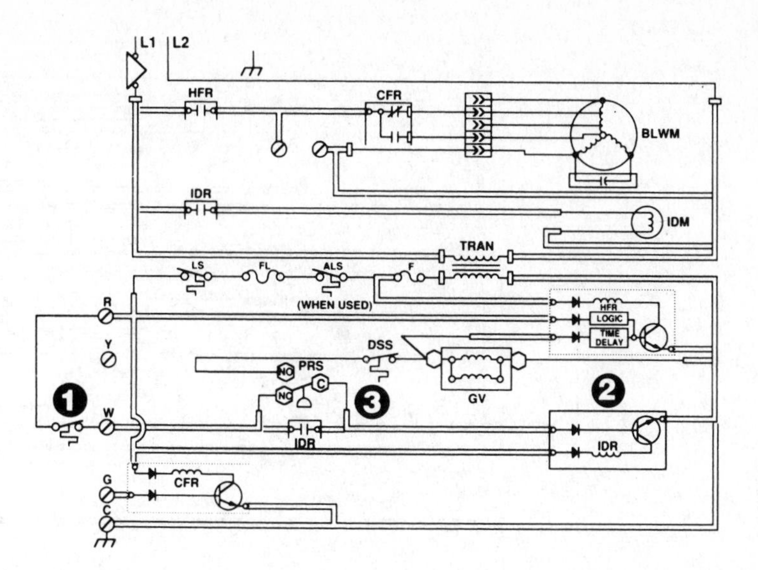

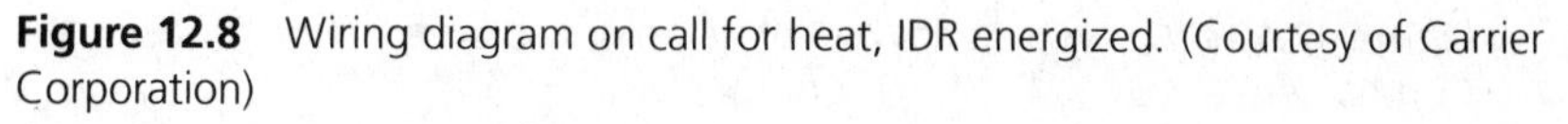

Figure 12.8 Wiring diagram on call for heat, IDR energized. (Courtesy of Carrier Corporation)

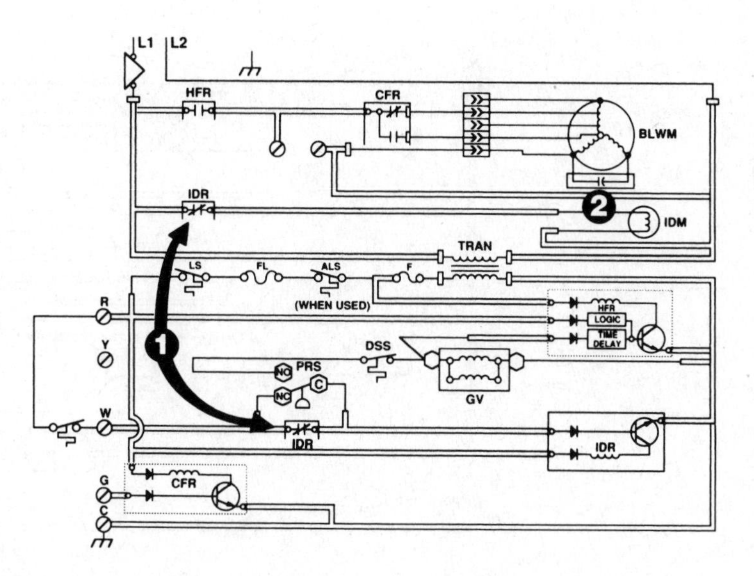

Figure 12.9 Wiring diagram on call for heat, IDM energized. (Courtesy of Carrier Corporation)

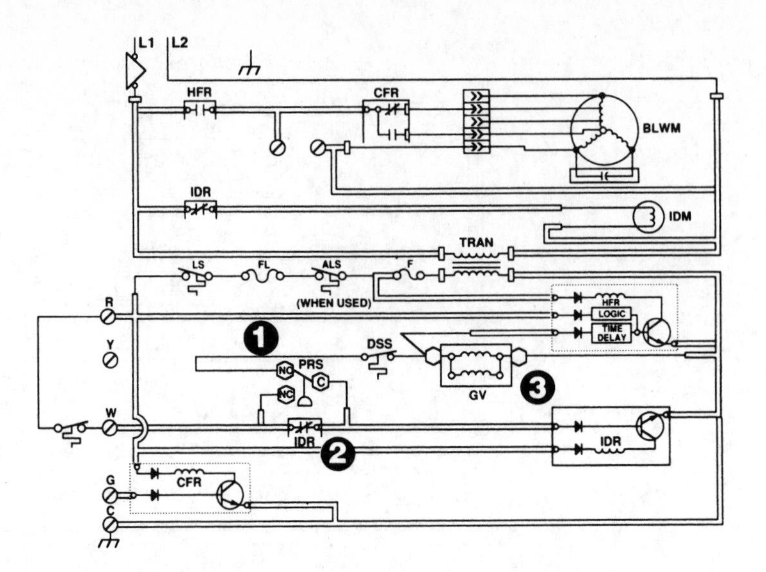

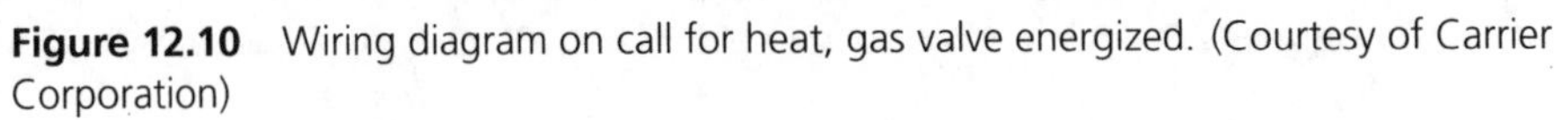
Figure 12.10 Wiring diagram on call for heat, gas valve energized. (Courtesy of Carrier Corporation)

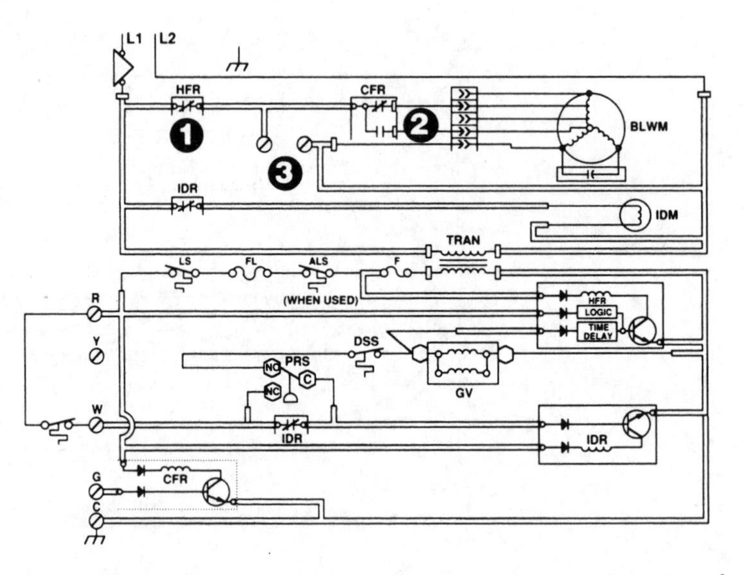

Figure 12.11 Wiring diagram on call for heat, fan motor energized. (Courtesy of Carrier Corporation)

3. If an electronic air filter is installed, power is available at the screw terminals. It is energized at the same time as the blower motor. Its load cannot exceed 1 A at 115 V.

Thermostat Satisfied (Figure 12.12) The furnace will continue to run until the thermostat is satisfied, at which point the gas valve is immediately de-energized, shutting off the burners. The blower motor continues to run for 120 seconds before shutting off. The blower-off-delay resistor on the circuit board can be clipped off to provide a 180-second run time.

Call for Cooling (Figure 12.13) On a call for cooling, the R-Y and R-G circuits are energized. The Y terminal is not used. It can be used, however, as a connection point for the outdoor condensing unit relay. The action that takes place when these circuits are energized is as follows:

1. The cooling blower relay is energized, closing the relay contacts that supply power to the blower motor speed circuit for cooling. The electronic air filter can also be activated at the same time.
2. The heating blower relay is also energized, closing its contacts, which provide a power path to the cooling blower relay.
3. Power is supplied to the condensing unit relay, which starts the compressor.

Continuous Blower Operation If continuous blower operation is used, the furnace operates as follows:

1. On cooling, the fan will run continuously at the higher cooling speed.
2. On heating, the fan will run continuously at the higher cooling speed, then on a call for heat, the blower shuts off for 45 seconds to allow the heat exchangers to warm up. After the warm-up period the blower will run at the speed selected for heating. When the thermostat is satisfied, the blower will continue to run for 120 or 180 seconds at the heating speed, then it will resume continuous operation at the higher cooling speed.

Figure 12.12 Thermostat satisfied. (Courtesy of Carrier Corporation)

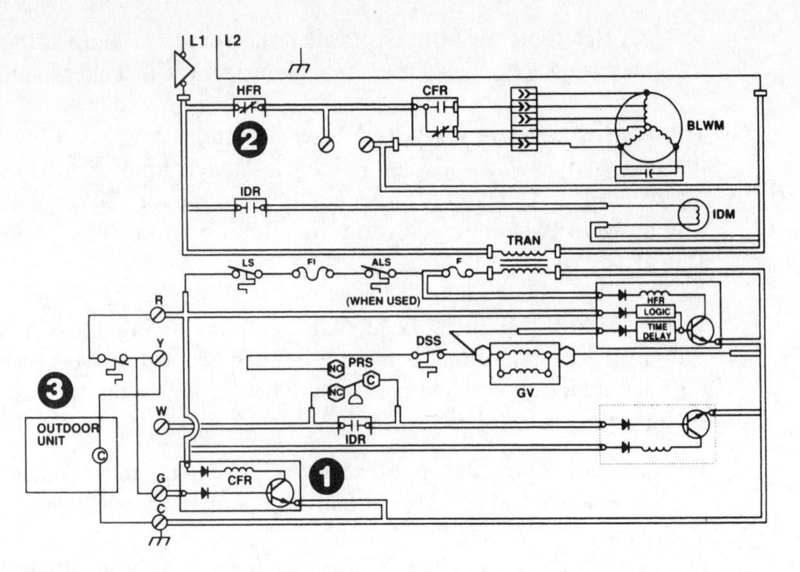

Figure 12.13 Wiring diagram on a call for cooling. (Courtesy of Carrier Corporation)

Characteristics of High-Efficiency Furnaces

The energy crisis has encouraged manufacturers of gas furnaces to design equipment that exceeds the usual 80% efficiency. The ability of high-efficiency furnaces (HEFs) to achieve better performance is based on two qualities:

1. *Lower flue gas temperature.* The furnaces extract more heat from the flue gas by lowering its temperature and bringing it nearer to ambient temperature before exhausting the flue gas.
2. *Condensation of the water vapor in the flue gas.* Condensation results in the recovery of the latent heat of vaporization. Almost 1000 Btu is released per pound of water condensed.

Components common to the various designs include the following:

1. *Induced draft fan.* This fan propels the flue gas mechanically through the heat-exchanger passages. On the pulse furnace, the induced draft fan serves only as a purge device and is used only in the precombustion and postcombustion cycles.
2. *Secondary heat-exchanger surface.* Such a surface permits removal of more heat from the flue gases than with only the normal primary heat exchanger. The basic heating surface reduces the flue gas temperature to around 500°F, and the secondary surface brings it down to around 150°F or lower.
3. *Condensate drain.* This drain is required to remove the products of combustion that condense at the lower flue gas temperatures. This condensate consists mainly of water vapor but can contain other substances, due to impurities in the gas, and can be corrosive. Therefore, it must be handled properly to prevent damage caused by improper piping or an inadequate disposal arrangement.

4. *Flue pipe design.* Due to the lower temperature and the reduced volume of flue gas, such materials as PVC plastic piping can be used for the vent stack. This feature has an advantage in the application of the equipment, simplifying the structural requirements.

One possible disadvantage of HEFs may be their limited capacities. Most of these units do not exceed 115,000 Btuh output. Due to the improved construction of most homes and the practice of more accurate furnace sizing, however, the smaller-capacity furnaces are adequate for many larger homes. The use of two furnaces is usually a good solution where more capacity is required.

Types of Designs

Generally speaking, the industry has created three distinctive HEF designs. All these designs have AFUE ratings in excess of 80%, meaning that the ratio of output to input is in the range of 80 to 97%. These designs include the following:

1. Recuperative design
2. Condensing design
3. Pulse design

Recuperative Design

The typical recuperative design operates in the range of 80 to 85% AFUE and is so called because it recoups some of the heat used for venting by employing a mechanical vent fan. The additional heat is often removed in an extra heat exchanger section that does not contain a burner. All vent gases from the other sections must pass through the extra section before being vented out of the furnace. Some manufacturers do not use an extra section but instead have a special design that causes the vent gases to remain longer in the heat exchanger. This design again extracts more heat and lowers the temperature of the vent gases. Figure 12.14 shows a typical recuperative design. Operation of this furnace is similar to that of the condensing design, described next.

Condensing Design

An illustration of a typical condensing design is shown in Figure 12.15. The figure lists the various components. With reference to the figure, this furnace operates as follows: Natural gas (or LP) is metered through a redundant gas valve (1) and is burned using monoport in-shot burners (2). Combustion takes place in the lower portion of the primary heat exchangers (3). The hot gases then pass into a collection chamber and through a stainless steel duct into a secondary heat exchanger (4). This heat exchanger has stainless steel tubes with turbulators and exterior aluminum fins. Combustion gases are drawn through the secondary heat exchanger by an induced draft fan (5) and exhausted to the vent stack (6). Moisture that condenses out of the flue gas, is collected in the drain assembly (7), and is directed to the floor drain.

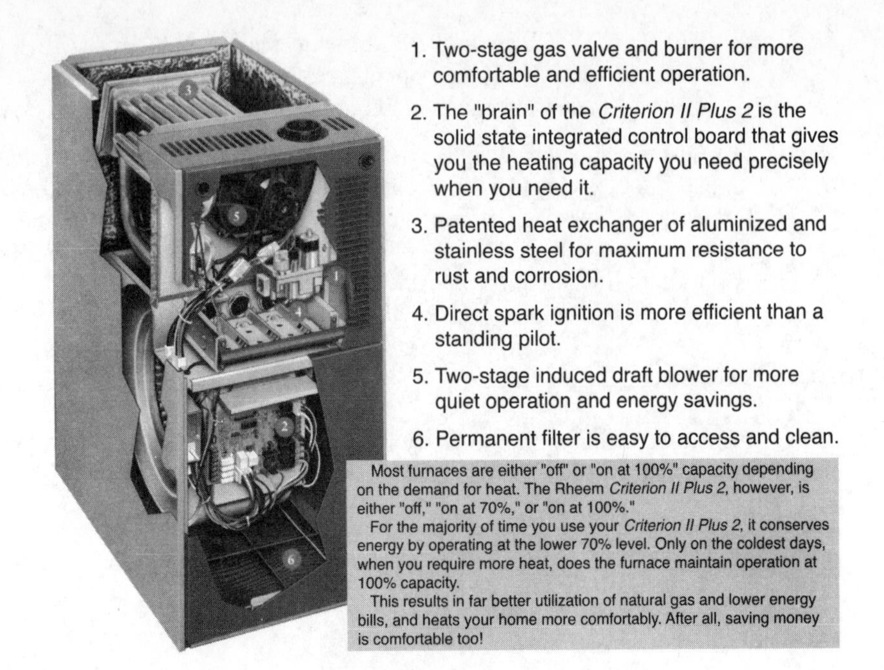

Figure 12.14 Typical recuperative design. (Courtesy of Rheem MFG. Co., Air Conditioning Division)

The multispeed direct-drive fan (8) delivers return air first over the secondary heating surface (4) and then over the primary surface (3) to the plenum outlet. Electronic ignition eliminates the need for a continuous pilot. The cabinet is insulated with fiberglass. Another illustration of a condensing design is shown in Figure 12.16.

Theory of Operation (Combustion) The efficiency of this furnace is obtained through three basic processes that are not found on standard-efficiency gas furnaces:

1. Very accurate control of the combustion process by metering of the volume of gas and fresh air entering the combustion chamber
2. The use of a secondary heat exchanger to absorb the residual heat from the flue gases that are leaving the primary exchanger
3. The introduction of air from outdoors to the combustion area of the furnace rather than from the immediate area of the furnace installation

1. REDUNDANT GAS VALVE
2. MONOPORT INSHOT BURNERS
3. ALUMINIZED, PRIMARY HEAT EXCHANGER
4. SECONDARY CONDENSING HEAT EXCHANGER
5. INDUCED DRAFT FAN
6. PVC VENT OUTLET
7. CONDENSATE DRAIN TRAP
8. MULTISPEED DIRECT-DRIVE MOTOR

Figure 12.15 Exposed view of condensing-type high-efficiency gas-fired furnace. (Courtesy of Bryant Air Conditioning)

Let's look closely at each of these processes and see how they work closely in unison to achieve the highest possible efficiencies.

Sequence of Operation (Combustion) Refer to Figure 12.17. As the furnace is firing, air is drawn through the combustion supply air tube (A) into the combustion chamber. The chamber is sealed (in fact, must be sealed) and operates at a slight negative pressure. A baffle ensures that each burner receives an adequate share of both primary and secondary combustion air. The ratio of gas to primary air is controlled by the conventional method of slotted shutters (B) between the gas orifice and burner tube venturi. The wider the shutter is opened, the greater the proportion of the mixture will be air. Each burner fires into a cell of the heat exchanger at an input of 20,000 or 25,000 Btuh,

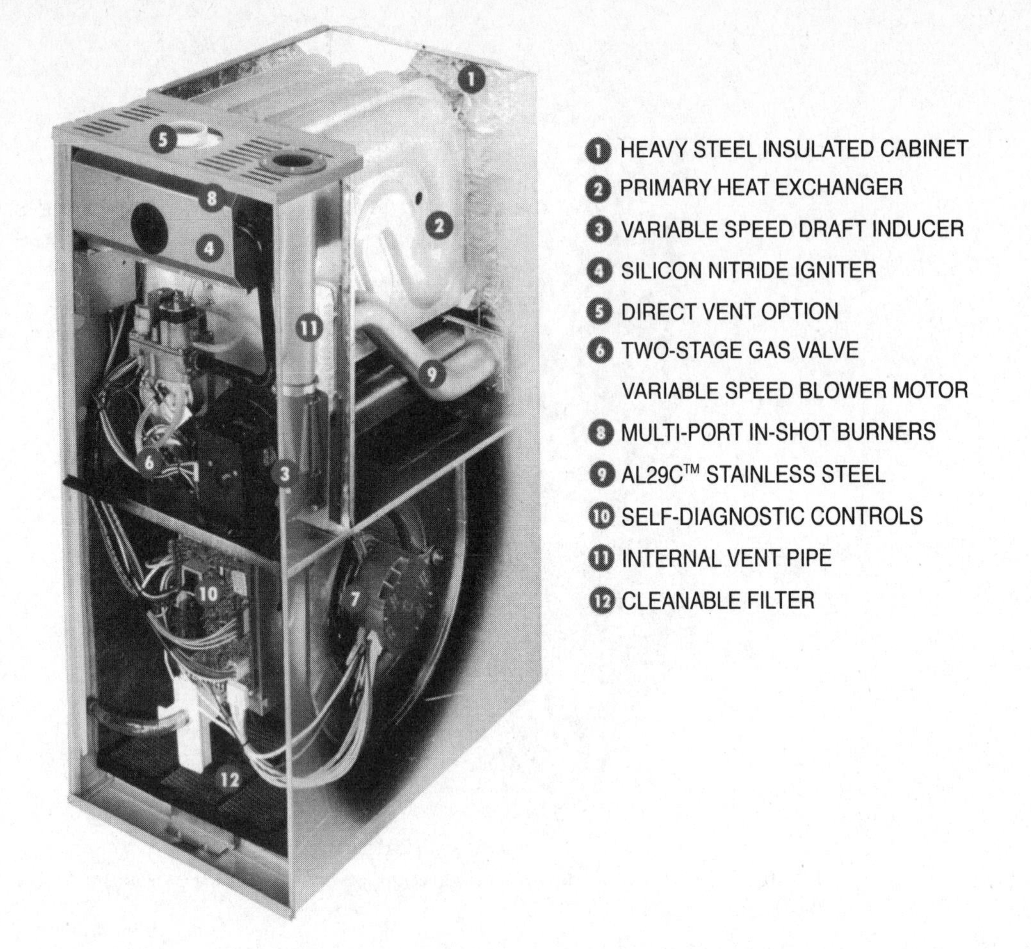

Figure 12.16 Cutaway view of condensing-type high-efficiency gas-fired furnace. (Courtesy of Trane Company Unitary Products Group)

depending on the capacity of the furnace. This is the primary heat exchanger (C) and is fabricated of aluminized steel, utilizing the weldless, one-piece process of construction. The operation of the furnace at this point is essentially the same as that of a conventional furnace. Airflow across the exchanger takes heat to the conditioned space, and the combustion products exit the main exchanger into the collector (D) through a series of baffles. These baffles slow the exit of the combustion gases so that more heat is drawn from them before they leave the primary heat exchanger.

The similarity in operation between a standard furnace and a condensing furnace ends here. The exiting flue gases would leave the conventional furnace and go up the flue at a temperature of 350 to 550°F, wasting a significant amount of heat. The condensing gas furnace utilizes this heat by routing the combustion gases to the secondary heat exchanger.

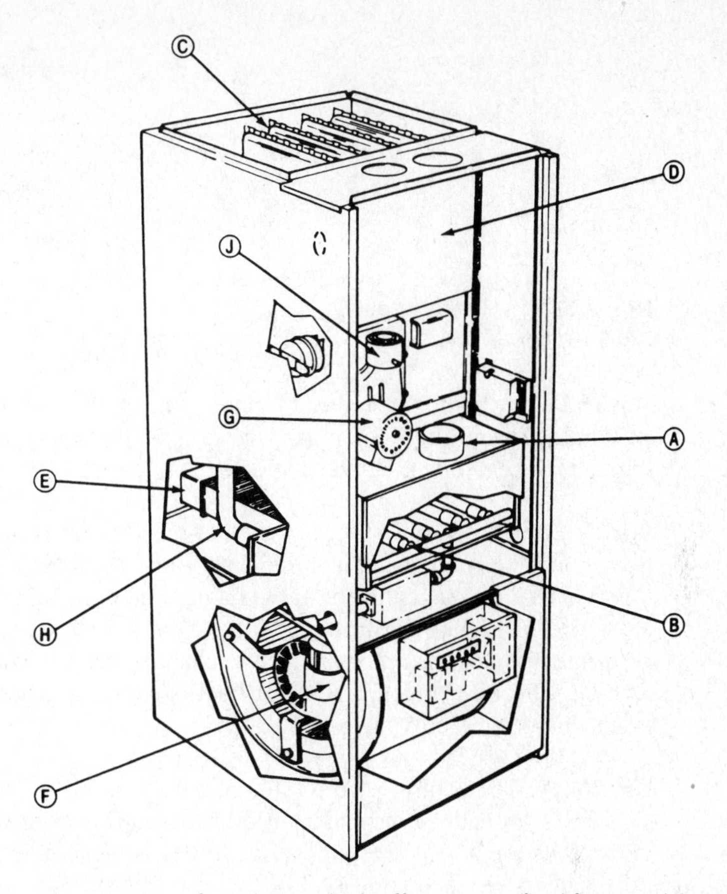

Figure 12.17 Cutaway view of condensing high-efficiency gas-fired furnace. (Courtesy of ARCOAIRE Air Conditioning & Heating)

This secondary exchanger (E) removes heat from the gases but also takes latent heat from the water vapor in the combustion products. This causes the water vapor to condense, hence the term *condensing furnace* and the need for a drainage system. The liquids that condense in the secondary exchanger are not all as harmless as water. Several dilute but powerful acids are also produced. This is the reason for the use of 29-4C stainless and high-temperature plastics in all areas of the furnace that are exposed to condensed combustion products. Even though these materials are expensive to buy and manufacture, they are essential to long product life and to the furnace's safety and reliability.

The combustion products now have little heat left in them. The remaining liquids from condensation are drained from the secondary exchanger to the drain trap/collector (F). The cooled (140 to 200°F) gases are drawn to the induced draft combustion blower (power ventor) (G) via a discharge tube (H) and through a restrictor. The restrictor controls the volume of gases drawn through the entire combustion system.

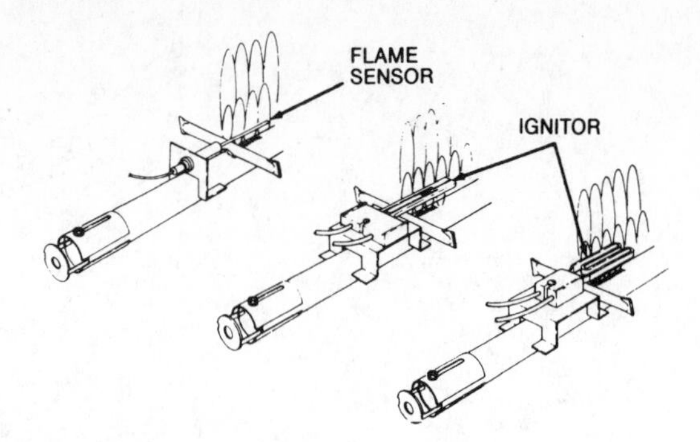

Figure 12.18 Slotted burners.
(Courtesy of Carrier Corporation)

Any remaining liquids are routed from the combustion blower to the drain trap/collector and drained from there to an external drain. The remaining gases exit the furnace through the vent pipe (J) to the outdoors.

Burners and Burner Orifices A slotted-tube burner is used in this furnace. These burners have certain flame requirements. Burner flame should run the length of the tube and be 3 to 4 in. in height. If the flame should lift from the burner, turn orange in color, or start rolling out from under the flash burner access shield after ignition, an adjustment is necessary, or the heat exchanger needs to be inspected and possibly cleaned. Checking the burner flame should give an indication of the need for further inspection (see Figures 12.18 and 12.19).

Gas Valve The control used on this furnace is a redundant combination gas valve. This design provides a manual main valve, servo-regulated automatic main diaphragm valve, plus the added safety of an additional in-line electromechanically operated redundant valve (see Figure 12.20).

Pressure Regulator A pressure tap has been provided on the gas valve (see Figure 12.20). Attach a pressure gauge fitting to this tap and check the outlet pressure. The pressure should be set as indicated on the rating plate. To adjust the pressure regulator, remove the cap covering the pressure regulator–adjusting screw. On valves with more than one adjusting screw, adjust only the screw marked "HI." Turn the adjusting screw out (counterclockwise) to decrease the pressure, and in (clockwise) to increase the pressure. Reinstall the cap securely.

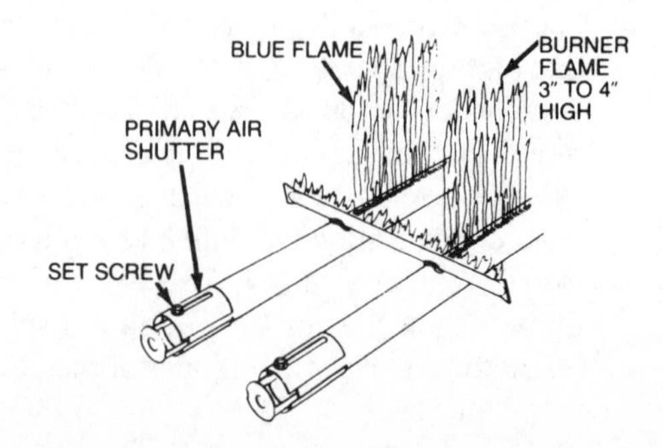

Figure 12.19 Burner assembly.
(Courtesy of ARCOAIRE Air Conditioning & Heating)

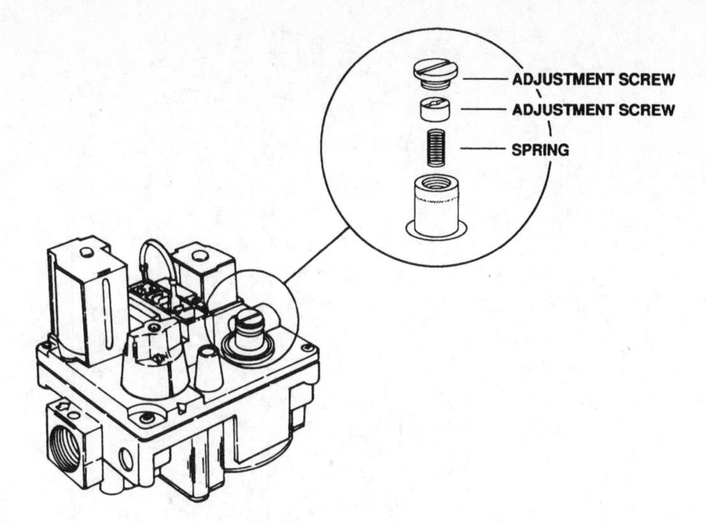

Figure 12.20 Redundant gas valve. (Courtesy of ARCOAIRE Air Conditioning & Heating)

Sequence of Operation (General) Let's follow the operation of the furnace through a demand for heat (refer to Figures 12.21 and 12.22). The room thermostat calls for heat, closing a circuit between terminals R and W. Inside the control center, voltage is applied to the heat relay coil (P), completing the W and C low-voltage circuit. The contacts close and supply line voltage to the induced draft motor.

The pressure switch (D) proves that the combustion blower is operating and closes on a rise in pressure differential across the combustion blower. This sends 24 V to the NC limit switch of the combination fan and limit control (E). If the limit is closed, voltage becomes available to a condensate safety sensor (G). This sensor ensures that the secondary heat exchanger is not filled with condensate and that the fresh-air intake and vent pipes are not blocked. In effect, it verifies airflow through the entire combustion system of the furnace.

Sequence of Operation (Electrical) Refer to Figures 12.21 and 12.22. The furnace requires an external 115-V power source of the ampacity stated on the nameplate. A junction box (F) for connection to line voltage is located at the upper left-hand front corner of the furnace. Line voltage is converted to 24-V control voltage by the furnace's transformer (A), located in the control center in the blower compartment. Neither line nor control voltage will be available unless the door interlock switch (B) is closed.

To operate the furnace as a heating unit only, a 24-V thermostat is the only external control required. This thermostat would be connected to the R and W terminals of the terminal strip (C). For operation with a heating and cooling unit or if continuous indoor blower operation is required, a combination heating/cooling thermostat is required. In this application R to G will energize the blower at high speed.

Terminal C is 24 V common; clock-type thermostats and an air-conditioning unit would require use of terminal C. Terminal Y is simply a binding post to connect the Y lead from the thermostat to the Y lead of the condensing unit and is not electrically

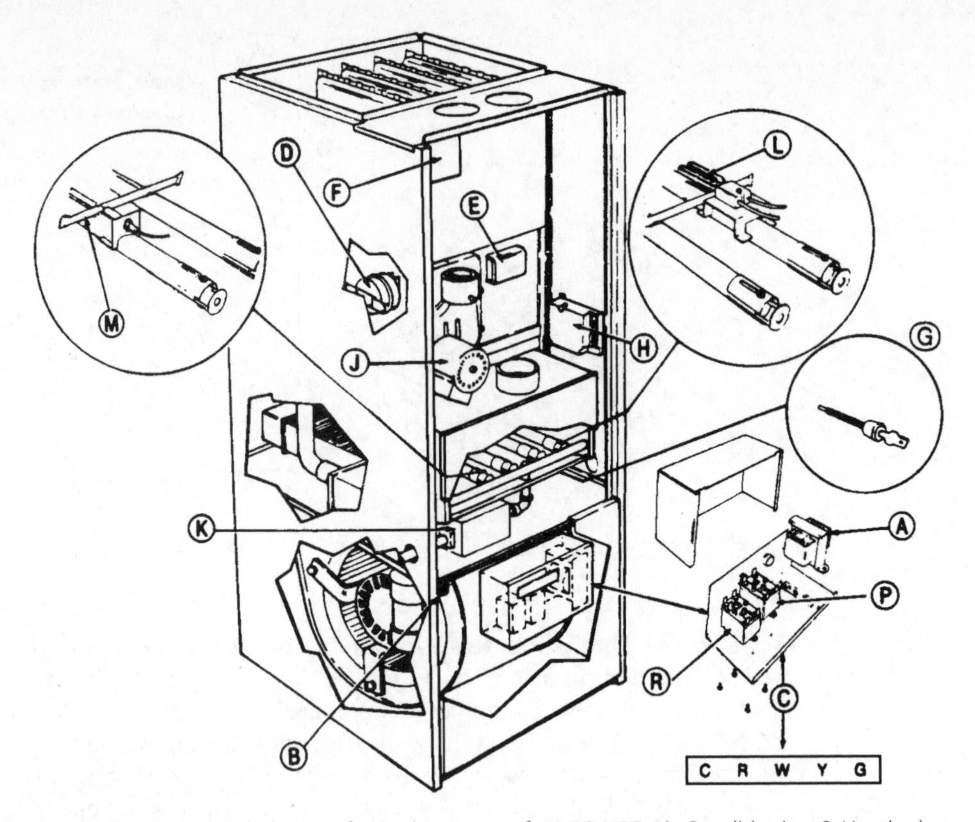

Figure 12.21 Exploded view of HEF. (Courtesy of ARCOAIRE Air Conditioning & Heating)

connected to the furnace controls in any way. Under no circumstances should the R and C terminals be connected to each other. Such action constitutes abuse and will damage the furnace controls. This damage will render the furnace transformer inoperable and will not be covered under warranty. The furnace transformer is capable of handling all internal furnace functions, and any recommended thermostat and any control voltage requirements of any matching approved split-system air-conditioning unit or heat pump. Heat pumps require the application of an "add-on" kit to interlock controls.

When this series of controls are all closed (the entire sequence occurs in a few seconds), 24 V is available to the ignition module (H), which then starts the ignition sequence. The module requires a 30-second prepurge from the time it receives power. This means that the combustion blower (J) will clear both heat exchangers and vent pipes of all gases (other than fresh air) before ignition can take place. **This is extremely important.** Igniting residual combustible vapors in a sealed combustion system can cause severe equipment damage.

When the prepurge cycle is complete, the ignition module (H) closes a 115-V circuit to the igniter (L) and allows approximately 15 seconds for the igniter to warm up. The igniter will now be glowing almost white-hot over the right-hand burner. The gas valve (K) receives 24 V from the module.

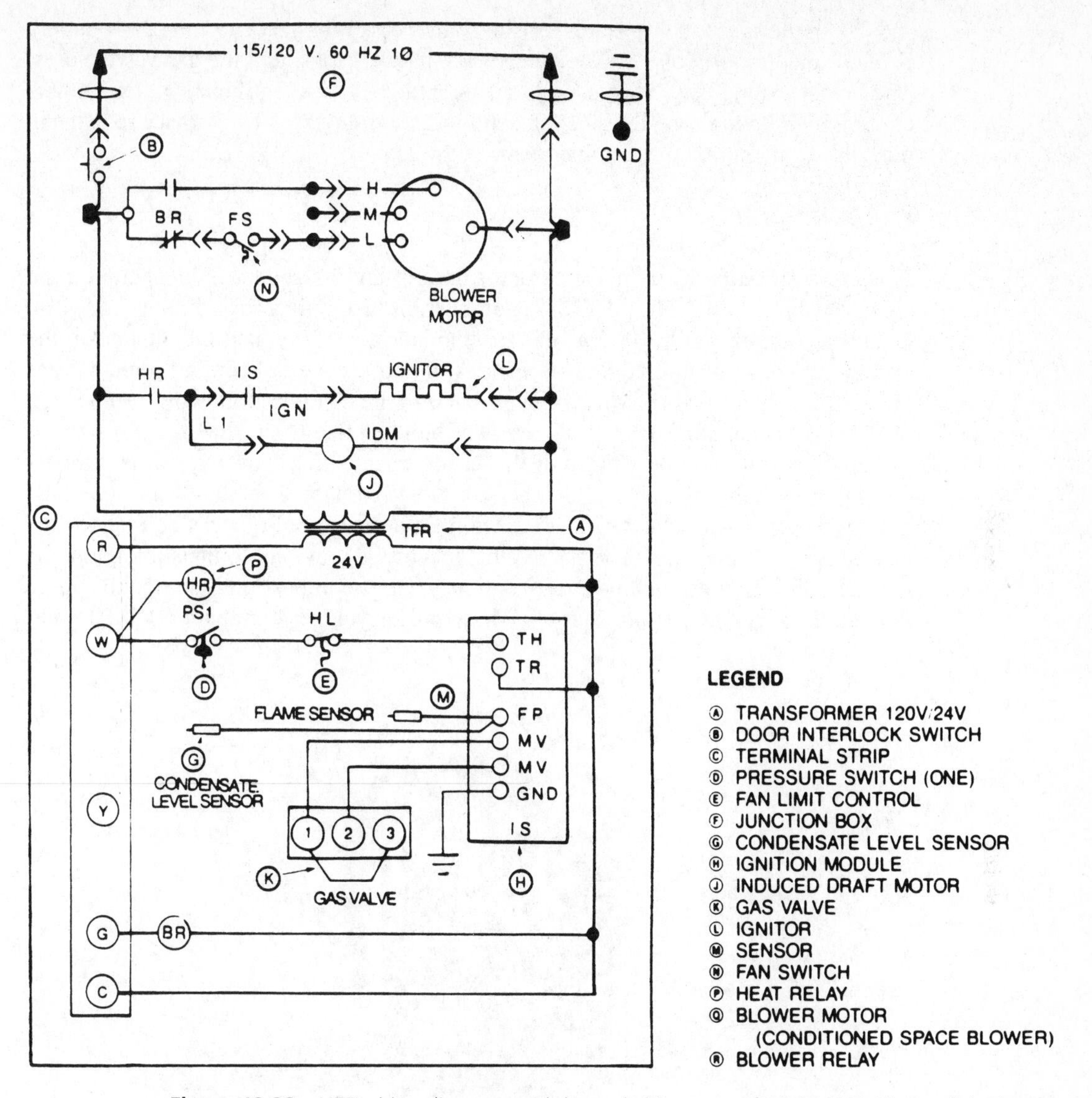

Figure 12.22 HEF wiring diagram, with legend. (Courtesy of ARCOAIRE Air Conditioning & Heating)

Ignition Occurs The sensor (M), located over the left-hand burner, must sense flame in 7 seconds or less; if it does not, the module will shut the gas valve off and retry the ignition sequence twice more before locking out. As the heat exchangers begin to heat up, the fan switch of the fan and limit control (N) closes, causing the main blower to run at heating speed (selected by the NC contacts of the blower relay), bringing air from the conditioned space across the secondary heat exchanger and the primary heat exchanger, then returning it to the conditioned space.

When the space thermostat is satisfied and the call for heat ends, the 24-V supply is interrupted to terminal W. The combustion blower stops, and the gas valve closes. The burner extinguishes. The main blower continues to run until the heat exchangers cool off and the fan switch of the fan and limit control opens. The furnace remains in this condition until there is another demand for heat.

Pulse Design

The essential parts of the pulse combustion unit and its sequence of operation are shown in Figures 12.23 and 12.24. Combustion takes place in a finned, cast-iron chamber. Note in the lower part of this chamber the gas intake, the air intake, the spark plug ignition, and the flame sensor. At the time of initial combustion the spark plug ignites the gas–air mixture. The pressure in the chamber following combustion closes the gas and air intakes. The pressure buildup also forces the hot gases out of the combustion chamber, through the tailpipe, into the heat-exchanger exhaust decoupler, and into the heat-exchanger coil. As the chamber empties after combustion its pressure becomes negative, drawing in a new supply of gas and air for the next pulse of combustion. The flame remnants of the previous combustion ignite the new gas–air mixtures, and the cycle continues. Once combustion is started the purge fan and the spark igniter are shut off. These pulse cycles occur about 60 to 70 times a second.

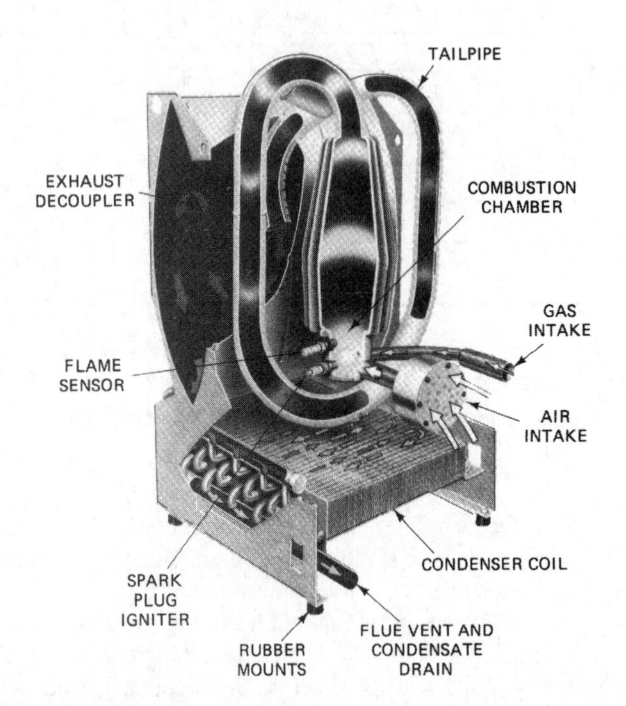

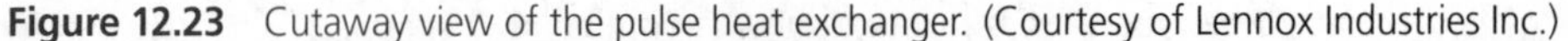

Figure 12.23 Cutaway view of the pulse heat exchanger. (Courtesy of Lennox Industries Inc.)

HOW A PULSE FURNACE WORKS:

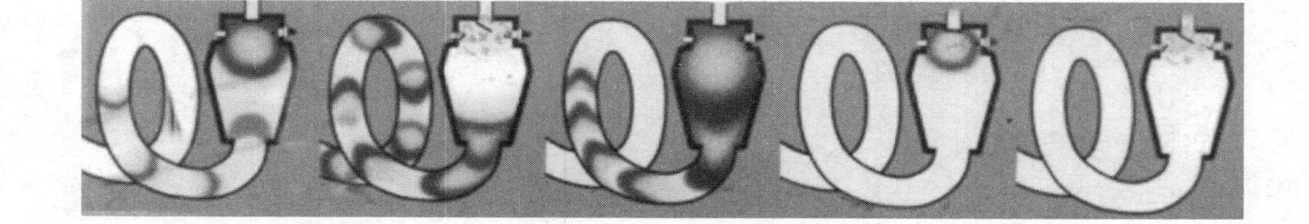

1 The heating process begins as air and gas enter the combustion chamber and mix near the spark igniter.

2 A spark creates initial combustion, similar to the way a spark plug works in your car. This causes a pressure build-up inside the combustion chamber.

3 The internal pressure relieves itself by forcing the products of combustion down a tailpipe and venting them outdoors.

4 As the combustion chamber empties, it creates a vacuum that prepares it for the next ignition. At the same instant, pressure "pulses" back from the end of the tailpipe.

5 The "pulse" returning from the tailpipe re-enters the combustion chamber, causing the new gas/air mixture in the chamber to ignite. As a result, the heating cycle is repeated at a rate of 60 to 70 times per second.

Figure 12.24 Sequence of pulse ignition cycle. (Courtesy of Lennox Industries Inc.)

The sequence of operation of the pulse furnace is as follows. When the room thermostat calls for heat it initiates the operation of the purge fan, which runs for 34 seconds. This is followed by the turning on of the ignition and the opening of the gas valve. The flame sensor provides proof of ignition and de-energizes the purge fan and spark ignition.

When the thermostat is satisfied, the gas valve closes and the purge fan is turned on for 34 seconds. The furnace continues to operate until the temperature in the plenum reaches 90°F. In case the flame is lost before the thermostat is satisfied, the flame sensor will try to reignite the gas–air mixture three or five times before locking out. Should there be a loss of either intake gas or air, the furnace will shut down automatically.

Note that combustion air is piped with the same PVC pipe as used for the exhaust gases. The pH range of the condensate liquids is from 4.0 to 6.0, which permits it to be drained into city sewers or septic tanks. The combustion mixture of gas and air are preset at the factory. No field adjustments are necessary.

Computerized Control of High-Efficiency Furnaces

Many features have been added to available furnaces as a result of the application of high technology. Here we describe the use of a computerized control system for the high-efficiency furnace. It is attractive from the standpoint of saving energy, improving occupant comfort level, and lowering the noise level of the installed equipment.

The furnace described has a 93.5 annual fuel utilization efficiency (AFUE). As an example of the energy it saves, assuming that it is installed in Cleveland, Ohio, the saving in a single heating season would be $230 on the gas bill and $90 on the electric bill as compared with a standard 78% AFUE furnace.

Primarily, we will review the control system, but let us first take a look at the major components, particularly those that differ from other high-efficiency furnaces. Figure 12.25 is a cutaway view of this furnace, showing the major components. Figure 12.26 shows the variable-speed inducer fan (2) and its speed controller (3). Figure 12.27 shows the high-pressure switch (1) and the low-pressure switch (2), which monitor combustion air flow during high- and low-fire operation, respectively.

Figure 12.28 shows the exterior of the miniature computer or microprocessor. It adjusts the burner input and the furnace blower speed to match the load. Figure 12.29 shows a view of the circuit board, with the cover removed. Figure 12.30 shows the LED indicator lights that signal fault codes to guide the service technician.

Figure 12.31 shows the Personality Plug, which lets the microprocessor know the size of the furnace to which it is connected. Figure 12.32 shows a series of switches that permit the service technician to supply input to the computer to check system operation. Figure 12.33 shows the rotary switch used in adjusting the airflow on cooling. At 400 cfm per ton, a 2-ton system requires 800 cfm. This would be accomplished by setting the switch on "2." Figure 12.34 shows the rotary control switch that is used to adjust the delay in the shutdown of the blower when there is a call for the heating mode. The information that follows shows various diagrams of the control arrangement and the sequence of operation.

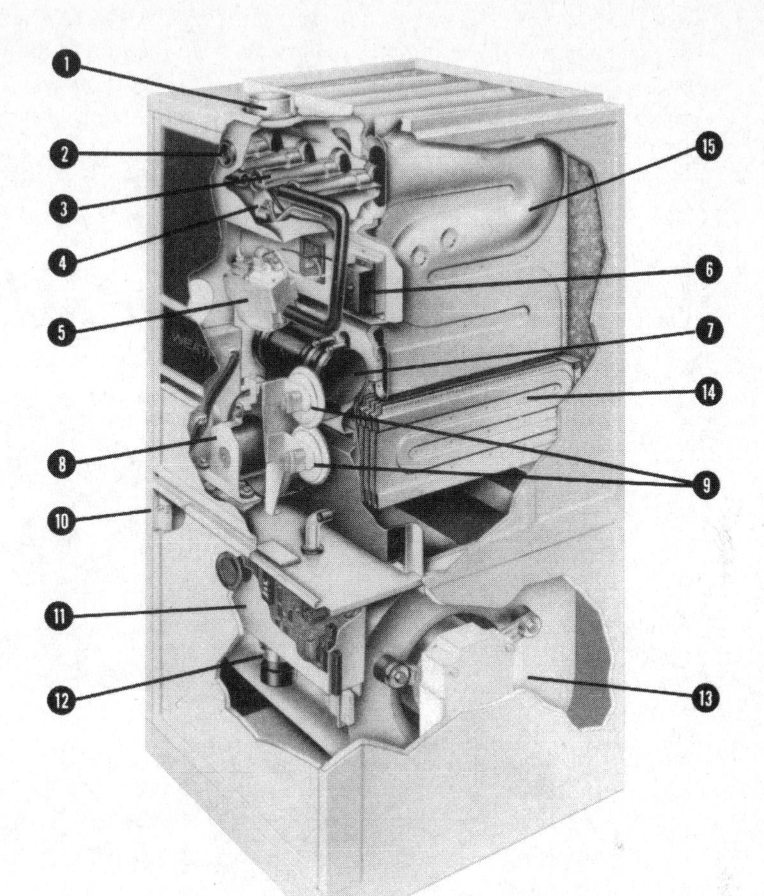

NOTE: The 58SXB Furnace is factory built for use with natural gas. The furnace can be field-converted for propane gas with a factory-authorized and listed accessory conversion kit.

NOTE: Control location and actual controls may be different than shown above.

1. Combustion-Air Pipe Connection
2. Sight Glass
3. Burner Assembly
4. Pilot
5. Two-Stage Gas Valve
6. Electronic Spark Ignition System
7. Vent Pipe Connection
8. Variable Speed Inducer Motor
9. Pressure Switches
10. Blower Door Safety Switch
11. Control Box
12. Condensate Trap
13. Blower Assembly With Variable-Speed Motor
14. Secondary Heat Exchanger
15. Primary Heat Exchanger

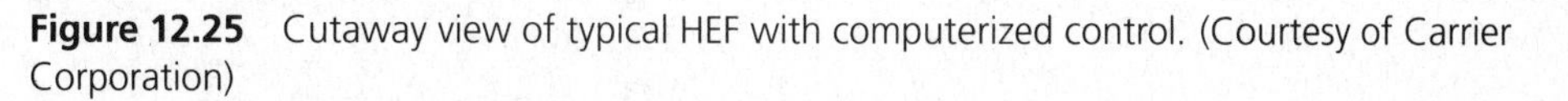

Figure 12.25 Cutaway view of typical HEF with computerized control. (Courtesy of Carrier Corporation)

Figure 12.26 Variable-speed condenser fan. (Courtesy of Carrier Corporation)

Figure 12.27 High/low pressure switch. (Courtesy of Carrier Corporation)

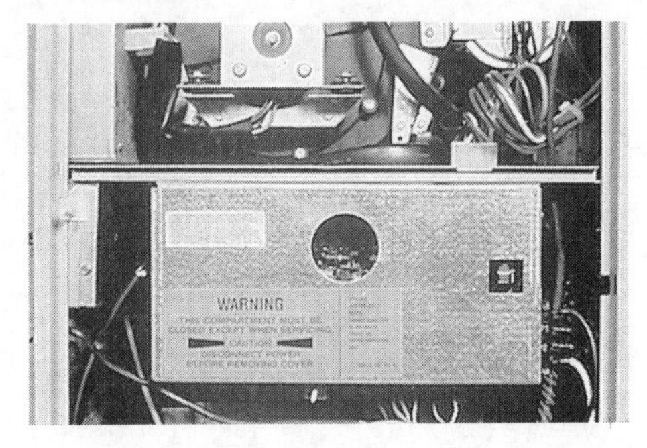

Figure 12.28 Computer or microprocessor. (Courtesy of Carrier Corporation)

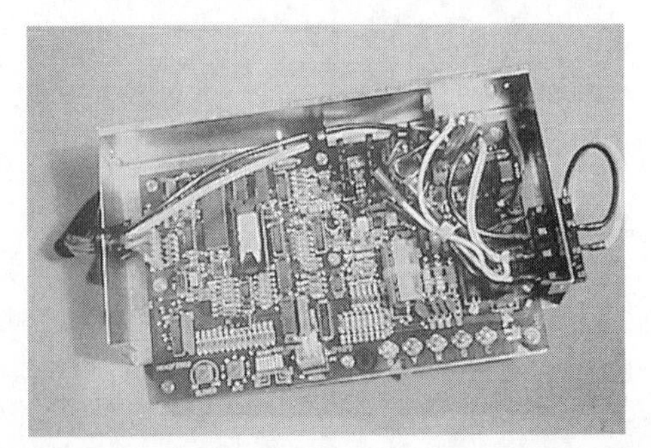

Figure 12.29 Circuit board. (Courtesy of Carrier Corporation)

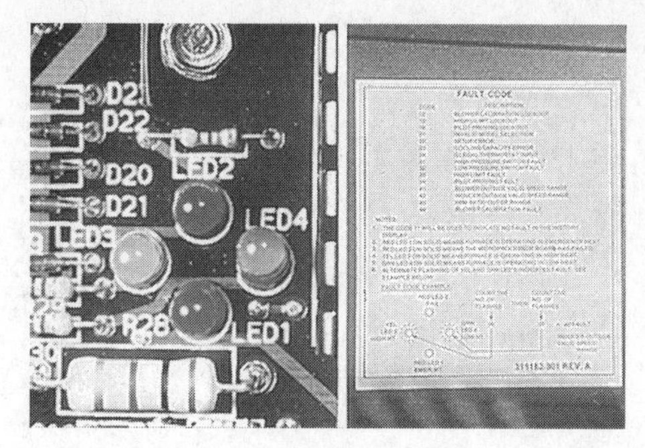

Figure 12.30 LED indicator lights. (Courtesy of Carrier Corporation)

Figure 12.31 Personality Plug. (Courtesy of Carrier Corporation)

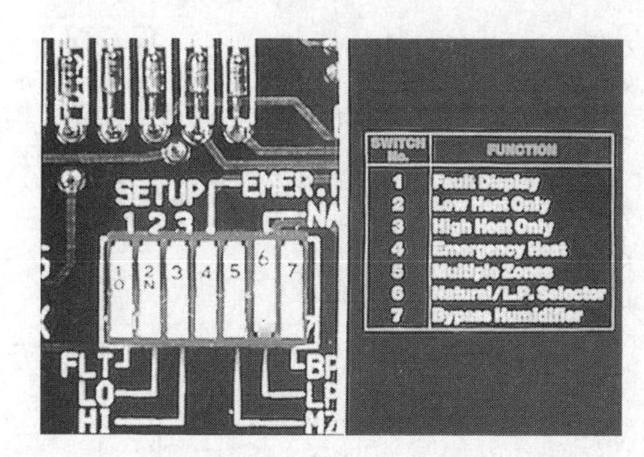

Figure 12.32 Computer switches. (Courtesy of Carrier Corporation)

Figure 12.33 Airflow switch. (Courtesy of Carrier Corporation)

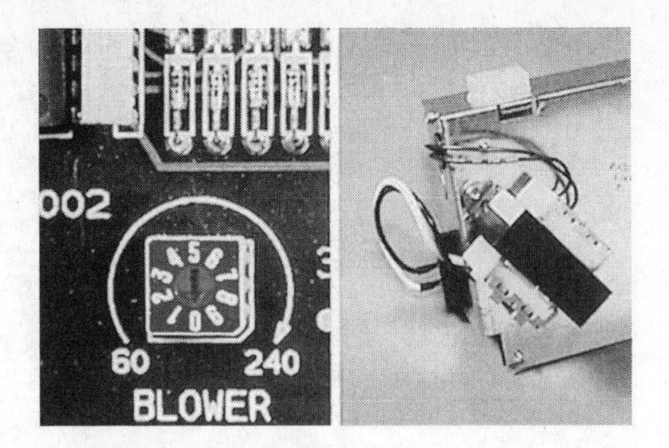

Figure 12.34 Blower control switch. (Courtesy of Carrier Corporation)

Power Supplied (Figure 12.35) On a call for heat, the blower door switch must be closed (1). Line-voltage power is supplied to the inducer motor controller (PCB1) and the blower motor controller (PCB2). Power is also applied to the transformer (2), which supplies low-voltage power to the microprocessor board (PCB3) and to the R terminal of the thermostat connection.

Call for Heat (Figures 12.36 and 12.37) Flow-sensing switches are energized (Figure 12.36). On a call for heat the R and W contacts on the thermostat close (1), supplying low-voltage power to the NC low- and high-flow-sensing pressure switches

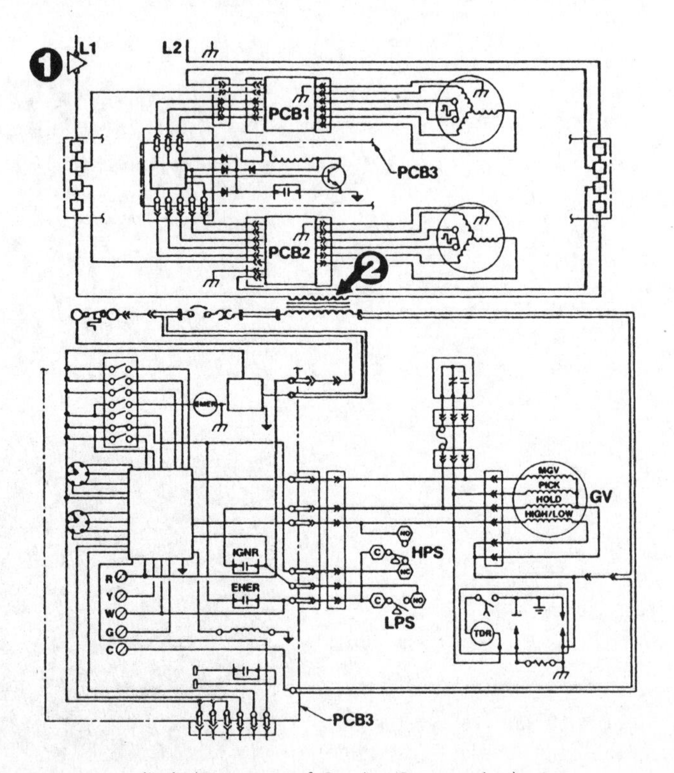

Figure 12.35 Power supplied. (Courtesy of Carrier Corporation)

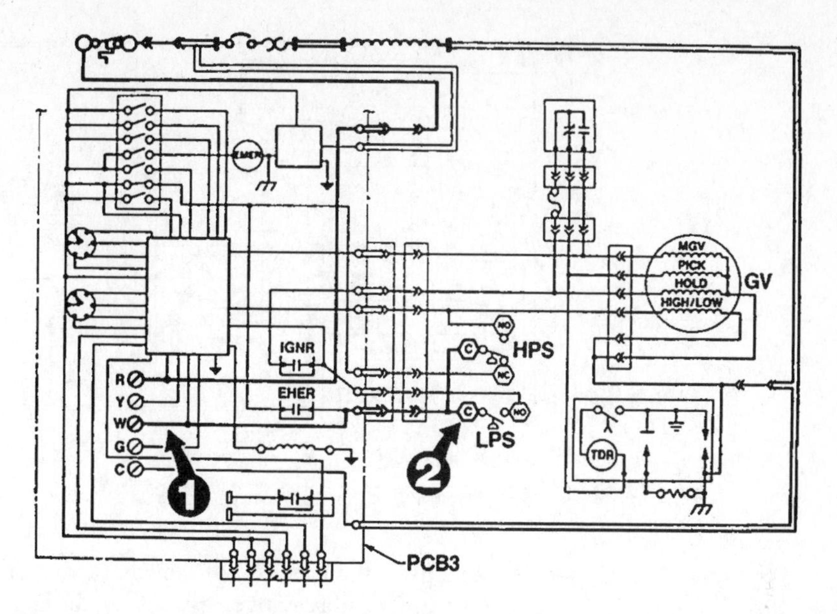

Figure 12.36 Call for heat, flow switch energized. (Courtesy of Carrier Corporation)

(2). The microprocessor signals the inducer motor to start (Figure 12.37), gradually bringing it up to speed.

Inducer Fan in Operation (Figure 12.38) With the inducer fan in operation, the NO contacts on the flow-sensing switches close (1), supplying power to the ignition relay contacts (2) and the high-fire solenoid in the gas valve (3). A signal is also sent to the microprocessor (4) to indicate that these actions have taken place.

Ignition Relay Energized (Figure 12.39) After the flow-sensing switch is made for 10 seconds, the microprocessor energizes the ignition relay (1). This supplies power to the holding coil of the gas valve.

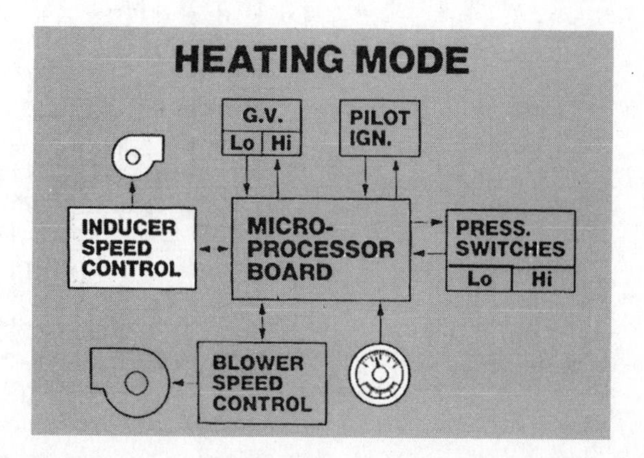

Figure 12.37 Microprocessor signals inducer motor to start. (Courtesy of Carrier Corporation)

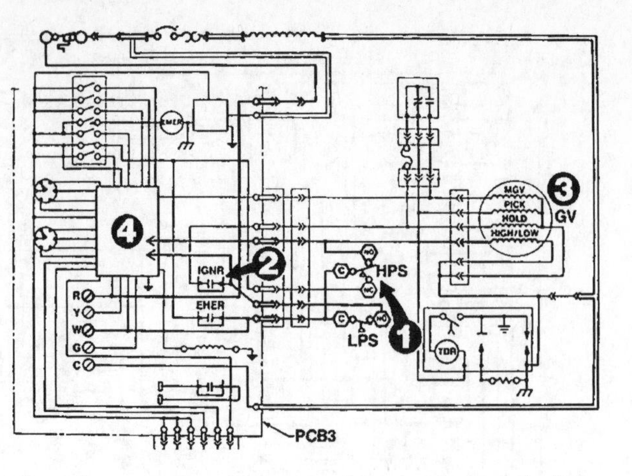

Figure 12.38 Inducer fan operation. (Courtesy of Carrier Corporation)

At the same time, the pick coil of the gas valve and the time-delay circuit of the spark generator are energized (2), through the NC contacts of the pilot assembly (3).

Gas Flows to Pilot (Figure 12.40) With the pick coil energized, gas flows to the pilot. After 10 seconds the time-delay circuit in the spark generator is energized (1), causing the pilot to be lighted by a spark.

Main Gas Valve Energized (Figure 12.41) Heat from the pilot flame causes the NO contacts to close and the NC contacts to open on the pilot assembly (1). This de-energizes the pick coil on the gas valve (2), energizes the main gas valve, and sends a flame-proving sig-

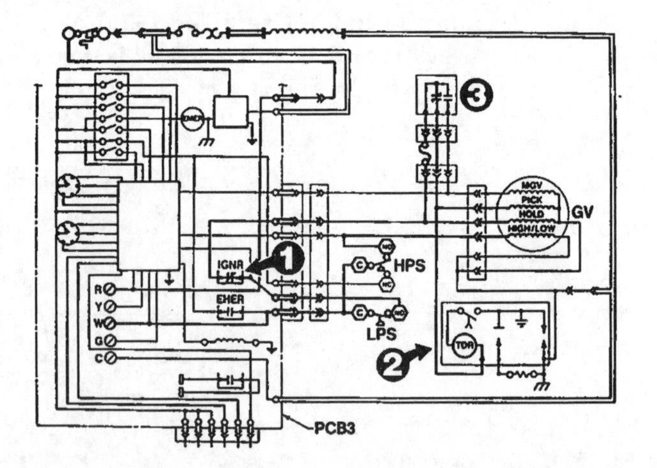

Figure 12.39 Ignition relay energized. (Courtesy of Carrier Corporation)

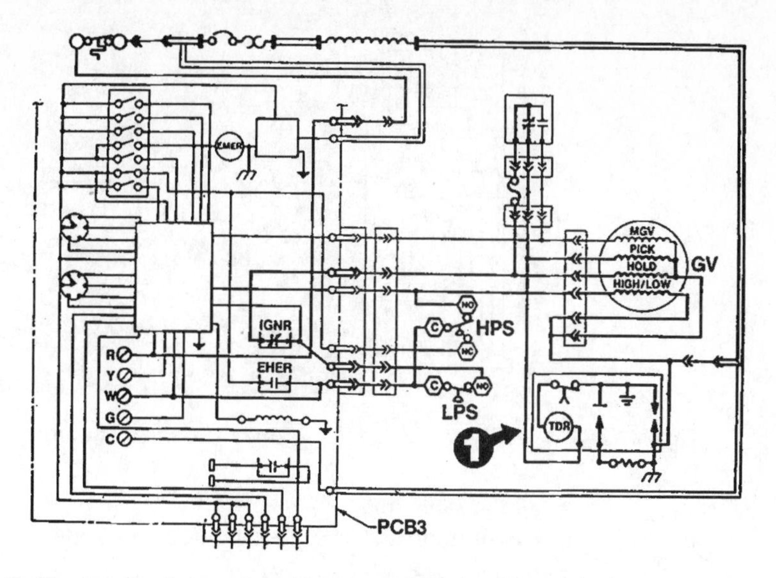

Figure 12.40 Gas flowing to pilot. (Courtesy of Carrier Corporation)

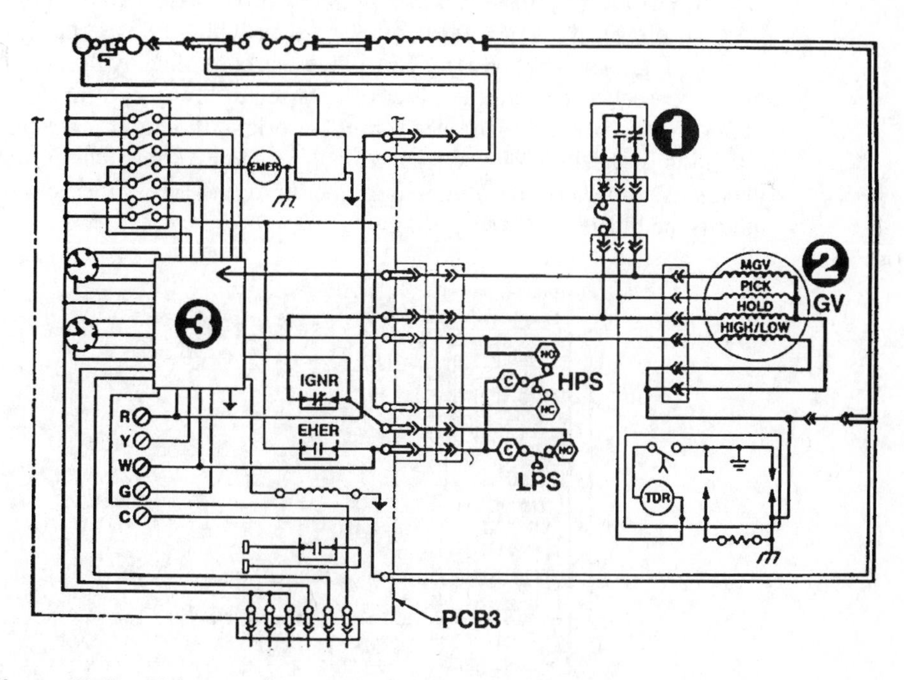

Figure 12.41 Main gas valve energized. (Courtesy of Carrier Corporation)

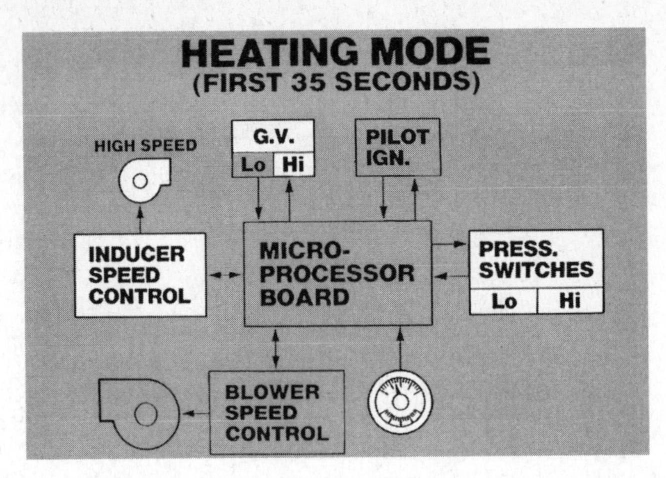

Figure 12.42 Heating mode, first 35 seconds. (Courtesy of Carrier Corporation)

nal to the microprocessor (3). The pilot flame then ignites the main burners in the high-fire mode.

Heating Mode (Figures 12.42 to 12.44) For the first 35 seconds of high-fire operation the blower does not run (Figure 12.42), which allows the heat exchanger to increase rapidly in temperature. This process is controlled by the computer. At the end of the 35-second period, the blower is started at low speed for 20 seconds (Figure 12.43). The computer then goes through a brief calibration procedure. It senses the condition of the air-supply system, calculates the rev/min value of the blower that is necessary to produce the required temperature rise across the heat exchangers, and adjusts the blower speed accordingly.

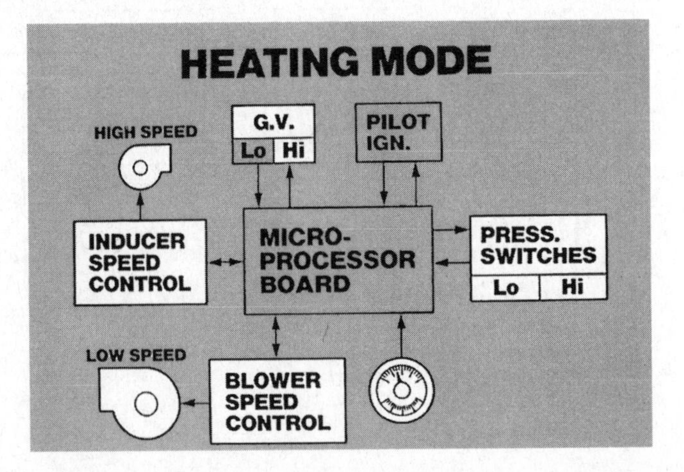

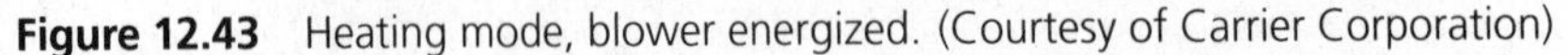

Figure 12.43 Heating mode, blower energized. (Courtesy of Carrier Corporation)

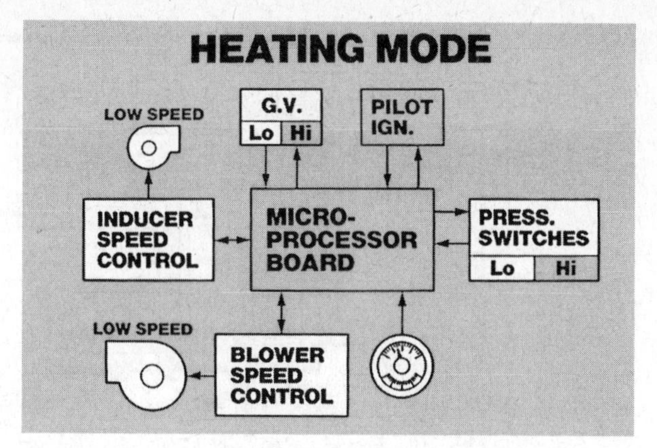

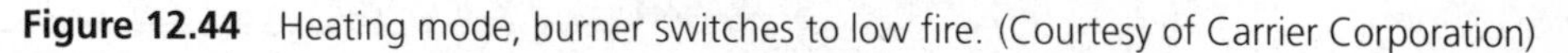
Figure 12.44 Heating mode, burner switches to low fire. (Courtesy of Carrier Corporation)

After 1 minute, which includes the high-fire operation and the calculation period, the burner switches to low fire (Figure 12.44). In this initial cycle the unit operates for 6 minutes. Then, if the thermostat is not satisfied, the system switches to high fire until the thermostat is satisfied. Whenever there is an interruption of power the system restarts in the initial-cycle condition. After the initial cycle the operation of the unit is controlled by the computer in accordance with inside and outside conditions.

Emergency Heat (Figures 12.45 and 12.46) In the emergency-heat mode the burner always operates at high fire. The blower operates at a high speed, higher than the air-conditioning speed (Figure 12.45). The setup switch (1) is closed when emergency heat is selected. On a call for heat the R-W circuit is energized, which causes the emergency heat relay (2) to energize. The emergency heat relay contacts close, starting the inducer

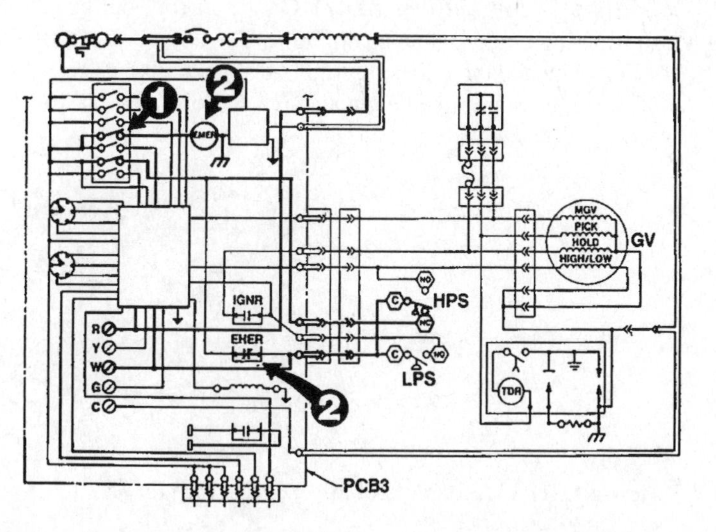

Figure 12.45 Emergency heat relay energized. (Courtesy of Carrier Corporation)

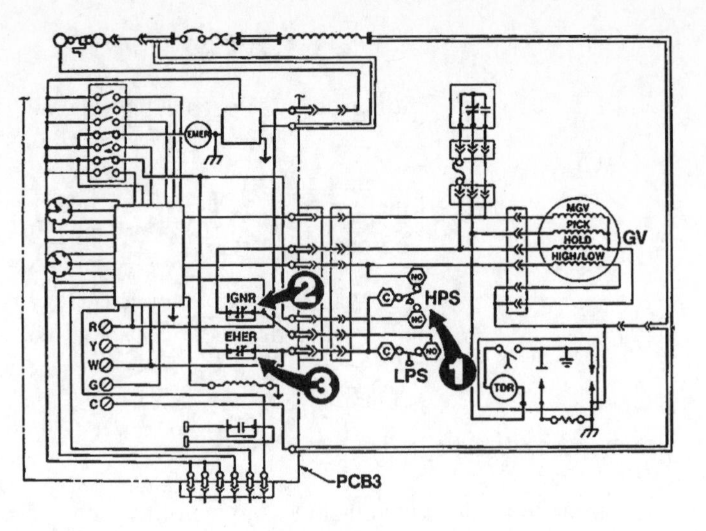

Figure 12.46 Emergency heat, burners energized. (Courtesy of Carrier Corporation)

motor and the blower motor. When the burners are lighted (Figure 12.46), low- and high-flow-sensing switches close (1). Power is supplied to the ignition relay (2), gas valve, and time delay of the spark generator. A holding path for the emergency-heat relay is established (3). The sequence of operation that follows is similar to normal operation except that the burner operates only in the high-fire mode and the blower only at high speed. The unit should be operated in the emergency heat mode only for relatively short periods of time.

Cooling Mode (Figure 12.47) On a call for cooling, the R-Y circuit is energized, starting the outdoor condensing unit. The computer starts the blower at low speed for 20 seconds and then goes through a brief calculation period and sets the rev/min value of the blower at a speed equivalent to 400 cfm per ton.

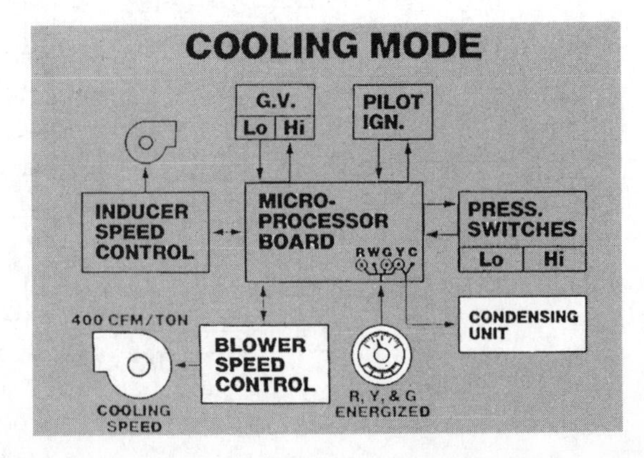

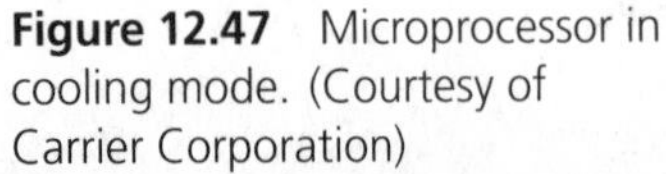
Figure 12.47 Microprocessor in cooling mode. (Courtesy of Carrier Corporation)

Honeywell Integrated Control System

The Honeywell integrated control system combines a number of traditional controls into a single package with few components. The system consists of the ST9120 furnace-control/fan-timer (Figure 12.48), the SV9501 SmartValve (Figure 12.49), and the Q3450 pilot burner (Figure 12.50.) All wiring between components is preassembled wiring harnesses with plug-in connectors. The hot surface igniter works on 24 V. Flame sensing is accomplished with a flame rectification rod. The ignition control module is contained within the gas valve. The module provides ignition and flame safety and emits a signal to the furnace-control/fan-timer to initiate a time-delay fan start.

Sequence of Operation for the Honeywell SmartValve Integrated System

1. On a call for heat the ST9120 furnace-control/fan-timer energizes the combustion air blower. Refer to Figure 12.51.
2. Once combustion-air flow is established, the air-proving switch makes, and 24-V power is sent to the SV9501 SmartValve.
3. The valve first checks for a flame signal. If there is none, the hot surface igniter is energized, and the pilot valve opens.
4. The pilot lights and the flame rod senses a flame.
5. The igniter shuts off. The main gas valve opens, the pilot lights the main burners, and the fan timer output from the valve is energized.

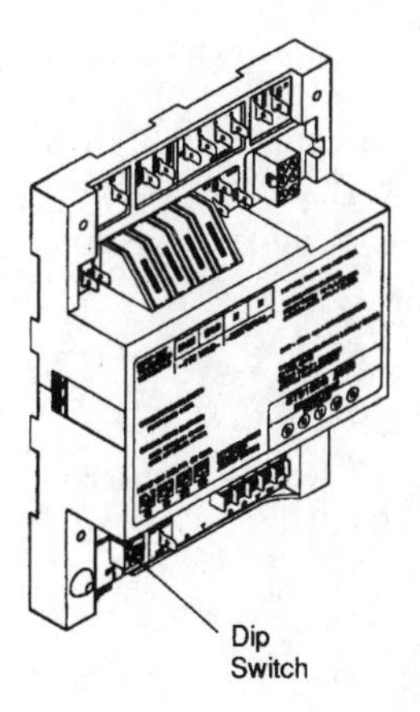

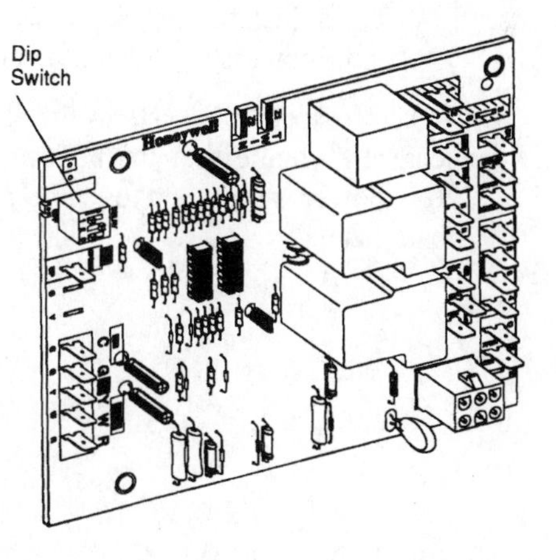

Figure 12.48 Honeywell Model ST9120 Furnace Control/Fan Timer. Right side has cover removed. (Courtesy of International Comfort Products Corporation, USA)

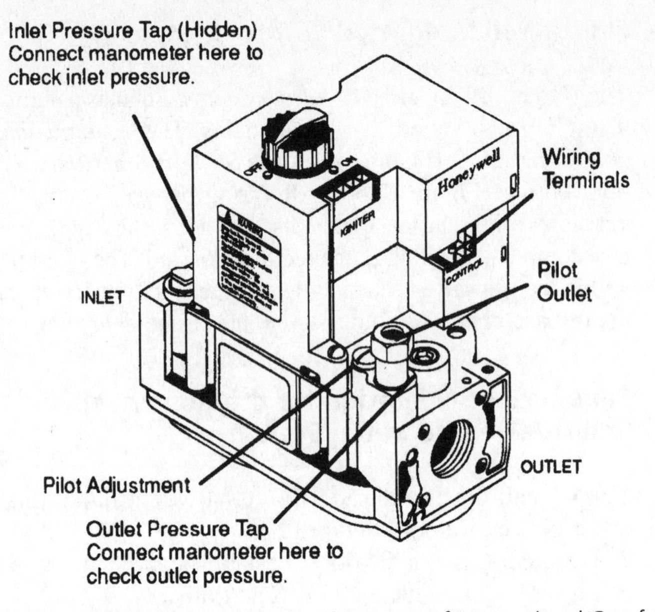

Figure 12.49 Honeywell SV9500 SmartValve. (Courtesy of International Comfort Products Corporation, USA)

6. When the fan timing is complete, the furnace circulating fan is energized at heat speed after the factory-set 30 seconds or the field-adjusted 60 seconds. Adjustment is via the dip switches located on the ST9120 (Figure 12.52.)
7. When the call for heat is satisfied, the ignition system is de-energized, and the gas valve shuts down.
8. The combustion air blower remains on through a 5-second postpurge time period before being de-energized.
9. A field-adjustable furnace fan off delay timing begins. When the factory-set 140-second timing is complete, the circulating fan is de-energized (Figure 12.51.)

In two-speed fan systems the electronic air cleaner, if connected to the ST9120, is energized when either the heat or cool speed of the furnace fan is energized, and 120-V power is provided to the humidifier circuit whenever the combustion air blower is energized.

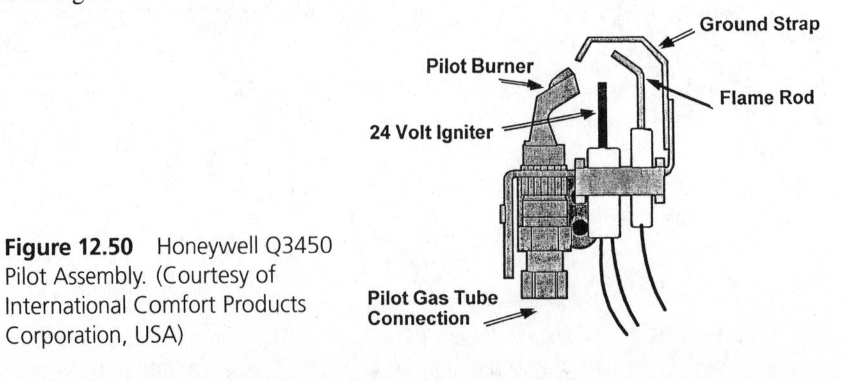

Figure 12.50 Honeywell Q3450 Pilot Assembly. (Courtesy of International Comfort Products Corporation, USA)

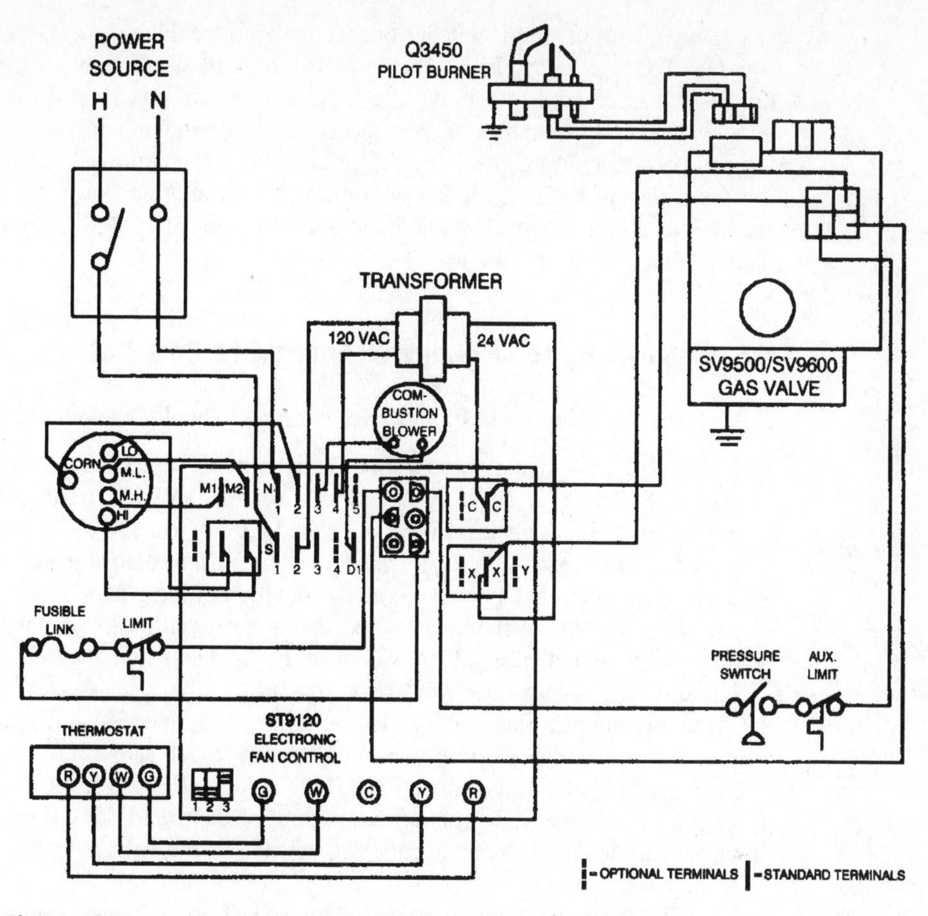

Figure 12.51 A diagram of the Honeywell SmartValve® System. (Courtesy of Honeywell Inc.)

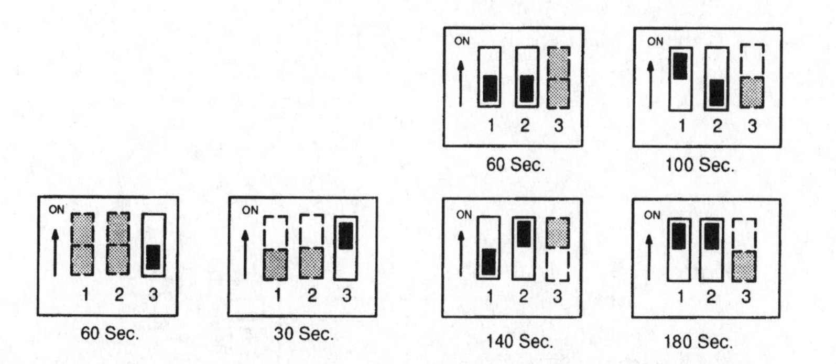

Figure 12.52 ST9120 fan control dip switch settings. (Courtesy of International Comfort Products Corporation, USA)

If any one of the limit switches opens, power to the thermostat and ignition system is cut, and the gas valve closes. The combustion air blower and the furnace circulating fan remain energized to remove the high-heat condition. Should the limit switch remake, the combustion blower remains on for a postpurge time period. The circulating fan powers down through its delay-off timing, after which normal operation resumes.

On a thermostat call for fan operation, the circulating fan is energized at heat speed. If a call for heat occurs, the fan remains running. If a call for cooling occurs, the fan switches to cool speed after a 4-second delay.

Troubleshooting the Honeywell SmartValve Integrated System

On a call for heat the combustion blower is running, but the main burner is not on.

1. If there is a pilot flame, measure the flame current. A reading of more than 25 mV on the SV9500 or more than 130 mV on the SV9501 indicates a defective SmartValve. Readings less than those stated require a check of the alignment of the pilot burner (Figure 12.53). The pilot brackets must remain at the 90° angle formed at the factory and must not be bent. The position of the ground electrode and flame rod are such that the flame envelops each at a 4:1 ratio. The proper ratio will produce a signal of at least 0.2 μA. There are no field adjustments. Replacement of the igniter/flame-rod assembly may be necessary.
2. If there is no pilot flame, check for voltage to the SmartValve (Figure 12.54). No power requires a check of upstream components such as the furnace control, pressure switch, and wiring connections. If the voltage is 20.4 V to 26.4 V, go to step 3.
3. Check for voltage to the igniter coming out of the SmartValve (Figure 12.55). No power requires replacement of the SmartValve. If the voltage is 20.4 to 26.4 V, go to step 4.
4. Visually verify that the igniter is glowing red. If not, replace the igniter/flame–rod assembly. If yes, go to step 5.
5. Check for gas supply to the valve and out of the pilot outlet of the valve. If the inlet gas pressure is greater than 5 in. W.C. but less than 3.2 in. W.C. at the outlet side, replace the SmartValve. If all gas pressures are correct, verify a clean orifice and then replace the igniter/flame-rod assembly.

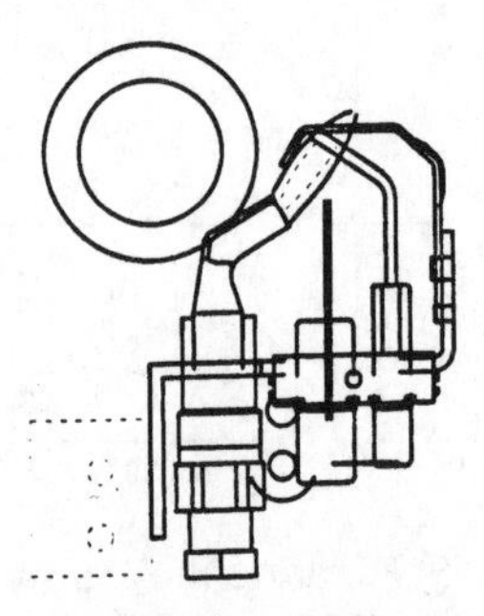

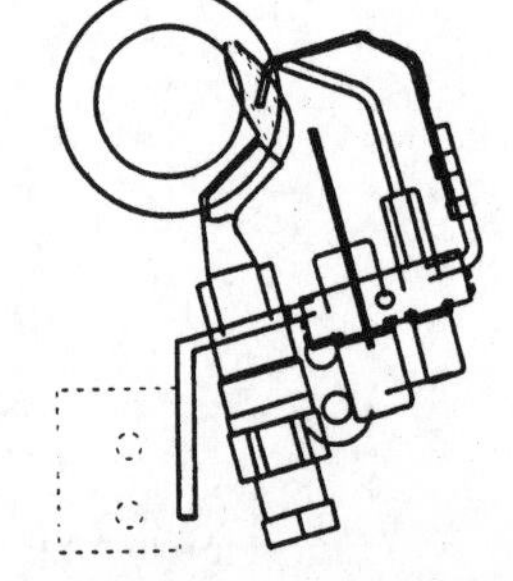

Figure 12.53 Alignment of the Q3450 pilot assembly. Note the flame impingement on the flame rod and grounding electrode. (Courtesy of International Comfort Products Corporation USA)

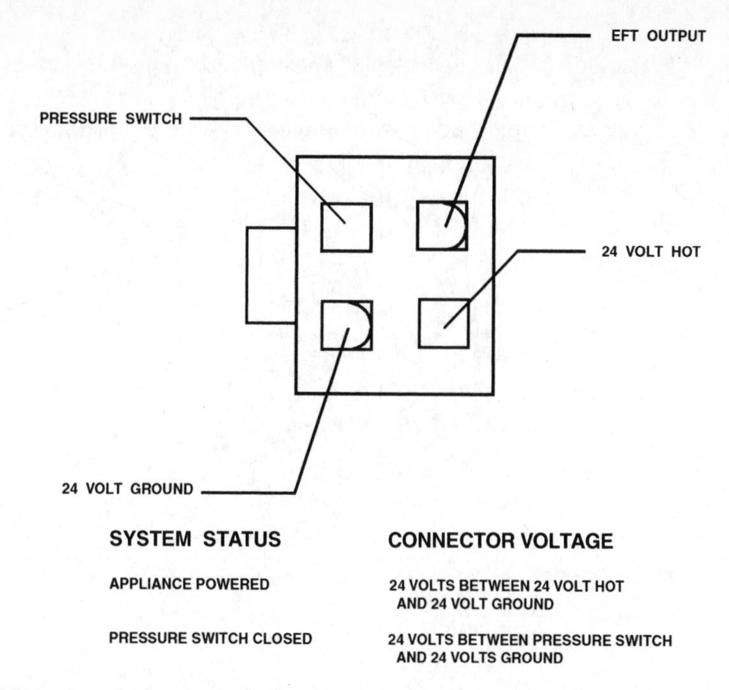

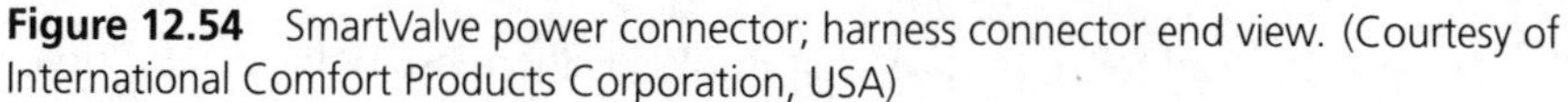

Figure 12.54 SmartValve power connector; harness connector end view. (Courtesy of International Comfort Products Corporation, USA)

Honeywell Two-Stage Integrated Control System

The Honeywell two-stage integrated control system is being used on many new high-efficiency furnaces because of its ability to provide part-load operation during milder winter temperatures. The system utilizes the Honeywell ST9162A furnace/fan control with the SV9540Q gas valve (Figure 12.56).

The sequence of operation is similar to that of the ST9120 with the following exceptions.

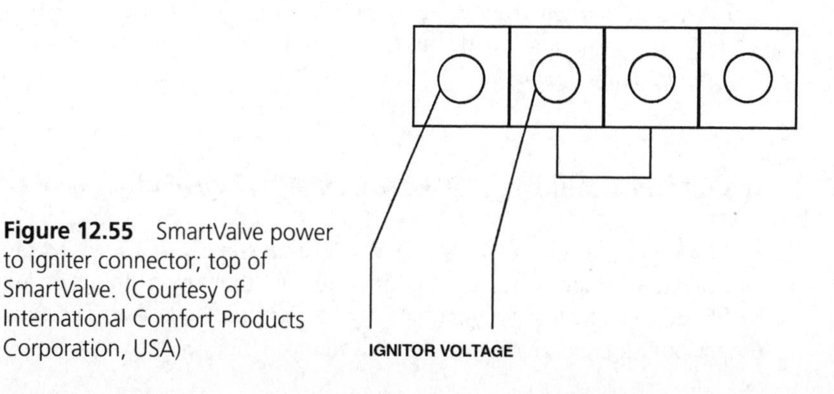

Figure 12.55 SmartValve power to igniter connector; top of SmartValve. (Courtesy of International Comfort Products Corporation, USA)

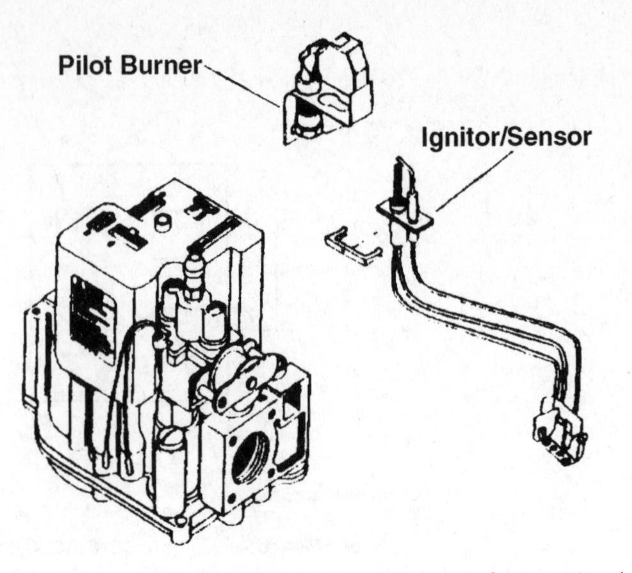

Figure 12.56 Honeywell SV9540Q Ignition System. (Courtesy of International Comfort Products Corporation, USA)

1. On a call for first-stage heating (*W*1 on thermostat), the combustion-air motor energizes at high speed, and the gas valve solenoid energizes at high fire. Timed from the opening of the main valve, the control waits for the selected Heat Fan On delay before de-energizing the high-fire solenoid, switching the combustion air motor to low speed, and energizing the furnace fan on low speed. The furnace remains on low fire until the thermostat is satisfied or calls for the second stage.
2. On a call for second-stage heating (*W*2 on thermostat), the control goes through the same light-off routine but remains at high fire and high speed at the end of the Fan On time delay.
3. With any change in a call from the *W*2 thermostat terminal, the control will switch between high and low fire without delay.
4. On a call for Fan On operation (G on thermostat), the control energizes the fan on low heating speed without delay. Any heating or cooling requests received during the Fan On operation will cause the fan to switch to the appropriate speed after the Fan On delay time expires.

Troubleshooting the ST9162A/SV9540Q Two-Stage System

The operation of the ST9162A/SV9540Q two-stage system can be checked with the use of two jumper wires for connection to the low-voltage terminal strip. The SV9540Q is both line voltage polarity and ground sensitive. The furnace must have proper polarity and grounding, or malfunctions can occur.

Checking the heating functions:

1. Jumper W1 to R or W1 and W2 to R on the control terminal strip.
2. Verify combustion blower startup.
3. Verify ignition system activation.
4. When the main burner lights, check the Heating Fan On time delay (usually 30 seconds.)
5. Verify heating speed fan operation.
6. Remove the jumper.
7. Check the postpurge time delay. The combustion blower will remain energized for 5 seconds.
8. Check the Heating Fan Off time delay (usually 140 seconds.)

If the furnace passes this testing sequence, the technician can assume that there are no problems with the ST9162A/SV9540Q system. Failure to pass does not mean a malfunction of the control system. Other components such as the combustion blower or the gas supply may be at fault.

Both the ST9162A furnace control and the SV9540Q gas valve contain sophisticated internal electronic components and have no user-serviceable components. Should a problem be verified with the device, it must be replaced.

The SV9540Q two-stage SmartValve has an LED indicator to help in troubleshooting:

1. If the LED is not lit, there is no power to the valve. Check 24-V output and line-voltage input at the ST9162A. Also check wiring harness connections.
2. A steady light is a normal indication whenever the system is powered.
3. Two flashes indicate a low-pressure switch stuck closed prior to start of cycle.
4. Three flashes indicate a low-fire switch remaining open after the combustion blower starts.
5. Four flashes indicate an open main or rollout limit switch.
6. Five flashes indicate flame signal sensed out of sequence.
7. Six flashes indicate a soft lockout from any ignition failure. After 5-minute delay, the control system will reset and initiate a new ignition sequence.
8. Eight flashes indicate a high-pressure switch stuck closed.
9. Nine flashes indicate a high-pressure switch stuck open.

"General 90" White–Rodgers 50A50 Total Furnace Control

The White–Rodgers 50A50 system consists of four components: the ignition system control, the hot-surface igniter, the flame sensor, and the gas valve (Figure 12.57) The General 90 furnace control manages the ignition sequence and controls the operation of the gas valve, igniter, combustion blower, and circulating blower. There are no internal user-serviceable components, and the control must be replaced if found defective. The hot surface igniter operates on 120 V ac. The gas valve operates on 24 V ac.

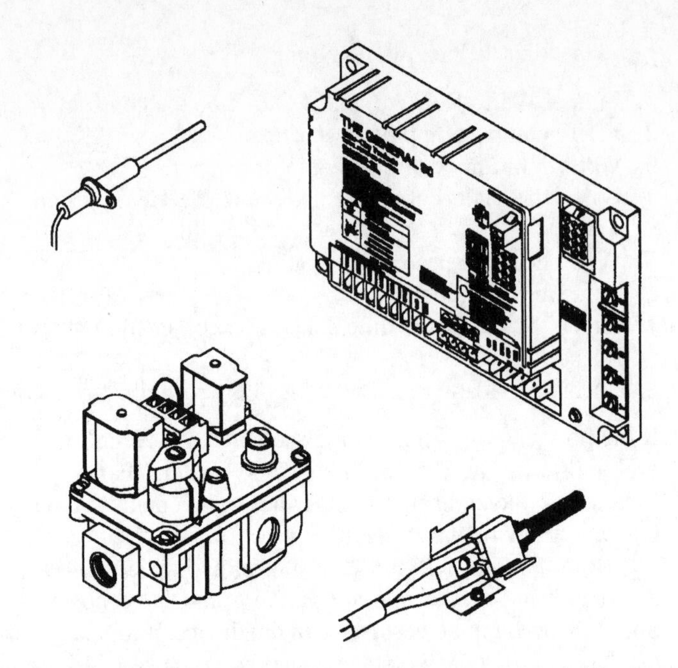

Figure 12.57 Components of the General 90 Total Furnace Control. (Courtesy of International Comfort Products Corporation, USA)

Sequence of Operation of a White–Rodgers 50A50 Furnace Control

Refer to Figure 12.58 for the sequence of operation

1. Line voltage feeds through the door switch when the access panel is in place.
2. The transformer provides 24 V across TH and TR.
3. 24 V is brought to the R terminal on the thermostat.
4. On a call for heat the thermostat closes from R to W and completes a circuit on the 50A50 control, causing the (K4) contacts to close.
5. 115 V energizes the inducer motor from the IND and IND/N terminals.
6. The pressure switch closes, proving the operation of the inducer motor. This prepurge lasts 30 seconds.
7. During the 30 seconds, the control does a self-test, making sure the flame rollout switch and the limit control are in the closed position.
8. Switch K6, located inside the control, completes the circuit to the hot-surface igniter across terminals IGN and IGN/N for 17 seconds.
9. After the 17 seconds of ignition, K7 and K8 close, energizing the main gas valve. The igniter remains energized for 4 seconds after the main gas valve is energized.
10. The main gas is proven by a flame rectification system that is built into the 50A50 control. (If the flame is not proven after 7 seconds, the main gas valve is de-energized and is returned to the prepurge/trial-for-ignition sequence six times, before locking out completely.)

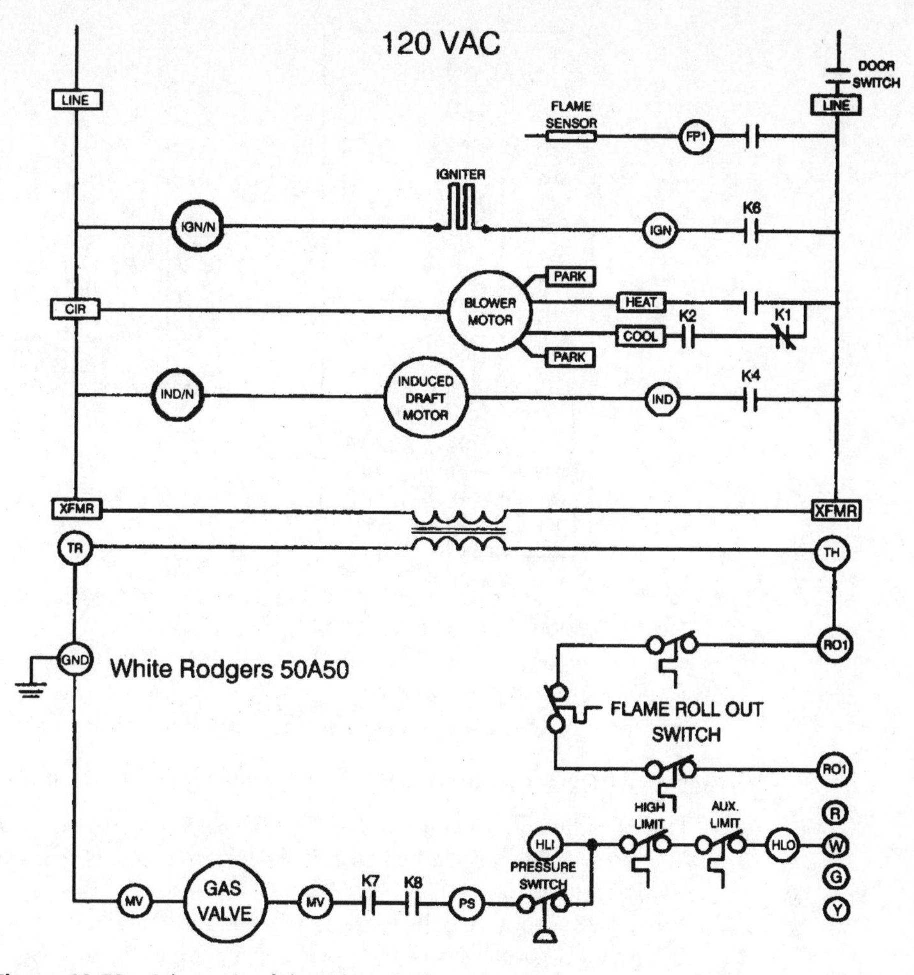

Figure 12.58 Schematic of the White–Rodgers 50A50 furnace control. (Courtesy of White–Rodgers Division, Emerson Electric Co.)

11. Energizing of the main valve starts the time-on delay for the blower circuit. The position of the dip switches (see Figure 12.59), located on the 50A50 control, determines the time that the fan is energized. Time-on settings are 15, 30, 45, or 60 seconds.
12. When the blower is energized the EAC and the HUM terminals bring 115 V to the electronic air cleaner and the humidifier circuits.
13. When the thermostat is satisfied, the 24-V supply is interrupted to the gas valve circuit. This causes the gas valve to close and K4 contacts to open, de-energizing the inducer motor.
14. The time-off delay begins on the blower motor circuit to de-energize the motor for the off cycle. The position of the dip switches, located on the 50A50 control, determines the time that the fan is de-energized. Time-off settings are 60, 90, 120, and 180 seconds.

HEAT "ON" DELAY

SW1	SW2	(SEC)
OFF	OFF	15
OFF	ON	30
ON	OFF	45
ON	ON	60

HEAT "OFF" DELAY

SW3	SW4	(SEC)
ON	ON	60
OFF	ON	90
ON	OFF	120
OFF	OFF	180

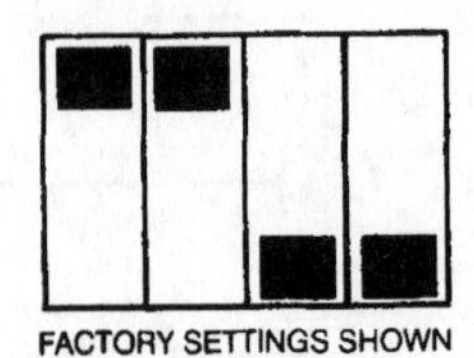

Figure 12.59 50A50 dip switch settings for Heat Fan On/Off time delay. (Courtesy of White–Rodgers Division, Emerson Electric Co.)

STUDY QUESTIONS

Answers to the study questions may be found in the sections noted in brackets.

12-1 Describe the control components for a 78% AFUE furnace. *[Components]*

12-2 What parts are energized when a 78% AFUE furnace is ready to start? *[Ready to Start]*

12-3 When a 78% furnace calls for heat, what part is energized first? *[Call for Heat]*

12-4 What is the normal number of pulses per second for a pulse furnace? *[Pulse Design]*

12-5 What is the pH range for the condensing liquid on an HEF? *[Pulse Design]*

12-6 Name four characteristics of HEFs. *[Characteristics of HEFs]*

12-7 Describe the piping used for combustion air on a HEF. *[Air for Combustion]*

12-8 How many speeds does the inducer motor have on a computerized HEF? *[Computerized Control]*

12-9 What is the function of the computer on an HEF? *[Computerized Control]*

12-10 Describe the action that takes place on a computerized HEF when calling for emergency heat. *[Computerized Control]*

CHAPTER 13

Components of Oil-Burning Furnaces

OBJECTIVES

After studying this chapter, the student will be able to:

- Identify the components of an oil burner
- Adjust an oil burner for best performance
- Provide service and maintenance for an oil burner

Oil-Burning Units

Oil must be vaporized to burn. There are two ways to vaporize oil:

1. *By heat alone.* For example, oil can be vaporized by heating an iron pot containing oil and igniting it with a flare or spark.
2. *By atomization and heating.* Oil is vaporized by forcing it under pressure through an orifice, causing it to break up into small droplets. The atomized oil is then ignited by an electric spark.

Types of Burners

There are two types of atomizing oil burners:

1. Low pressure
2. High pressure

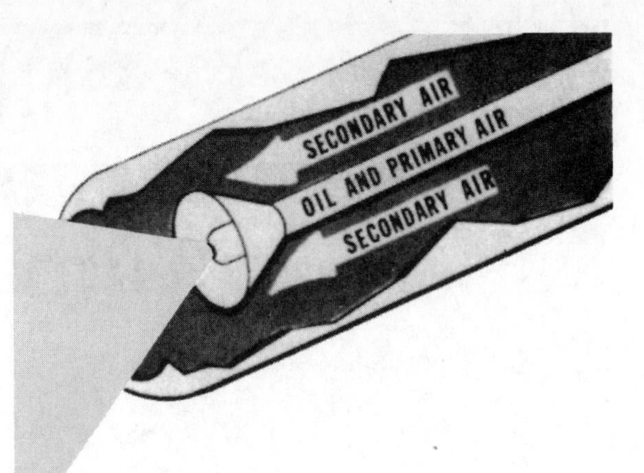

Figure 13.1 Low-pressure gun-type oil burner. (Courtesy of Carrier Corporation)

Low-Pressure Burner

In a low-pressure gun-type burner (Figure 13.1), oil and primary air are mixed prior to being forced through an orifice or nozzle. A pressure of 1 to 15 psig on the mixture, plus the action of the orifice, causes atomization of the oil. Secondary air is drawn into the spray mixture after it is released from the nozzle. Electric spark ignition is used to light the combustible mixture.

Although the low-pressure burner has achieved some success, it has been replaced to a great extent by the high-pressure burner. Higher pressure produces better atomization and high efficiencies.

High-Pressure Burner

A high-pressure gun-type burner (Figure 13.2) forces oil at 100 psig pressure through the nozzle, breaking the oil into fine, mistlike droplets. Some models require operating oil pump pressures in the 140–200 psig range. The atomized oil spray creates a low-pressure area into which the combustion air flows. Combustion air is supplied by a fan through vanes, creating turbulence and complete mixing action.

The high-pressure gun burner is the most popular domestic burner. It is simple in construction, relatively easy to maintain, and efficient in operation. Therefore, the balance of this unit concentrates exclusively on providing information relative to the high-pressure burner.

High-Pressure Burner Components

The high-pressure oil burner is an assembled unit, whose component parts are made by a few well-known manufacturers. The parts are mass produced, low in cost, and readily available for servicing requirements.

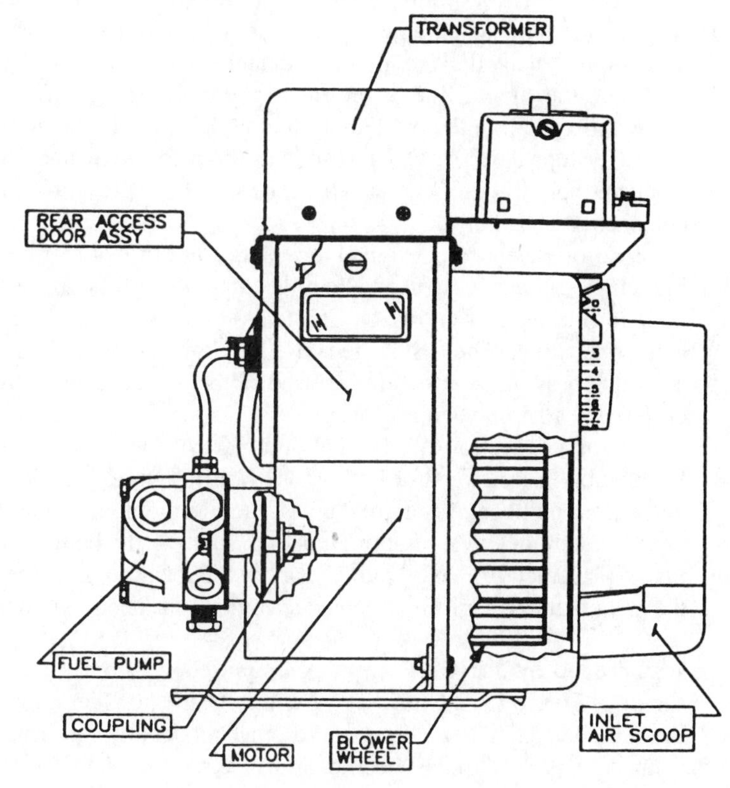

Figure 13.2 Component parts of a high-pressure oil burner. (Courtesy of R.W. Beckett Corporation)

Power Assembly

The power assembly consists of the motor, fan, and fuel pump. The nozzle assembly consists of the nozzle, the electrodes, and parts related to the oil–air mixing action. The ignition system consists of the transformer and electrical parts. The motor drives the fan and fuel pump. The fan forces air through the blast tube to provide combustion air for the atomized oil. The fuel pump draws oil from the storage tank and delivers it to the nozzle.

The oil–air mixture is ignited by an electric spark formed between two properly positioned electrodes in the nozzle assembly. The spark is created by a transformer that increases the 120-V primary power supply to a 10,000-V secondary supply to form a low-current spark arc.

Fuel Pumps All fuel pumps are of the rotary type, using cams, gears, or a combination of both. The principal parts of the fuel pump in addition to the gears are the shaft seal, the pressure-regulating valve, and the automatic cutoff valve. The shaft seal is necessary, since the pump is driven by an external source of power. The pressure-regulating valve has an adjustable spring that permits regulation of the oil pressure. The pump actually delivers more oil than the burner can use. The excess oil is dumped back into the supply line or returns to the tank. All pumps also have an adjustment screw for regulating the pressure of the oil delivered to the nozzle. The automatic cutoff valve stops the flow of oil as soon as the pressure drops. Thus, when the burner is stopped, the oil is quickly cut off to the nozzle to prevent oil from dripping into the combustion chamber.

Fuel-oil pumps are designed for single-stage or two-stage operation (Figure 13.3). The single-stage unit is used when the supply of oil is above the burner, and the oil flows to the pump by gravity. The two-stage unit is used when the storage tank is below the burner. The first stage is used to draw the oil to the pump. The second stage is used to provide the pressure required by the nozzle. The suction on the pump should not exceed a 15-in. vacuum.

Piping connections to the fuel pump are of two types: single-pipe and two-pipe. Installation instructions for both are shown in Figure 13.4. Various pump models are made by a number of manufacturers. Some pumps have clockwise rotation; others have counterclockwise rotation. It is important when changing a pump to use an identical replacement, since oil pumps vary in the location of connections. Always refer to the manufacturer's information and specification sheets for connection details.

Test Procedure for Oil Pumps A simplified test procedure for oil burner pumps is shown in Figure 13.5. Note that the connections shown are for a specific model pump. Similar connections can be made to other models by referring to the manufacturer's connection details on the specification sheets.

Motor The motor supplies power to rotate the pump and fan. Usually, a split-phase motor (with a start and a run winding) is used (Figure 13.6). This type of motor provides enough torque (starting power) to move the connected components. In the event

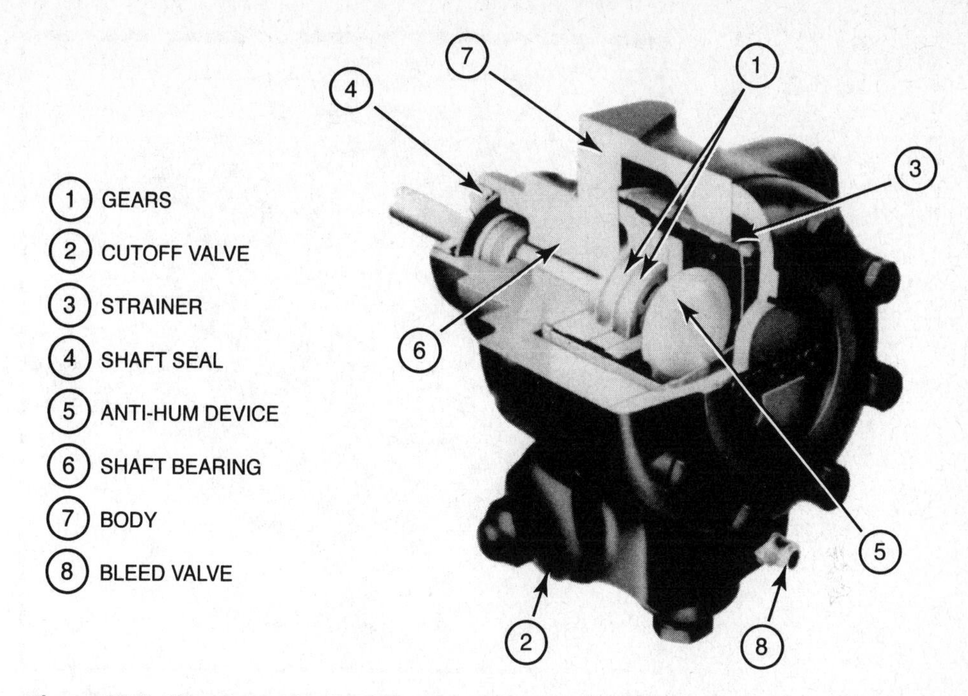

Figure 13.3 Two-stage oil pump. (Courtesy of Suntec Industries, Inc.)

that a motor fails, it must be replaced by substituting one with the same horsepower, direction of rotation, mounting dimension, revolutions per minute, shaft length, and shaft diameter.

Multiblade Fan and Air Shutters The multiblade (squirrel cage) centrifugal fan delivers air for combustion. It is enclosed within the fan housing. The inlet to the fan has an adjustable opening so that the amount of air volume handled by the fan can be manually controlled. The outlet of the fan delivers the combustion air through the blast tube of the burner (Figure 13.2).

Shaft Coupling The shaft coupling connects the motor to the fuel pump (Figure 13.7). This coupling:

- Provides alignment between the pump and motor shafts
- Absorbs noise that may be created by the rotating parts
- Is strongly constructed to endure the starting and stopping action of the motor

Static Disk, Choke, and Swirl Vanes As shown in Figure 13.8, the static disk, which is located in the center of the draft tube, causes the air from the fan to build up velocity at the inside surface of the tube. The choke is located at the end of the tube

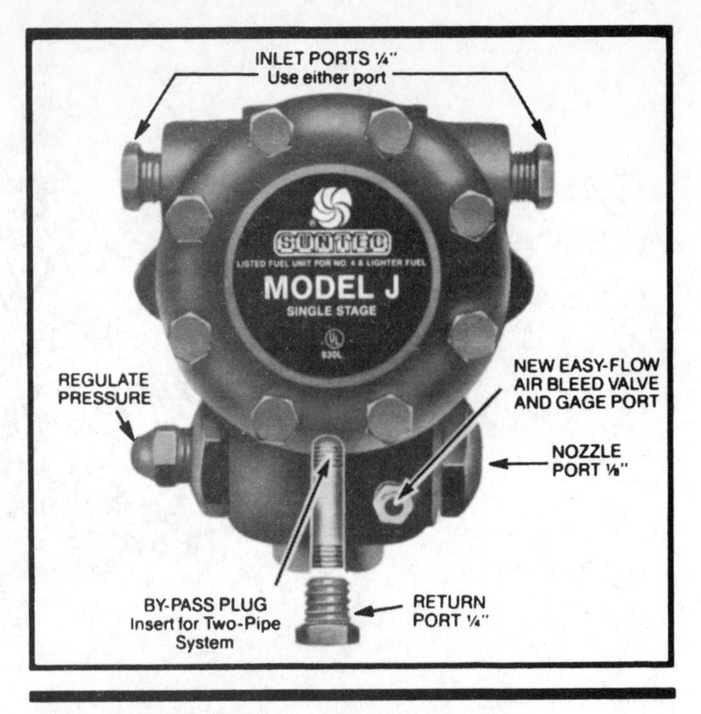

ONE-PIPE SYSTEM • INLET LINE ONLY

Fuel units are shipped without by-pass plug installed. Verify that no one has installed the by-pass plug. For line lengths under 50 feet, use 3/8" O.D. copper tubing. For line lengths 50 to 100 feet, use 1/2" O.D. copper tubing. "J" and "H" models are not recommended for lift above two feet, max 2" Hg inlet vacuum, except for the J2-F (Fig. 2).

TWO-PIPE SYSTEM • INLET AND RETURN LINE

Remove internal by-pass plug from plastic bag and insert as shown in illustration. Tighten securely. For recommended line sizes, refer to charts on reverse side. Maximum operating vacuum at fuel unit for the "J" model is 12" Hg. The Model "H" should not be used where inlet vacuum exceeds 20" of Hg at 1725 and 17" of Hg at 3450 rpm.

Figure 13.4 Oil pump showing connection locations. (Courtesy of Suntec Industries, Inc.)

and restricts the area, thus further increasing the air velocity. The swirl vanes located near the choke make the leaving air turbulent, which assists the mixing action.

Nozzle Assembly

The nozzle assembly consists of the oil feed line, the nozzle, electrodes, and transformer connections (Figure 13.9). The assembly serves to position the electrodes in

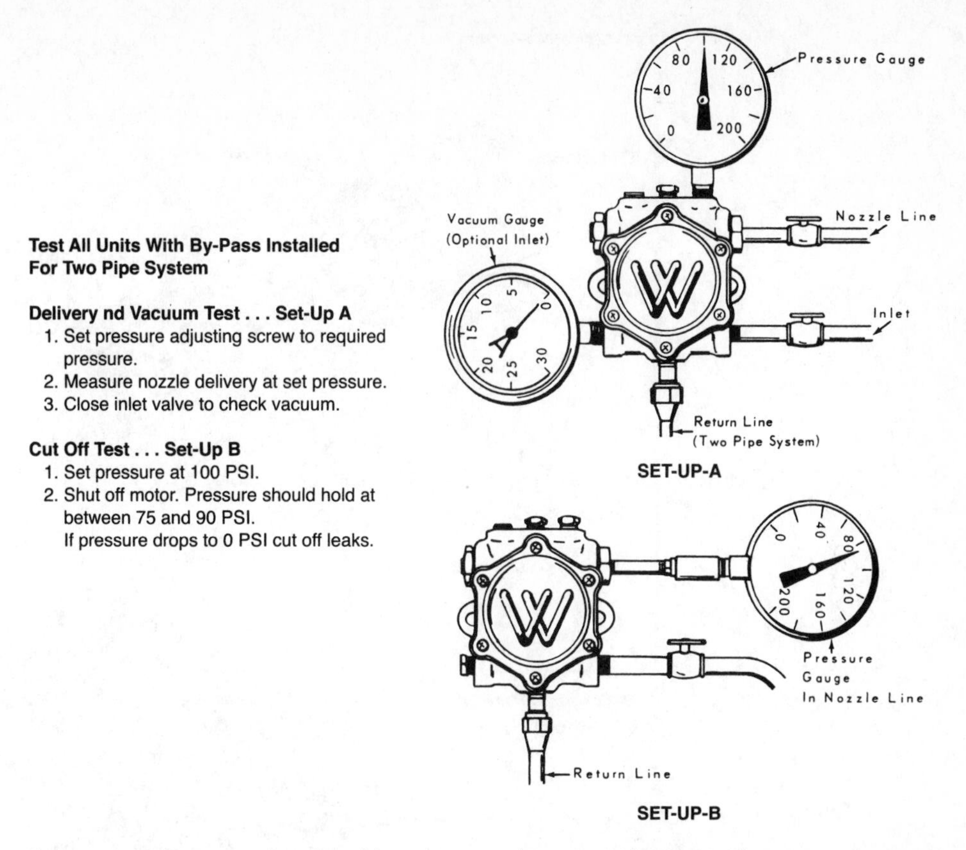

Figure 13.5 Test procedure for oil burner pump. (Courtesy of Webster Electric Company, Inc.)

Figure 13.6 Oil burner motor. (Courtesy of Essex Group, Controls Division, Steveco Products, Inc.)

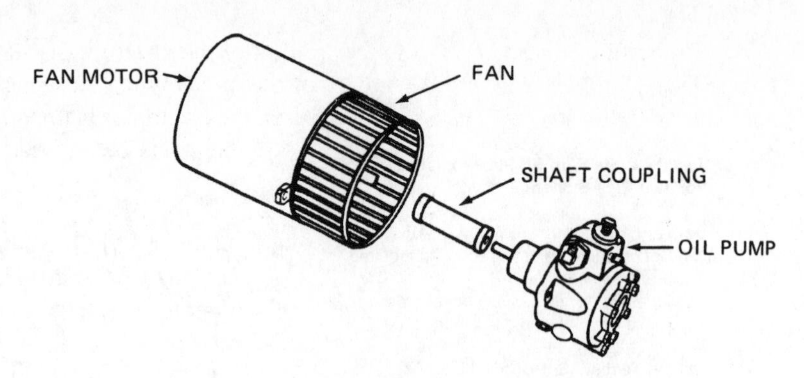

Figure 13.7 Shaft coupling. (Courtesy of Carrier Corporation)

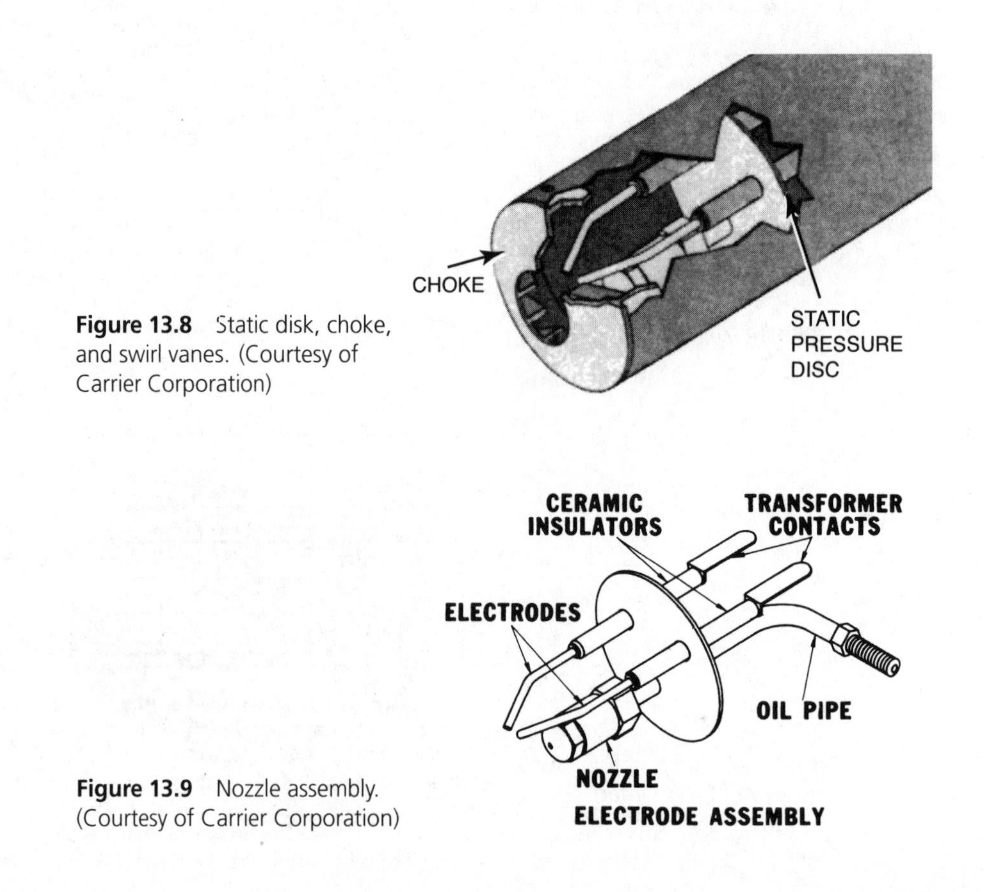

Figure 13.8 Static disk, choke, and swirl vanes. (Courtesy of Carrier Corporation)

Figure 13.9 Nozzle assembly. (Courtesy of Carrier Corporation)

respect to the nozzle opening and provides a mounting for the high-potential electrical leads from the transformer. The assembly is located near the end of the blast tube.

Electrode Position The correct position of the electrodes is shown in Figure 13.10. The electrodes must be located out of the oil spray but close enough for the spark to arc into the spray. The spark gap between the electrodes is important. If the electrodes are not centered on the nozzle orifice, the flame will be one-sided and will cause carbon to form on the nozzle.

Nozzle A nozzle construction is shown in Figure 13.11. The purpose of the nozzle is to prepare the oil for mixing with the air. The process is called *atomization.* Oil first enters the strainer, which has a mesh that is finer in size than the nozzle orifice in order to catch solid particles that could clog the nozzle.

From the strainer, the oil enters slots that direct the oil to the swirl chamber. The swirl chamber gives the oil a rotary motion when it enters the nozzle orifice, thus shaping the spray pattern. The nozzle orifice increases the velocity of the oil. The oil leaves in the form of a mist or spray and mixes with the air from the blast tube. Because of the fine tolerances of the nozzle construction, servicing is usually impractical. A defective or dirty nozzle is normally replaced.

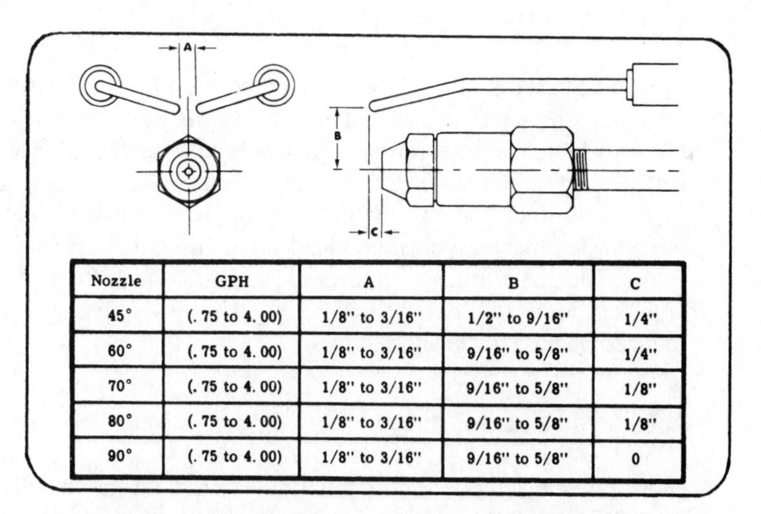

Nozzle	GPH	A	B	C
45°	(.75 to 4.00)	1/8" to 3/16"	1/2" to 9/16"	1/4"
60°	(.75 to 4.00)	1/8" to 3/16"	9/16" to 5/8"	1/4"
70°	(.75 to 4.00)	1/8" to 3/16"	9/16" to 5/8"	1/8"
80°	(.75 to 4.00)	1/8" to 3/16"	9/16" to 5/8"	1/8"
90°	(.75 to 4.00)	1/8" to 3/16"	9/16" to 5/8"	0

Recommended Electrode Settings. NOTE: Above 4.00 GPH, it may be advisable to increase dimension C by ⅛" to insure smooth starting. When using double adapters: (1) Twin ignition is the safest and is recommended, with settings same as above. (2) With single ignition, use the same A and B dimensions as above, but add ¼" to dimension C. Locate the electrode gap on a line midway between the two nozzles.

Figure 13.10 Electrode positioning. (Courtesy of Delavan Corporation)

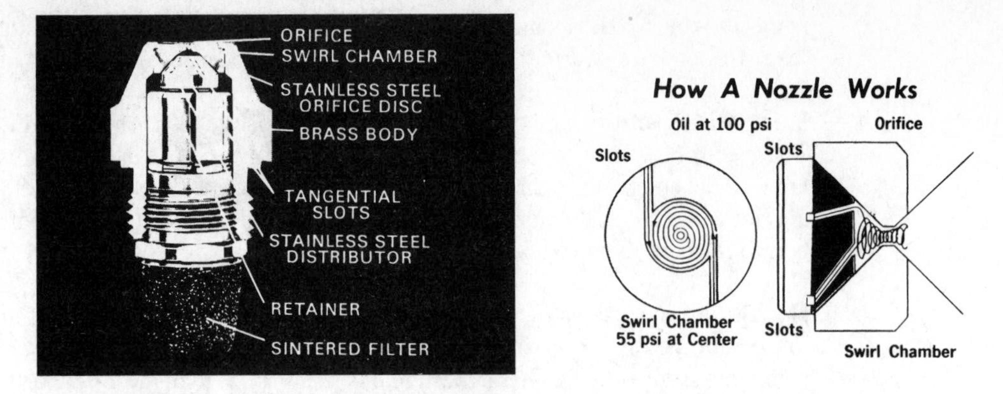

Figure 13.11 Nozzle construction. (Courtesy of Delavan Corporation)

Spray Patterns The requirements for nozzles vary with the type of application. Nozzles are supplied with two variations: the shape of the spray and the angle between the sides of the spray. There are three spray shapes, as shown in Figure 13.12: hollow (H), semihollow (SH), and solid (S). The hollow and the semihollow are most popular on domestic burners because they provide better efficiencies when used with modern combustion chambers.

The angle of the spray must correspond to the type of combustion chamber (Figure 13.13). An angle of 70° to 90° is usually best for square or round chambers. An angle of 30° to 60° is best for long, narrow chambers.

Ignition System All high-pressure burners have electric ignition. The power is supplied by a step-up transformer connected to two electrodes. A transformer is shown in Figure 13.14. The transformer supplies high voltage, which causes a spark to jump between the two electrodes. The force of the air in the blast tube causes the spark to arc (or bend) into the oil–air mixture, igniting it. Ceramic insulators surround the electrodes where they are close to metal parts. Insulators also serve to position the electrodes. The transformer* increases the voltage from 120 V to 10,000 V but reduces the amperage (current flow) to about 20 mA. This low amperage is relatively harmless and reduces wear on the electrode tips.

Oil Storage and Piping

Most cities have codes and regulations governing the installation and piping of oil tanks to which the installation mechanic must adhere. There are two tank locations:

1. Outside (underground)
2. Inside (usually in the basement)

An outside tank can be 550, 1000, or 1500 gal in capacity. An inside tank usually has a capacity of 275 gal. Tanks must be approved by Underwriters Laboratories Inc. (UL).

*Because of its high voltage, caution should be used in checking the transformer.

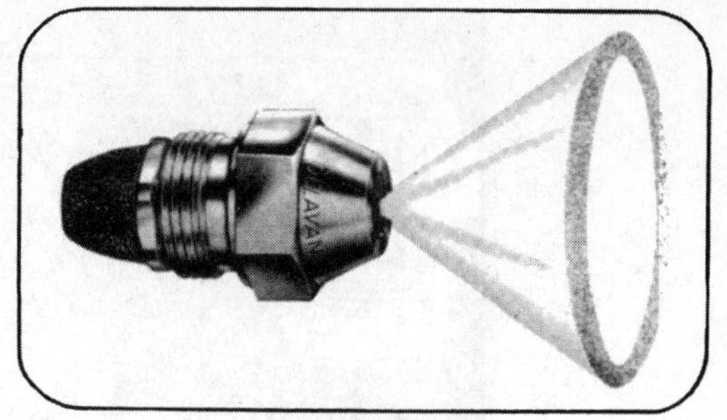

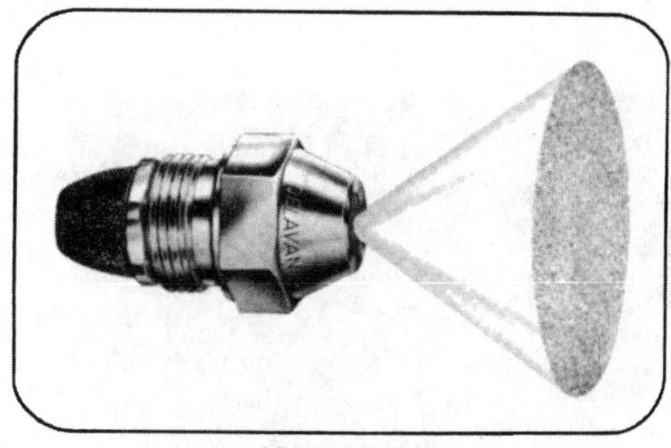

Figure 13.12 Spray patterns. (Courtesy of Delavan Corporation)

Outside Underground Tank As shown in Figure 13.15, an outside underground tank should be installed at least 2 ft below the surface. The fill pipe should be 2 in. iron pipe size (IPS) and the vent pipe 1¼ in. IPS. The vent pipe must be of a greater height than the fill pipe. The bottom of the tank should slope away from the end with the suction connection. A slope of 3 in. over its length is adequate. Piping connections to the tank should use swing connections to prevent breakage of the pipe when movement occurs. Oil-suction and return lines are usually made of ⅜-in.-outside-diameter (OD) copper tubing, positioned within 3 in. of the bottom of the tank.

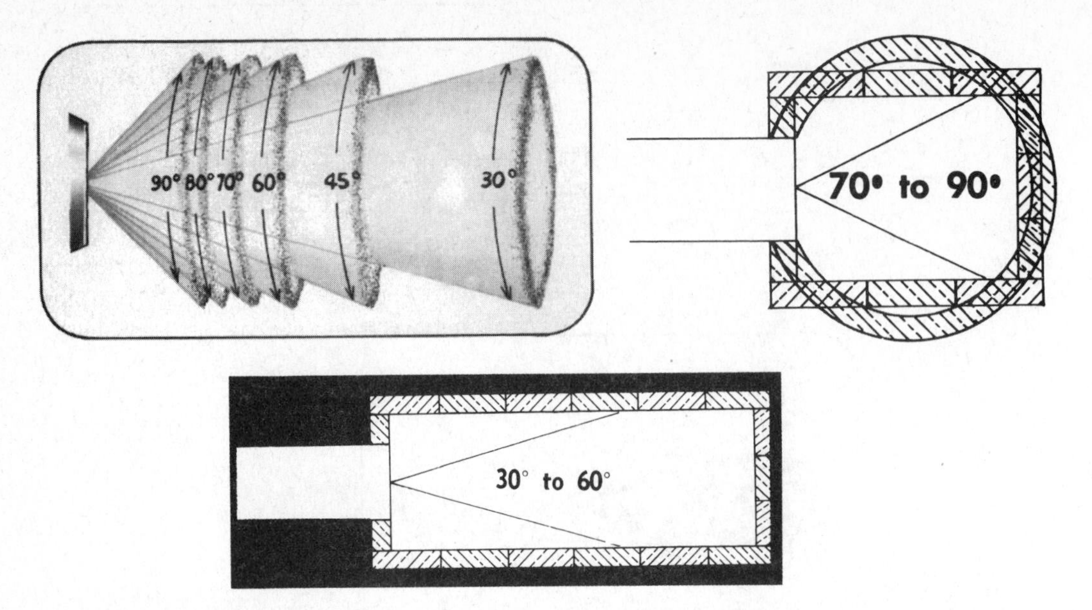

Figure 13.13 Relationship of spray angle to combustion chamber shape. (Courtesy of Delavan Corporation)

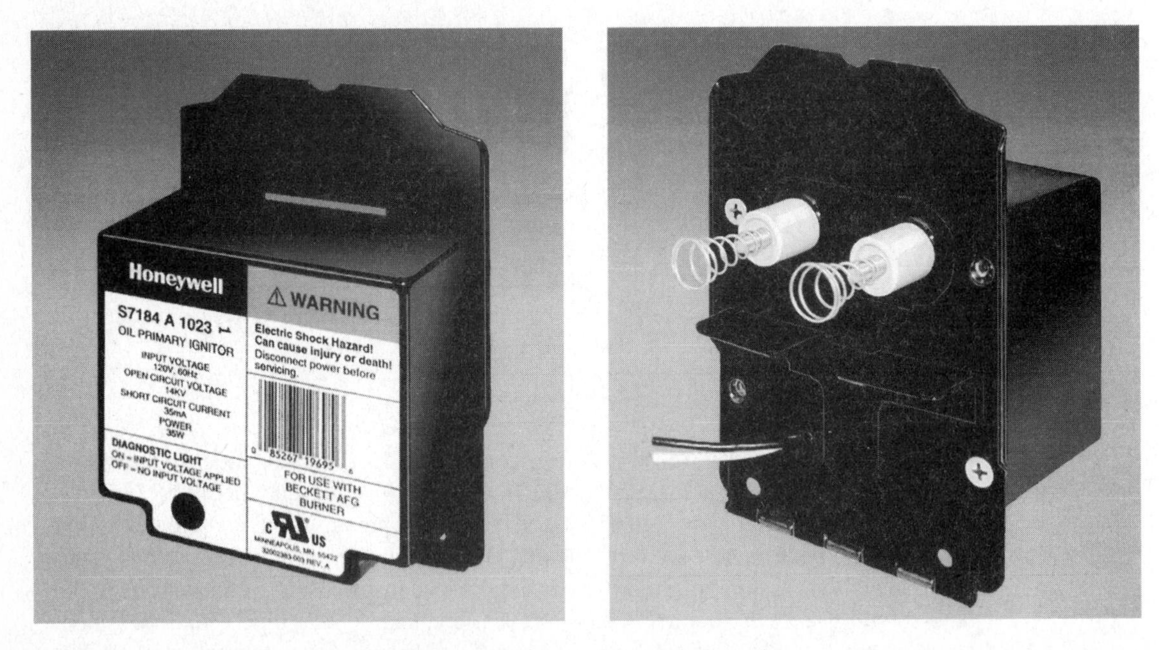

Figure 13.14 Transformer. (Courtesy of Honeywell, Inc.)

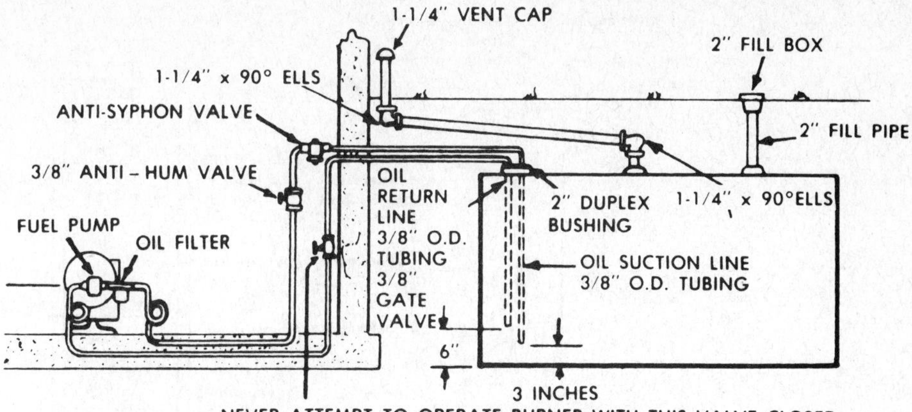

TWO STAGE PUMP (2 PIPE)

DISTANCE TANK BOTTOM BELOW PUMP (FEET)	MAX. RUN LENGTH (FT.) INCLUDES LIFT		DISTANCE TANK BOTTOM BELOW PUMP (FEET)	MAX. RUN LENGTH (FT.) INCLUDES LIFT	
	3/8" O.D. TUBING	1/2" O.D. TUBING		3/8" O.D. TUBING	1/2" O.D. TUBING
1	65	100	9	45	100
2	63	100	10	42	100
3	60	100	11	40	100
4	58	100	12	37	100
5	55	100	13	35	100
6	53	100	14	32	100
7	50	100	15	30	100
8	48	100	16	27	100

Figure 13.15 Outside underground tank installation. (Courtesy of Heil–Quaker)

A UL-approved suction-line filter should be used. A globe valve should be installed between the tank and the filter, and check valve should be installed in the suction line at the pump to keep the line filled at all times. A suitable oil-level gauge should be installed. The tank should be painted with at least two coats of tar or asphaltum paint.

Inside Tank An inside tank installation is shown in Figure 13.16. Many of the regulations and installation conditions for inside tanks are the same as those for outside tanks. The fill pipe should be 2 in. IPS and the vent pipe, 1¼ in. IPS. An oil-level gauge should be installed. The oil piping is usually ⅜-in.-OD copper tubing. A filter should be installed in the oil-suction lines.

Inside tank installation varies in some ways from outside tank installation. The slope for the tank bottom should be about 1 in. and directed toward the oil-suction connection. Two globe-type shutoff valves should be used, one at the tank and one at the burner. The valve at the tank should be a fuse type. The oil line between the tank and the burner should be buried in concrete. The tank should be at least 7 ft. from the burner.

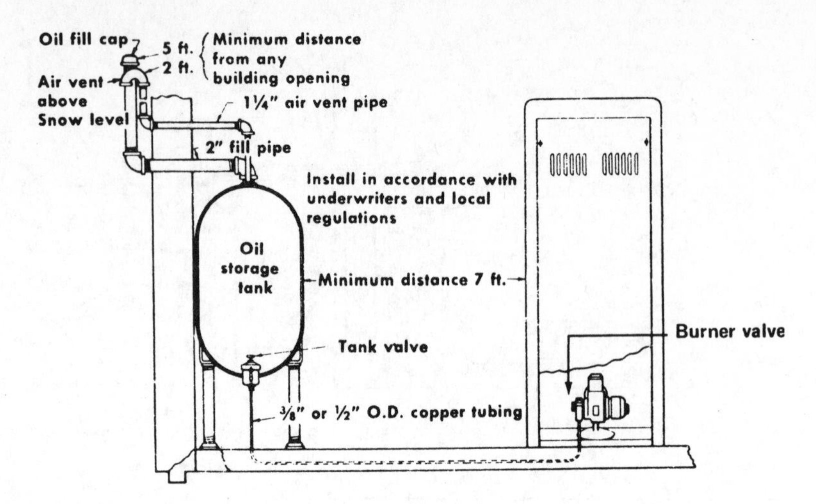

Figure 13.16 Inside tank installation. (Courtesy of Webster Electric Company, Inc.)

Accessories

Two important accessories are the combustion chamber and the draft regulator. The combustion chamber is placed in the lower portion of the heat exchanger, surrounding the flame on all sides with the exception of the top. A barometric damper is used as a draft regulator. The draft regulator is placed in the flue pipe between the furnace and the chimney.

Combustion Chambers

The purpose of a combustion chamber is to protect the heat exchanger and to provide reflected heat to the burning oil. The reflected heat warms the tips of the flame, assuring complete combustion. No part of the flame should touch the surface of the combustion chamber. If the combustible mixture does touch, the surface will be cooled and incomplete combustion will result. It is essential that the chamber fit the flame. The nozzle must be located at the proper height above the floor. The bottom area of the combustion chamber is usually 80 in.^{2}/gal for nozzles between 0.75 and 3.00 gpm, 90 in.^{2}/gal for nozzles between 3.50 and 6.00 gpm, and 100 in.^{2}/gal for nozzles 6.50 gpm and higher.

Three types of material are generally used for combustion chambers:

1. Metal (usually stainless steel)
2. Insulating firebrick
3. Molded ceramic

Metal is used in factory-built, self-contained furnaces and is used without backfill (material in space between chamber and heat exchanger). Insulated firebrick is used for conversion burners to fit almost any type of heat-exchanger shape and uses fiberglass backfill. The molded ceramic chamber is prefabricated to the proper shape and size and uses fiberglass backfill (Figure 13.17).

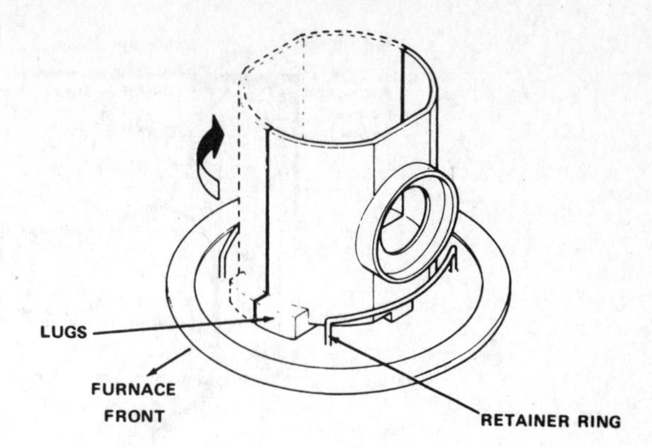

Figure 13.17 Molded ceramic combustion chamber. (Courtesy of Luxaire, Inc.)

Draft Regulators

A draft regulator maintains a constant draft over the fire, usually −0.01 to −0.03 in. W.C. Too high a draft causes undue loss of heat through the chimney. Too little draft causes incomplete combustion. A draft regulator (Figure 13.18) consists of a small door in the side of the flue pipe. The door is hinged near the center and controlled by adjustable weights. Building air is admitted to the flue pipe as required to maintain a proper draft over the fire.

Maintenance and Service

Annual maintenance of oil burner equipment is essential to good operation. Service procedures are outlined later in this section (see Figure 13.19).

Burner Assembly

1. Clean fan blades, fan housing, and screen.
2. Oil motor with a few drops of SAE No. 10 oil.
3. Clean pump strainer.
4. Adjust oil pressure to 100 psig. (Some models require operating pressures in the 140–200 psig range.)
5. Check oil pressure cutoff.
6. Conduct combustion test and adjust air to burner for best efficiency.

Nozzle Assembly

See Figures 13.20 and 13.21.

1. Replace nozzle.
2. Clean nozzle assembly.

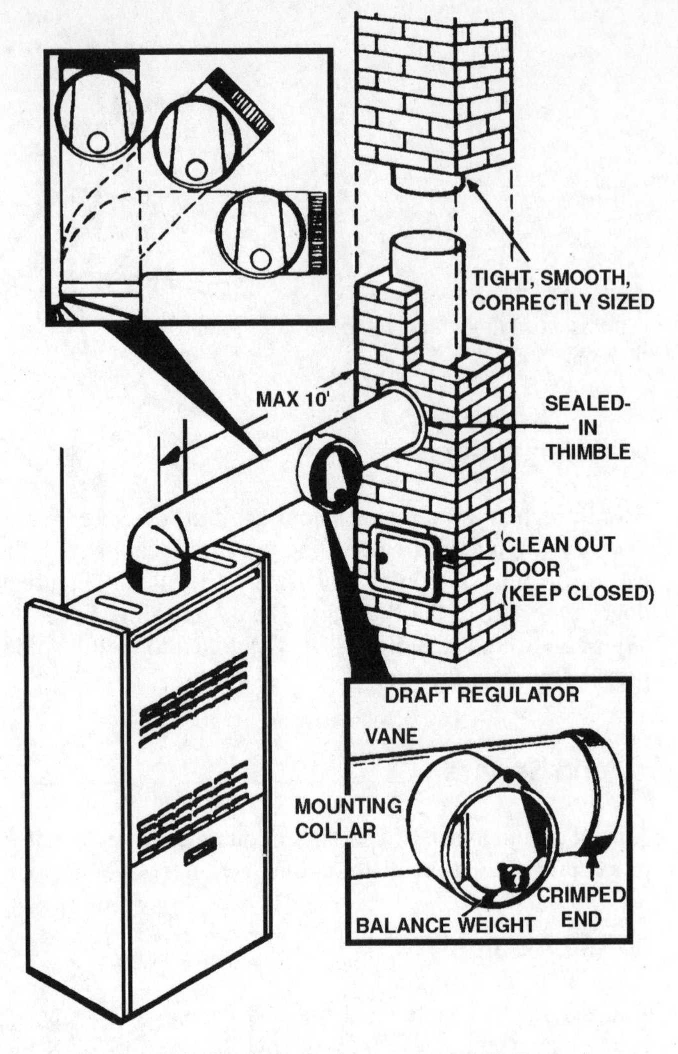

Figure 13.18 Application of draft regulator. (Courtesy of Heil–Quaker)

3. Check ceramic insulators for hairline cracks, and replace if necessary.
4. Check location of electrodes, and adjust if necessary.
5. Replace cartridges in oil-line strainers.

Ignition System and Controls

1. Test transformer spark.
2. Clean thermostat contacts.

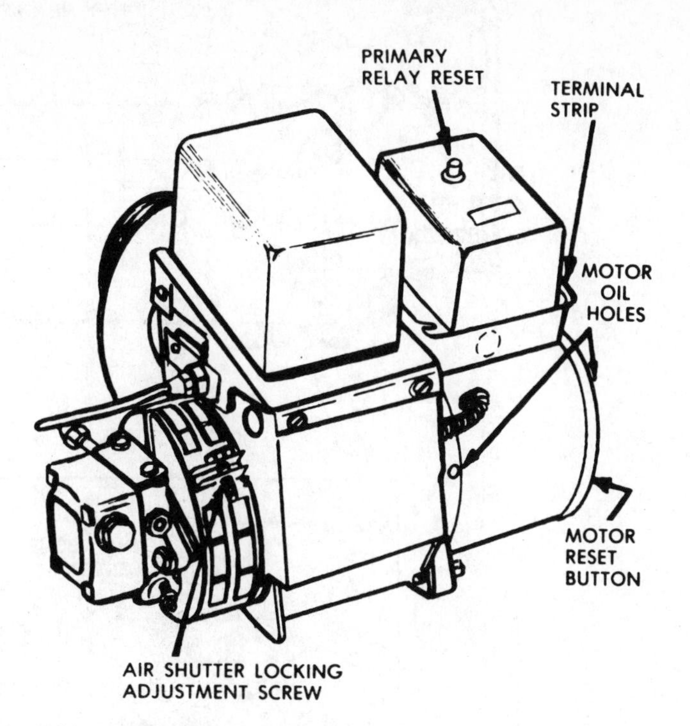

Figure 13.19 Oil burner. (Courtesy of Heil–Quaker)

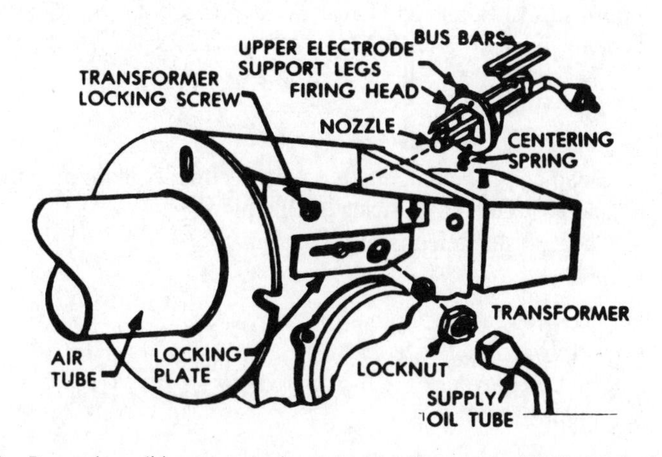

Figure 13.20 Removing oil burner nozzle assembly. (Courtesy of Heil–Quaker)

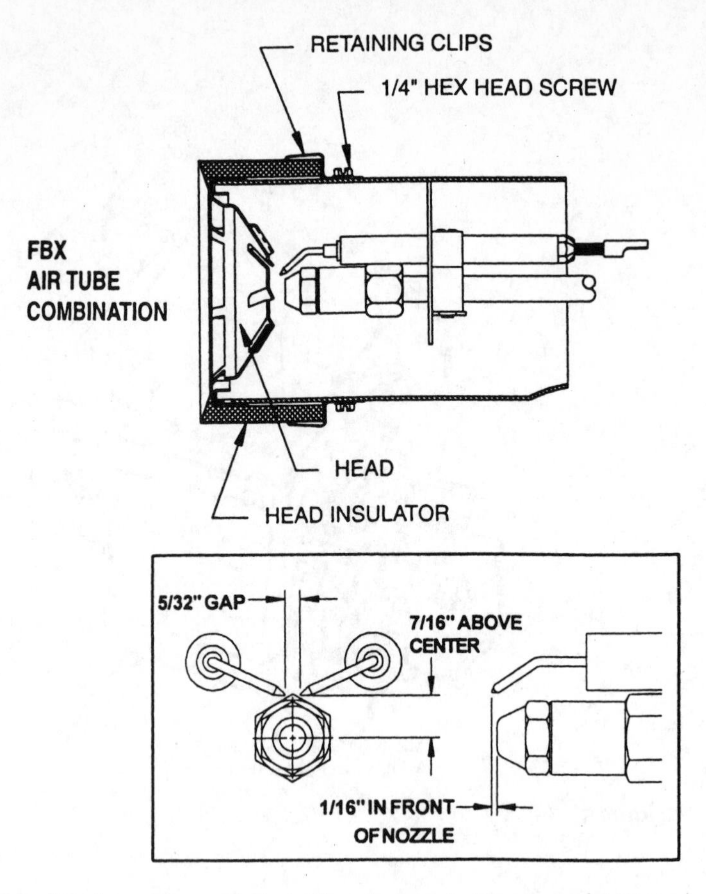

Figure 13.21 Nozzle position and electrode adjustment. (Courtesy of R.W. Beckett Corporation) See Figure 13.10 for settings based on different nozzle angles.

3. Clean control elements that may become contaminated with soot, especially those that protrude into furnace or flue pipe.
4. Check system electrically.

Furnace

See Figure 13.22.

1. Clean combustion chamber and flue passages.
2. Clean furnace fan blades.
3. Oil fan motor.
4. Replace air filter.

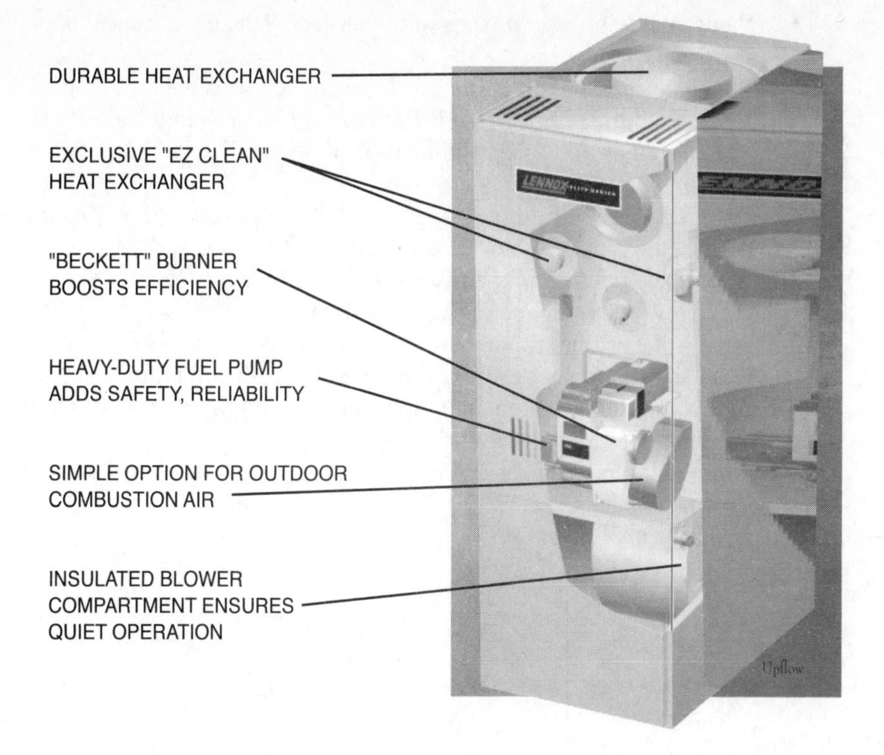

Figure 13.22 Components of a typical forced oil-fired warm air furnace. (Courtesy of Lennox Industries Inc.)

After this work is completed, run the furnace through a complete cycle and check all safety controls (Chapter 14). Clean up the exterior of the furnace and the area around the furnace.

Caution: If the unit runs out of oil or if there is an air leak in the lines, the fuel pump can become air-bound and not pump oil. To correct this situation, the air must be bled from the pump and replaced with oil. On a one-pipe system, this is done by loosening or removing the plug on the port opposite the intake. Start the furnace and run it until oil flows out of the opening; then turn it off and replace the plug. The system can then be put back into operation. A two-pipe pump is considered to be self-priming. If it does fail to prime, however, follow the procedure just described.

STUDY QUESTIONS

Answers to the study questions may be found in the sections noted in brackets.

13-1 How is oil vaporized on a pressure-operated gun-type oil burner? *[Oil-Burning Units]*

13-2 Name and describe two types of oil burners. Which is the most popular? *[Types of Burners]*

13-3 On a high-pressure oil burner, is the air mixed with the oil before or after passing through the nozzle? *[High-Pressure Burner Components]*

13-4 Are the electrodes that ignite the oil located in or out of the oil spray? *[Electrode Position]*

13-5 Describe the application of single- and two-stage oil pumps. *[Fuel Pumps]*

13-6 What is the type of motor used on a gun-type oil burner? *[Motor]*

13-7 What is the purpose of the nozzle on an oil burner? *[Nozzle]*

13-8 What governs the shape of flame used on the oil burner? *[Spray Patterns]*

13-9 What is the normal capacity of an inside oil tank? *[Oil Storage and Piping]*

13-10 What size of copper tubing is used to connect the oil burner? *[Inside Tank]*

13-11 What are the maintenance and service requirements of an oil burner? *[Maintenance and Service]*

13-12 What action is necessary in case the burner runs out of oil and fails to start? *[Maintenance and Service]*

CHAPTER 14

Oil Furnace Controls

OBJECTIVES

After studying this chapter, the student will be able to:

- Identify the various controls and circuits on an oil-fired furnace
- Determine the sequence of operation of the controls
- Service and troubleshoot the oil heating unit control system

Use of Oil Furnace Controls

Many of the oil furnace controls are similar to those of gas furnaces. The main difference is in the control of the fuel-burning equipment. The major components controlled are:

- Oil burner (including ignition)
- Fan
- Accessories

As with gas furnaces, certain variations must be incorporated in the oil furnace control system to comply with various furnace models. A downflow furnace, for example, needs an auxiliary limit control and a timed fan start control, which are not required on an upflow furnace.

Circuits

The control circuits for an oil-burning furnace consist of:

- Power circuit
- Fan circuit
- Ignition circuit
- Oil burner circuit
- Accessory circuits

Primary Control

Several of the control circuits are combined in a single primary control to simplify control construction and field wiring. The primary control is a type of central control assembly that supplies power for the ignition and oil burner circuits. The primary control actuates the oil burner and provides a safety device that stops the operation of the burner if the flame fails to ignite or is extinguished for any reason. This safety device prevents any sizable quantity of unburned oil from flowing into the furnace, thus reducing the possibility of an explosion. Two types of oil primary control sensors are commonly used:

1. Bimetallic sensor
2. Cad-cell sensor

Bimetallic Sensor The bimetallic sensor type of oil primary control is often called a *stack relay* because it is normally placed in the flue pipe between the heat exchanger and the barometric damper. It includes a relay that permits the low-voltage (24-V) thermostat to operate the line-voltage (120-V ac) oil burner. The internal wiring for a bimetallic sensor type of primary control is shown in Figure 14.1. The entry of the power supply (120-V ac) is shown at the top of the diagram. Line-voltage power is connected to the oil burner and to the transformer located in the primary control. The balance of the control operates on low voltage.

Figure 14.2 shows the installation of the bimetallic sensor (pyrotherm detector) in the furnace flue. Figure 14.3 shows the mechanical action that takes place when the bimetal expands as it senses heat from the flame. The normal position of the switches in this control are cold contacts NC, hot contacts NO. When the bimetal is heated, the cold contacts break, and the hot contacts make.

When the thermostat calls for heat (see Figure 14.1) the cold contacts are made, and the safety switch heater is energized. If the flame fails to light and the cold contacts remain closed, the safety switch heater will "warp out" (open) the safety switch. The heaters and switch constitute a type of heat relay. With current flowing through this circuit, about 90 seconds is required to trip (open) the safety switch. If the burner produces flame and heats the bimetallic element, the switching occurs, opening the cold contacts and closing the hot contacts. This action causes the current to bypass the safety switch heater, and the burner continues to run.

A stack relay type of primary control is shown in Figure 14.4. Line-voltage power is connected across terminals 1 and 2, and the oil burner motor is connected across 2 and 3. Constant ignition is also connected across terminals 2 and 3. Intermittent ignition is connected across terminals 2 and 4. The thermostat is connected to terminals T and T (or W and B).

Two types of ignition are used in the field: constant and intermittent. With constant ignition the electrodes spark continuously. Intermittent ignition operates only when the burner is started. With constant ignition, both the oil burner motor and the ignition transformer are wired across terminals 2 and 3. Bimetallic stack relays are found on many furnaces now in the field. The new units, however, are usually

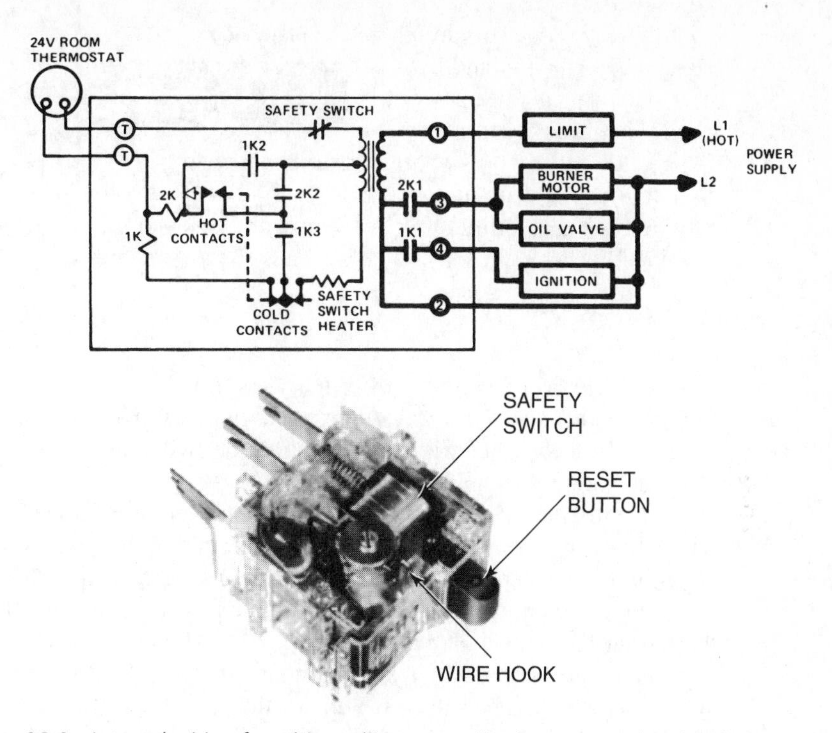

Figure 14.1 Internal wiring for a bimetallic sensor oil primary control with safety switch. (Courtesy of Honeywell Inc.)

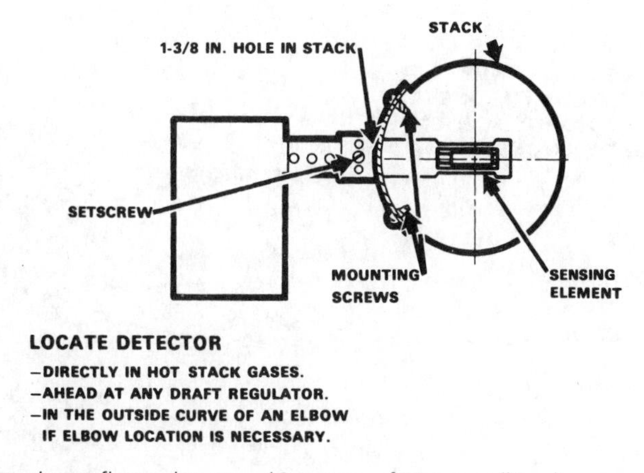

Figure 14.2 Pyrotherm flame detector. (Courtesy of Honeywell Inc.)

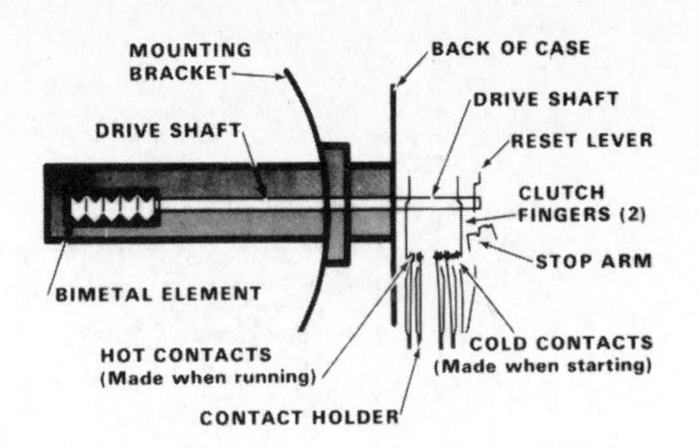

Figure 14.3 Pyrotherm flame detector operation. (Courtesy of Honeywell Inc.)

equipped with a cad-cell type of primary control. *For the service technician's protection, it is important that the hot line of the power supply be brought to the number 1 terminal.* In accordance with good practice, the switching action of this control should be in the hot line.

If the thermostat calls for heat, 120-V ac power must be available across terminals 2 and 3 on a constant-ignition system. If intermittent ignition is used, though, 120-V ac power must also be available across terminals 2 and 4. After the thermostat has been removed, the thermostat terminals can be jumped to simulate calling for heat by the thermostat.

The control should be tested without supplying heat to the bimetallic element. When this is done, the safety switch should shut off the power to the burner in about 90 seconds. If heat is supplied to the bimetallic element, the burner should continue to run. Sometimes these controls get "out of step"; that is, the cold contacts do not remake when the bimetal cools down. This situation should be corrected by following

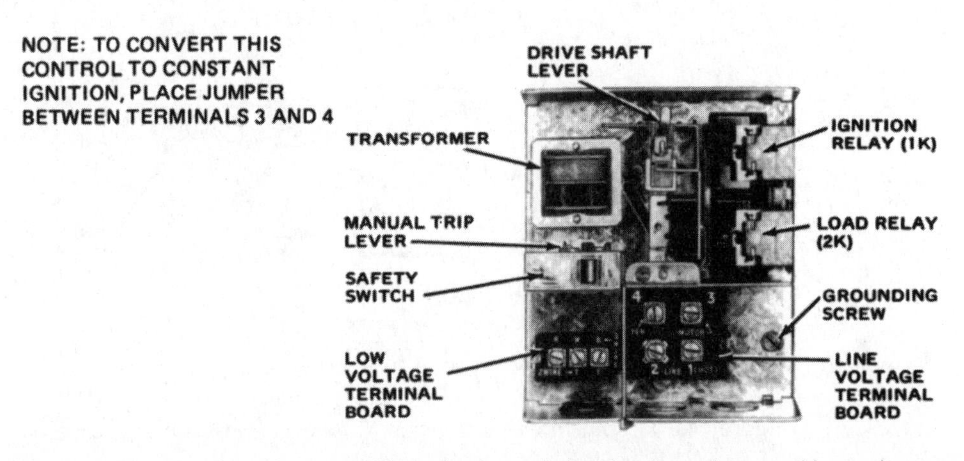

Figure 14.4 Typical primary control (pyrotherm type) showing line-voltage and low-voltage electrical terminals. (Courtesy of Honeywell Inc.)

the restepping procedure shown in Figure 14.5. To test the contact terminals, jumper across the cold contacts, as shown in Figure 14.6. If the bimetal is cold and contacts cannot be restepped, replace the primary control.

Stack relays have some built-in limitations. They are slow-acting due to the thermal lag of the bimetal. Wiring must be completed in the field, so performance cannot be fully checked by the manufacturer. The bimetal element is exposed to the products of combustion and requires more maintenance than the cad-cell detector. Stack relays are no longer used in new installations. They are, however, used for the replacement of defective units. All new residential oil heating systems utilize cad-cell primary controls.

Cad-Cell Sensor A cad cell is shown in Figure 14.7, and its location is shown in Figure 14.8. The cad-cell primary control has one feature that is not present in the bimetallic sensor. If the cad cell senses stray light (light not from the flame) before the thermostat calls for heat, the burner cannot be started by the thermostat. This is a protective arrangement. Therefore, when testing the operation of the cad-cell primary, block off all light that might reach the cad cell.

The cad cell itself can be tested with a suitable ohmmeter. The ohmmeter must be capable of reading up to 100,000 Ω. In the absence of light, the cad cell should have a resistance of about 100,000 Ω. In the presence of light, its resistance should not exceed 1500 Ω.

To test the primary control, start the burner and within 30 seconds jump the cad-cell terminals with a 1000- to 1500-Ω resistor. If the burner continues to run, the cad-cell primary is operating correctly. In actual operation, a poor or inadequate oil burner flame can be a problem. The safety arrangement on a cad-cell primary control usually stops the burner in about 30 seconds if the cad cell does not sense an adequate flame.

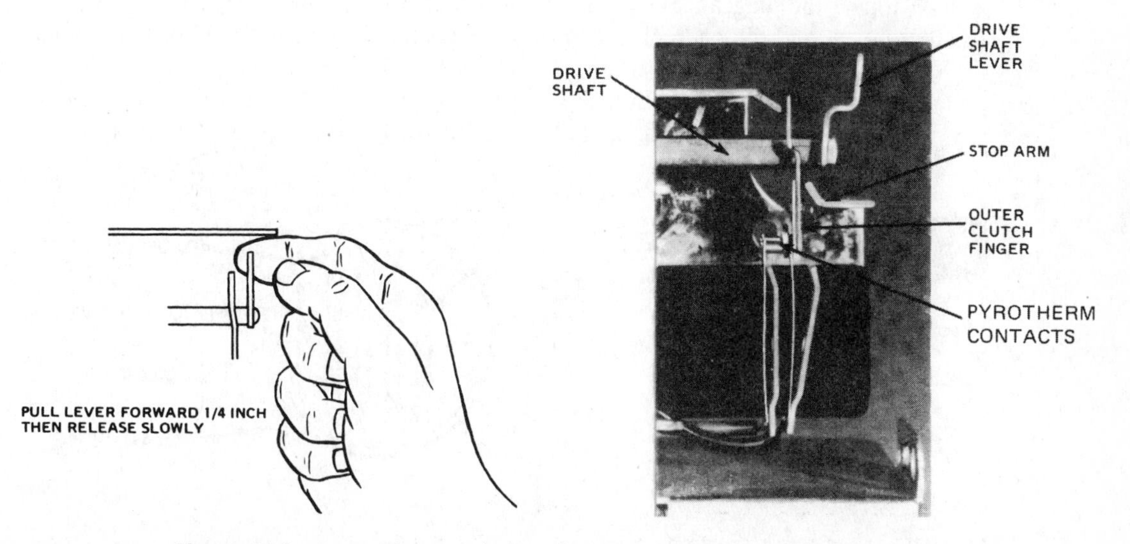

Figure 14.5 Restepping pyrotherm relay control. (Courtesy of Honeywell, Inc.)

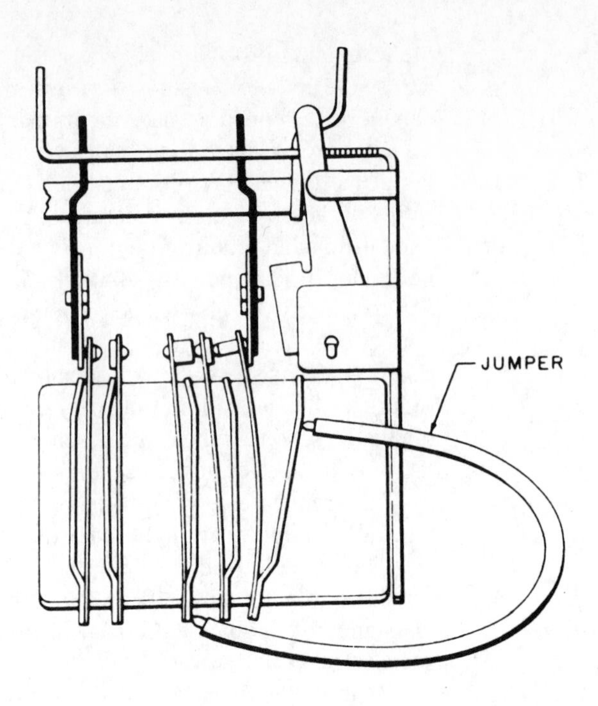

Figure 14.6 Use of jumper to test cold contacts on pyrotherm relay control. (Courtesy of Honeywell Inc.)

Electrical Response The ability of the cad cell to change electrical resistance when exposed to different intensities of light makes possible the operation of a safety control circuit. Figure 14.9 shows the internal wiring connections for the primary cad-cell types of primary control as well as the internal circuiting. The power (120-V ac) connection to the burner and primary control transformer are shown at the top of the diagram. The balance of the control is low voltage (24-V ac). The thermostat and the cad cell are shown connected to the right side of the primary control (see Figure 14.10b).

When the thermostat calls for heat, the oil burner starts and the safety switch heater is energized. If the flame fails to ignite, the heater warps out the safety switch

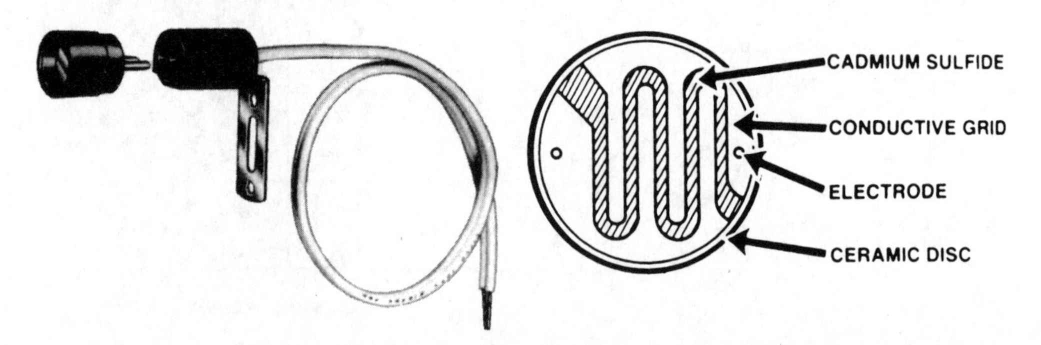

Figure 14.7 Cad-cell assembly and details of face. (Courtesy of Honeywell Inc.)

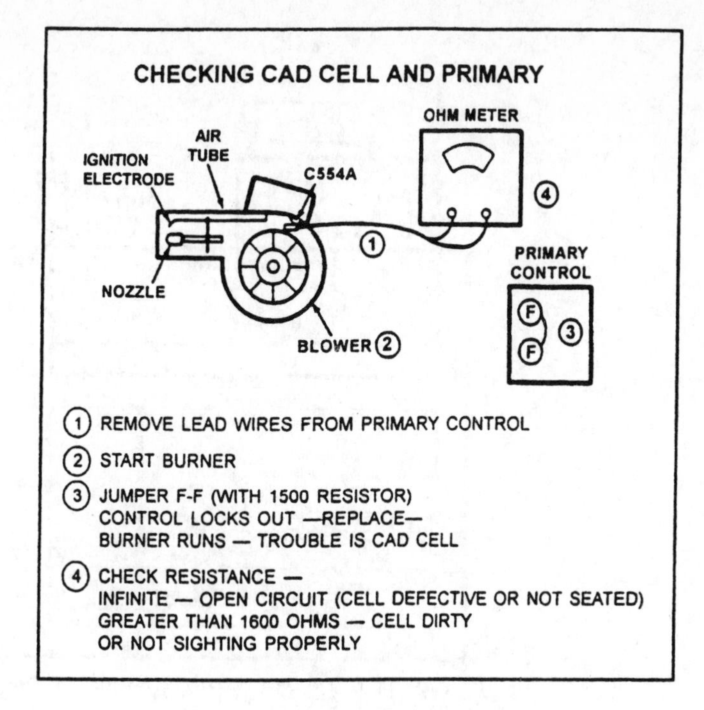

CAD CELL TROUBLESHOOTING

OHMMETER READING	CAUSE	ACTION
0 ohms	Short circuit.	Check for pinched cad cell leadwires.
Less than 1600 ohms but not 0	Cad cell and application are operating correctly.	None.
Over 1600 ohms but not infinite.	Dirty or defective cell, improper sighting, or improper air adjustment.	1. Clean cell face and recheck 2. Check flame sighting 3. Replace cell and recheck. 4. Adjust air band to get good reading
Infinite resistance.	Open circuit.	Check for improper wiring, loose cell in holder, or defective cell.

Figure 14.8 Cad-cell location in oil burner assembly. (Courtesy of R.W. Beckett Corporation)

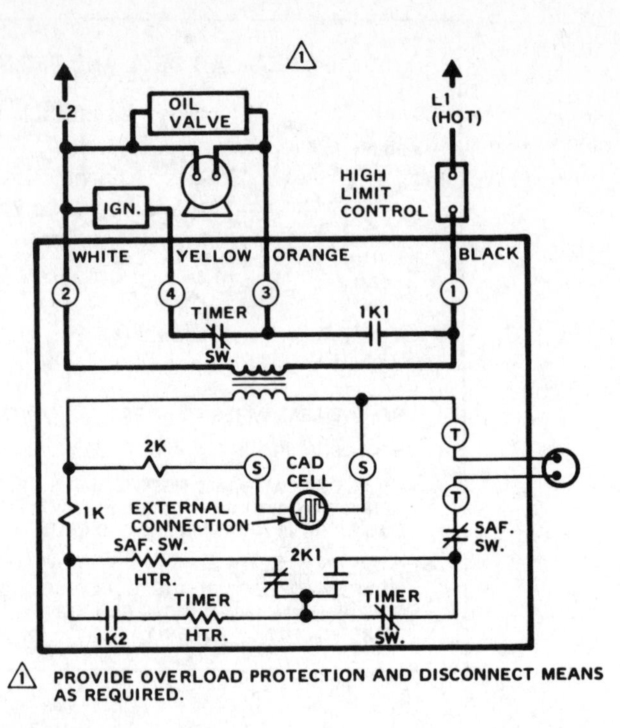

Figure 14.9 Wiring for cad-cell oil primary control, with intermittent ignition. (Courtesy of Honeywell Inc.)

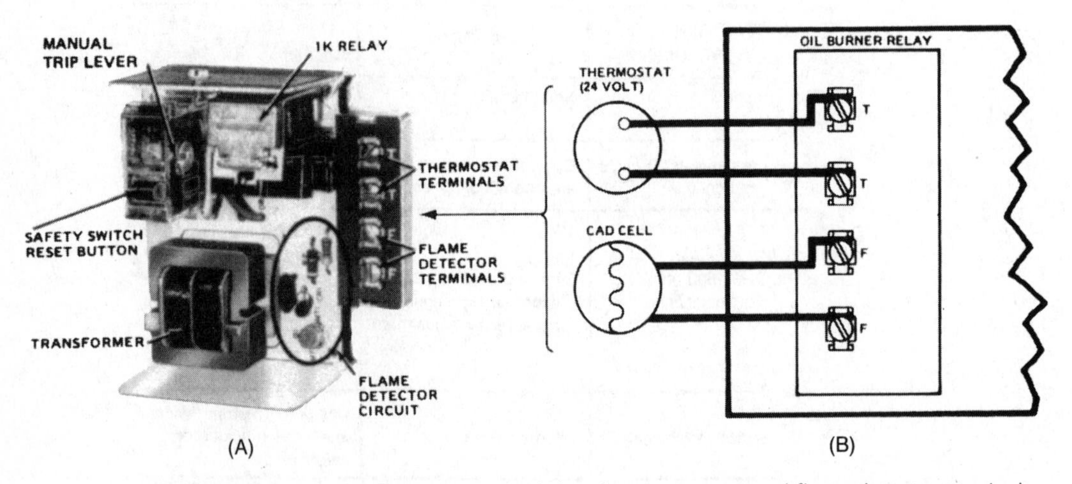

Figure 14.10 Cad-cell primary control showing thermostat and flame detector terminals. (Courtesy of Honeywell Inc.)

in about 30 seconds, turning off the burner. If the flame is produced, it is sensed by the cad cell. This causes the safety switch heater to be bypassed and the burner stays on.

The cad-cell primary control power supply (120-V ac) is wired to the black (hot) and white (neutral) connections. The oil burner motor is wired to the white and orange connections. The thermostat is connected to terminals T and T. The cad cell is connected to terminals S and S (or F and F), as shown in Figure 14.10.

Options Available There are two important options available on the cad-cell primary control:

1. Constant or intermittent ignition
2. Variations in motor control arrangement

On constant ignition, both the oil burner motor and the ignition transformer are wired across the white and orange connections. On intermittent ignition, the motor is wired across white and orange, and the ignition transformer is wired across white and yellow.

Solid-State Sensor Solid-state technology is based on manipulating semiconductors to perform different functions in an electronic circuit. The solid-state circuit, which replaces the sensitive relay in cad-cell primaries, consists of two resistors, one capacitor, and two solid-state switches. One of these is a bilateral switch, and the other is a Triac, as shown in Figure 14.11.

The Triac acts as an NO switch. It conducts current between terminals A and B after a triggering current is applied to the gate terminal. It continues to conduct after the triggering current is removed until current between A and B drops to zero. At this time, it becomes nonconducting, and another triggering signal is required before it will conduct again.

The bilateral switch is frequently used to trigger a Triac. When only terminals 1 and 2 are used, it acts as a resistor until a certain voltage, called the *breakover voltage,*

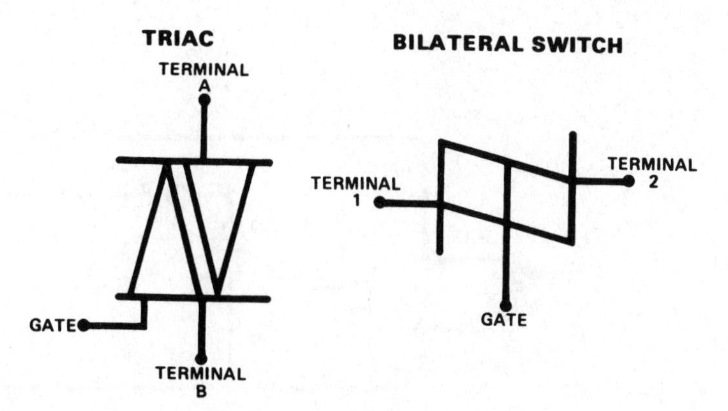

Figure 14.11 Solid-state switches. (Courtesy of Honeywell Inc.)

is reached. This voltage is very small, on the order of 8 V. When the breakover voltage is passed, the resistance of the switch collapses and current flows. When the bilateral switch is used as a three-terminal device, current is applied to the gate terminal to change the breakover voltage.

In Figure 14.12, which shows a solid-state flame-sensing circuit, the resistors and the capacitor in the cad-cell primary are not, strictly speaking, solid-state devices, but they are used extensively in solid-state circuits. A resistor simply opposes the flow of current. In a dc circuit, a capacitor blocks the flow of current. In an ac circuit, the capacitor stores current on one half-cycle and releases it on the reverse half-cycle. Resistor R1 in Figure 14.12 determines the voltage drop across the F-F terminals, and resistor R2 protects the solid-state components from abnormally high voltages. The capacitor provides an extra burst of current each half-cycle to ensure that the Triac is triggered when necessary. The solid-state flame-sensing circuit is less affected by ambient light hitting the cad cell than is a comparable sensitive relay.

Sequence of Operation The schematic of an intermittent-ignition model of a cad-cell primary control is shown in Figure 14.13. When the line switch is closed, the internal transformer is powered, whether or not the thermostat is calling for heat. A line-voltage thermostat can be used with this control if terminals T-T on the primary are jumpered. In this case, the primary and cad cell will not be powered until the thermostat calls for heat.

The solid-state flame-sensing circuit is designed so that the Triac will conduct current only when cad-cell resistance is high, indicating no flame. If cad-cell resistance is high, the voltage across the bilateral switch exceeds the breakover voltage, causing it to conduct and trigger the Triac. The capacitor provides a current pulse each half-cycle to make sure the breakover voltage is exceeded.

On flame failure, cad-cell resistance increases until the Triac is energized. The safety switch heats until the bimetal warps enough to break the circuit and stop the burner. The primary control must be reset before the burner can be restarted. If there is a power failure during a call for heat, the burner shuts down safely and automatically. The system automatically returns to normal operation when power is restored.

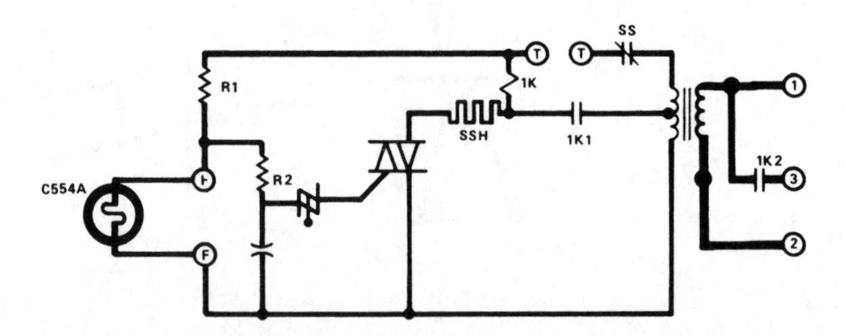

Figure 14.12 Solid-state flame-sensing circuit. (Courtesy of Honeywell Inc.)

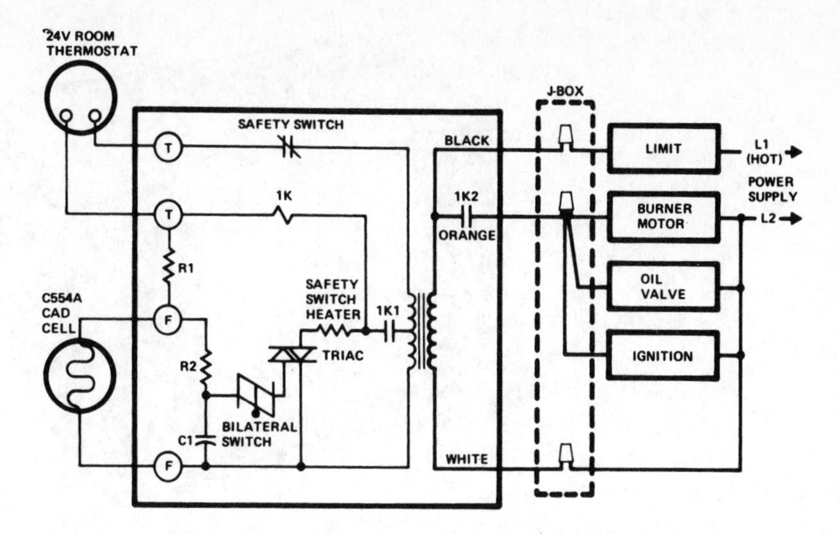

Figure 14.13 Wiring diagram for solid-state cad-cell primary. (Courtesy of Honeywell Inc.)

The Development of Modern Primary Controls

Advances in electronics technology have made an exciting impact in the oil heat industry. These developments have enabled control manufacturers to incorporate high-performance features in residential primary controls. Residential controls are now being made available that have features that were previously found only with the expensive primary control units for commercial or industrial applications. Features like valve-on delay (prepurge), burner motor-off delay (postpurge), and interrupted duty ignition are becoming universal. Preignition, limited reset and recycle, and alarm contacts are also helping to make today's controls advanced and powerful tools. This section discusses some of these important improvements and includes a discussion of the Honeywell R7184 series control (Figure 14.14), the most recent entry into the marketplace.

The following are some of the key developments:

Relays

Microcontrollers

Timers

Flame sensing

LED indicators

Self-checking and system checking

Advanced features

Relays have become smaller and more reliable. It is now common to see several relays inside each control, so that the motor, igniter and valve may all be controlled by separate relays. Solid-state relays are increasingly being used instead of the traditional electromechanical relays.

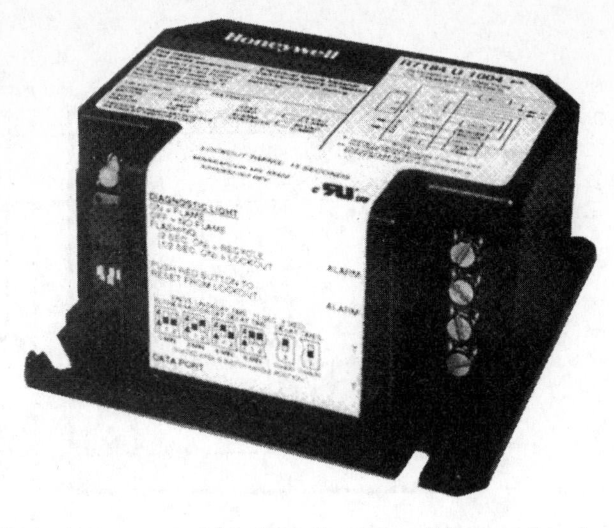

Figure 14.14 Basic primary control function. (Courtesy of Honeywell Inc.)

Microcontrollers (small computer chips designed specifically to control other electronics) have increased the ability to control the burner's components separately and with more intelligence. A microcontroller's software code can easily add control functions and make complex decisions, eliminating the need for large decision-making circuits. Circuits are now needed mostly to provide processor and 24-V power, to convert sensor signals, and to drive relays.

Timers have developed from being controlled by bimetal switches to being controlled by semiconductors. Now, many different timers can all be running at the same time. Timings are less affected by temperature, vibration, light, line voltage, and the like. They can be accurate to within fractions of a second, instead of varying by 10% to over 50% for some traditional controls.

Flame sensing has progressed from stack-mounted bimetal switches to cadmium sulfide sensors (cad cells) and ultraviolet sensors. These sensors can more quickly and accurately sense the flame, and they also have the ability to sense the varying brightness of a flame. Software programming can monitor the rise and fall of a cad-cell resistance to interpret its signal better. For instance, the cad-cell signal can be smoothed out to help prevent nuisance lockouts. The Honeywell R7184 can display the cad-cell resistance, so an ohmmeter is not required.

Note: For proper operation, it is important that the cad-cell resistance be below 1600 Ω.

LED indicators offer an increasing amount of diagnostic information such as recycling, flame status, or different lockout modes. Indicators like these are making it easier than ever to know what was happening before the technician arrived at the installation.

Self-checking and system checking to ensure a safe, proper starting and operation of the burner are now standard procedures in microcontroller-based controls.

Advanced features, both new and borrowed from more expensive controls, are being added to residential controls. In the near future, look for increased use of the reset button, special pump-priming procedures, redundant safety features, brownout protection, advanced recycle methods, and other features that will make controls better and the technician's job easier.

A handy R7184 quick reference guide for technicians is included in Figure 14.15.

Power Circuit

Following is the key to the schematic diagrams used in the study of the power circuit and other circuits described in this chapter:

BK	Black-color wire	FC	Fan control
BL	Blue-color wire	FM	Fan motor
G	Green-color wire	FR	Fan relay
OR	Orange-color wire	L	Limit
R	Red-color wire	LA	Limit auxiliary
W	White-color wire	L_1 L_2	Line 1, line 2 of power supply
Y	Yellow-color wire	S	Cad-cell terminal
C	24-V common connection	T	Thermostat terminal
CC	Compressor relay	TFS	Timed fan start

In the power circuit shown in Figure 14.16, 120-V ac power is supplied to the primary control. A fused disconnect is placed in the hot side of the line. A limit control is wired in series with the hot side of the power supply (L_1) and the black connection on the primary control. The transformer is a part of the primary control.

Fan Circuit

The simplest type of fan circuit consists of a fan control in series with a single-speed fan motor, as shown in Figure 14.17 . In a multiple-speed fan motor the switch on the subbase of the thermostat changes the fan speed, as shown in Figure 14.18.

On downflow and horizontal furnaces, a timed fan start (TFS) is used to start the fan, and the fan control (FC) stops it, as shown in Figure 14.19. An auxiliary limit control is used in addition to the regular limit control and is wired in series with the primary control.

Ignition and Oil Burner Circuits

Ignition and oil burner circuits are shown in Figure 14.20. The ignition transformer is wired in parallel with the oil burner motor on a constant-ignition system. The burner is wired to the orange and white connections on the primary control.

Beckett TECHNICIAN'S QUICK REFERENCE GUIDE

The following service procedures will help you become familiar with the R7184 series primary controls. For control operation, please refer to the basic control functions described below. For further information, wiring instructions, and troubleshooting, please refer to the Honeywell R7184 Installation Instructions (Honeywell 69-1233).

PRIMING THE PUMP

1. Initiate a call for heat.
2. While the ignition is on, press and release the reset button (hold ½ sec. or less). If the control has not locked out since its most recent complete heat cycle, the lockout time will be extended to 4 minutes (45 sec. in earlier units), and the ignition will remain on for the entire heat cycle.
3. Bleed the pump until all froth and bubbles are purged. If prime is not established within the extended lockout time, the control will lock out. Press the reset button to reset the control and to return to step 2. **Note:** The reset button can be held for 30 seconds at any time to reset the control's lockout counter to zero and send the control to standby.
4. Repeat steps 2 and 3, if needed, until the pump is fully primed and the oil is free of bubbles. Then terminate the call for heat, and the control will resume normal operation.

RESETTING FROM RESTRICTED LOCKOUT

If the control locks out three times in a row without a complete heat cycle between attempts, the lockout becomes restricted in order to prevent repetitious resetting by the homeowner. To reset, hold down the reset button for 30 seconds (until the LED flashes twice).

DISABLE FUNCTION

Any time the motor is running, press and hold the reset button to disable the burner. The burner will remain off as long as the button is held and will return to standby when released.

LED INDICATOR KEY

LED	STATUS
On	Flame sensed
Off	Flame not sensed
Flashing (½ sec. on, ½ sec. off)	Lockout / Restricted Lockout
Flashing (2 sec. on, 2 sec. off)	Recycle

CAD CELL RESISTANCE CHECK

While the burner is firing, and after the ignition has been turned off, press and release the reset button (hold ½ sec. or less) to check the cad cell resistance. The LED will flash 1 to 4 times, depending on the cad cell resistance (see the chart at the right). For proper operation, it is important that the cad cell resistance is below 1600 Ohms.

LED FLASHES	CAD CELL RESISTANCE
1	0 – 400 Ohms
2	400 – 800 Ohms
3	800 – 1600 Ohms
4	≥ 1600 Ohms

R7184 SERIES CONTROL FEATURES

MODEL	ADVANCED FEATURES
R7184A1000	Interrupted duty ignition microprocessor-based control.
R7184B1016	All features of the R7184A1000 plus a 15 second valve-on delay (prepurge).
R7184P1031*	All features of the R7184B1016 plus a burner motor-off delay (postpurge): field selectable ½, 2, 4, or 8 minutes.
R7184P1049 / R7184U*	All features of the R7184P1031 plus dry alarm contact terminals (30VAC, 2A max). Prepurge and postpurge can be field disabled together by a DIP switch setting.

*Postpurge timings may be different.

TYPICAL SEQUENCE OF OPERATION

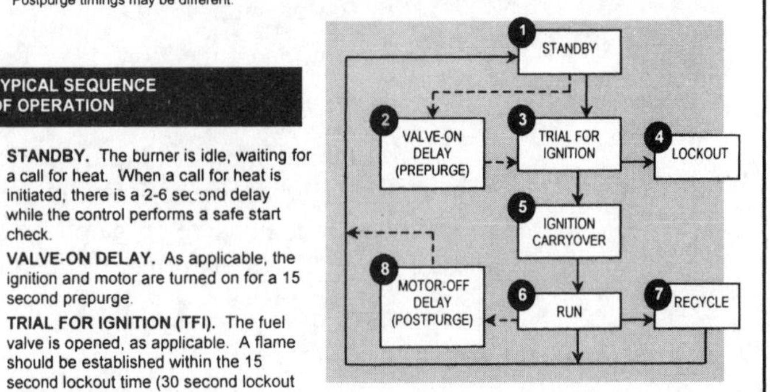

1. **STANDBY.** The burner is idle, waiting for a call for heat. When a call for heat is initiated, there is a 2-6 second delay while the control performs a safe start check.
2. **VALVE-ON DELAY.** As applicable, the ignition and motor are turned on for a 15 second prepurge.
3. **TRIAL FOR IGNITION (TFI).** The fuel valve is opened, as applicable. A flame should be established within the 15 second lockout time (30 second lockout time is available).
4. **LOCKOUT.** If flame is not sensed by the end of the TFI, the control shuts down on safety lockout and must be manually reset. If the control locks out three times in a row, the control enters restricted lockout. Follow the instructions to the left to reset the control.
5. **IGNITION CARRYOVER.** Once flame is established, the ignition remains on for 10 seconds to ensure flame stability. It then turns off.
6. **RUN.** The burner runs until the call for heat is satisfied. The burner is then sent to burner motor-off delay, as applicable, or it is shut down and sent to standby.
7. **RECYCLE.** If the flame is lost while the burner is firing, the control shuts down the burner, enters a 60 second recycle delay, and then repeats the ignition steps outlined above. If the flame is lost three times in a row, the control locks out to prevent cycling with repetitious flame loss caused by poor combustion.
8. **BURNER MOTOR-OFF DELAY.** If applicable, the fuel valve is closed and the burner motor is kept on for the selected postpurge time before the control returns the burner to standby.

Figure 14.15 A handy R7184 quick reference guide for technicians. (Courtesy of R.W. Beckett Corporation)

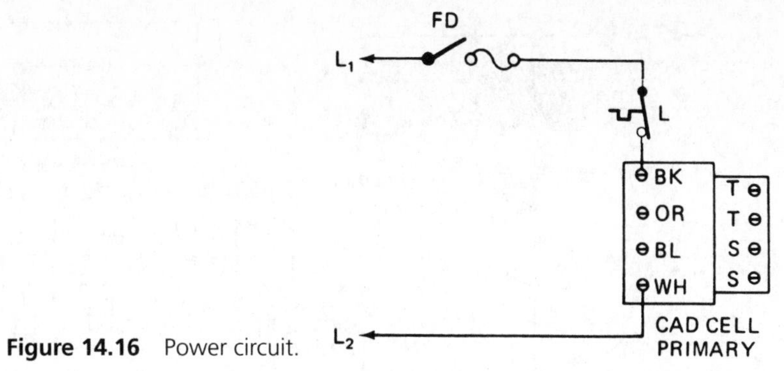

Figure 14.16 Power circuit.

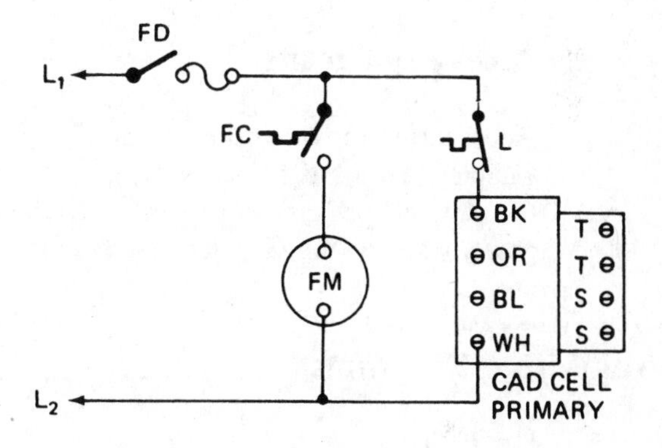

Figure 14.17 Fan circuit with single-speed fan motor.

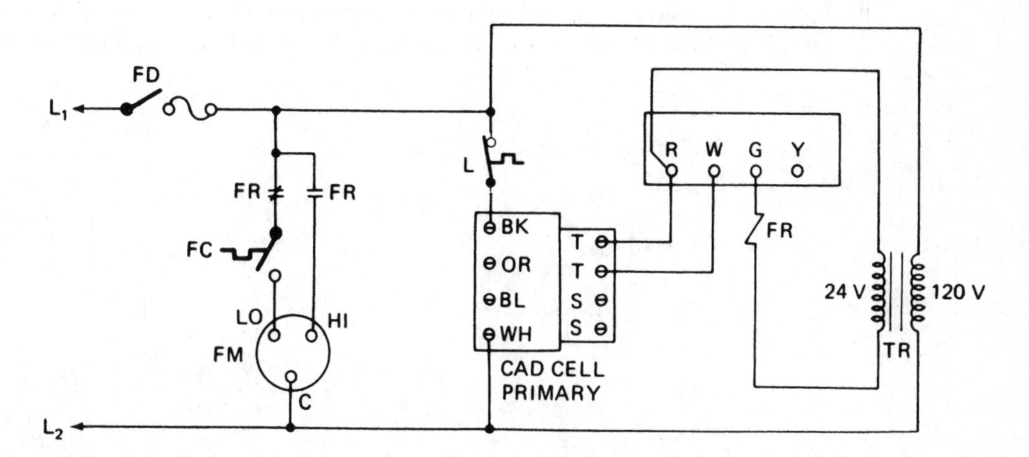

Figure 14.18 Fan circuit with multispeed fan motor.

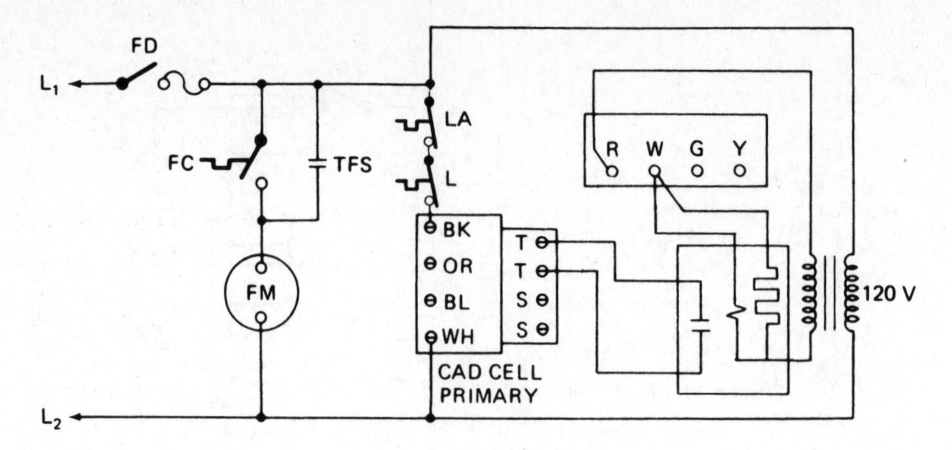

Figure 14.19 Fan circuit for downflow and horizontal furnaces.

Accessory Circuits

Accessories for oil furnaces can be added in a manner similar to those shown in Chapter 11 for gas furnaces. Because the transformer for heating is contained within the cad-cell primary, a separate transformer is usually required for control circuit power when cooling is added (see Figures 14.18 and 14.19).

Typical Wiring Diagrams

Upflow

Figure 14.21 shows a typical connection diagram for an upflow oil furnace. Note that on this unit most of the wiring connections are brought to a common junction box.

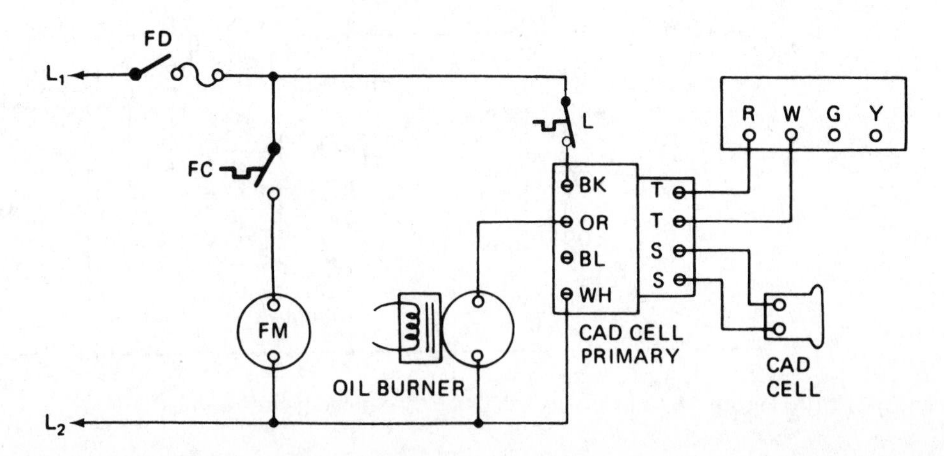

Figure 14.20 Ignition and oil burner circuits.

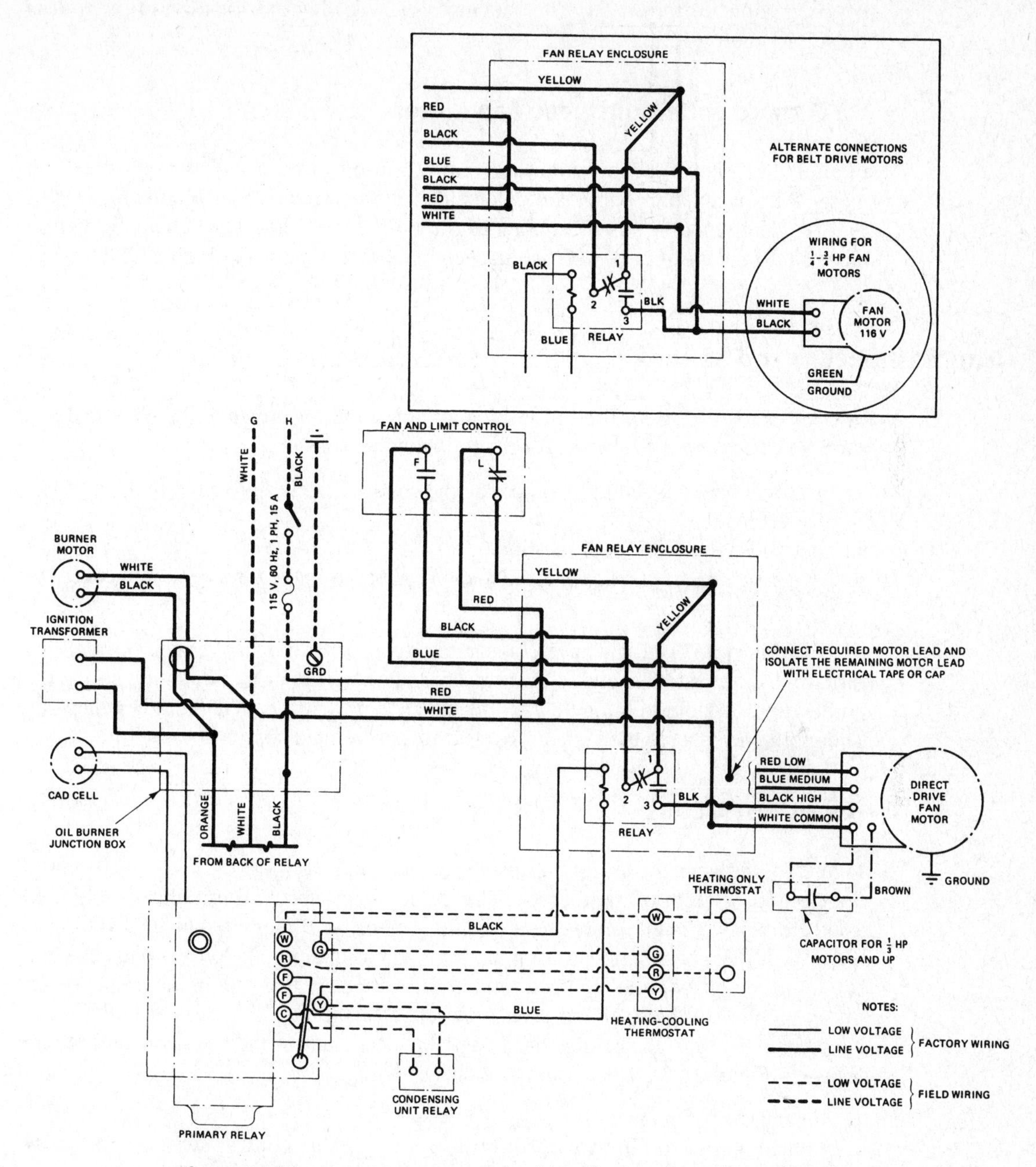

Figure 14.21 Heat–cool connection wiring diagram for upflow oil furnace. (Courtesy of ARCOAIRE Air Conditioning & Heating)

Because both the fan and limit controls are of line-voltage type, a jumper can be used between common terminals. The schematic wiring diagram for the upflow unit is shown in Figure 14.22.

Oil Furnace with Multispeed Fan Motor

Figure 14.23 shows a connection diagram for an oil furnace with a multispeed fan motor. All wiring connections are made either in the primary control junction box or in a separate junction box that includes the indoor fan relay. The schematic wiring diagram for the oil furnace with multispeed fan motor is shown in Figure 14.24.

Troubleshooting and Service

In troubleshooting an oil-fired furnace, attention should be given to the electrical and fluid flows through the furnace, for the following reasons:

- Electrical power is required to operate the control system, ignite the fuel, and energize the loads.
- Fuel oil is required for combustion.
- Air is required for combustion and to convey heat from the furnace to the spaces being heated.

Electricity, oil, and air have a separate circuit or path of movement through the furnace. Each must be supplied in the proper place, in the proper quantity, and at the proper time. Troubleshooting the fuel oil and air supply was covered in other chapters. Therefore, the concern here is chiefly electrical power and control.

Use of Test Meters

Figure 14.25 shows the use of various test meters in measuring voltage and current in the electrical system of the furnace. The technique in electrical troubleshooting is to "start from power" and follow the availability of power through the control system to the load. If the power supply is stopped for any reason along the proper path, the reason for the stoppage should be determined and corrected. If power in the proper quantity is supplied to the load and it does not operate, the load device is at fault. Seven locations where meter readings are taken along the path of the electrical current are shown in Figure 14.25. Locations to be checked are:

1. Voltage supply to the unit
2. Voltage across the limit control to determine if this control switch is properly closed
3. Voltage across the thermostat to be certain that it is calling for heat
4. Voltage at the oil burner motor and ignition transformer to be certain that 120 V ac is being supplied

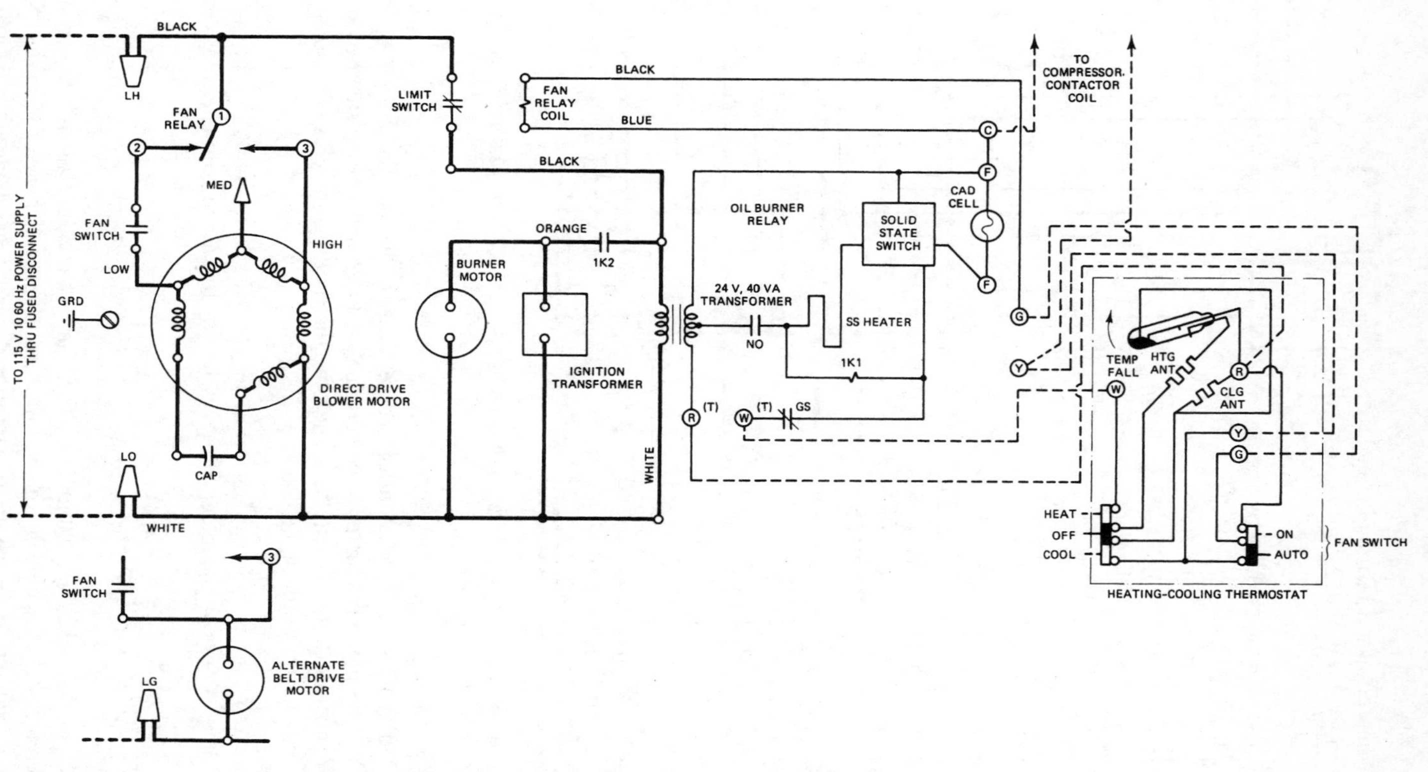

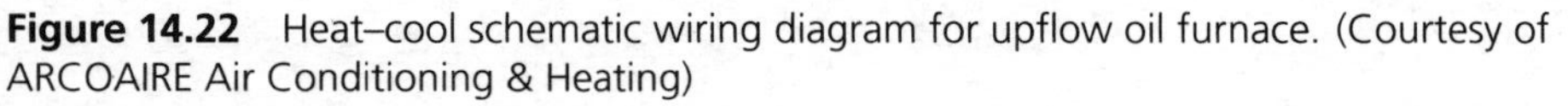

Figure 14.22 Heat–cool schematic wiring diagram for upflow oil furnace. (Courtesy of ARCOAIRE Air Conditioning & Heating)

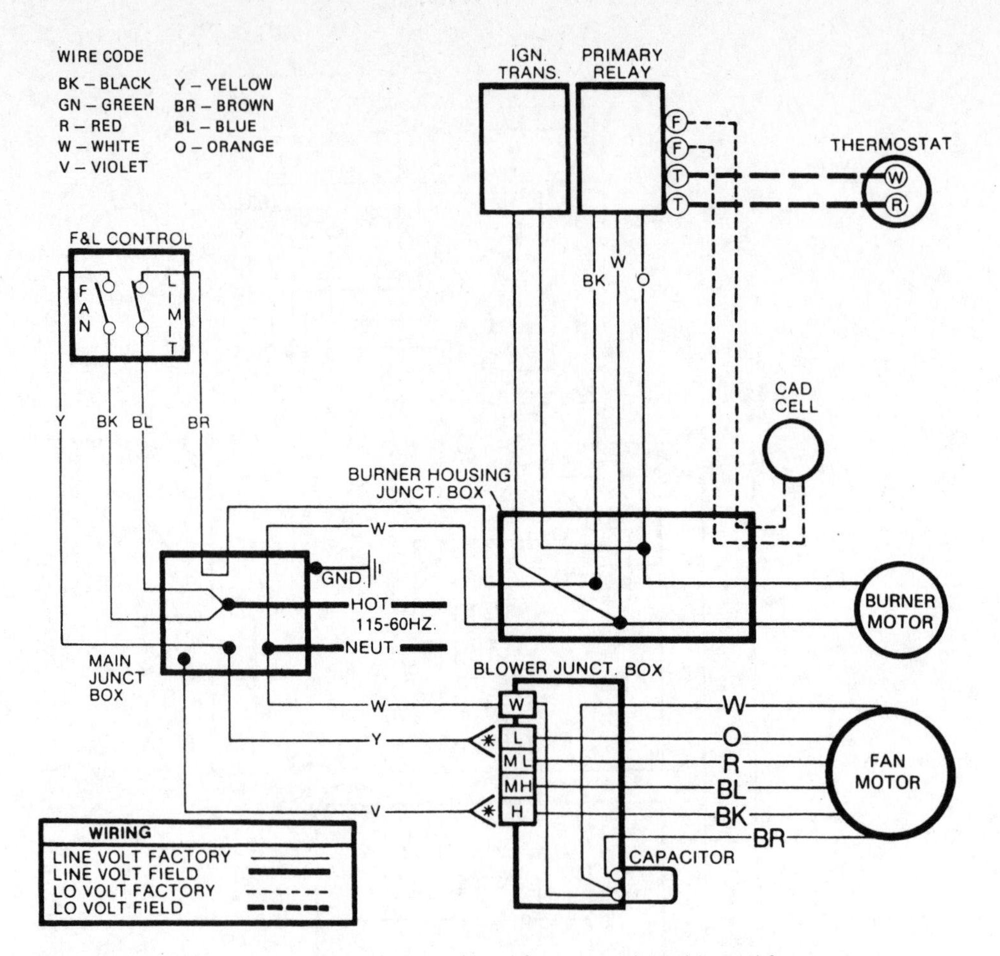

Figure 14.23 Connection wiring diagram for oil furnace with multispeed fan motor. (Courtesy of Heil–Quaker)

5. Voltage across the fan control to determine that it is calling for fan operation
6. Voltage at the fan motor to be certain that 120 V ac is being supplied
7. Amperage to the fan motor to measure the running amperes (should agree with the data on the motor nameplate)

In meter readings 2, 3, and 5, a voltmeter is used to test a switch. If the switch is open, a voltage shows on the meter. If the switch is closed, the voltmeter reads zero.

Testing Components

The most complex element in an oil-fired furnace control system is the primary control. Therefore, it is this element that has the greatest potential need for service. Specific testing procedures are necessary to determine if the primary control is functioning properly.

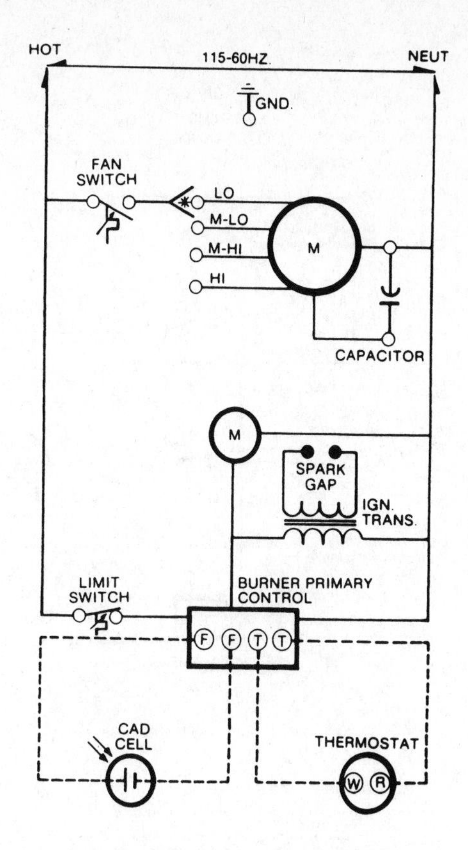

Figure 14.24 Schematic wiring diagram for oil furnace with multispeed fan motor. (Courtesy of Heil–Quaker)

Other Servicing Techniques

Most oil-burning furnaces have a manual test overload on the oil burner motor and a reset-type safety switch on the primary control. After an overload, these devices must be manually reset. The tripping of an overload switch is an indicator of a problem that must be found and corrected. Resetting the overload switches may start the equipment, but unless the problem is resolved, nuisance tripout will continue to recur.

Regardless of the complaint, much time can often be saved by first checking the following routine conditions:

- Does the tank contain fuel?
- Is power being delivered to the building?
- Are all hand-operated switches closed?
- Are all hand valves in the oil supply line open?
- Are all limit controls in their normal (closed) position?
- Is the thermostat calling for heat?
- Are all overload switches closed?

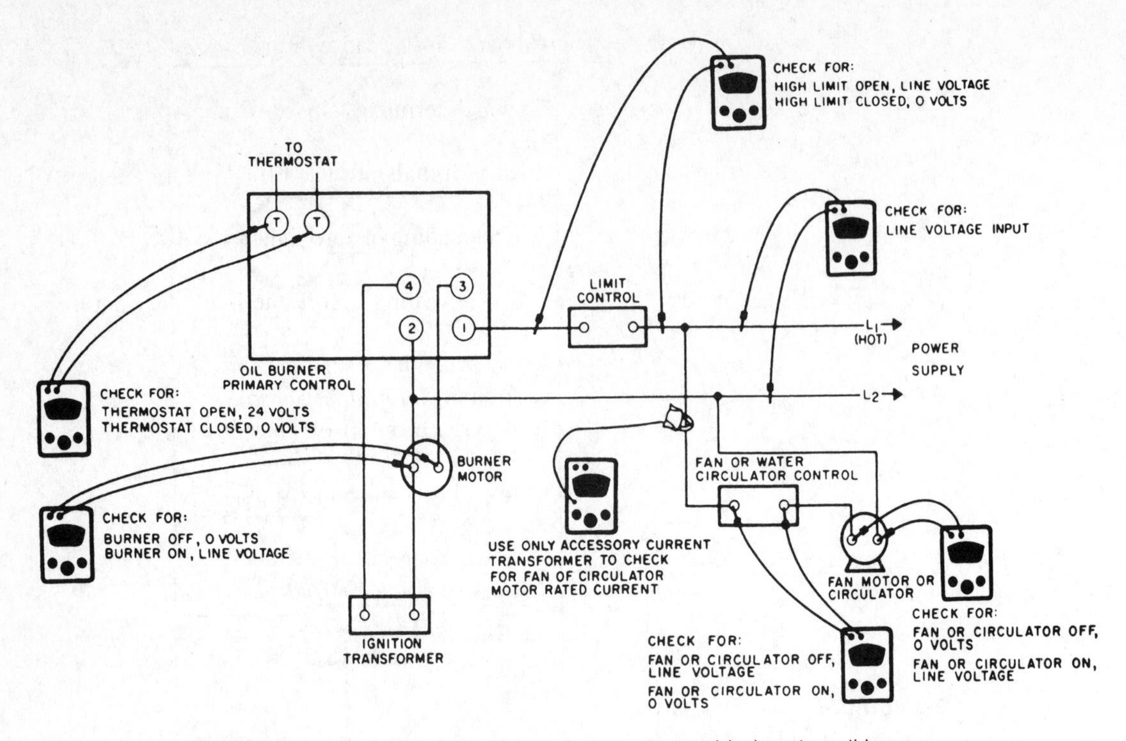

Figure 14.25 Use of various electrical meters in troubleshooting oil burner systems. (Courtesy of Honeywell Inc.)

Low voltage can cause problems such as low oil pump speed, motor burnouts from overload, and burners that fail to operate because relays do not pull in. The power supply voltage should be checked while the greatest power usage is occurring. Low voltage should never be overlooked as a source of service problems. When the service to the building is at fault, the local power company should be contacted. If the problem is within the building, the services of an electrician may be required to solve the problem.

STUDY QUESTIONS

Answers to the study questions may be found in the sections noted in brackets.

14-1 What are the various circuits in an oil burner control system? *[Circuits]*

14-2 What are the two types of primary controls? *[Primary Control]*

14-3 On the stack relay, are the cold contacts normally open or closed when starting? *[Bimetallic Sensor]*

14-4 What are the component materials of a cad cell? *[Cad-Cell Sensor]*

14-5 What is the resistance of a cad cell in the absence of light? *[Cad-Cell Sensor]*

14-6 To which terminals of the primary control is the cad cell connected? *[Cad-Cell Sensor]*

14-7 How many low-voltage terminals are on the cad-cell primary control? *[Cad-Cell Sensor]*

14-8 On the cad-cell primary control, which terminal is used for the hot side of the power supply? *[Cad-Cell Sensor]*

14-9 On the bimetal stack relay, which terminals are used for the power supply? *[Bimetallic Sensor]*

14-10 In testing a switch with a voltmeter, a reading of zero indicates what position of the switch? *[Use of Test Meters]*

14-11 On a stack relay, if the flame fails, how long before the burner will stop on safety? *[Bimetallic Sensor]*

14-12 What is the resistance of the cad cell when it senses light? *[Cad-Cell Sensor]*

14-13 Describe the solid-state cad-cell circuit. *[Solid-State Sensor]*

14-14 Which control requires the most service on an oil burner control system? *[Testing Components]*

14-15 In a solid-state controller, how many relays can be incorporated? Explain a few. *[Modern Primary Controls]*

14-16 Why is the modern primary control flame-sensing method preferred over the stack-mounted bimetal switch? *[Modern Primary Controls]*

CHAPTER 15

Electric Heating

OBJECTIVES

After studying this chapter, the student will be able to:

- Identify the various types of controls and circuits used on an electric furnace
- Determine the sequence of operation of the controls
- Service and troubleshoot the electric furnace control system

Conversion of Electricity to Heat

An electric heating furnace converts energy in the form of electricity to heat. The conversion takes place in resistance heaters. Electric furnaces differ from gas or oil furnaces in that no heat exchanger is required. Return air from the space being heated passes directly over the resistance heaters and into the supply-air plenum. The amount of heat supplied depends on the number and size of the resistance heaters used. The conversion of electricity to heat (the heat equivalent for 1 W of electrical power) takes place in accordance with the following formula:

$$1 \text{ W} = 3.415 \text{ Btu}$$

Electric Furnace Components

Figure 15.1 shows an electric forced warm air furnace (upflow model). Return air is brought into the blower compartment through the filters, then passed over the heating elements and sent out into the distribution system. An electric furnace requires no flue. All the heat produced is used in heating the building. Input is equal to output. Thus, it operates at 100% efficiency. Major components, excluding the controls, are:

- Heating elements
- Fan and motor assembly

TYPICAL ELECTRIC FURNACE FEATURES

BUILT-IN COOLING COIL COMPARTMENT — Slide-in type for easier conversion to summer cooling. Accommodates 1½, 2, 2½, and 3 ton air conditioner cooling coils. See note on reverse side on heat pump installation.

CONTROLS — On demand from the wall thermostat, the heating elements are energized by electrical contactors. The 15 thru 30 KW versions have the blower motor interlocked with each stage for safety. Easily two staged.

LIMIT SWITCH — Thermal snap disc in each heating element shuts off power automatically if system air temperature becomes excessive.

BUILT-IN TRANSFORMER — Provides power supply for heating and optional cooling controls.

BLOWER RELAY — Provides automatic blower speed change-over to meet heating and cooling air delivery requirements.

BRANCH CIRCUIT FUSING — Factory installed in models rated over 48 amps.

HEATING ELEMENTS — Nickel-chrome wire with individual fusible links for long life. Entire assembly slides out for easy maintenance.

MOTOR — Multi-speed for both heating and cooling.

BLOWER — Heated air is quietly circulated by large volume centrifugal blower that is matched to the electrical heating system for efficiency. Slides out for easy maintenance.

FILTERS — Twin permanent type slide out from front for easy cleaning on all models except Models EFC5 and EFC10.

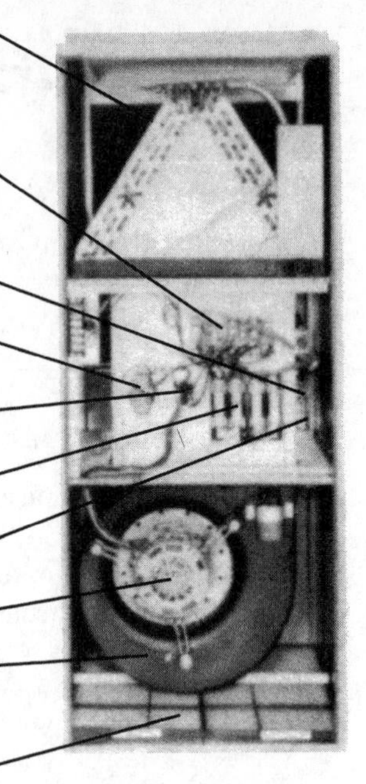

Figure 15.1 Electric forced warm air furnace, upflow model. (Courtesy of Bard Manufacturing Company)

- Furnace enclosure
- Accessories such as filters, humidifier, and cooling (optional)

Following is the key to the schematic diagrams used in this chapter:

AC	Auxiliary contacts	L	Limit control
C	24-V common connection	L_1, L_2	Line 1, line 2 of power supply
FC	Fan control	OT	Outdoor thermostat
FD	Fused disconnect	R	Red wire, thermostat connection
FL	Fuse link	SQ	Sequencer
FM	Fan motor	TFS	Timed fan start
FU	Fuse	TR	Transformer
HR	Heat relay	W	White wire, thermostat connection

Heating Elements

The heating elements are made of Nichrome, a metal consisting chiefly of nickel and chromium. Heating elements are rated in kilowatts of electrical power consumed (1 kW = 1000 W). A typical heating element assembly is shown in Figure 15.2. The Nichrome wire is supported by insulating material placed to provide a minimum amount of air resistance and a maximum amount of contact between the air and the resistance heaters.

Each element assembly contains a thermal fuse and a safety limit switch (Figure 15.3). The safety limit switch is shown in schematic form in Figure 15.4. The limit switch is usually set to open at 160°F and to close when the temperature drops to 125°F. The thermal fuse, a backup for the safety limit switch, is set to open at a temperature slightly higher than that of the limit switch. In addition to the safety controls in the element assembly, each assembly is fused where it connects to the power supply. Triple safety protection is thereby provided.

Electric furnaces have various individual numbers and sizes of heating element assemblies, depending on the total capacity of the equipment. Elements are placed on the line (power supply) in stages in order not to overload the electrical system on startup. In most units, the maximum size of an element energized or de-energized at one time is 5 kW (17,075 Btuh). Figure 15.4 shows how the heater sequencer operates.

Fan-and-Motor Assembly

The fan-and-motor assembly is similar to that used on a gas or oil furnace. Fans may be either direct-drive or belt-drive. There is a trend toward the use of multispeed direct-drive fans, since they facilitate adjustment of the airflow by changing the blower

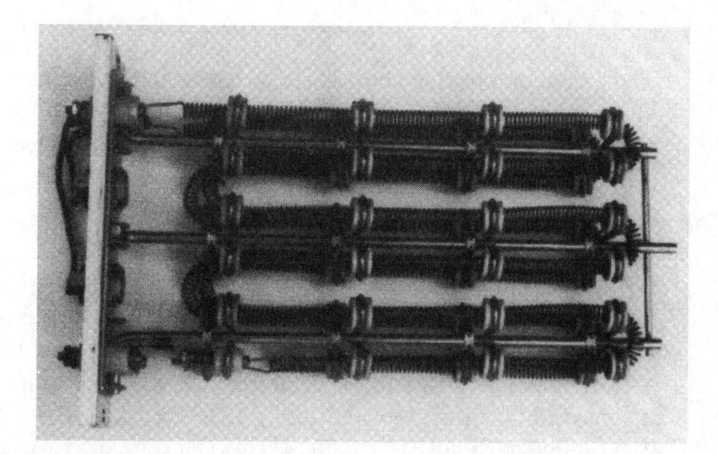

Figure 15.2 Typical heating element.

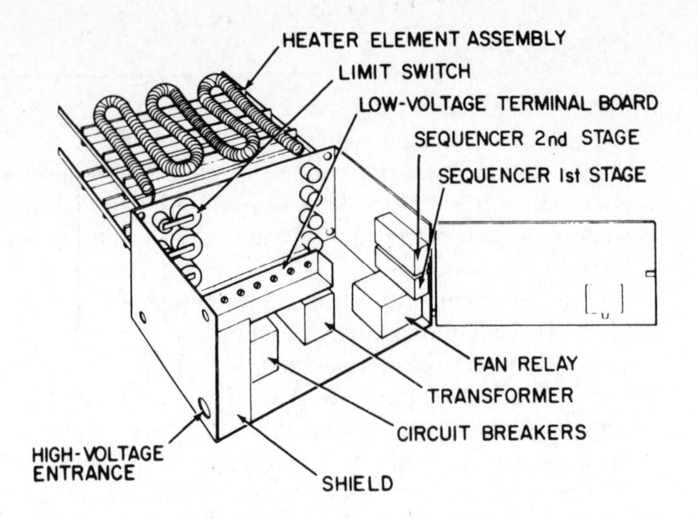

Figure 15.3 Heater element assembly with circuit breakers. (Courtesy of Bryant Air Conditioning)

speed. This is usually necessary when cooling is added, since larger air quantities and, consequently, higher speeds are required.

Enclosures

The exterior of the casing is similar to that of a gas or oil furnace but without the flue pipe connection. The interior is designed to permit the air to flow over the heating elements. The section supporting the heating elements is usually insulated from the exterior casing by an air space.

Accessories

Filters, humidifiers, and cooling are added to an electric heating furnace in a manner similar to that used for gas and oil furnaces. Electric heating can therefore provide all the related climate control features provided by other types of fuel.

Power Supply

The power supply for an electric furnace is 208/240-V, single-phase, 60-Hz. This power is supplied by three wires: two are hot, and one is neutral. Fused disconnects are placed in the hot lines leading to the furnace. The National Electrical Code® limits to 48 A the amount of service in a single circuit. If the running current exceeds this amount, additional circuits and fused disconnects must be provided. The fuses for a 48-A service must not exceed 60 A. This is a National Electrical Code® requirement, specifying that fuses should not exceed 125% of full-load amperes ($48 \times 1.25 = 60$).

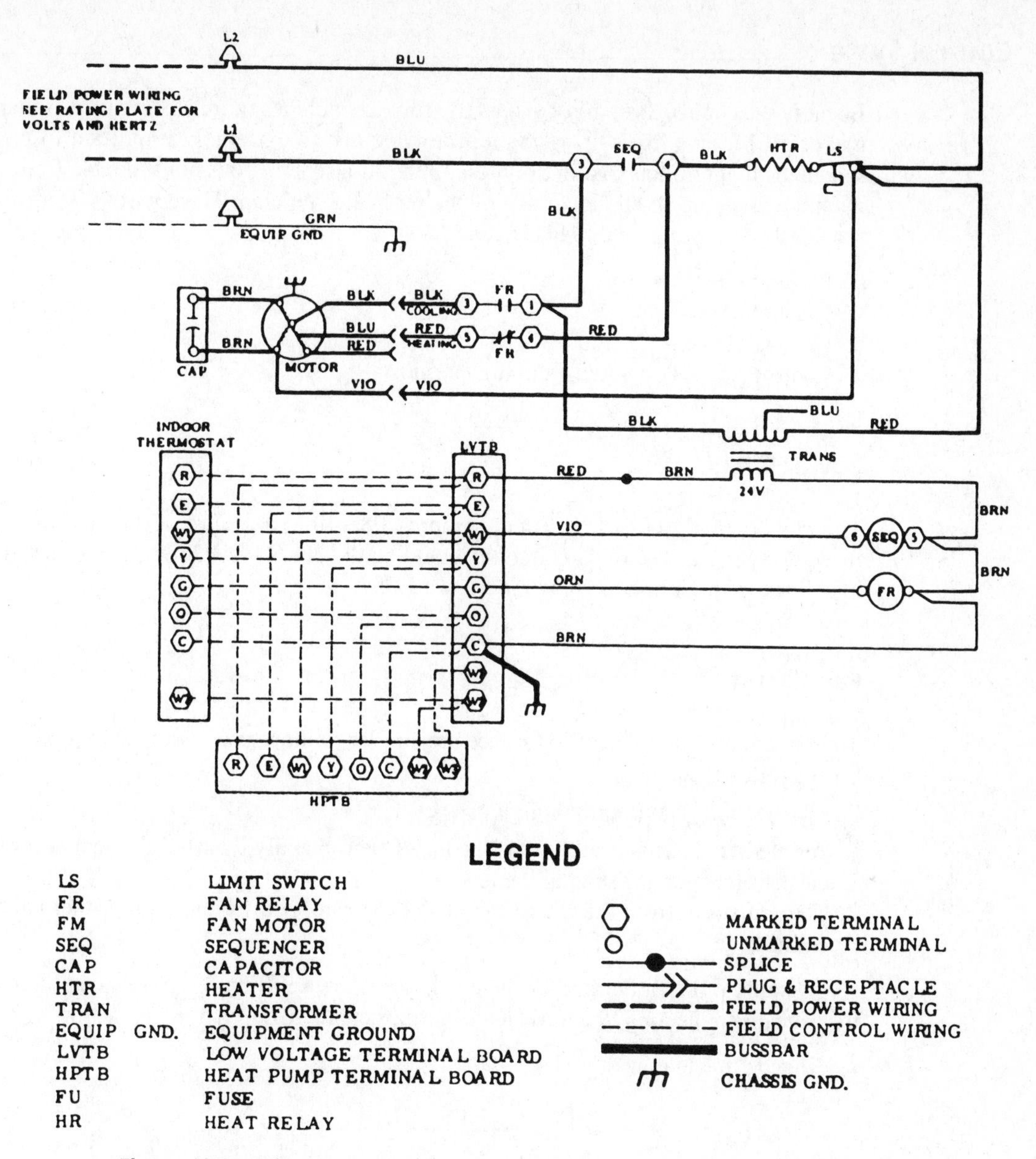

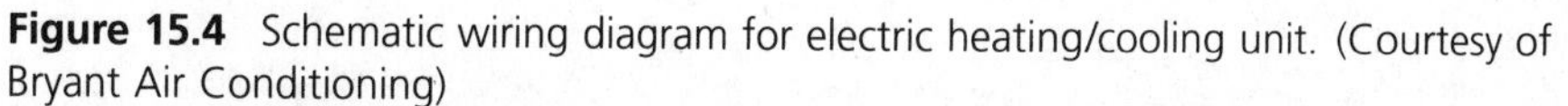

Figure 15.4 Schematic wiring diagram for electric heating/cooling unit. (Courtesy of Bryant Air Conditioning)

All wiring must be enclosed in conduit with proper connectors. Because 208/240 V is considerably more dangerous than lower voltages, every possible protection must be provided. The National Electrical Code® also requires that the furnace be grounded. The ground wire in the power supply is provided for this purpose.

Control System

Because the control system for an electric furnace includes more electrical parts than a gas or oil furnace, the wiring is more involved. It is extremely important to use a schematic diagram to assist in troubleshooting. If a schematic is not available from the manufacturer, one should be drawn by the service technician. The control system consists of the following electrical circuits:

- Power circuit
- Fan circuit
- Heating element circuits
- Control circuits operated from the thermostat

Power Circuit

The power supply consists of one or more 208/240-V ac sources, directed through fused disconnects to the load circuits, including the 208/240-V/24-V transformer. The load lines are both hot, as shown in Figure 15.5.

Fan Circuit

The fan circuits on an electric furnace are similar to those used with gas or oil except for the following changes:

1. The fan motor is usually 208/240 V ac.
2. A timed fan start is used to start the fan. This is usually a part of the sequencer (an electrical device for staging the loads).
3. The fan starts from the timed fan start but is stopped by the regular thermal fan control.

Details of the fan circuit arrangement are shown in Figure 15.6. When the thermostat calls for heating, *R* and *W* in the thermostat are made, energizing the low-voltage

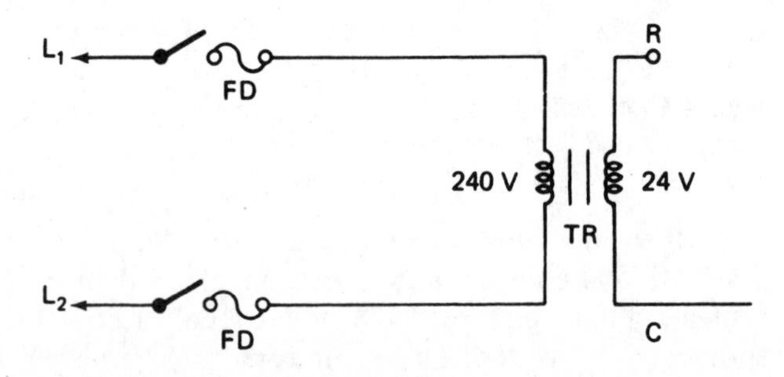

Figure 15.5 Power circuit showing hot load lines.

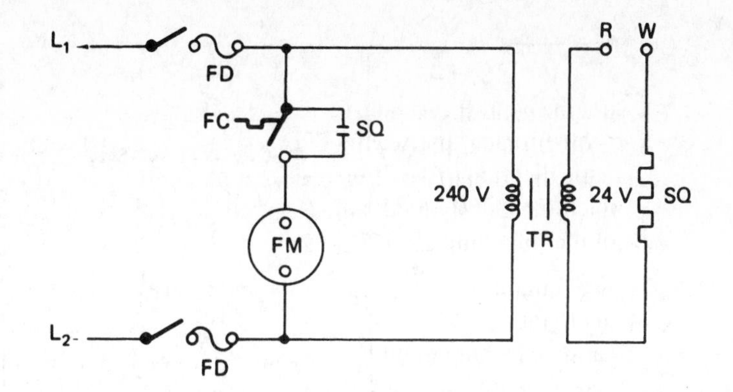

Figure 15.6 Typical fan circuit.

heater in the sequencer. After approximately 45 seconds, the fan starts to operate. Just as soon as the temperature leaving the furnace rises to the fan control cut-in temperature, that control makes. It has no effect on the fan, however, since it is already running.

Approximately 45 seconds after the thermostat is satisfied, the timed fan start switch opens. It has no effect on the fan, though, since the fan control is made. When the leaving air temperature drops to the cutout setting of the fan control, the fan stops.

Heating Element Circuits

The heating element assemblies are connected in parallel to the power supply. These assemblies are turned on by the sequencer or heat relays so that the load is gradually placed on the line. Each heating element has a sequencer switch wired in series with the element to start its operation. Figure 15.7 shows the fan and the heating elements operated by a sequencer.

Control Circuits

When the thermostat calls for heating, R and W make, energizing the low-voltage heater (SQ) in the sequencer. In approximately 45 seconds, the switch to the fan (SQ_1) and the first heating element (SQ_2) are made simultaneously. Approximately 30 seconds later, the switch to the second heater element (SQ_3) is made. If additional heating elements are used, there is a comparable 30-second delay before each succeeding element is operated.

A two-stage thermostat is used on some systems. This thermostat has two mercury bulb switches, one for each stage. The first stage is set to make at a temperature a few degrees higher than the second stage. The first stage controls the fan and some of the heating elements. The second stage controls the balance of the heating elements.

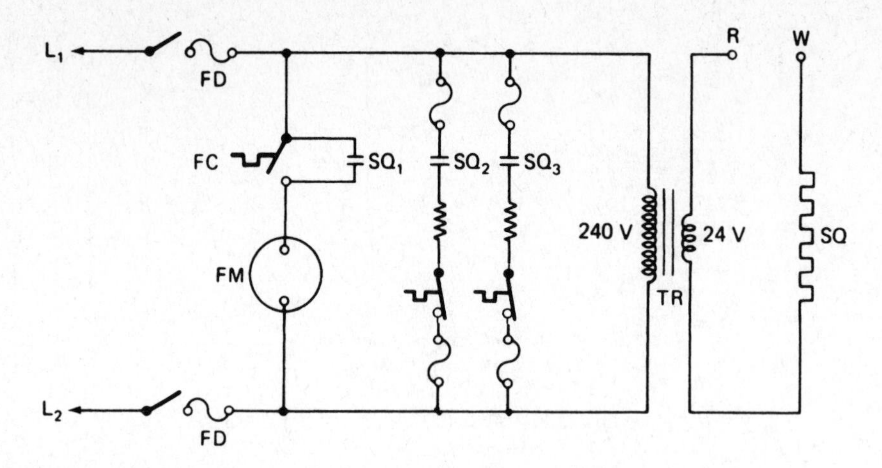

Figure 15.7 Fan and heating elements operated by a sequencer.

Referring to Figure 15.8, when the first stage of the thermostat calls for heat, R and W_1 make, energizing the heater SQ in the sequencer. This action switches on the first element and the fan, then the second element is energized in 30 seconds. If the room temperature continues to drop to the setting of the second stage of the thermostat, R and W_2 are made, energizing the heat relay heater (HR). In approximately 45 seconds the switch to the third heating element (HR_1) is made, operating the final stage of heating.

The second-stage thermostat can also be an outdoor thermostat rather than a part of the first-stage thermostat assembly. The low-voltage control circuit for this type of installation is shown in Figure 15.9.

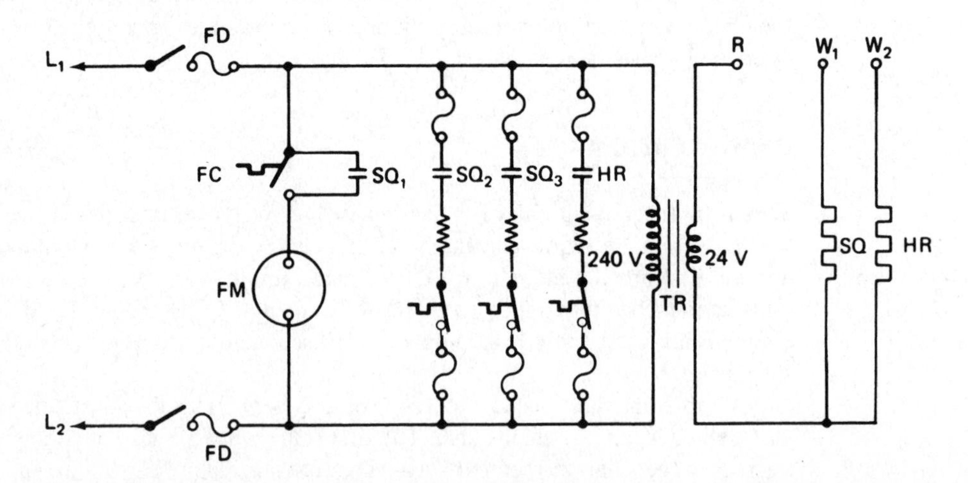

Figure 15.8 Thermostat assembly in two-stage heating.

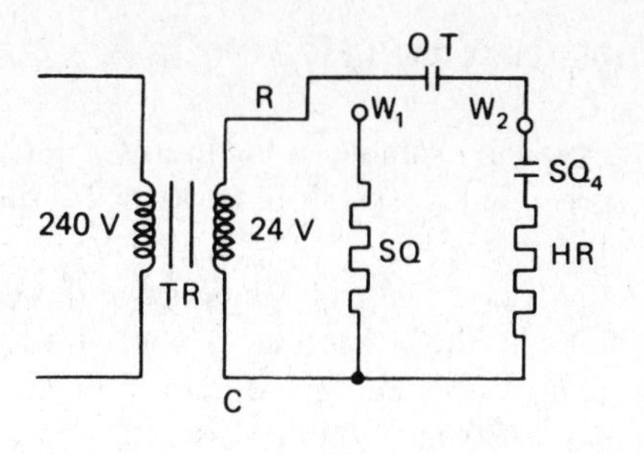

Figure 15.9 Use of outdoor thermostat.

The first-stage (standard) thermostat operates the first-stage heating in the usual manner. A switch on the sequencer (SQ_4), an auxiliary switch, is also made in stage 1 to permit stage 2 to operate. When the outside temperature reaches the setting of the outdoor thermostat (OT), the second-stage heating is operated in a manner similar to that shown in Figure 15.8. Separate heat relays for each load can be used as an alternative to using the sequencer for staging the loads on an electric heating furnace. These relays function much like the sequencer, as shown in Figure 15.10.

When R and W make, the heater on heat relay 1 (HR_1) is energized along with the fan relay (FR). In this arrangement the fan starts immediately, and in about 30 seconds heat relay 1 switch (HR_1) closes, operating the first heater element. At the same time, the other HR_1 switch in the 24-V circuit closes, energizing the heater on heat relay 2. In approximately 30 seconds, the two switches (HR_2) in heat relay 2 close. One operates heating element 2, and the other energizes the heater in heat relay 3. In approximately 30 seconds switch HR_3 closes, operating heating element 3. The fan and all heating elements are thereby placed on the line in sequence, as shown in Figure 15.10.

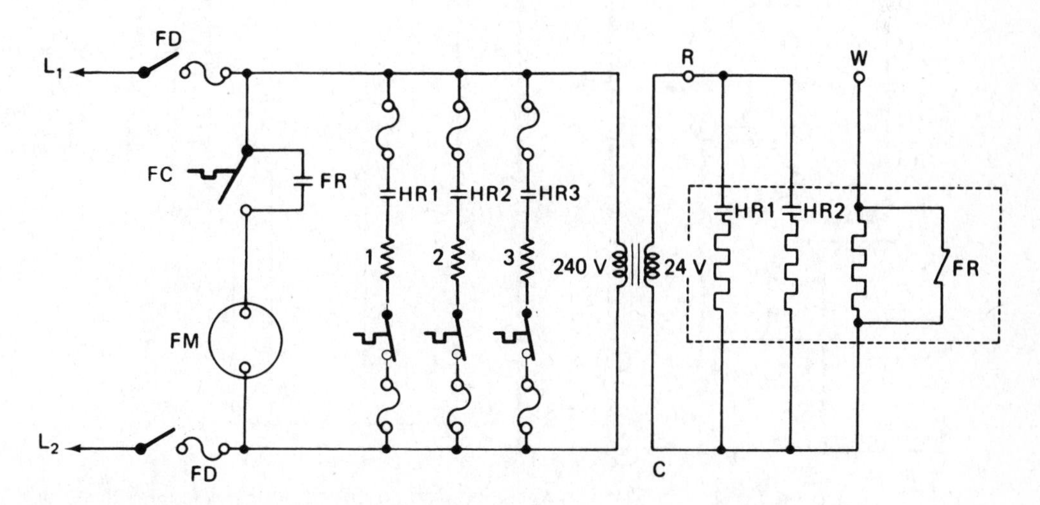

Figure 15.10 Use of heat relays. (single-stage thermostat)

Use of Sequencers and Heat Relays

A typical example of the use of sequencers to stage the loads on an electric furnace is shown in Figures 15.11 and 15.12. The furnace shown in Figure 15.11 has two heating elements.

This unit uses a single-stage thermostat. When R and W are made in the thermostat, the fan immediately starts on low speed. At the same time, the heater in the sequencer is energized, and the heater elements are operated in stages at approximately 30-second intervals.

In reference to Figure 15.11, the set of NO contacts of energized sequencer SEQ (between 1 and 2) closes and completes the circuit through the set of NC contacts of indoor fan relay FR (between 5 and 4) to the low-speed tap of indoor fan motor FM. The fan motor starts instantly. The circuit to heater elements HTR 1 is also completed, and these elements are energized.

After a short built-in time delay that prevents both heater elements from energizing simultaneously, the set of NO contacts of energized sequencer SEQ (between 3 and 4) closes and completes the circuit to heater elements HTR 2, and these elements are energized.

The fan motor and the energized heater elements remain on until the room temperature rises to a point above the heating control setting of the room thermostat. At this point the electrical connection between thermostat terminal R to terminal W (or W 1) opens. This open circuit de-energizes sequencer coil SEQ. The electric heating cycle is now off until there is another demand for heating by the room thermostat.

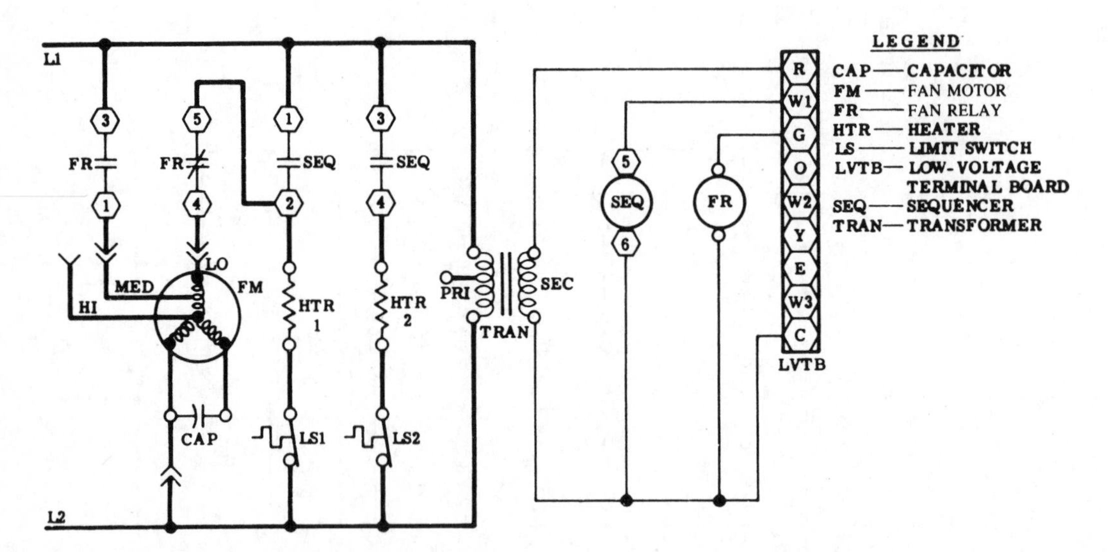

Figure 15.11 Typical schematic wiring diagram for electric heat sequencer. (Courtesy of Bryant Air Conditioning)

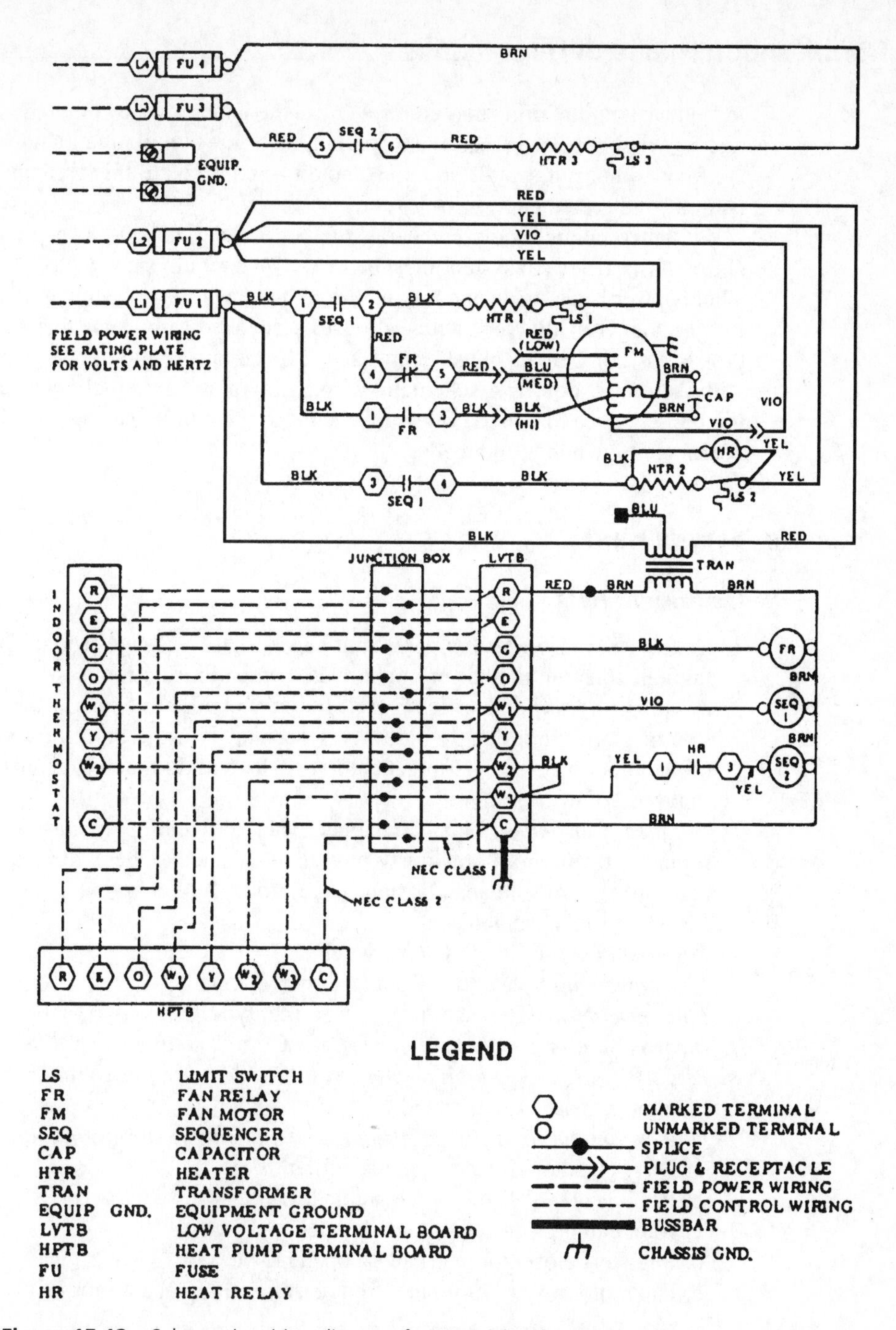

Figure 15.12 Schematic wiring diagram for 12–20 kW heating/cooling unit. (Courtesy of Bryant Air Conditioning)

Troubleshooting and Service

In troubleshooting, it is good practice to use the nature of the complaint as a key to the area in which service is required. For example, if the complaint of a homeowner is that the fan will not run, troubleshooting should be confined to this area until the problem is found and corrected.

If the complaint is more general, such as no heat, then a thorough and systematic check of the electrical system must be made. This requires an electrical check, starting where power is available and tracing through the electrical system to determine where it is no longer available. Switches must be checked to be sure that they are in their proper positions. Loads must be checked to be certain that they operate when supplied with the proper power. A schematic wiring diagram is extremely helpful in determining the sequence in which power travels through the unit and for checking the correct operation of switches and loads.

Service Hints

Insufficient Heat

1. *System or building incomplete.* Your new furnace will not produce proper comfort until all construction has been completed and all insulation is in place.
2. *Building not properly winterized.* If a drafty condition exists, we suggest installation of storm doors and windows, as needed. If walls or ceiling are excessively cold, proper insulation (if practical to install) will do much to reduce fuel bills and improve comfort.
3. *Furnace overloaded.* This can happen when a dwelling is enlarged (by adding on rooms or opening up previously unused attic space). Check the required heat load against the furnace capacity, and make proper and economical recommendations for solving this problem.
4. *Power supply turned off.* Close switch.
5. *Low power supply voltage.* Contact power company.
6. *Main fuse blown.* Replace fuse with correct type.
7. *Incorrect thermostat anticipator setting.* Correct setting.
8. *Fan operating at too high a speed, resulting in a low-temperature rise.* Reconnect fan motor to next lower speed.
9. *Unit cycling on limit controls because of inadequate air circulation.*
   a. Dirty air filter—clean or replace filter.
   b. Fan wheel blades dirty—clean blades.
   c. Duct dampers closed—open dampers.
   d. Registers closed or restricted—open registers.
   e. Fan motor nonoperational—find reason and replace motor.

Other causes of insufficient heat could be open heater element or elements, blown element fuses, blown fan motor and transformer fuses, blown thermal link fuses, shorted contactor control, shorted wiring, shorted transformer, or loose terminal connections. Any of these malfunctions requires the services of a technician or electrician.

Rooms Too Hot—or Some Too Cold

1. *Thermostat located incorrectly.* Read the section in the thermostat installation instructions on locating the thermostat, and relocate it as necessary.
2. *System out of balance.* Readjust dampers.
3. *Registers blocked.* Check carefully to make sure that rugs or furniture are not covering or blocking discharge or return-air registers.
4. *Air passages blocked.* Check to see that return-air passages are not blocked. Remove obstructions such as fallen insulation.

Fan/Motor noisy Check fan assembly for loose bolts, then lubricate motor bearings.

Checking Power

It is essential that the proper power be supplied to the equipment and to the load devices in the furnace. Manufacturers' data on the nameplate indicate the proper voltage. Most equipment can be operated within a range of 10% above or below the rated voltage. If power is not available within these limits, the equipment will not operate properly.

Checking Switches

1. *Voltmeter.* This test can be used only when power is on and the voltmeter leads are placed across the two terminals of the switch. When the switch is closed, the voltmeter reads 0. When the switch is open, the voltmeter reads the voltage in the circuit.
2. *Ohmmeter. This test is used only when the power is off.* The leads of the ohmmeter are placed across the two terminals of the switch. The switch must be disconnected from the circuit. A 0 (zero) reading indicates that the switch is closed. An ∞ (infinity) reading indicates that the switch is open.
3. *Jumper.* This test is used with the power on. If a jumper across the two terminals of the switch operates the load, the switch is open.

Checking Loads

1. *Voltmeter.* Test with power on. Determine if proper voltage is available at the load. With proper voltage the load should operate.
2. *Ammeter.* Test with power on.

The ammeter test is used to check the current used by the total furnace or any one of its load components (Figure 15.13). The meter readings are compared with the data on the nameplate. This test offers an excellent means of checking the current through the sequencer to be certain that the elements are being staged on at the proper time. The jaws of the clamp-on ammeter are placed around one of the main power supply lines.

Figure 15.13 Ammeter used to read current draw of electric furnace.

STUDY QUESTIONS

Answers to the study questions may be found in the sections noted in brackets.

15-1 What is the principal difference between the electric heating furnace and other types? *[Conversion of Electricity to Heat]*
15-2 What is the heat equivalent of 1 W? *[Conversion of Electricity to Heat]*
15-3 What is the efficiency of an electric furnace? *[Electric Furnace Components]*
15-4 What is the cutout temperature on an electric furnace limit control? *[Heating Elements]*
15-5 What is the maximum-size heating element? *[Heating Elements]*
15-6 What voltage is supplied to an electric furnace? *[Power Supply]*
15-7 What are the voltages on a control circuit transformer on an electric furnace? *[Power Circuit]*
15-8 Are the heating elements connected in series or parallel? *[Heating Element Circuits]*
15-9 What is the purpose of the sequencer? *[Heating Element Circuits]*
15-10 The outdoor thermostat is used in place of which type of room thermostat? *[Heating Element Circuits]*
15-11 Which instrument is used to check the flow of current through the heating elements? *[Checking Loads]*
15-12 What are the causes of insufficient heat? *[Service Hints]*
15-13 What causes rooms to be too hot? *[Service Hints]*
15-14 Should the fan always operate when the heating elements are on? *[Fan Circuit]*

CHAPTER 16

Estimating the Heating Load

OBJECTIVES

After studying this chapter, the student will be able to:

- Utilize heating load calculation procedures described in the Air Conditioning Contractors of America's *Manual J*

Overview of Heating Loads

The primary purpose of a heating system is to provide comfort conditions with the expenditure of a minimum amount of energy. This process requires that the heating system be properly sized to fit the building requirements. Most heat load calculations are derived from ASHRAE information published by the Air Conditioning Contractors of America (ACCA) in *Manual J,* which has become the standard of the residential heating industry.

Manual J provides an orderly procedure for calculating the total load of a structure, which is then used as the basis for selecting furnace size. The *Manual J* calculation also provides the loads of the individual rooms, which can be used to design the duct system using ACCA's *Manual D.* This chapter deals only with the use of *Manual J.* Selecting the correct furnace size is very important, since an undersized unit will not handle the load, whereas an oversized unit with excess capacity can result in poor control, inefficiency, excessive operating costs, and other problems.

Typical Calculation

To illustrate how to use *Manual J,* a typical calculation is shown. An abbreviated floor plan is given in Figure 16.1. For simplicity, the outside walls and partitions are shown as single lines. In actual practice the outside dimensions of the house would be used

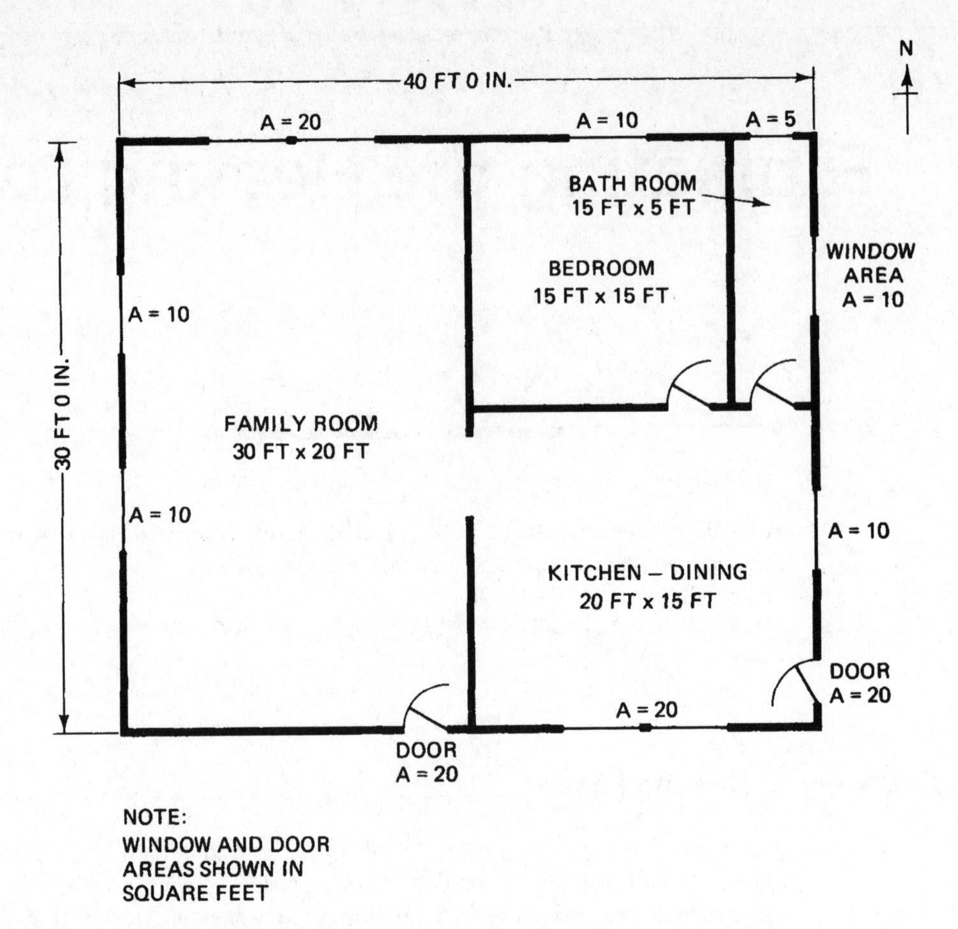

Figure 16.1 Floor plan.

for the "entire house" calculation. The inside dimensions of the outside walls would be used in calculating the individual room loads.

Determining Outside Design Temperature

In the example shown, the city selected from Table 16.1 is Detroit, Michigan, which has a winter design temperature of +6°F. This is not the lowest temperature reached in the area, but 97½% of the time the temperature will be above the +6°F outdoor design temperature.

Selecting Inside Design Temperature

The recommended selection is 70°F.

Table 16.1 Outdoor design conditions for representative cities in Georgia and Michigan

		Winter	
Location	*Latitude (Degrees)*	*97½% Design (db)*	*Heating D.D. Below 65°F*
Georgia			
Atlanta	33	+22	2990
Augusta	33	+23	2410
Savannah	32	+27	1850
Michigan			
Detroit	42	+6	6290
Grand Rapids	42	+5	6890
Lansing	42	+1	6940

Source: Courtesy of Air Conditioning Contractors of America.

Determining Design Temperature Difference

Design temperature difference is the difference between the inside design temperature of 70°F and the outside design temperature of 6°F:

$$70°F - 6°F = 64°F$$

Rounding this figure to the nearest 5°F interval, we use the value 65°F. (This is done to comply with the 5°F temperature increments shown in the tables.)

Determining Construction Numbers

The construction numbers for windows, doors, walls, ceilings, and floors are given in Table 16.2. These numbers represent the materials used in construction of these various exposures. Referring to the excerpts from Table 16.2, note that each type of construction is represented by a number and a letter. A "double-pane window" with a "metal frame" is therefore identified as construction number "3C."

Determining Heat-Transfer Multipliers

The heat-transfer multiplier (HTM) for each construction number is based on the design temperature difference. For example, for a double-pane window, the construction number 3C has an HTM value of 47.1 at a winter temperature difference of 65°F. A summary of the construction numbers and HTM values for the example is shown in Figure 16.2.

Recording Construction Numbers

The construction numbers and the HTM values are recorded in the appropriate locations on the data sheet (see Figure 16.3).

Table 16.2 Heat-transfer multipliers (heating)

			Winter T.D.		
			60	*65*	*70*
				–HTM–	
No. 3 double-pane window					
Clear Glass					
A. Wood frame			33.1	35.8	38.6
B. T.I.M. frame*			36.5	39.6	42.6
C. Metal frame			43.5	(47.1)	50.8
No. 11 metal doors					
A. Fiberglass core			35.4	(38.4)	41.3
B. Fiberglass core and storm			22.0	23.9	25.7
C. Polystyrene core			28.2	30.6	32.9
No. 12 wood frame exterior walls with sheathing and siding or brick, or other exterior finish					
	Cav. insul.	*Sheathing*			
A.	None	½ in. GYPSUM BRD. (R. 0.5)	16.3	17.6	19.0
B.	None	½ in. ASPHALT BRD. (R. 1.3)	13.0	14.1	15.2
C.	R-11	½ in. GYPSUM BRD. (R. 0.5)	5.4	5.8	6.3
D.	R-11	½ in. ASPHALT BRD. (R. 1.3)	4.8	(5.2)	5.6
No. 16 ceiling under ventilated attic space or unheated room					
A. No insulation			35.9	38.9	41.3
B. R-7 insulation			7.2	7.8	8.4
C. R-11 insulation			5.3	5.7	6.2
D. R-19 insulation			3.2	(3.4)	3.7
No. 19 floors over an unheated basement, enclosed crawl space					
A. Hardwood floor + no insul.			9.4	10.1	10.9
B. Hardwood floor + R11 insul.			2.4	(2.6)	2.8
C. Hardwood floor + R13 insul.			2.3	2.5	2.7

*T.I.M. Thermally improved metal frame.
Source: Courtesy of Air Conditioning Contractors of America

Determining Infiltration

For the sample calculation, referring to Table 16.3, we see that the number of winter air changes per hour for this 1200-ft^2 residence of average construction is 1.0 air change per hour. Using procedure A in Table 16.3, we obtain an HTM value for infiltration of 84.7. This value is recorded on the data sheet.

	Const. No.	*HTM*
A. Determine outdoor design temperature + 6°F db (Table 1) Detriot, Mich.		
B. Select inside design temperature +70°F db		
C. Design temperature difference: 70°–6° = 64°F (For convenience, use 65°F.)		
D. Windows: all rooms—clear glass, double pane, metal frame (no storm); Table 2	3C	47.1
E. Doors: Metal, fiberglass core, no storm; Table 2	11A	38.4
F. First-floor walls: basic frame constructions, plastic vapor barrier, R-11 insulation, 1/2-in. asphalt brd. (R1.3) sheating, and face brick; Table 2	12D	5.2
G. Ceiling: basic construction, under vented attic, with R-19 insulation; Table 2	16D	3.4
H. Floor: Hardwood plus R-11 insulation; over enclosed unheated crawl space; Table 2	19B	2.6

Note: Ceilings are 8 ft.; all duct work is located in the unheated crawl space and is covered with R-4 insulation (7A, multiplier 0.10).

Figure 16.2 Assumed design conditions and construction (heating). (Courtesy of Air Conditioning Contractors of America)

Determining Duct-Loss Multiplier

For the example shown, refer to Table 16.4; based on supply-air temperatures below 120°F and the duct with R-4 insulation located in the crawl space, the duct-loss multiplier is 0. 10 (10%). This value is recorded on the data sheet (see Figure 16.3).

Recording Exposed Areas

The exposed areas for the entire house and individual rooms are recorded on the data sheet (see Figure 16.3).

Multiplying Exposed Areas by HTM Factors

The exposed areas are multiplied by the HTM factors to obtain the heat loss through each exposure, and the individual exposure losses are added to obtain a subtotal. The duct loss is added to the subtotal. For the example shown, the total Btuh loss for the entire house is 32,745 Btuh (29,768 + 2977). For one individual room (the family room) the load is 15,327 Btuh (13,934 + 1393). The same procedure applies to the other rooms in the house.

1	Name of Room					Entire House			1 FAMILY ROOM			2 BEDROOM	
2	Running Ft Exposed Wall					40+30+40+30=140'			20+30+20=70'			15'	
3	Room Dimensions, Ft					40'0" x 30'0"			30' x 20'			15' x 15'	
4	Ceiling Ht, Ft		Directions Room Faces			8'	NEWS		8'	NWS		8'	N
	Type of Exposure		Const. No.	HTM		Area or Length	Btuh		Area or Length	Btuh		Area or Length	Btuh
				Htg	Clg		Htg	Clg		Htg	Clg		Htg
5	Gross Exposed Walls and Partitions	a	12D			1120			560			120	
		b											
		c											
		d											
6	Windows and Glass Doors (Htg)	a	3C	47.1		95	4475		40	1884		10	471
		b											
		c											
		d											
7	Windows and Glass Doors (Clg)	North											
		E & W or NE & NW											
		South or SE & SW											
8	Other Doors		11 A	38.4		40	1536		20	768			
9	Net Exposed Walls and Partitions	a	12 D	5.2		985	5122		500	2600		110	572
		b											
		c											
		d											
10	Ceilings	a	16 D	3.4		1200	4080		600	2040		225	
		b											
11	Floors	a	19B	2.6		1200	3120		600	1560		225	
		b											
12	Infiltration . . . Calc.			84.7		135	11,435		60	5082		10	
13	Sub Total Btuh Loss = 6 + 8 + 9 + 10 + 11 + 12						29,768			13,934			
14	Duct Btuh Loss			10%			2,971			1,393			
15	Total Btuh Loss = 13 + 14						32,745			15,327			
16	People @ 300 and Appliances 1200												
17	Sensible Btuh Gain = 7 + 8 + 9 + 10 + 11 + 12 + 16												
18	Duct Btuh Gain				%								
19	Total Sensible Gain = 17 + 18												

Figure 16.3 Sample of worksheet for *Manual J* load calculation. (Courtesy of Air Conditioning Contractors of America)

Table 16.3 Infiltration evaluation of winter air changes per hour

Floor Area	*900 or Less*	*900–1500*
Best	0.4	0.4
Average	1.2	(1.0)
Poor	2.2	1.6

Best, average, and poor categories are based on the quality of the structure. *For example, Average* would include a plastic vapor barrier, caulking, weather-stripping, exhaust fans dampered, combustion air from inside, intermittent ignition, and a flue damper.

Procedure A—winter infiltration, HTM calculation.

1. Winter infiltration CFM
   1.0 AC/H × 9600 ft.^{3} volume × 0.0167
   = 160 ft.^{3}/min
2. Winter infiltration Btuh
   1.1 × 160 ft.^{3}/min × 65°Winter TD
   = 11,440 Btuh
3. Winter infiltration HTM
   11,440 Btuh ÷ 135 total window and door area
   = 84.7 HTM

Cubic volume of house = floor area × ceiling height. In example: 1200 × 8 = 9600 ft.^{3}

Source: Courtesy of Air Conditioning Contractors of America.

Computer Programs

A number of computer programs are available for calculating heating and cooling loads. Generally speaking, there are two types:

1. Those that relate directly to the information published by ASHRAE
2. The Right-J Program, which relates directly to *Manual J,* published by ACCA

A designer using the Right-J Program should determine that the program follows the most recent edition of *Manual J.* A number of important changes have been made

Table 16.4 Duct loss multipliers

Supply Air Temperatures Below 120°F	*Duct Loss Multipliers*	
Duct Location and Insulation Value	*Winter Design Below 15°F*	*Winter Design Above 15°F*
Enclosed in unheated space		
Crawl space or basement–none	0.20	0.15
Crawl space or basement–R2	0.15	0.10
Crawl space or basement–R4	(0.10)	0.05
Crawl space or basement–R6	0.05	0.00

Source: Courtesy of Air Conditioning Contractors of America.

in the calculation procedures in recent years. Of primary importance is the calculation of infiltration. Tests have shown that when compared with newer structures, some older types of construction have high infiltration factors, which can practically double the total load.

The computer programs are fast and accurate, provided that good data are supplied and the operator has sufficient training in operating the computer. It is advisable to compare computer results with manual calculations when using the computer programs initially as a check for possible errors.

The printouts from the computer programs are useful in making presentations or submitting information to building authorities. The summaries indicate not only the final load calculations but also the construction factors used. Another feature of the computer programs is the ease in making "what-if" (prediction) calculations. For example, it might be of interest to compare various amounts of insulation to determine the most cost-effective thickness to use. This would confirm that the increased initial cost is justified based on the savings in operating costs. It is then a simple matter to recalculate the load based on assumed changes in construction factors.

STUDY QUESTIONS

Answers to the study questions may be found in the sections noted in brackets.

16-1 What is the primary purpose of the heating system? *[Overview of Heating Loads]*

16-2 Why is *Manual J* considered a standard of the industry for calculating residential heating loads? *[Overview of Heating Loads]*

16-3 Are the outside or inside dimensions of a house used in calculating a heating load? *[Typical Calculation]*

16-4 Which table is used to determine the outside design temperature? *[Typical Calculation]*

16-5 Which table is used to select the construction factors? *[Typical Calculation]*

16-6 Give an example of determining the design temperature difference. *[Typical Calculation]*

16-7 Give an example of selecting the heat-transfer multiplier (HTM). How is it used? *[Typical Calculation]*

16-8 What is infiltration, and what is its unit of measurement? *[Typical Calculation]*

16-9 What factors are multiplied to determine the heat loss through an exposure? *[Typical Calculation]*

16-10 What is the construction number for hardwood floors with no insulation over an unheated basement? Also, what is the HTM factor for a winter design temperature difference of 65°F? *[Determining Heat Transfer Multipliers]*

CHAPTER 17

Evaluating a Heating System

OBJECTIVES

After studying this chapter, the student will be able to:

- Evaluate the furnace selection and the air distribution system for an existing residence
- Determine the input value of fuel used for a heating system
- Measure the air quantity actually being circulated in a forced warm air heating system

Diagnosing the Problem

To diagnose a problem and then decide what is needed to correct it, a service technician must first determine whether the difficulty is a system problem or a mechanical problem. Some typical system problems are:

- Drafts
- Uneven temperature
- Not enough heat

Drafts are currents of relatively cold air that cause discomfort when they come in contact with the body. Drafts may be caused by a downflow of cold air from an outside wall or window. Uneven temperatures can cause discomfort and are generally due to the following:

1. Different temperatures in a room near the floor and near the ceiling (sometimes called *stratification*) (Figure 17.1)
2. Different temperatures in one room compared with another
3. Different temperatures on one floor level compared with another

Uneven room temperatures may be caused by incorrect supply diffuser locations. Cold outside surfaces of a room should be warmed by properly located supply-air out-

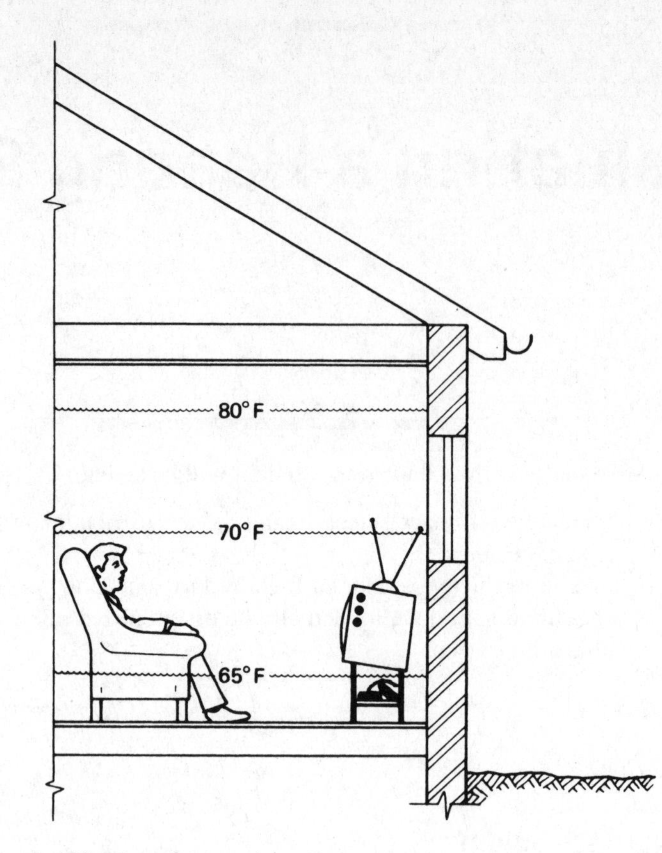

Figure 17.1 Stratification of room air.

lets. Temperature differences among rooms may be caused by improper balancing of the system due to the lack of duct dampers in the branch run to each outlet. Balancing is the process of regulating the flow of air into each room to produce even temperatures. Not enough heat may be caused by too small a furnace or improper fuel input to the unit to match the heating load of the house.

Other system problems can be solved or improved by the service technician. Other problems, however, are built into the design of the system and can be corrected only by redesign or replacement of major components.

Evaluating the System

Three items should be considered:

1. Furnace (energy source and efficiency)
2. Fan
3. Air distribution systems (supply and return)

Two questions relate to the furnace:

1. Has the proper size been selected?
2. Is the furnace adjusted to produce its rated output?

In Chapter 16 a method was given for determining the heating load of a building. Any losses that occur, such as duct loss or ventilation air, are added to the total room loss to determine the total required heating load. The furnace output rating should be equal to, but not greater than 15% higher than, the total required heating load. If the furnace is too small, it will not heat properly in extreme weather. If it is too large, the "off" cycles will be too long, and the result will be uneven heating.

To determine if a furnace is producing its rated capacity, it is necessary to check:

1. Fuel input
2. Combustion efficiency

Energy Source

Gas Many manufacturers void their warranty if the gas input is not adjusted to within 2% of the rated input of the furnace, as shown on the furnace nameplate (see Figure 17.2). To determine input, proceed as follows:

1. Obtain the BTU/ft^3 rating of the gas from the local utility.*
2. Turn off all other appliances and operate the furnace continuously during the test.
3. Measure the length of time it takes for the furnace to consume 1 ft^3 of gas.

For example, if it takes 36 seconds to use 1 ft^3 of gas (according to the 1-ft^3 dial on the meter), and the gas rating is 1000 Btu/ft^3, the input is 1000 Btu/ft^3 $\times$ 100 ft^3 = 100,000 Btuh. The 100 is found on the "36 seconds" line in the 1-ft^3 dial column in

*Consult the local gas company for this information.

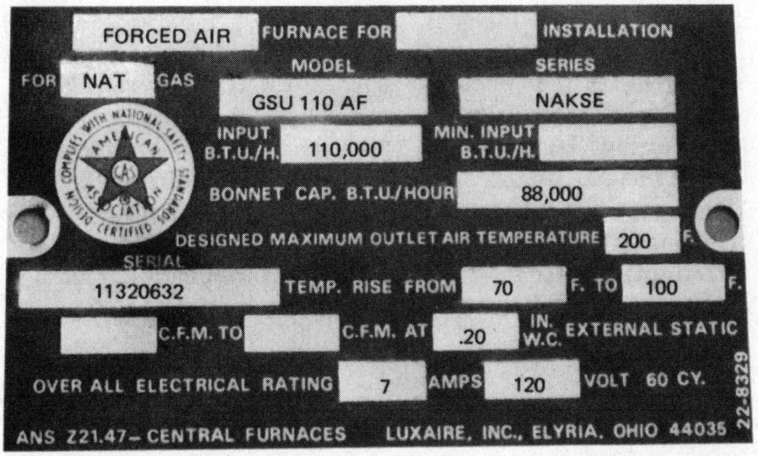

Figure 17.2 Typical gas furnace nameplate.

Figure 17.3. If the furnace is rated at 100,000 Btuh input, the usage will be within requirements.†

Another method of computing the gas flow to a furnace or other appliance is by utilizing the following formula:

$$\text{cubic feet of gas} = \frac{3600 \times \text{dial on the gas meter}}{\text{number of seconds for one revolution}}$$

$$= \frac{3600 \times 1\ \text{ft}^3\ \text{dial}}{36\ \text{seconds/revolution}} = 100\ \text{ft}^3$$

$$\text{furnace input} = 100\ \text{ft}^3 \times 1000\ \text{Btu/ft}^3 = 100{,}000\ \text{Btuh}$$

Some meters have low volume dials other than 1 ft^3. Using the table shown in Figure 17.3 and the method described, a similar check on the gas input can be made. If

†Input rating for a gas furnace is given on the furnace nameplate.

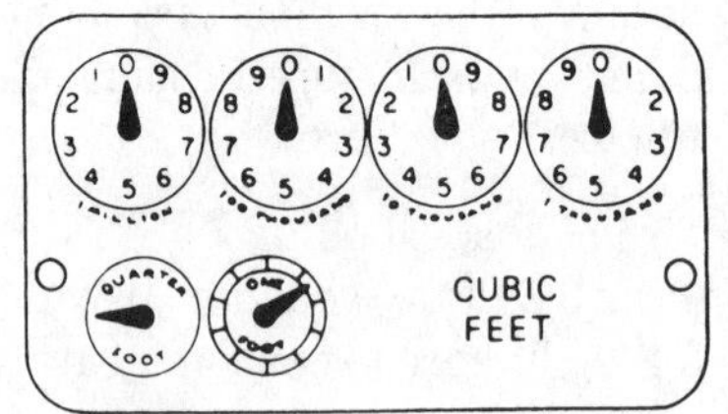

THE DIAL IS MARKED AS TO HOW MUCH GAS IS MEASURED FOR EACH REVOLUTION. USING THE NUMBER OF SECONDS FOR ONE REVOLUTION, AND THE SIZE OF THE TEST DIAL, FIND THE CUBIC FEET OF GAS CONSUMED PER HOUR FROM THE TABLE

SECONDS FOR ONE REV.	SIZE OF TEST DIAL ¼ CU.FT.	½ CU.FT.	1 CU.FT.	2 CU.FT.	5 CU.FT.
10	90	180	360	720	1800
11	82	164	327	655	1636
12	75	150	300	600	1500
13	69	138	277	555	1385
14	64	129	257	514	1286
15	60	120	240	480	1200
16	56	113	225	450	1125
17	53	106	212	424	1059
18	50	100	200	400	1000
19	47	95	189	379	947
20	45	90	180	360	900
21	43	86	171	343	857
22	41	82	164	327	818
23	39	78	157	313	783
24	37	75	150	300	750
25	36	72	144	288	720
26	34	69	138	277	692
27	33	67	133	267	667
28	32	64	129	257	643
29	31	62	124	248	621
30	30	60	120	240	600
31	–	–	116	232	581
32	28	56	113	225	563
33	–	–	109	218	545
34	26	53	106	212	529
35	–	–	103	206	514

SECONDS FOR ONE REV.	SIZE OF TEST DIAL ¼ CU.FT.	½ CU.FT.	1 CU.FT.	2 CU.FT.	5 CU.FT.
36	25	50	100	200	500
37	–	–	97	195	486
38	23	47	95	189	474
39	–	–	92	185	462
40	22	45	90	180	450
41	–	–	–	176	439
42	21	43	86	172	429
43	–	–	–	167	419
44	–	41	82	164	409
45	20	40	80	160	400
46	–	–	78	157	391
47	19	38	76	153	383
48	–	–	75	150	375
49	–	–	–	147	367
50	18	36	72	144	360
51	–	–	–	141	355
52	–	–	69	138	346
53	17	34	–	136	240
54	–	–	67	133	333
55	–	–	–	131	327
56	16	32	64	129	321
57	–	–	–	126	316
58	–	31	62	124	310
59	–	–	–	122	305
60	15	30	60	120	300

Figure 17.3 Gas meter measuring dials and table for determining gas input based on dial readings. (Courtesy of Bard Manufacturing Company)

the input is not correct, it should be changed to the furnace input rating by adjusting the gas-pressure regulating valve or by changing the size of the burner orifices (see the manufacturer's instructions for details).

Oil The input rating to an oil furnace can be checked by determining the oil burner nozzle size and measuring the oil pressure. Nozzles are rated for a given amount of oil flow at 100 pounds per square inch gauge (psig) oil pressure. One gallon of grade No. 2 oil contains 140,000 Btu. The oil burner nozzle is examined and the input flow is read on the nozzle in gallons per hour (gal/h). Thus a nozzle rated at 0.75 gal/h operating with an oil pressure of 100 psi would produce an input rating of 105,000 Btu (0.75 gal/h × 140,000 Btu). The oil pressure is measured and adjusted, if necessary, to 100 psig within 3%. Figure 17.4 illustrates a typical oil furnace nameplate and oil burner nozzle.

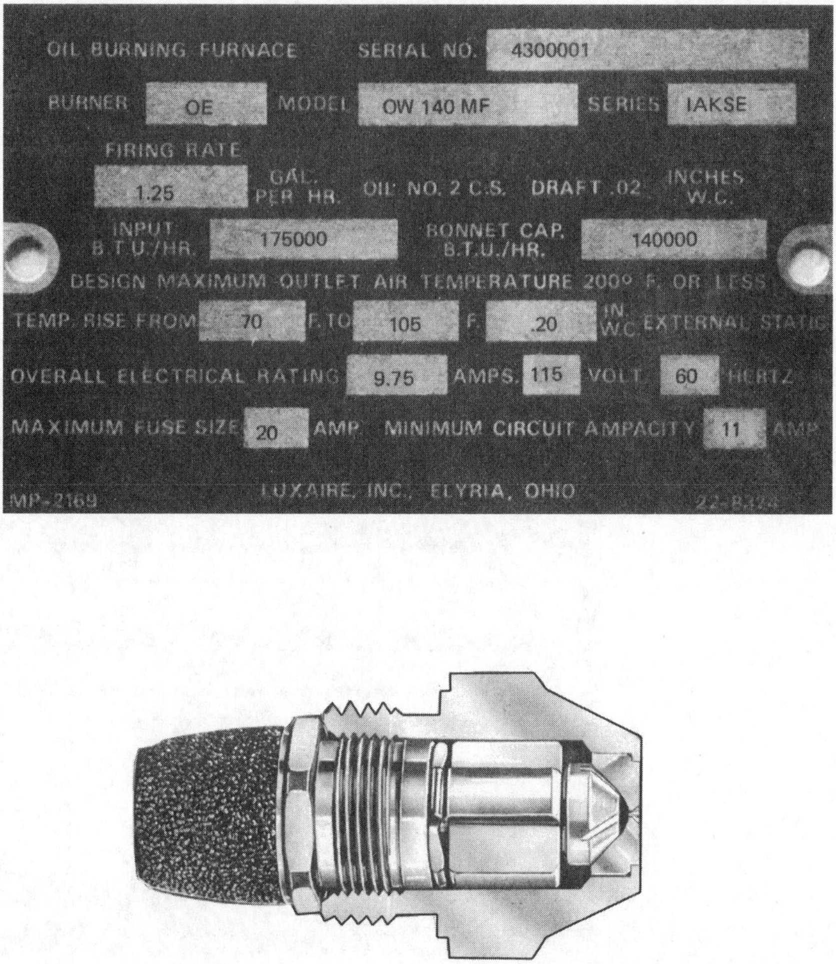

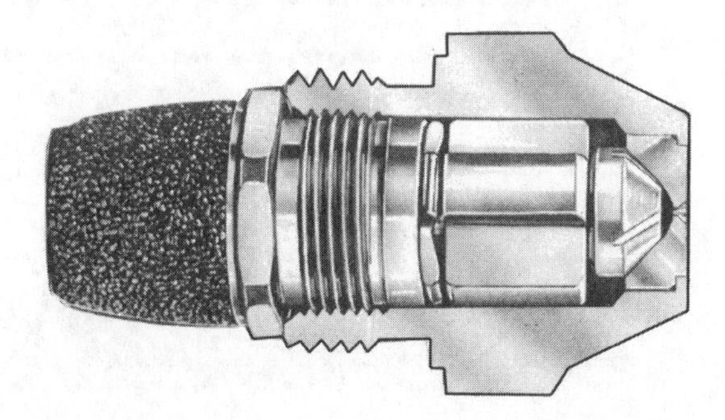

Figure 17.4 Typical oil burner nozzle and oil furnace nameplate. (Courtesy of Delavan Corporation)

Electricity For an electric furnace the input is usually rated in watt-hours (Wh) or kilowatt-hours (kWh) (1000 Wh = 1 kWh of electricity consumed). Watts (W) can be converted to Btu by multiplying by 3.4131 (1 W = 3.4131 Btu). For example, if an electric furnace is rated at 30 kWh, the input is 102,393 Btuh (30 kWh × 1000 h/kWh × 3.4131 Btu). Figure 17.5 illustrates a typical electric furnace nameplate.

Efficiency

For both gas and oil furnaces, the output rating is reduced from the input rating by an efficiency factor. For an electric furnace, though, the input is equal to the output, since there is no heat loss up the chimney. The general formula is

$$\text{Btuh output} \div \text{Btuh input} = \text{efficiency factor}$$

The output rating of both gas and oil furnaces is shown on the nameplate as a percentage of input. The efficiency of the installed furnace is usually lower, however. A service technician should perform an efficiency test to determine if the rated efficiency is actually being maintained. The method of determining and adjusting combustion efficiency for oil furnaces is covered in Chapter 13. Gas furnace adjustments are covered in Chapter 5. In most cases where the combustion efficiency is low, the technician can adjust the fuel burner so that the furnace may again approach the design efficiency. Having measured the efficiency of the furnace, the technician can calculate the actual output from the formula given. For example, if a gas furnace is operating at 80% efficiency** and has an

**Manufacturer's rating.

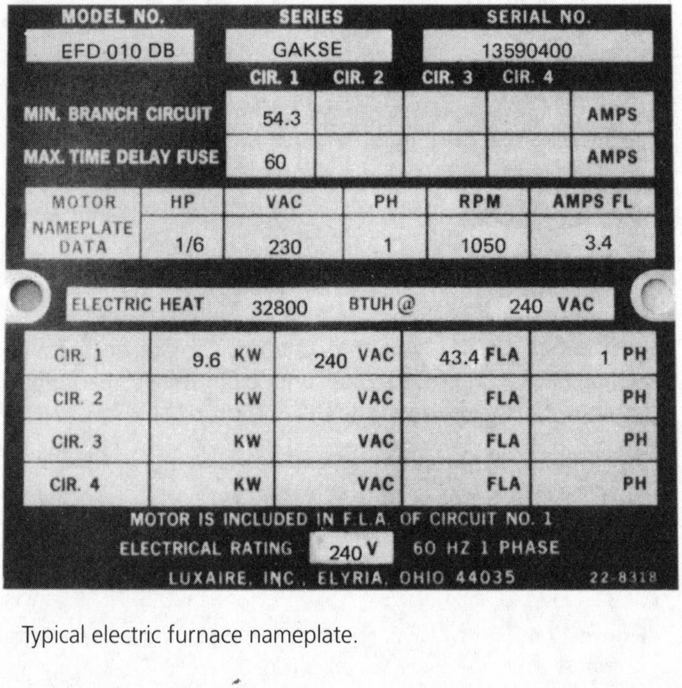

MODEL NO.	SERIES	SERIAL NO.
EFD 010 DB	GAKSE	13590400

	CIR. 1	CIR. 2	CIR. 3	CIR. 4
MIN. BRANCH CIRCUIT	54.3			AMPS
MAX. TIME DELAY FUSE	60			AMPS

MOTOR NAMEPLATE DATA	HP	VAC	PH	RPM	AMPS FL
	1/6	230	1	1050	3.4

ELECTRIC HEAT 32800 BTUH @ 240 VAC

CIR. 1	9.6 KW	240 VAC	43.4 FLA	1 PH
CIR. 2	KW	VAC	FLA	PH
CIR. 3	KW	VAC	FLA	PH
CIR. 4	KW	VAC	FLA	PH

MOTOR IS INCLUDED IN F.L.A. OF CIRCUIT NO. 1
ELECTRICAL RATING 240 V 60 HZ 1 PHASE
LUXAIRE, INC., ELYRIA, OHIO 44035 22-8318

Figure 17.5 Typical electric furnace nameplate.

input of 100,000 Btuh, the output (bonnet capacity) is 80,000 Btuh (100,000 Btuh × 0.80 efficiency). An electric furnace with 30 kWh (102,393 Btuh) input operates at 100% efficiency and has an output of 102,393 Btuh (102,393 Btuh × 1.00 efficiency).

It is seldom desirable to select a furnace with a rated output equal to the heat loss of the house. If this were done, the furnace could be too small due to system losses or system inefficiency. Other losses can occur that are not accounted for in the heat-loss calculations, and combustion efficiencies may not always be achieved. For these reasons it is good practice to add 10 or 15% to the calculated heat loss to determine the required furnace output.

Fans

The fan produces the movement of air through the furnace, absorbing heat from the heat exchanger surface and carrying it through the distribution system to the areas to be heated. Fan factors involve:

- Air volume
- Static pressure
- Causes of poor air distribution

Air Volume For the system to operate satisfactorily, the fan must deliver the proper air volume. Most furnaces permit some flexibility (variation) in the air volume capacity of the furnace. A furnace having an output of 80,000 Btuh may be able to produce 800, 1200, and 1600 cfm of air volume, depending on the requirements. Systems designed for heating only usually require less air than systems designed for both heating and cooling. The proper air volume for heating is usually determined by the required temperature rise. The temperature rise is the supply-air temperature minus the return-air temperature at the furnace. Systems used for heating only should be capable of a temperature rise of 85°F. Systems designed for heating and cooling should be capable of a minimum temperature rise of 70°F.

To find the quantity of air the furnace is actually circulating, a service technician measures the temperature rise. One thermometer is placed in the return-air plenum and the other is placed in the supply-air duct. While the furnace is operating continuously the readings are taken and the temperature rise (difference) computed. Thermometer locations for checking temperature rise are shown in Figure 17.6.

For a typical heating-only application, the return-air temperature is 65°F, and the supply-air temperature is 150°F, indicating an 85°F temperature rise (150°F − 65°F = 85°F). For a typical heating/cooling system, the return-air temperature is 65°F, and the supply air temperature is 135°F, indicating a temperature rise of 70°F (135°F − 65°F = 70°F). The air volume circulated is proportionately different for the two types of systems. To determine air volume the following formula is used:

$$\text{cfm} = \frac{\text{Btuh output of furnance}}{\text{temperature rise (°F)} \times 1.08}$$

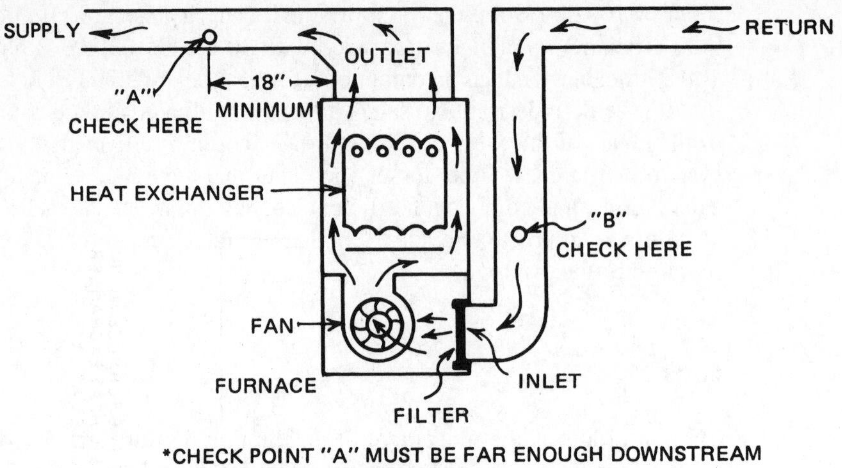

Figure 17.6 Thermometer locations for checking temperature.

Thus if the temperature rise on an 80,000-Btu-output furnace is 85°F, the air volume is

$$\text{cfm} = \frac{80{,}000 \text{ Btuh}}{80°\text{F} \times 1.08} = 870$$

If the temperature rise on an 80,000-Btuh-output furnace is 70°F, the air volume is

$$\text{cfm} = \frac{80{,}000 \text{ Btuh}}{70°\text{F} \times 1.08} = 1058$$

Many furnaces have two-speed fans, with the low speed frequently used for heating and the high speed for cooling. Adjustments in the fan speed can be made on most furnaces to regulate the air volume to meet the requirements of the heating system. This topic is discussed further in Chapter 4.

Static Pressure A manufacturer rates the fan air volume of a furnace to produce a quantity of air at a certain external static pressure. *Static pressure* is the resistance to airflow offered by any component through which the air passes. *External static pressure* is the resistance of all components outside the furnace itself. Thus if a unit is rated to supply 600 cfm at an external static pressure of 0.20 in., it means that the total resistance offered by supply ducts + return ducts + supply diffusers + return grilles must not exceed 0.20 in. of static pressure. A system external and internal static pressure diagram is shown in Figure 17.7.

Static pressure is measured in inches of water column that the pressure of the fan is capable of raising on a water gauge *manometer* (Figure 17.8). A manometer is an instrument for measuring pressure of gases and vapors. The pressures in residential systems are small, so an inclined tube manometer is used to increase the accuracy of the readings

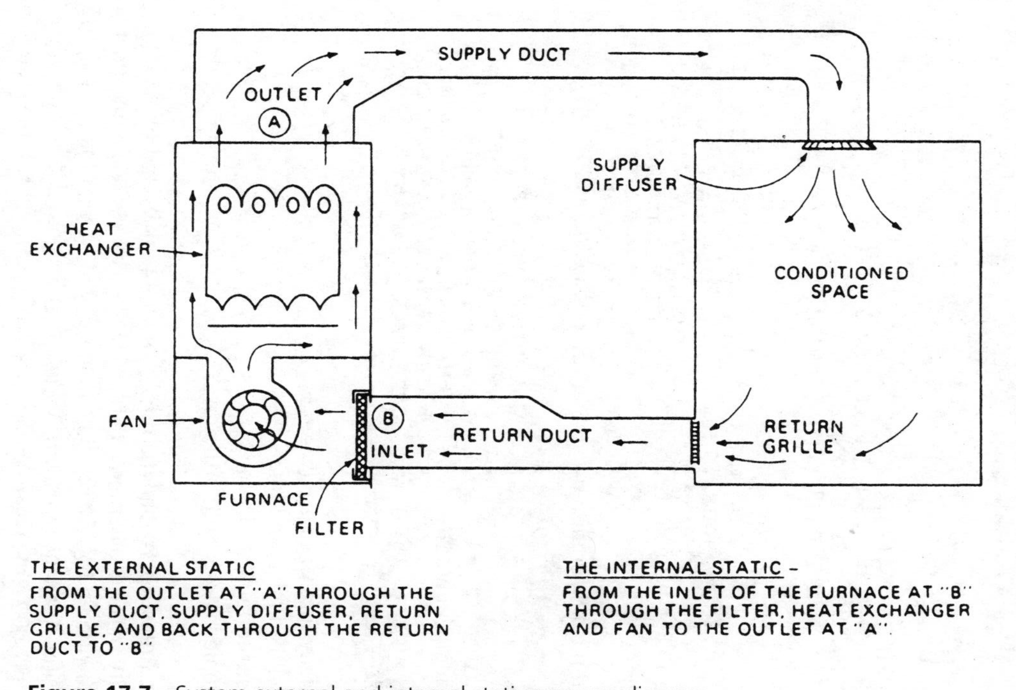

Figure 17.7 System external and internal static pressure diagram.

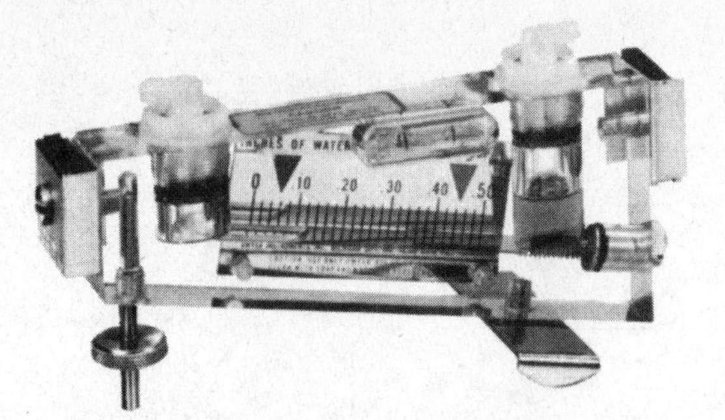

MODEL 170 INCLINED-TYPE PORTABLE MANOMETER, WITH BUILT-IN LEVELING AND MAGNETIC CLIPS. RANGE: 0–0.50 IN. WATER (COURTESY, DWYER INSTRUMENTS, INC.)

SERIES 1222-8-D FLEX-TUBE U-TUBE MANOMETER, MAGNETIC MOUNTING CLIPS AND RED GAUGE OIL INCLUDED. RANGE: 8-IN. WATER (4–0–4) (COURTESY, DWYER INSTRUMENTS, INC.)

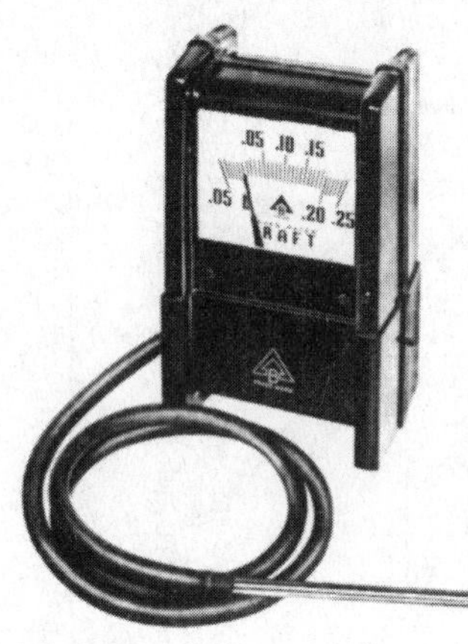

MODEL MZF, DRY-TYPE. SUPPLIED WITH 5-IN. DRAFT TUBE AND 9 FT OF RUBBER TUBING. RANGE +0.05 TO 0.25 IN. WATER (COURTESY, BACHARACH INSTRUMENT, COMPANY.)

Figure 17.8 Types of manometers.

taken. A manometer indicates the air pressure delivered by the fan above atmospheric pressure. To illustrate how small these fan readings are, 1 atmosphere (14.7 psi) is equal to 408 in. W.C. Two-tenths of an inch (0.20 in. W.C.) of pressure is 1/2040 atmosphere.

A great deal of care must be taken in designing an air distribution system to stay within the rated external static pressure of the furnace. If the system resistances are too high, the amount of airflow of the furnace is reduced. All manufacturers' ratings of external static pressure are based on clean filters. Dirty filters can cause reduced air volume and poor heating. A table showing the relationship between increased external static pressure and decreased air volume (cfm) is given in Figure 17.9.

The greater the air volume and static pressure of a furnace, the larger the horsepower of the motor required to deliver the air. Therefore, units used for heat pumps, which operate at higher air volumes and higher static pressures, require higher-horsepower

AIR DELIVERY—(CFM) FOR DIRECT DRIVE BLOWERS

MODEL NUMBER	BLOWER SIZE	MOTOR H.P.	BLOWER SPEED	EXTERNAL STATIC PRESSURE—INCHES WATER COLUMN													
				0.1		0.2		0.3		0.4		0.5		0.6		0.7	
				CFM	TEMP. RISE	CFM	TEMP. RISE	CFM	TEMP. RISE	CFM	TEMP. RISE	CFM	TEMP. RISE	CFM	TEMP. RISE	CFM	TEMP. RISE
GUH060A012	11.8x8	½	HIGH	1609	32.4	1550	33.6	1491	35.0	1431	36.4	1365	38.2	1280	40.7	1202	43.4
			MED.	1294	40.3	1268	41.1	1231	42.4	1188	43.9	1131	46.1	1070	48.9	992	52.6
			LOW	1045	49.9	1021	51.1	995	52.4	960	54.3	924	56.5	875	59.6	820	63.6
GUH075A012	11.8x8	½	HIGH	1609	40.1	1550	41.6	1491	43.3	1431	45.1	1365	47.3	1280	50.4	1202	53.7
			MED.	1294	49.9	1268	50.9	1231	52.4	1188	54.3	1131	57.1	1070	60.3	992	65.1
			LOW	1045	61.8	1021	63.2	995	64.9	960	67.2	924	69.8	875	73.8	820	78.7
GUH100A016	11.8x10.6	¾	HIGH	1642	52.4	1568	54.9	1505	57.2	1441	59.7	1359	63.3	1270	67.8	1170	78.5
			MED. HI	1621	53.1	1557	55.3	1487	57.9	1415	60.8	1357	63.4	1252	68.7	1152	74.7
			MED. LOW	1465	58.7	1399	61.5	1343	64.1	1285	67.0	1197	71.9	1122	76.7	1030	83.6
			LOW	1335	64.5	1295	66.4	1249	68.9	1207	71.3	1152	74.7	1084	79.4	1010	85.2
GUH125A016	11.8x10.6	¾	HIGH	1810	58.8	1750	60.8	1675	63.5	1628	65.4	1577	67.5	1470	72.4	1371	77.6
			MED. HIGH	1802	59.0	1730	61.5	1655	64.3	1575	67.6	1478	72.0	1380	77.1	1262	84.3
			MED. LOW	1685	63.1	1625	65.5	1565	68.0	1490	71.4	1403	75.8	1308	81.4	1188	89.6
			LOW	1562	68.1	1515	70.2	1460	72.9	1390	76.6	1313	81.0	1220	87.2	1103	96.5
GUH125A020	11.8x10.6	¾	HIGH	2160	48.5	2085	50.0	2010	52.1	1955	53.6	1890	55.4	1820	57.6	1680	62.4
			MED. HIGH	1891	55.4	1837	57.0	1784	58.7	1729	60.6	1675	62.6	1582	66.2	1490	70.3
			MED. LOW	1777	59.0	1728	60.6	1680	62.4	1614	64.9	1549	67.6	1495	70.1	1441	72.7
			LOW	1631	64.2	1596	65.7	1562	67.1	1505	69.6	1448	72.4	1389	75.4	1330	78.8

Figure 17.9 Manufacturer's data showing external static pressures. (Courtesy of ARCOAIRE Air Conditioning & Heating)

motors than do units used for heating only. The speed, or revolutions per minute (rev/min), of the fan wheel must often be increased for cooling. The wet cooling coil installed external to the basic furnace has a resistance to the airflow in addition to ductwork and grilles. The total external static pressure of a cooling system may be as high as 0.50 in. W.C. static pressure. Following are two typical performance ratings for a unit that can be applied to either heating only or heating and cooling:

Use	*Air Volume (CFM)*	*External Static (in. W.C.)*	*Motor HP*
Heating only	700	0.20	⅛
Heating and cooling	800	0.50	¼

Causes of Poor Air Distribution One of the first jobs of a service technician in evaluating the distribution system is to determine whether it has been designed for heating only or for heating and cooling. If the air volume is found to be too low, it can be caused by one or more of the following:

1. Incorrect fan speed
2. Closed or partially closed dampers
3. Dirty filters
4. Incorrectly sized ducts
5. High-pressure-drop duct fittings

If improper fan speed is the problem, the technician can check on the possibility of speeding it up. This may require a larger motor. The increased speed must be kept within permissible noise levels. If closed dampers are the problem, adjustments can be made. Dampers are installed for balancing the system (regulating the airflow to each room) (Figure 17.10). Misadjustments should be corrected and, if necessary, the system rebalanced. If dirty filters are the problem, the owner should be advised of proper preventive maintenance measures. If duct sizes are the problem, some improvement can be made by speeding up the fan. There are limitations to improving this condition without a major redesign, however.

High-pressure-drop duct fittings are a problem that can often be corrected by minor changes in the ductwork. The pressure drop in duct fittings is usually given in terms of the equivalent length of straight duct of the same size having an equal pressure drop.

Figure 17.11 shows pressure drops in terms of equivalent lengths (EL) for various duct fittings. Abrupt turns and restrictions should be avoided whenever possible. When ductwork is placed in an unheated attic, crawl space, or garage, insulated ductwork should be used. Various types of flexible ducts and methods of insulating ducts are illustrated in Figure 17.12.

Supply-Air Distribution Systems

In general, there are three types of supply-air distribution systems:

1. Radial
2. Perimeter loop
3. Trunk duct and branch

Selection of a system depends on the house construction and room arrangement. Each system has its advantages for certain types of applications.

Radial A radial system consists of a number of single pipes running from the furnace to the supply-air outlets. In these systems the furnace is located near the center of the house so that the various runouts (supply-air ducts) will be as nearly equal in length as possible. This system helps provide warm floors and is used with both crawl-space construction and concrete-slab floors.

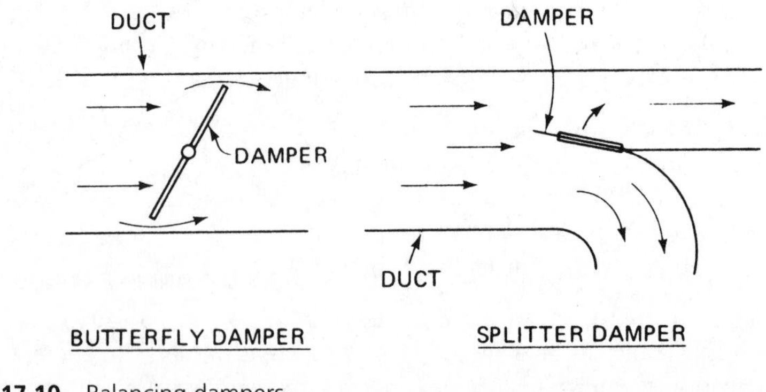

Figure 17.10 Balancing dampers.

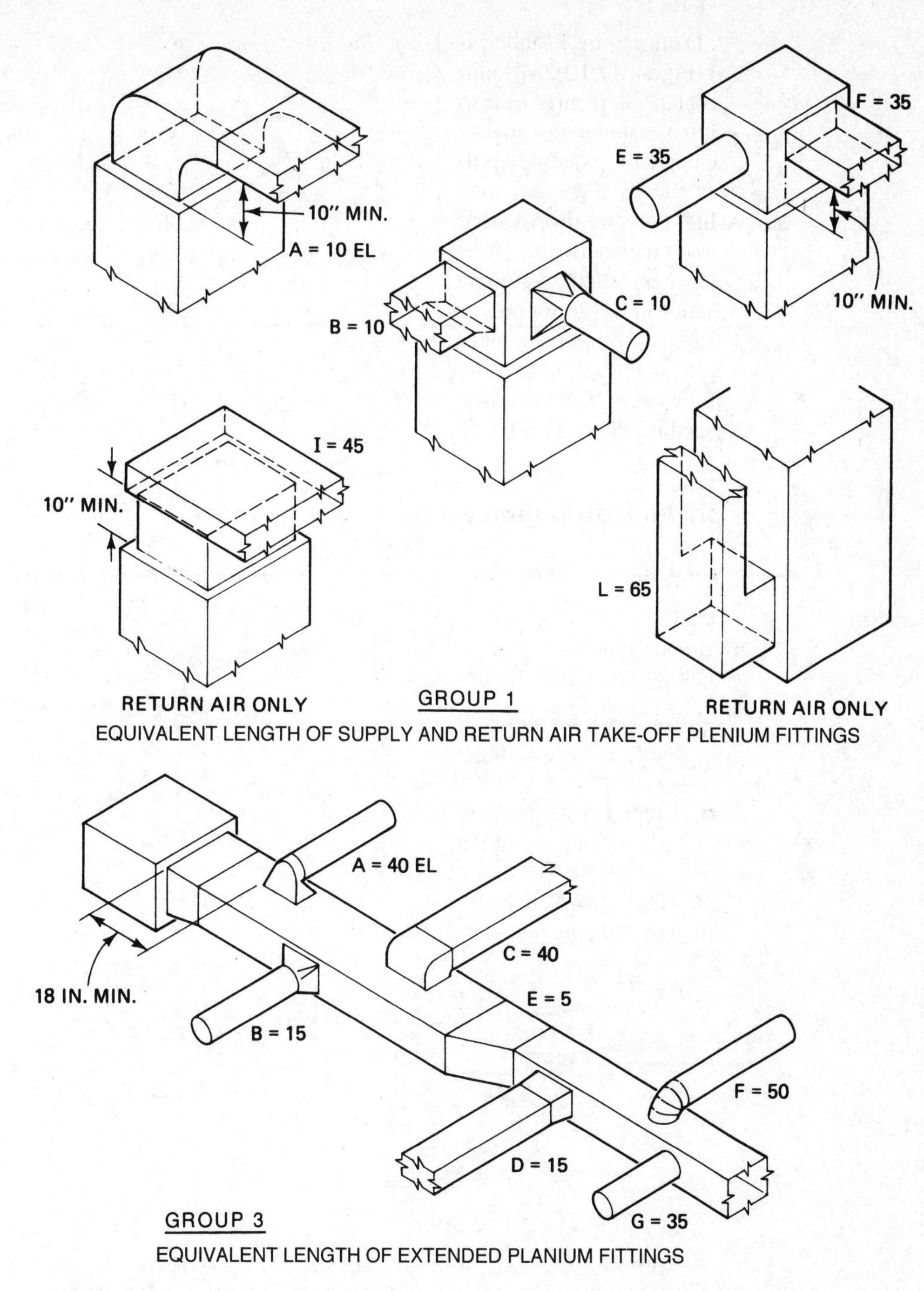

Figure 17.11 Supply and return fittings. (Courtesy of Air Conditioning Contractors of America)

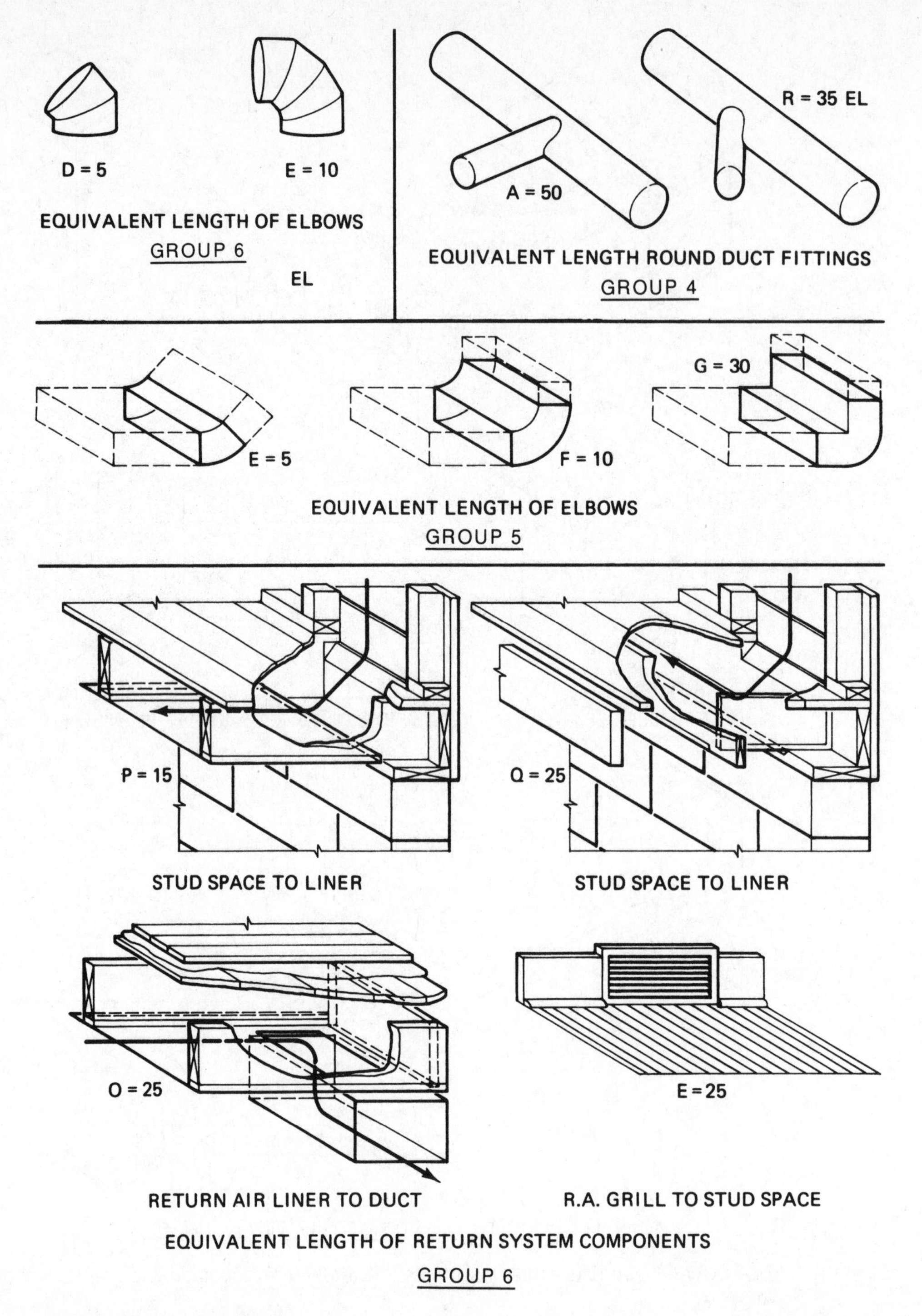

Figure 17.11 Continued

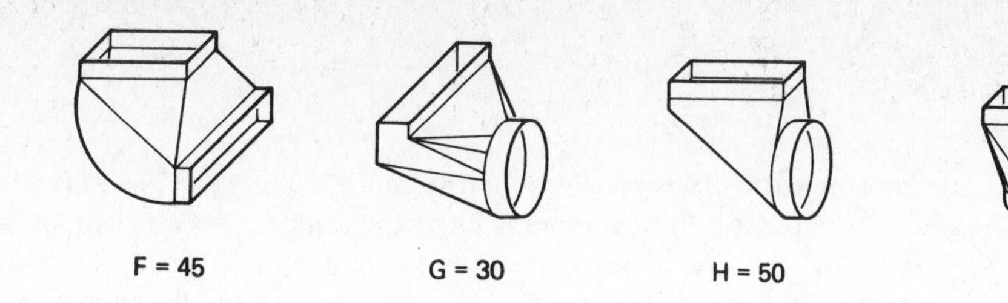

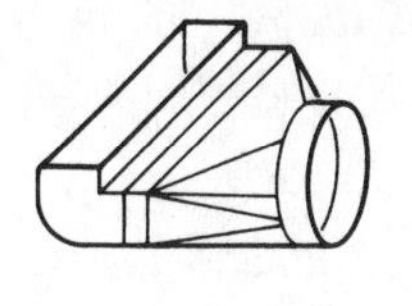
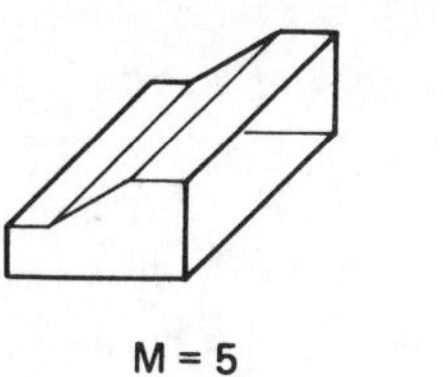
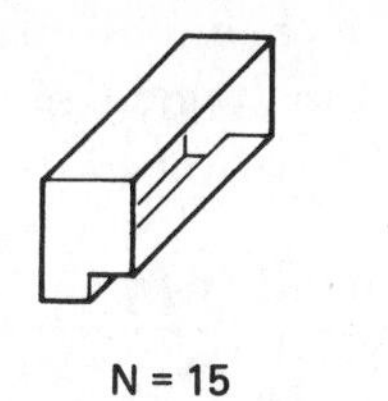
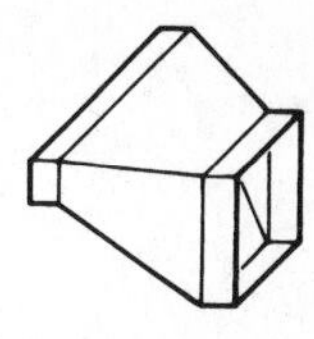
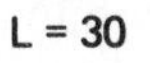

M = 5   N = 15   O = 15

EQUIVALENT LENGTH OF BOOT FITTINGS
GROUP 7

Figure 17.11 Continued

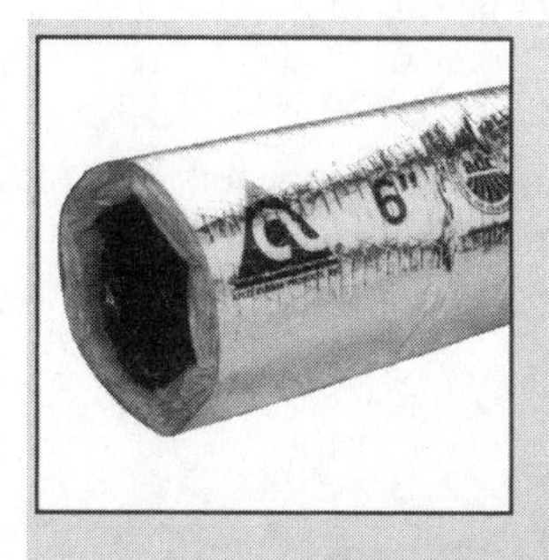

Our premium insulated duct has a double ply, metalized polyester vapor barrier with Tri-Directional reinforcements.	
UPC #030	Double ply polyester core. R-Value = 4.2*
UPC #036	Double ply polyester core. R-Value = 6.0*
UPC #031	Double ply polyester core. R-Value = 8.0* (Not available in 20″ dia.)
UPC #037	CPE Polymeric core. R-Value = 4.2*
UPC #039	CPE Polymeric core. R-Value = 6.0*

This durable, all purpose product has a double ply, grey polyester vapor barrier with spiral reinforcements.	
UPC #070	Double ply polyester core. R-Value = 4.2*
UPC #076	Double ply polyester core. R-Value = 6.0*
UPC #077	CPE Polymeric core. R-Value = 4.2*
UPC #079	CPE Polymeric core. R-Value = 6.0*

This product has a tough black polyethylene vapor barrier with scrim reinforcement.	
UPC #080	Double ply polyester core. R-Value = 4.2*
UPC #086	Double ply polyester core. R-Value = 6.0*
UPC #088	Double ply polyester core. R-Value = 8.0* (Not available in 20″ dia.)
UPC #087	CPE Polymeric core. R-Value = 4.2*

* R-Values are classified by Underwriters Laboratories Inc., in accordance with the ADC Flexible Duct Performance and Installation Standards (1991) using ASTM C - 518 at installed wall thickness on flat insulation only.

Figure 17.12 Various types of flexible ducts. (Courtesy of ATCO Rubber Products, Inc.)

Perimeter Loop The perimeter-loop system (Figure 17.13) is a modification of the radial system. A single duct running along the perimeter (outer edge) of the house supplies air to each supply-air diffuser located on the floor above. Radial ducts connect the perimeter loop to the furnace. The runout ducts from the furnace to the loop are larger and fewer than those in the radial system. This system lends itself for use with slab-floor construction. The perimeter loop supplies extra heat along the perimeter of the house, warming the floor and heating the exterior walls of the house. The greatest heat loss in a slab floor is near its perimeter.

Extended Plenum The trunk-and-branch system is the most versatile system. It permits the furnace to be located in any convenient location in the house. Large ducts carry the air from the furnace to branches or runouts going to individual air outlets. The trunk duct (large duct from furnace) may be the same size throughout its entire length if the length does not exceed 25 ft. This is called an *extended plenum* (Figure 17.14). On large systems the trunk duct should be reduced in size, after branches remove a portion of the air, so that it tapers toward the end (Figure 17.15). A trunk duct can be used for basement installations or overhead installations of ductwork. It can be used for both supply and return systems.

Diffuser Placement The placement of supply-air diffusers, or registers, depends to some extent on the climate of the area in which the house is located. In northern climates, where cold floors can be a problem and where outside exposures must be thor-

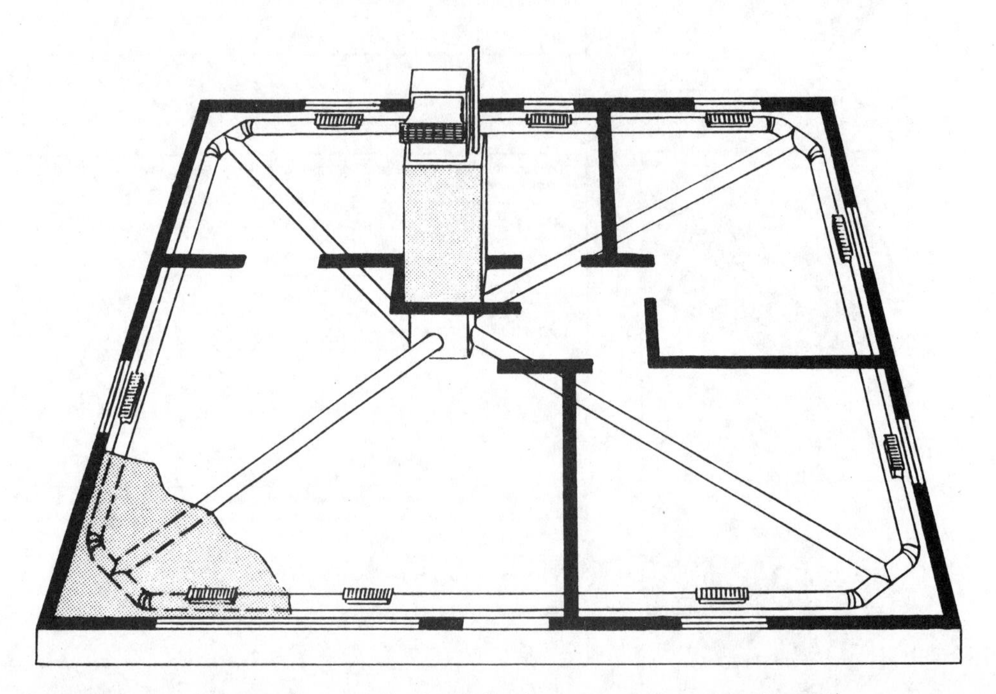

Figure 17.13 Perimeter-loop supply-air distribution system. (Courtesy of Carrier Corporation)

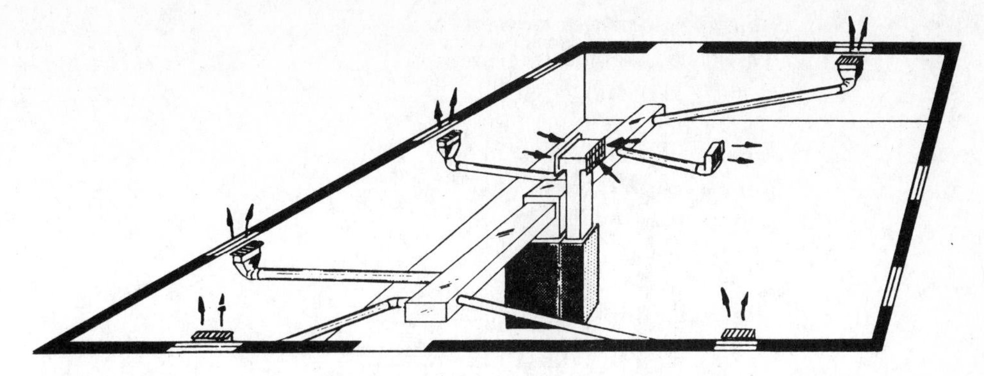

Figure 17.14 Extended-plenum air-supply distribution system. (Courtesy of Carrier Corporation)

oughly heated, a perimeter location of the supply-air diffusers is best. The register can be located in the floor adjacent to each exposed wall or in the baseboard. In many southern climates, where cooling air distribution is more important than heating air distribution, diffusers can be located on an inside wall as high as 6 ft above the floor. In extremely warm climates, the supply-air outlet can be placed in the ceiling. Figures 17.16 and 17.17 show and describe various locations for the placement of air-supply

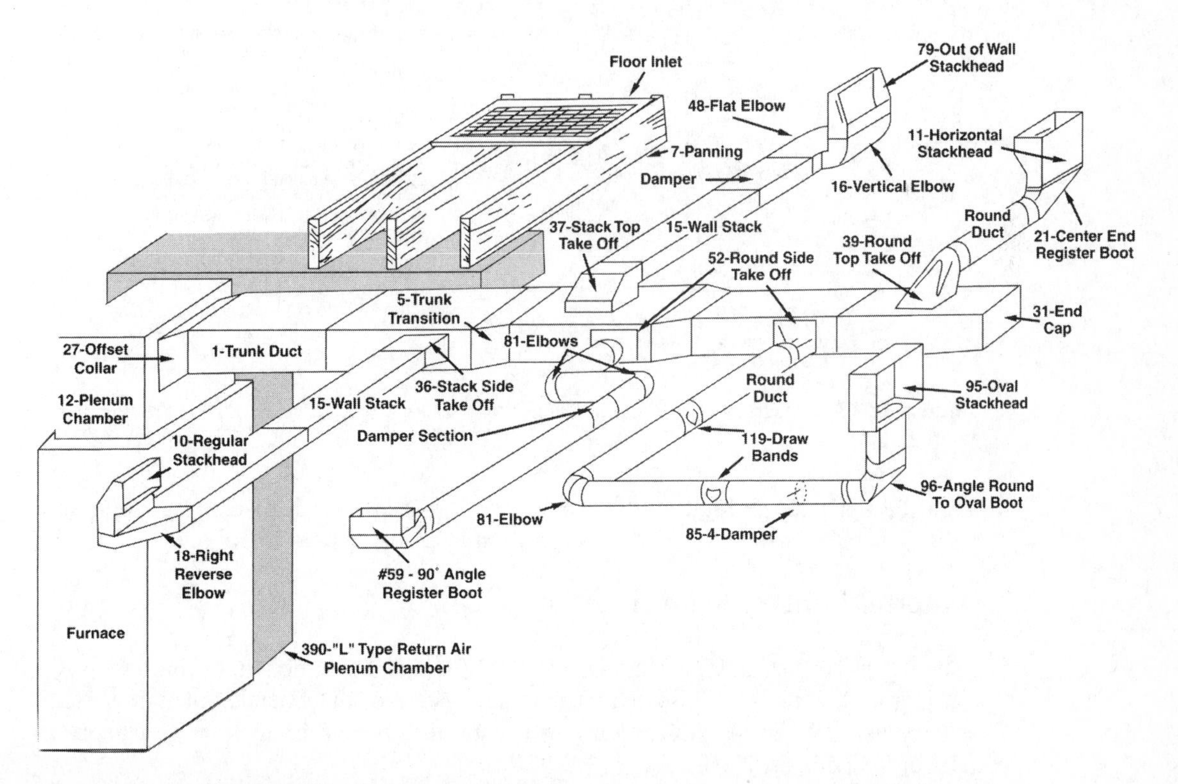

Figure 17.15 Reducing trunk duct used in supply- or return-air distribution systems.

[A] REGISTERS SET TO DIRECT AIR UPWARD ALONG THE WALL AT AS WIDE AN ANGLE AS POSSIBLE

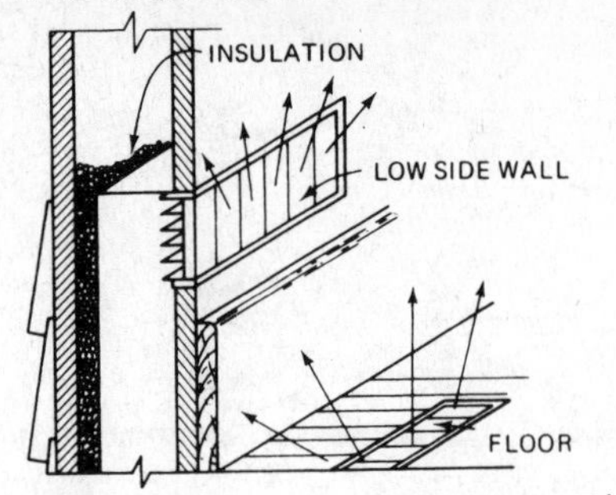

FLOOR OR LOW SIDEWALL PERIMETER OUTLETS[A]

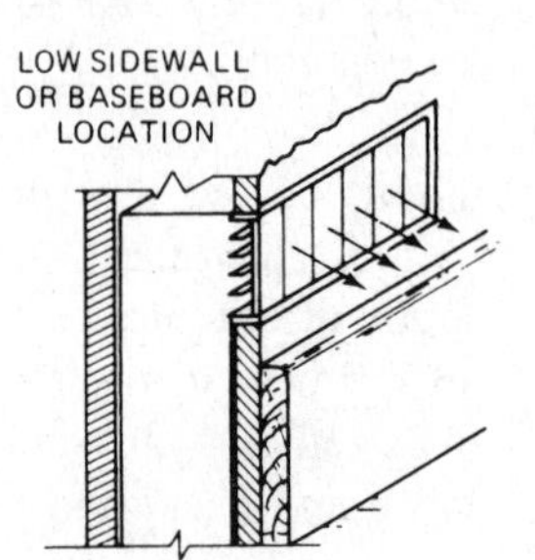

[A] VERTICAL BARS WITH ADJUSTABLE DEFLECTION, OR FIXED VERTICAL BARS WITH DEFLECTION TO RIGHT AND LEFT NOT EXCEEDING ABOUT 22 DEG. FOR LOW SIDEWALL LOCATION, THE DEFLECTION FOR HORIZONTAL, MULTIPLE VANE REGISTERS SHOULD NOT EXCEED 22 DEG. FOR BASEBOARD LOCATIONS, THE DEFLECTION FOR HORIZONTAL, MULTIPLE VANE REGISTERS SHOULD NOT EXCEED ABOUT 10 DEG.

RECOMMENDED TYPE OF BASEBOARD AND LOW SIDEWALL INSTALLATION ON WARM WALL[A]

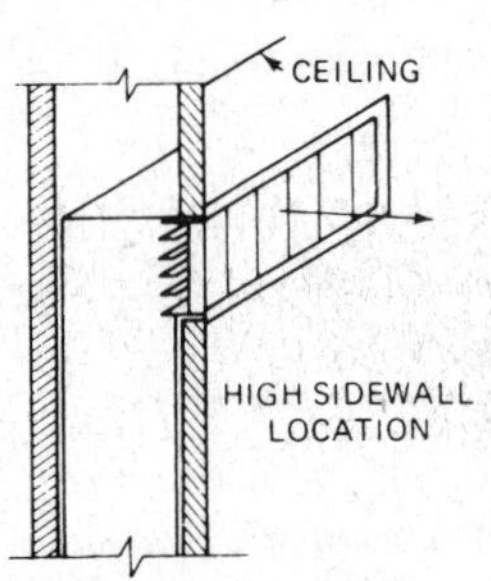

[A] HORIZONTAL VANES, IN BACK OR FRONT, TO GIVE DOWNWARD DEFLECTIONS NOT TO EXCEED 15 TO 22 DEG.

RECOMMENDED TYPE OF HIGH SIDEWALL INSTALLATION ON WARM WALL[A]

Figure 17.16 Placement of air-supply distribution outlets. (By permission from the ASHRAE Handbook)

distribution outlets. Figure 17.18 shows various types of diffusers, registers, and grilles, some of which are used for supply outlets and some for return inlets.

Return-Air Distribution Systems

The return-air distribution system is usually of simple construction, having less pressure drop than the supply system. The radial system or the extended-plenum system can be used. Where possible, return air is carried in boxed-in (enclosed) joist spaces (Figures 17.15 and 17.19).

GENERAL CHARACTERISTICS OF OUTLETS

GROUP	OUTLET TYPE	OUTLET FLOW PATTERN	SIZE DETERMINED BY
1	CEILING AND HIGH SIDEWALL	HORIZONTAL	MAJOR APPLICATION – HEATING OR COOLING
2	FLOOR REGISTERS, BASEBOARD AND LOW SIDEWALL	VERTICAL, NONSPREADING	MAXIMUM ACCEPTABLE HEATING TEMPERATURE DIFFERENTIAL
3	FLOOR REGISTERS, BASEBOARD AND LOW SIDEWALL	VERTICAL, SPREADING	MINIMUM SUPPLY VELOCITY DIFFERS WITH TYPE AND ACCEPTABLE TEMPERATURE DIFFERENTIAL
4	BASEBOARD AND LOW SIDEWALL	HORIZONTAL	MAXIMUM SUPPLY VELOCITY SHOULD BE LESS THAN 300 FPM

GROUP	MOST EFFECTIVE APPLICATION	PREFERRED LOCATION
1	COOLING	NOT CRITICAL
2	COOLING AND HEATING	NOT CRITICAL
3	HEATING AND COOLING	ALONG EXPOSED PERIMETER
4	HEATING ONLY	LONG OUTLET – PERIMETER, SHORT OUTLET – NOT CRITICAL

Figure 17.17 General characteristics of air-supply distribution outlets. (By permission from the *ASHRAE Handbook*)

Grille Placement In northern climates, where supply-air diffusers are located along the perimeter, return-air grilles (returns) are usually placed near the baseboard on inside walls. Because the room temperature is not affected by the location of returns, though, they may be placed high on the inside walls to prevent drafts. In southern climates, where high inside wall outlets or ceiling diffusers are used, returns can be placed in any convenient location on inside walls. Short-circuiting of the supply air directly into the return, however, must be avoided. Whereas supply-air outlets (one or more) are placed in every room to be heated, connecting rooms with open doorways can share a common return. Returns are not placed in bathrooms, kitchens, or garages. Air for these rooms must be taken from some other area. The general rule is that the total duct area for the return-air system must be equal to the total duct area for the supply-air system. Figure 17.20 shows possible locations for the placement of air-return grilles in the return-air system.

Provision for Cooling

When cooling as well as heating is supplied from the same outlets, it is necessary to direct the supply air upward for cooling. Cold air is heavier than warm air and tends to puddle (collect near the floor). To raise cold air, baseboard diffusers should have adjustable vanes or baffles. This upward movement can be accomplished by using air deflectors (Figure 17.21).

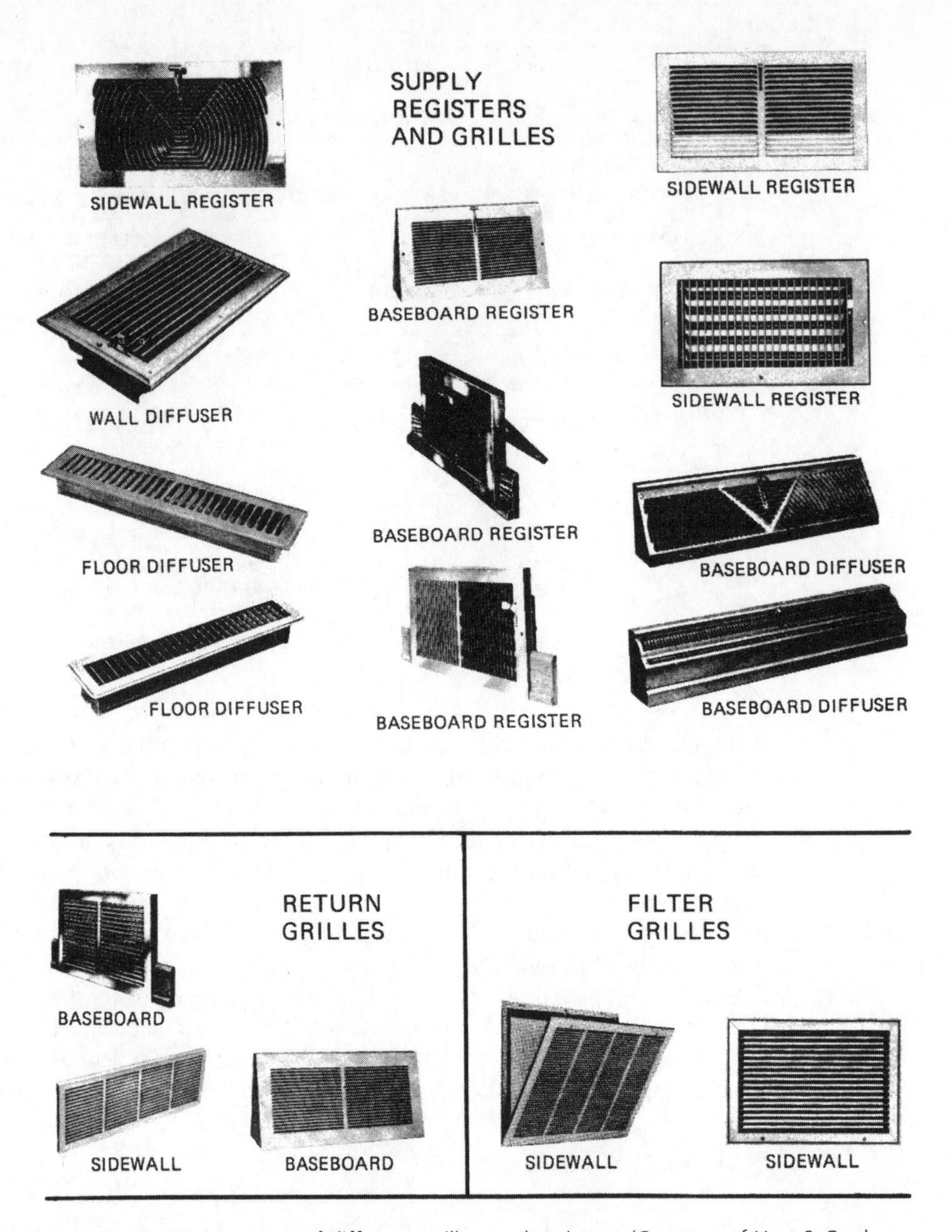

Figure 17.18 Various types of diffusers, grilles, and registers. (Courtesy of Hart & Cooley, Inc.)

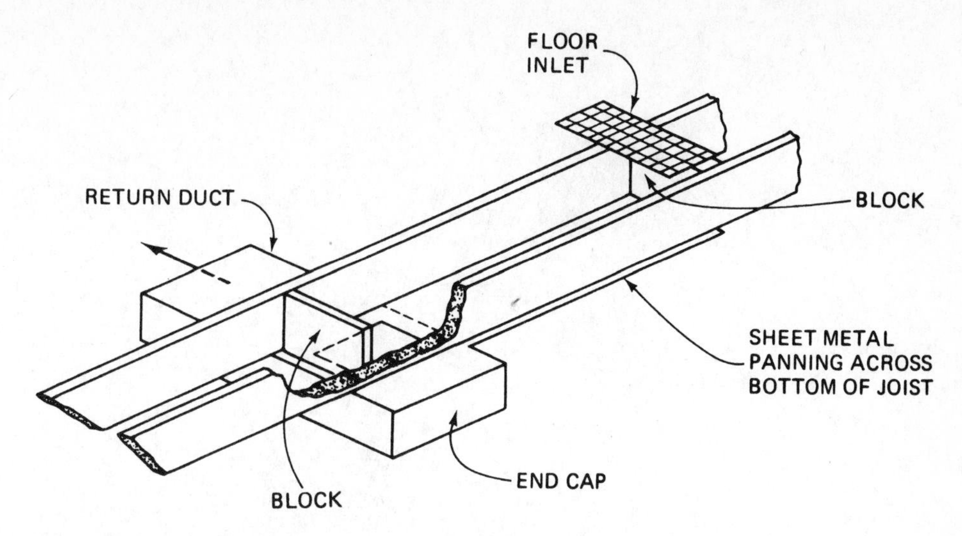

Figure 17.19 Boxed-in joist spaces in return-air distribution system.

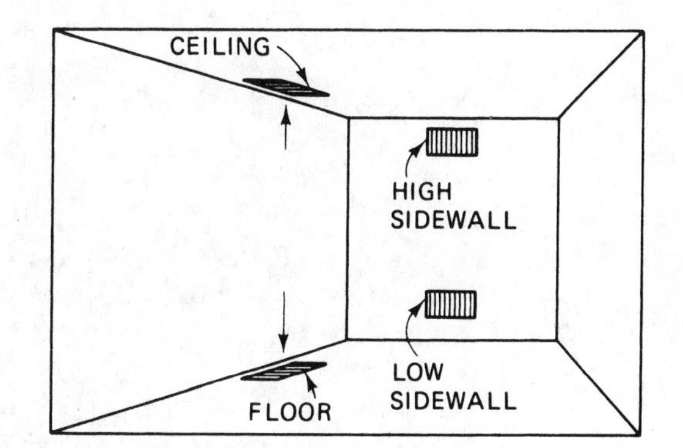

Figure 17.20 Placement of air-return grilles in return-air distribution system.

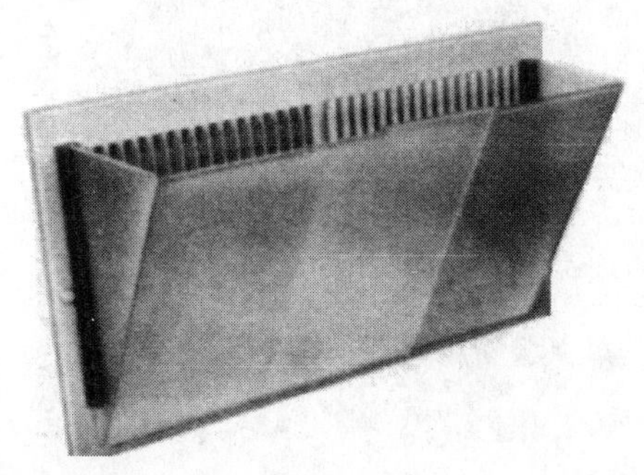

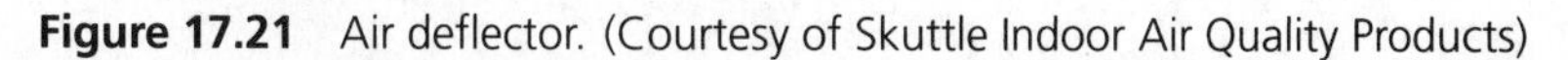

Figure 17.21 Air deflector. (Courtesy of Skuttle Indoor Air Quality Products)

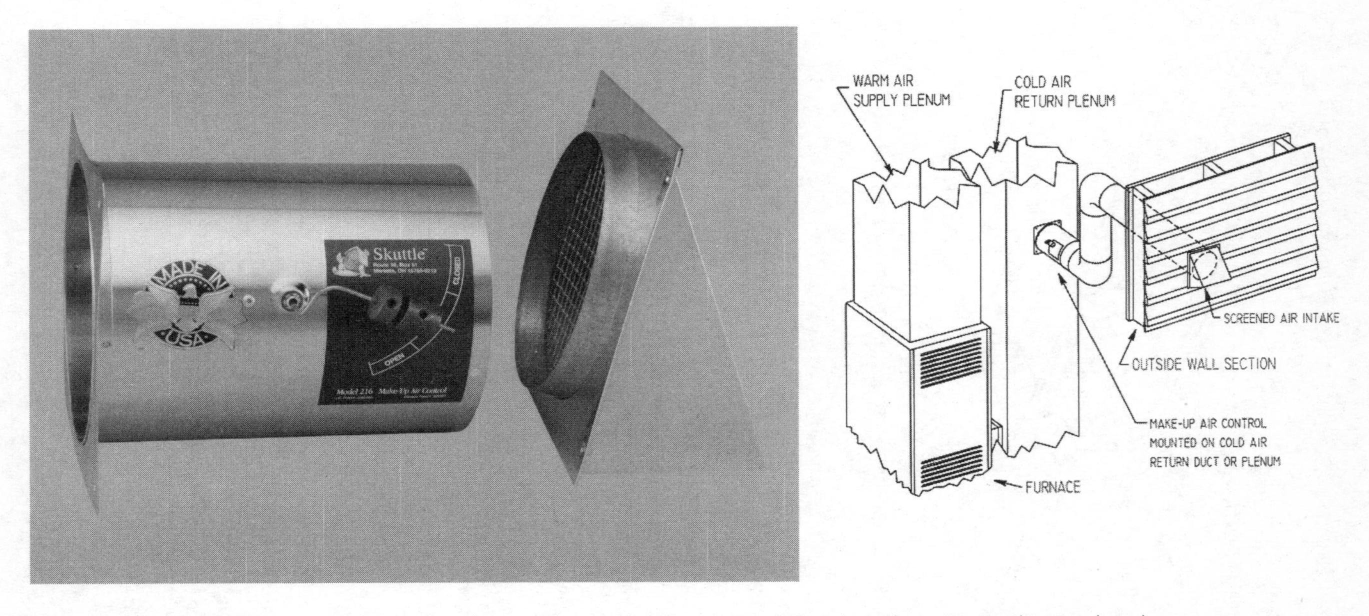

Figure 17.22 Barometric and thermal makeup air controls. (Courtesy of Skuttle Indoor Air Quality Products)

Other Provisions

- Dampers must be provided in each branch supply run from the furnace to balance the system.
- All ducts running in unconditioned areas such as attics or garages must be insulated.
- Ventilation air should be connected directly to the furnace or return-air plenum, not to the intermediate return-air ductwork.
- Ducted outside air can be made available to the furnace for ventilation air and/or combustion air (see Figure 17.22).

STUDY QUESTIONS

Answers to the study questions may be found in the sections noted in brackets.

17-1 What are some common problems of residential heating systems from a comfort standpoint? *[Diagnosing the Problem]*

17-2 Name three types of uneven temperatures that cause discomfort. *[Diagnosing the Problem]*

17-3 Name three parts of the system in which to look for comfort problems. *[Diagnosing the Problem]*

17-4 What is the method of determining whether a gas furnace is operating at its rated input? *[Energy Source: Gas]*

17-5 Which dial on the gas meter is used for timing the rated input of a furnace? *[Energy Source: Gas]*

17-6 What is the method of rating the flow of oil burner nozzles? *[Energy Source: Oil]*

17-7 What is the Btu equivalent for 1 W of electricity? *[Energy Source: Electricity]*

17-8 Give the formula for determining the efficiency of a furnace. *[Efficiency]*

17-9 Give the formula for determining the cubic feet per minute of air circulated. *[Air Volume]*

17-10 Define external static pressure in terms of system resistance to the flow of air. *[Static Pressure]*

17-11 How is static pressure measured? What units are used? *[Static Pressure]*

17-12 How does external static pressure affect the flow of air? *[Static Pressure]*

17-13 Is the volume of air required for cooling less than, equal to, or greater than the volume or air required for heating? *[Static Pressure]*

17-14 What are the five principal causes of poor air distribution? *[Causes of Poor Air Distribution]*

17-15 Describe the three types of supply-air distribution systems. *[Supply-Air Distribution Systems]*

17-16 Define the equivalent length rating of a duct fitting. *[Supply-Air Distribution Systems]*

17-17 What is the best location for return-air grilles? *[Supply-Air Distribution Systems]*

CHAPTER 18

Installation Practice

OBJECTIVES

After studying this chapter, the student will be able to:

- Evaluate the quality of a heating installation in a residence
- Determine the proper wire size and fuse size for electrical circuits
- Determine the proper gas pipe size for a residential gas furnace installation
- Evaluate the venting requirements for a residential gas furnace installation

Evaluating Construction and Installation

When a service technician is called on to correct a system complaint, there are two areas to be evaluated.

1. *Building construction.* Does the heating system conform to the building requirements?
2. *Furnace installation.* Is the equipment installed properly?

System complaints involve such conditions as drafts, uneven temperatures, cold floors, and other conditions that cause discomfort, even when the equipment may be operating well mechanically. Some of these complaints are due to the design of the system, which cannot be modified without considerable expense. Other system problems can be improved, if not fully corrected, when a service technician understands the cause.

Building Construction

The types of building construction that require special treatment for best heating results include:

- Structures with basements
- Structures over crawl spaces

- Concrete-slab construction
- Split-level structures

Structures with Basements These structures should have some heat in the basement to produce warm floors on the level above. This practice should be followed whether or not the basement is finished for a recreation room. The structure should also be designed so that the basement can be properly heated at reasonable cost. The recommended construction for a properly heated basement is shown in Figures 18.1 and 18.2.

The following are desirable construction features of structures with basements:

- Subsoil below floor is well drained.
- A moisture barrier is placed below the floor and between the outside walls and the ground.
- Insulation is placed in the joist around the perimeter of the structure.

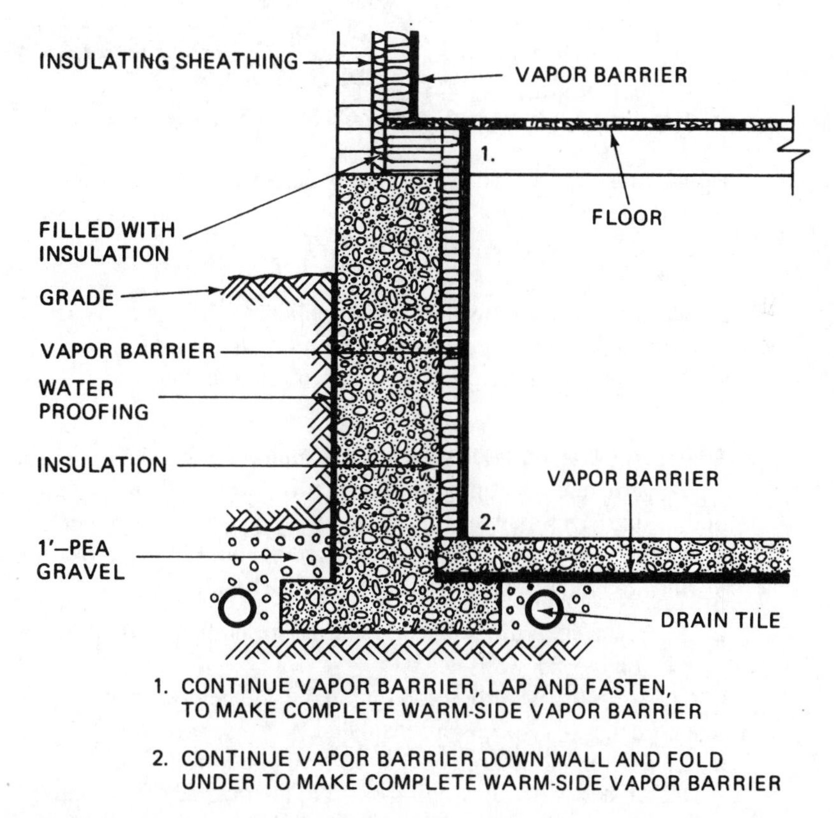

Figure 18.1 Application of insulation in basement of existing structure. (Courtesy of the Detroit Edison Company)

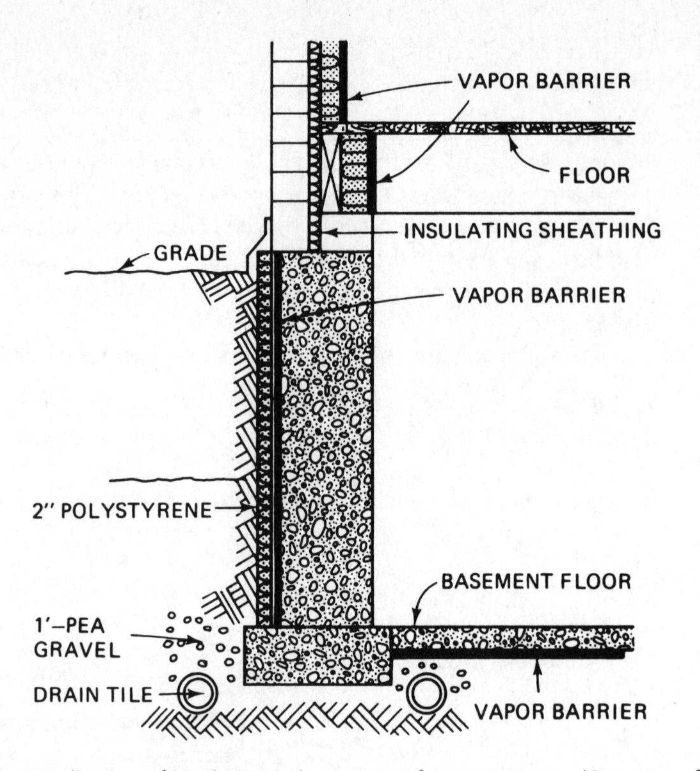

Figure 18.2 Application of insulation in basement of new structure. (Courtesy of the Detroit Edison Company)

- Vaporproof insulating board is applied to the inside basement walls of existing structures and on the outside of new structures.
- Any piping that pierces the vapor barrier is properly sealed.

Structures Over Crawl Spaces It is important that the crawl space be heated in this type of structure. Heating produces warm floors on the level above. The best construction is similar in many ways to basement construction. Where crawl spaces are built above the bare ground a moisture barrier should be used (see Figure 18.3). The following are features of crawl-space construction:

- A moisture barrier is placed over the ground and extended upward a minimum of 6 in. on the sidewalls.
- The walls are waterproofed on the outside below grade.
- Perimeter joist spaces are insulated.
- On existing structures insulation is applied to the inside of the crawl-space walls, and on new structures it is applied to the outside of the walls.
- Ventilation of the crawl space is provided in summer only.
- Dampers permit adjustment of the heat feeding the crawl-space area.

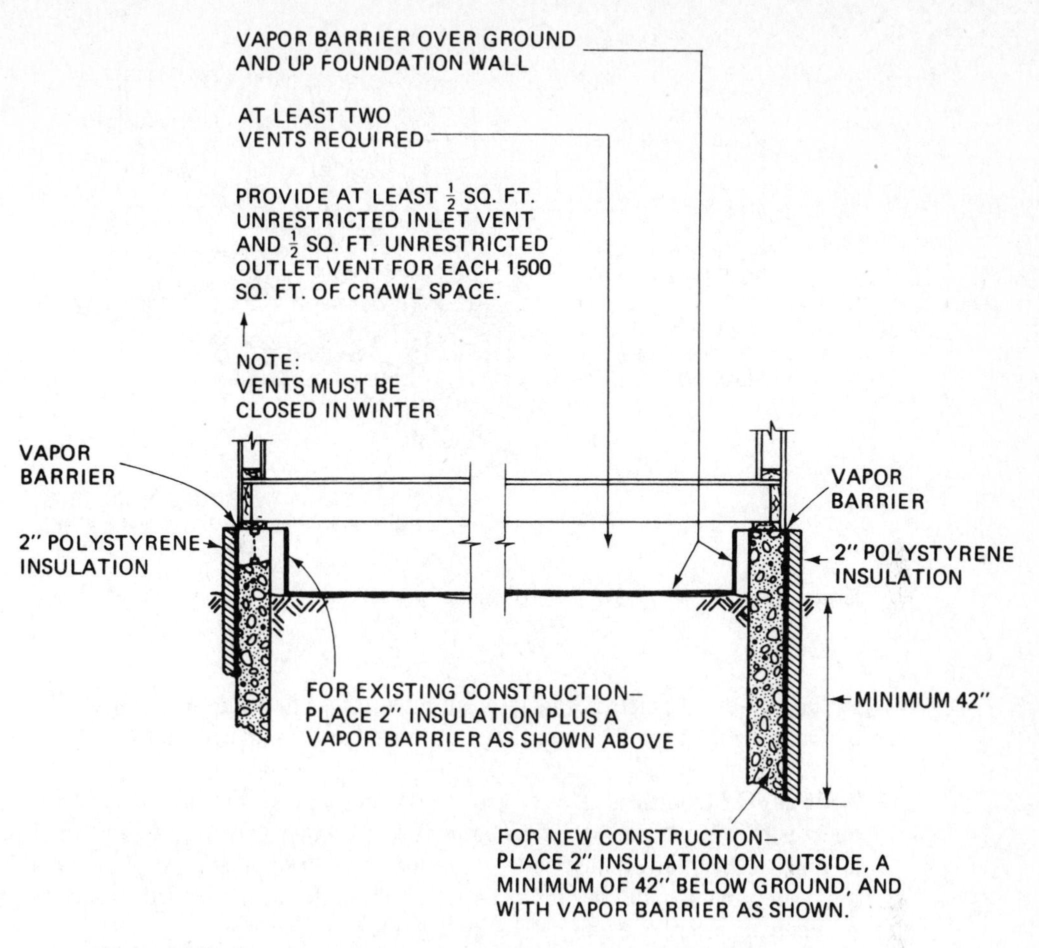

Figure 18.3 Recommended crawl-space construction and ventilation.

Concrete-Slab Construction It is important to warm the floor in concrete-slab construction, particularly around the perimeter, where greatest heat loss occurs. This can be done by placing the heat-distributing ducts in the concrete floor. Some of the items included in this construction and shown in the illustrations are as follows:

- Edge insulation is installed around the perimeter of the slab, as shown in Figure 18.4.
- Ducts, embedded in the floor, supply perimeter heating (Figure 18.5).
- A concrete pit of proper size is poured below the furnace for the supply-air plenum chamber (Figure 18.6).
- A moisture barrier is placed over the ground before the concrete slab is poured.
- Feeder ducts slope downward from the perimeter to the plenum pit.

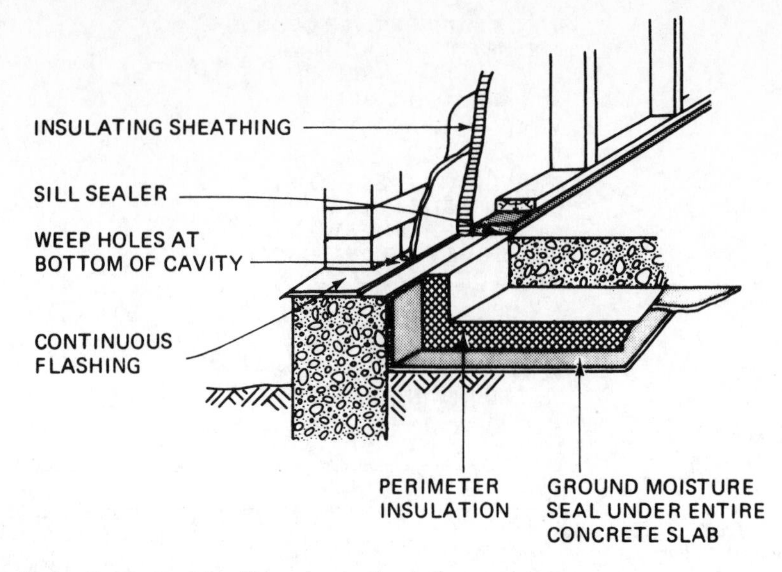

Figure 18.4 Concrete slab with perimeter insulation.

- Ducts are constructed of waterproof materials with waterproof joints.
- Dampering of individual outlets is provided at the register location.

Split-Level Structures This type of construction presents a problem in balancing the heat distribution to provide even heating because each level has its own heat-loss characteristics. Each level must be treated separately from an air-distribution and balancing standpoint. Continuous fan operation is strongly recommended. In large structures of this type, a separate unit can be installed for each section.

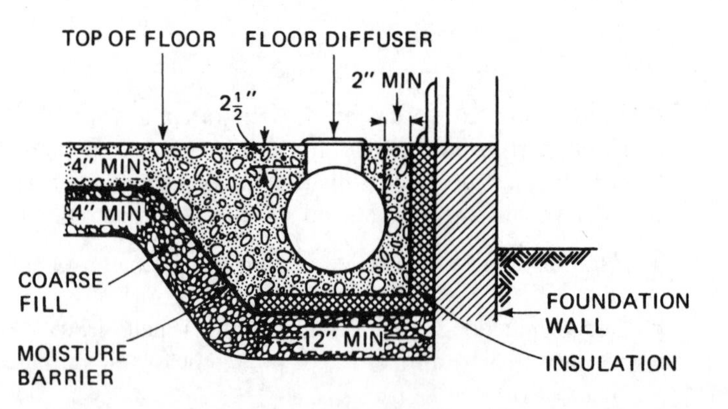

Figure 18.5 Cross section of slab construction containing perimeter duct made of one type of material. (By permission from the *ASHRAE Handbook*)

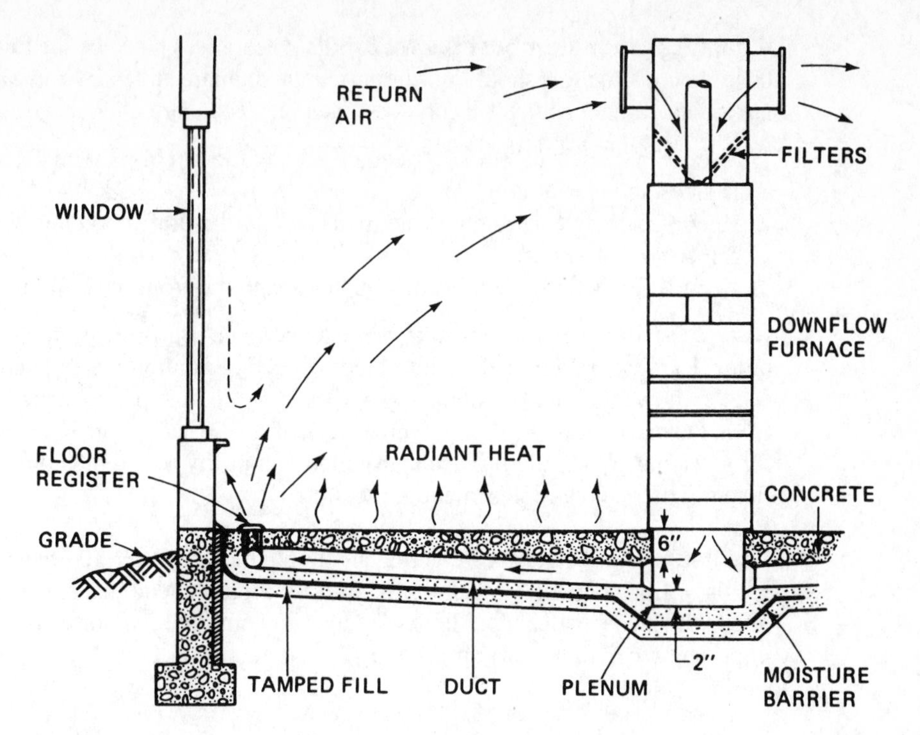

Figure 18.6 Construction of feeder ducts to plenum pit. (Courtesy of Carrier Corporation)

Furnace Installation

When a furnace installation is inspected for the first time, the service technician should determine if minimum standards have been met in the original installation. The owner should be advised of any serious faults that may affect safety and performance. Some installation conditions of importance are

- Clearances from combustible materials
- Circulating air supply
- Air for combustion, draft hood dilution, and ventilation
- Vent connections
- Electrical connections
- Gas piping

A service technician should comply with the local codes and regulations that govern installations. If there are no local codes, the equipment should be installed in accordance with the recommendation made by the National Board of Fire Underwriters, the American National Standards Institute (ANSI 2223.1), and the American Standards Association (ASA 221.30).

Clearances from Combustible Materials Clearances between the furnace and combustible construction should not be less than standard unless permissible clearances are indicated on the attached furnace nameplate. Standard clearances are as follows:

1. Keep 1 in. between combustible material and the top of plenum chamber, 0 to 6 in. at sides and rear of unit.
2. Keep 9 in. between combustible material and the draft hood and vent pipe in any direction.
3. Keep 18 in. between combustible material and the front of the unit.

Accessibility clearances take precedence over fire protection clearances (minimum clearances). Figure 18.7 shows recommended minimum clearances in a confined space. Allow at least 24 in. at the front of the furnace if all parts are accessible from the front. Otherwise, allow 24 in. on three sides of the furnace if the back must be reached for servicing. When the furnace is installed in a utility room, the door must be of sufficient size to allow replacement of the unit.

Circulating Air Supply Circulating air supply may be 100% return air or any combination of fresh outside air and return air. It is recommended that return-air plenums be lined with an acoustical duct liner to reduce any possible fan noise. This is particularly important when the return-air grille is close to the furnace.

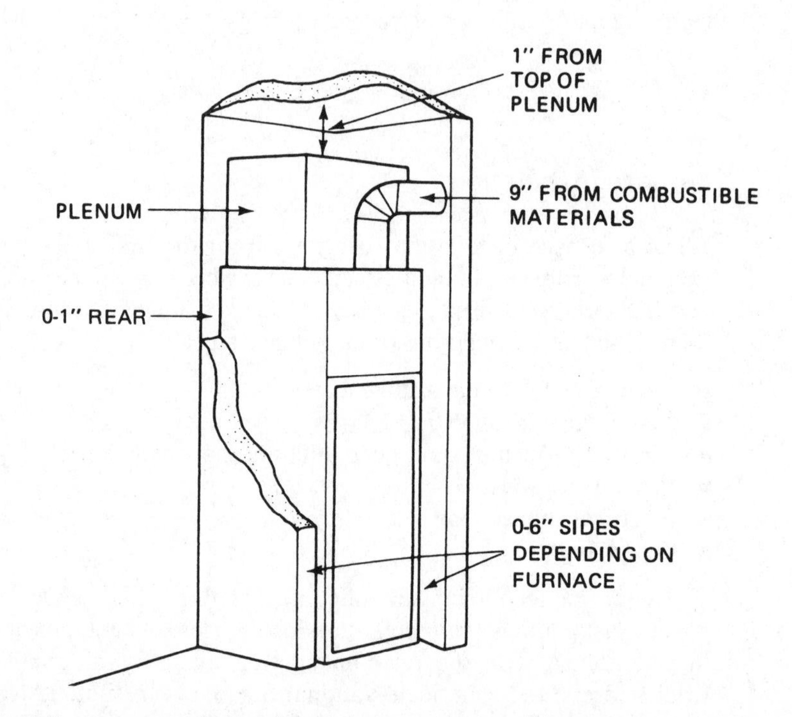

Figure 18.7 Recommended minimum clearances in a confined space.

All duct connections to the furnace must extend outside the furnace closet. Return air must not be taken from the furnace room or closet. Adequate return-air duct height must be provided to allow filters to be removed and replaced. All return air must pass through the filter after it enters the return-air plenum.

Air for Combustion, Draft Hood Dilution, and Ventilation Air for combustion, draft hood dilution, and ventilation differ somewhat for two types of conditions:

1. Furnace in a confined space
2. Furnace in an unconfined space

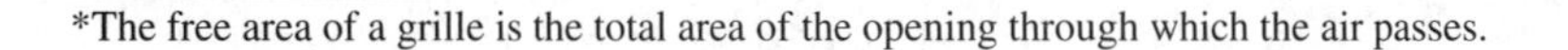

Confined Space If the furnace is located in a confined space, such as a closet or small room, provisions must be made for supplying combustion and ventilation air (Figures 18.8 and 18.9). Two properly located openings of equal area are required. One opening should be located in the wall, door, or ceiling above the relief opening of the draft diverter. The other opening should be installed in the wall, door, or floor below the combustion air inlet of the furnace. The total free area* of each opening must be at least 1 in.^{2} for each 1000 Btuh input. It is recommended that the two per-

*The free area of a grille is the total area of the opening through which the air passes.

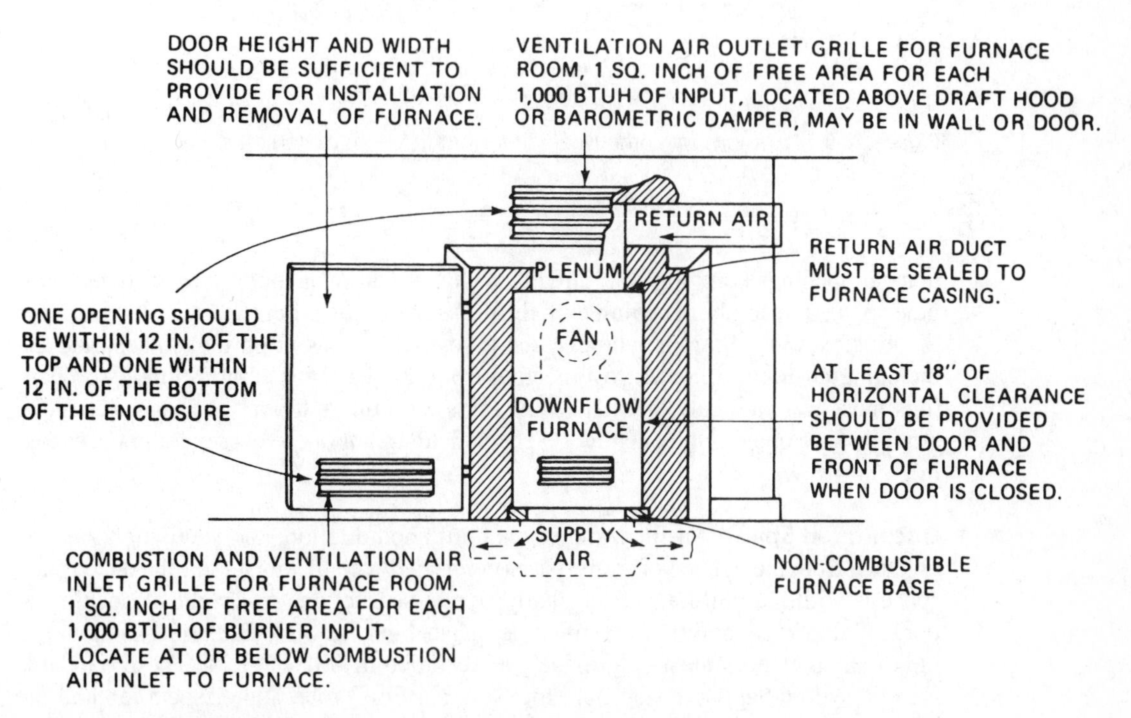

Figure 18.8 Provisions for combustion and ventilation air for furnace closet.

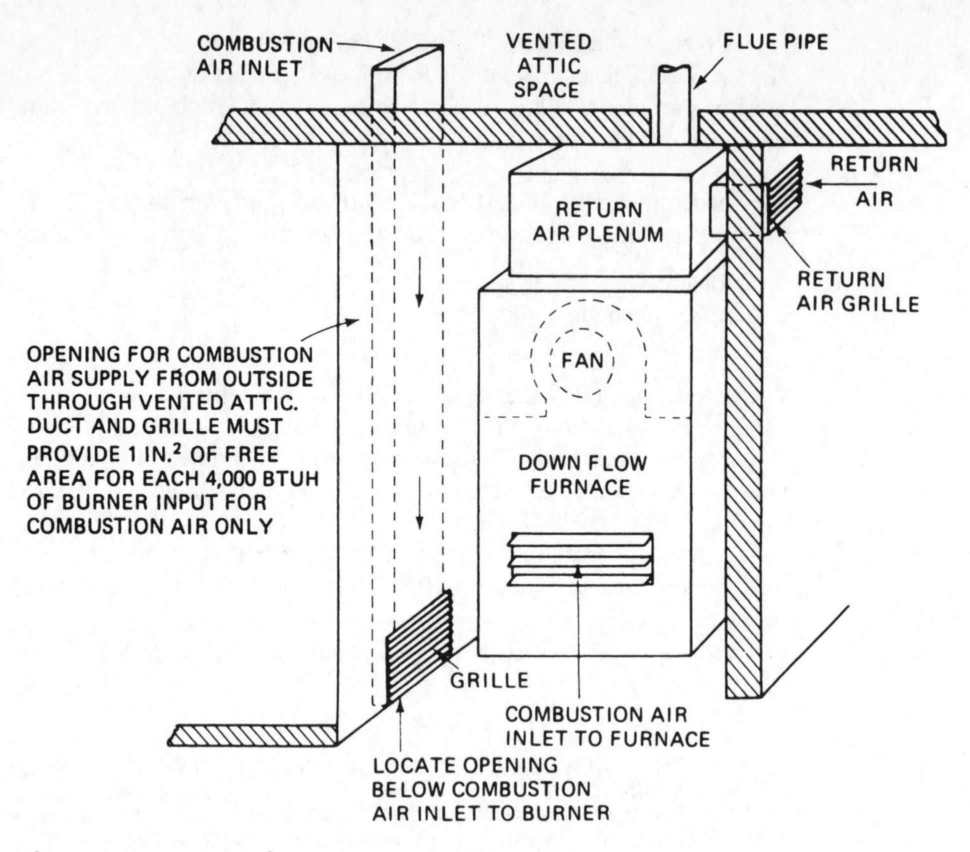

Figure 18.9 Provisions for combustion air for furnace from vented attic space.

manent openings communicate directly with an additional room(s) of sufficient volume so that the combined volume of all spaces meets the criteria.

In closet installations where space is restricted, it is important to separate the incoming air used to prevent overheating from that used for combustion. Two openings must communicate directly or by ducts with the outdoors or spaces (crawl or attic) that are open to the outside. A schematic diagram of the air separation is shown in Figure 18.9.

Unconfined Space Air for combustion, draft hood dilution, and ventilation must be obtained from the outside or from spaces connected with the outside. If the unconfined space is within a building of unusually tight construction, a permanent opening or openings having a total free area of not less than 1 in.^{2} per 4000 Btuh of total input rating of all appliances must be provided, as specified in ASIA 221.30. These standards are adopted and approved by both the National Fire Protection Association and the National Board of Fire Underwriters, NFPA No. 54.

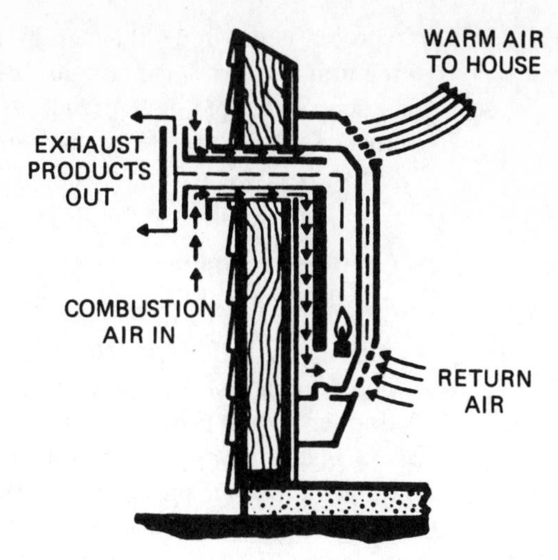

Figure 18.10 Installation of direct-vented room heater. (Used by permission of the copyright holder, American Gas Association)

A direct-vented space heating system vents directly through the outside wall, as shown in Figure 18.10. Direct-vented systems draw their combustion air from the outdoors and exhaust the product of combustion in the same manner. In most cases, conditioned air is not used for combustion, and makeup air is not required. Direct-vent systems may be mechanically controlled draft or natural draft. Most direct-vent furnaces will have a fan section that can be attached to the return air to aid in air circulation.

Installation in Crawl Space Codes in most communities no longer allow the installation of horizontal furnaces in crawl spaces. When a replacement is necessary, a new location must be found for the furnace.

Electrical Wiring Service All electrical wiring and connections should be made in accordance with the National Electrical Code® and with any local ordinances that may apply. Some of the important items to be observed in providing electrical service are as follows:

1. A separate 120-V power circuit properly fused should be provided, with a disconnect service readily accessible at the furnace.
2. The fuse size and the wire size are determined by the National Electrical Code®. The wire size and fuse size are based on 125% of the nameplate-rating, full-load amperes (FLA). FLA is the manufacturer's running current rating when the motor is operated at peak load. The minimum branch circuit from the building service to the furnace should be AWG No. 14 wire protected by a 15-A circuit breaker or fuse.
3. All replacement wire used within the combustion area of a furnace should be of the same type and size as the original wire and rated for 105°C.
4. Unless a circuit breaker is used, the fuse should be of the time-delay type.
5. Strain-relief connectors should be used at the entrance and exits of junction boxes.

6. Control circuit wire (for 24 V) should be AWG No. 18 wire with 105°C temperature rating. Low-voltage circuits (24 V) shall not be placed in any enclosure, compartment, or outlet box with line-voltage circuits.
7. Furnace shall be provided with a nameplate, voltage, and amperage ratings. Figure 18.11 shows typical furnace specifications.

Gas Piping A recommended gas piping arrangement at the furnace is shown in Figure 18.12. Some of the characteristics of good gas piping are as follows:

1. Piping should include a vertical section to collect scale and dirt.
2. A ground joint union should be placed at the connection to the furnace.
3. A drip leg should be installed at the bottom of the vertical riser.
4. The manual shutoff valve must be located external to the furnace casing.
5. Natural gas service pressure from the meter to the furnace is 7 in. W.C. (4.0 oz). At the furnace regulator it is then reduced to 3.5 in. (2.0 oz) to the burners. Liquid petroleum (LP) gas is furnished with a tank pressure regulator to provide 11 in. W.C. (6.3 oz) to the furnace burner.

Natural Gas Pipe Sizing Adequate gas supply must be provided to the gas furnace. The installation contractor is usually responsible for providing piping from the gas source to the equipment. Therefore, if other gas appliances are used in the building, the entire gas piping layout must be examined from the gas meter to each piece of equipment to determine the proper gas piping sizes for the systems.

If additional gas requirements are being added to an existing building, it may also be advisable to be certain that the gas meter supplied is adequate. This can be done by advising the gas company of the total connected load. The utility company will provide the necessary service at the meter.

General Procedures The procedure for determining the proper gas piping sizes from the meter is as follows:

1. Sketch a layout of the actual location of the gas piping from the meter to each gas appliance. Indicate on the sketch the length of each section of the piping diagram. Indicate on the sketch the Btuh input required for each gas appliance. See Figure 18.13 for an example. Note in this diagram that there are three appliances on the system.
2. Convert the Btuh input of each gas appliance to cubic feet per hour by dividing the Btuh input by the Btu/ft^3 rating of the gas used. The example shown uses natural gas with a rating of 1000 Btu/ft^3. Thus, for the appliances shown in Figure 18.13, the cubic feet of gas requirement for each appliance is as follows:

Outlet	*Appliance*	*Input (Btuh)*	*Natural Gas (ft^3/h*
A	Water heater	30,000	30
B	Range	75,000	75
C	Furnace	136,000	136

		NATURAL GAS				
	Model No.	NDGK040CF	NDGK050AF	NDGK075AF NDGK075CF	NDGK100AG NDGK100CG	NDGK125AK NDGK125CK
Capacity:	Input (Btuh)	40,000	50,000	75,000	100,000	125,000
	Heating capacity (Btuh)	38,000	48,000	70,000	91,000	114,000
	Heating capacity (Btuh) (Ca.)	37,154	46,034	67,648	88,174	108,937
	D.O.E. A.F.U.E.%	94.7	94.8	93.2	91.4	91.1
	C.A. seasonal eff.%	81.8	83.5	84.2	83.4	81.0
	Temp. rise (°F)	20–50	20–50	40–70	50–80	35–65
Flue size/type		2 in./PVC	2 in./PVC	2 in./PVC	2 in./PVC	2 in./PVC
Gas piping size		½ in.	½ in.	½ in.	½ in.	½ in.
Burners (number)		2	2	3	4	5
Electrical data	Volts-Ph-Hz.	115/60/1				
	(F.L.A.)	8.0 A	8.0 A	8.0 A	10.8 A	11.1 A
Transformer size (VA)		40				
Filter	Size (in.)	(2) 15 × 20 × 1				
Data	Type	Washable				
Cooling	Max. CFM @ .5 ESP	1300	1200	1230	1380	1965
Capacity	Nominal Tons	3	3	3	3½	5
Weight	Net (Lbs.)	180	180	190	215	242
	Shipping (Lbs)	202	200	210	240	265

Figure 18.11 Typical furnace specifications.

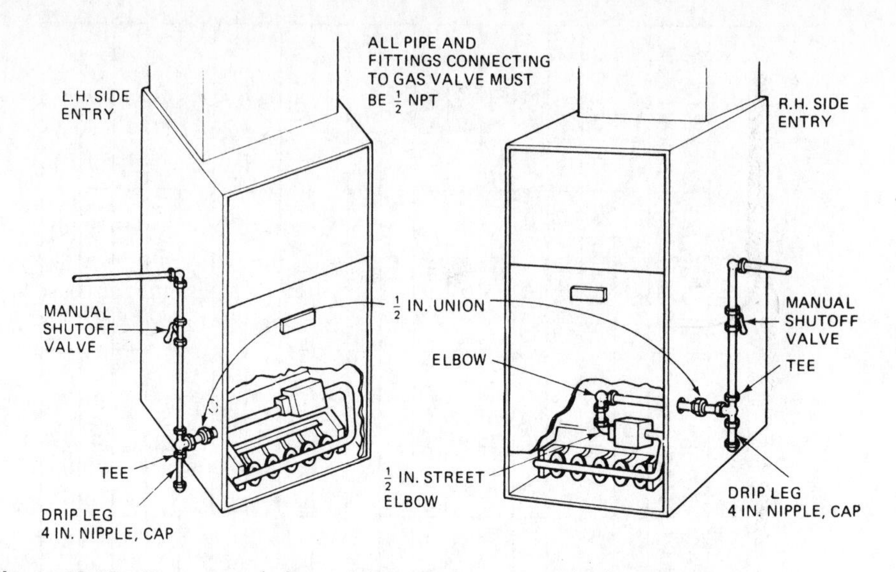

Figure 18.12 Recommended gas piping arrangement at the furnace.

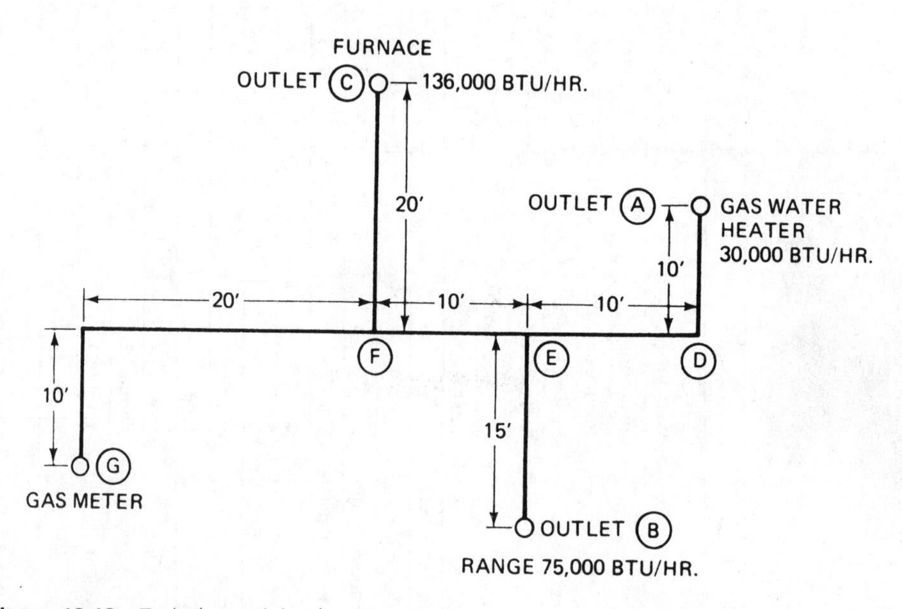

Figure 18.13 Typical gas piping layout.

3. Determine the amount of gas required in cubic feet per hour for each section of the piping system, as shown on the layout.
4. Using the appropriate figure (Figures 18.14, 18.15, 18.18, 18.20, and 18.21), select and record the proper pipe sizes.

Most residential gas piping systems are sized using Figure 18.14 because most natural gas is delivered from the meter at a pressure of 0.5 psig and a specific gravity of 6/10 (0.6). The specific gravity refers to the relative weight of 1 ft^3 of the gas as compared with the weight of an equal quantity of air.

From Figure 18.14 the maximum length of pipe to the appliance or the greatest distance from the meter is used for sizing the entire piping system. Thus, in Figure 18.13 the length of pipe to outlet (A) is 60 ft, and this number is used for all pipe sizes. Using the preceding data with the piping layout in Figures 18.13 and 18.14 determine the pipe sizes as follows:

1. Because 60 ft is the greatest distance from the meter, point A to point G (10 ft + 10 ft + 10 ft + 20 ft + 10 ft), use the 60 "length of pipe" line in Figure 18.14.
2. Staying on the 60 "length of pipe" line, move down the column until the cubic feet per hour of the gas carried in AD (30) is covered. A ½-in. pipe, which can carry up to 50 ft^3/h, is used.
3. DE carries the same 30 ft^3/h, so it is also ½ in.
4. BE carries 75 ft^3/h, so it requires ¾-in. pipe, which carries up to 105 ft^3/h.
5. EF carries 105 (30 + 75) ft^3/h, so it is also ¾ in.

SAMPLE PROBLEM STEPS

	Nominal Iron Pipe Size, Inches	Length of Pipe, Feet													
		10	20	30	40	50	(60)	70	80	90	100	125	150	175	200
	¼	32	22	18	15	14	12	11	11	10	9	8	8	7	6
	⅜	72	49	40	34	30	27	25	23	22	21	18	17	15	14
2,3	½	132	92	73	63	56	50	46	43	40	38	34	31	28	26
4,5	¾	278	190	152	130	115	105	96	90	84	79	72	64	59	55
6	1	520	350	285	245	215	195	180	170	160	150	130	120	110	100
7	1¼	1050	730	590	500	440	400	370	350	320	305	275	250	225	210
	1½	1600	1100	890	760	670	610	560	530	490	460	410	380	350	320
	2	3050	2100	1650	1450	1270	1150	1050	990	930	870	780	710	650	610
	2½	4800	3300	2700	2300	2000	1850	1700	1600	1500	1400	1250	1130	1050	980
	3	8500	5900	4700	4100	3600	3250	3000	2800	2600	2500	2200	2000	1850	1700
	4	17,500	12,000	9700	8300	7400	6800	6200	5800	5400	5100	4500	4100	3800	3500

For specific gravity figure, check your local utility company.

Figure 18.14 Maximum capacity of pipe in cubic feet of gas per hour for gas pressures of 0.5 psig or less and a pressure drop of 0.3 in. W.C. (based on 0.60 specific gravity gas). (Used by permission of the copyright holder, American Gas Association)

6. CF carries 136 ft^3/h, so it requires 1-in. pipe, which carries up to 195 ft^3/h.
7. FG carries 241 (136 + 105) ft^3/h, so it requires 1¼-in. pipe, which carries up to 400 ft^3/h.

Pipe Section	*Total-ft³/h Natural Gas Carried in Section*	*Iron Pipe Size (in.)*
AD	30	½
DE	30	½
BE	75	¾
EF	105 (30 + 75)	¾
CF	136	1
FG	241 (136 + 105)	1¼

When using specific gravities other than 0.60, use Figure 18.15. For example, assume that the natural gas has a specific gravity of 0.75. From Figure 18.15 the multiplier is 0.90. This multiplier would be used to redetermine the cubic feet per hour of gas handled by each part of the piping system.

When the 0.90 multiplier is applied to the preceding example, the results are as follows:

Pipe Section	*ft³/h Determined in Example*	*ft³/h Using Multiplier of 0.90*	*Iron Pipe Size (in.)*
AD	30	27 (0.90 × 30)	½
DE	30	27 (0.90 × 30)	½
BE	75	68 (0.90 × 75)	¾
EF	105	94 (0.90 × 105)	¾
CF	136	122 (0.90 × 136)	1
FG	241	217 (0.90 × 241)	1¼

Commercial/Industrial Sizing The volume of gas to be provided must be determined directly from the manufacturer's input ratings for the equipment being

Specific Gravity	Multiplier	Specific Gravity	Multiplier
.35	1.31	1.00	.78
.40	1.23	1.10	.74
.45	1.16	1.20	.71
.50	1.10	1.30	.68
.55	1.04	1.40	.66
.60	1.00	1.50	.63
.65	.96	1.60	.61
.70	.93	1.70	.59
.75	.90	1.80	.58
.80	.87	1.90	.56
.85	.84	2.00	.55
.90	.82	2.10	.54

Figure 18.15 Multipliers to be used only with Figures 18.14 and 18.16 when applying different specific gravity factors. (Used by permission of the copyright holder, American Gas Association)

EQUIVALENT LENGTH OF STRAIGHT PIPE									
SCREWED FITTING		½	(¾)	1	1¼	1½	2	2½	3
90 ELL		1.55	2.06	2.62	3.45	4.02	5.17	6.16	7.67
45 ELL		0.73	0.96	1.22	1.61	1.88	2.41	2.88	3.58
TEE		3.10	4.12	5.24	6.90	8.04	10.3	12.3	15.3
GAS COCK		0.36	0.48	0.61	0.81	0.94	1.21	1.44	1.79

Figure 18.16 Equivalent lengths in feet, computed on the inside diameter of schedule 40 steel pipe.

installed. In industrial gas pipe sizing, the measured length and the equivalent length of fittings and valves are used to arrive at the total equivalent length (TEL), which is then used for sizing the pipe. See Figure 18.16 and the example given.

Note: The "equivalent length" given for a fitting or gas cock is the resistance to flow that could be experienced in a section of straight pipe the same size. In the example the resistance to flow of the ¾-in. 90° ell is 2.06 ft, which is equal to the resistance to flow in 2.06 ft of ¾-in. straight pipe (as shown in Figure 18.16).

Example 18–1

From Figure 18.16 and the layout in Figure 18.17, the calculation for the TEL would be as follows, using the ¾-in. steel pipe.

$$\begin{aligned} \text{Measured length} &= 38 \text{ ft} \\ \text{Fittings} &= 8.10 \text{ ft} \\ \text{Gas cock} &= 0.48 \text{ ft} \\ \text{TEL} &= 46.58 \text{ ft} \end{aligned}$$

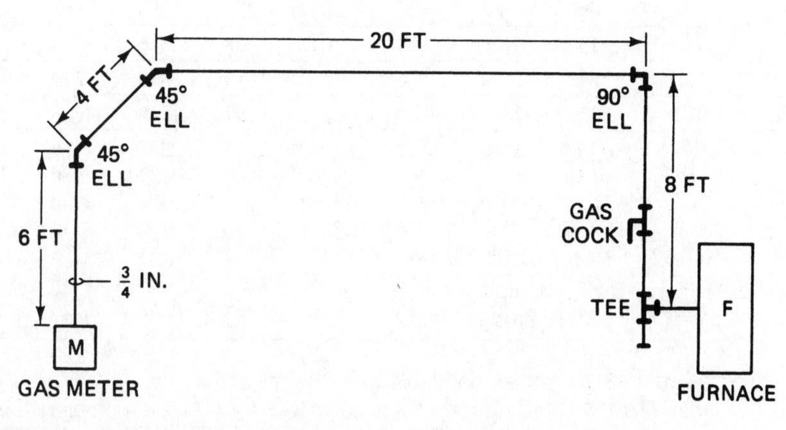

Figure 18.17 Sample piping layout, gas meter to furnace.

The simplified method for sizing gas piping for commercial and industrial applications is described next.

1. Measure the length of pipe from the gas meter to the most remote outlet. This is the only measurement necessary. To find the TEL, simply add 50% to the measured length.
2. Determine the Btuh required for each piece of equipment. Divide these figures by the heating value of the gas. For example:

$$\frac{\text{Btuh}}{\text{heating value}} = \text{ft}^3/\text{h}$$

$$\frac{75{,}000}{1000} = 75\ \text{ft}^3/\text{h}$$

3. Determine the volume (ft^3/h) of gas that each section of the piping will carry.
4. Use the measured length of pipe and volume of gas in each section and outlet. Use Figure 18.18 for sizing.

Example 18–2 Determine the necessary pipe size for each section and outlet of Figure 18.19. The natural gas has a specific gravity of 0.6 and a heating value of 1000 Btu/ft^3. The gas pressure is 8 in. W.C. with a maximum allowable pressure drop of 0.5 in. W.C. in the piping.

Nominal Iron Pipe Size Inches	Total Equivalent Length of Pipe in Feet										
	50	100	150	200	250	300	400	500	1000	1500	2000
1	284	195	157	134	119	108	92	82	56	45	39
1¼	583	400	322	275	244	221	189	168	115	93	79
1½	873	600	482	412	386	331	283	251	173	139	119
2	1681	1156	928	794	704	638	546	484	333	267	229
2½	2680	1842	1479	1266	1122	1017	870	771	530	426	364
3	4738	3256	2615	2238	1983	1797	1538	1363	937	752	644
3½	6937	4767	3828	3277	2904	2631	2252	1996	1372	1102	943
	9663	6641	5333	4565	4046	3666	3137	2780	1911	1535	1313
	17,482	12,015	9649	8258	7319	6632	5676	5030	3457	2776	2376
	28,308	19,456	15,624	13,372	11,851	10,738	9190	8145	5598	4496	3848
	58,161	39,974	52,100	27,474	24,350	22,062	18,883	16,735	11,502	9237	7905
	105,636	72,603	58,303	49,900	44,225	40,071	34,296	30,396	20,891	16,776	14,358
	167,236	114,940	92,301	78,998	70,014	63,438	54,295	48,120	33,073	26,559	22,731

Figure 18.18 Pipe sizing table for pressures under 1 lb. Approximate capacity of pipes of different diameters and lengths in cubic feet per hour with pressure drop of 0.5 in. W.C. and 0.6 specific gravity. (Used by permission of the copyright holder, American Gas Association)

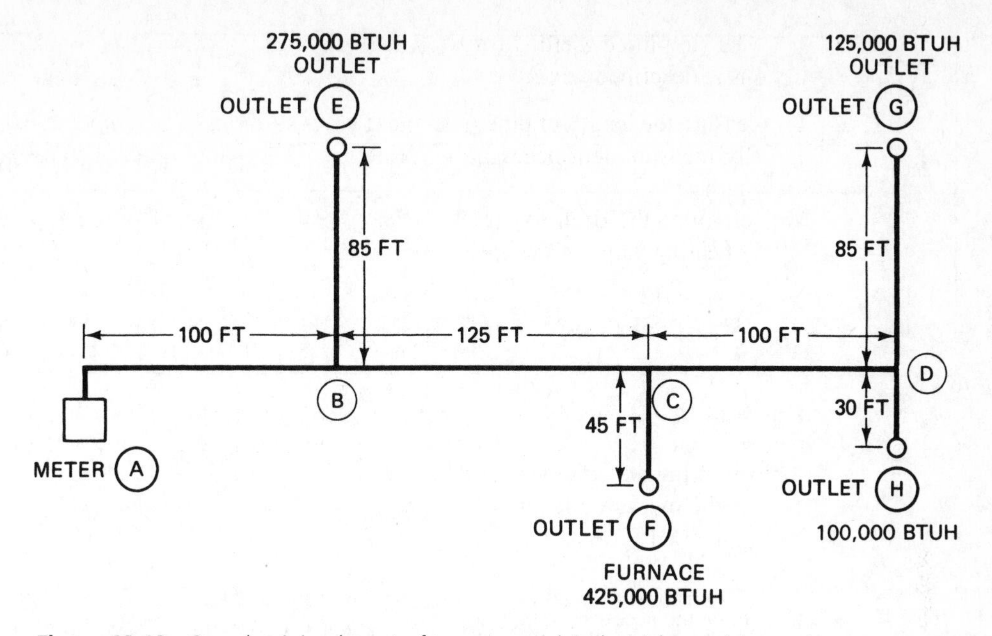

Figure 18.19 Sample piping layout of a commercial–industrial gas piping system.

Solution

1. The measured length of the piping from the meter (A) to the most remote outlet (G) is 100 ft + 125 ft + 100 ft + 85 ft = 410 ft.
2. Increase the measured length by 50%: 410 ft + 205 ft = 615 ft. The value 615 ft is then the TEL that is used to determine the pipe sizes from Figure 18.17.
3. When the calculated TEL falls between two columns in Figure 18.18, the larger TEL is used. In this case 615 ft falls between 500 and 1000; therefore, the 1000 column is used.

Main Sections	*Btuh*	*Flow in Section ft^3/h*	*Nominal Iron Pipe Size (in.)*
AB	925,000	925	3
BC		925 − 275 = 650	3
CD		650 − 425 = 225	2
Branches	***Btuh***	***Flow in Branch ft^3/h***	***Nominal Iron Pipe Size (in.)***
BE	275,000	275	2
CF	425,000	425	2½
DH	100,000	100	1¼
DG	125,000	125	1½

Liquid Petroleum Gas Pipe Sizing Liquid petroleum (LP) gas is popular in areas where natural gas is in short supply. Trailers and recreation vehicles (RVs) use LP gas

Nominal (I.D.) Iron Pipe Size, Inches	Length of Pipe, Feet											
	10	20	30	40	50	60	70	80	90	100	125	150
½	275	189	152	129	114	103	96	89	83	78	69	63
¾	567	393	315	267	237	217	196	185	173	162	146	132
1	1071	732	590	504	448	409	378	346	322	307	275	252
1¼	2205	1496	1212	1039	913	834	771	724	677	630	567	511
1½	3307	2299	1858	1559	1417	1275	1181	1086	1023	976	866	787
2	6221	4331	3465	2992	2646	2394	2205	2047	1921	1811	1606	1496

Figure 18.20 Maximum capacity of pipe in thousands of Btuh of undiluted liquefied petroleum gases (at 11 in. W.C. inlet pressure, based on a pressure drop of 0.5 in. WC.). (Used by permission of the copyright holder, American Gas Association)

for refrigeration as well as for heating. Figures 18.20 and 18.21 are used to calculate the pipe or tubing sizes necessary to supply LP gas to the furnace.

Venting

Venting of natural draft furnaces is accomplished with the temperature difference between the vent gases inside the flue and the surrounding air. The greater this temperature difference, the greater the push up the chimney. Sufficient draft is required also to move the moisture that is produced in the combustion process. This moisture is normally in the vapor state but can condense and form liquid water while in the chimney. This condensate can be very corrosive. It can attack mortar, vent connectors, and

Outside Diameter, Inch	Length of Tubing, Feet									
	10	20	30	40	50	60	70	80	90	100
⅜	39	26	21	19	—	—	—	—	—	—
½	92	62	50	41	37	35	31	29	27	26
⅝	199	131	107	90	79	72	67	62	59	55
¾	329	216	181	145	131	121	112	104	95	90
⅞	501	346	277	233	198	187	164	155	146	138

Figure 18.21 Maximum capacity of semirigid tubing in thousands of Btuh of undiluted liquefied petroleum gases (at 11 in. W.C. inlet pressure, based on a pressure drop of 0.5 in. W.C.). (Used by permission of the copyright holder, American Gas Association)

heat exchangers. Condensation can be avoided by maintaining the vent gases at temperatures high enough to keep the chimney warm.

Temperatures of Various Furnaces

Conventional furnaces are designed to maintain the vent gas temperature above 400°F. Because most vent condensation occurs at approximately 126°F, the gases will not condense in a properly constructed and sized chimney. Standard furnaces with electric pilot ignition, forced draft, or induced draft still operate at above 400°F.

Recuperative furnaces with a seasonal efficiency of 80%+ reach this efficiency by removing additional heat from the flue gases. This brings the vent gas temperature down to approximately 180°F. Oversized or poorly constructed chimneys may reduce that temperature to 126°F and cause condensation.

Condensing furnaces with a seasonal efficiency of 90%+ operate with a vent temperature of approximately 120°F. Condensate removal is designed into this furnace. Therefore, conventional vent piping and chimneys are not used.

Masonry Venting

When used for flue gas venting, masonry chimneys must be lined with a fireclay tile, and the tile liner must be installed with tight joints. Because the liner does not touch the brick or block of the chimney, it produces a vacuum bottle–like insulating effect. The lining is designed to heat up quickly and remove very little heat from the flue gases. If the tile is not installed correctly, gases will escape to the brick or block, cool rapidly, and condense. The remaining gases within the liner will not be able to keep the liner warm and will also condense. At the cooler temperature they will not maintain a sufficient push up the chimney and will spill into the building through the diverter. An oversized chimney requires more heat to keep the liner warm than is available in the flue gas, which causes the same problem.

Type B Venting

The type B vent consists of an outer galvanized steel pipe and an inner aluminum pipe with an airspace between the two pipes (Figure 18.22). Each manufacturer of type B vents produces a unique system, and various manufacturers' components must not be interchanged. Installation instructions must be followed with each system as to joint connections, securing, and vent termination height. Generally, sheet metal screws should not be used.

The ideal installation is to have a correctly sized type B vent for each appliance. Usually, though, a furnace and water heater are combined to be vented through one stack. If the water heater has a low Btu rating and operates frequently while the furnace

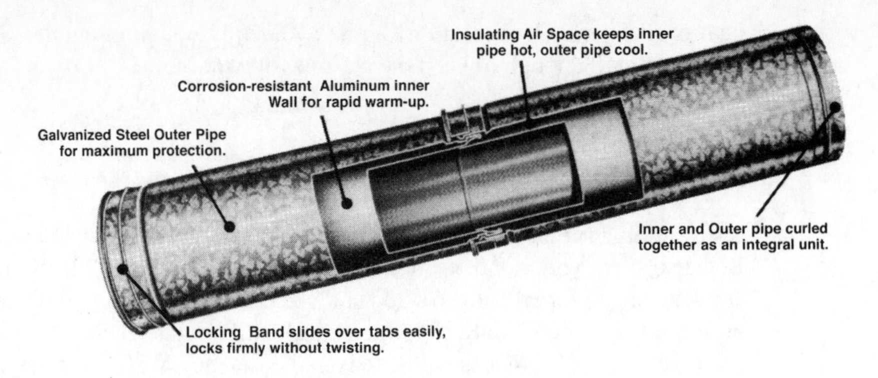

Figure 18.22 Metlvent® type B double-wall insulated gas vent. (Courtesy of Hart & Cooley, Inc.)

is not operating, there may be insufficient heat in the vent gas to avoid condensation. This is the reason to size a common vent with great accuracy.

Plastic Venting

Furnaces that operate with flue gases that do or may condense require a noncorrosive vent system. Condensing furnaces, with flue gas temperatures at 120°F, generally are vented with low-temperature polyvinyl chloride (PVC) or chlorinated polyvinyl chloride (CPVC) plastic pipe and fittings.

As shown in Figure 18.23, horizontal pipe should be supported every 3 ft, with the first support as close to the furnace as possible. The induced draft fan and/or furnace cabinet should not support the weight of the flue pipes. The piping should be pitched upward from the furnace ¼-in. per foot to allow condensate to drain back into the furnace condensate removal system. Vertical pipe should not extend more than 30 ft above the appliance.

All PVC piping must be prepared for gluing by using a purple cleaner–primer. Any other pipe preparation will not hold the pipe together in the long run and is not allowed by code.

Roof vent outlets are installed through a rubber flashing such as is used for plumbing stacks (Figure 18.24). Care should be taken to avoid locations where the wind can cause downdrafts. Any vent piping passing through unconditioned areas, such as an attic, garage, or crawl space, must be insulated to prevent excess condensation in the pipe.

Sidewall venting requires vent termination kits as designed by the manufacturer. An open pipe should not be allowed to face into the prevailing winds (Figure 18.25). Sidewall vents may not terminate within 10 ft of the property line.

The manufacturer's installation instructions must be followed for installation with reference to air intakes. If the furnace is operated without outside combustion air, the vent termination must not be any closer than 4 ft from any opening into the building. When outside combustion air is used, the manufacturer's instructions must be followed.

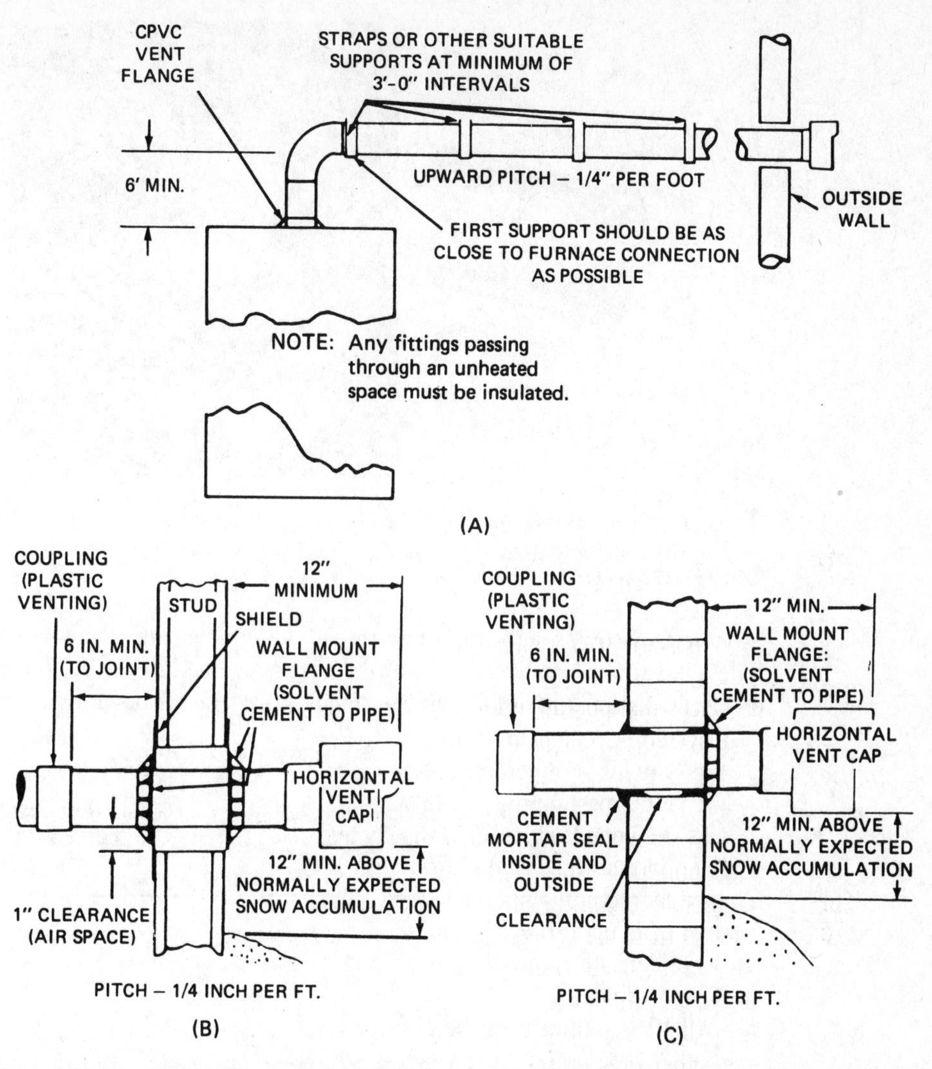

Figure 18.23 Plastic pipe venting: (a) support and horizontal pitch required; (b) installation through combustible walls; (c) installation through noncombustible walls. (Courtesy of the Trane Company Unitary Products Group)

Condensate Removal System

To remove the condensate on 90+ furnaces, the individual manufacturer's drain kits must be used (Figure 18.26). All kits provide a means of trapping the condensate to assure that no vent gases pass into the building or into the circulating airstream. It is imperative that no additional or substitute traps be installed. Some kits also include a switch that will open the heating control circuit if the trap should become overfilled with condensate.

Figure 18.24 Rubber roof flashing.

Condensate Neutralizer

Some communities will not allow the acidic condensate to enter the sewer system unless it is first neutralized. Neutralizer kits are available as an option from many manufacturers. Installation of a condensate neutralizer kit is shown in Figure 18.27.

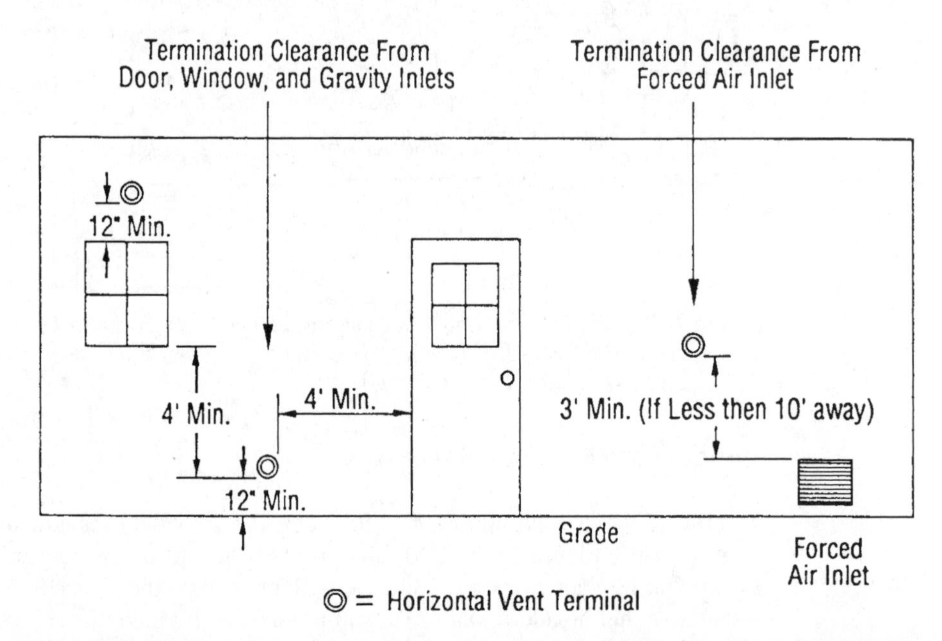

Figure 18.25 Vent termination clearances to openings in buildings. (Courtesy of ARCOAIRE Air Conditioning & Heating)

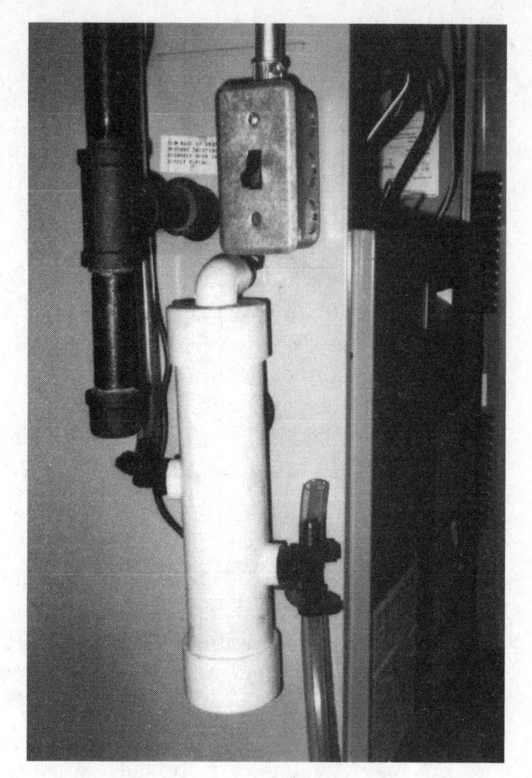

Figure 18.26 Typical condensate trap kit.

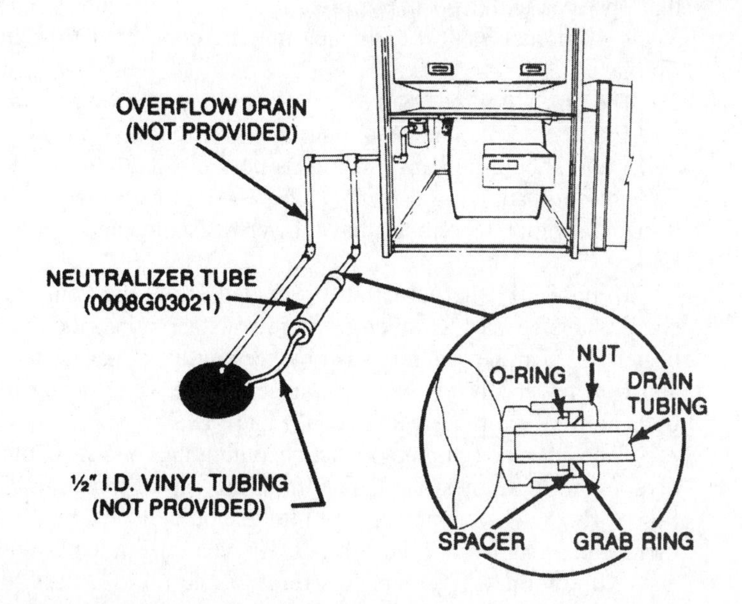

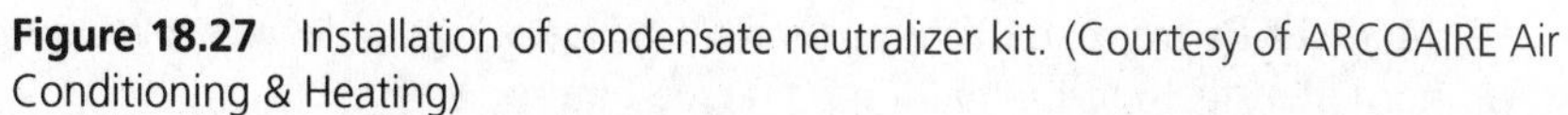

Figure 18.27 Installation of condensate neutralizer kit. (Courtesy of ARCOAIRE Air Conditioning & Heating)

Vent Sizing Categories

Gas appliances are separated into four categories for venting purposes.

1. Category 1 appliances do not produce positive vent static pressure, operate at over 275°F vent gas temperatures, and rely on the heat content of combustion products to vent. Even if these furnaces are equipped with an induced draft blower, the blower is not strong enough to pressurize the vent. The new 78–80% furnaces and the midefficiency 80–90% furnaces fall into this category.
2. Category 2 is for natural draft appliances with a flue gas temperature of less than 275°F. At the time of this writing, furnaces of this type are not being manufactured.
3. Category 3 appliances rely on the heat content of the combustion products and mechanical or other means to vent. Recuperative furnaces fall into this category.
4. Category 4 includes the highest-efficiency furnaces that are designed to produce condensation in the heat exchanger.

General Considerations for Category 1 Appliances Correct sizing of the furnace, based on calculated load, will extend the running time, thereby decreasing the potential for condensation in the vent pipe. If the vent is oversized, the flue gases will move too slowly, and condensation will result. When the vent is too small, it will not carry the total volume of the flue gases.

Two or more category 1 appliances may be connected to a single vent or chimney. A category 1 appliance must not be connected to any portion of a vent system that operates under positive pressure. Positive pressure would result with category 3 and 4 appliances. No appliance may be connected to a chimney flue serving a fireplace.

The pipe from the appliance to the chimney or vent is called the *vent connector* (see Figure 18.28). The vent connector must be the same size as the appliance vent collar and should be kept as short as possible. Adding an elbow creates resistance. For example, adding a 6-in. 90° elbow would be the equivalent of adding 20 ft of horizontal pipe; 45° elbows have lower resistance and can be utilized in most vent runs.

To minimize the potential for condensation, the vent connector to the furnace should have at least as much or more resistance than the connector from the water heater. This can be accomplished by keeping the water heater vent connector shorter and with fewer elbows, and by connecting the water heater to the vertical vent at a point above the furnace connection (Figure 18.28).

The vent pipe (chimney) must be a minimum of 5 ft in total height from the draft diverter of the highest appliance up to the cap of the vent. This is dimension H in Figure 18.28. The connecting tee fitting should always be the same size as the common vent, as shown in Figure 18.29. The vent pipe must extend at least 3 ft, by code, above the point where it passes through the roof. Unless the vent termination is approved by the manufacturer for reduced clearances, the vent must end 2 ft above any portion of the building within 10 ft (see Figure 18.30).

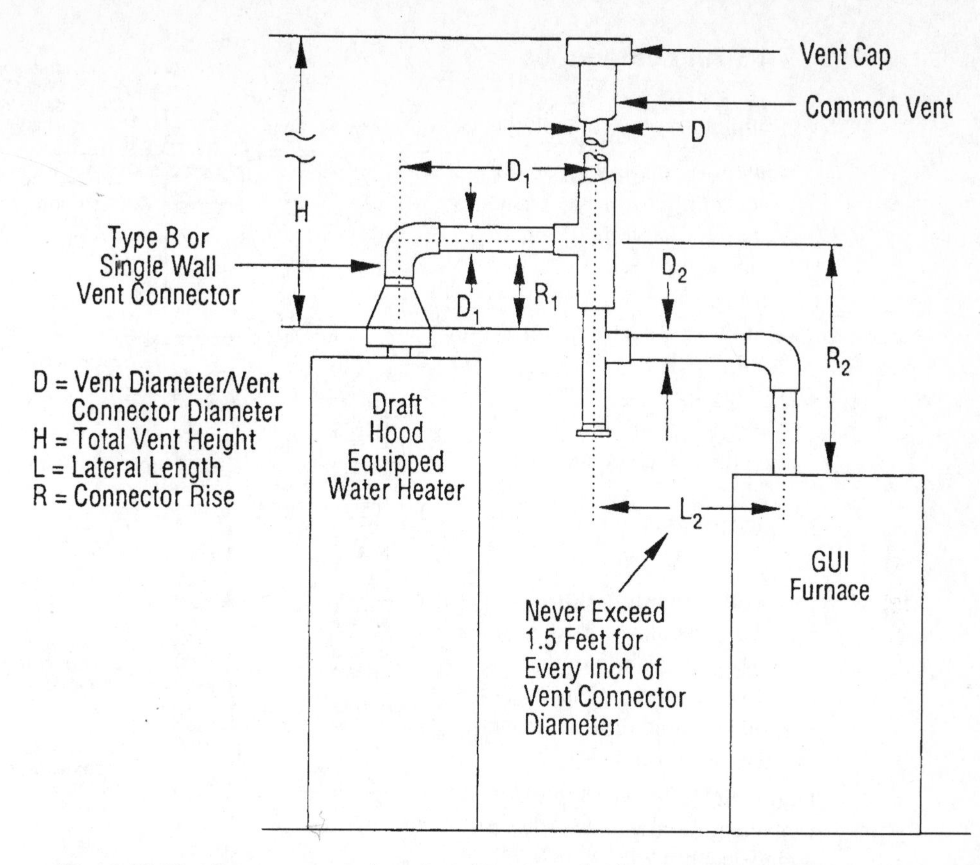

Figure 18.28 Common vertical venting. (Courtesy of ARCOAIRE Air Conditioning & Heating)

Category 3 Appliance Venting Midefficiency furnaces and recuperative furnaces that produce a positive pressure in the vent must have vent connections that are gastight to prevent leakage into the conditioned spaces. Furnaces are usually vented separately, but special designs may allow some common venting with water heaters. Special venting materials or aluminum piping is required because of the potential for condensation in the vent. Category 3 furnaces cannot be vented through a B vent.

Category 4 Condensing Furnace Venting PVC plastic may be specified for venting systems that will not exceed 160°F. For higher-temperature conditions, up to 210°F, CPVC may be required. These furnaces may not be connected to a common vent and must have a properly sized single vent directly to the outdoors. For sizing purposes, the manufacturer's required calculations must be strictly followed regarding vent length and number of elbows. Most residential category 4 furnaces have vent sizes in the 2- to 3-in. range. Figure 18.31 shows typical installation requirements.

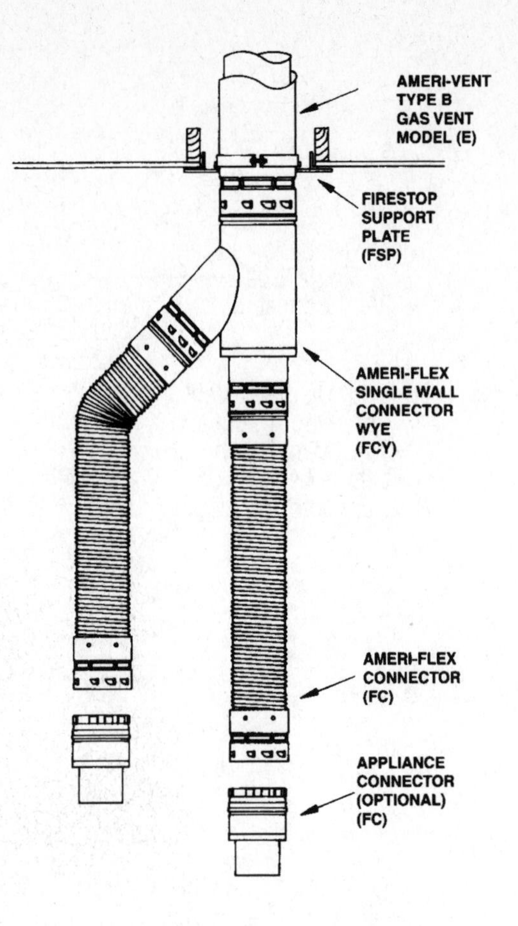

Figure 18.29 Correct use of vent tee. (Courtesy of American Metal Products)

Vent Sizing Tables

Category 1 appliance vent sizing is normally the responsibility of the installer. Category 3 and 4 sizing is the responsibility of the manufacturer. Vent systems can be divided into two basic types:

1. Single-appliance systems
2. Multiple-appliance systems

Separate sizing tables must be used for single and multiple applications. In addition, separate sizing tables must be used for B vents and masonry chimneys.

Figure 18.32 shows a condensed sizing table for type B double-wall vents with single-wall metal vent connectors serving a single category 1 appliance. The first column is vent height from the appliance draft diverter to the top (cap) of the vent. The second column is any lateral (horizontal) dimension between the draft diverter and the cap. The numbers across the top represent the size of the vent connector and the vent

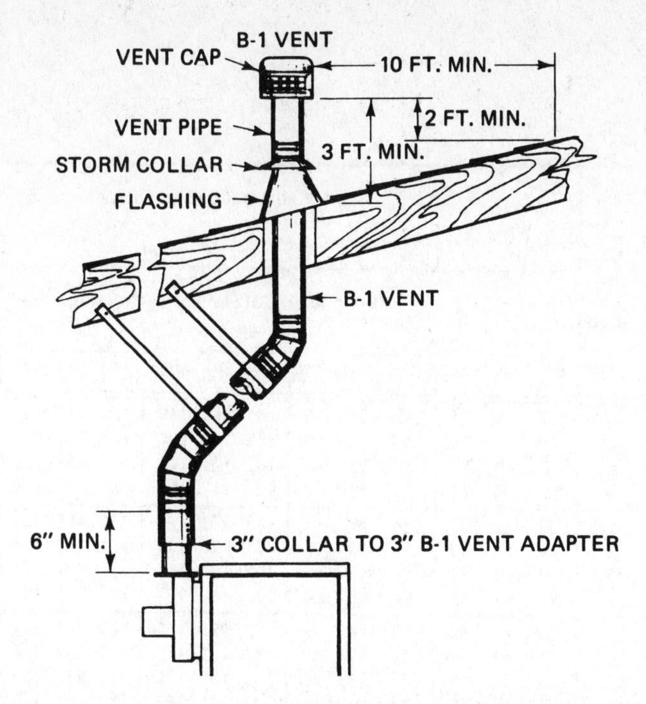

Figure 18.30 Recommendations for installation of type B vents.

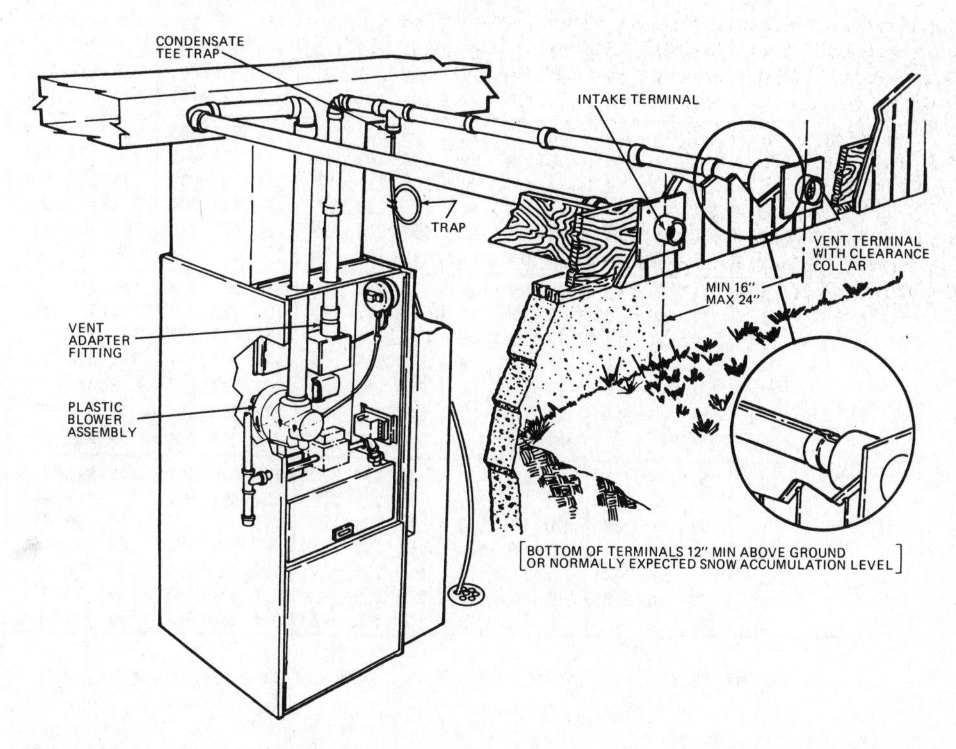

Figure 18.31 Typical installation requirements. (Courtesy of ARCOAIRE Air Conditioning & Heating)

Height H (ft)	Lateral L (ft)	Vent and Connector Diameter - D (inches)																	
		3"			4"			5"			6"			7"			8"		
		Appliance Input Rating in Thousands of BTU Per Hour																	
		FAN		NAT	FAN		NAT	FAN		NAT	FAN		NAT	FAN		NAT	FAN		NAT
		Min	Max	Max	Min	Max	Max	Min	Max	Max	Min	Max	Max	Min	Max	Max	Min	Max	Max
6	0	38	77	45	59	151	85	85	249	140	126	373	204	165	522	284	211	695	369
	2	39	51	36	60	96	66	85	156	104	123	231	156	159	320	213	201	423	284
	4	NR	NR	33	74	92	63	102	152	102	146	225	152	187	313	208	237	416	277
	6	NR	NR	31	83	89	60	114	147	99	163	220	148	207	307	203	263	409	271
8	0	37	83	50	58	164	93	83	273	154	123	412	234	161	580	319	206	777	414
	2	39	56	39	59	108	75	83	176	119	121	261	179	155	363	246	197	482	321
	5	NR	NR	37	77	102	69	107	168	114	151	252	171	193	352	235	245	470	311
	8	NR	NR	33	90	95	64	122	161	107	175	243	163	223	342	225	280	458	300
10	0	37	87	53	57	174	99	82	293	165	120	444	254	158	628	344	202	844	449
	2	39	61	41	59	117	80	82	193	128	119	287	194	153	400	272	193	531	354
	5	52	56	39	76	111	76	105	185	122	148	277	186	190	388	261	241	518	344
	10	NR	NR	34	97	100	68	132	171	112	188	261	171	237	369	241	296	497	325
15	0	36	93	57	56	190	111	80	325	186	116	499	283	153	713	388	195	966	523
	2	38	69	47	57	136	93	80	225	149	115	337	224	148	473	314	187	631	413
	5	51	63	44	75	128	86	102	216	140	144	326	217	182	459	298	231	616	400
	10	NR	NR	39	95	116	79	128	201	131	182	308	203	228	438	284	284	592	381
	15	NR	NR	NR	NR	NR	72	158	186	124	220	290	192	272	418	269	334	568	367
20	0	35	96	60	54	200	118	78	346	201	114	537	306	149	772	428	190	1053	573
	2	37	74	50	56	148	99	78	248	165	113	375	248	144	528	344	182	708	468
	5	50	68	47	73	140	94	100	239	158	141	363	239	178	514	334	224	692	457
	10	NR	NR	41	93	129	86	125	223	146	177	344	224	222	491	316	277	666	437
	15	NR	NR	NR	NR	NR	80	155	208	136	216	325	210	264	469	301	325	640	419
	20	NR	NR	NR	NR	NR	NR	186	192	126	254	306	196	309	448	285	374	616	400
30	0	34	99	63	53	211	127	76	372	219	110	584	334	144	849	472	184	1168	647
	2	37	80	56	55	164	111	76	281	183	109	429	279	139	610	392	175	823	533
	5	49	74	52	72	157	106	98	271	173	136	417	271	171	595	382	215	806	521
	10	NR	NR	NR	91	144	98	122	255	168	171	397	257	213	570	367	265	777	501
	15	NR	NR	NR	115	131	NR	151	239	157	208	377	242	255	547	349	312	750	481
	20	NR	NR	NR	NR	NR	NR	181	223	NR	246	357	228	298	524	333	360	723	461
	30	NR	NR	NR	NR	NR	NR	NR	NR	NR	NR	NR	NR	389	477	305	461	670	426
50	0	33	99	66	51	213	133	73	394	230	105	629	361	138	928	515	176	1292	704
	2	36	84	61	53	181	121	73	318	205	104	495	312	133	712	443	168	971	613
	5	48	80	NR	70	174	117	94	308	198	131	482	305	164	696	435	204	953	602
	10	NR	NR	NR	89	160	NR	118	292	186	162	461	292	203	671	420	253	923	583
	15	NR	NR	NR	112	148	NR	145	275	174	199	441	280	244	646	405	299	894	562
	20	NR	NR	NR	NR	NR	NR	176	257	NR	236	420	267	285	622	389	345	866	543
	30	NR	NR	NR	NR	NR	NR	NR	NR	NR	315	376	NR	373	573	NR	442	809	502

Figure 18.32 Vent tables. Capacity of type B double-wall vents with single-wall metal connectors serving a single category 1 appliance. (Courtesy of Gas Appliance Manufacturers Association)

in inches diameter. This size must never be smaller than the vent collar on the furnace. The main body of the table represents the Btu input rating of the appliance in thousands (i.e., 38 = 38,000 Btu). Figures are shown for both natural draft and fan assisted draft appliances.

To use the table for a 120,000-Btuh natural draft appliance, first determine the height and any lateral offset for the vent system. Look for the appropriate row in columns 1 and 2. In that row move toward the right until you reach a number under a column "NAT-Max" that is 120 or larger. From that point, move up to establish the vent size.

For a fan-assisted induced draft appliance, follow the same instruction, but look for the input rating to fall between a set of minimum and maximum columns. The minimum is important to avoid condensation due to an oversized chimney.

Figure 18.33 shows a condensed common vent table for type B double-wall vents with single-wall connectors serving two or more category 1 appliances. To establish the common vent size, determine the vent height and the types of appliances connected (fan-assisted or natural draft) and their combined Btu input. In the row for the height, move to the right until you reach the combined Btuh input under the appropriate appliance combination column. Move to the top to establish the common vent size.

Figure 18.34 shows a condensed common vent table for a masonry chimney with single-wall metal connectors serving two or more category 1 appliances. Whenever a category 1 furnace is removed from a common venting system, particular attention must be paid to the sizing of the venting for the remaining appliance, such as a water heater. Without the larger appliance, the common vent is too large and may cause condensation and insufficient draft. Many codes require that the remaining appliance be vented through a chimney liner, which may be a type B vent or a flexible liner, as shown in Figure 18.35.

Vent Height H (ft)	Common Vent Diameter - D (inches)																				
	4"			5"			6"			7"			8"			9"			10"		
	Combined Appliance Input Rating in Thousands of Btu Per Hour																				
	FAN +FAN	FAN +NAT	NAT +NAT	FAN +FAN	FAN +NAT	NAT +NAT	FAN +FAN	FAN +NAT	NAT +NAT	FAN +FAN	FAN +NAT	NAT +NAT	FAN +FAN	FAN +NAT	NAT +NAT	FAN +FAN	FAN +NAT	NAT +NAT	FAN +FAN	FAN +NAT	NAT +NAT
6	92	81	65	140	116	103	204	161	147	309	248	200	404	314	260	547	434	335	672	520	410
8	101	90	73	155	129	114	224	178	163	339	275	223	444	348	290	602	480	378	740	577	465
10	110	97	79	169	141	124	243	194	178	367	299	242	477	377	315	649	522	405	800	627	495
15	125	112	91	195	164	144	283	228	206	427	352	280	556	444	365	753	612	465	924	733	565
20	136	123	102	215	183	160	314	255	229	475	394	310	621	499	405	842	688	523	1035	826	640
30	152	138	118	244	210	185	361	297	266	547	459	360	720	585	470	979	808	605	1209	975	740
50	167	153	134	279	244	214	421	353	310	641	547	423	854	706	550	1164	977	705	1451	1188	860
100	175	163	NR	311	277	NR	489	421	NR	751	658	479	1025	873	625	1408	1215	800	1784	1502	975

Figure 18.33 Vent tables. Common vent capacity of type B double-wall vents with single-wall metal connectors serving two or more category 1 appliances. (Courtesy of Gas Appliance Manufacturers Association)

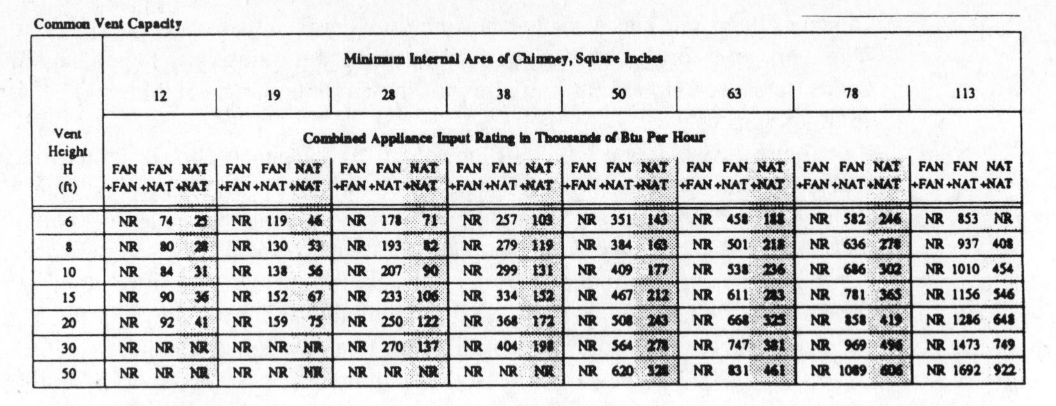

Common Vent Capacity

Vent Height H (ft)	Minimum Internal Area of Chimney, Square Inches: 12			19			28			38		
	Combined Appliance Input Rating in Thousands of Btu Per Hour											
	FAN +FAN	FAN +NAT	NAT +NAT	FAN +FAN	FAN +NAT	NAT +NAT	FAN +FAN	FAN +NAT	NAT +NAT	FAN +FAN	FAN +NAT	NAT +NAT
6	NR	74	25	NR	119	46	NR	178	71	NR	257	103
8	NR	80	28	NR	130	53	NR	193	82	NR	279	119
10	NR	84	31	NR	138	56	NR	207	90	NR	299	131
15	NR	90	36	NR	152	67	NR	233	106	NR	334	152
20	NR	92	41	NR	159	75	NR	250	122	NR	368	172
30	NR	NR	NR	NR	NR	NR	NR	270	137	NR	404	198
50	NR	NR	NR	NR	NR	NR	NR	NR	NR	NR	NR	NR

Vent Height H (ft)	Minimum Internal Area of Chimney, Square Inches: 50			63			78			113		
	Combined Appliance Input Rating in Thousands of Btu Per Hour											
	FAN +FAN	FAN +NAT	NAT +NAT	FAN +FAN	FAN +NAT	NAT +NAT	FAN +FAN	FAN +NAT	NAT +NAT	FAN +FAN	FAN +NAT	NAT +NAT
6	NR	351	143	NR	458	188	NR	582	246	NR	853	NR
8	NR	384	163	NR	501	218	NR	636	278	NR	937	408
10	NR	409	177	NR	538	236	NR	686	302	NR	1010	454
15	NR	467	212	NR	611	283	NR	781	365	NR	1156	546
20	NR	508	243	NR	668	325	NR	858	419	NR	1286	648
30	NR	564	278	NR	747	381	NR	969	496	NR	1473	749
50	NR	620	328	NR	831	461	NR	1089	606	NR	1692	922

Figure 18.34 Vent tables. Common vent capacity of masonry chimney flue with single-wall metal connectors serving two or more category 1 appliances. (Courtesy of Gas Appliance Manufacturers Association)

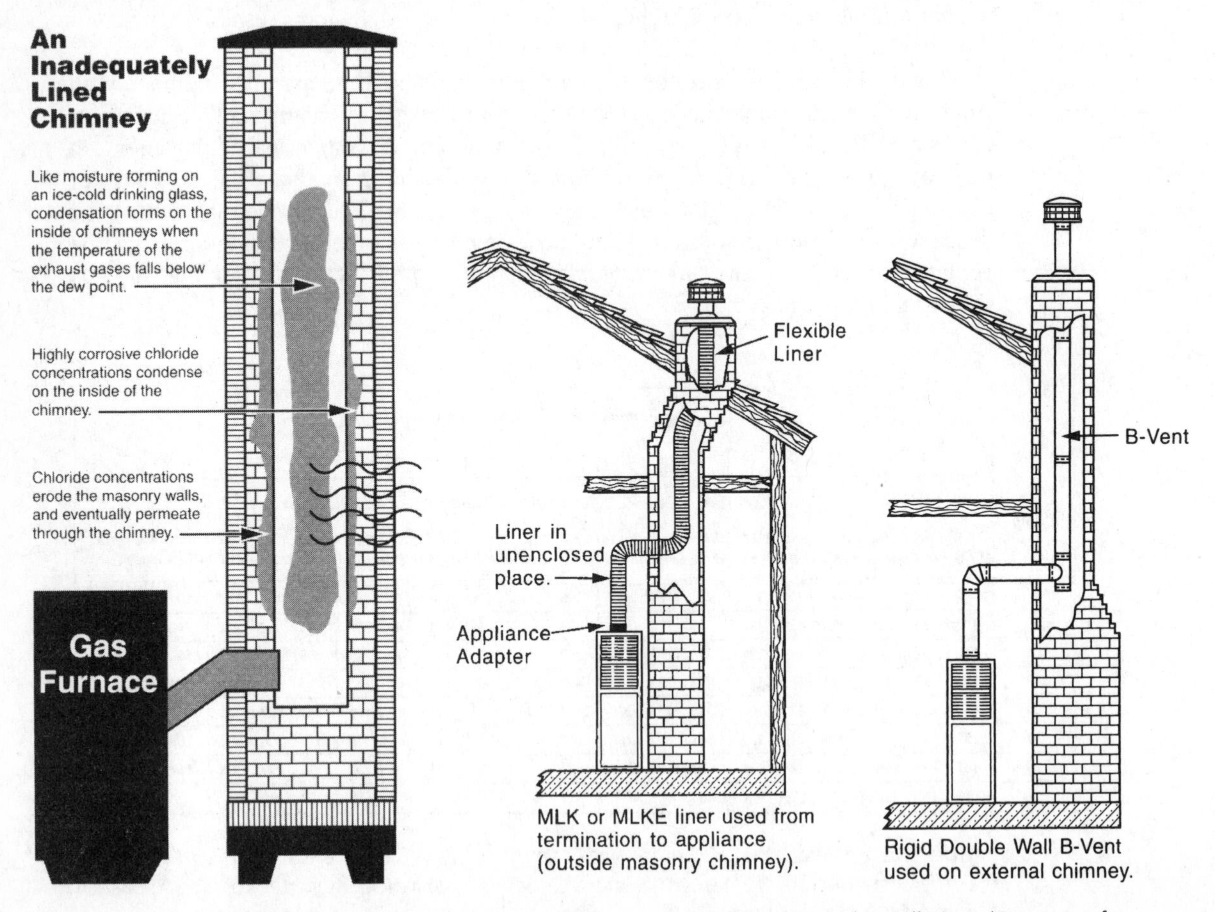

Figure 18.35 Flexi-Liner chimney liner components and typical installation. (Courtesy of Flex-L International, Inc.)

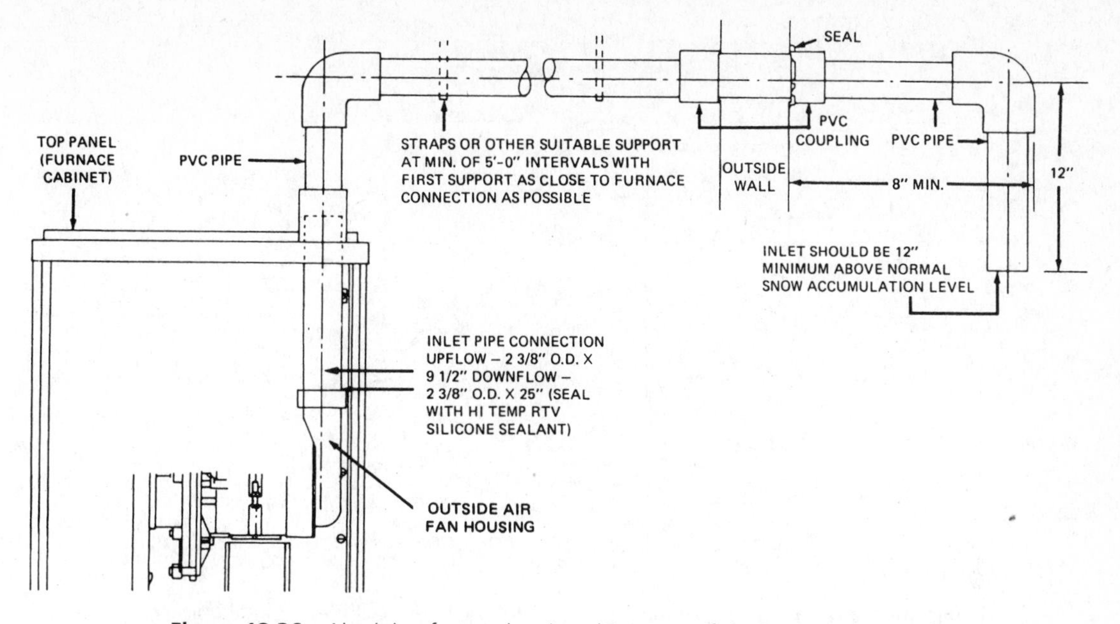

Figure 18.36 Air piping for combustion. (Courtesy of the Trane Company Unitary Products Group)

Combustion Air

Many high-efficiency furnaces are specifically designed to use outside air for combustion, whereas others may have an accessory kit to convert for outside combustion air. Outside air is required when the furnace is installed in a closet, a small mechanical room, a tightly constructed building, or in a corrosive environment. Some manufacturing processes, chemical storage facilities, stables, and beauty parlors are corrosive environments. With some manufacturers, installing the combustion air kit allows termination of the vent pipe closer to a window or other openings into the building.

Combustion air is normally provided through PVC piping. The manufacturer's instruction must be followed to obtain the correct size. Most pipe sizes are 1.5 to 3 in. in diameter. The outside termination of the combustion air piping must be weatherproof, since the piping is pitched down toward the furnace and there is no provision to eliminate water in the pipe. All piping must be secured as shown in Figure 18.36 to avoid any sagging, movement, or pressure on furnace components. The outdoor termination should be located within 24 in. of the vent termination but be at least 14 in. away from it (see Figure 18.37).

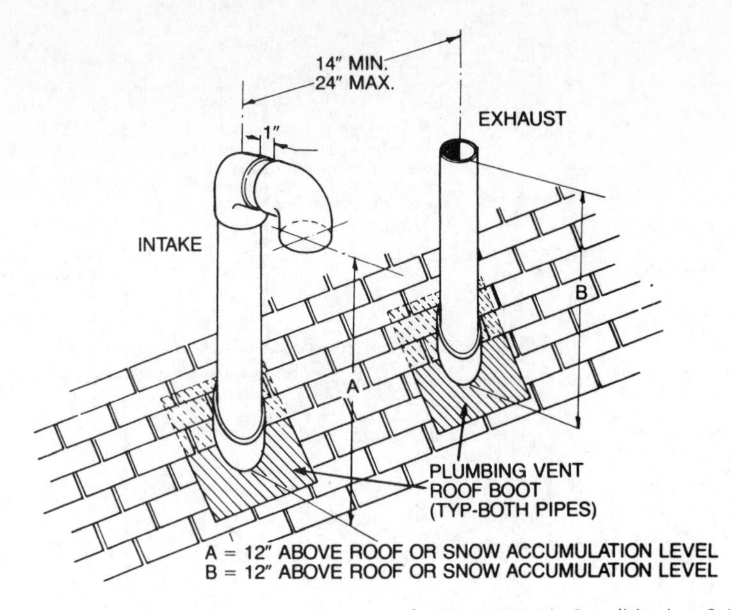

Figure 18.37 Roof vent termination. (Courtesy of ARCOAIRE Air Conditioning & Heating)

Codes

There are various codes enforced for furnace, duct, piping, and venting installations, including the *BOCA* National Mechanical Code, the Uniform Mechanical Code (UMC), the National Fire Protection Association (NFPA), and others. Always determine which code is being enforced in your local community before doing any work. A person licensed to do mechanical work is not normally allowed to perform electrical and plumbing work.

Review Problem

Using the sketch in Figure 18.38, assume the following loads:

Outlet A: 50,000

Outlet B: 60,000

Outlet C: 150,000

Determine the pipe sizes for natural gas using a pressure drop of 0.5 in. W.C. and 0.60 specific gravity gas.

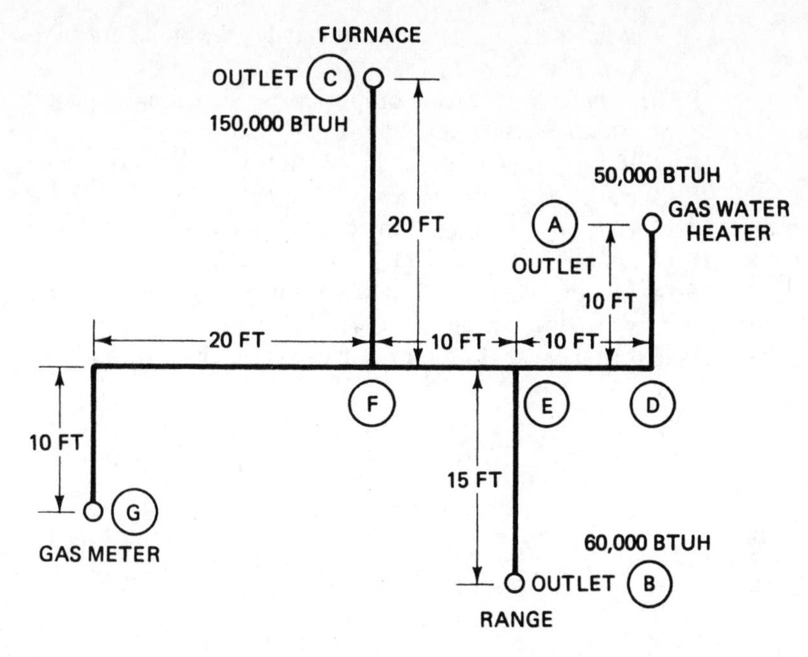

Figure 18.38 Gas piping layout from Figure 18.13 with new loads.

STUDY QUESTIONS

Answers to the study questions may be found in the sections noted in brackets.

18-1 What are the desirable basement, crawl-space, and concrete-slab construction features? *[Building Construction]*

18-2 Name three rules that apply to clearances from materials. *[Clearances from Combustible Materials]*

18-3 Name one important rule to be followed in supplying air for combustion. *[Air for Combustion, Draft Hood Dilution, and Ventilation]*

18-4 Name three (of the five) items of recommended practice for reconnecting gas to the furnace. *[Gas Piping]*

18-5 Briefly describe the procedure for determining the proper gas piping sizes from the meter to the various gas appliances. *[Gas Pipe Sizing]*

18-6 What is the natural gas pressure usually delivered from the meter? *[Natural Gas Pipe Sizing]*

18-7 What is the specific gravity of natural gas in most areas? *[Natural Gas Pipe Sizing]*

18-8 What design gas pipe pressure drop is considered good practice for residential gas piping? *[Natural Gas Pipe Sizing]*

18-9 What is the gas pressure delivered to the piping for LP gas? *[Liquid Petroleum Gas Pipe Sizing]*

18-10 What is the pressure drop used for sizing and piping using LP gas? *[Liquid Petroleum Gas Pipe Sizing]*

18-11 If the furnace is placed in a confined space, what special provisions need to be made? *[Air for Combustion, Draft Hood Dilution, and Ventilation]*

18-12 Why is vent sizing so important? *[Venting]*

18-13 Describe a type B vent. *[Type B Venting]*

18-14 List two types of plastic vent pipes and state the maximum temperature for which they are approved. *[Plastic Venting]*

18-15 List the types of furnaces in each category. *[Vent Sizing Categories]*

CHAPTER 19

Heating System Maintenance and Customer Relations

OBJECTIVES

After studying this chapter, the student will be able to:

- Determine the maintenance requirements for a heating system
- Instruct the owner on proper care and operation of the system

Importance of Proper Maintenance

All mechanical equipment requires proper maintenance to maintain its original efficiency. Heating systems are no exception. With proper care, a heating system will provide good performance for many years. In addition to providing periodic maintenance, the service technician can often upgrade the system when new technology is available. The maintenance items that should receive attention include the following:

1. Testing and adjusting the fuel burning unit
2. Cleaning the air passages of the system
3. Servicing the fan–motor assembly
4. Rebalancing the system
5. Adjusting the thermostat anticipator and fan control
6. Testing the power and safety controls
7. Cleaning the heat exchangers
8. Changing air filters
9. Servicing accessories
10. Upgrading the equipment wherever possible

Testing and Adjusting

Information on combustion is covered in Chapter 3. The use of various types of combustion test instruments is supplied in Chapters 5, 8, and 13. It is important to adjust the fuel-burning unit to produce the highest efficiency that can be maintained continuously. The firing rate of the fuel and the supply of combustion air are the principal adjustments.

The firing rate of the fuel should be set to match the input requirements of the furnace indicated on the nameplate. For example, on a gas furnace, if the input rating is 100,000 Btuh, and the heating value of the gas is 1000 Btu/ft^3, the rate at which gas should be supplied is 100 ft^3/h. This can be tested by turning the thermostat up and timing the gas flow through the meter. If the input is incorrect, an adjustment can be made at the gas pressure regulator or to the size of the burner orifices. On an oil burner, the size of the nozzle and the oil pressure determine the firing rate.

Gas Furnace

Observe the flame. It should be a soft blue color without yellow tips. Adjust the primary air if necessary. If the flame will not clean up when adjusted, remove the burners and clean them both inside and outside. Measure the gas pressure (Figure 19.1). It should be 3.5 in. W.C. for natural gas and 11.0 in. W.C. for LP gas. Measurements are made at the manifold pressure tap with main burners operating and with other gas-burning appliances in operation. Check the thermocouple. Under no load, it should generate 18 to 30 mV dc. When under load, it should generate at least 7 mV dc.

Oil Furnace

Clean the burner (Figure 19.2) and run a complete combustion test. Replace the oil burner nozzle if necessary. Adjust the air volume for maximum efficiency. Replace the oil-line filter. Clean the bimetallic element on the stack relay (if used). Observe the

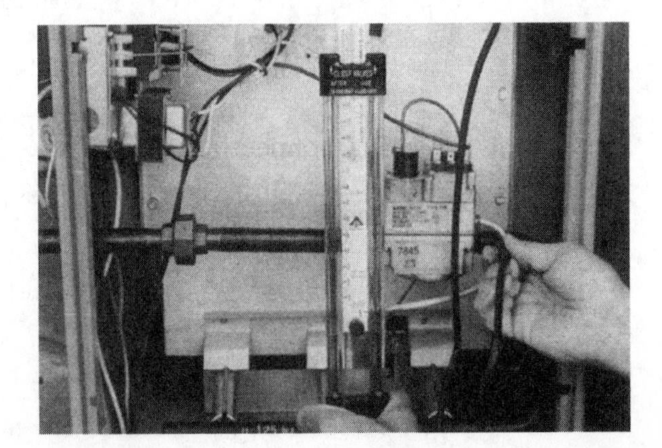

Figure 19.1 Measuring manifold gas pressure.

Figure 19.2 Typical oil burner assembly. (Courtesy of Wayne Combustion Systems)

flame after the burner is restarted. The flame should be yellow in color and centered in the combustion chamber.

At this point a final adjustment should be made using proper test instruments. Unless otherwise specified in the appliance manufacturer's instructions, the unit should be set as follows: After allowing 10 minutes for warm-up, air should be set so that the smoke number is no greater than 1. Less than No. 1 smoke is desired. (*Note:* Occasionally, a new heating unit requires more time than this to burn cleanly due to the oil film on heating surfaces.) Carbon dioxide measured in the stack (ahead of draft control) should be at least 8% for oil rates 1.0 gal/h or less and 9% for oil rates over 1.0 gal/h. The unit should be started and stopped several times to assure good operation.

Check the oil pressure (Figure 19.3). It should be 100 psig during operation and 85 psig immediately after shutdown. Some models require operating pressures in the 140–200 psig range. Motor life will be increased by proper oiling. Use a few drops of nondetergent oil at both motor oil holes twice each year. The line filter cartridge

Figure 19.3 Checking the oil pressure. (Courtesy of Sundstrand Hydraulics)

should be replaced every year to avoid contamination of the fuel unit and atomizing nozzle. Finally, the area around the heating unit should be kept clean and free of any combustible materials, especially papers and oily rags.

Electric Furnace

Set the thermostat to call for heat. Check by reading the incoming power amperage to make certain that all heater elements are operating. On the three- and four-element furnaces that are equipped with a fused disconnect block, both legs of the heating elements are provided with cartridge fuses internal to the furnace. Two elements and the blower are grouped together on a 60-A fuse. On three-element furnaces, a single element is fused at 30 A.

Cleaning Air Passages

If construction work has taken place in the area or if the occupant is negligent in keeping the filters clean, there may be a layer of airborne dust on the outside surface of the heat exchangers. This surface needs to be kept clean. Particularly on high-efficiency furnaces, debris will collect on the external surface of the secondary heat exchanger. Stiff bristle brushes can be used to dislodge foreign material, and a commercial vacuum cleaner can be used to remove dirt and lint.

On high-efficiency furnaces it may be necessary to remove the blower to reach the secondary heat exchanger for cleaning, as shown in Figure 19.4. Brush strokes must be in the direction of the finned surface to avoid damage to the fins. Inspect and clean the blades of the fan wheel using a brush and vacuum. Care should be taken not to dislodge balance weights (clips) that may be on the fan wheel.

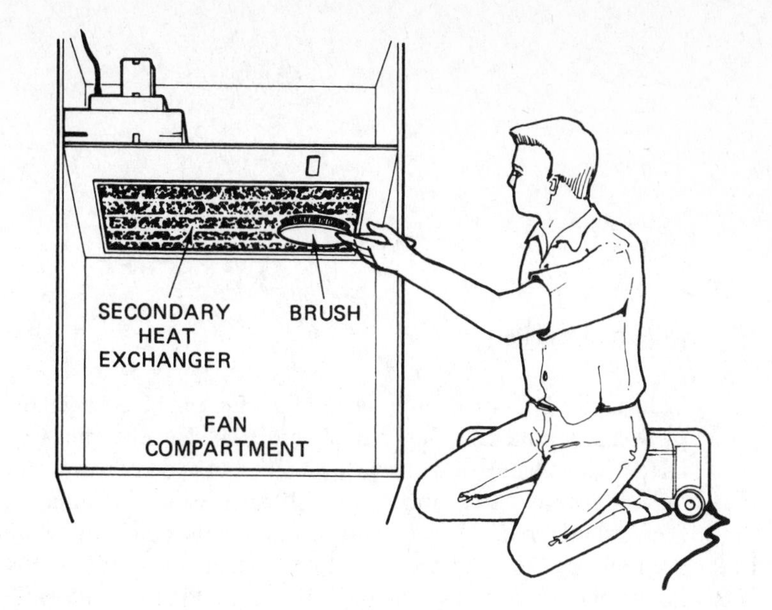

Figure 19.4 Cleaning secondary heat exchanger.

Servicing Fan-Motor Assembly

The fan–motor assembly requires periodic attention. The unit may be belt-driven, as shown in Figure 4.14, or direct-driven, as shown in Figure 4.15. Most direct-drive units have multiple-speed arrangements. Some motors and bearings require lubrication, and some do not. Oil or grease should be applied in accordance with the manufacturer's instructions.

Clean out the blades of the fan wheel. Motors that require oil should be oiled twice a year. If there is more than ⅛-in. end play on the shaft, move the thrust collar closer to the bearing. Furnace fans are designed to rotate in one direction. It is therefore important to determine proper rotation at the time of installation or when changing the fan motor (Figure 19.5). If it becomes necessary to reverse the rotation, follow the instructions on the motor terminal block for reversing the lead wires. The directions are usually found under the cover where the lead wires enter the motor.

It is important to determine if the proper amount of air is being supplied by the fan–motor assembly. The easiest way to determine the volume of air handled by the furnace is to measure the temperature rise through the furnace. The temperature rise plus the output rating of the furnace provides the information necessary for calculating the cubic feet per minute (cfm) handled by the fan:

$$\text{cfm} = \frac{\text{output Btuh}}{\text{temperature rise} \times 1.08}$$

If the air volume handled by the fan does not comply with the manufacturer's requirement, the fan speed must be adjusted.

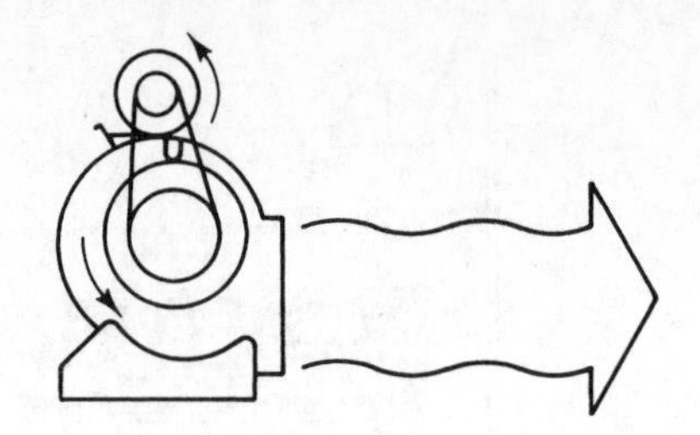

Figure 19.5 Determining fan rotation.

Belt-Driven Fans

If the unit is belt-driven, the speed of the fan can be changed by adjusting the variable-pitch motor pulley, as shown in Figure 19.6. To adjust fan speed, loosen the setscrew in the outer flange outward to decrease fan speed or inward to increase speed. The outer flange must be rotated by half-turns to avoid damage to the threads, but the flange should not be rotated outward so that the belt is riding on the hub rather than on the flanges of the pulley. After each adjustment of the motor pulley, lock the outer flange in place by tightening the setscrew on the flat of the pulley hub. Line up the fan and motor pulleys by using a straightedge across two outer edges of the fan pulley, and adjust the pulley on the motor shaft so that the belt is parallel with the straightedge. Tighten the setscrew securely on the flat of the motor shaft.

Most fan motors rotate at 1725 rev/min. The following table shows how the revolutions per minute can be increased or decreased by changing either the diameter of the motor pulley or the fan pulley.

$$\text{rev/min of equipment} = \frac{\text{rev/min of motor} \times \text{diameter of motor pulley}}{\text{diameter of equipment pulley}}$$

For example, if the motor pulley is 2.00 in. in diameter and the fan pulley is 6 in., the fan will turn at 575 rev/min. If the motor pulley is increased in diameter to 3.00 in., the revolutions per minute will increase to 862. It should be noted that an increase in fan revolutions per minute increases the amperage draw on the motor. Some procedures relating to belt installation, alignment, and adjustments of belt-driven fans are shown in Figure 19.7.

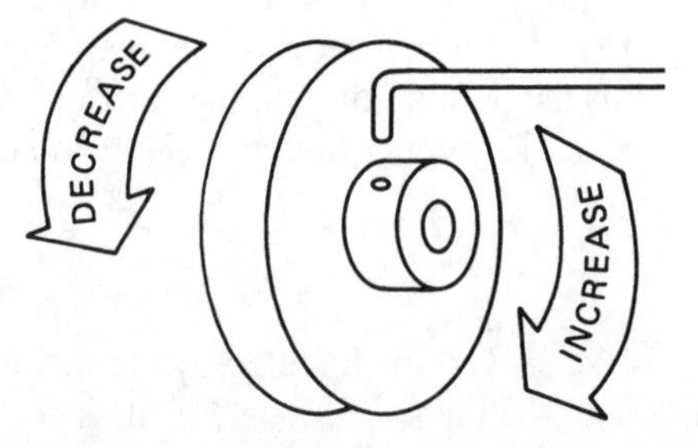

Figure 19.6 Adjustment of variable-speed pulley.

WIPE OFF PULLEYS AND BELT with a clean rag to get rid of all oil and dirt. Dirt and grease are tough abrasives that cause the belt to wear out faster, throwing it out of balance and shortening its life.

INSTALL V-BELT in pulley grooves by loosening the belt take-up or the adjusting screw on the motor. Do not "roll" or "snap" the belt on the pulleys; this causes much more strain than the pulleys should have. Be sure the belt doesn't "bottom" in the pulley grooves.

ALIGN BOTH PULLEYS AND SHAFTS by moving the motor on its motor mount. You can do this "by eye," but you're a lot safer if you hold a straight edge flush against the blower pulley, then move the motor until the belt is absolutely parallel to the straight edge.

HERE'S A SHORT-CUT WAY TO CHECK PULLEY ALIGNMENT: Sight down the top of the belt from slightly above it. If the belt is straight where it leaves the pulley and does not bend, you can bet that the alignment is reasonably good.

CHECK BELT TENSION before proceeding further. Remember that a V-belt "rides" the inside of the pulley faces. Since the sides of the belt wedge in the pulleys, the V-belt does not have to be tight. It should be as loose as possible without slipping in the pulley grooves.

USE THIS RULE-OF-THUMB FOR ADJUSTING BELT TENSION: Using the belt take-up or motor adjusting screw, tighten the belt until the slack side can be depressed about ¾" for each foot of span between the pulleys. WARNING: EXCESSIVE BELT TENSION IS THE MOST FREQUENT CAUSE OF BEARING WEAR AND RESULTING NOISE.

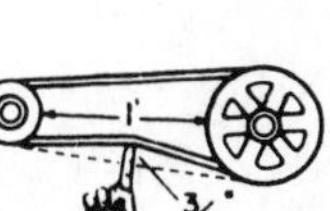

Figure 19.7 Removal, alignment, and adjustment of V-belts. (Courtesy of Lau Industries/Conaire Sales Division)

EQUIPMENT SPEED (REV/MIN) FOR MOTORS TURNING 1725 REV/MIN

Diameter of motor Pulley (in.)	*Diameter of Pulley on Equipment (in.)*					
	5	*(6)*	*7*	*8*	*9*	*10*
1.25	431	359	308	270	240	216
1.50	518	431	370	323	288	259
1.75	604	503	431	377	335	302
(2.00)	690	(575)	493	431	383	345
2.25	776	646	554	485	431	388
2.50	862	719	616	539	479	431
2.75	949	791	678	593	527	474
3.00	1035	862	739	647	575	518
3.25	1121	934	801	701	623	561
3.50	1208	1006	862	755	671	604
3.75	1294	1078	924	809	719	647
4.00	1380	1150	986	862	767	690

Direct-Drive Fans

Direct-drive fans (see Figure 19.8A) are more popular than the belt-driven type because they require less maintenance. Furnaces that are equipped for air conditioning are supplied with a multispeed motor. Multispeed motors have several connections in the motor windings, which allow the selection of different speeds. Most electric motors have a wiring diagram attached to the motor casing or to the underside of the plate that covers the motor terminals. The motor connections shown in Figure 19.8B are for a three-speed, capacitor-start motor.

When cooling is added to an existing furnace, the evaporator coil adds an additional pressure drop in the distribution system. In most cases, cooling also requires more air than heating. If the furnace does not already have a multispeed motor for the fan, it is advisable to install one. The fan speed for heating is usually slower than that used for cooling. Some high-efficiency furnaces, however, require a faster fan speed than that used for cooling. A combination fan relay and transformer (Figure 19.9) can be installed to provide a means of automatically changing the speed of the fan motor from heating to cooling.

The use of the fan relay and transformer as applied to single-speed fans is shown in Figure 19.10A, and that for two-speed fans is shown in Figure 19.10B.

(A) DIRECT-DRIVE FAN

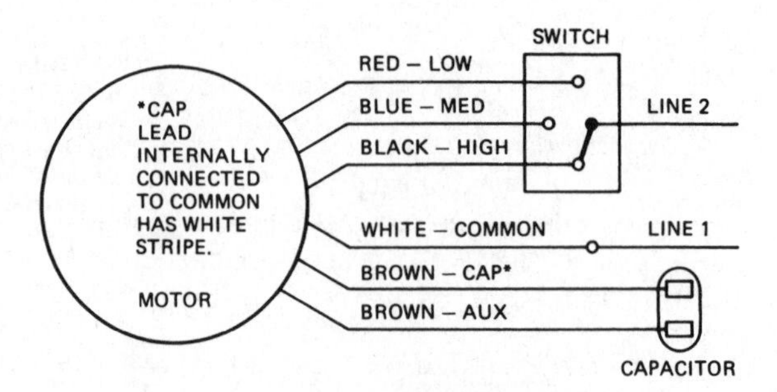

(B) CONNECTIONS ON THREE-SPEED, CAPACITOR-START MOTOR.

Figure 19.8 Direct-drive multispeed fan with capacitor-start motor. ((A) Courtesy of Conaire Division, Philips Industries, Inc.; (B) courtesy of Universal Electric)

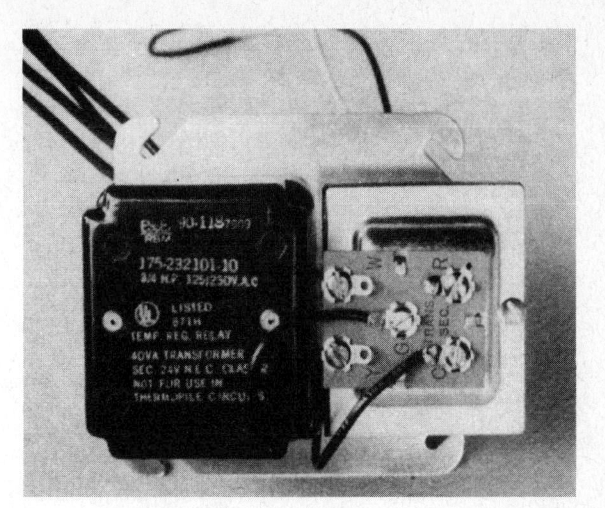

Figure 19.9 Combination fan relay and transformer.

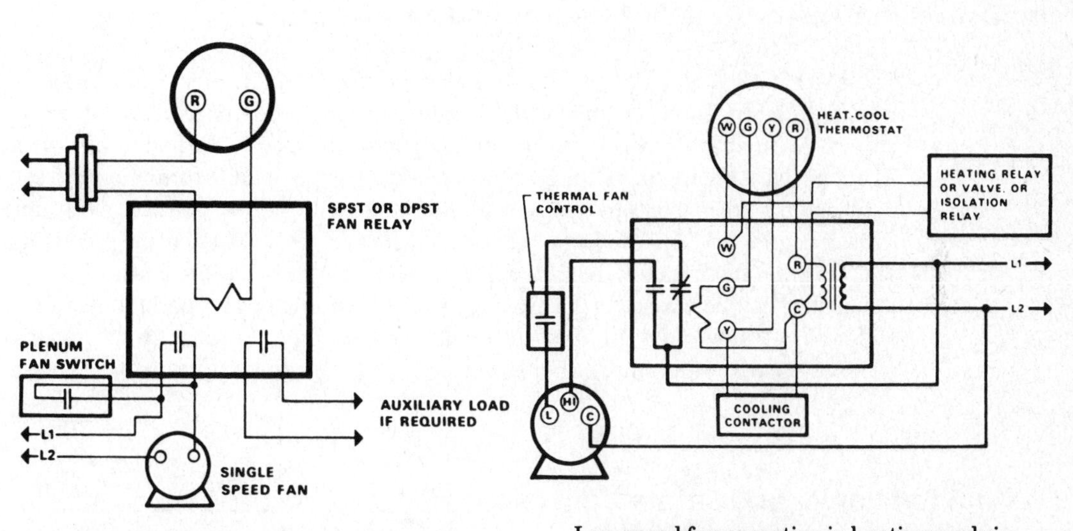

In the system shown, a call for cooling makes R-G in the thermostat and completes the circuit through the fan relay coil. The relay pulls in to energize the system fan. The plenum fan switch, for controlling the fan in the heating mode, is wired in parallel with the normally open fan relay contacts.

(A) SINGLE SPEED FAN

Low speed fan operation in heating mode is maintained through a normally closed contact in the fan relay. When the fan relay coil is energized on a call for cooling by the thermostat, this contact opens. This cuts off power to the low speed fan, and the normally open contacts close to energize the high speed windings of the fan motor.

(B) TWO SPEED FAN

Figure 19.10 Application of fan relay and transformer. (Courtesy of Honeywell Inc.)

Rebalancing the System

After a forced warm air system has been in use for a period of time, some rooms may be heating better than others. Rebalancing the system is therefore in order. Air balancing is performed by adjusting the branch-duct dampers to produce uniform temperatures throughout the building (Figure 17.10). The procedure is as follows:

1. Place a thermometer in each room at table height.
2. Open all supply and return air duct dampers.
3. Open all baffles and register dampers.
4. Set the thermostat to call for heat.
5. Adjust the dampers while the furnace is running to produce uniform temperatures in all the rooms.

The balancing should be done during weather cold enough to permit continuous operation of the heating equipment for a substantial period of time. Continuous fan operation with the furnace cycling at the rate of 8 to 10 times per hour provides good conditions for balancing.

Adjusting the Thermostat Anticipator and Fan Control

The thermostat anticipator is adjusted to control the length of the operating cycle. Raising the amperage setting of the anticipator lengthens the cycle. Lowering the amperage setting of the anticipator shortens the cycle. Refer to the section "Thermostat Anticipator Setting" in Chapter 10 for detailed information for setting anticipators. When continuous fan action is not used, the fan control on an upflow furnace starts the fan when the bonnet temperature reaches its cut-in point (Figure 19.11). If the fan is delivering air that is too cool for comfort, the cut-in point of the fan control can be raised. It is advantageous to have the longest possible fan-on time and still not blow cold air. The differential of the fan control (cut-in temperature minus cutout temperature) should be great enough to prevent short cycling of the fan.

Testing the Power and Safety Controls

It is important to use a separately fused branch electrical circuit containing a properly sized fuse or circuit breaker for the furnace. A means for disconnecting the furnace must be located within sight and be readily accessible. Check the supply voltage to be certain that the proper power is being delivered to the furnace. To check the supply voltage to determine that it is adequate, the following procedure is recommended:

1. On a gas-fired forced-air furnace, remove the transformer. Then, check the power supply leads in the junction box (Figure 19.12).

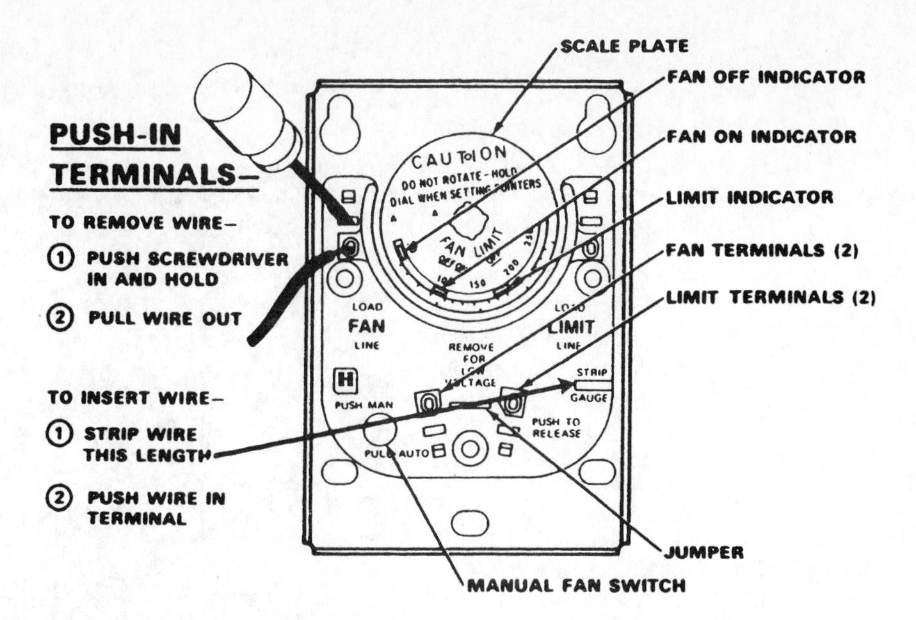

Figure 19.11 Fan control adjustment. (Courtesy of Honeywell Inc.)

2. On an oil-fired forced-air furnace, remove the cad-cell control from the top of the oil burner assembly. Then, check the power supply leads in the junction box (Figure 19.13).

Check to be sure that all wiring connections are tight and that wiring insulation is in good condition. With a thermometer, check the cutout point of the limit control by shutting down the fan. The bonnet temperature must not exceed 200°F.

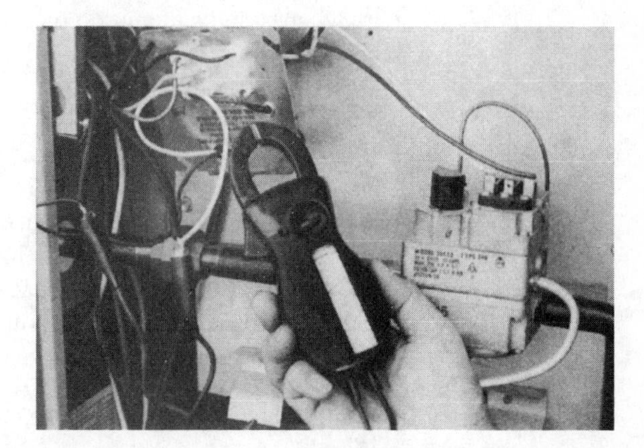

Figure 19.12 Checking the power supply on a gas-fired furnace.

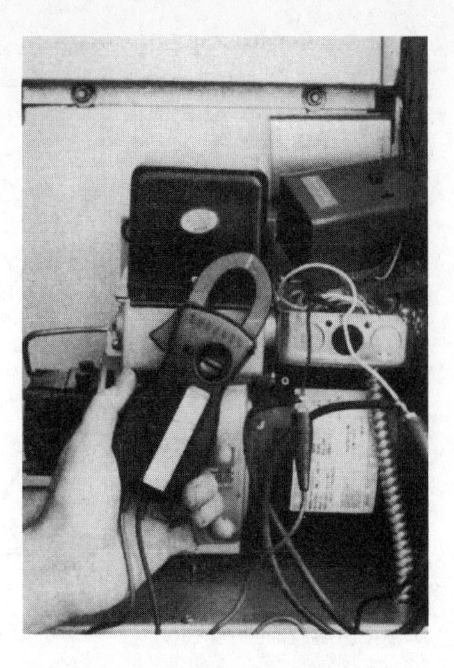

Figure 19.13 Checking the power supply on an oil-fired furnace.

Safety Check of the Limit Control

The limit control shuts off the combustion control system and energizes the circulating-air fan motor if the furnace overheats. The recommended method of checking the limit control is gradually to block off the return air after the furnace has been operating for at least 5 minutes. As soon as the limit has proven safe, the return-air opening should be unblocked to permit normal air circulation. By using this method to check the limit control, it can be established that the limit is functioning properly and is fail-safe if there is a motor failure. (Downflow or horizontal furnaces have a manual reset limit switch located on the fan housing.) On high-efficiency furnaces, there are many sensing switches and safeguard controls that must be checked to be certain that the equipment will run properly.

Safety Check of the Flow-Sensing Switch

1. Turn off 115-V power to the furnace. Remove the control door and disconnect the inducer motor lead wires from the inducer printed-circuit board.
2. Turn on 115-V power to the furnace. Close the thermostat switch as if making a normal furnace start. The pilot should light and then cycle off and on. If the main burners do not light, the flow-sensing switch is functioning properly.
3. Turn off 115-V power to the furnace. Reconnect the inducer motor wires, replace the control door, and turn on 115-V power.

Safety Check of the Draft Safeguard Switch

The purpose of this control is to permit the safe shutdown of the furnace during certain blocked-flue conditions.

1. Disconnect power to the furnace and remove the vent pipe from the furnace outlet collar. Be sure to allow time for the vent pipe to cool down before removing.
2. Restore power to the furnace. Allow the furnace to operate for 2 minutes, then block (100%) the flue outlet. The furnace should cycle off within 2 minutes.
3. Reconnect the vent pipe. Wait 5 minutes and then reset the draft safeguard switch.

Cleaning the Heat Exchanger

The only time it is necessary to disassemble the furnace and clean the interior surface of the heat exchanger is when the fuel-burning device is out of adjustment and sooting occurs in the flue passages. On some conventional furnaces, it is possible to use a brush to dislodge the soot (Figure 19.14) and a vacuum to remove it. On high-efficiency furnaces, two heat exchangers may require cleaning. A typical location of these heat exchangers is shown in Figure 19.15.

Changing Air Filters

Clean or replace air filters as often as necessary to permit proper airflow. Replacement filters should be similar in material, thickness, and size to those removed. The air filters must be clean when testing any system for air volume. A dirty filter creates an unnecessary restriction, thereby reducing air volume.

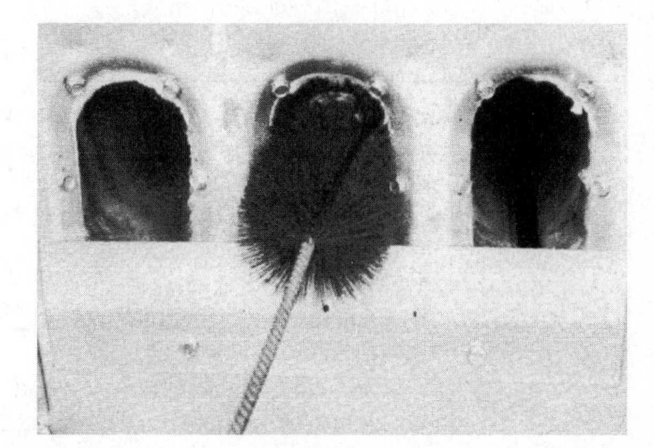

Figure 19.14 Cleaning a heat exchanger.

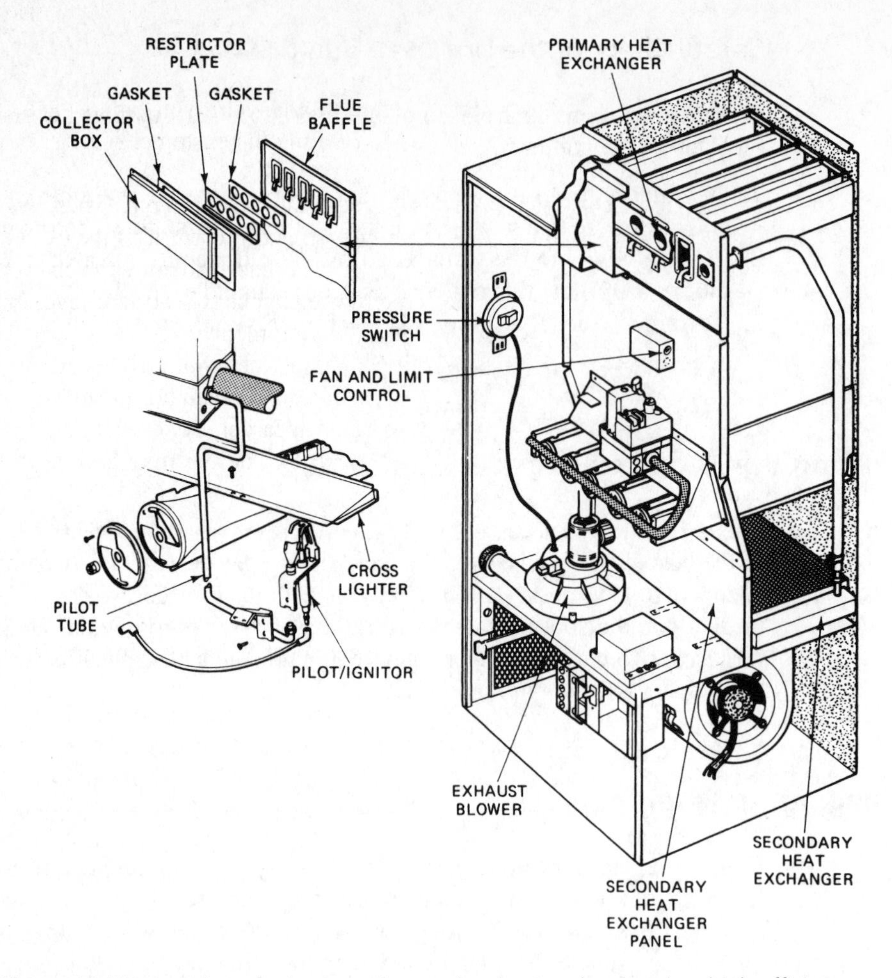

Figure 19.15 Location of primary and secondary heat exchangers in a high-efficiency furnace.

Servicing Accessories

Clean and service accessories in accordance with the manufacturer's instructions. Humidifiers may require cleaning more than once a year, depending on water conditions. The condenser surface on condensing units used for cooling must be kept clean. Condenser fans usually require lubrication.

Upgrading the Furnace

The owner should be advised of any advanced technology that will improve efficiency and comfort. These items include programmable thermostats and electronic ignition.

Customer Relations

You can have a toolbox full of the latest tools, yet if you cannot "get along" with your customers, what good are the tools? Treating the customer right is every bit as important as having the right tools.

Why are customer relations so important? Simply put, when you complete your service call, write out your service report, and present it to the customer, you expect payment. Gas valves do not write checks. Furnaces do not write checks. Motors do not write checks. People—customers—write checks. Creating a favorable attitude is as much a part of your job as repairing the equipment.

Consider the customers' frame of mind when you arrive. Perhaps they have been without heat for several hours. Perhaps they have heard horror stories about service people who make excessive charges. Perhaps an earlier service call failed to solve the problem. When you come up the driveway, the customer may have already made a case against you.

Your first job is to make a favorable impression. Here are a few of the things you can do to get off to a good start:

1. *Radiate value.* Your truck should be clean and orderly. Your tools should be contained in a suitable box, easy to transport. Your instruments should be in proper cases, ready to use.
2. *Polish up your self-image.* Your appearance should be that of a professional. You should be confident of your ability and show it. Your smile should radiate success.
3. *Develop good work habits.* Plan your work. Use accurate information. Refer to the instruction manual whenever necessary. Seek the advice of the manufacturer if necessary. Exercise good time management. If you have undesirable work habits that hold you back, correct them. Make yourself more valuable to your customer as well as to yourself.

Remember, sometimes customers appear to be difficult, but for the most part, they are not inherently rude. Something else may be going on in their life that has them upset. Whatever it may be, it is part of your job as service person to show empathy and understanding and do the right thing for your customer. Becoming a service technician is one of the most exciting, rewarding careers you can choose. Put your heart into it. Develop your "people skills" and you will be rewarded.

Instructions for Equipment Service

Discuss with the owner the features of the installation, how the system operates, and the service necessary to maintain the equipment. It is important to leave a copy of the detailed service report with the customer. The proper instructions for maintenance and care shown in Figures 19.16 and 19.17 will be a helpful guide for the customer.

Warranty

The manufacturer's limited warranty (Figure 19.18) and the contractor's warranty should be supplied to the owner. Although the manufacturer does not include labor in replacing a defective part during the first year, usually the contractor's warranty does. Manufacturers often offer an extended warranty on parts such as heat exchangers.

MAINTENANCE AND CARE FOR THE OWNER OF A GAS-FIRED FURNACE

Warning:

1. TURN OFF ELECTRICAL POWER SUPPLY TO YOUR FURNACE BEFORE REMOVING ACCESS DOORS TO SERVICE OR PERFORM MAINTENANCE.
2. When removing access doors or performing maintenance functions inside your furnace, be aware of sharp sheet metal parts and screws. BE EXTREMELY CAREFUL WHEN HANDLING PARTS OR REACHING INTO THE UNIT.

AIR FILTER

The air filter should be checked at least every 6 to 8 weeks and changed or cleaned whenever it becomes dirty.

The size of the air filter varies, depending on the furnace model and size. When replacing your furnace filter, always use the same size and type of filter that was originally supplied.

Warning: NEVER OPERATE YOUR FURNACE WITHOUT A FILTER IN PLACE.

Failure to heed this warning may result in damage to the furnace blower motor. An accumulation of dust and lint on internal parts of your furnace can cause a loss of efficiency and, in some cases, fire.

When inspecting, cleaning, or replacing the air filter in your furnace, refer to the appropriate following procedures that apply to your particular furnace.

1. Turn OFF the electrical supply to the furnace. Remove the control and fan access doors, respectively.
2. Gently remove the filter and carefully turn the dirty side up (if dirty) to avoid dislodging dirt from the filter.
3. If the filter is dirty, wash it in a sink or bathtub or outside with a garden hose. Always use cold water, and use a mild liquid detergent if necessary. Then, allow the filter to air dry.
4. Reinstall the clean filter with the crosshatch binding side facing the furnace fan. Be sure that the filter retainer is under the flange on the furnace casing.
5. Replace the fan and control access doors, and restore electrical power to your furnace.

COMBUSTION AREA AND VENT SYSTEM

The combustion area and vent system should be visually inspected before each heating season. An accumulation of dirt, soot, or rust can result in loss of efficiency and improper performance. Accumulations on the main burners can result in the burners firing out of normal time sequence. This delayed ignition is characterized by an especially loud sound that can be quite alarming.

Caution: If your furnace makes an especially loud noise when the main burners are ignited by the pilot, shut down your furnace and call your service person.

1. Turn OFF the electrical supply to your furnace and remove the access doors.
2. Carefully inspect the gas burner and pilot areas for dirt, rust, or scale. Then, inspect the flue connection area and flue pipe for rust.

Caution: If dirt, rust, soot, or scale accumulations are found, call your service person.

Figure 19.16 Typical gas-fired furnace maintenance and care instructions.

MAINTENANCE AND CARE FOR THE OWNER OF AN OIL-FIRED FURNACE

SAFETY INFORMATION

Warning: These instructions are intended to aid you in the safe operation and proper maintenance of your oil furnace. Read these instructions thoroughly before attempting to operate the furnace. The purpose of WARNINGS and CAUTIONS in these instructions is to call attention to the possible danger of personal injury or equipment damage; they deserve careful attention and understanding. If you do not understand any part of these instructions, contact a qualified licensed service person for clarification.

1. Do not attempt to light the burner manually with a match or other flame.
2. Do not attempt to start the burner with oil or oil vapors in the combustion chamber.
3. Do not operate the furnace without all fan doors and compartment covers securely in place.
4. Do not attach any kind of device to the flue or vent.
5. Always shut off electrical power to the furnace before attempting any maintenance.

HOW YOUR SYSTEM OPERATES

As the flame from the oil burner warms the heat exchanger, the fan switch will start the fan operation. Warm air should now gently circulate from the supply diffusers throughout the dwelling and return to the furnace through return air grille(s).

When the temperature of the circulating air reaches the temperature setting of the thermostat, the oil burner will stop operation, the heat exchanger will cool, and the fan operation will soon be interrupted.

OPERATING INSTRUCTIONS

Burner operation

1. Set the room thermostat above room temperature.
2. Turn the electric switch on. The burner should start automatically.
3. After burner starts, reset the room thermostat for desired temperature.

If Burner Does Not Start

1. Check the fuse in the burner circuit.
2. Make sure the room thermostat is set above the room temperature.

Warning: An explosion or flash fire can occur if the combustion chamber is not free of oil when using the primary control reset button. Make sure the combustion chamber is free of oil before using the reset button.

3. Wait 5 minutes in order to allow the control to cool so that it will recycle. Reset the primary control.
4. If the burner still does not start, call your service person.

Note: CAD CELL MAY BE EXPOSED TO DIRECT ARTIFICIAL LIGHT OR SUNLIGHT, WHICH MAY ENTER THROUGH THE BURNER AIR-CONTROL BAND.

ANNUAL MAINTENANCE

1. Before the heating season, lubricate the burner motor using SAE No. 20 motor oil.

Caution: Do not oil more than necessary. Follow the instructions attached to the fan housing.

2. Have the burner, heating unit, and all controls checked by your service contractor to assure proper operation during the heating season.

TO CHANGE OIL FILTER

1. Turn electric switch off.
2. Close the oil-line valve at tank.
3. Carefully remove the lower section of the filter assembly containing the oil filter cartridge.
4. Drain excess oil from the oil line into a container; then dispose of the oil and the old cartridge.
5. Place a new filter cartridge in the filter assembly and screw it back into place.
6. Loosen the "bleed" screw on top of oil cartridge assembly, allowing air to be purged. Tighten the bleed screw.
7. Open valve in oil line at tank.

THINGS YOU MAY DO

1. **Warning:** Disconnect the main power to the unit before attempting any maintenance.
2. Keep the air filters clean.

Caution: DO NOT OPERATE YOUR SYSTEM FOR EXTENDED PERIODS WITHOUT FILTERS. ANY RECIRCULATED DUST PARTICLES WILL BE HEATED AND CHARRED BY CONTACT WITH THE FURNACE HEAT EXCHANGER. THIS RESIDUE WILL SOIL CEILINGS, WALLS, DRAPES, CARPETS, AND OTHER HOUSEHOLD ARTICLES.

LUBRICATION INSTRUCTIONS

Highboy Units

In order to oil the motor and fan, it is necessary to slide the fan out of the furnace. Remove the fan panel on the front of the furnace and slide the fan section out.

Counterflow Units

In order to oil the motor and fan, it is necessary to remove the upper inner panel of the fan compartment. Follow the instructions on the label.

Lowboy and Horizontal Units

In order to oil the motor and fan, it is necessary to remove only the fan compartment door. The fan is positioned within the compartment to make the fan bearings accessible for oiling.

Proper belt tension is important and may be checked by depressing the belt at a point halfway between the pulleys approximately 1 in. Belt tension may be adjusted by turning the adjustment screw attached to motor base to raise or lower motor.

Figure 19.17 Typical oil-fired furnace maintenance and care instructions.

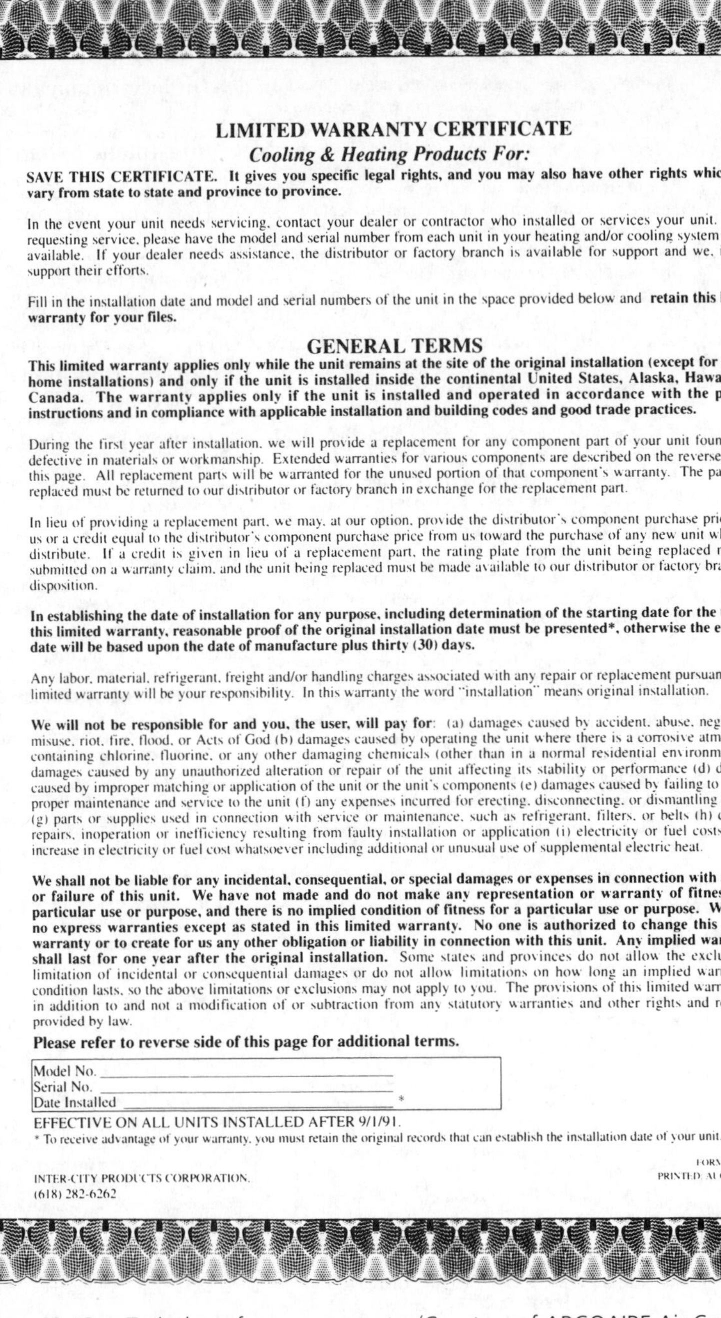

LIMITED WARRANTY CERTIFICATE

Cooling & Heating Products For:

SAVE THIS CERTIFICATE. It gives you specific legal rights, and you may also have other rights which may vary from state to state and province to province.

In the event your unit needs servicing, contact your dealer or contractor who installed or services your unit. When requesting service, please have the model and serial number from each unit in your heating and/or cooling system readily available. If your dealer needs assistance, the distributor or factory branch is available for support and we, in turn, support their efforts.

Fill in the installation date and model and serial numbers of the unit in the space provided below and **retain this limited warranty for your files.**

GENERAL TERMS

This limited warranty applies only while the unit remains at the site of the original installation (except for mobile home installations) and only if the unit is installed inside the continental United States, Alaska, Hawaii, and Canada. The warranty applies only if the unit is installed and operated in accordance with the printed instructions and in compliance with applicable installation and building codes and good trade practices.

During the first year after installation, we will provide a replacement for any component part of your unit found to be defective in materials or workmanship. Extended warranties for various components are described on the reverse side of this page. All replacement parts will be warranted for the unused portion of that component's warranty. The part to be replaced must be returned to our distributor or factory branch in exchange for the replacement part.

In lieu of providing a replacement part, we may, at our option, provide the distributor's component purchase price from us or a credit equal to the distributor's component purchase price from us toward the purchase of any new unit which we distribute. If a credit is given in lieu of a replacement part, the rating plate from the unit being replaced must be submitted on a warranty claim, and the unit being replaced must be made available to our distributor or factory branch for disposition.

In establishing the date of installation for any purpose, including determination of the starting date for the term of this limited warranty, reasonable proof of the original installation date must be presented*, otherwise the effective date will be based upon the date of manufacture plus thirty (30) days.

Any labor, material, refrigerant, freight and/or handling charges associated with any repair or replacement pursuant to this limited warranty will be your responsibility. In this warranty the word "installation" means original installation.

We will not be responsible for and you, the user, will pay for: (a) damages caused by accident, abuse, negligence, misuse, riot, fire, flood, or Acts of God (b) damages caused by operating the unit where there is a corrosive atmosphere containing chlorine, fluorine, or any other damaging chemicals (other than in a normal residential environment) (c) damages caused by any unauthorized alteration or repair of the unit affecting its stability or performance (d) damages caused by improper matching or application of the unit or the unit's components (e) damages caused by failing to provide proper maintenance and service to the unit (f) any expenses incurred for erecting, disconnecting, or dismantling the unit (g) parts or supplies used in connection with service or maintenance, such as refrigerant, filters, or belts (h) damage, repairs, inoperation or inefficiency resulting from faulty installation or application (i) electricity or fuel costs or any increase in electricity or fuel cost whatsoever including additional or unusual use of supplemental electric heat.

We shall not be liable for any incidental, consequential, or special damages or expenses in connection with any use or failure of this unit. We have not made and do not make any representation or warranty of fitness for a particular use or purpose, and there is no implied condition of fitness for a particular use or purpose. We make no express warranties except as stated in this limited warranty. No one is authorized to change this limited warranty or to create for us any other obligation or liability in connection with this unit. Any implied warranties shall last for one year after the original installation. Some states and provinces do not allow the exclusion or limitation of incidental or consequential damages or do not allow limitations on how long an implied warranty or condition lasts, so the above limitations or exclusions may not apply to you. The provisions of this limited warranty are in addition to and not a modification of or subtraction from any statutory warranties and other rights and remedies provided by law.

Please refer to reverse side of this page for additional terms.

Model No. ______________________________
Serial No. ______________________________
Date Installed ___________________________ *

EFFECTIVE ON ALL UNITS INSTALLED AFTER 9/1/91.

* To receive advantage of your warranty, you must retain the original records that can establish the installation date of your unit.

INTER-CITY PRODUCTS CORPORATION.
(618) 282-6262

FORM 7650-149
PRINTED AUGUST 1991

Figure 19.18 Typical gas furnace warranty. (Courtesy of ARCOAIRE Air Conditioning & Heating)

STUDY QUESTIONS

Answers to the study questions may be found in the sections noted in brackets.

19-1 Name 10 items that are included under furnace maintenance. *[Importance of Proper Maintenance]*

19-2 Which parts of the furnace require adjustment? *[Importance of Proper Maintenance]*

19-3 How should the firing rate of the furnace be determined? *[Testing and Adjusting]*

19-4 How is the air quantity determined? *[Servicing Fan-Motor Assembly]*

19-5 What is the relation of the amount of air required for cooling to the amount required for heating? *[Direct-Drive Fans]*

19-6 In balancing the airflow, where should the thermometers be situated? *[Rebalancing the System]*

19-7 How often should the furnace cycle per hour? *[Adjusting the Thermostat Heat Anticipator and Fan Control]*

19-8 How often should the fan motor be lubricated? *[Servicing Fan–Motor Assembly]*

19-9 What is the proper color for a gas flame? *[Gas Furnace]*

19-10 How frequently should oil burners be serviced? *[Oil Furnace]*

19-11 What is the maximum time that heat exchangers are under warranty? *[Warranty]*

19-12 What are the important areas of concern in customer relations? *[Customer Relations]*

CHAPTER 20

Energy Conservation

OBJECTIVES

After studying this chapter, the student will be able to:

- Recommend construction factors to reduce building exposure losses
- Determine the value of retrofit measures to reduce the use of energy required for heating
- State the types of auxiliary equipment or modifications that are available for energy conservation

Energy Conservation Measures

Figure 20.1 shows the extent to which space heating contributes to the nation's use of energy. Nearly 18% of the total energy consumed is required for heating. The addition of domestic hot water and air conditioning increases the proportion of energy used to about 25% of the total national consumption. Although heating is an essential element of our existence, only recently has a great deal of attention been directed to conserving the valuable natural resources required for heating.

A number of ways to conserve energy are discussed in other chapters, including the use of high-efficiency furnaces and heat pumps, and the improved use of controls. This chapter is devoted primarily to describing how the building envelope can be improved to save energy. Included also are other conservation measures for specific applications. This information is divided into the following three areas:

1. Reducing exposure losses
2. Reducing outside air infiltration
3. Using other retrofit measures

Although heating-installation and service personnel are not concerned directly with the construction of the building, they are frequently called on to offer recommendations for reducing the size of the heating load and for improving comfort levels. An

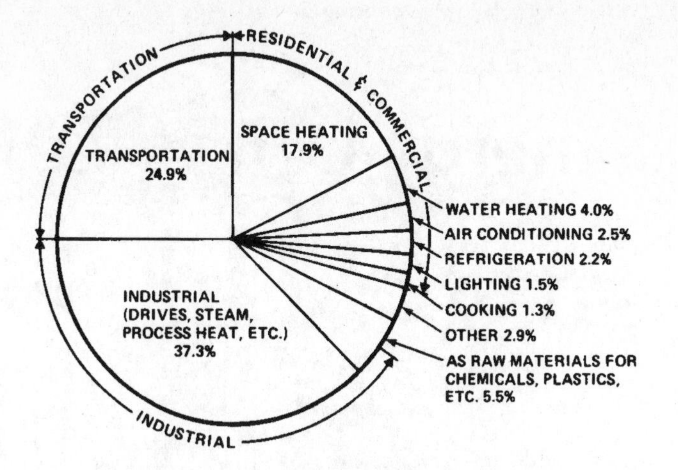

Figure 20.1 Factors contributing to the national use of energy. (Courtesy of The Dow Company)

understanding of the construction factors and other modifications to improve performance can be a valuable asset.

Reducing Exposure Losses

Figure 20.2 shows the exposure losses for a typical well-built house. The construction shown includes fiberboard sheathing, insulated doors, dual-glazed windows, R-19 ceiling insulation, and R-11 wall insulation. Variations in the size and shape of the house and its window area will alter the heat-loss distribution. Typical losses are as follows:

Exposure	*Percent of Total Heat Loss*
Frame walls	17
Ceiling	5
Basement walls	20
Basement floor	1
Windows	16
Doors	3
Air leakage	38

Some reduction in heat loss can be accomplished by adopting the options shown in Figure 20.3.

Insulation The insulating values of various types of materials are commonly given in terms of R-factors, where *R* represents the resistance of the material or materials to thermal loss—that is, the R-value is the ability of the insulation to slow the transfer of heat. Higher R-values represent more insulating ability.

Figure 20.4 shows the amount of either *batts* or *loose fill* material required to achieve various insulating values. For example, 6 in. of glass mineral fiber batts are

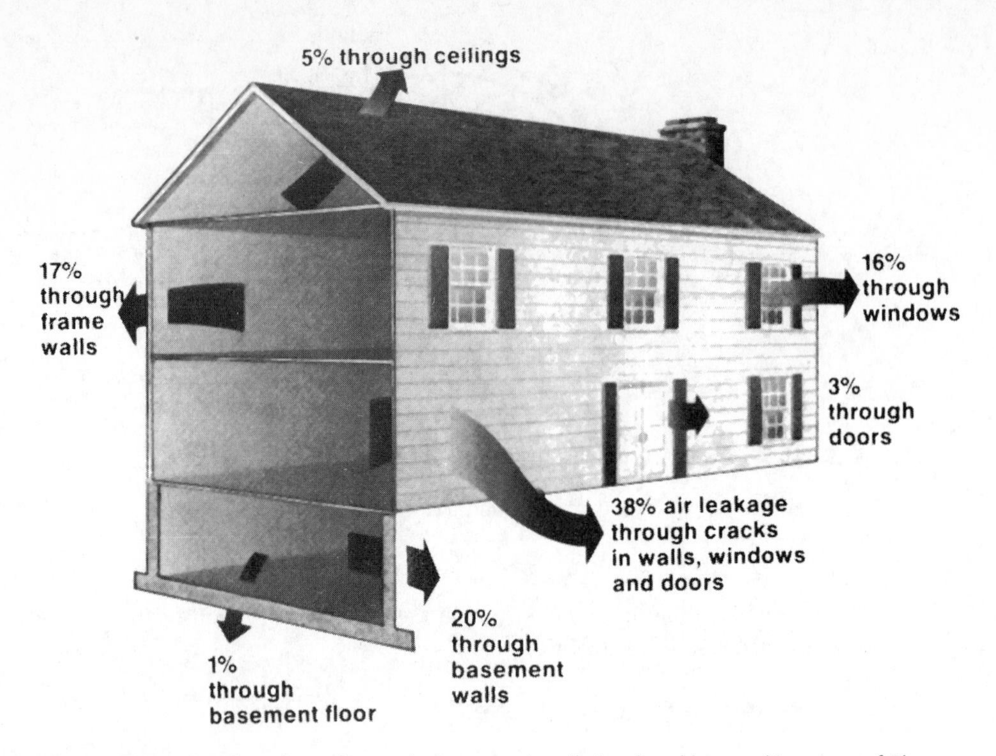

Figure 20.2 Heat loss through a typical conventionally insulated home. (Courtesy of The Dow Company)

OPTION	APPROXIMATE SAVINGS ON HOME HEAT LOSS
CEILING: RAISE INSULATION FROM R-19 TO R-30	2 %
WALL (4" STUD): RAISE INSULATION FROM R-11 TO R-13	1 %
WALL (6" STUD): RAISE INSULATION FROM R-11 TO R-19	5 %
WINDOWS: CHANGE FROM DUAL GLAZING TO TRIPPLE GLAZING	5 %
SHEATHING: SWITCH FROM CONVENTIONAL FIBERBOARD TO 1" OR R-4.41 SHEATHING AND INSTALL FROM ROOF LINE TO SILL PLATE	14 %
SHEATHING: SWITCH FROM CONVENTIONAL FIBERBOARD TO 1" OR R-5.41 SHEATHING AND INSTALL FROM ROOF LINE TO FROST LINE	24 %

Figure 20.3 Thermal improvement options. (Courtesy of The Dow Company)

BATTS OR BLANKETS

INSULATION VALUE	GLASS MINERAL FIBER	ROCK MINERAL FIBER
	INCHES	INCHES
R-11	3½	3
R-13	3⅝	3½
R-19	6	5½
R-22	6½	6
R-26	8¼	7
R-30	9½	8½
R-38	12	11

LOOSE FILL (POURED OR BLOWN)

INSULATION VALUE	** GLASS MINERAL FIBER		** ROCK MINERAL FIBER		*** CELLULOSE FIBER
	INCHES	BAGS/1000 SQ. FT.	INCHES	BAGS/1000 SQ. FT.	INCHES
R-11	5	11	3¾	23	3
R-13	6	13½	4¾	28	3½
R-19	8¾	20	6½	38	5⅛
R-22	10	22	7½	45	6
R-26	12	27	9½	56	7
R-30	13¾	30	10¼	62	8⅛
R-38	17½	40	12¾	77	10¼

**Blown mineral fiber insulation has both a depth and density relationship. R-values depend upon both number of inches and bags per 1,000 square feet of insulation.

***According to the National Cellulose Insulation Manufacturers Association, blown cellulose insulation shall have an R-value that is greater than or equal to 3.70 per inch of thickness with a density not to exceed 3 pounds per cubic foot.

Figure 20.4 R-values related to insulation thickness. (Courtesy of the U.S. Department of Energy)

needed to produce R-19 thermal resistance. Using loose fill, 8¾ in. is required to produce the same R-value. The number of bags of loose fill required per 1000 ft^2 to provide the various thicknesses is also shown. A comparison of the various types of insulation with reference to such qualities as ease of application, fire resistance, and vapor barrier requirements is given in Figure 20.5.

Where to Insulate Refer to Figure 20.6 for recommendations as to where to insulate for a more energy efficient home. Also included are the suggested amounts of insulation given in R-factors and inches. Figure 20.7 shows an example of a state energy code, in this case Michigan. Note that other insulation options are available that meet the code requirements. Always check with the local official to assure compliance with all local ordinances and code interpretations. Code requirements are shown for several cities, along with the measures that need to be followed in order to meet the code.

Cellulose insulation has been produced and installed in new and existing homes for more than 40 years. This insulation is nontoxic and noncarcinogenic, and requires no special health warning labels. Cellulose insulation conforms to the space where it is

Type	*Comments*	*Application*
BATTS/BLANKETS Preformed glass fiber or rock wool with or without vapor barrier backing	Fire resistant, moisture resistant, easy to handle for do-it-yourself installation, least expensive and most commonly available.	Unfinished attic floor, rafters, underside of floors, between studs
RIGID BOARD 1. Extruded polystyrene bead 2. Extruded polystyrene 3. Urethane 4. Glass fiber	All have high R-values for relatively small thickness. 1,2,3: Are not fire resistant, require installation by contractor with ½ in. gypsum board to ensure fire safety. 3: Is its own vapor barrier; however, when in contact with liquid water, it should have a skin to prevent degrading. 1,4: Require addition of vapor barrier. 2: Is its own barrier.	Basement walls, new construction frame walls; commonly used as an outer sheathing between siding and studs
LOOSE FILL (POURED IN) 1. Glass fiber 2. Rock wool 3. Treated cellulose fiber	All easy to install, require vapor barrier bought and applied separately. Vapor barrier may be impossible to install in existing walls. 1, 2, 3: Fire resistant, moisture resistant.	Unfinished attic floor, uninsulated existing wall
LOOSE FILL (BLOWN IN) 1. Glass fiber 2. Rock wool 3. Treated cellulose fiber	All require vapor barrier bought separately; all require space to be filled completely. Vapor barrier may be impossible to install in existing walls. 1,2, 3: Fire resistant, moisture resistant. 3: Fills up spaces most consistently. When blown into closed spaces, has slightly higher R-value.	Unfinished attic floor, finished attic floor, finished frame walls, underside of floors

Figure 20.5 Insulation materials chart. (Courtesy of the U.S. Department of Energy)

applied—even around wires, electrical boxes, and pipes. Cellulose qualifies as a Class I material and has a flame spread of less than 25 as determined by ASTM-E 119. The fire retardants in cellulose do not deteriorate, evaporate, sublime, leach out, or otherwise disappear over time. This insulation has long been regarded as superior to other fiber insulation materials in sealing the building envelope against air infiltration and sound travel. Low-dust cellulose for blown installation is available. This material produces

Where To Insulate For A More Energy-Efficient Home

CATHEDRAL CEILINGS

Product	Thickness	Use
R-38C	(10 1/4" Thick)	For 2 x 12 Joists
R-30C	(8 1/4" Thick)	For 2 x 10 Joists

ATTICS/FLAT CEILINGS

NEW CONSTRUCTION/UNINSULATED ATTICS

Product	Thickness	Use
R-38	(12" Thick)	Faced — With Vapor Barrier
R-30	(9 1/2" Thick)	Faced — With Vapor Barrier

ADDING TO EXISTING INSULATION

Product	Thickness	Use
R-30	(9 1/2" Thick)	Unfaced — No Vapor Barrier
PINKPLUS® R-25	(8 3/4" Thick)	Unfaced — Featuring MIRAFLEX™ fiber Perforated Poly-Wrap — No Vapor Barrier
R-25	(8" Thick)	Unfaced — No Vapor Barrier
R-19	(6 1/4" Thick)	Unfaced — No Vapor Barrier

EXTERIOR WALLS

Product	Thickness	Use
R-21	(5 1/2" Thick)	For 2 x 6 Walls
PINKPLUS R-19	(6 1/4" Thick)	Faced — For 2 x 6 Walls Perforated Poly-Wrap — With Vapor Barrier
R-19	(6 1/4" Thick)	For 2 x 6 Walls
R-15	(3 1/2" Thick)	For 2 x 4 Walls
PINKPLUS R-13	(3 1/2" Thick)	For 2 x 4 Walls Perforated Poly-Wrap — Kraft-Faced One Side With Vapor Barrior
R-13	(3 1/2" Thick)	For 2 x 4 Walls
FOAMULAR® R-5	(1" Thick rigid foam)	For exterior walls — No Vapor Barrier
FanFold R-1.5	(3/8" Thick rigid foam)	For exterior wall residing applications — No Vapor Barrier

Unless otherwise noted, Kraft-Faced insulation or a 4- to 6-mil polyethylene vapor barrier should be used.

INTERIOR WALLS & FLOORS/CEILINGS: SOUND CONTROL

Product	Thickness	Use
QuietZone™	(3 1/2" Thick acoustic insulation)	For 2 x 4 Walls — Kraft-Faced One Side
PINKPLUS R-13	(3 1/2" Thick)	For 2 x 4 Walls Perforated Poly-Wrap — Kraft-Faced One Side
R-13	(3 1/2" Thick)	For 2 x 4 Walls Unfaced — No Vapor Barrier
R-11	(3 1/2" Thick)	For 2 x 4 Walls Unfaced — No Vapor Barrier

FLOORS

Product	Thickness	Use
PINKPLUS R-19	(6 1/4" Thick)	Perforated Poly-Wrap — Kraft-Faced One Side — With Vapor Barrier
R-19	(6 1/4" Thick)	Faced — With Vapor Barrier

CRAWLSPACE WALLS

Product	Thickness	Use
PINKPLUS R-25	(8 3/4" Thick)	Unfaced — Featuring MIRAFLEX fiber Perforated Poly-Wrap — No Vapor Barrier
R-19	(6 1/4" Thick)	UnFaced — No Vapor Barrier

BASEMENT WALLS

Product	Thickness	Use
R-21	(5 1/2" Thick)	For 2 x 6 Walls
PINKPLUS R-19	(6 1/4" Thick)	For 2 x 6 Walls Perforated Poly-Wrap — Kraft-Faced One Side With Vapor Barrier
R-19	(6 1/4" Thick)	For 2 x 6 Walls
PINKPLUS R-13	(3 1/2" Thick)	For 2 x 4 Walls Perforated Poly-Wrap — Kraft-Faced One Side With Vapor Barrier
R-13	(3 1/2" Thick)	For 2 x 4 Walls
R-11	(3 1/2" Thick)	For 2 x 4 Walls
FOAMULAR INSULPINK™ R-7.5	(1 1/2" Thick rigid foam)	For Basement Walls No Vapor Barrier

Figure 20.6 Where to insulate for efficiency. (Courtesy of Owens Corning)

virtually no visible dust. Wet-spray cellulose, which is installed in wall cavities and covered by sheet rock, is one of the fastest growing insulation products in new construction. Cellulose insulation saves money by making homes more energy efficient.

Attic Ventilation Related to the use of attic insulation is the need for adequate attic ventilation. In summer the attic temperature can be excessive unless the attic is well ventilated. Two means are available: (1) natural ventilation and (2) the use of an exhaust fan. The selection depends on the attic construction and the feasibility of providing the necessary openings.

Michigan State Energy Code

Insulation Recommendations

CATHEDRAL CEILINGS R-30C
ATTICS / FLAT CEILINGS R-30
EXTERIOR 2 X 6 WALLS ..... R-21 OR R-19
EXTERIOR 2 X 4 WALLS ..... R-15 OR R-13
FLOORS OVER CRAWL SPACE R-19
FLOORS OVER UNHEATED SPACE .... R-19 OR R-11
BASEMENT WALLS (HEATED SPACE) R-11

Measures to Meet the Code

- Install kraft-faced R-15 (3 1/2") batts or R-13 (3 1/2") batts in 2x4 wall framing, or R-19 (6 1/4") in 2x6 framing. Unfaced batts may be used when a separate vapor retarder is applied.
- Ceilings can be insulated with R-30 (9 1/2") kraft-faced batts. Install attic ventilation of 1 square foot of vent for every 300 square feet of ceiling area. Where possible, place half the area at the eaves and the other half near the roof ridge, using a ridge vent, roof vents or gable end vents. Use eave baffles to maintain air flow at the eave.
- Cathedral ceilings can be framed in with 2x10 rafters and insulated with friction fitted R-30C (8 1/4") kraft-faced insulation. Use baffles at the eave if the rafters are notched. Make sure the clearance between the roof deck and insulation is maintained during installation. Do not inset staple the facing flanges. If required to hold the insulation in place in irregular spaced cavities, face staple the flanges. Install 1 square foot of vent for every 300 square feet of cathedral area.
- Floors can be insulated with R-19 (6 1/4") kraft-faced insulation with the facing installed adjacent to the sub-floor.
- For crawlspaces install 4- or 6-mil polyethylene on the ground. Install at least 2 vents with 1 square foot for every 1500 square feet of floor. Provide cross ventilation by placing openings on opposite ends of the space.

Code Requirements

MICHIGAN ENERGY CODE (ASHRAE 90A)
RESIDENTIAL 1, 2 FAMILY BUILDINGS REQUIRED PROVISIONS[1]

City	HDD	Wall[5]	Attic	Floor[6]	Slab R-Value[4]	
		Uo	Uo	Uo	No Heat	Heated
Bay City	6776	.176	.039	.08	5.5	7.7
Detroit	6240	.182	.042	.08	5.2	7.3
Flint	6975	.174	.038	.08	5.6	7.8
Grand Rapids	7026	.173	.038	.08	5.7	7.9
Houghton	9288	.150	.033	.08	7.2	9.6
Lansing	6337	.181	.041	.08	5.2	7.4
Traverse City	7765	.166	.034	.08	6.2	8.5

1. Always check with your local code office to assure compliance with all local ordinances and code interpretations.
2. Other insulation options to meet the code are available.
3. Wall calculations assume a window performance of U-.67 or better.
4. 2x4 wall is assumed to have an R-3 foam sheathing applied on the exterior side.
5. Windows, doors and all forms of non-opaque walls are included in the wall Uo given above.
6. Insulation shall extend downward from the top of the slab for at least 24" or downward to the bottom of the slab, then horizontally beneath the slab for a total distance of at least 24".

Figure 20.7 Michigan State Energy Code. (Courtesy of Owens Corning)

The following two methods of providing *natural ventilation* can be used. In making a selection, the important element to consider is the amount of net vent area each will provide. As noted in Figure 20.8, the recommended net vent area is dependent on the method used.

Natural Ventilation Method	*Recomended Net Vent Area*
Continuous ridge vents plus soffit vent	1.5 in.^{2}/ft.^{2} of ceiling area
Gable, roof, or turbine vents plus soffit vent	3.0 in.^{2}/ft.^{2} of ceiling area

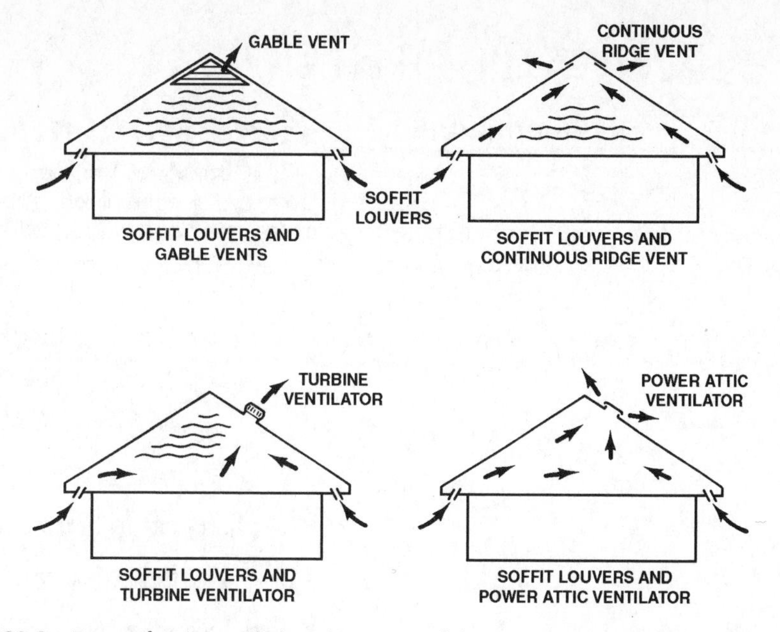

Figure 20.8 Types of attic ventilation. (Courtesy of the U.S. Department of Energy)

Powered ventilation consists of a thermostatically controlled vent placed near the peak of the roof in the center of the attic. If the house is large (over 2000 ft^2) or has a T- or L-shaped configuration, more than one is needed. The fan capacity of the ventilator should provide a minimum of 1.5 cfm per square foot of ceiling area. Soffit vents, which provide intake air, should have a minimum of 80 in. of net free area for each 100 cfm of ventilator capacity. The thermostat should be set to turn on at 100°F and off at 85°F.

Attic Ventilation Channels Installing attic ventilation channels between the rafters or trusses provides an unobstructed air channel through the insulation. Poor airflow increases moisture and reduces the insulation's efficiency. Figure 20.9 provides details related to construction and application.

Vapor Barriers To protect the insulation against condensation, a vapor barrier is recommended. An excellent example of energy-efficient wall construction is shown in Figure 20.10. Note the position of the polyethylene vapor barrier and the use of sheathing with a high R-factor.

The U-factor Thus far the discussion of insulation has centered around the R-factor, the resistance of a material to the flow of heat. A more useful factor in determining heat load is the *U-factor.* The U-factor is the reciprocal of the R-factor; thus,

$$U = \frac{1}{R}$$

Attic Ventilation Channels

Installing attic ventilation channels between the rafters or trusses provides an unobstructed air channel through the insulation. Without attic ventilation channels air cannot flow freely from the soffit to the exhaust vents. Poor air flow reduces insulation efficiencies and accelerates problems due to moisture.

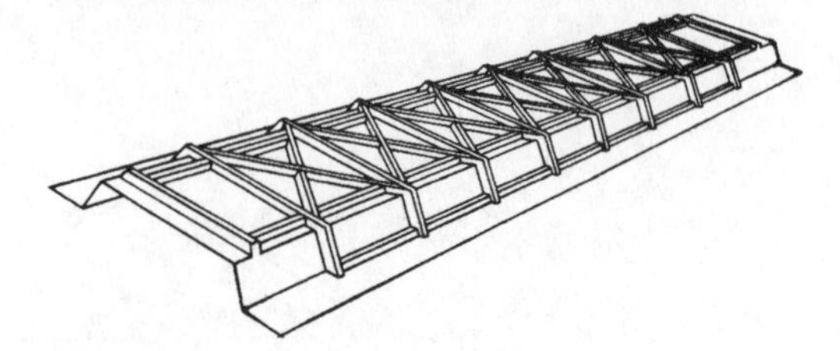

- **Full Joist Coverage for Maximum Air Flow**
- **Wide Flanges for Easy Stapling**
- **Models Designed Specifically for 16" and 24" Applications**

MODEL NO.	JOIST SPACING	WIDTH	LENGTH	AIR CHANNEL DEPTH	FREE VENT
PV1448	16" O.C.	14"	48"	1.4"	15 sq."
PV2248	24" O.C.	22"	48"	1.4"	26 sq."

ProVent Flame Spread: Class 1 Building Material
(ASTM E84-91a, ANSI 2.5, NFPA 255, UBC 42-1, UL 723).
Material of Construction: High Impact Polystyrene

- **Cooler Attic in Summer**
- **Improves Insulation Effectiveness Year Around**
- **Drier Attic in Winter**
- **Helps Prevent Ice Dam Formation**

One ventilation channel per rafter or truss cavity is recommended.

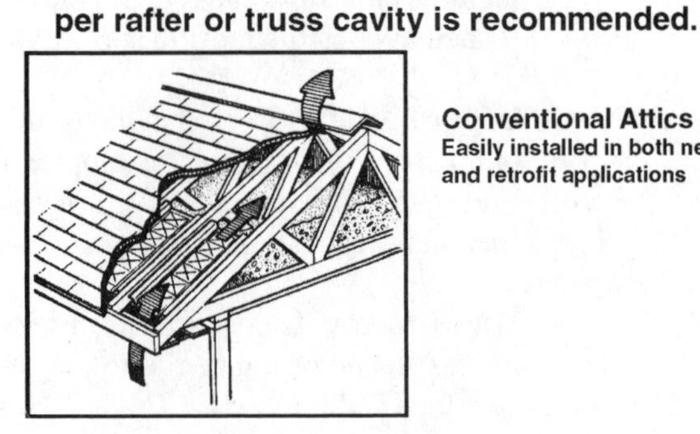

Conventional Attics
Easily installed in both new and retrofit applications

Figure 20.9 Attic ventilation channels. (Courtesy of ADO Products)

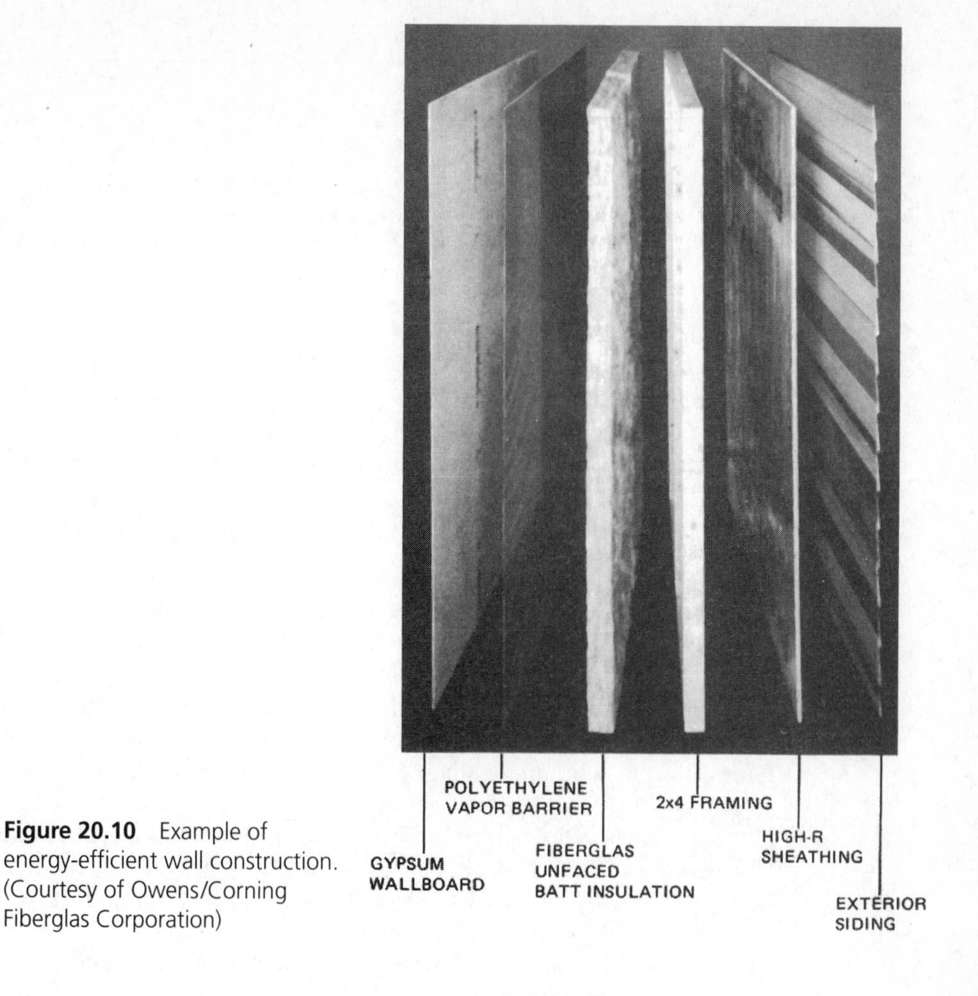

Figure 20.10 Example of energy-efficient wall construction. (Courtesy of Owens/Corning Fiberglas Corporation)

The U-factor is an overall heat-transfer coefficient. It represents the amount of heat in Btuh units that will flow through a given material or materials per square foot of surface area per degree of difference in temperature. For example, if the R-value is R-13, the U-factor is

$$U = \frac{1}{R} = \frac{1}{13} = 0.08 \text{ Btuh/ft}^2/^\circ\text{F}$$

For an exposed wall, the R-factor is the sum of all the R-factors of the individual components in the wall. The following tabulation illustrates the determination of the R-factor for a wall composed of various components. Note that this wall has a total R-factor of 15.18. This amount represents the resistance of the wall to the flow of heat through it.

Material or Surface	*R-Factor*
Exterior surface resistance	0.17
Bevel-lapped siding, ½ × 8 in.	0.81
Fiberboard insulating sheathing, ¾ in.	2.10
Insulation, mineral wool batts, 3½ in.	11.00
⅜-in. gypsum lath and ⅜-in. plaster	0.42
Interior surface resistance	0.68
Total R-factor	15.18

In most northern climates it is good practice to construct walls with an R-factor of 15.0 or greater, and ceilings with a factor of 25.0 or greater. Windows should be constructed with double glass, or single glass with a storm window.

The corresponding U-factor, or heat-transfer coefficient, in the preceding example, where the R-factor for the exposed wall is 15.18, is determined as follows:

$$U = \frac{1}{R} = \frac{1}{15.18} = 0.07 \text{ Btuh/ft}^2/^\circ\text{F}$$

With this information, the effectiveness of adding insulation can be evaluated in terms of heat savings.

Energy-Efficient Windows Windows with built-in thermal barriers reduce the heat loss through the glass and the framing, as shown in Figure 20.11. Note that the features listed are in groups dealing with the frame, the sash, and the glazing.

Figure 20.12 describes the selection of efficient windows for homes in the central climates. Because the climate varies in different areas, the central climate zone and Chicago have been chosen as an example. In Figure 20.12, the benefits of high-performance windows are discussed. The Energy Star and the NFRC label are also explained. Point 3 suggests that annual energy costs for a typical house in the area be compared. Point 4 suggests the use of a computer program such as RESFEN to estimate and compare energy costs. RESFEN is a computer program for calculating the annual heating and cooling energy use and costs due to fenestration systems. RESFEN also calculates their contribution to peak heating and cooling loads. It is available at no charge on the World Wide Web at *windows.lbl.gov/software*. This program takes into account the climate zone, house design, and utility rates.

Figure 20.13 provides a set of guidelines for comparing window properties in the central climate zone. This zone is a mixed climate requiring both heating and cooling. Discussed are the U-factor, solar heat gain coefficient (SHGC), visible transmittance (VT), and air leakage (AL).

Figure 20.14 deals with the comparison of window performance in Chicago, Illinois. Six different glazing care studies are given, with a description of the glazing and its effect on the U-factor, SHGC, and VT. From these data, the annual energy cost has been calculated for heating and cooling.

This information was prepared by John Carmody and Kerry Haglund, University of Minnesota, and Dariush Arasteh, Lawrence Berkley National Laboratory, and is

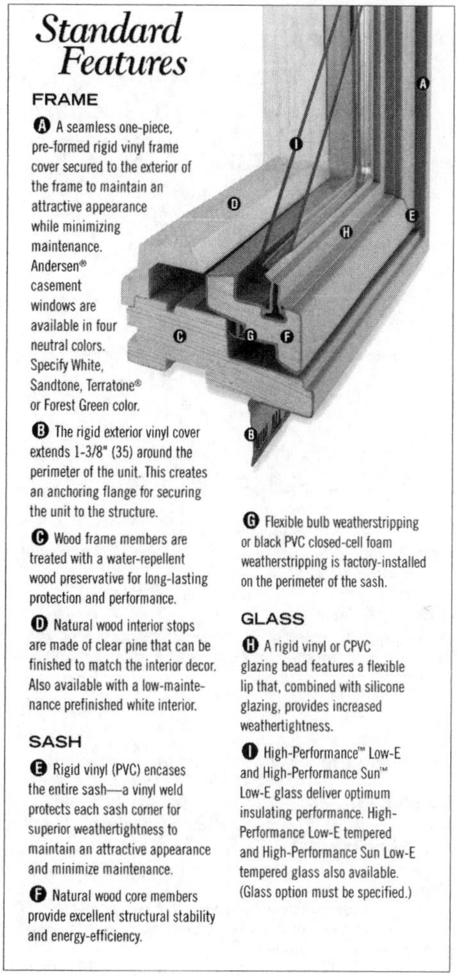

Standard Features

FRAME

(A) A seamless one-piece, pre-formed rigid vinyl frame cover secured to the exterior of the frame to maintain an attractive appearance while minimizing maintenance. Andersen® casement windows are available in four neutral colors. Specify White, Sandtone, Terratone® or Forest Green color.

(B) The rigid exterior vinyl cover extends 1-3/8" (35) around the perimeter of the unit. This creates an anchoring flange for securing the unit to the structure.

(C) Wood frame members are treated with a water-repellent wood preservative for long-lasting protection and performance.

(D) Natural wood interior stops are made of clear pine that can be finished to match the interior decor. Also available with a low-maintenance prefinished white interior.

SASH

(E) Rigid vinyl (PVC) encases the entire sash—a vinyl weld protects each sash corner for superior weathertightness to maintain an attractive appearance and minimize maintenance.

(F) Natural wood core members provide excellent structural stability and energy-efficiency.

(G) Flexible bulb weatherstripping or black PVC closed-cell foam weatherstripping is factory-installed on the perimeter of the sash.

GLASS

(H) A rigid vinyl or CPVC glazing bead features a flexible lip that, combined with silicone glazing, provides increased weathertightness.

(I) High-Performance™ Low-E and High-Performance Sun™ Low-E glass deliver optimum insulating performance. High-Performance Low-E tempered and High-Performance Sun Low-E tempered glass also available. (Glass option must be specified.)

Figure 20.11 Construction of a high-performance window. (Courtesy of Andersen Windows, Inc.)

SELECTING EFFICIENT WINDOWS FOR HOMES IN THE CENTRAL CLIMATES

In recent years, low-E coatings and other improvements have revolutionized window performance. This fact sheet is intended to help you select energy efficient windows in a new home or replace windows in an existing home.

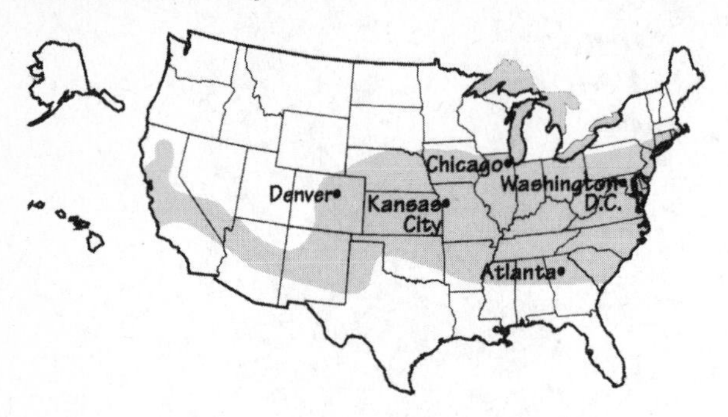

Central Climate Zone — Heating and Cooling

Benefits of High Performance Windows

Cooling and Heating Season Savings
Low-E coatings and other technologies can significantly reduce winter heat loss and summer heat gain.

Improved Daylight and View
New glazings with spectrally selective low-E coatings can reduce solar heat gain significantly with a minimal loss of visible light (compared to older tints and films).

Improved Comfort
In both summer and winter occupant comfort is increased; window temperatures are more moderate and there are fewer cold drafts.

Reduced Condensation
Frame and glazing materials that resist heat conduction do not become cold and result in less condensation.

Reduced Fading
Coatings on glass or plastic films within the window assembly can significantly reduce the ultraviolet (UV) and other solar radiation which causes fading of fabrics and furnishings.

Lower Mechanical Equipment Costs
Using windows that significantly reduce solar heat gain means that cooling equipment costs may be reduced.

Look for the Energy Star

The Department of Energy (DOE) and the Environmental Protection Agency (EPA) have developed an Energy Star designation for products meeting certain energy performance criteria.

Look for an Energy Star for the Central Zone (heating and cooling).

Look for the NFRC Label

The National Fenestration Rating Council (NFRC) has developed a window energy rating system based on whole product performance. The NFRC label provides the only reliable way to determine the window properties and to compare products.

For recommended window properties in this region, see the following page.

Compare Annual Energy Costs for a Typical House in Your Region

The annual energy use from computer simulations for a typical 2000-square-foot house in your region can be compared for different window options. See the following pages for examples.

Estimate and Compare Annual Energy Costs for Your House

Using a computer program such as RESFEN to compare window options is the only method of obtaining reasonable estimates of the heating and cooling costs for your climate, house design, and utility rates. See the next page for information on RESFEN.

Figure 20.12 Selecting efficient windows. (Courtesy of the Efficient Windows Collaborative, a project of the Alliance to Save Energy)

Guidelines for Comparing Window Properties in the Central Climate Zone

The Central Climate Zone is mainly a mixed climate requiring both heating and cooling.

U-FACTOR

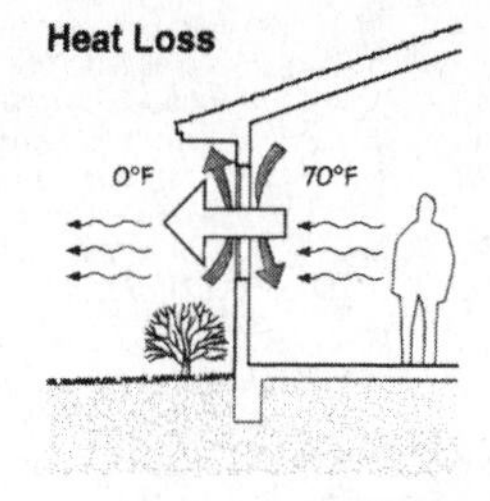

The rate of heat loss is indicated in terms of the U-factor (U-value) of a window assembly. The insulating value is indicated by the R-value which is the inverse of the U-value. The lower the U-factor, the greater a window's resistance to heat flow and the better its insulating value.

Recommendation: Select windows with a U-factor of 0.40 or less. The larger your heating bill, the more important a low U-factor becomes.

SOLAR HEAT GAIN COEFFICIENT (SHGC)

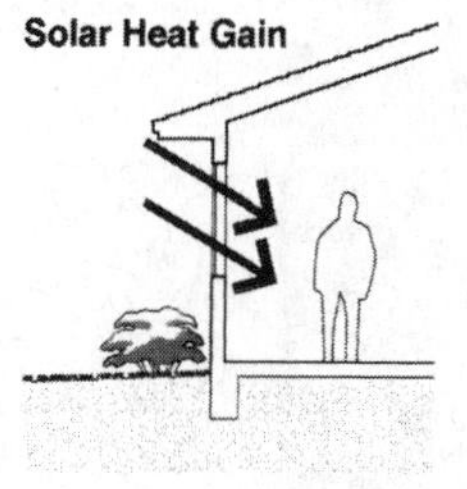

The SHGC is the fraction of incident solar radiation admitted through a window. SHGC is expressed as a number between 0 and 1. The lower a window's solar heat gain coefficient, the less solar heat it transmits.

Recommendation: If you have significant air conditioning costs or summer overheating problems, look for SHGC values of 0.40 or less. If you have moderate air conditioning requirements, select windows with a SHGC of 0.55 or less. While windows with lower SHGC values reduce summer cooling and overheating, they also reduce free winter solar heat gain. Use a computer program such as RESFEN to understand heating and cooling trade-offs.

VISIBLE TRANSMITTANCE (VT_{glass} and VT_{window})

The visible transmittance (VT) is an optical property that indicates the amount of visible light transmitted. VT is expressed as a number between 0 and 1. The higher the VT, the more daylight is transmitted. A high VT is desirable to maximize daylight.

Recommendation: A window with VT*glass* above 0.70 (for the glass only) is desirable to maximize daylight and view. This translates into a VT*window* above 0.50 (for the total window including a wood or vinyl frame).

AIR LEAKAGE (AL)

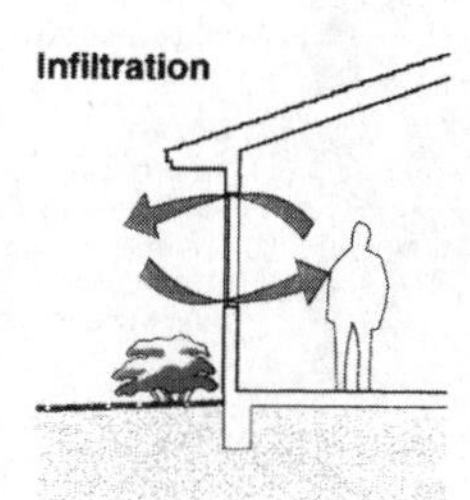

Heat loss and gain occur by infiltration through cracks in the window assembly. AL is expressed in cubic feet of air passing through a square foot of window area. The lower the AL, the less air will pass through cracks in the window assembly. While many think that AL is extremely important, it is not as important as U-factor and SHGC.

Recommendation: Select a window with an AL of 0.30 or below (units are cfm/sq ft).

Figure 20.13 Guidelines for comparing window properties. (Courtesy of the Efficient Windows Collaborative, a project of the Alliance to Save Energy)

Comparing Window Performance in Chicago, Illinois

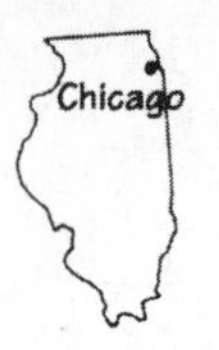

The annual energy performance figures shown here were generated using RESFEN for a typical 2000 sq. ft. house with 300 sq. ft. of window area (15% of floor area). The windows are equally distributed on all four sides of the house and include typical shading (interior shades, overhangs, trees and neighboring buildings). The heating system is a gas furnace with air conditioning for cooling. The figures below are based on typical energy costs for this location (natural gas, $0.50/therm and electricity, $0.122/kWh). U-factor, SHGC, and VT are for the total window including frame.

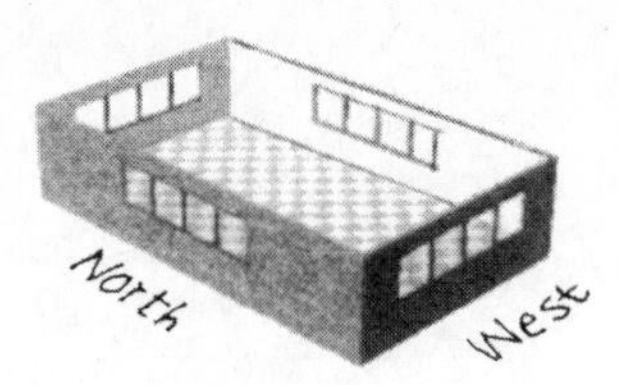

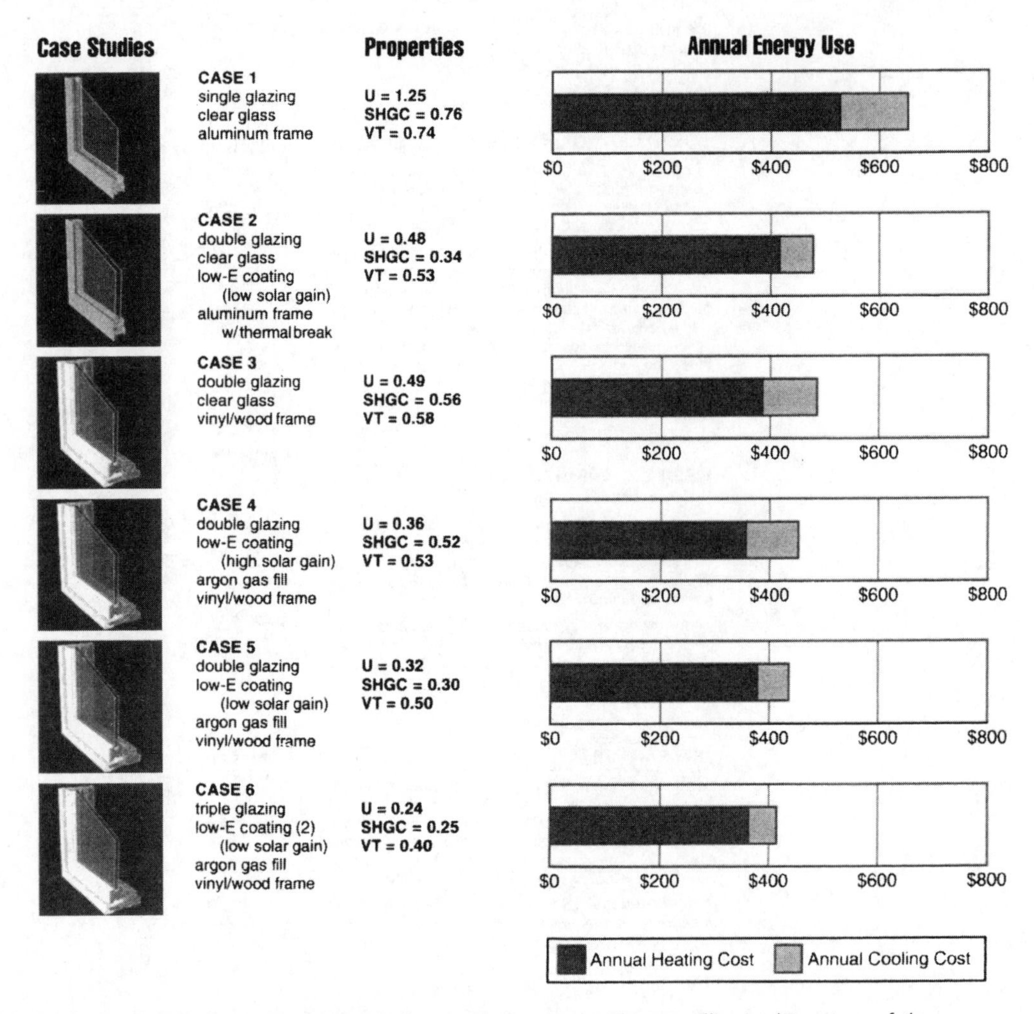

Figure 20.14 Comparing window performance in Chicago, Illinois. (Courtesy of the Efficient Windows Collaborative, a project of the Alliance to Save Energy)

provided by the Efficient Windows Collaborative, a project of the Alliance to Save Energy. More information can be obtained at *www.efficientwindows.org*.

Energy-Efficient Doors Energy-efficient doors are available. These are doors that incorporate either polystyrene or polyurethane insulation and a thermal barrier to reduce heat loss through the framing. One door manufacturer's rating chart gives the overall R-factor for various types of door construction (Figure 20.15). With the use of a full polyurethane core, the R-factor is 15.15, which is almost equal to the wall factor of 15.18 calculated for the exposed wall in the preceding example. With such a door a storm door is virtually unnecessary. In fact, a storm door might be counterproductive because of the tendency to leave the primary door open at times to avoid the inconvenience of opening and closing two doors.

Weatherstripping Weatherstripping and caulking around doors, windows, and other wall penetrations stop the entry of outside air into a building. Even if no noticeable drafts are present, air movement can be occurring. A typical 3-ft-wide door with a ⅛-in. separation at the threshold may not produce a noticeable draft, but, nonetheless, 4½ in.^{2} of open area exists. This is equivalent to a 1³⁄₁₆-in.-diameter hole (see Figure 20.16).

Reducing Air Leakage

There are many ways to tighten a structure to reduce air infiltration. It is first esssential to find out where leakage may occur. There are two main sources of leakage: (1) through building materials and (2) at intersections where two building

	R factor
Solid core wood door	2.90*
Stile and rail wood door	2.79*
Hollow core wood door	2.18*
Steel door with polystyrene core	7.14
Steel door with ¾" urethane "honeycomb" core	8.42
Storm door (aluminum)	1.84**
THERMA-TRU door (with a full polyurethane core door)	15.15

**R-factor reference according to ASHRAE (American Society of Heating, Refrigeration and Air Conditioning Engineers) figures.*
***Based on engineering calculations.*

Figure 20.15 R-factors for various types of exterior doors. (Courtesy of Pease Company, Ever-Strait Division)

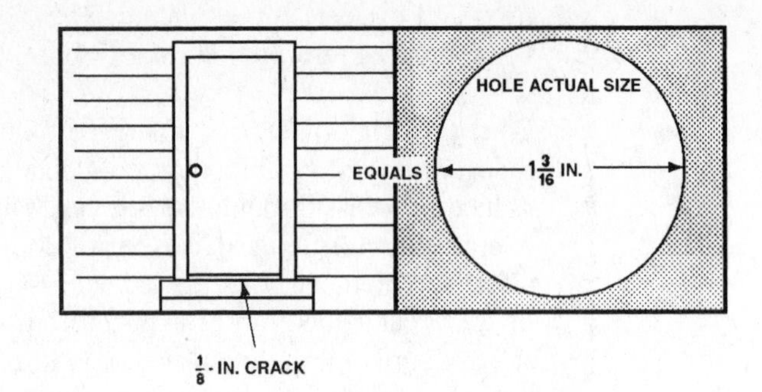

Figure 20.16 Door clearance of ⅛ in. is equivalent to area of 1³⁄₁₆-in. hole.

components meet. It is a simple matter to install a vapor barrier under plasterboard when a new house is constructed. The outside of the framed in house can also be wrapped with a nonperforated continuous fiber, such as Tyvek (Figure 20.17). Possible air and water leakage sources are listed, as are possible interior moisture sources. The house wrap material has microscopic pores that are large enough for water vapor to pass through, yet small enough to resist air and water penetration.

Some examples of intersections are floors with walls, walls with ceilings, chimneys with floors and walls, and penetrations in the wall, floor, or ceiling to accommodate fixtures for plumbing, electrical switches and outlets, or light switches. Air leakage can also occur through bypasses from the basement and the house interior to the attic. These bypasses are most often found in wall cavities and around chimneys, vents, and flue pipes. Wherever a leak is found, it must be eliminated.

Determine the leakage sites. Inspect the house to determine remedial measures to tighten the construction. Some retrofit measures that can be used include the following:

1. Seal holes around pipes and wire.
2. Seal cracks around beams in open beam ceilings.
3. Seal cracks around heating registers and exhaust fan cutouts.
4. Seal cracks around edges of fireplaces and mantelpieces.
5. Seal cracks inside and around built-in cabinets and bookshelves.
6. Seal holes in recessed fluorescent light fixtures.
7. Seal gaps beneath baseboards and molding and behind electric baseboard heaters.
8. Install a fireplace seal-tight damper in the chimney flue.
9. Weatherstrip and caulk leaky windows and doors and seal cracks around the window and door frames.

Air-to-air Heat Exchangers Cutting down infiltration and reducing air leakage can create nearly airtight homes. This can cause problems unless some form of ventilation is provided. Some problems caused by airtight houses are shown in Figure 20.18.

The air-to-air heat exchanger is a device for recovering heat from exhaust air and using it to heat air supplied from the outside for ventilation. Figure 20.19 shows how

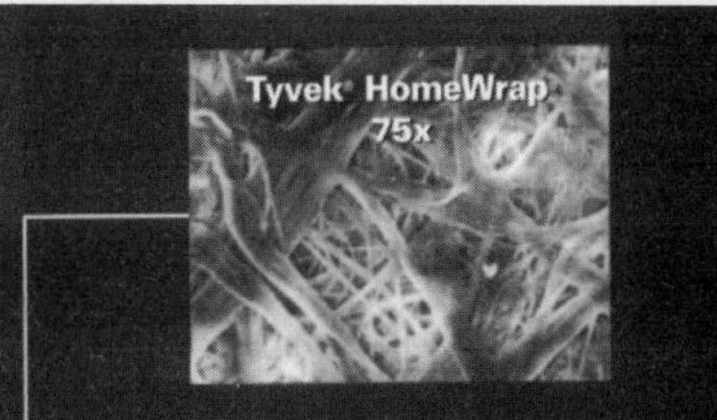

The technology behind Tyvek® has no equal:

DuPont Tyvek® is a non-perforated continuous microfiber web with microscopic pores large enough for moisture vapor to pass through, yet small enough to resist air and liquid water penetration. It's micropores are 1 to 10 microns, smaller than a human hair.

Perforated housewraps rely on mechanical pin-perforations to provide vapor permeability. With perforated membranes there are not enough pin holes to allow moisture to escape properly, yet the mechanical holes are large enough to let bulk water in. Holes are about 500 to 600 microns.

Balance of Performance:	Tyvek® HomeWrap®
Air Penetration Resistance air-ins test [cfm/ft² @ 75 PA]	**HIGHEST** air penetration protection [.007]
Bulk Water Holdout (AATCC-127) hydrostatic head [cm]	**HIGHEST** water holdout [210]
Moisture Vapor Permeance (ASTM E96) vapor transmission [perms]	**HIGH** breathability, lets moisture escape [58]
Durability	**EXCELLENT** tear strength, good wet strength

Start with the assumption that water and moisture always will penetrate walls. Whether from inside sources or from outside weather conditions, water and moisture inevitably find ways to get inside walls.

A "forgiving" wall is one that is able to manage water and moisture, increasing the capacity of a wall to dry out before damage can occur.

Possible Air Leakage:

1. Electrical, plumbing penetrations through top and bottom wall plates, both inside and outside.

2. Gaps between sheathing.

3. Interior partition walls that connect to floor or attic.

4. Joints between sole plate and subfloor along exterior walls.

5. Gaps in drywall and top and bottom wall plates.

6. Uninsulated walls near unheated spaces like the garage or crawl spaces.

7. Wherever two framing members meet, forming a crack.

8. Around windows and doors.

Possible Water Leakage:

1. Poorly fitted and flashed doors and windows.

2. Inadequate overhangs.

3. Wind-driven rain forced through any crack or crevice, by higher outside pressure to lower inside pressure.

4. Plumbing or electrical protrusions not properly sealed.

Interior Moisture Sources:

1. Showers and baths.

2. Dishwashers.

3. Improperly vented exhaust fans.

4. Green or wet framing lumber.

5. Cooking.

6. Plants.

• Wind easily penetrates the more than 2000 feet of cracks and joints in an average size home.

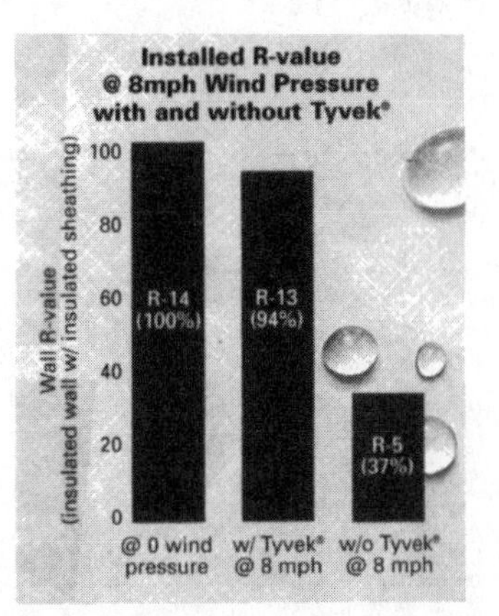

An 8 mph wind, (average across the U.S.) on an unwrapped wall system with a 1/4-inch fanfold loses almost 50% of its installed R-value.

Figure 20.17 Performance of Tyvek® HomeWrap®. (Courtesy of Tyvek Corporation)

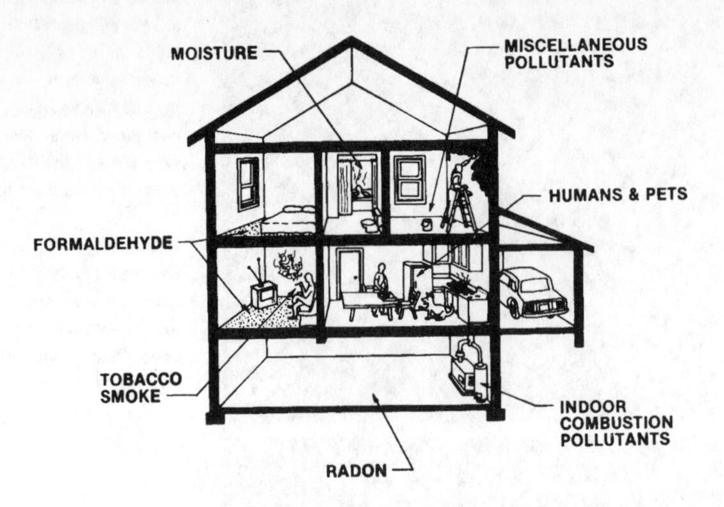

Figure 20.18 Sources of indoor pollution. (Courtesy of the U.S. Department of Energy)

this heat-recovery unit operates. In winter, fresh, dry, cold air is forced through the multipath heat exchanger core, and stale, warm indoor air is forced through in the opposite direction. The fresh-air temperature rises 80% or more of the indoor–outdoor temperature difference. The rate of airflow can be regulated by the homeowner to meet individual needs. During the colder winter months, some moisture in the exhaust air can freeze when subjected to low outside temperatures. The unit includes automatic defrost to ensure proper operation during severe winter conditions. A humidistat can be provided to switch the fan speed to control indoor humidity levels.

Figure 20.20 illustrates the application of an air-to-air heat-exchanger system in a typical home. The best mounting location is usually the basement adjacent to other

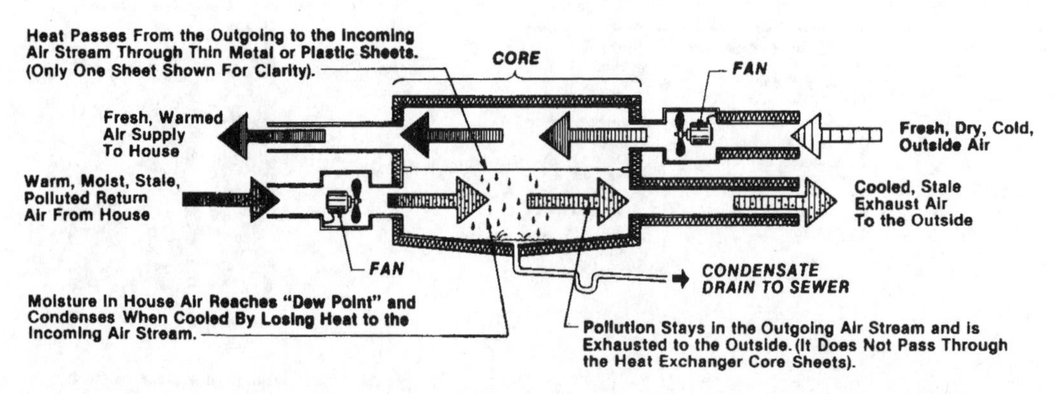

Figure 20.19 Operation of air-to-air exchanger. (Courtesy of the U.S. Department of Energy)

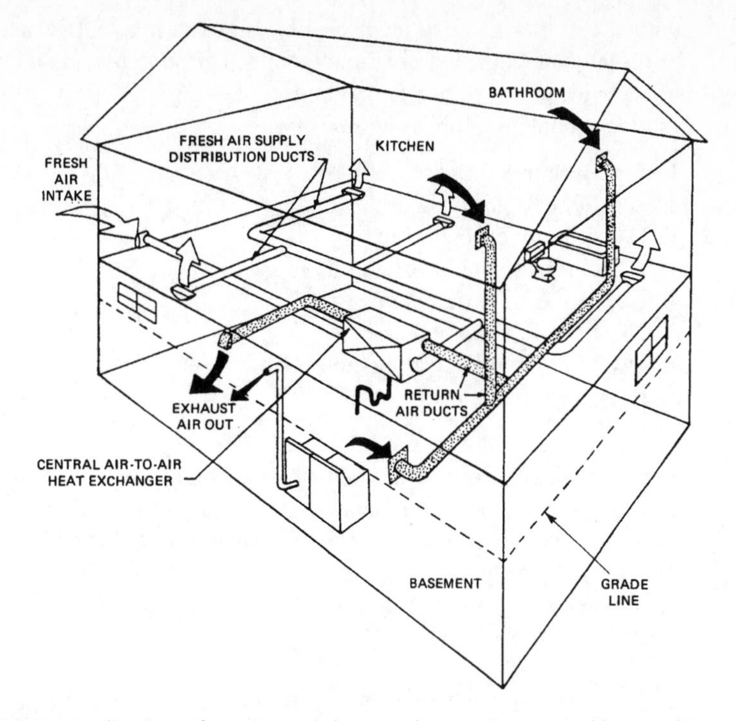

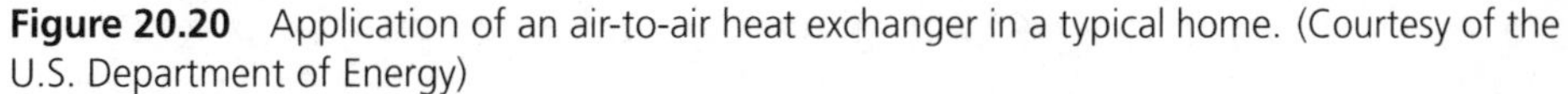

Figure 20.20 Application of an air-to-air heat exchanger in a typical home. (Courtesy of the U.S. Department of Energy)

mechanical equipment, or in the laundry area. Select a location with the best relationship to the ductwork and as close as possible to the outside wall. A provision for a condensate drain will have to be made, a plumbing stack, fixture, or floor drain should be handy. The unit may be mounted between or below the floor joists, on wall brackets, or on the floor.

Costs can be substantially reduced when installing a central, ducted system in new construction or retrofitting an existing system if the proper planning takes place. Exchanger manufacturers should be consulted in the design phase of the project to facilitate this planning process. Ductwork layout and design must be planned for, and many problems can be avoided if the system is well integrated into the house design.

Using Other Retrofit Measures

There is a continuing effort among manufacturers to offer new products to reduce energy consumption. Some of these products fit specific requirements, whereas others have more general applications. In the following section various measures that have unique features are described. The list is not complete, because of the dynamic nature of research and development in this field. The student should be constantly on the alert for new ideas and products that conserve energy. The topics that are discussed in this section include the following:

1. Fireplace treatment
2. Use of heat-pump water heaters
3. Use of more efficient lighting
4. Alterations to the heating unit
5. Use of programmed thermostats
6. Use of solar heating
7. Use of alternative heating equipment
8. Improved maintenance

Fireplace Treatment The open fireplace can be a source of considerable heat loss. To prevent this, it is recommended that wood-burning fireplaces be installed with separate outside air intake vents for combustion air. This provision replaces the normal drawing of already heated air from within the house. Glass doors on the unit also channel radiant heat into the room while preventing warm air from escaping up the chimney, as shown in Figure 20.21.

Use of Heat Pumps for Hot Water Heat-recovery units that use rejected energy for heating water are now available on many residential cooling and heat-pump units.

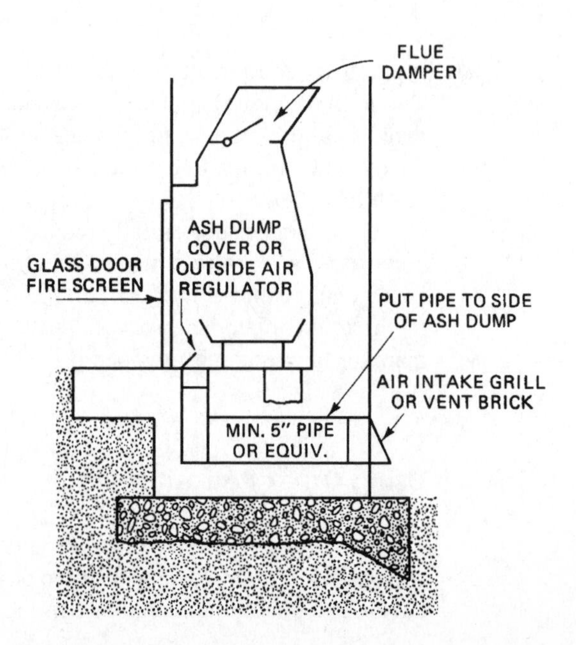

Figure 20.21 Example of an energy-efficient fireplace.

These units are valuable for supplementing the conventional water heater. Complete-package heat-pump water heaters are also available. They operate year-round in the heating mode to supplement the standard gas or electric units (see Chapter 24 for information on heat pumps).

Use of More Efficient Lighting Energy-efficient fluorescent lighting is recommended whenever appropriate. Fluorescent lights produce from 55 to 92 lumens of light per watt, compared with 15 to 25 lumens per watt for incandescent light. A *lumen* is a measure of the flow of light. Fluorescent lights have a longer life than incandescent bulbs. The average life of a fluorescent bulb is around 20,000 hours compared with less than 1000 hours for incandescent bulbs.

Alterations to the Heating Unit Alterations to the heating unit to improve efficiency were discussed in earlier chapters and should be consulted for complete information. Important measures to consider for retrofitting include:

1. Replacing the unit with a high-efficiency furnace (see Chapter 12)
2. Sizing the furnace to fit the load (see Chapter 16)
3. Installing electronic ignition (see Chapter 11)

Use of Programmable Thermostats The use of an automatic temperature setback thermostat is one example of a system modification that can be made (Figure 20.22). A 5 to 10% savings in fuel can be realized by lowering the temperature in the space 5°F for 8 hours each night. A savings of 9 to 16% can be realized by lowering the space temperature 10°F for 8 hours at night. Tables showing night setback savings for various parts of the country are shown in Figure 20.23.

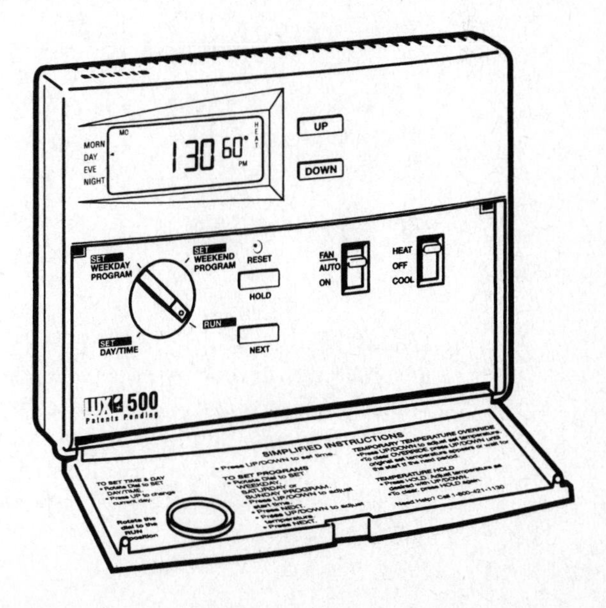

Figure 20.22 Seven-day programmable thermostat. (Courtesy of LUX Products Corporation)

City	①	②	③	④
ALBUQUERQUE, NM	12%	24%	16%	16%
ATLANTA, GA	15%	27%	12%	19%
ATLANTIC CITY, NJ	12%	23%	13%	20%
BILLINGS, MT	10%	20%	9%	16%
BIRMINGHAM, AL	15%	28%	12%	17%
BOISE, ID	11%	22%	8%	15%
BOSTON, MA	11%	22%	13%	20%
BUFFALO, NY	10%	20%	14%	22%
BURLINGTON, VT	9%	18%	14%	22%
CHARLESTON, SC	16%	29%	13%	19%
CHEYENNE, WY	10%	19%	12%	17%
CHICAGO, IL	11%	21%	13%	20%
CINCINNATI, OH	12%	24%	12%	19%
CLEVELAND, OH	10%	21%	13%	21%
COLORADO SPRINGS, CO	11%	22%	11%	16%
COLUMBUS, OH	11%	22%	12%	19%
CORPUS CHRISTI, TX	17%	30%	10%	15%
DALLAS, TX	15%	28%	9%	14%
DENVER, CO	11%	22%	10%	17%
DES MOINES, IA	11%	20%	12%	19%
DETROIT, MI	11%	21%	13%	22%
DODGE CITY, KS	12%	23%	9%	15%
GREENSBORO, NC	14%	25%	12%	19%
HOUSTON, TX	16%	30%	9%	14%
INDIANAPOLIS, IN	11%	22%	12%	19%
JACKSON, MS	16%	30%	%	17%
JACKSONVILLE, FL	17%	30%	%	17%
KANSAS CITY, MO	12%	23%	10%	16%
LAS VEGAS, NV	15%	27%	%	11%
LITTLE ROCK, AR	15%	27%	10%	16%
LOS ANGELES, CA	15%	30%	20%	27%
LOUISVILLE, KY	13%	24%	%	18%
MADISON, WI	10%	19%	%	19%
MEMPHIS, TN	15%	26%	11%	17%
MIAMI, FL	18%	30%	11%	17%
MILWAUKEE, WI	10%	19%	13%	19%
MINNEAPOLIS, MN	9%	18%	12%	20%
NEW ORLEANS, LA	16%	30%	11%	17%
NEW YORK, NY	12%	23%	13%	20%
OKLAHOMA CITY, OK	14%	26%	11%	16%
OMAHA, NE	11%	20%	12%	19%
PHILADELPHIA, PA	11%	24%	13%	20%
PHOENIX, AZ	16%	30%	%	11%
PITTSBURGH, PA	11%	22%	13%	20%
PORTLAND, ME	10%	19%	15%	21%
PORTLAND, OR	13%	24%	11%	20%
PROVIDENCE, RI	11%	21%	16%	24%
ROANOKE, VA	12%	24%	12%	19%
SALT LAKE CITY, UT	11%	21%	10%	16%
SAN DIEGO, CA	16%	30%	25%	33%
SAN FRANCISCO, CA	14%	26%	14%	19%
SEATTLE, WA	12%	24%	16%	23%
SIOUX FALLS, SD	10%	19%	11%	18%
SPOKANE, WA	11%	20%	10%	18%
SPRINGFIELD, MA	11%	20%	13%	20%
ST. LOUIS, MO	12%	23%	11%	18%
SYRACUSE, NY	11%	20%	13%	21%
WASHINGTON, DC	13%	25%	13%	20%
WILMINGTON, DE	12%	23%	10%	20%

① 10° * SINGLE HEATING SETBACK 70° TO 60° 8 HOURS/DAY

② 10° * DOUBLE HEATING SETBACK 70° TO 60° 8 HOURS/DAY, TWICE/DAY

③ 5° SINGLE COOLING SETUP 75° TO 80° 11 HOURS/DAY

④ 5° DOUBLE COOLING SETUP 75° TO 80° 9 HOURS/DAY, 7 HOURS/NIGHT

*SAVINGS FOR A 5° HEATING SETBACK ARE AT LEAST $\frac{1}{2}$ OF SAVINGS FOR A 10° SETBACK.

ACTUAL SAVINGS DEPEND ON YOUR HOME, GEOGRAPHIC LOCATION, NUMBER OF SETBACKS AND AMOUNT OF SETBACK DEGREES.

Figure 20.23 Percentage of heating/cooling energy savings with thermostat setback in average home. (Courtesy of Honeywell Inc.)

Use of Solar Heating In the 1970s there was a tremendous push in a great variety of areas to solve the energy crisis. Federal legislation provided an income tax credit for money spent on qualifying energy conservation expenditures.

The maximum allowance was $300, based on a 15% credit on the first $2000 of expenditures. Eligible items included insulation, thermal windows and doors, more efficient replacement burners, non-pilot-light ignition systems, and automatic setback thermostats.

Federal tax credit was also provided for the installation of renewable energy source systems up to a maximum of $2200, based on 30% of the first $2000, and 20% of the next $8000 of expenses. Approved items included active and passive solar systems, and wind and geothermal devices.

Presently, ways of reducing energy costs are again being evaluated, and solar heating could well provide some answers.

Solar heating is used by homeowners principally for:

1. Heating space
2. Heating domestic water
3. Heating swimming pool water

From an economic standpoint, domestic hot-water solar heating produces the greatest return for the average homeowner, since hot water is needed year-round. This type of heating permits utilization of the summer solar radiation, which is greater than that available in winter for most parts of the country. Figure 20.24 shows and describes what the manufacturer calls a user-friendly, do-it-yourself solar water heating system. This particular package consists of a series of preassembled components ready to be connected to the collector and water tank. The heat-exchanger package (Figure 20.25) is not affected if the collector or tank needs to be repaired or replaced.

Figure 20.26 shows the heat exchanger and typical installations of the passive-phase-change solar system. This system relies on the change of refrigerant from the liquid to vapor to liquid phases. The system is quiet and maintenance free, if properly installed by a qualified refrigeration service contractor. In the passive system, the heat exchanger must always be installed higher than the collector in order that the condensed liquid refrigerant can return to the collector by gravity.

As the sun's rays fall on the collector plate, phase change takes place in the collector, since the liquid refrigerant immediately boils and changes to a vapor. The hot vapor moves by its own vapor pressure to the heat exchanger, where the reverse phase change takes place. The hot vapor gives up its heat to the domestic water, condenses, and returns to to the bottom of the collector by gravity.

Potable domestic water is circulated through the secondary circuit from the finned copper coil in the heat exchanger to the water storage tank. A differential control causes a circulating pump to start when the temperature of the vapor in the heat exchanger is 5° above the water storage tank temperature. The pump will run only when it is economical to store heat.

Under favorable conditions of sunlight, two 18-ft^2 collector modules will heat approximately 60 gallons of water per collector per day from 65°F to 120°F or even to 150°F. A larger model of the passive heat exchanger provides for heating water for two different purposes from one phase-change collector source. It is particularly intended for heating domestic water and a hot tub at the same time.

Collectors and heat exchangers are installed only in matched groups and are UL recognized or UL listed. All refrigeration and other codes must be adhered to.

SOLAR WATER HEATING SYSTEM

FOR THE HANDY HOMEOWNER

- **Easy Installation** – completely pre-assembled pump and heat exchanger package and collectors.
- **Easy Maintenance** – built of many standard, readily available components for ease of replacement by the homeowner.
- **Low Cost** – unlike many systems, the heat exchanger package is not built onto or into the storage tank. Therefore the storage tank may be readily replaced years later at a minimum cost without replacing other parts of the solar system.
- **Easy Mounting** – the collector yard mount #9354 eliminates roof mounting and provides more convenient access to collectors.

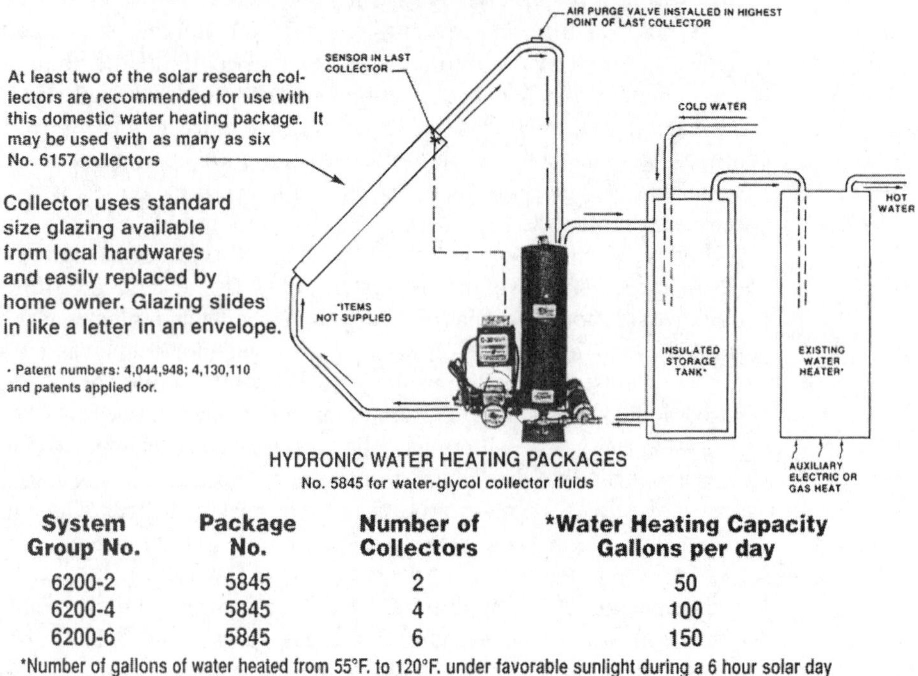

System Group No.	Package No.	Number of Collectors	*Water Heating Capacity Gallons per day
6200-2	5845	2	50
6200-4	5845	4	100
6200-6	5845	6	150

*Number of gallons of water heated from 55°F. to 120°F. under favorable sunlight during a 6 hour solar day

Figure 20.24 User-friendly solar water-heating system. (Courtesy of Solar Research)

In places where the heat exchanger cannot be installed above the solar collector, the active-phase-change solar sytem is recommended (Figure 20.27). The phase-change operation is identical to that of the passive unit; however, the active system differs in that the condensed liquid is returned to the collectors by a pump, which is controlled by an electric float. Power used by the return pump is very small, since only

SOLAR HYDRONIC WATER HEATING PACKAGE

Efficient-Proven-Simplified Installation

This unique, closed system, solar water heating package consists of 8 components all pre-assembled and tested at the factory so that there is a minimum amount of hook up work required in the field.

The plumbing contractor, or handy home owner, connects the package to the collector (not included) and to a readily available standard insulated storage tank (not included).

Electrical wiring of control and pumps has been done at the factory. Therefore, it is only necessary to extend a sensor wire to the collector and plug in the wall cord of the water heating package.

Since the package is not built onto or into the storage tank, the storage tank may be readily replaced, when required, at a minimum cost without replacing other parts of the system.

Although primarily intended to serve as a booster for the existing hot water heater it can supply nearly all the hot water needs of a family under optimum conditions. The system works in all climates and must have an approved collector fluid to protect against freezing and corrosion.

PART No. 5845

PATENT NUMBERS
4,044,948 AND 4,130,110

The package consists of the following components all hooked together to save work in the field.

1. Expansion tank containing heat exchanger
2. Differential control
3. Primary pump
4. Secondary pump
5. Relief valve
6. Fill port
7. Check port
8. Drain valve

Included: Instructions

Not Included: Collectors and Storage Tank

OVERALL HEIGHT	H = 30 in.
OVERALL WIDTH	W = 23 In.
NET WEIGHT	= 54 lb.
SHIPPING WEIGHT	= 65 lb.

Figure 20.25 Solar hydronic water-heating package. (Courtesy of Solar Research)

about one-eighth as much liquid needs to be pumped as is used in a hydronic (water-based) heating system.

Use of Alternative Heating Equipment In many locations, favorable fuel rates can be obtained if the supply may be interrupted. To allow for continuous conditioning of the building, a dual-fuel furnace can be utilized. Solid-fuel coal or wood is burned in the primary heat exchanger. The secondary fuel is either natural gas or No. 2 fuel oil. If there should ever be a need to do so, the oil and gas burners can be easily exchanged. The natural gas burner can also be converted to LP gas.

<u>PASSIVE</u> PHASE CHANGE SOLAR SYSTEMS

While lower in cost and most highly efficient, the passive system is also the most simple, quiet and trouble-free of our phase change systems. When properly installed, no regular maintenance is required. There is very little to get out of order and the basic system usually operates many years and, in fact, almost indefinitely without repairs.

IT IS THEREFORE RECOMMENDED THAT THE PASSIVE SYSTEM BE USED WHEREVER POSSIBLE.

However, in the passive system the heat exchanger must always be installed higher than the top of the collector in order that condensed liquid refrigerant in the heat exchanger can return to the collector by gravity. In spite of this requirement, the passive system can be used in a high percentage of cases, especially if some thought is given to various possibilities.

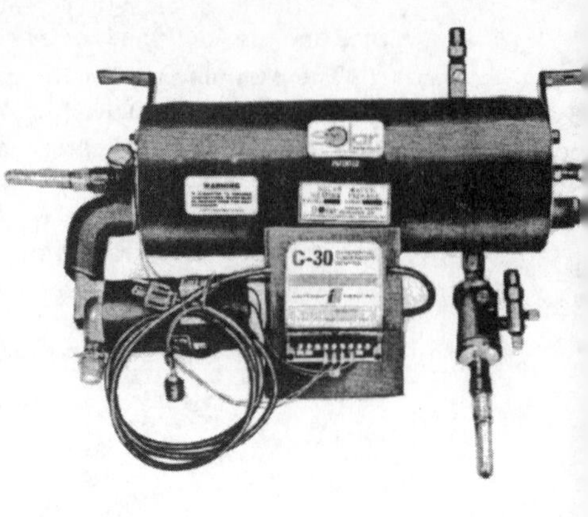

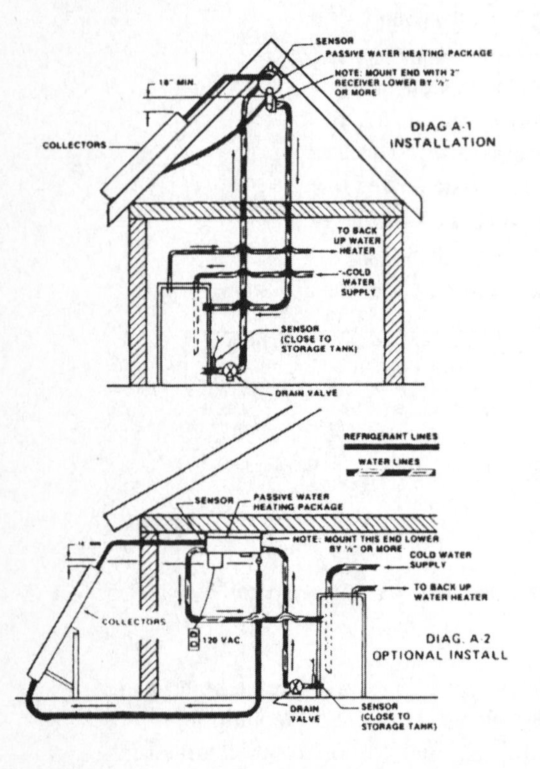

MODEL NO.	5835	6020	5953	6018	6027
Overall Height	19 in.	19 in.	19 in.	30 in.	19 in.
Overall Width	31 in.	35 in.	55 in.	14 in.	45 in.
Net Weight	39 lbs.	45 lbs.	55 lbs.	39 lbs.	50 lbs.
Shipping Weight	44 lbs.	50 lbs.	60 lbs.	44 lbs.	60 lbs.
Horz. or Vert	Horz.	Horz.	Horz.	Vert.	Horz.

Under favorable conditions of sunlight, two 18 sq. ft. collector modules will heat approximately 60 gallons of water per collector day from 65°F to 120° or even to 150°. With the No. 5835 or No. 6018 passive heat exchangers a maximum of four 18 sq. ft. collectors may be used. Six 18 sq. ft. collectors may be used with the No. 6020 passive heat exchanger. With the No. 5953 passive heat exchanger, a maximum of twelve 18 sq. ft. collectors may be used.

Collectors and heat exchangers are sold only in matched groups.

Solar Research collector plates, heat exchangers and receivers are UL recognized or UL listed.

Installation is to be made by a qualified refrigeration service contractor according to instructions furnished. All refrigeration and other codes must be adhered to.

Figure 20.26 Passive-phase-change solar system. (Courtesy of Solar Research)

ACTIVE PHASE CHANGE SOLAR SYSTEMS

ACTIVE SYSTEMS ARE RECOMMENDED FOR USE WHEN IT IS IMPRACTICAL TO USE THE PASSIVE SYSTEM. THAT IS, ACTIVE SYSTEMS ARE USED ONLY WHERE THE HEAT EXCHANGER CANNOT BE INSTALLED AT A HIGHER LEVEL THAN THE TOP OF THE COLLECTORS.

These systems are very similar to our phase change passive systems. Hot vapor moves passively by its own vapor pressure from the collector (as in the passive system) to the heat exchanger where it condenses as it gives up its heat. However, the active system differs from the passive in that the condensed liquid is then returned back to the collectors by a pump, controlled by an electric float.

Therefore, the active return system makes it possible to install the heat exchanger almost anywhere, without regard to position of the collector. The heat exchanger may be either lower, or at the same level as the collectors.

WARRANTY: 5-year limited Pro Rata warranty on parts manufactured by Solar Research.
1-year limited warranty on components such as controls, motors, pumps, valves, etc.

COLLECTORS
REFRIGERANT LINES
WATER LINES
COLD WATER
SENSOR
HOT WATER
RR 5951 WATER HEATING PACKAGE
SENSOR
INSULATED STORAGE TANK

Part No. 5951 may be used with 36, 72 or 108 square feet of Solar Research collector surface. It is supplied from the factory as a complete pre-tested unit consisting of heat exchanger, receiver, float assembly with sight glass, liquid refrigerant pump, protective strainer, check valve, motor and special differential control. Components are designed to provide a minimum of engineering in the field.

OVERALL HEIGHT	=	34 in.
OVERALL WIDTH	=	28 in.
NET WEIGHT	=	93 lb.
SHIPPING WEIGHT	=	110 lb.

Installation must be made by a qualified refrigeration service engineer and all codes pertaining to the use of refrigerants must be followed.

Figure 20.27 Active-phase-change solar system. (Courtesy of Solar Research)

Improved Maintenance

Regular maintenance is important on any heating system. If maintenance is neglected, the efficiency and reliability of the system will decrease. Firing devices should be cleaned and serviced on a regular basis. Throwaway filters should be replaced when dirty. Motors should be oiled. Where efficiency can easily be measured, as on an oil-fired installation, this should be done at least once a year to assure good performance.

Distribution systems should be kept clean. Thermostats should be kept in calibration. Ductwork should be balanced to provide an even distribution of heat. The thermostat location should be changed or the thermostat replaced if it does not provide adequate temperature control. Proper maintenance is essentially good housekeeping practice, which results in improved performance.

Many new types of energy-saving devices are becoming available on the market because of the urgent need for saving fuel. Anyone involved in the heating business should carefully evaluate these innovations, since fuel should be saved by all practical means.

STUDY QUESTIONS

Answers to the study questions may be found in the sections noted in brackets.

20-1 What percentage of the total energy consumed in the United States is used for heating? *[Energy Conservation Measures]*

20-2 What does an R-factor represent in terms of insulating value? *[Insulation]*

20-3 What is the largest factor in the thermal resistance of a properly constructed wall? *[Insulation]*

20-4 What should the R-factor be for a ceiling, in keeping with good practice? *[Insulation]*

20-5 An R-factor of 13 is equivalent to what U-factor? *[Insulation]*

20-6 What percentage of the heat loss of an average house is due to air leakage? *[Reducing Exposure Losses]*

20-7 What is the density of expanded polystyrene in pounds per cubic foot? *[Insulation]*

20-8 What is the increase in insulating value if 4 in. of insulation is changed to 12 in.? *[Insulation]*

20-9 What is the recommended net vent area for continuous ridge vents? *[Attic Ventilation]*

20-10 What R-factor is practical for a polystyrene core door? *[Energy-Efficient Doors]*

20-11 How is an air-to-air heat exchanger used? *[Air-to-Air Exchangers]*

20-12 What is the best method of reducing heated-air losses from a fireplace? *[Fireplace Treatment]*

20-13 What types of lighting are most energy efficient for residences? *[Use of More Efficient Lighting]*

CHAPTER 21

Indoor Air Quality

OBJECTIVES

After studying this chapter, the student will be able to:

- Define the elements that make up polluted air
- Make a survey of a building to determine the conditions causing air pollution
- Know the procedures required to control pollution
- Make recommendations for improving indoor air quality

Introduction

Air is a relatively thin layer of mixed gases that surrounds the earth. This atmosphere consists primarily of nitrogen (78%) and oxygen (21%). Air in its pure form is invisible, odorless, and tasteless and is required to support life. Oxygen is the active element in the air that is inhaled and is supplied to the bloodstream to carry out the metabolic process essential to life. Oxygen is converted to carbon dioxide when it is exhaled. Anyone confined to a closed space must be provided with adequate ventilation. The supply of fresh outside air is as important as exhausting the stale air to maintain normal respiratory functions.

Air Contamination

Normal air contains certain foreign material. These *permanent atmospheric impurities* enter the air as a result of natural processes such as wind erosion, sea-wave action, and volcanic eruption. Human-produced contamination can come from a number of sources, including power plants, industrial factories, transportation, construction, and agriculture.

Since the earth is a closed system, these contaminants stay within the earth's atmosphere and must be accepted and endured or modified to rid them of undesirable qualities. As these impurities accumulate, the detrimental ones become a problem. To maintain health and safety, air quality both inside occupied buildings and outside must be controlled to meet certain standards.

Studies have shown:

1. People spend 80–90% of their time indoors.
2. 50% of illness is directly related to or aggravated by the air people breathe.
3. A person breathes 200,000 liters of air daily.
4. There are up to 30 million pollutants in 1 ft^3 of air.
5. Only 25% of the dust and dirt in the air can be seen with the naked eye.
6. The microscopic portion of the particulate is what poses a problem from the health standpoint.
7. Homes are making people sick due to tight construction. Tight construction saves energy but traps contaminated air in and keeps outside air out.

Because the heating and air conditioning industry is highly responsible for indoor climate, designers and technician must, therefore, provide equipment and services to maintain a healthy indoor environment.

A distinction needs to be made between the terms *contaminant* and *pollutant.* The word *contaminant* is used to indicate unwanted constituents that may or may not be detrimental to human health. The term *pollutant* is used to indicate contaminants that do present a health risk.

Particle Size

Air pollution is not easily detectable. It can escape our normal senses. If we could see the impurities in the air we breathe, it would often be unacceptable. Contaminants can be classified roughly as follows:

1. Dust, fumes, and smokes that are solid particulate matter, although smoke often contains liquid particles
2. Mists, fogs, and smokes that are suspended liquid particles
3. Vapor and gases

These contaminants differ a great deal in size, as shown in Figure 21.1. Only a few are visible to the human eye. Note that the scale at the top of Figure 21.1 is in microns (μm). A micron is one-millionth of a meter, or one-thousandth of a millimeter. For comparison, human hairs have a size range between 25 and 300 μm.

Particle size is important in determining the type of filter to be used to capture the contamination. Particles 10 μm or larger are visible to the naked eye. Smaller particles are visible only in high concentrations such as smoke or clouds. Hygienists are concerned primarily with particles 2 μm or smaller, since this is the range of sizes most likely to be retained in the lungs. Particles larger than 8 to 10 μm are separated and retained by the upper respiratory tract and are rapidly cleared by swallowing or coughing. The size ranges for a few common pollutants are: pollen, 10 to 100 μm; bacteria,

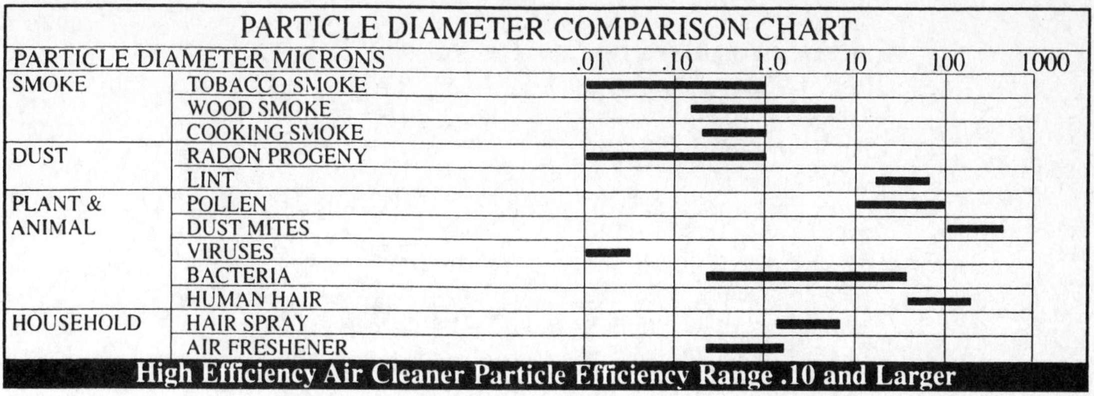

PARTICLE DIAMETER COMPARISON CHART

PARTICLE DIAMETER MICRONS		.01	.10	1.0	10	100	1000
SMOKE	TOBACCO SMOKE						
	WOOD SMOKE						
	COOKING SMOKE						
DUST	RADON PROGENY						
	LINT						
PLANT & ANIMAL	POLLEN						
	DUST MITES						
	VIRUSES						
	BACTERIA						
	HUMAN HAIR						
HOUSEHOLD	HAIR SPRAY						
	AIR FRESHENER						
High Efficiency Air Cleaner Particle Efficiency Range .10 and Larger							

Figure 21.1 Sizes and characteristics of airborne solids and liquids. (Courtesy of Skuttle Mfg. Co.)

0.4 to 50.0 μm; tobacco smoke, 0.01 to 1.0 μm; and viruses, 0.01 to 0.06 μm. Removal equipment needs to be selected to deal with these contaminants.

Indoor Air Quality

Figure 21.2 lists 27 examples of pollutants that we are often exposed to in our homes. Some, like tobacco, smoke, and paint fumes, are very obvious; however, most are present at lower intensities and are therefore not as obvious. Whatever their intensity, they are still injurious to health. Although all the occupants of a home are adversely affected, children and the elderly are the most susceptible.

Studies have shown that the levels of indoor pollutants may be 2–10 times higher than outdoor levels. The most effective way to improve indoor air quality is to reduce or eliminate the sources of pollution. One of the effective methods is the application of appropriate air filters.

Figure 21.3 is a filter selection guide. The lower section shows the sizes of various pollutants and the group into which each falls: 1, 2, 3 or 4. This manufacturer's listing shows products by group, and gives each filter's range of effectiveness to assist in the selection of the proper product. For example, to filter out a group 3 pollutant, a Micro-Particle filter could be used. To filter out a group 4 pollutant, a Micro-Kontrol™ filter could be used.

HVAC Equipment Designed to Improve Indoor Air Quality

A number of manufacturers have developed equipment that can be applied to climate control systems to greatly improve the indoor air quality. Three major pieces of equipment that meet this requirement are as follows:

1. Quality media air filter (Figure 21.4)
2. Advanced electronic air cleaner (Figure 21.5)
3. Packaged energy recovery ventilator (Figure 21.7)

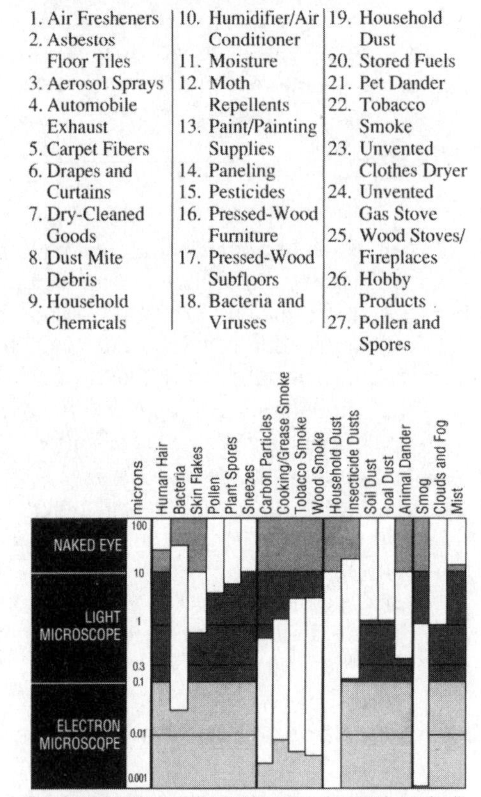

The air you breathe contains a wide variety of contaminants in all shapes and sizes. The Honeywell F29 effectively captures most dirty air particles, making your air cleaner and healthier.

Figure 21.2 Pollutants in the air we breathe. (Courtesy of Honeywell Inc.)

FILTER SELECTION GUIDE

Percent Efficiency Arrestance (A.S.H.R.A.E. Standard 52-76 Arrestance)

50 60 70 80 90 92 97 98 99

GROUP ONE
U-Trims®
Everlast™
Washable Panels
Electrostatics
Kontrol Seal™
Foam
Polyester
Fiberglass
Auto Rolls
Odor Kontroller® (flat)

Percent Efficiency Atmospheric Dust Spot (A.S.H.R.A.E. Standard 52.1-1992 Atmospheric)

20 30 40 50 60 70 80 90 95 98

GROUP TWO
Maxi-Pleat® HP
Maxi-Pleat® TP
Dust Kontroller®
Change-A-Pleat™
Electrostatics
Kontrol Seal™
Dust Patroller®
Odor Kontroller® (pleated)

Percent Efficiency .3 D.O.P. (Mil-standard 282)

10 20 30 40 50 60 80

GROUP THREE
Micro-Particle
Mega-Pak™
Kontrol-Pak™
Airo-Flo™
Kontrol-Flo™
Dust Patroller®
Turbo 2000™
H.E.L.P.A.®

GROUP FOUR
Micro-Kontrol™
Mega-Pak™

The positioning of Air Kontrol's products, by name, on the above chart is intended to give each its range of filtration effectiveness and assist in the selection of the proper product. This simple way to classify air filters is according to their efficiency (ASHRAE Standard 52.1-1992) as the four groups above do. The user of this catalog will notice that each product listed above under its performance group has been identified by the group number throughout.

FILTER SELECTION GUIDE PARTICLE SIZE

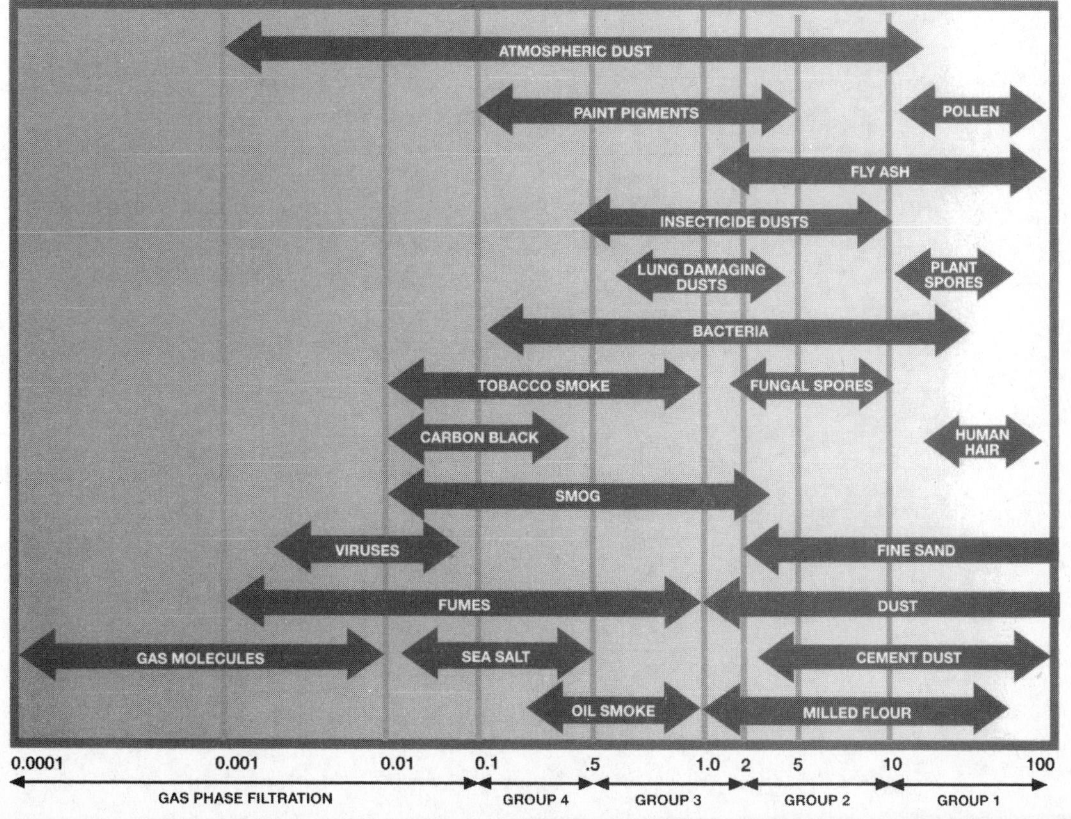

Figure 21.3 Filter selection guide. (Courtesy of Air Kontrol)

Figure 21.4 High-capacity, high-efficiency air filters. (Courtesy of Skuttle Indoor Air Quality Products)

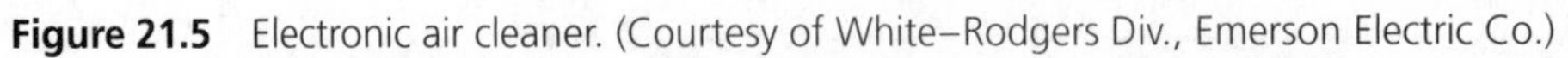

Figure 21.5 Electronic air cleaner. (Courtesy of White–Rodgers Div., Emerson Electric Co.)

Media Air Filter

The media air filter removes as much as five times the amount of contaminants as the typical furnace filter. Ordinary filters trap only the largest particles. These filters trap some particles as small as 0.5 μm, removing such contaminants as whole-grain pollen and plant spores. Some of the other features include the following:

1. They keep the heating surfaces, fan blades, and air passages clean for more efficient operation.
2. The filters are approximately 6 in. deep and are constructed of pleated media paper. With proper maintenance they can last up to a year.
3. They can be replaced with an electronic air filter to upgrade the system.

Electronic Air Cleaner

There are a number of types of electronic air filters, but even the best ones produce some ozone. The maximum ozone level should be in the range 0.005 to 0.020 ppm, which is well below the limits established by the EPA. Electronic air cleaners perform up to 30 times better at cleaning the air than typical furnace filters. They remove irritating allergens such as pollen and dust mites and even microscopic particles such as bacteria, viruses, and tobacco smoke as small as 0.01 μm. Some of the other features include the following:

1. They can save 10 to 15% in operating cost by keeping the parts of the HVAC system clean.
2. There is no cost for filter replacement. Filters are washed and restored to their original efficiency.
3. Home furnishings stay clean longer, effecting additional savings.

High-Efficiency Portable Room HEPA Filter

To this point we have been dealing with equipment that is installed; however, many situations that require something portable, which can be moved from room to room as needed. The F29, shown in Figure 21.6, is a HEPA (high-efficiency particulate air) filter. It is the most efficient media filter available. It is capable of removing 99.97% of common airborne contaminants.

Because the HEPA filter is portable, it can go where it is needed the most. This unit weighs only 15.5 lb and is virtually maintenance free. The filter media lasts up to 3 years. A solid-state three-speed switch controls a quiet fan. This unit can be used in homes, hospitals, schoolrooms, and motel rooms. A 16-ft by 16-ft room will experience nine air changes per hour, an 18-ft by 18-ft room seven air changes per hour, and a 20-ft by 20-ft room five air changes per hour.

Energy Recovery Ventilator

This unit is an accessory that can be added to a forced warm air heating system to supply and control the introduction of outside air and to exhaust stale air. It is a type of *air-*

High Efficiency HEPA Filter Portable Room Air Cleaner

- Easily portable — weighs a lightweight 15.5 pounds. A built-in carrying handle makes it easy to move the unit wherever clean air is needed.
- HEPA (High Efficiency Particulate Air) filter is a mimimum 99.97% efficient at removing common airborne contaminants.
- Virtually maintenance-free – HEPA filters require no cleaning, only periodic replacement.
- 75 square foot (extended surface area) HEPA filter lasts up to 3 years in normal use.
- Exclusive, patented 360° airflow pattern provides maximum air cleaning in all areas of the room.
- Solid-state 3-speed switch controls a quiet fan circulating air at 125/250/300 cubic feet per minute (CFM).
- Energy-efficient motor draws only 350 watts at highest fan speed, plugs into any standard 120 volt outlet.
- Replaceable, activated charcoal prefilter controls smoke, odors and other gases.
- Honeywell (2) year limited warranty.

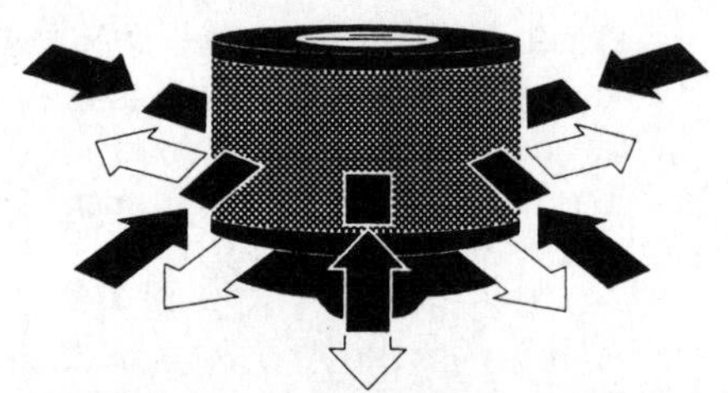

In goes the bad air, out comes the good air...An exclusive, paten airflow pattern draws dirty air from a 360° radius, then gently blow clean air at floor level distributing it throughout the room.

Figure 21.6 Portable HEPA room air cleaner. (Courtesy of Honeywell Inc.)

to-air heat exchanger. Figure 21.7 shows a diagrammatic view of the outside-air and the exhaust-air connections to the heat recovery ventilator. Frequently, dedicated air ducts are run to the bathrooms and laundry rooms to provide exhaust from these areas.

Figure 21.8 is another example of an energy recovery ventilator (ERV). Two models are available that handle 22,000 ft^3 and 35,000 ft^3 of interior home volume. This

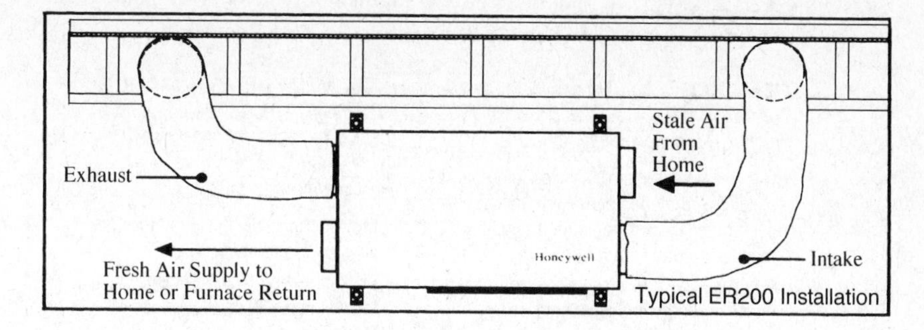

Figure 21.7 Energy recovery ventilator (ERV). (Courtesy of Honeywell Inc.)

unit is designed for dependable year-round operation, continuosly expelling stale air while simultaneously introducing fresh air. The core performs all the moisture and temperature recovery with no moving parts.

An automatic defrost system prevents ice buildup in cold weather. The blowers are specially designed to move air quietly. Washable filters are easily accessible.

Figure 21.9 shows the energy recovery core. During the heating season, this unit transfers heat from the outgoing air to the inbound fresh air. During the cooling season it lowers the temperature of the warm outside air as it passes through the core. This unit accomplishes approximately 50% moisture recovery during the heating season. It also helps control excess moisture during the cooling season.

Figure 21.10 describes the three ERV operating controls: the Detector (option 1), the Electro (option 2), and the Basic (option 3). The UL-approved systems operate on 120 V with a maximum power of 150 to 225 W, drawing 1.3 to 1.9 A.

A new series of three air system models has been developed by Broan-NuTone. Figure 21.11 shows these units. The GSFH1K for HEPA filtration only; the GSVH1K for HEPA filtration and fresh air ventilation; and the CSHH3K for HEPA filtration, fresh air ventilation and heat recovery. Also shown in 21.11 are the air system controls.

Figure 21.12 shows a typical basement installation of a system. For maximum performance and ultimate whole house air cleaning (and fresh air distribution), the air handler furnace fan should run continuously.

Figure 21.13 shows the internal construction of the three models. Detailed specifications include the maximum purification area, total air flow, operating voltage, power consumption, dimensions and recommended installation locations. A complete installed system is shown in Figure 21.14.

Additional Indoor Air-Quality-Devices

Figures 21.15, 21.16 and 21.17 are air purification systems that use ultraviolet (UV) germicidal light. UV light has been used since 1936 in controlling airborne pathogens such as viruses and bacteria and is used in many applications, such as in water treat-

text continued on p. 473

Continuous, year-round operation requires reliable and user-friendly equipment. Broan's dependable GUARDIAN ERV Systems meet the demand.

RELIABLE *design and quality Broan construction help ensure long system life and trouble-free operation.*

- Our proven enthalpic core technology, with specially developed permeable material, performs all moisture and temperature recovery without moving parts.
- Automatic defrost system prevents ice build-up on the enthalpic core, for efficient operation even in cold weather. Also, system design helps prevent pressure drops under most conditions. So you don't have to worry about the system drawing fumes from the chimney or spillage from the furnace into the house.

MAXIMUM PERFORMANCE *is something you can count on from GUARDIAN Systems thanks to Broan's technological leadership in home ventilation products.*

- Blowers specially designed to move air quietly and efficiently.
- Exceptional recovery efficiency allows for maximum energy savings.

EASY TO MAINTAIN *design helps you keep system operating at peak efficiency with a minimum of effort.*

- Washable foam filters are easily accessible for your convenience.

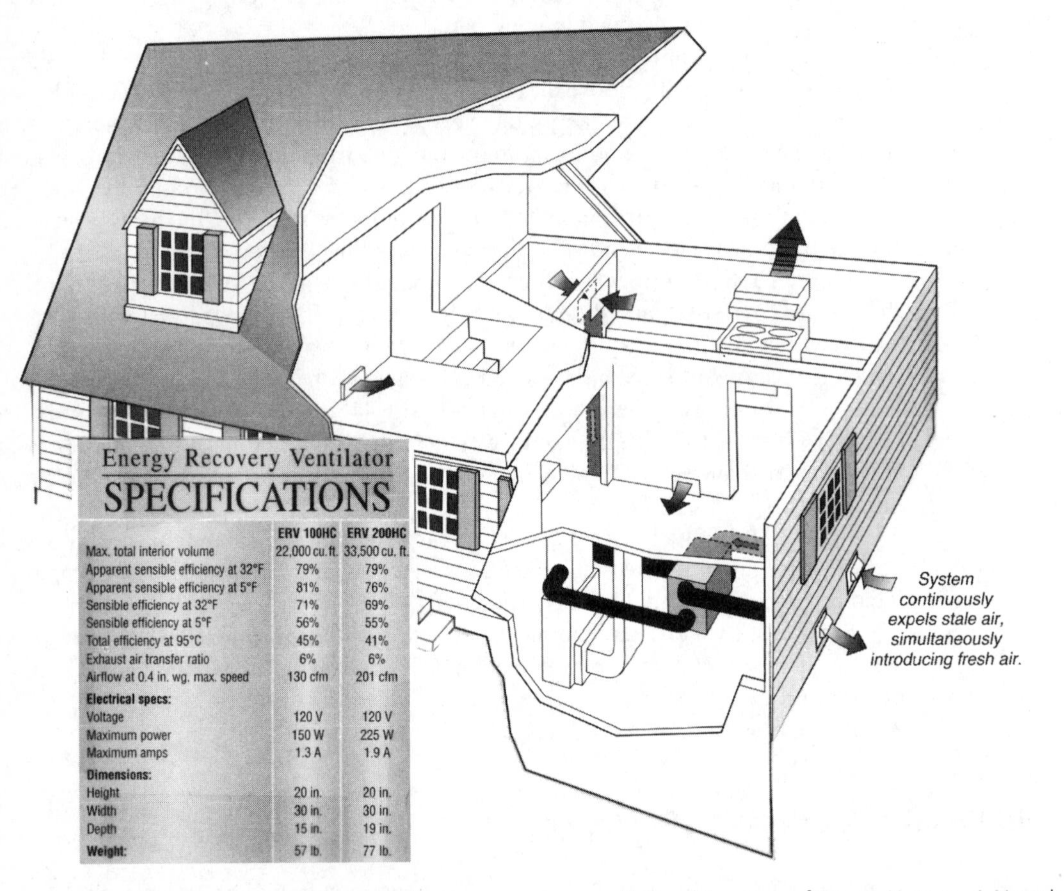

Energy Recovery Ventilator
SPECIFICATIONS

	ERV 100HC	ERV 200HC
Max. total interior volume	22,000 cu. ft.	33,500 cu. ft.
Apparent sensible efficiency at 32°F	79%	79%
Apparent sensible efficiency at 5°F	81%	76%
Sensible efficiency at 32°F	71%	69%
Sensible efficiency at 5°F	56%	55%
Total efficiency at 95°C	45%	41%
Exhaust air transfer ratio	6%	6%
Airflow at 0.4 in. wg. max. speed	130 cfm	201 cfm
Electrical specs:		
Voltage	120 V	120 V
Maximum power	150 W	225 W
Maximum amps	1.3 A	1.9 A
Dimensions:		
Height	20 in.	20 in.
Width	30 in.	30 in.
Depth	15 in.	19 in.
Weight:	57 lb.	77 lb.

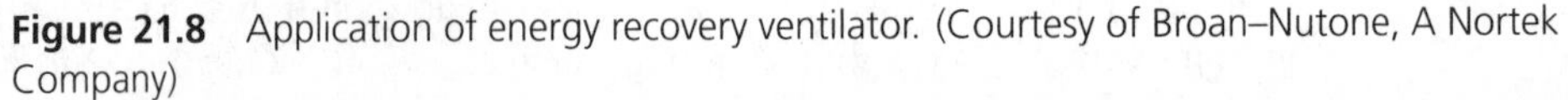

Figure 21.8 Application of energy recovery ventilator. (Courtesy of Broan–Nutone, A Nortek Company)

Maximum Heat Recovery

During the heating season, this system transfers heat from outgoing stale air to inbound fresh air. During the cooling season, it lowers the temperature of warm outside air as it passes through our enthalpic core.

Effective Moisture Transfer

Approximately 50% moisture recovery helps prevent your home from drying out during the heating season, while eliminating window condensation under most conditions. It also helps control excess moisture during the cooling season, reducing air conditioner demands.

Figure 21.9 Energy recovery module. (Courtesy of Broan–Nutone, A Nortek Company)

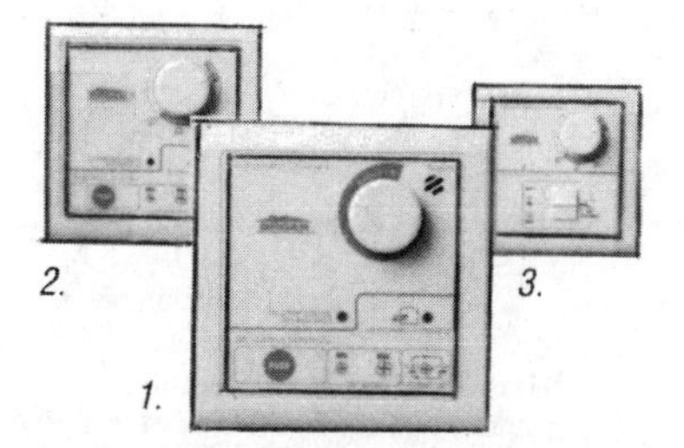

CONVENIENT ERV CONTROLS

Your choice of three wall controls makes routine operation of the ERV easy. Central location makes it convenient.

1. The Detector (VT3W)

- Sensor detects increases in pollutants. Adjusts speed setting automatically.
- Features three operating modes.
- Includes a maintenance light to let you know when it's time to clean the filter.

2. The Electro (VT2W)

- Dehumidistat automatically shifts unit to high speed whenever humidity exceeds preset comfort level.
- Same operating modes and maintenance light as the Detector.

3. The Basic (VT1W)

- Features MIN and MAX settings.
- Automatic dehumidistat control.

EASY OPERATION

- Simple speed control switch on side of unit lets you run the system on an as-needed basis to keep operating costs to a minimum.
- Optional dehumidistat automatically switches system to high speed when humidity level rises above your preselected comfort setting, for perfect control of excess moisture.

CHOICE OF SPEEDS

- Control switch on side of unit lets you decide when to run the system and what air flow speed to use, according to moisture control requirements.
- Optional dehumidistat for automatic control of excess moisture.

LOW MAINTENANCE

- Quality built for dependable, trouble-free operation.
- Foam filters are easily accessible and completely washable for maximum convenience.

All GUARDIAN Systems are UL and HVI-certified, helping assure proper, reliable operation.

Figure 21.10 Controls and operation. (Courtesy of Broan–Nutone, A Nortek Company)

CHOOSE FROM 3 MODELS

HEPA Filtration - GSFH1K

- Designed exclusively to filter indoor air and eliminate microscopic particles
- Improves indoor air quality and helps reduce the symptoms associated with respiratory problems
- Recirculates air throughout the house
- **Provides 270 CFM of filtered air.**
- Installation kit included

HEPA Filtration Fresh Air Ventilation - GSVH1K

- Combines HEPA filtration and ventilation
- Filters indoor air and eliminates microscopic particles
- Eliminates excess humidity during the cold seasons and gaseous pollutants by exhausting stale air and replacing it with fresh filtered air
- Relieves the symptoms associated with respiratory problems
- **Offers 70-105 CFM of fresh air**
- **Provides a total of 270 CFM of fresh/filtered air.**
- Installation kit included

HEPA Filtration Fresh Air & Heat Recovery - GSHH3K

- In addition to the benefits of the HEPA Filtration Fresh Air Ventilation System (above), this model has a heat recovery system that preheats air from outside before circulating it throughout the home, resulting in increased comfort
- Recirculation mode available
- **Offers 70-105 CFM of fresh air**
- **Provides a total of 270 CFM of fresh/filtered air**
- Installation kit included

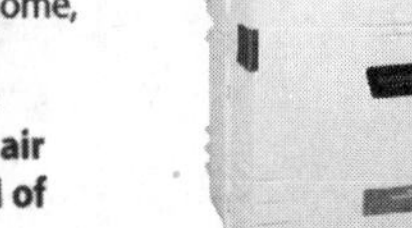

STANDARD ON ALL MODELS

- ***True HEPA filter included with each unit***
- ***Up to 2,500 sq. ft. capacity***
- ***Whole-house HEPA air purification***
- ***Two pre-filters included***
- ***Easy installation***
- ***2-year warranty***

GuardianPlus™ Air Systems are the only whole-house filtration devices on the market to feature an electronic, low voltage wall mounted control switch for convenient operating flexibility.

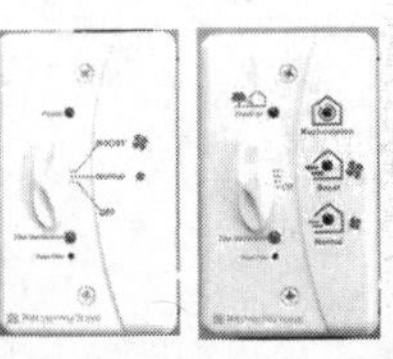

- Remote control of the systems operating modes.
- All electric, low voltage device.
- LED indicates systems' operation mode and signals filter maintenance.

Figure 21.11 Three air system models. (Courtesy of Broan–Nutone LLC)

Three Basic Strategies according to the US Environmental Protection Agency (EPA) to reduce indoor air pollutants are: Source Control, Ventilation, and Air Filtration.

Strategy I - Source Control

Source control is the elimination of the indivual sources of polution.

Strategy II - Ventilation

Ventilation is the introduction of outside air into the home, and the exhausting of contaminated air from the homes.

Strategy III - Filtration

Filtration is the process of removing airborne particles, dust, etc., using different filtration devices. Without filtration, we are recycling the same airborne particles and bacteria over and over.

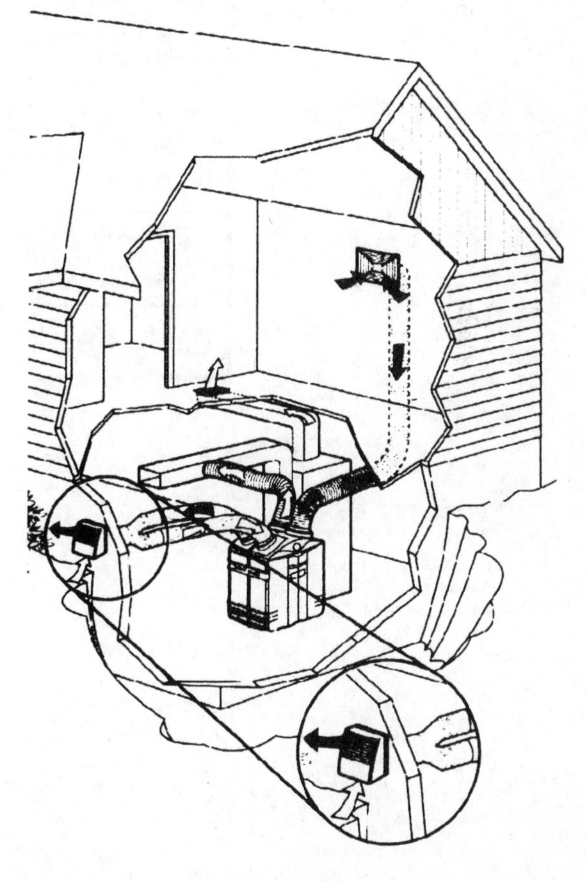

Figure 21.12 Typical basement installation (Courtesy of Broan–Nutone LLC)

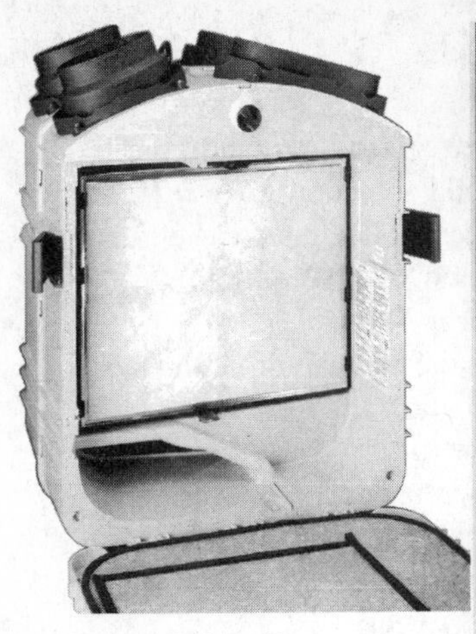

The HEPA filtration model GSFH1K provides HEPA filtration throughout the house. It picks up contaminated air and filters out it's particulat impurities.

	HEPA Filtration GSFH1K	HEPA Filtration Fresh Air Ventilation GSVH1K	HEPA Filtration Fresh Air & Heat Recovery GSHH3K
Maximum purification area	2,500 sq ft	2,500 sq ft	2,500 sq ft
Rate of air flow (fresh air) CFM	N/A	70-105 CFM	70-105 CFM
Total air flow CFM	270	270	270
Diameter of air exhaust and fresh air outlet vents	n/a	5 or 6 in.	5 or 6 in.
Diameter of indoor air ports	8 in.	8 in.	8 in.
Voltage (AC)	120 v	120 v	120 v
Power	170 W	218 W	232 W
Dimensions: Height Length Depth	29.0 in 22.9 in 22.9 in	29.4 in 22.9 in 22.9 in	29.4 in 22.9 in 22.9 in
Weight (approx.)	35 lbs	42 lbs	48 lbs
Installation kit	Included	Included	Included
Shipping Weight	53	55	57
Wall Control	Included	Included	Included
Installation location	Basement, attic, garage, crawl space	Basement, attic, garage, crawl space	Basement, attic, garage, crawl space

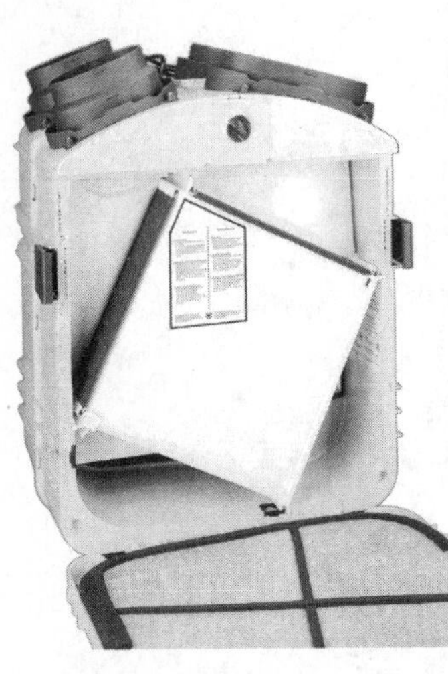

The model GSHH3K combines HEPA filtration and fresh air with the added benefit of heat recovery.

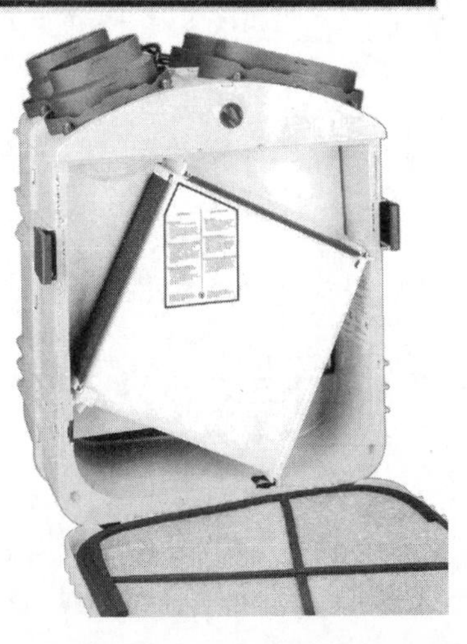

The HEPA filtration model GSVH1K combines the filtration of the GSFH1K with added ventilation. It expels stale contaminated air and replaces it with clean, fresh outside air.

Figure 21.13 Internal construction and specifications for the three models. (Courtesy of Broan–Nutone LLC)

Figure 21.14 Complete installation of a Guardian Plus™ System. (Courtesy of Broan–Nutone LLC)

Model 9700PP
12VDC. Portable for Auto, RV or Marine
43cfm blower

Photo-Catalytic Portable Room Air Purifier

- Dimensions: 3"H X 12"W X 18"L
- Electrical Usage: 110VAC
- Design Air Flow: 147cfm variable speed blower
- Unit Capacity: Handles up to 400 sq. ft. room
- Integrates into your forced-air heating & cooling system
- One year limited warranty

Figure 21.15 Photo-Catalytic air purifier. (Courtesy of Elite Environmental Products)

High-Efficiency Inline Air Cleaner with Germicidal Photo-Catalytic Light

- Dimensions: 9.25"W X23"H X 29" L
- Electrical Usage: 110VAC
- Filter: (1) pleated media filter
- 18" Germicidal Photo-Catalytic Lamp
- For systems up to 5 Tons
- Integrates into your forced-air heating & cooling system
- Three year limited warranty (1 year lamps)

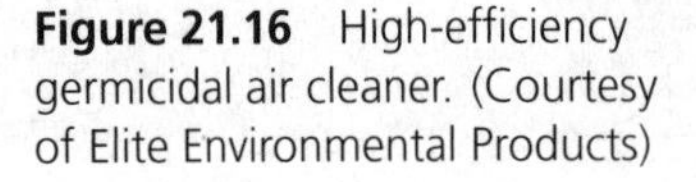

Figure 21.16 High-efficiency germicidal air cleaner. (Courtesy of Elite Environmental Products)

Evaporator Coil Germicidal Air Purifier

Model 2218

- Supply side model sterilizes coil mold
- Dimensions: (2) 6"H X 8.5"W X 18"Probe
- Electrical Usage: 110VAC
- For systems up to 2.5 Tons
- (2) 18" germicidal photo-catalytic lamps
- Integrates into your forced-air heating & cooling system
- Three year limited warranty (1 year lamps)

Figure 21.17 Evaporator germicidal air purifier. (Courtesy of Elite Environmental Products)

ment plants, hospital surgical rooms, dairies, and residential homes. When microscopic bioaerosols are exposed to the germicidal light, they are sterilized.

Figure 21.15 shows a Photo-Catalytic portable room air purifier. The model 9700PP operates on 12 V dc and can be used in automobile, RV, or marine applications. The model 9500PP has a design airflow of 147 cfm and can handle up to a 400-ft^2 room. It can also be integrated into a forced-air heating/cooling system.

Figure 21.16 shows a high-efficiency in-line air cleaner with a germicidal Photo-Catalytic light. The model 9000 operates on 110 V ac and is suitable for heating/cooling systems up to 5 tons. The unit consists of a filter rack, a pleated media filter, and an 18-in. germicidal lamp.

Figure 21.17 shows the Model 2218 evaporator coil germicidal air purifier. The two 18-in. lamps are mounted in the heating/cooling system so that they sterilize coil mold. This unit handles sytems up to 2.5 tons. All models are UL approved.

STUDY QUESTIONS

Answers to the study questions may be found in the sections noted in brackets.

21-1 What is the difference between *contaminated* air and *polluted* air? *[Air Contamination]*

21-2 How small a particle can be seen with the naked eye? *[Particle Size]*

21-3 What is the most effective way to improve indoor air quality? *[Indoor Air Quality]*

21-4 How efficient is the HEPA filter? *[High-Efficiency Portable Room HEPA Filter]*

21-5 How small a particle can be trapped by the media air filter? *[Media Air Filter]*

21-6 How small a particle can be trapped by the electronic air cleaner? *[Electronic Air Cleaner]*

21-7 What is the function of an energy recovery ventilator? *[Energy Recovery Ventilator]*

21-8 What percentage of illness is directly related to or aggravated by the air people breathe? *[Air Contamination]*

21-9 What is the purpose of the energy recovery core? *[Energy Recovery Ventilator]*

21-10 How does a germicidal light aid in air purification? *[Additional Indoor Air-Quality Devices]*

CHAPTER 22

Zoning

OBJECTIVES

After studying this chapter, the student will be able to:

- Improve the performance of a heating system using zoning
- Select and install the proper devices to form an operating zone control system
- Program and adjust the system to achieve maximum comfort and efficiency

Introduction

The availability of high-technology components has made it possible to upgrade residential and small commercial heating and cooling systems. A series of "comfort systems" have evolved all the way from the "potbelly" stove to the "electronic marvels" that can greatly enhance the way we live. To mention a few of these innovations:

1. The high-efficiency furnace, designed to conserve energy, covered in Chapter 12
2. The heat pump, designed to utilize the same apparatus for both heating and cooling, covered in Chapter 24
3. Provision for improved indoor air quality to provide a healthier environment, covered in Chapter 21
4. The application of an integrated comfort system that includes individual zone control, electronic balancing, and computerized monitoring, covered in this chapter

Homeowners and small commercial users are increasingly demanding more functions from the climate control system. These include heating, cooling, humidity control, improved air cleaning, fresh-air regulation, zoned temperature control, and energy conservation. In this chapter we pay particular attention to the use of an integrated system that supplies greater comfort and efficiency through the use of zoned control.

Zoning

Zoning is a type of system design in which the air supply to various areas of the building is separated and controlled individually. The air supply to each zone is fitted with an adjustable zone damper that is controlled by a room thermostat located in the zone. The selection of areas for each zone can be determined by considering the usage or exposure orientation or both. For residential applications the areas for zoning are usually selected based primarily on use, due to 24-hour occupancy. For small commercial installations the building orientation is often the controlling factor.

Why is zoning so important? It is because load factors differ in various parts of the building. To maintain comfort conditions in each important area, the system must provide individual zone control. If only one thermostat is used, the service person will balance the system for one set of conditions and find that at another time, rebalancing is required. Controlling each zone according to its needs is essential in maintaining maximum customer satisfaction. For example, in a residence the following zoning may prove favorable:

Zone 1: kitchen, dining room, utility room

Zone 2: family room and living room

Zone 3: bedrooms

A desirable characteristic of a zoned system is that it often permits the use of a smaller conditioning unit. Because all zones will not be calling for peak capacity at the same time, conditioning only the part of the building that needs it at the time saves energy.

A recommended procedure for planning a zoned system for a new residence is as follows:

1. Decide on the areas to be included in each zone.
2. Determine the peak heating and cooling demands and peak cfm values for each zone.
3. Determine the heating and cooling equipment size and performance required for the structure, based on the demands for the building and not the sum of the individual zone loads.
4. Lay out and size the supply-air ducts and locate the comfort zone dampers. Each zone must be large enough to handle its peak requirements.
5. Decide where to locate the thermostats for each zone in accordance with good practice.
6. Proceed with the installation arrangements.

Figures 22.1 to 22.6 illustrate various applications of zoned systems.

Figure 22.1 shows a three-zone, split-level system. Each zone has a thermostat and zone damper controlling the air flow to that zone.

Figure 22.2 shows zoned control for a bilevel house: one zone controlling the first-floor living area, another zone controlling the second-floor bedrooms, and a third zone controlling the downstairs recreation room.

Figure 22.3 shows a five-zone radial system with the ductwork located in the slab.

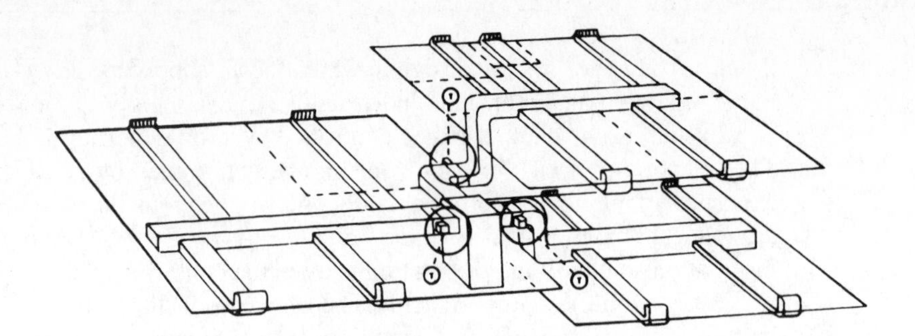

3-ZONE SPLIT LEVEL System in a typical split level house is shown with a damper and thermostat controlling the flow of air through the trunk duct to each zone. Installation requires a 3-Zone Mastertrol, 3 dampers, with thermostats.

Figure 22.1 Three-zone split-level system. (Courtesy of Honeywell Inc.)

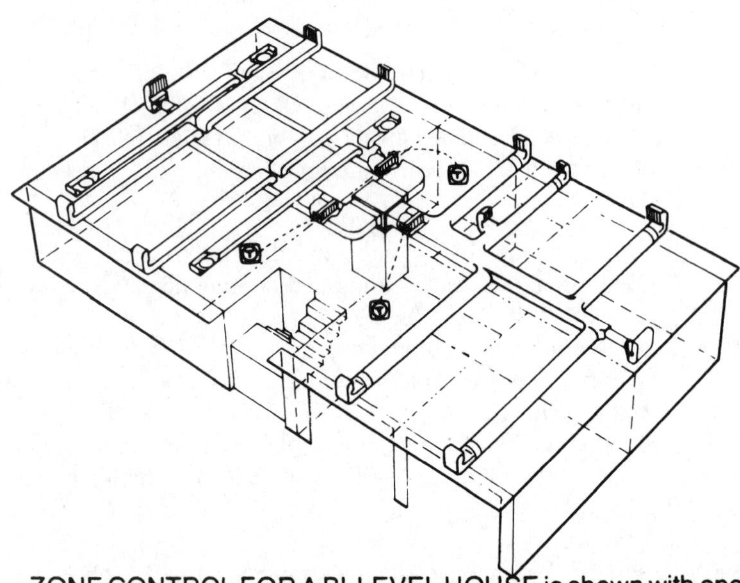

ZONE CONTROL FOR A BI-LEVEL HOUSE is shown with one thermostat and damper controlling the upstairs living area zone, same for the bedroom zone, and same for downstairs recreation area. A three zone Mastertrol Panel, 1 40 VA transformer, plus three dampers and thermostats is the only material required for the installation.

Figure 22.2 Zone control for a bilevel house. (Courtesy of Honeywell Inc.)

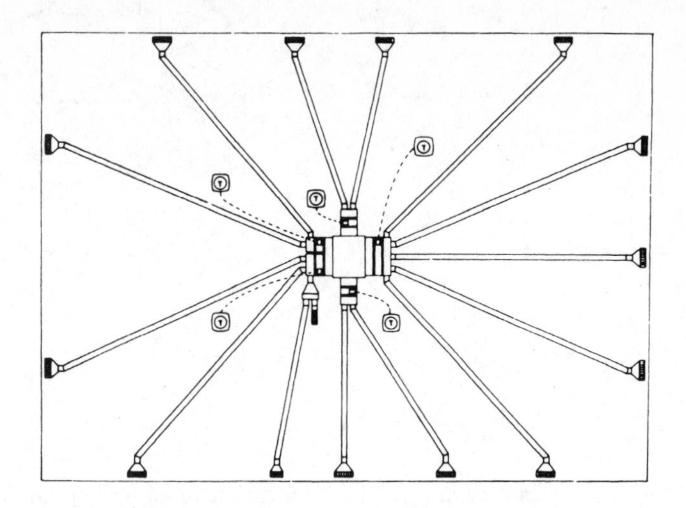

ROOM BY ROOM TEMPERATURE CONTROL is shown in this 5-zone radial system, which may be located either overhead or in slab.

Figure 22.3 Five-zone radial system. (Courtesy of Honeywell Inc.)

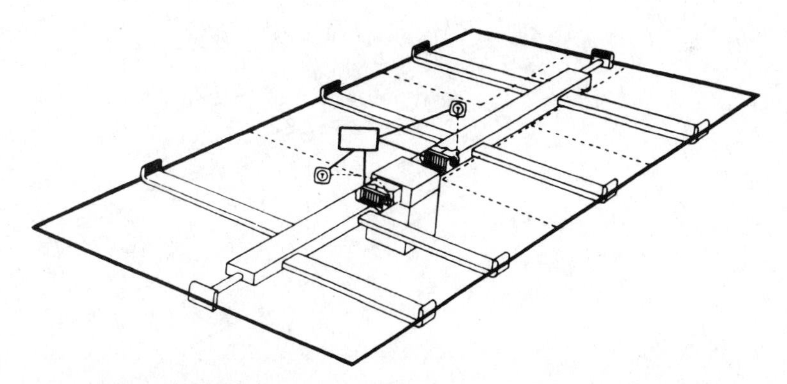

2-ZONE RANCH HOUSE is shown with a thermostat in each zone controlling a corresponding zone damper. For a Mastertrol installation, the following material would be required: One 2-Zone Mastertrol Panel, 2 AOBD Dampers with thermostats, and 1 40 VA Transformer.

Figure 22.4 Two-zone ranch house. (Courtesy of Honeywell Inc.)

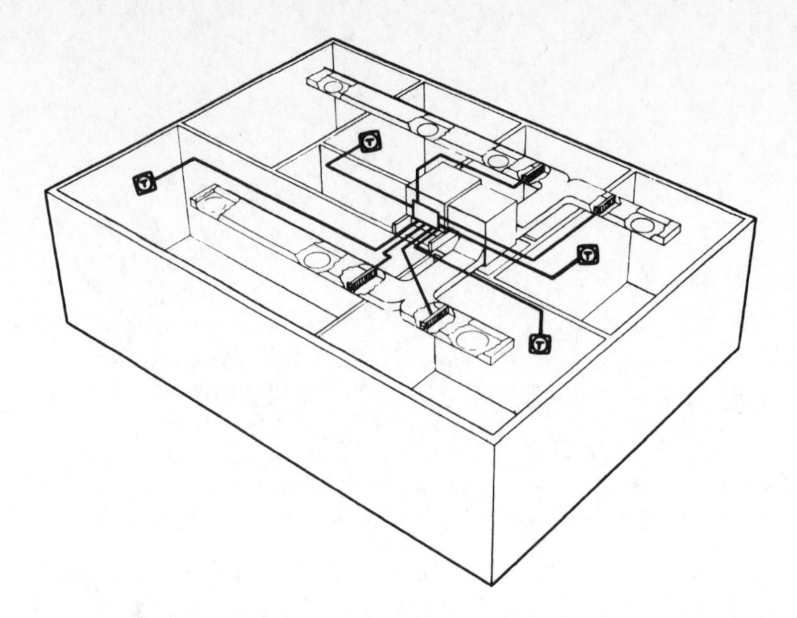

A 4-ZONE SYSTEM for a professional office or home is shown with the air to each room or zone controlled by a damper and thermostat. Installation requires a 4-Zone Mastertrol. 4 dampers and thermostats, and one 40VA transformer.

Figure 22.5 Four-zone system. (Courtesy of Honeywell Inc.)

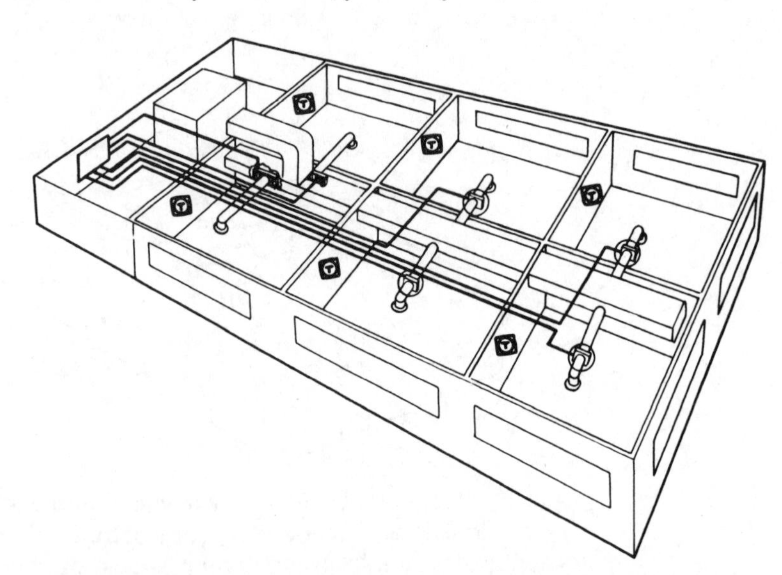

ROOM BY ROOM COMFORT CONTROL may be installed on most any new or existing heating-cooling system by controlling the flow of air at each outlet with automatic square to round transition dampers.

Figure 22.6 Individual-room zone control system. (Courtesy of Honeywell Inc.)

Figure 22.4 shows a two-zone ranch house system with separate zones for the living and sleeping areas.

Figure 22.5 shows a four-zone system for a small office building or home. Here both building use and orientation were taken into consideration in planning the zoning.

Figure 22.6 shows an individual room control system with a separate thermostat and damper for each room.

When considering zoning for an existing system, the designer must carefully examine the supply ductwork. The ductwork must be readily accessible for the installation of the zone dampers. Ideally, the trunk ducts should be split to supply the required zoning areas. It may be necessary on some jobs to use motorized dampers at the outlets to provide zoning. Often, zoning will help a system that is slightly undersized, since all zones seldom call for peak capacity at the same time. Zoning should not, however, be expected to correct a poorly designed duct system. The following are important considerations in planning any zoned system:

1. The airflow over the heating or cooling heat exchangers must not be reduced below limits set by the manufacturer.
2. Air velocities must be held below levels that could produce objectionable noise.
3. Some excess pressure–relieving arrangement needs to be provided to prevent airflow problems when some zones are calling for conditioned air and the rest are shut off.

Relieving Excess Air Pressure

When a number of zones are satisfied and their dampers shut off, the remaining open zones may be undersized to handle the minimum required airflow from the HVAC unit. To correct this condition, some type of modification needs to be made. One solution is to oversize the ducts for individual zones so that the duct for the active zone, when others are shut off, may be large enough to handle the required air supply.

Another method of solving the problem, which is usually the preferred solution, is to provide a bypass arrangement. Bypassing some of the air allows the total air handled by the conditioner to be held above minimum requirements.

For a retrofit installation, the limitations of an existing system are greater and require more ingenuity on the part of the planner. With proper study, however, zoning can be applied to most systems to improve performance.

It is recommended that the following formula be used in determining bypass airflow:

$$\underset{\text{(cfm)}}{\textit{air handler}} - \underset{\text{(cfm)}}{\textit{smallest zone peak}} - \underset{\text{(cfm)}}{\textit{leakage of all closed dampers}} = \underset{\text{(cfm)}}{\textit{bypass air flow}}$$

There are a number of ways to provide the bypass:

1. Divert the air to another space inside the building, such as an unoccupied room, utility room, hallway, or basement area. This is usually the preferred arrangement (see Figure 22.7).

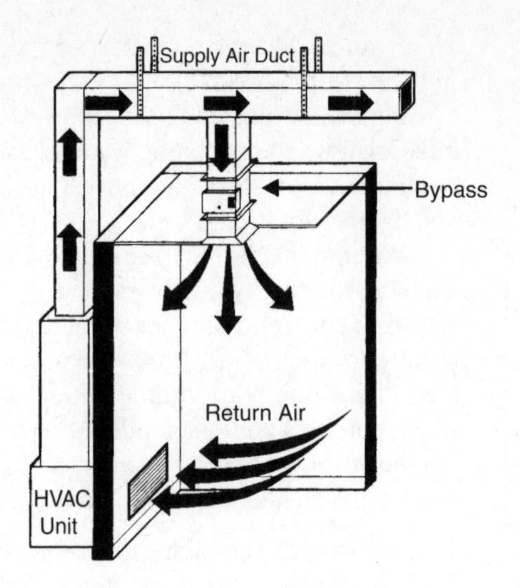

Figure 22.7 Using bypassed air as a supply source. (Courtesy of Carrier Corporation)

2. Return the bypassed air to the intake side of the HVAC unit (see Figure 22.8). This system works well when the zone or zones calling for conditioned air can be supplied in a reasonable amount of time. With prolonged operation, though, the high-limit temperature may be reached on heating, and the low-limit temperature may be reached on cooling. Either of these conditions can cause the equipment to shut down or short-cycle, which is unsatisfactory.

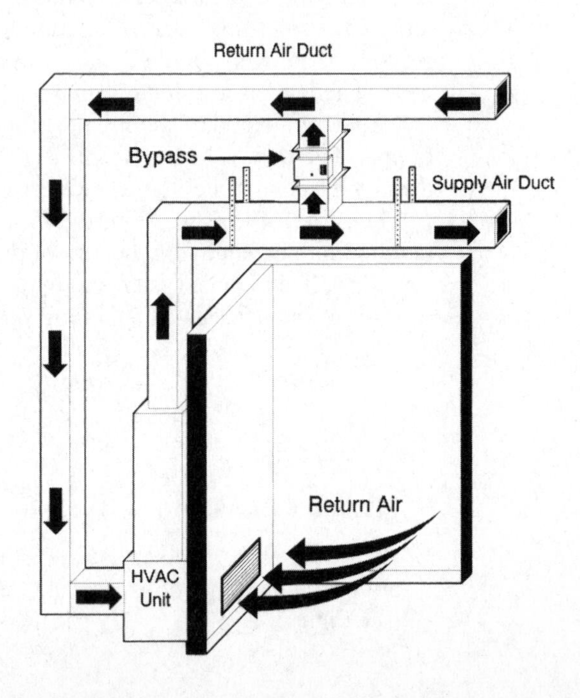

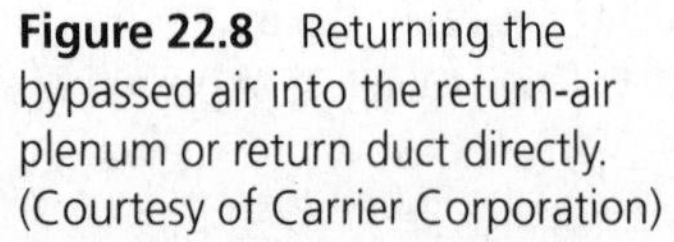

Figure 22.8 Returning the bypassed air into the return-air plenum or return duct directly. (Courtesy of Carrier Corporation)

3. Undersize the zoned dampers so that there is some leakage past the damper even during normal operation. For example, a 22 × 8 zone damper can be placed in a 24 × 8 zone duct. This arrangement permits a leakage of air past the damper location when the damper is closed. This may be satisfactory on some jobs, but the performance is difficult to predict.
4. Oversize each zone duct to handle 60 to 70% of the air handler cfm and omit any bypass arrangement. This design can be used only on new and relatively small installations.

Types of Zone Dampers

There are various types of automatic zone dampers available. They are selected to comply with the configuration of the ductwork and the method of control. The types are as follows:

1. Opposed-blade damper for rectangular ducts (Figure 22.9)
2. Round duct damper (Figure 22.10)
3. Multivalve supply-register damper (Figure 22.11)
4. Square ceiling-diffuser damper (Figure 22.12)
5. Round ceiling-diffuser damper (Figure 22.13)
6. Floor diffuser damper (Figure 22.14)

Motorized dampers are regulated automatically by the zone thermostats. Some dampers can be fully modulated, and others are available only for two-position operation. Either the rectangular or the round duct damper can be used in the fresh-air supply duct if regulation is desired.

Automatic Bypass Dampers

There are two types of automatic bypass dampers:

1. *Barometric static pressure relief damper:* requires no electrical connections and operates on low system pressures, usually in a range below 0.5 in. S.P. (static pressure)

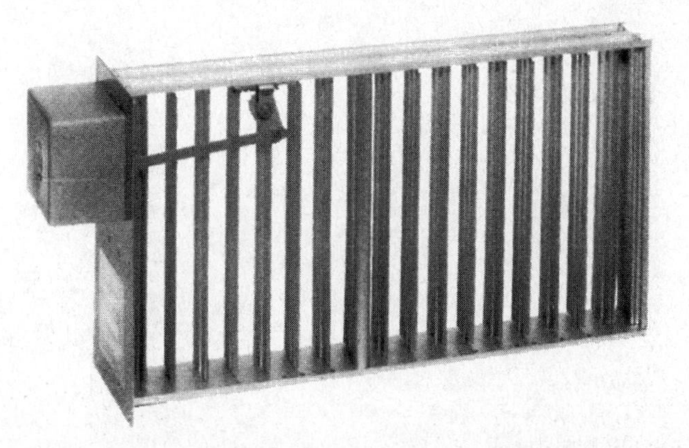

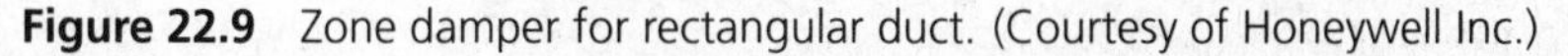

Figure 22.9 Zone damper for rectangular duct. (Courtesy of Honeywell Inc.)

Figure 22.10 Zone damper for round duct. (Courtesy of Honeywell Inc.)

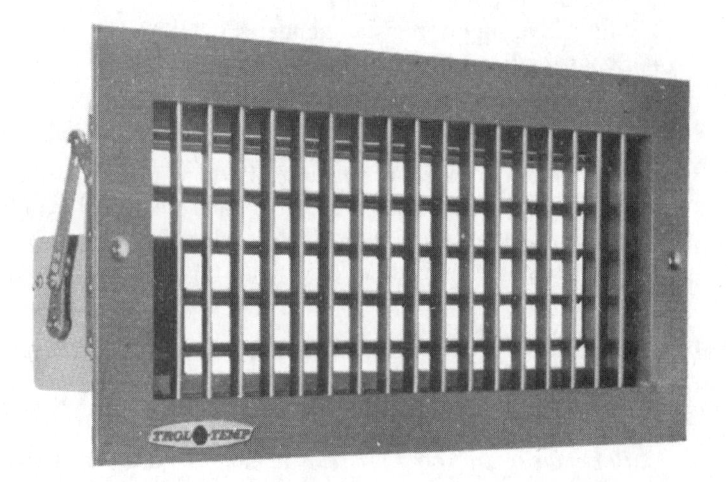

Figure 22.11 Multivalve horizontal register damper. (Courtesy of Honeywell Inc.)

Figure 22.12 Square ceiling-diffuser damper. (Courtesy of Honeywell Inc.)

Figure 22.13 Round ceiling-diffuser damper. (Courtesy of Honeywell Inc.)

2. *Motorized static pressure relief damper:* operates in response to a pressurestat located in the supply duct

A single-blade bypass damper is shown in Figure 22.15.

Types of Zoning Systems

A number of manufacturers provide zoning equipment and controls to simplify the installation of a zoned system. These arrangements differ in many respects, such as the number of zones handled, methods of control, and types of adjustments provided. These systems are briefly reviewed here so that the HVAC technician can choose the system that best fits the job. The following systems are presented:

1. Honeywell Trol-A-Temp zone control system
2. Carrier Comfort Zone system
3. California Economizer Digitract zone system
4. Enerstat Zone Control System 2

Honeywell Trol-A-Temp Zone Control System

The Trol-A-Temp zone control system is a series of integrated devices designed to control a multizone HVAC system (Figure 22.16). The control center is the Mastertrol automatic balancing system (MABS) (Figure 22.17). This device is capable of controlling two or four zones. For systems requiring five or more zones, the TotalZone series of control panels is used.

Figure 22.14 Floor diffuser damper. (Courtesy of Honeywell Inc.)

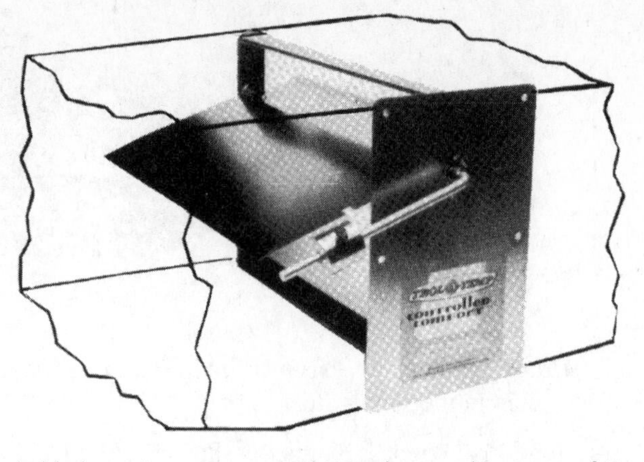

Figure 22.15 Single-blade static pressure–regulating damper. (Courtesy of Honeywell Inc.)

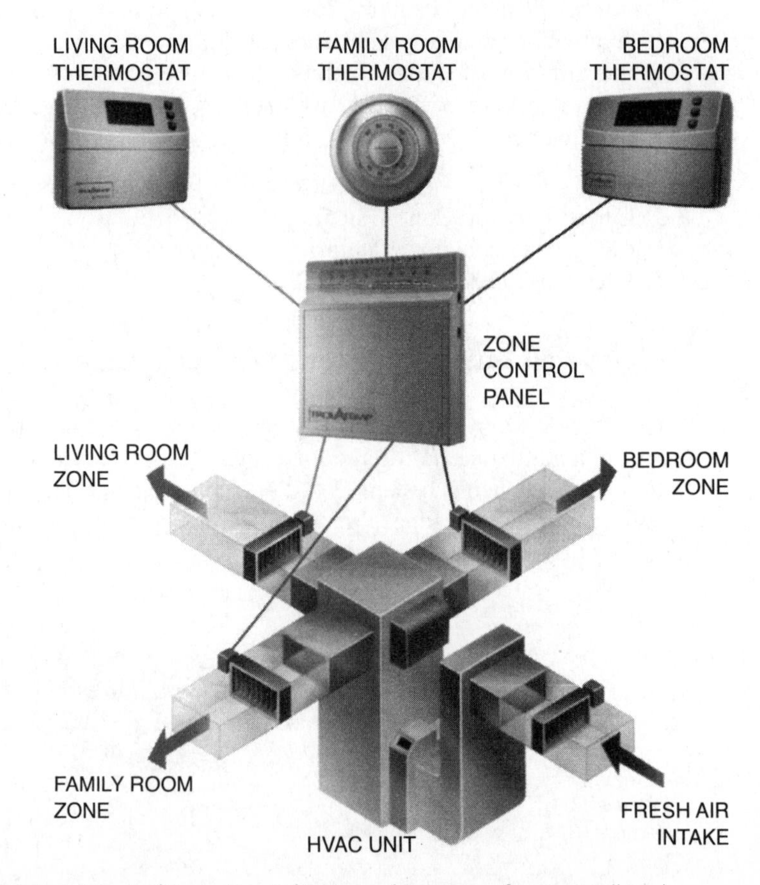

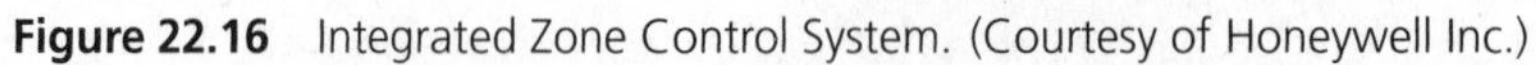

Figure 22.16 Integrated Zone Control System. (Courtesy of Honeywell Inc.)

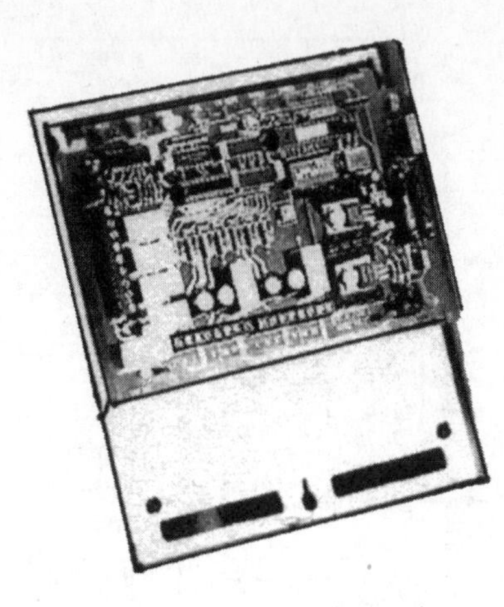

Figure 22.17 Mastertrol Automatic Balancing System, MABS EZ Zone control panel. (Courtesy of Honeywell Inc.)

The MABS EZ-2 and EZ-4 are microprocessor-based forced-air zone-control panels that provide automatic or manual changeover control of single- and two-stage heating and cooling and heat pumps with or without auxiliary heat. A typical wiring diagram for a Mastertrol zone panel with four thermostats and a two-stage heating/cooling system is shown in Figure 22.18.

The MABS EZ panel uses a standard heating cooling thermostat for each zone to allow any zone to call for heating or cooling. When one zone calls for heating while another calls for cooling, the MABS EZ accepts the first call. Once that first call is satisfied or after 20 minutes from initiation of the opposite call, the MABS switches over to the other mode of conditioning. If neither call is satified in the 20 minutes, the system switches back and forth until both calls are satisfied.

In the standard Mastertrol system the zone dampers operate in two positions, fully open when calling for conditioned air and fully closed when satisfied. When a call is satisfied, the system enters a purge mode. This holds open the damper for the last zone calling and purges into that zone. When all zones are satisfied, all zone dampers go to a fully open position. The purge time can be set for 3½ or 2 minutes using dip switch 1. The ON setting provides 2 minutes.

Once all zones are satisfied, each thermostat fan switch may be used to control fan operation. If circulation is desired, the fan switch is placed in the ON position. The fan runs, and the dampers are closed to the zones where the fan switch is set to AUTO.

The MABS EZ panel can control up to two stages of heating and two stages of cooling using a single-stage thermostat. The panel uses a timer to initiate the second stage, which is adjustable from 5 to 30 minutes using dip switches 5, 6, and 7 on the panel. See Figure 22.19 for the stage-timer configuration.

If a call for heating or cooling is not satisfied within the time set with the dip switches, the panel energizes the second stage of heating or cooling. Both stages remain on until the call is satisfied.

Additional controls can be used, depending on the requirements of the installation.

1. Night setback can be provided in any zone by replacing the standard thermostat with a 7-day time clock and programmable thermostat.

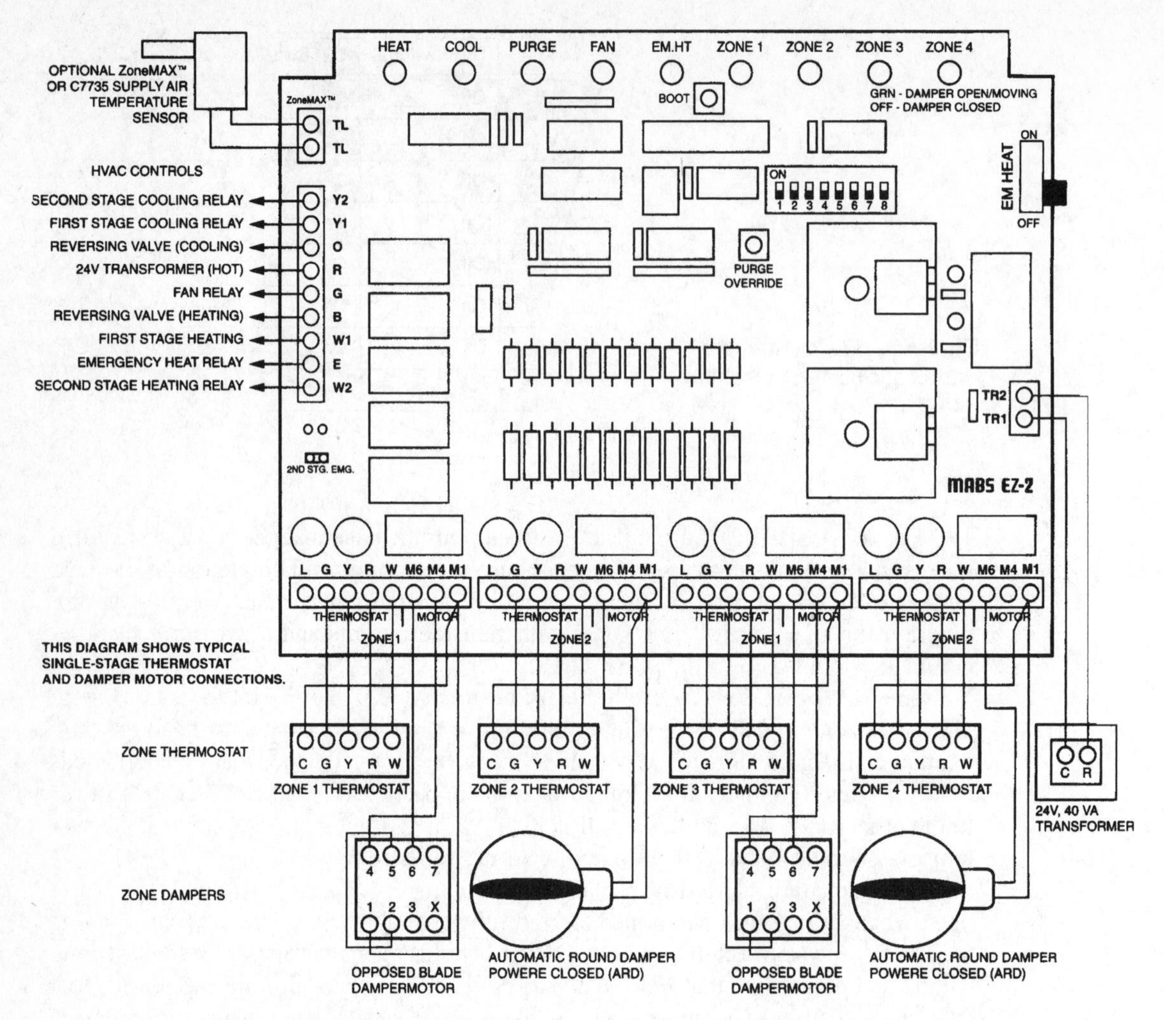

Figure 22.18 Typical wiring diagram for the Mastertrol panel with four thermostats and a two-stage heating/cooling system. (Courtesy of Honeywell Inc.)

2. A duct-mounted temperature probe that monitors the supply-air temperature can be provided to control capacity or to prevent overheating or coil icing. The ZoneMAX Sensor attaches to the TL terminals at the upper left corner of the panel. If the system trips because a high temperature of 160°F (71°C) or a low temperature of 40°F (4°C) was exceeded the heating or cooling equipment shuts down, and the fan purges the conditioned air from the plenum. The ZoneMAX Sensor resets and allows normal operation when the supply-air temperature falls 10° in heating mode or rises 10° in cooling mode. The 10° will prevent short-cycling of the equipment. If one of the limits is tripped, the appropriate LED (Figure 22.20) will flash continuously until it resets. The ZoneMAX Sensor wiring must be at least 12 in. away from any line voltage wiring to ensure correct operation.

Minutes	DIP Switch 5	DIP Switch 6	DIP Switch 7
5	Off	Off	Off
8	On	Off	Off
10	Off	On	Off
12	On	On	Off
15	Off	Off	On
20	On	Off	On
25	Off	On	On
30	On	On	On

Figure 22.19 Two-stage timer configuration. (Courtesy of Honeywell Inc.)

The MABS EZ features several LED status indicators to aid troubleshooting (Figure 22.20). The LEDs are located at the top of the panel and are visible through a window in the cover. If none of the lights are lit, the built-in thermal circuit breaker may be tripped. The breaker protects against shorts in the thermostat and damper wiring. It does not protect against shorts in the HVAC equipment.

		Status		
LED	Color	Lighted	Not Lighted	Flashing
Heat	Red	Heat call	Not in heat call	ZMS high temper-ature limit tripped
Cool	Green	Cool call	Not in cool call	ZMS low temper-ature limit tripped
Purge	Amber	Purge mode	Not in Purge mode	–
Fan	Green	Fan only call	No fan only calls	–
Em Heat	Red	Emergency heat switch on or ther-mostat in emergency heat	Not in emergency heat mode	–
Zone 1,2, 3,4	Green	Damper open or moving	Damper closed	–

Figure 22.20 LED status indicators. (Courtesy of Honeywell Inc.)

A heat-pump or multistage thermostat is not required for the MABS EZ panels because of the internal timer configuration. A heat-pump thermostat is used only when emergency heat control is desired from the thermostat. An emergency heat switch is located on the side of the panel and can be activated without removing the cover.

Normal fan operation is initiated by the HVAC equipment. The fan can be set to come on from the panel with a call for heat. Set dip switch 8 to the ON position to activate this mode.

Carrier Comfort Zone System

The Carrier Comfort Zone system is designed primarily for residential application. The basic system provides automatic control for four zones with heating only or with heating and cooling. It is recommended for applications requiring 5 tons (2000 cfm) or smaller with a maximum 1 in. W.G. inlet static pressure at the zone dampers. A typical system layout is shown in Figure 22.21.

Each zone is controlled by an individual zone thermostat operating a fully modulating zone damper. All wiring is run to a comfort-zone center that is located in any easily accessible location. In the standard system an automatic barometric bypass damper is provided to relieve excess air pressures when a reduced number of zones are calling for conditioned air.

A four-zone controller is placed in zone 1. This zone is considered the most occupied zone of the house, usually the living room or family room. This controller not only controls zone 1 but also provides the programming facility for setting up schedules for all zones. In the standard system remote room sensors are provided for zones 2, 3, and 4. A duct-temperature sensor is located between the bypass damper and the heating unit. The standard wiring diagram is shown in Figure 22.22. The wiring for single-stage heating is shown in Figure 22.23.

Accessories The accessories available include:

1. Home access module (HAM)
2. Smart sensor
3. Smart sensor power pack
4. Outdoor air temperature sensor
5. Outdoor air damper
6. Motorized bypass damper

The home access module (HAM) is used to detect Comfort Zone system errors. The HAM will report that the system is operating correctly, or if an error does occur, it will automatically call an Air Conditioning Service (ACS) dealer or the ACS National Response Center. The HAM must be programmed via report logger or PC command center software prior to installation.

The smart sensor and smart sensor power pack are used together. A smart sensor can be used in place of a remote room sensor. It provides the ability to view and adjust the setpoint in the zone where the sensor is located.

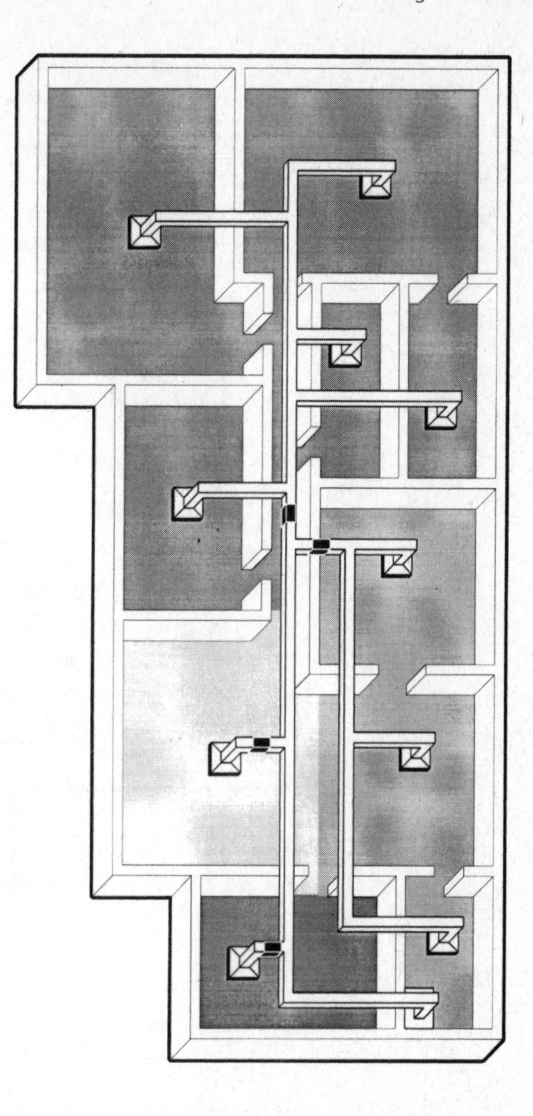

Figure 22.21 Typical four-zone Comfort Zone residential system. (Courtesy of Carrier Corporation)

The motorized bypass damper can be used in place of the barometric bypass damper. It senses the position of all zone dampers and opens the bypass, as required, to relieve duct pressure. The outside air damper, activated by the outside air temperature sensor, controls the outside air entering the building. This control operates on an economizer basis, using outside air to cool when temperature conditions permit it. An accessory wiring diagram is shown in Figure 22.24.

Programming and Operating the System The Comfort Zone controller provides the facility for programming the system. From this device the unique demands of each zone can be set for different times of the day and week. The principal parts of this controller are shown in Figure 22.25.

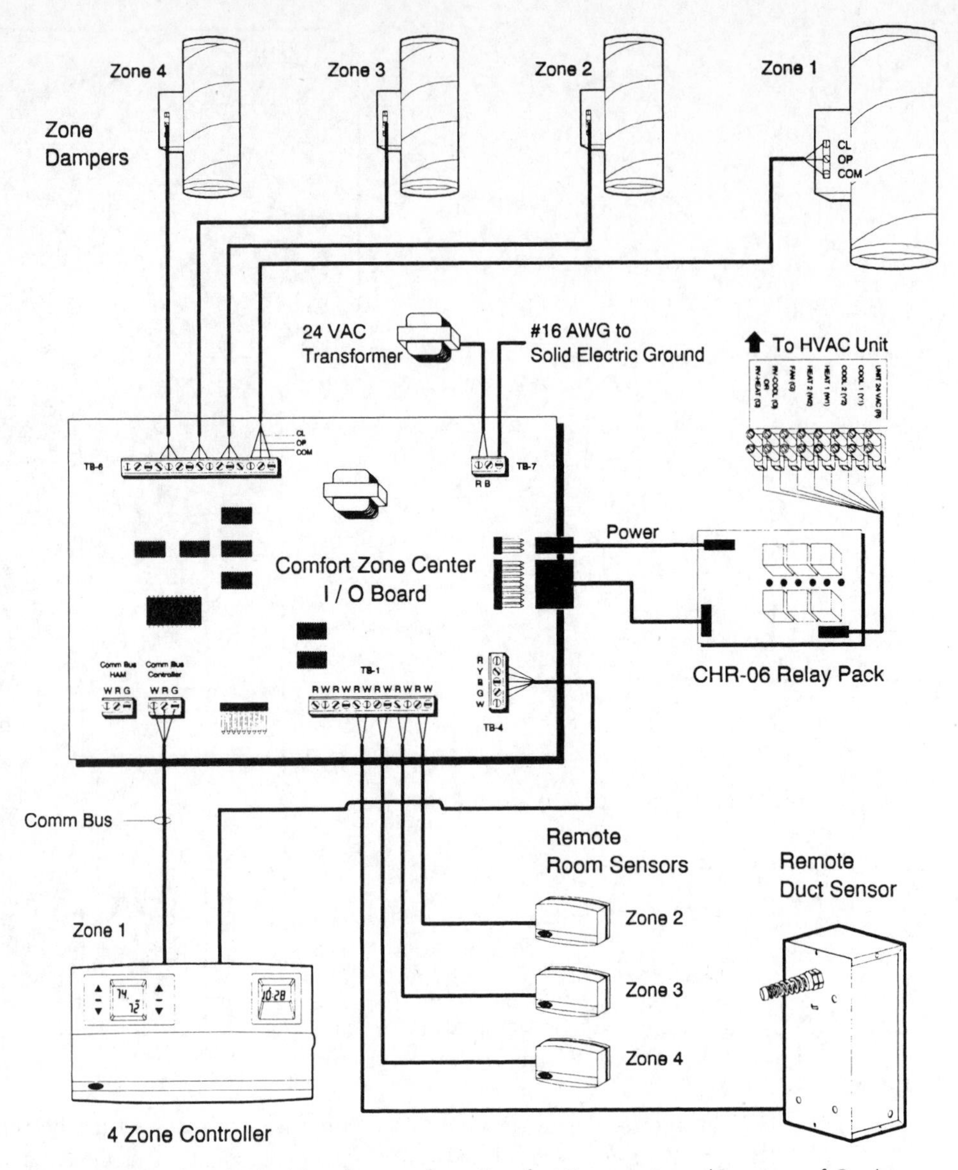

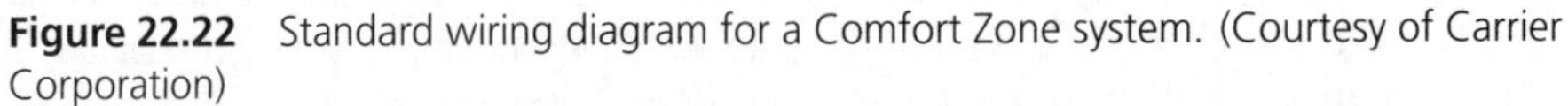
Figure 22.22 Standard wiring diagram for a Comfort Zone system. (Courtesy of Carrier Corporation)

The controller display indicates the zone setpoints, zone temperatures, and programming information for adjusting the setpoints. The clock display indicates the current time and day. During programming, it shows the start times and weekly periods. The programming adjustment buttons are used to program weekly periods and start times for each zone and to set the clock. The system switches provide for selection of

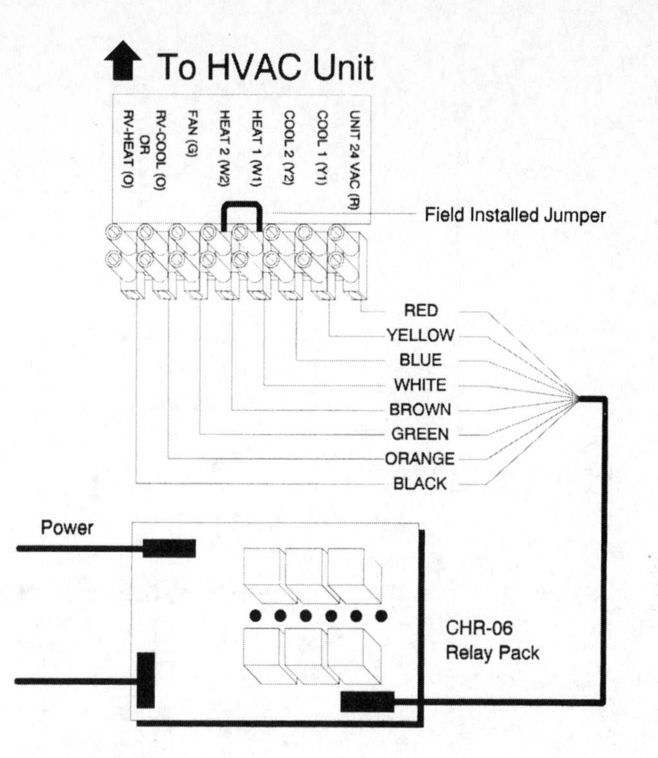

Figure 22.23 Wiring diagram for a single-stage heating system. (Courtesy of Carrier Corporation)

heat, cool, fan, and emergency heat operation. The zone selector dial provides for the selection of each zone and the vacation mode, for programming. It also has an optional selection area for use in installation and service.

DIGITRACT Comfort Control Systems

The DIGITRACT system by California Economizer, Inc. is designed to provide an inexpensive and easy-to-install two-zone system for two-story and ranch-style homes. Many second-floor rooms overheat both in the winter and in the summer due to the gravity flow of warm air. Often, the typical solution is to install two separate HVAC units, one for each floor; however, installing one unit and the DIGITRACT zoning system (Figure 22.26) reduces the equipment, installation, and maintenance costs, and allows for smaller equipment sizing. The normal division in a ranch home is a zone for the living quarters and a zone for the sleeping quarters. This is especially useful in cooling, since the cooling equipment can be sized to supply only the maximum load in one zone, which saves energy as compared with conditioning an entire building at one time.

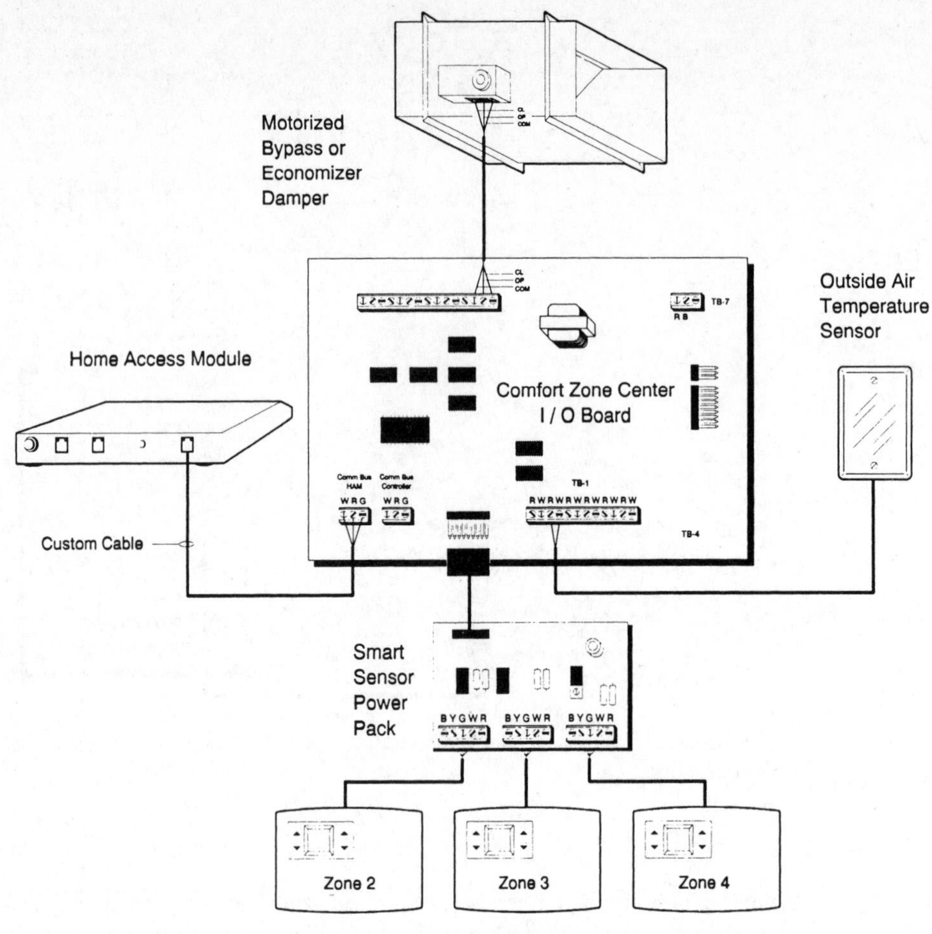

Figure 22.24 Accessory wiring diagram. (Courtesy of Carrier Corporation)

The system consists of:

1. A system controller for either a gas heating/electric cooling or heat-pump application (Figure 22.27)
2. Round or rectangular zone dampers
3. Two standard or programmable thermostats
4. Optional barometric bypass
5. Optional capacity control

To maximize the operation of the DIGITRACT system the manufacturer recommends properly sized ductwork, along with a bypass damper and a capacity control device. The duct system works at normal static pressure and velocity without excessive

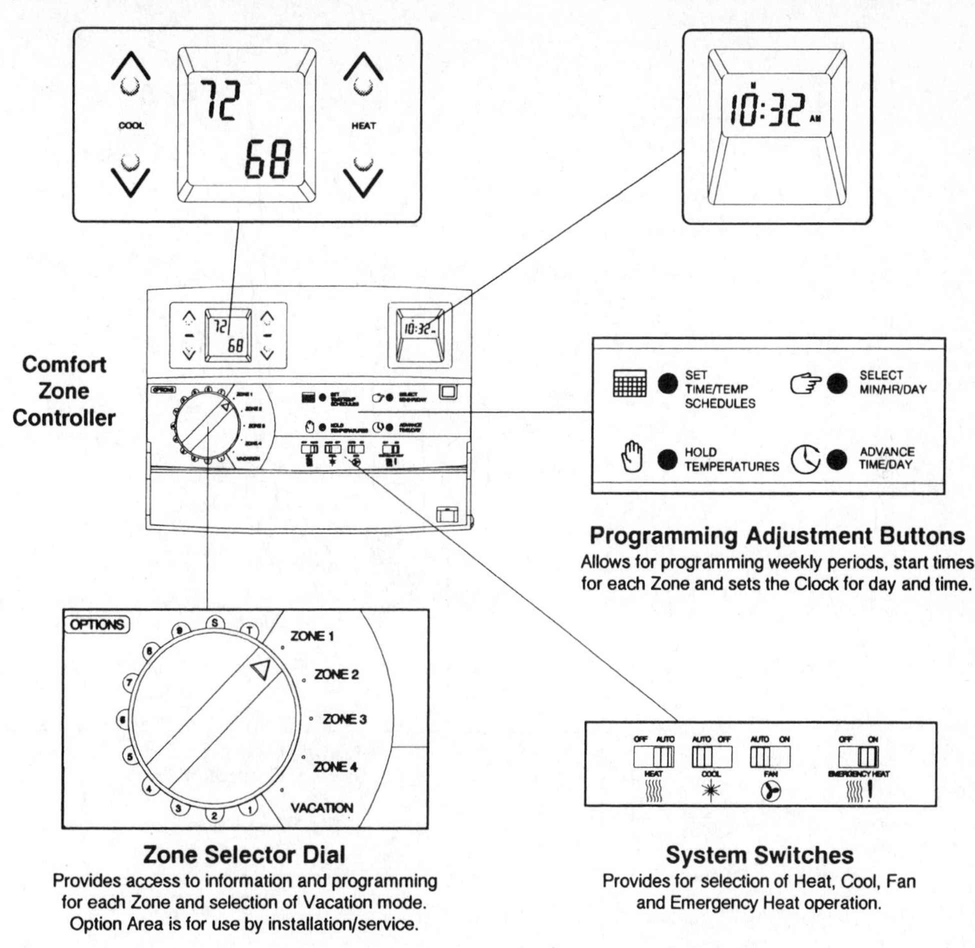

Figure 22.25 Comfort Zone controller. (Courtesy of Carrier Corporation)

noise. The capacity control cycles the compressor and gas valve to avoid coil freeze-up and overheating of the heat exchanger.

Another duct option is to oversize the trunk ducts and branch runs to allow each zone to handle at least 70% of the total system cfm. The use of a bypass damper or capacity control thus may be eliminated. Care must be taken to select supply registers and return grills large enough to avoid air noises. This type of system is best used where air delivery in each zone is roughly equal.

There are two types of DIGITRACT system controllers available. One is used with gas heating/electric cooling (GE) systems (Figure 22.27), the other is for heat

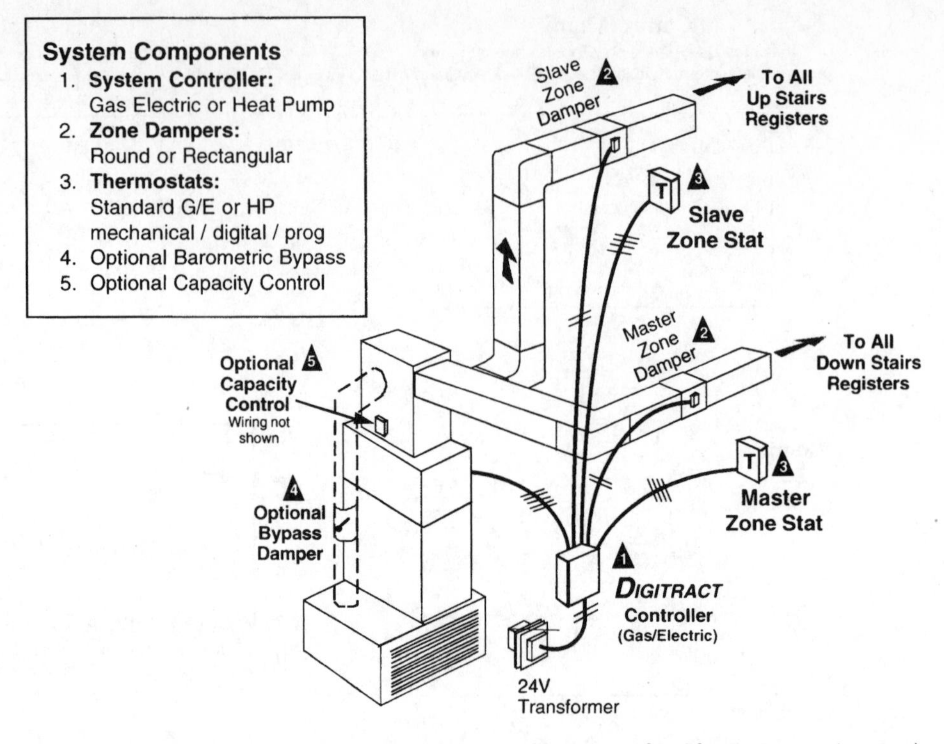

Figure 22.26 DIGITRACT Comfort Control System. (Courtesy of California Economizer, Inc.)

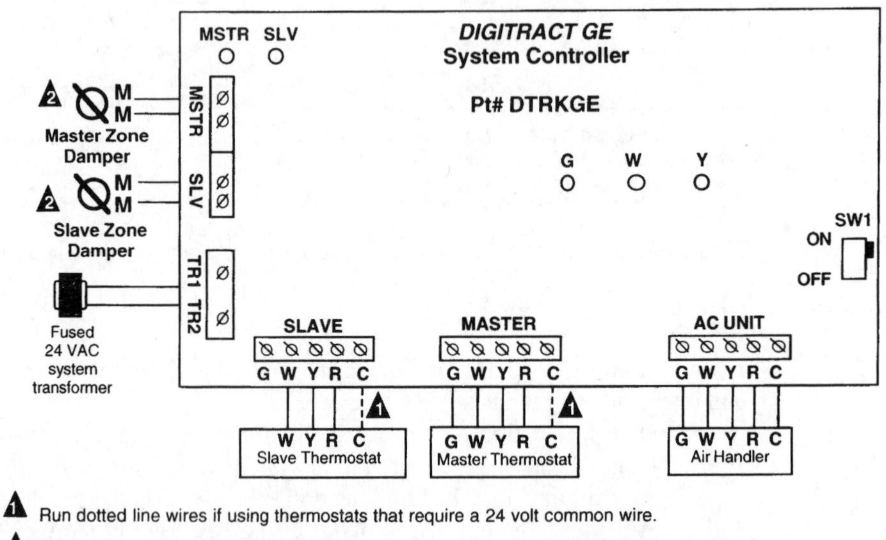

Figure 22.27 DIGITRACT system controller for gas heating/electric cooling application. (Courtesy of California Economizer, Inc.)

pumps. The gas/electric controller provides for automatic changeover, whereas the heat-pump controller is manual changeover.

The DIGITRACT system requires two thermostats, a master and a slave. For gas/electric units the master and slave thermostat must be provided with a cool–off–heat–auto switch for automatic changeover operation and a fan auto-on switch. A standard heat–off–cool thermostat may be used but will require manual changeover from heating to cooling. Only the master zone thermostat can activate continuous fan operation. For heat-pump applications both thermostats must be of heat-pump design with a W2 terminal for electric supplemental heat. Programmable thermostats of the appropriate design may be used.

DIGITRACT Sequence of Operation

1. No zones are calling:
   When neither thermostat is calling for heating or cooling, both dampers are open and the furnace or compressor is off. During this time, unconditioned air can be circulated through both zones by turning the fan switch on the master thermostat to ON.
2. One zone is calling:
   When only one zone is calling, the controller turns on either heating or cooling as called for, and the damper for the zone not calling is powered closed. The MSTR or SLV light on the upper left of the controller is lit corresponding to the zone that is *not* calling.
3. Two zones calling for the same mode:
   When both zones are calling for the same mode of operation, the unit turns on for the mode called, heating or cooling, and both dampers remain open.
4. Two zones calling for different modes:
   If the thermostats are calling for different modes, the furnace or compressor runs the mode of the first thermostat calling and closes the damper of the zone calling in the other mode. After the first thermostat is satisfied, the unit shuts off for 5 minutes, then the dampers change positions and the furnace or compressor starts for the still-calling thermostat. Both heating and cooling needs can be addressed automatically, as required by each zone. If standard heat–off–cool thermostats are used, the the homeowner must manually switch between heat and cool.
5. Lock out and purge:
   After all calls are satisfied, new calls will be locked out for 5 minutes. During this period, the damper of the last calling zone is open and the other zone is closed. If the master thermostat fan switch is in the ON position during this time, any residual conditioned air will be purged into the open zone.

Status Lights There are two damper operation lights and three mode lights on the controller (Figures 22.27 and 22.28). The damper lights, marked MSTR and SLV, indicate that the corresponding damper is closing or closed. The mode light G indicates that the furnace blower is on for cooling or manual operation. During heating, the fan is powered by the fan control, rather than the system controller, and the G light is off. W indicates the furnace is on, and Y indicates the air conditioning is on. The two lights may be

LIGHT	COLOR	FUNCTION
G	Green	Blower fan on
W	Red	Furnace on
Y	Yellow	Air conditioner on
MSTR	Red	Master zone damper close
SLV	Red	Slave zone damper close

Figure 22.28 Status light description table. (Courtesy of California Economizer, Inc.)

on even if at times the equipment is not running. This is because the capacity controller has temporarily shut down the unit to prevent overheating or a coil freeze-up.

Leaving-Air Temperature Control (TRLAT) The TRLAT is a single-stage capacity control that monitors the temperature of the air leaving the the supply plenum. During low-load cooling operation, the TRLAT senses the lower air temperature and cycles the compressor to prevent coil freeze-up. In the heating mode, the TRLAT senses increased leaving-air temperature and cycles the gas valve to prevent overheating. With the use of dip switches, the cooling cutout temperature can be set inside the TRLAT for 41°, 44°, 47°, or 50°F. The heating cutout can be set for 125°, 140°, 150°, or 160°F. For heat pumps the cutout is 118°F and not changeable. The TRLAT is wired between the DIGITRACT system controller (AC unit terminals) and the furnace terminal board (Figure 22.29). It is physically mounted in the supply-air duct, as far away from the evaporator coil as possible. It must not be mounted past the takeoff for the bypass damper (Figure 22.30).

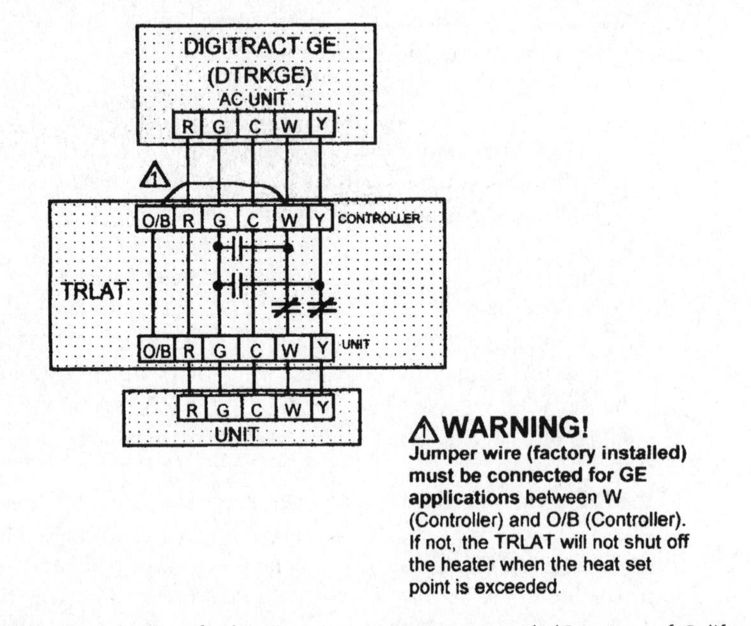

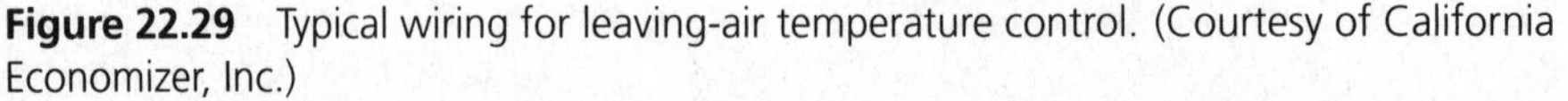
Figure 22.29 Typical wiring for leaving-air temperature control. (Courtesy of California Economizer, Inc.)

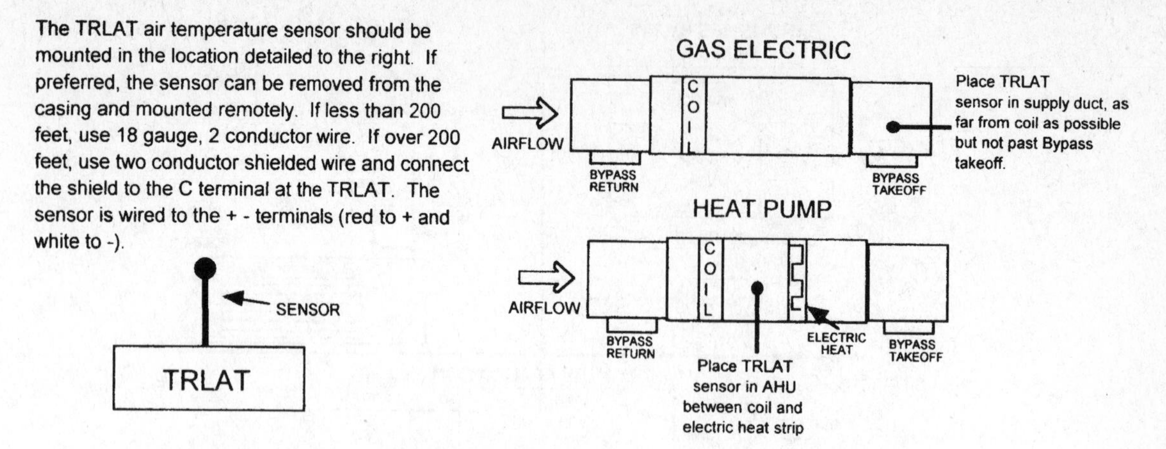

Figure 22.30 Sensor placement. (Courtesy of California Economizer, Inc.)

Bypass Damper A bypass damper is used to provide constant air delivery across the evaporator coil and through the heat exchanger. The bypass is a barometric damper and must be sized for the nominal cfm of the system (Figure 22.31) up to a maximum of 2000 cfm for a 5-ton unit. The goal of the damper is to modulate open, allowing excess air to flow from the supply side to the return side of the air-handling unit (Figure 22.32). In this way, excessive pressure cannot build up in the duct system when one of the zone dampers is closed. Maintaining constant duct pressure prevents intermittent air-velocity noise at the registers.

Since it is not motorized, the barometric bypass damper must be adjusted carefully.

1. Set all thermostats for no call and turn the fan switch to ON.
2. Adjust the counterweight (Figure 22.32) to balance the damper so that it is just ready to open. The damper should stay in the closed position.
3. Set one of the thermostats to call for heating or cooling. The bypass damper should open. If it does not, readjust the counterweight.
4. Set the second thermostat to call for the same heating or cooling as in step 3. The bypass damper should shut. If it does not, readjust the counterweight.

SYSTEM SIZE	DAMPER SIZE	PART #	L	W	NOMINAL CFM
UP TO 2.5 TONS	9 INCH	101ABBD09	11"	12"	1000
3 TONS	10 INCH	101ABBD10	12"	13"	1200
4 TONS	12 INCH	101ABBD12	14"	15"	1600
5 TONS	14 INCH	101ABBD14	16"	17"	2000

Figure 22.31 Bypass selection chart. (Courtesy of California Economizer, Inc.)

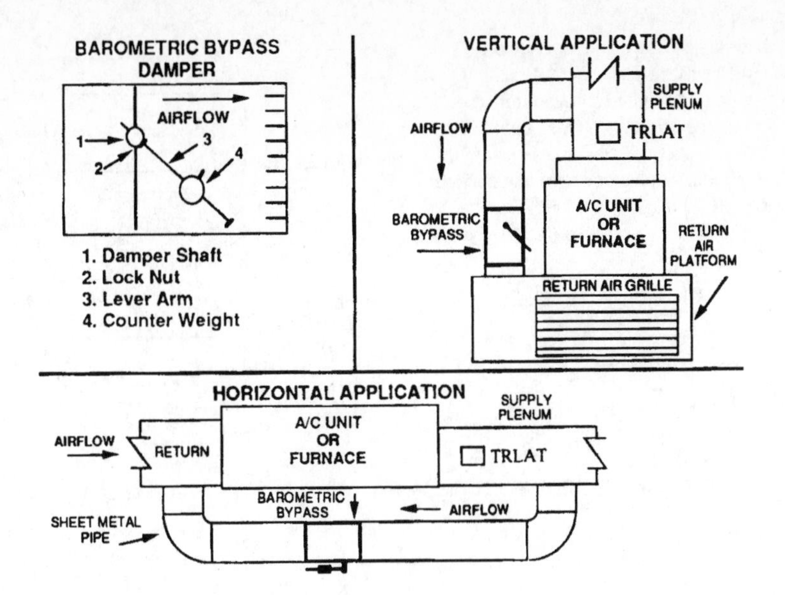

Figure 22.32 Bypass damper applications. (Courtesy of California Economizer, Inc.)

The DIGITRACT Zone Control System is used for two zone residential or small commercial applications. For larger homes or offices the System 1000 Zone Control allows a single HVAC unit to be controlled by up to seven zone thermostats (Figure 22.33).

Enerstat Zone Control System 2

Enerstat offers zone control systems for 2-, 5-, and 10-zone applications. We shall describe only the 2-zone system. Information on the other systems can be obtained from the manufacturer or from its sales outlets. System 2 is designed to condition only the occupied half of the residence at one time. It is suggested that the living area be selected for one zone and the bedroom area for the other. Some of the important features included in the system are as follows:

1. Either solid-state programmable or standard thermostats can be used
2. Solid-state logic panel
3. Logic panel can be used with heat pumps and gas, oil, or electric furnaces
4. Will control one- or two-stage heating-cooling system
5. Auto or manual heating/cooling changeover
6. Continuous or intermittent fan operation in both zones
7. Anti-short-cycle protection
8. Two-position or modulating damper operation
9. Bypass damper not required for most systems
10. LED (light) indication on all relays

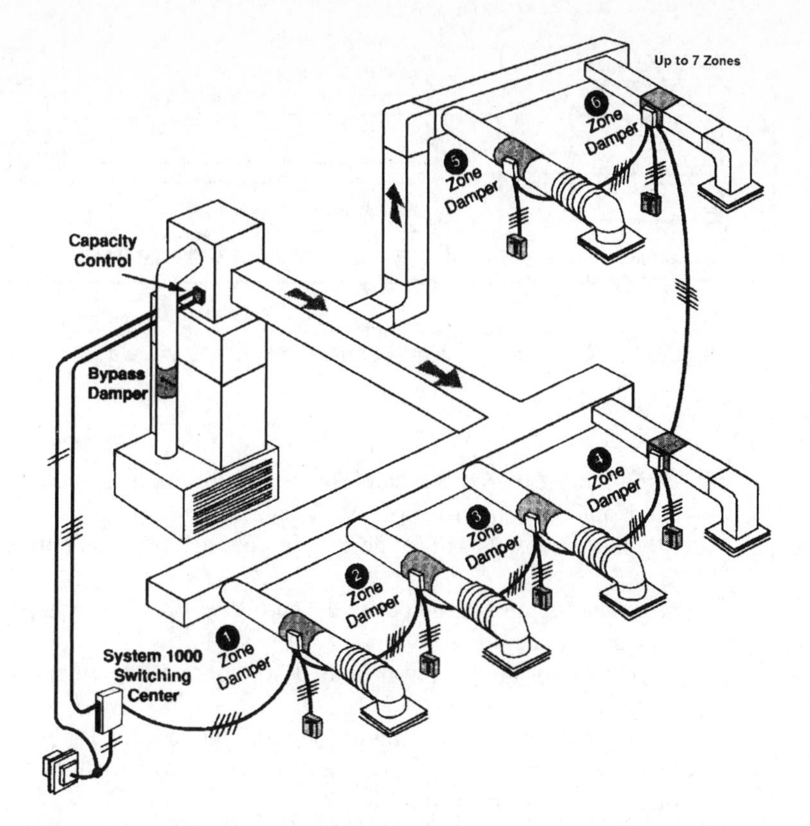

Figure 22.33 California economizer system 1000. (Courtesy of California Economizer, Inc.)

Description of the System Two programmable setback/set-up or conventional thermostats are required to control the two-zone damper motors. The thermostats are programmed or set manually to deliver the required amount of heat or cooling to each area. The system can be set in either manual or automatic heating/cooling changeover. If set in the automatic position, the first thermostat to call for heating or cooling establishes the mode of the system. Short-cycling protection is provided to prevent rapid changeover. A typical system drawing is shown in Figure 22.34.

Sequence of Operation If either the zone 1 or zone 2 thermostat calls for heating or cooling, the fan will start unless delayed by the fan limit control. The heating and cooling equipment cycles as needed. The operating sequence for the zone dampers is as follows:

1. If the zone 1 thermostat calls for heating or cooling, the zone 1 damper remains open, and the zone 2 damper is driven closed.

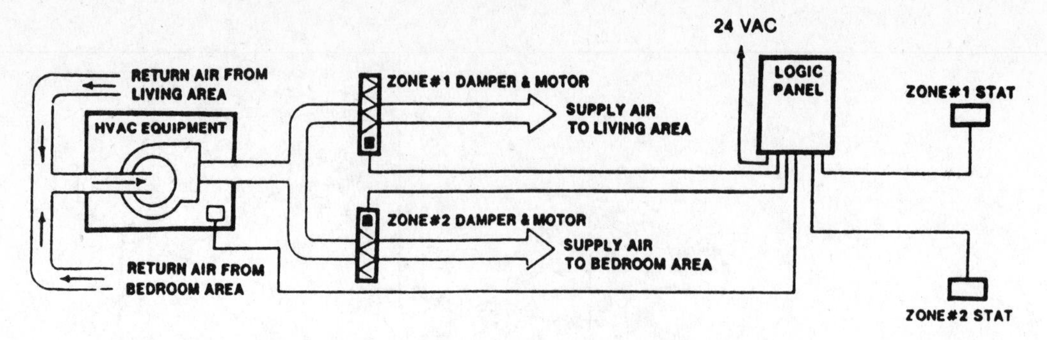

Figure 22.34 System drawing. (Courtesy of Enerstat–Valera Corp.)

2. If the zone 2 thermostat calls for heating or cooling, the zone 2 damper remains open, and the zone 1 damper is driven closed.
3. If both thermostats call for heating or cooling at the same time, both zone dampers open.
4. If a power failure should occur, the spring return on the damper motors will operate to open both dampers.
5. The system can be programmed to operate in either manual or automatic changeover.
6. If the thermostats are calling for opposite modes (one for heating, one for cooling), the zone calling first is served first. The other zone will be served 5 minutes after the first zone stops calling.

Schematic Diagram of Connections to the Logic Panel Figure 22.35 shows the wiring connections from the logic panel to various components, for both the gas/oil and heat-pump systems. A key to the terminal designations is shown, as are a series of notes that apply to the operation of these systems.

Design of the Air Distribution System

1. It is recommended that the duct for each zone be sized for approximately 75% of the total air-supply cfm. This will increase the outlet velocity for the zone that is calling only a small amount. The important result is that the cfm of air across the cooling coil is kept above minimum requirements.
2. Often, all that is necessary is to increase the branch size from 6 in. round to 7 in. round and to increase the size of the supply grilles from 8 in. × 4 in. to 12 in. × 4 in.
3. If the condition exists that one zone is considerably larger than the other, it is recommended that some leakage be provided past the larger zone damper when it is in closed position (about 20%) to release excess air pressure.
4. Balance dampers are recommended on all branch runs to facilitate balancing.

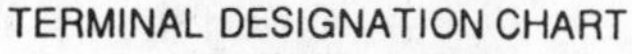

TERMINAL DESIGNATION CHART
(for Heat-Cool systems)

W	- Heating (one stage only)
W1	- First stage Heating
W2	- Second stage Heating
Y	- Cooling (one stage only)
Y1	- First stage Cooling
Y2	- Second stage Cooling
G	- Fan
RC	- 24 Vac Hot (cooling)
RH	- 24 Vac Hot (heating)
X	- 24 Vac common
D1	- Damper zone #1
D2	- Damper zone #2

TERMINAL DESIGNATION CHART
(for Heat Pump systems)

AUX	- Auxiliary Heat
EH	- Emergency Heat
HP1	- Heat Pump 1st stage
HP2	- Heat Pump 2nd stage
Y	- Compressor (heating/cooling)
G	- Fan
O	- Changeover (cooling)
B	- Changeover (heating)
RC	- 24 Vac Hot (cooling)
RH	- 24 Vac Hot (heating)
X	- 24 Vac Common
D1	- Damper zone #1
D2	- Damper zone #2

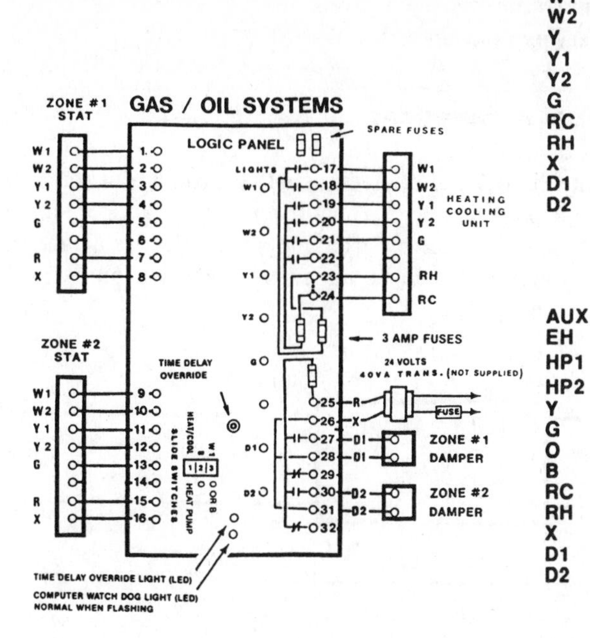

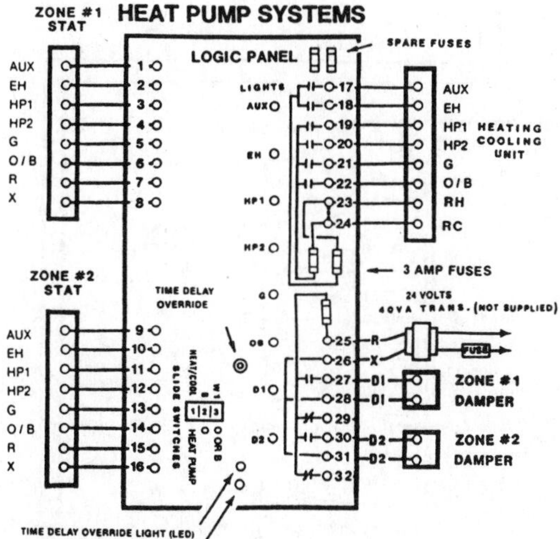

NOTES

1) Logic panel has built-in minimum on, minimum off time delays and may not appear to respond to a signal from the thermostats for several minutes.

2) All heating and cooling equipment must supply it's own 24 Vac power. If separate transformers for Heating and Cooling are used, remove jumper between 23 and 24 (RH-RC).

3) If only two zone dampers are required, contractor should furnish and install a 40VA control transformer and a Buss SSU box with a 0.5 Amp fuse.

4) If system 2 is installed in a residence, one stat should be located in the living area (not in a hall). The other stat should be located in the master bedroom (not in the hall).

5) It is normal for D2 light (LED) to come on when power is connected to panel even if there are no calls from the thermostat.

6) Damper motors are powered closed and spring return open. Lights D1 or D2 will be lit if damper is open.

Up to 3 motors may be connected in parallel on each zone (must use 75VA transformer).

Damper motor power requirements: 24Vac at 0.5 Amp, 12VA.

Logic Panel power requirements: 24Vac at 0.6 Amp, 15VA. (add 12VA for each damper)

Figure 22.35 Logic panel connections. (Courtesy of Enerstat–Valera Corp.)

Preparing the Logic Panel for Operation Refer to the manufacturer's instructions for:

1. Setting up the logic panel
2. Overriding the compressor time delay during testing
3. Use with special heat-pump equipment

Special Zone Damper Applications

1. In a few applications, individual dampers must be installed in some or all of the branch runs. The logic panel is set up to handle up to three damper motors per zone. Additional dampers may be added by installing an auxiliary relay and transformer.
2. If needed, an additional zone can be added by installing an extra thermostat, damper, transformer, and relay.

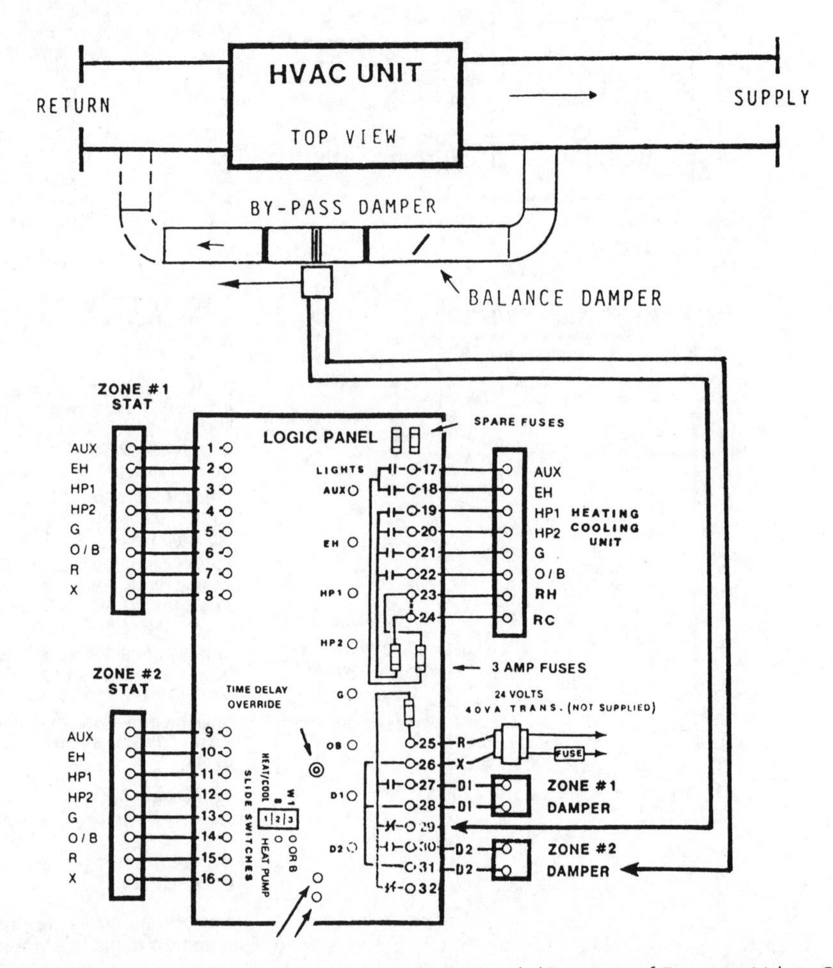

Figure 22.36 Bypass damper connections to logic panel. (Courtesy of Enerstat–Valera Corp.)

Bypass Dampers If one zone is substantially larger than another, it may be advisable to install a bypass damper to relieve the air pressure. Provision is made in the logic panel for this addition. High- and low-limit discharge thermostat should also be installed. It is recommended that the low limit be set at 50°F and the high limit at 150°F. The wiring connections to the logic panel are shown in Figure 22.36.

Two-speed Fan Operation Two DPST 24-V ac relays are required to control a two-speed fan. If one zone damper is open, the fan will run at low speed. If both zone dampers are open, the fan will run at high speed. If the unit is equipped with two-stage cooling, only one stage will come on when a single zone is calling. See Figure 22.37 for wiring connections of a two-speed fan to the logic panel.

Auxiliary Zones An auxiliary zone is used to serve a light-load area, such as a basement or single room. It has no direct control of the heating or cooling equipment. The branch duct to the auxiliary zone can be taken off ahead of or after the existing zone damper. Wiring diagrams for the auxiliary zone controls are shown in Figure 22.38.

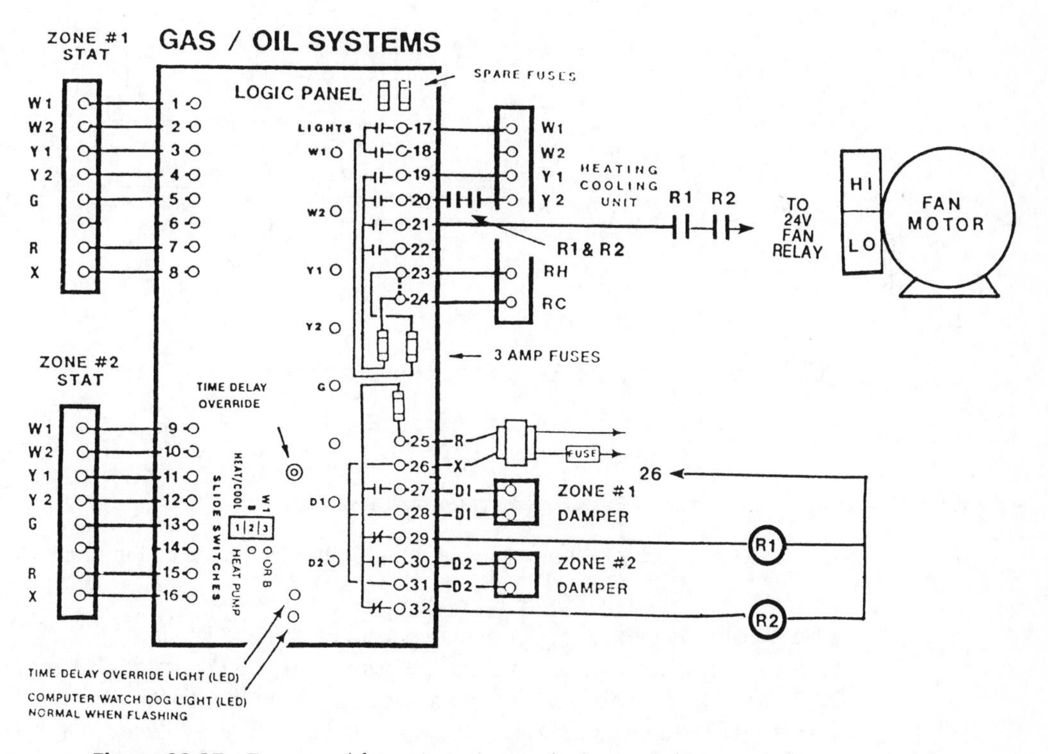

Figure 22.37 Two-speed fan connections to logic panel. (Courtesy of Enerstat–Valera Corp.)

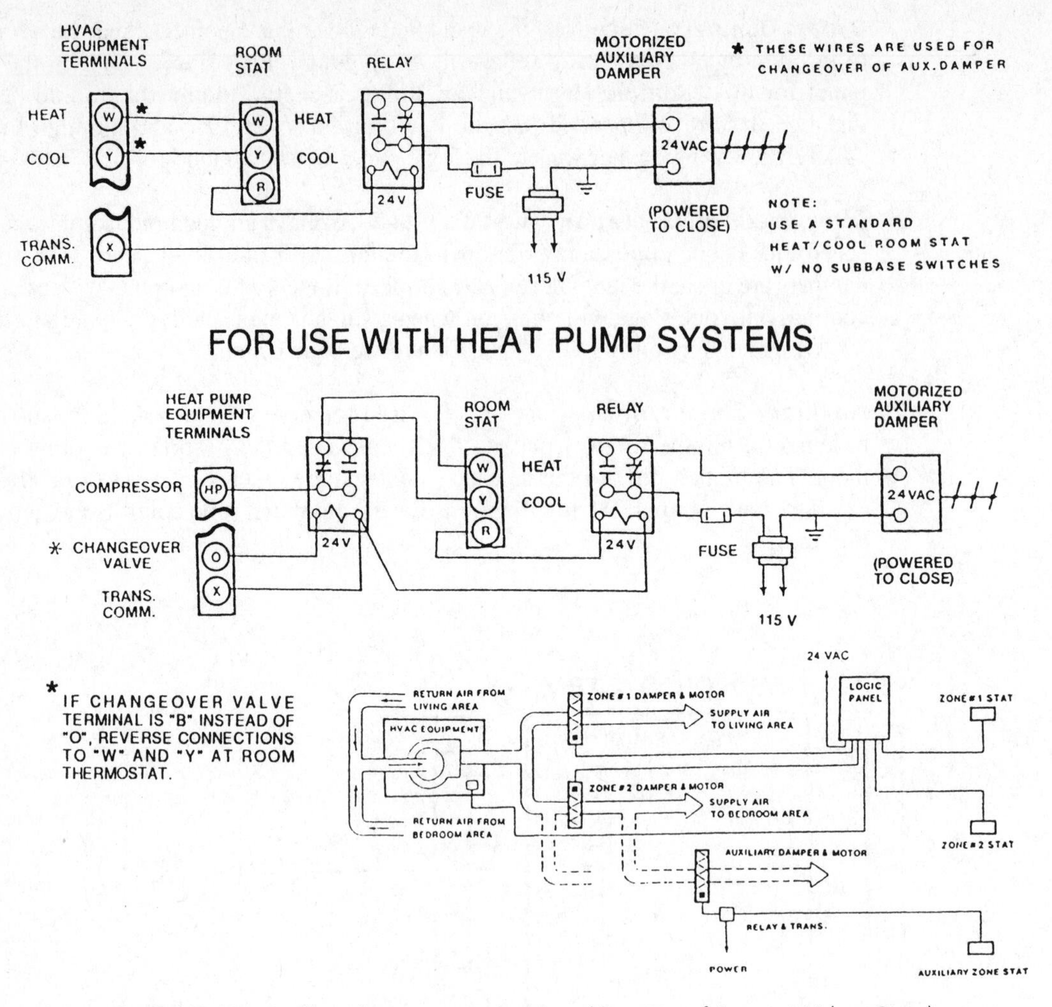

Figure 22.38 Auxiliary damper control wiring. (Courtesy of Enerstat–Valera Corp.)

STUDY QUESTIONS

Answers to the study questions may be found in the sections noted in brackets.

22-1 What is meant by *zoning? [Zoning]*
22-2 What is the procedure for planning a zoned system? *[Zoning]*
22-3 For a two-zone system, how is the residence usually divided? *[Zoning]*
22-4 How is excess air pressure controlled? *[Relieving Excess Air Pressure]*
22-5 Describe the types of zone dampers available. *[Types of Zone Dampers]*
22-6 Describe the types of automatic bypass dampers. *[Automatic Bypass Dampers]*

22-7 Describe the steps in adjusting a barometric bypass damper. *[(DIGITRACT) Bypass Damper]*

22-8 What accessories are available for the Trol-A-Temp system? *[(Honeywell) Accessory Equipment]*

22-9 What accessories are available for the Comfort Zone system? *[(Carrier) Accessories]*

22-10 Describe the essential parts of the Comfort Zone system controller. *[(Carrier) Programming and Operating the System]*

CHAPTER 23

Hydronic Heating

OBJECTIVES

After studying this chapter, the student will be able to:

- Determine the proper design for a residential hot water heating system
- Select the system components required for an installation
- Troubleshoot a system and solve problems

Introduction

In certain areas of the country there is a preference for using hot water for residential heating. Modern hot water heating systems require an understanding of hydronic technology, such as the physics of heat transfer, piping circuitry, control operation, and combustion characteristics. The hydronic heating system uses boiler-heated water as a heat-transfer medium. The hot water is pumped through the radiation located in the spaces to be heated. The comfort received from this type of system is highly desirable.

The following is a brief review of some of the features of hydronic heating:

1. Heat is supplied along the outside wall where the heat loss is greatest. The outside walls and window areas are heated, reducing the radiation loss from the occupants to these cold surfaces.
2. Air rises by convection from the baseboard heat source toward the ceiling, then it circulates through the room, returning along the floor to the bottom of the baseboard (Figure 23.1). This gentle movement of warm air, without drafts, adds to the comfort of the system.
3. The mass of hot water being heated creates a "flywheel" effect. The room temperature heats up and cools down slowly, preventing sudden changes in room temperature that can be uncomfortable and preventing the "cold 70" condition common to many forced-air heating systems, in which the thermostat is satisfied but the occupants are uncomfortable.

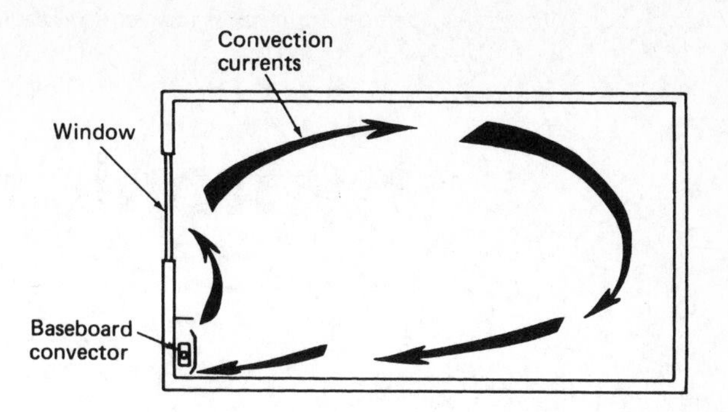

Figure 23.1 Convection air current.

4. Panel-type radiant heating is designed to supply heat by circulating the warm water through an arrangement of piping, either installed in the floor, subfloor, or ceiling, as shown in (Figure 23.2).
5. Should the occupants require additional humidification or desire summer cooling, a separate air system needs to be provided. A cooling system designed specifically for homes with hydronic heat can supply adequate cooling through ceilings or side walls.

Types of Systems

Five common piping arrangements are employed in a residential hot water heating system: the series loop, one pipe or single main, two-pipe reverse return, two-pipe direct return, and radiant panel. These piping arrangements are described next.

A series-loop baseboard system is shown in Figure 23.3. A single pipe, called a main, enters directly into a section of baseboard, exits at the opposite end, and then continues on to the next section. Thus the water flowing through one section of baseboard will flow through all sections of baseboard. A variation of the series loop is the zone control system, in which two or more loops are piped in parallel, and each is controlled individually with a thermostat and a zone valve.

Both the series-loop and the zone control systems are arranged so that the circulator operates when any zone calls for heating. The boiler burner is also cycled "on"

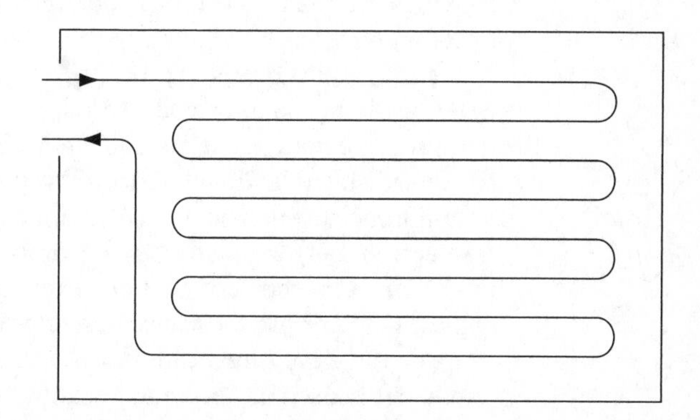

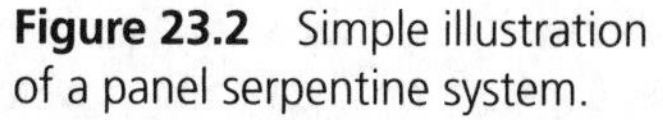
Figure 23.2 Simple illustration of a panel serpentine system.

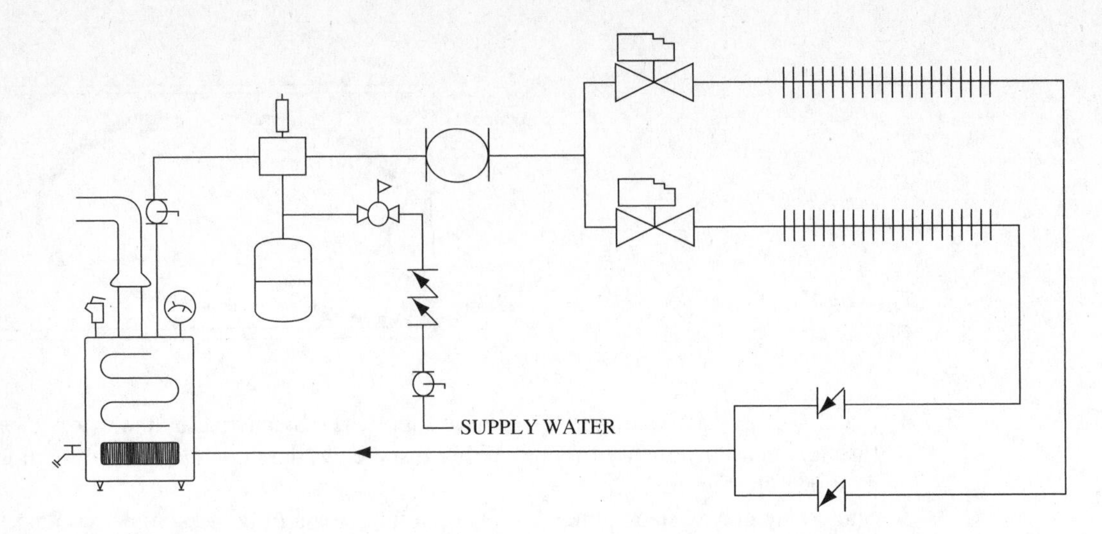

Figure 23.3 Series-loop baseboard system with two zones. (Refer to Legend on page 524)

when any one of the zones calls for heat. Boiler water temperature is maintained by an immersion aquastat that operates the gas valve or oil burner. A high-limit aquastat will override the burner operation should the water temperature exceed its limit setting. The supply water main is pitched upward from the boiler. The return main is pitched downward to the boiler. Air vents are provided for each convector and at the end of the supply main.

A simple one-pipe or single-main system utilizes a single pipe or main that is run from the boiler outlet, alongside all the radiation, and back to the boiler return connection. The system uses a special tee, called a *diverter fitting,* to control the water flow through the individual sections of radiation, as shown in Figure 23.4.

When the radiation branches are connected the diverter fitting is installed on the supply end of the main, and a standard tee is used to connect the return. For proper location of the tees on the main, check with the manufacturer of the radiation. In cases where the radiation has to be located below the main, two diverter tees must be used, and they should be spaced the same distance apart as the length of the radiation.

The two-pipe reverse-return system uses one pipe or main from the boiler to supply hot water to the radiation and a second pipe or main to return the water to the boiler. The hot water supplied to the first radiation off the main is the last to return to the boiler, as can be seen by following the path of the piping shown in Figure 23.5. This piping design is said to be self-balancing. The main frequently decreases in size as each section is taken off and increases in size as additional return piping is added on the path back to the boiler. It is recommended that even though the piping is self-balancing, balancing valves should be used, because of pump sizing and circulator velocity.

The two-pipe direct-return layout is shown in Figure 23.6. The direct-return system is rarely used in residential heating because of the balancing problems that could

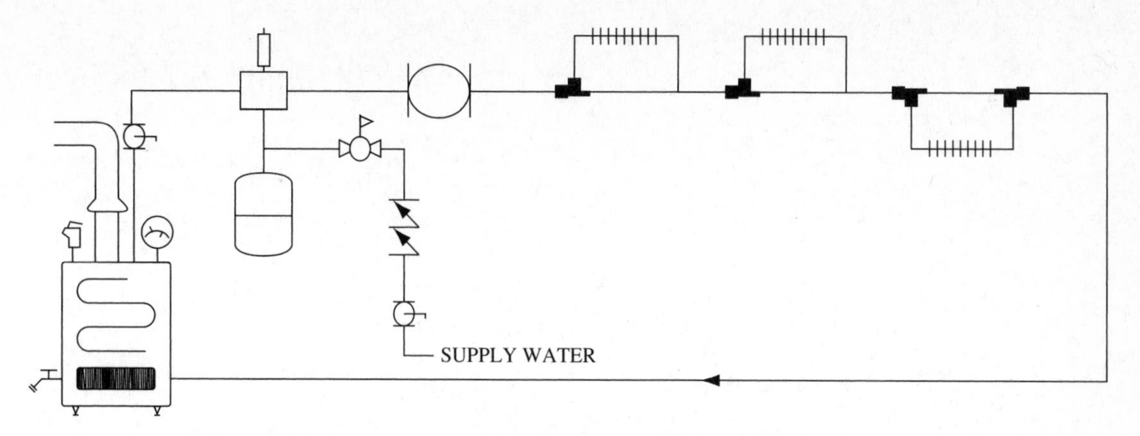

Figure 23.4 One-pipe or single-main system with diverter fitting.

arise. The piping follows the principle "first in" and "first out," as shown in the figure, and is run as short as possible. To prevent short-circuiting of the water, balancing valves are used on each section of radiation. The water temperature in the main is the same to all sections of the radiation or convectors. The temperature drop through the various sections is determined by the balancing valve setting. The manufacturer's recommendations for a balancing valve should be followed because of the critical flow rate required in each section of radiation. The restriction to water flow caused by the balancing valve also affects the temperature output from the radiation. On commercial projects the general contractor's engineer will specify the balance valve setting.

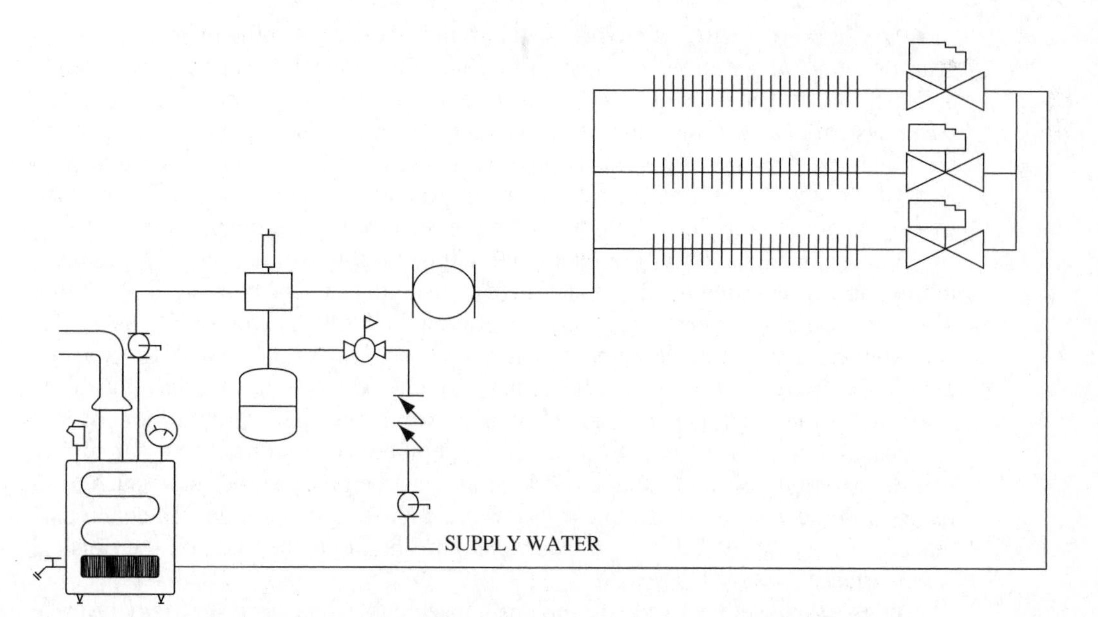

Figure 23.5 Two-pipe reverse-return system showing balancing valves.

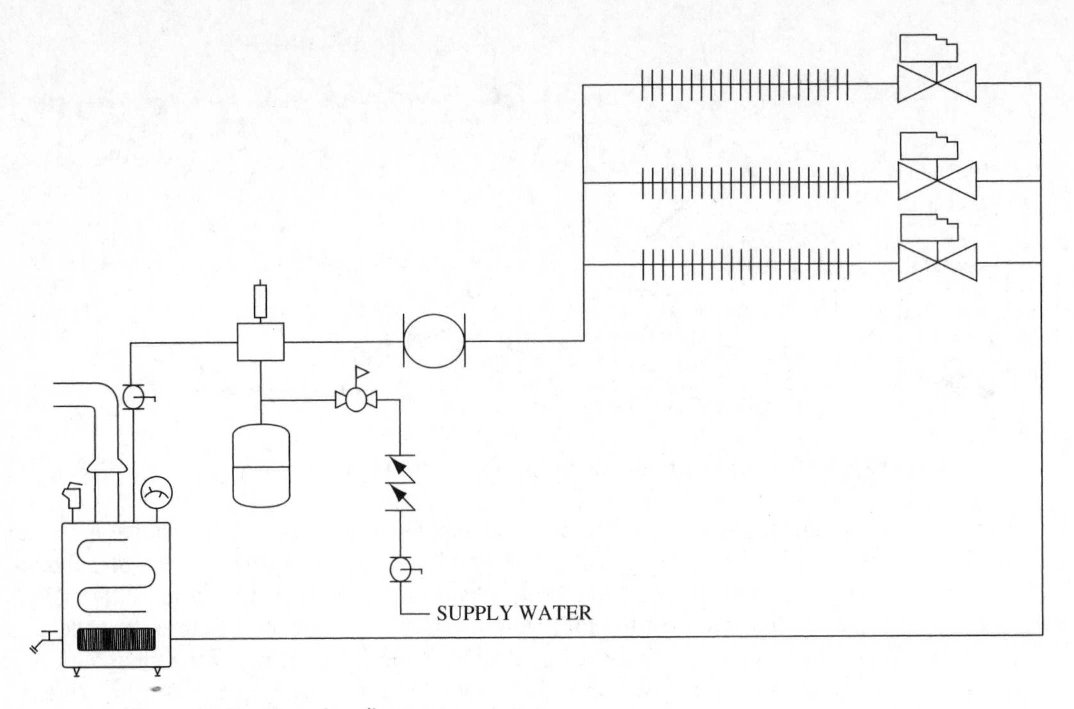

Figure 23.6 Two-pipe direct-return system.

Panel hot-water systems are installed in floors or ceilings, utilizing a supply and a return main to which all the coils are connected. Balancing valves and vents on each coil section can then be made accessible in a floor pit or in the ceiling of a closet. The panel system is shown in Figure 23.7. This type of design uses panels or serpentine coils. *Serpentine* is the term used for pipes routed back and forth in an S-arrangement. A similar system can be designed for snow removal. Freeze protection can be provided by using a glycol antifreeze solution as the transfer medium instead of pure water.

It is recommended that the technician contact the manufacturer of equipment, tubing, and accessories for assistance before installing or servicing a panel system. The floor surface temperature should be between 85°F (30°C) and 90°F (38°C.) The water temperature in the tubing below the floor should not exceed 140°F (90°C), and a mixing valve is required to vary the temperature based on the insulation value of the floor. For instance, a higher supply temperature would be required for a floor covered with carpet than for a floor with the tubing embedded in concrete or for piping between subfloor joists. Special consideration must be given to tile floors such as in bathrooms and the kitchen. A thin gypsum-based underlayment or poured concrete is placed on top of the subfloor and made smooth. The tile is then laid on the finished floor surface.

Years ago, panel heat was installed using steel and copper piping. Many of these systems failed when the corrosive action of the concrete created hard-to-repair leaks in

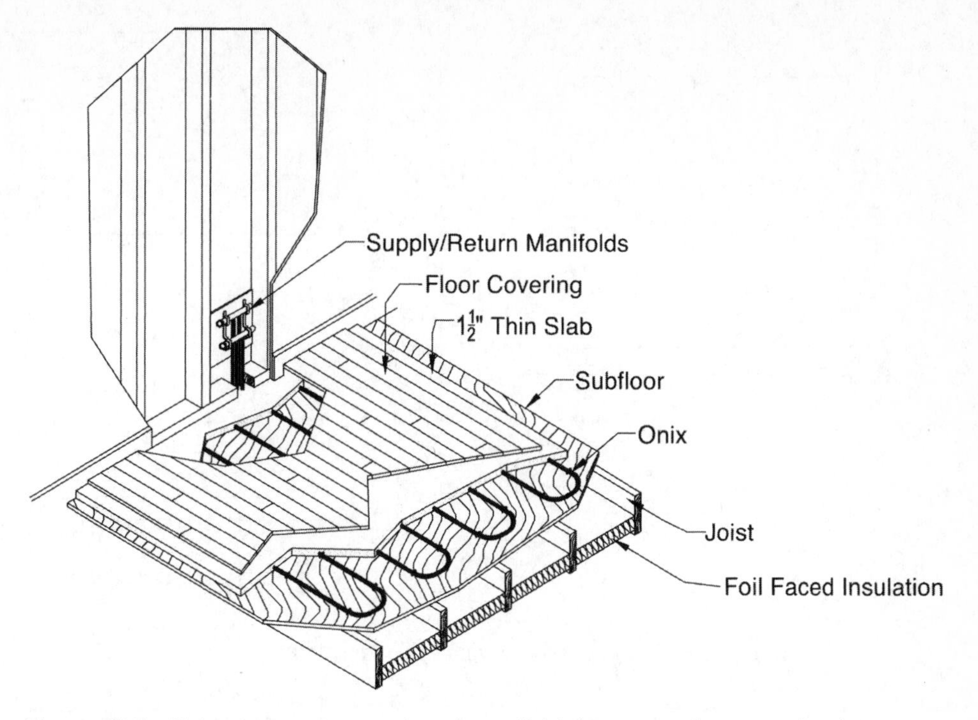

Figure 23.7 Piping arrangement and trunk manifold. (Courtesy of Watts Radiant)

the buried pipe and fittings. Today, cross-linked polyethylene (PEX) or multicomposite tubing is used for panel-heat applications because it is extremely tough and bendable. This tubing is available in various lengths and is normally supplied in 160- to 200-ft lengths with pipe diameters ranging from ½ to 1 in.

Primary/Secondary System

The piping layout shown in Figure 23.8 illustrates a primary/secondary system. The benefit is that heated water from the boiler blends with the return water from the radiation. The basic system shown is designed to handle the heating load of three different areas. There are four circulators involved in this type of system. Circulator A is the primary circulator and maintains a constant temperature of approximately 120°F (55°C) between the boiler and the secondary loops. The blending of return water and boiler water reduces the possibility of thermal shock to the boiler on a call for heat in any of the secondary units. It also prevents the flue gases from condensing in the boiler, which could cause severe damage over time.

The primary circulator operates continuously while the secondary loops demand hot water. The temperature in the primary loop may rise as the demand increases, but it will never fall below the preset primary loop temperature.

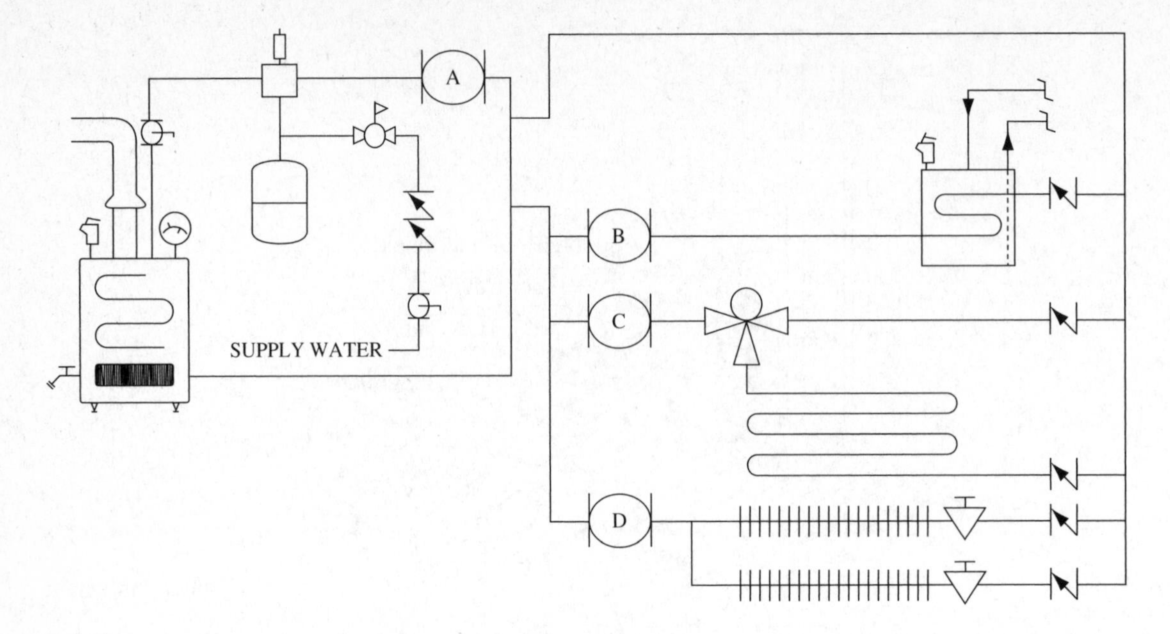

Figure 23.8 Primary/secondary pumps and piping.

Circulators B, C, D are secondary circulators, providing hot water to the various zones. Circulator B supplies hot water to the domestic hot-water tank. Circulator C supplies hot water to the panel-heat section, maintaining a pipe temperature not to exceed 120°F (50°C) in the floor. This temperature setting is maintained by a three-way control valve.

Circulator D supplies hot water to different sections of baseboards or convectors. If the radiation is in a different room, balancing valves are recommended to obtain an even temperature. These are shown in Figure 23.8. Check valves are installed in each of the zones to prevent backflow.

The secondary circulators run independently on the demand from the zone they serve. These circulators utilize the compression tank on the primary loop and should always pump away from the supply tee.

Selecting the Equipment for a Typical System

We now consider some of the design aspects of hot water heating systems, lay out a typical residential system, and select suitable equipment. The following are the steps involved:

1. Determine the heating load.
2. Sketch the piping layout.
3. Select the circulator pump.

4. Select the boiler.
5. Select the radiation.
6. Select the accessories.
7. Select the control system.

Determining the Heating Load

Refer to Chapter 16 for calculating the heating load for the building. This information provides not only the total load in Btuh under design conditions but also the loads for individual rooms.

For our typical house we shall use the one-floor layout shown in Figure 23.9. We assume a full basement with space for the boiler and the piping. The individual loads for the rooms are as follows:

Room	Btu Heat Loss
Bedroom #1	8,431
Bath	2,614
Kitchen	13,601

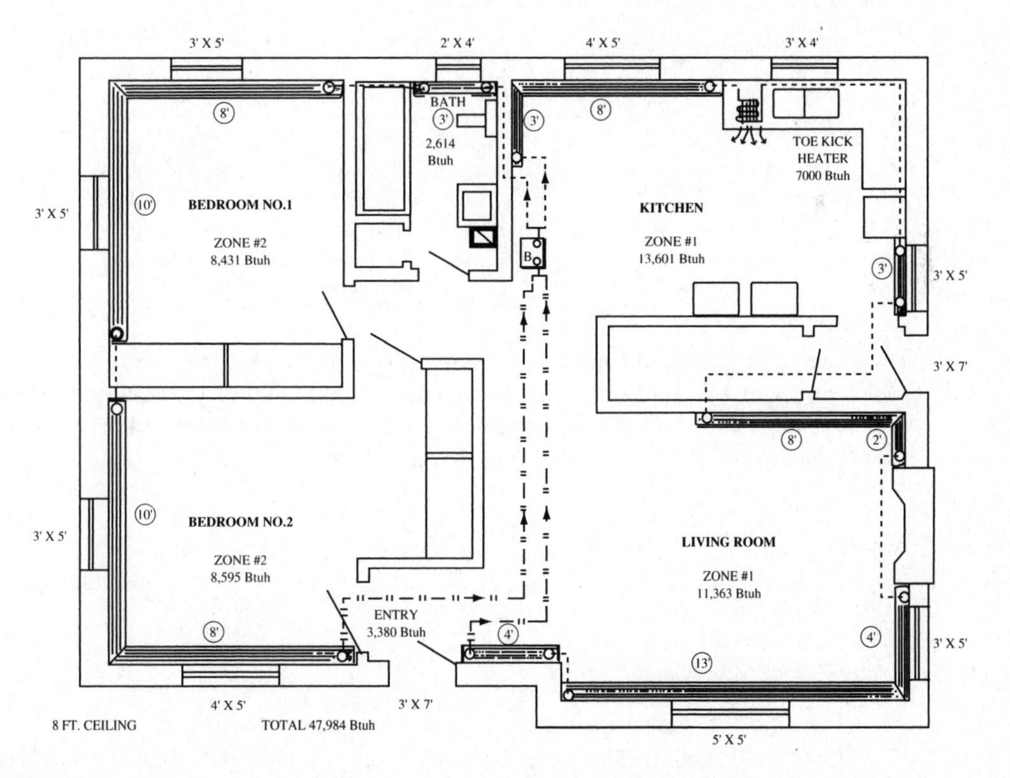

Figure 23.9 House plan showing boiler and piping. Goes up if plan is turned.

Living Room	11,363
Entry	3,380
Bedroom #2	8,595
Entire House	47,984

Sketching the Piping Layout

For our example we shall use a two-zone series-loop system using baseboard fin-tube radiation (Figure 23.10). The heating elements will be installed along the outside wall, to offset as much heat loss as possible. In this type of arrangement, the main trunk piping is run within the baseboard enclosure, along with sufficient fin tube to match the load requirements. The total length of pipe and fin tube measures 102 ft for the bedroom zone and 128 ft for the living zone. Because each circuit has 12 ft of calculated pipe length contained in the same boiler manifolds, the total length for the house is 218 ft (102 + 128 − 12 = 218).

Sizing the Circulator and Trunk Lines

Zone Selection

Zone 1			*Zone 2*		
Kitchen	13,601	Btuh	Bathroom	2,614	Btuh
Living room	11,363	Btuh	Bedroom #1	8,431	Btuh
Entry	3,380	Btuh	Bedroom #2	8,595	Btuh
Total	28,344	Btuh	Total	19,640	Btuh
Total length of radiation required for zone 1 is 65 linear ft.			Total length of radiation required for zone 2 is 61 linear ft.		

The Btu capacity of a hot water heating system is equal to the gallons of water circulated per minute (gpm) times the temperature drop (TD) of water, times a conversion factor of 500, which is the product of the pounds of water per gallon (8.33) times the number of minutes in an hour (60). In formula form,

$$\text{Btu} = \text{gpm} \times \text{TD} \times 500 \qquad \text{gpm} = \frac{\text{Btu}}{\text{TD} \times 500}$$

For our system example,

$$\text{circulator gpm} = \frac{47{,}984 \text{ Btu}}{20 \times 500} = 4.8 \text{ gpm}$$

Note: A TD of 20°F between the boiler supply outlet and the boiler return is considered good practice.

The pressure in feet of head that the pump needs to overcome must be determined:

$$1 \text{ ft of head} = 0.433 \text{ psi (pounds per square inch)}$$

$$1 \text{ psi} = 2.31 \text{ ft of head}$$

The example house plan is divided into two zones, as indicated. The boiler will have a 1-in. supply and return manifold, as that size falls within the range available on the chart in Figure 23.10. Each zone will have a ¾-in. baseboard series-loop system, because that size is commercially the most available and the least expensive. The baseboard information in Figure 23.10 indicates that with an average water temperature of 180°F (85°C) each foot of ¾-in. baseboard will emit 580 Btuh per linear foot. In the house plan, the kitchen requires more heat than the available wall space allows, so a toe-kick heater is selected with a maximum output of 7000 Btuh to supplement the baseboard. The 7000 Btuh is subtracted from the total heat loss of 47,984 Btuh. The remaining loss of 40,984 Btuh is divided by 580 Btuh per foot of baseboard. Thus 71 ft of ¾-in. baseboard will be needed.

The baseboard specifications are shown in Figure 23.10. The locations of the radiation units, including the toe-kick heater, are shown on the house plan. If ½-in. baseboard is required, use the appropriate chart to give the heat emission rate for that size. Most systems, however, use ¾-in baseboard.

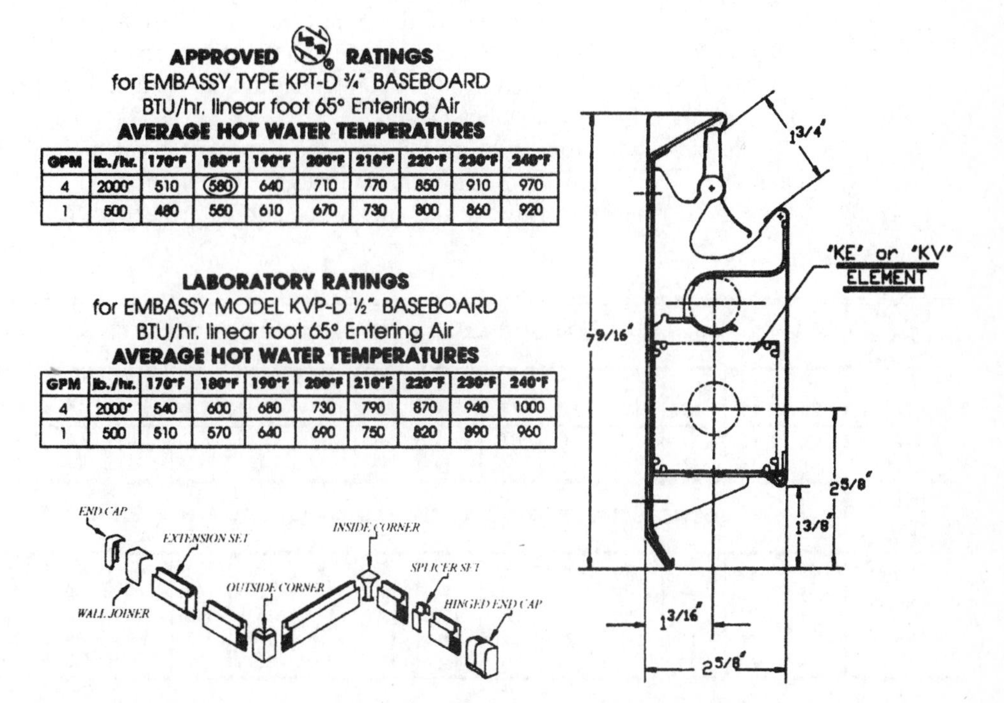

APPROVED RATINGS
for EMBASSY TYPE KPT-D ¾" BASEBOARD
BTU/hr. linear foot 65° Entering Air
AVERAGE HOT WATER TEMPERATURES

GPM	lb./hr.	170°F	180°F	190°F	200°F	210°F	220°F	230°F	240°F
4	2000*	510	580	640	710	770	850	910	970
1	500	480	550	610	670	730	800	860	920

LABORATORY RATINGS
for EMBASSY MODEL KVP-D ½" BASEBOARD
BTU/hr. linear foot 65° Entering Air
AVERAGE HOT WATER TEMPERATURES

GPM	lb./hr.	170°F	180°F	190°F	200°F	210°F	220°F	230°F	240°F
4	2000*	540	600	680	730	790	870	940	1000
1	500	510	570	640	690	750	820	890	960

Figure 23.10 Baseboard radiation and enclosure. (Courtesy of Embassy Industries, Inc.)

Circuit or Trunk Pipe Size	GALLON PER MINUTE CAPACITY OF CIRCUIT															
½"	2.3	2.0	1.9	1.8	1.7	1.7	1.6	1.5	1.5	1.4	1.3	1.2	1.2	1.1	0.9	0.8
¾"	5.0	4.3	4.1	3.8	3.7	3.6	3.4	3.2	3.1	2.9	2.8	2.6	2.4	2.2	2.0	1.8
1"	9.6	8.3	7.7	7.3	7.0	6.8	6.5	6.3	5.9	5.7	5.5	5.0	4.6	4.3	3.8	3.4
1¼"	..	18	17	16	15	15	14	14	13	12	11	11	9.7	9.0	8.3	7.3
Available Head in Ft. of Water	**TOTAL LENGTH OF CIRCUIT (AS MEASURED ON PIPING LAYOUT)**															
4	35	45	50	60	65	70	75	80	90	100	110	130	150	180	220	290
5	45	60	65	70	80	90	95	100	120	130	140	160	190	230	290	360
6	55	70	80	90	100	110	120	130	140	160	180	200	240	290	350	450
7	65	90	100	110	120	130	140	150	170	190	210	240	290	340	420	540
8	75	100	110	130	140	150	160	180	200	220	250	290	330	400	490	620

Figure 23.11 Table for sizing the circulator. (Courtesy of the Hydronics Institute, Inc.)

We now need to determine the available head of the system by referring to Figure 23.11. The 1-in. supply and return manifolds are checked for the total flow of 4.8 gpm. Locate the 1-in. trunk size and then move to the right to a number equal to or larger than 4.8 gpm. Follow that column down to a number equal to or greater than the total piping of 218 ft. Follow that row to the left to obtain the available head of 7 ft of water.

Repeat this step for the individual zones using ¾-in. pipe. In Figure 23.12 move from the ¾-in. circuit pipe size to the right to a number equal to or larger than the 2.9 gpm for the living zone. Follow that column down to a number equal to or greater than the 128-ft length of that zone to obtain the available head of 5 ft of water. For the bedroom zone, the available head is 4 ft.

The circulator will be selected based on the highest head calculated, 7 ft, using the chart shown in Figure 23.13. This chart relates the circulator head available to overcome the resistance to flow, to the amount of hot water needed in gpm. For pump

Circuit or Trunk Pipe Size	GALLON PER MINUTE CAPACITY OF CIRCUIT															
½"	2.3	2.0	1.9	1.8	1.7	1.7	1.6	1.5	1.5	1.4	1.3	1.2	1.2	1.1	0.9	0.8
¾"	5.0	4.3	4.1	3.8	3.7	3.6	3.4	3.2	3.1	2.9	2.8	2.6	2.4	2.2	2.0	1.8
1"	9.6	8.3	7.7	7.3	7.0	6.8	6.5	6.3	5.9	5.7	5.5	5.0	4.6	4.3	3.8	3.4
1¼"	..	18	17	16	15	15	14	14	13	12	11	11	9.7	9.0	8.3	7.3
Available Head in Ft. of Water	**TOTAL LENGTH OF CIRCUIT (AS MEASURED ON PIPING LAYOUT)**															
4	35	45	50	60	65	70	75	80	90	100	110	130	150	180	220	290
5	45	60	65	70	80	90	95	100	120	130	140	160	190	230	290	360
6	55	70	80	90	100	110	120	130	140	160	180	200	240	290	350	450
7	65	90	100	110	120	130	140	150	170	190	210	240	290	340	420	540
8	75	100	110	130	140	150	160	180	200	220	250	290	330	400	490	620

Figure 23.12 Table for sizing the circulator. (Courtesy of the Hydronics Institute, Inc.)

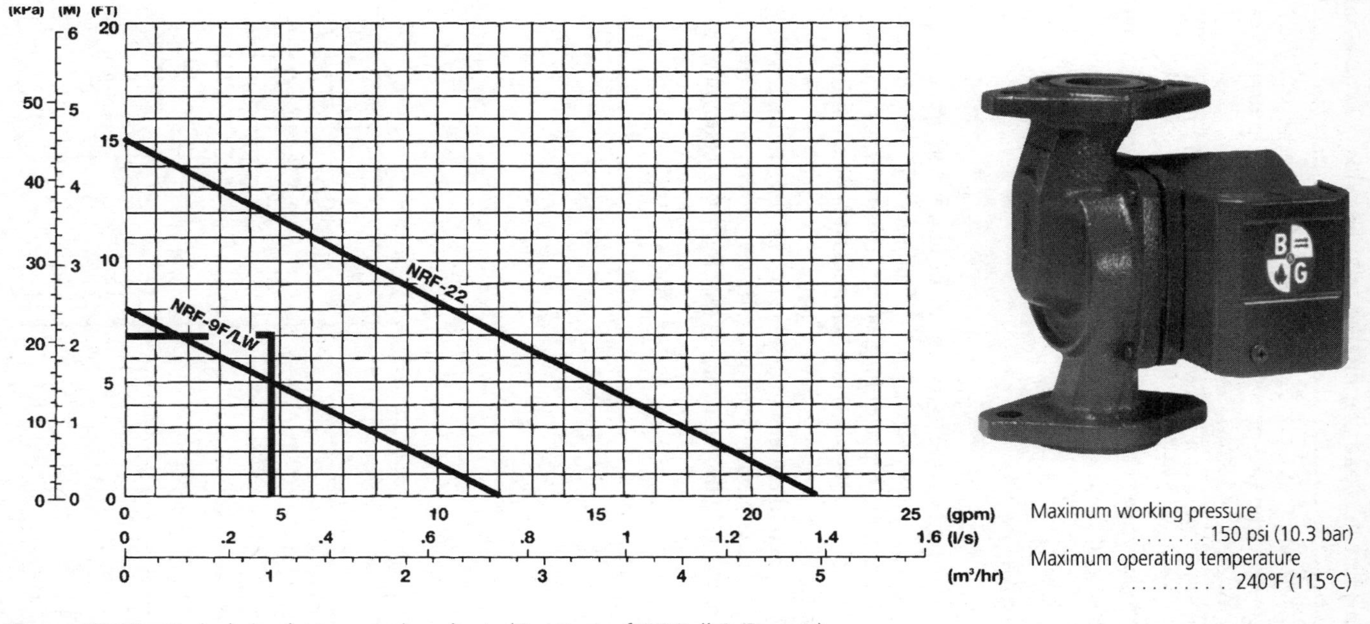

Figure 23.13 Typical circulator pumping chart. (Courtesy of ITT Bell & Gossett)

selection it is assumed that both zones are calling for heat at the same time; therefore, from our previous calculation, the total heat loss of the building, 47,984 Btuh, is divided by 10,000, resulting in 4.8 gpm. A line drawn up from 4.8 gpm at the bottom of the chart and a line drawn over from 7 ft on the left of the the chart, intersect at a point between the two pump curves. The NRF-22 circulator would thus surpass the system resistance to flow.

The piping and control application is shown in Figure 23.14. This diagram shows a boiler-and-piping arrangement for a two-zone system. A circulator could be used in place of each zone valve and accomplish the same results. Manufacturer's recommendations should always be consulted before altering a system.

Boiler Ratings

Residential boilers are rated in accordance with procedures set up by the U.S. Department of Energy (DOE). Manufacturers follow these guidelines to calculate and publish the heating capacity for a boiler. This heating capacity rating is the useful heat output available to supply space heating and is given in Btuh or in MBh (thousands of Btu per hour).

Some manufacturers also give a net IBR rating, which is the DOE heating output capacity minus a pipe-loss allowance. When a boiler is selected, the net IBR output capacity must be equal to or slightly greater than the heating load for the structure.

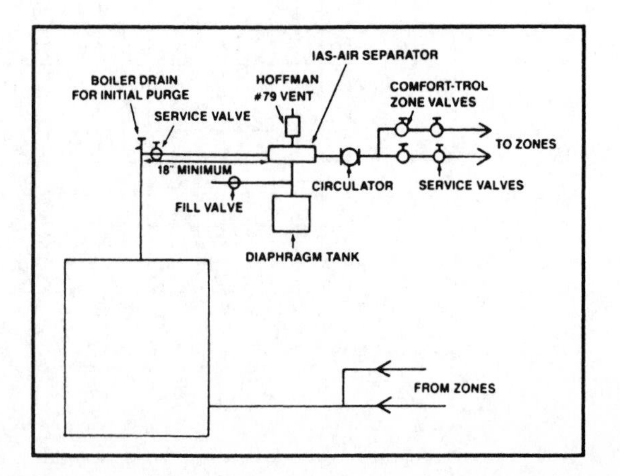

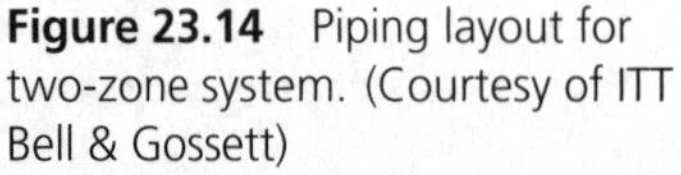

Figure 23.14 Piping layout for two-zone system. (Courtesy of ITT Bell & Gossett)

Boiler Installation Codes

ANSI–ASME requires code compliance with the National Fuel Gas Code ANSI Z 223.1 and the National Electrical Code® and NFPA-70. Installation must also conform to standards for controls and safety devices as per ANSI/ASME CSD-1.

In Canada, the installation must be in accordance with standards CAN/CGA-B149(.1 or .2) codes for gas-burning equipment, and with standard C.S.A. C22.1 Canadian Electrical Code Parts 1 and 2, and or local codes.

The gas-fired hot water boiler shown in Figure 23.15 is designed so it can be installed in a limited space such as is found in condominiums, townhouses, or homes without basements. A safety vent switch is installed in the flue passage just before the vent stack diverter. In case of flue blockage, which causes high temperatures, the vent switch will shut off gas flow. This boiler provides an annual fuel utilization efficiency (AFUE) of 83%.

A. Vent damper
B. Drafthood
C. Blocked vent switch
D. Deluxe insulated jacket
E. Iron nipple joins sections
F. Pressure temperature gauge
G. Durable cast iron sections for long life
H. Pinned heating surface for maximum heat extraction
I. Concealed step opening gas valve (SmartValve™ shown)
J. Flame roll-out switch
K. Stainless steel burners
L. Brand name quality controls
M. Self lubricating circulator never needs oil

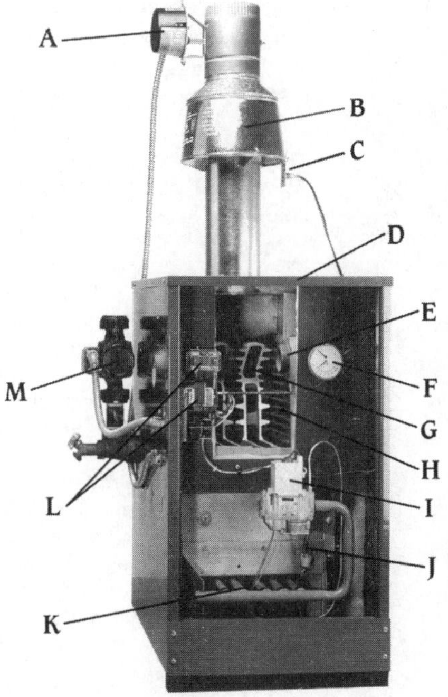

Figure 23.15 Typical gas-fired cast-iron hot-water boiler. (Courtesy of Burnham Corporation)

Figure 23.16 illustrates a standard oil-fired, cast-iron boiler. It is supplied as a package, with circulator pump, oil burner, and completely wired controls. It has an output efficiency of 80%. This boiler features horizontal sections and a specially designed air separator. It is very compact, with a height of only 48½ in. with heating capacities ranging from 84 to 303 MBh.

The pulse-design hot-water boiler maintains an AFUE of above 90%. The Multi-Pulse series features a welded heat exchanger that incorporates a cast stainless steel pulse combustion chamber and a welded-steel pressure vessel with spiral stainless steel fire tubes (Figure 23.17).

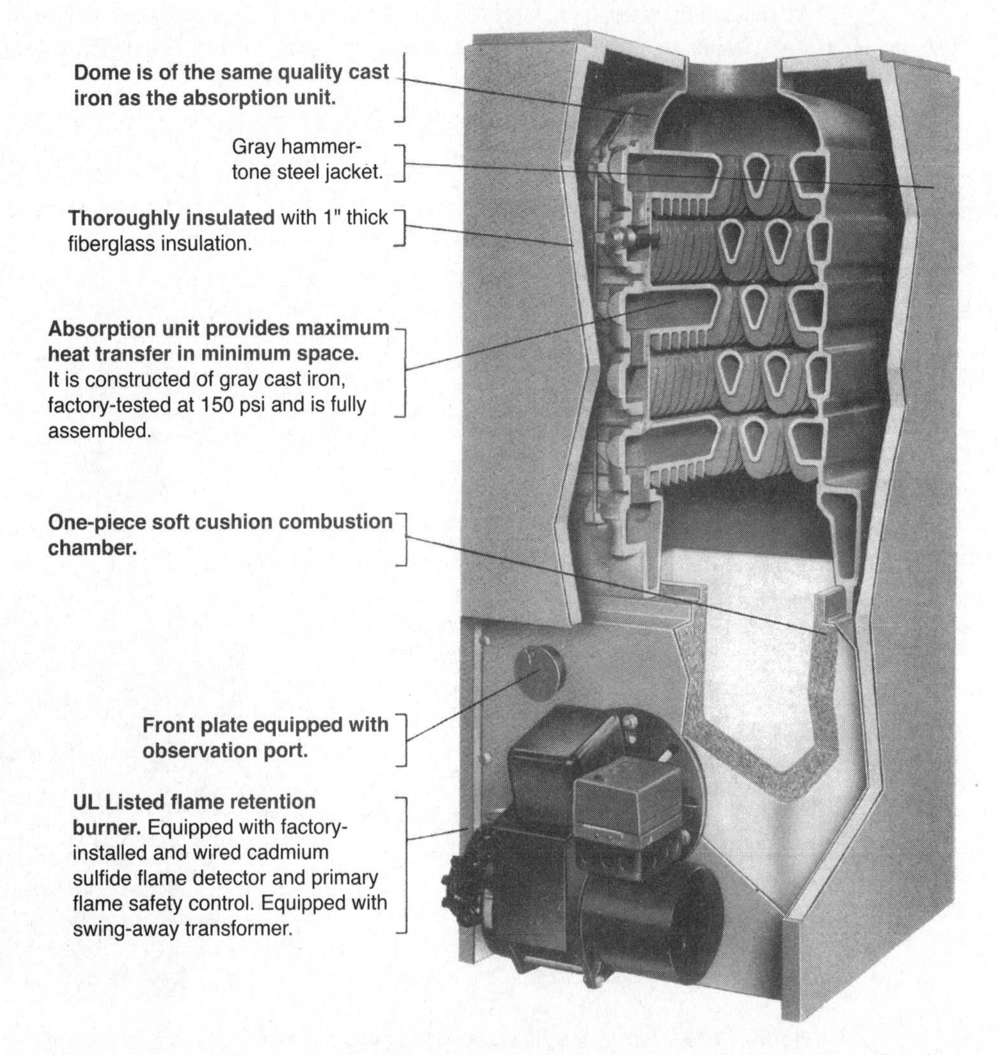

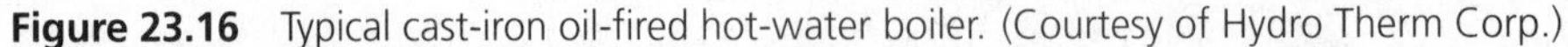

Figure 23.16 Typical cast-iron oil-fired hot-water boiler. (Courtesy of Hydro Therm Corp.)

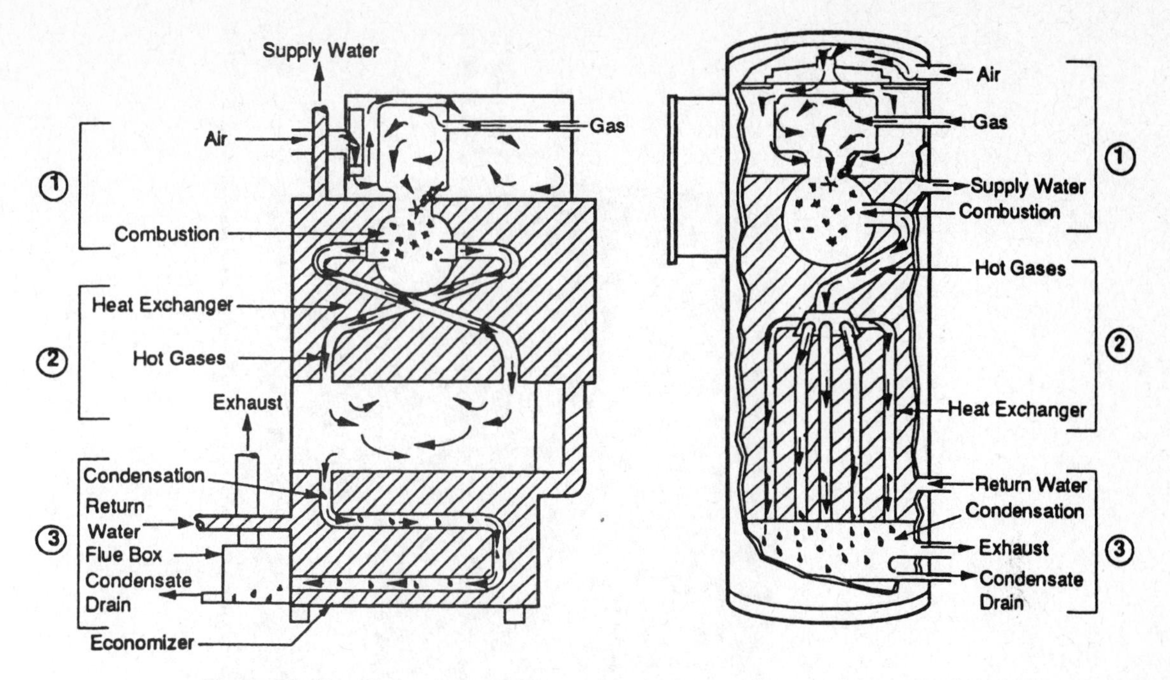

Figure 23.17 Cutaway view of the multipulse heat exchanger. (Courtesy of HydroTherm Corp.)

The pulse boiler is an advanced technology, requiring no burners, pilot, flue, or chimney. The boiler uses 100% outside air supplied through 1½- to 2-in. schedule 40 PVC pipe. The fuel–gas mixture is initially ignited by a spark plug. Thereafter, it is ignited by the previous cycle's heat. The products of combustion gases are vented through a 1½- to 2-in. chlorinated polyvinyl chloride (CPVC) pipe. Pipe can be routed either through an unused, insulated chimney, a roof, or an outside wall. Condensed water vapor from the flue gases is removed through a ¼-in. ID pipe.

Boiler Combustion-Air Requirements

All requirements are set by NFPA through the National Fuel Gas Code No. 54-1984 ANS1 Z223.1. Check the local code before replacing or installing any new equipment. For example, an unconfined space, such as an open basement, must have 50 ft^3 of room volume per 1000 BTUs of gas input. As another example, a boiler with 100,000 Btu/h input would require 5000 ft^3 of room volume.

For boilers with a sealed combustion chamber, the process is a simple concept. Fresh air for combustion is drawn directly into the combustion chamber from the outdoors. Exhaust gases are vented directly outside through a second pipe. Both the intake of fresh air and the venting of exhaust gases are powered by a fan. The Boiler Plus Series venting is shown in Figure 23.18.

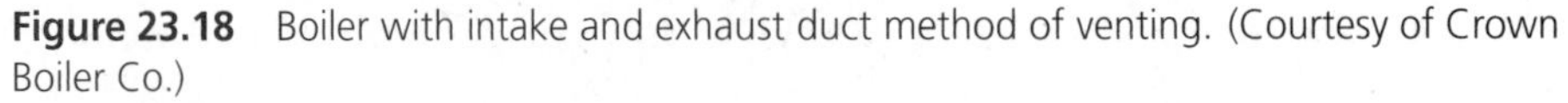

Figure 23.18 Boiler with intake and exhaust duct method of venting. (Courtesy of Crown Boiler Co.)

Selecting the Boiler

Boilers are available in various sizes and can be gas or oil fired. They can be cast iron, steel or copper. The cast-iron boilers can be vertical or horizontal, depending on the deflection of the flue passage. Each cast-iron section of the boiler has pins or fingers protruding out of the body of the section. Heat is extracted from the flue gases by these cast-iron pins and transferred by conduction to the water.

In the steel fire-tube design, water surrounds a group of steel tubes through which the hot combustion gases pass. A spiral pipe design in the fire tubes helps increase heat transfer.

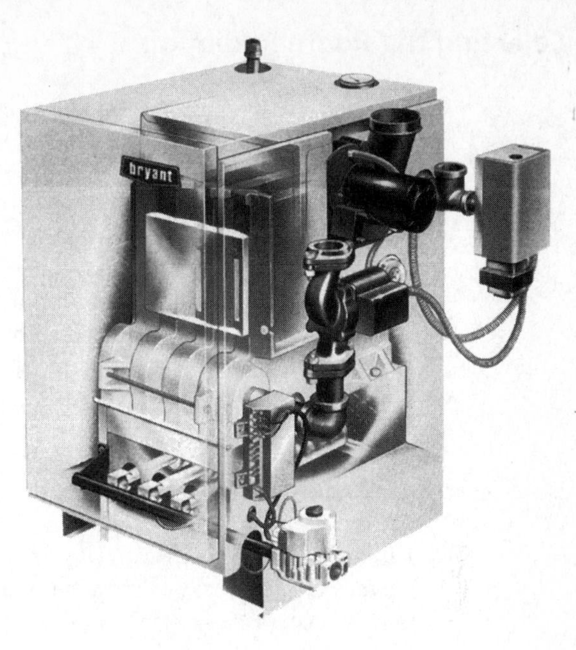

Figure 23.19 Package gas-fired boiler. (Courtesy of Carrier Corporation)

In the copper water tube or copper fin-tube boiler the water piping is contained inside a copper funnel, where it is exposed to flue gases. Copper boilers are small and can directly heat domestic water or pools. Tube boilers require constant water circulation while being fired to prevent thermal stress damage.

A standard gas-fired cast-iron boiler is illustrated in Figure 23.19. Most residential boilers of this type are packaged with a circulator pump and oil burner and have prewired controls. Oil-fired boilers deliver heat with energy efficiencies up to 84.4% AFUE.

The contoured internal vanes guide the flow of water through the heat exchanger for more efficiency. Thermal pins on the internal walls of the heat exchanger increase heat transfer efficiency.

Boiler Installation, Operation, and Maintenance

A boiler must be installed in accordance with national and local codes. These codes apply to the fuel supply, the venting requirements, and the safety provisions. The manufacturer's instructions shipped with the equipment should be followed carefully regarding the location and setting of the boiler, installation of water piping and hydronic components, wiring the boiler, providing combustion air requirements, startup procedures, and maintenance.

Selecting the Room Radiation

Three different types of room radiation can be applied: baseboard fin tube, convectors, and radiators. Baseboard fin tube is an enclosure approximately 8 in. tall containing bare piping and finned piping, as shown in Figure 23.10. It replaces the wooden baseboard where it is installed. For most residences, fin-tube radiation is most popular. It spreads out the supply of heat over a large area and is usually more attractive than convectors or radiators. In rooms where baseboard space is at a premium, however, such as a bath or kitchen, it may be advisable to use convectors or radiators. Convectors are larger enclosures that contain either multiple rows of finned piping or a cast-iron heating element. Radiators are multiple tall sections of heating elements, with legs, placed directly on the floor near exterior walls.

Selecting the Accessories

Although most boiler packages are supplied completely assembled, some system components are selected and installed on the job. The design of the piping and the placement of accessories is critical to the operation of the system.

Location of the Circulator Pump

On the demand for heat the circulator's job is to move hot water from the boiler to the baseboards and to return the cooler water back to the boiler to be reheated. The pump should ideally be located on the supply side of the system, pumping away from the point of no pressure change. This is further explained under the heading "Plain Steel Compression Tanks." The location of the replacement water feed line and compression tank connection are critical, and the arrangements shown in Figure 23.20 should be followed.

Circulator Pump

Old-style circulators have a bearing assembly that holds the pump shaft. The motor spins at 1750 rev/min riding on two sleeve bearings. This is an exterior bearing seal, as shown in Figure 23.21. This pump is specifically designed to be easily serviced in the field with a replaceable motor, bearing assembly, coupler, seal, and impeller. The motor and sleeve bearings are to be lubricated only with the oil type recommended by manufacturer specifications.

New-style circulators (Figure 23.22) currently used on all new hot-water boilers have a wet rotor with a water system seal. The pump works well with water temperatures up to 225°F (107°C) but should not exceed 230°F (110°C). The speed of the new wet rotor pump is between 2650 and 3400 rev/min. The stator is a one-piece, high-nickel stainless steel and is isolated from the system's water, thus maintaining precision bearing alignment for longer life.

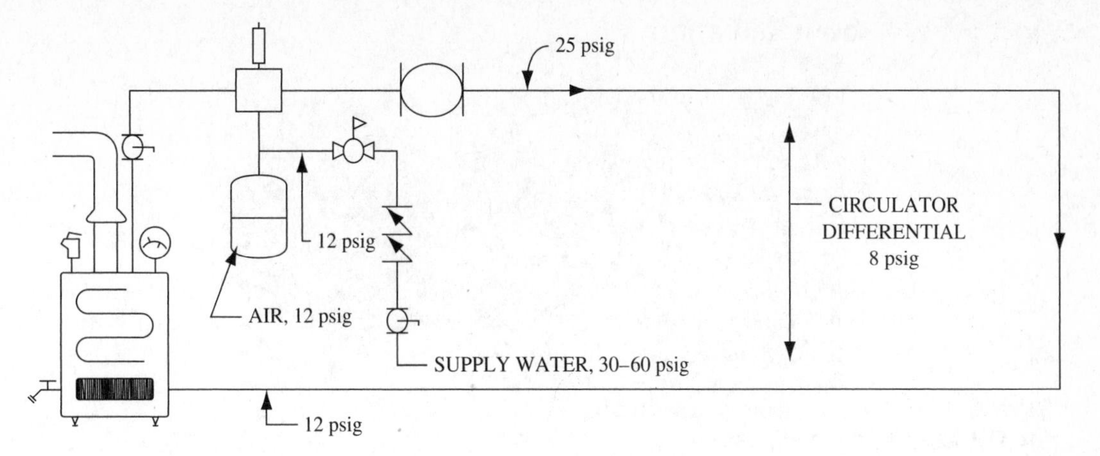

Figure 23.20 Proper location of circulator in a residential hot-water system.

Location of Valves

In Figure 23.23, note that shutoff valves are used in the system to isolate the boiler and accessories. Should service be required, it can be performed without draining the system. Valves are also used to provide a proper path for filling the system. The supply line furnishing city water has three valves: a shutoff valve, a check valve, and a pressure-regulating valve. This pressure is sufficient to supply water to the radiation located on the second floor of a residence. The check valve is required by code to prevent contaminating the water supply in case the system pressure should exceed the city water pressure. The bypass valve shown is used to modulate or reduce the supply water to a zone by allowing water to short-circuit to the return line. This hand-balancing adjustment can be automated with a modulating three-way zone valve.

Filling the System

One good way to fill the system and at the same time purge the air is to open valve A, shown in Figure 23.23, close valve B, and open the drain valve. This arrangement uses city water pressure to fill the system. Care must be taken to return the system from city water pressure to 12 psig when the automatic feed is placed in control. Most boilers are shipped complete with a pressure/temperature relief valve and a boiler water temperature/altitude gauge, which measures water pressure and its equivalent altitude in feet.

Air-Purging Arrangement

The piping must be filled with a continuous supply of water. Air pockets must be eliminated. The high points of the system must have vents for removing any air that may be

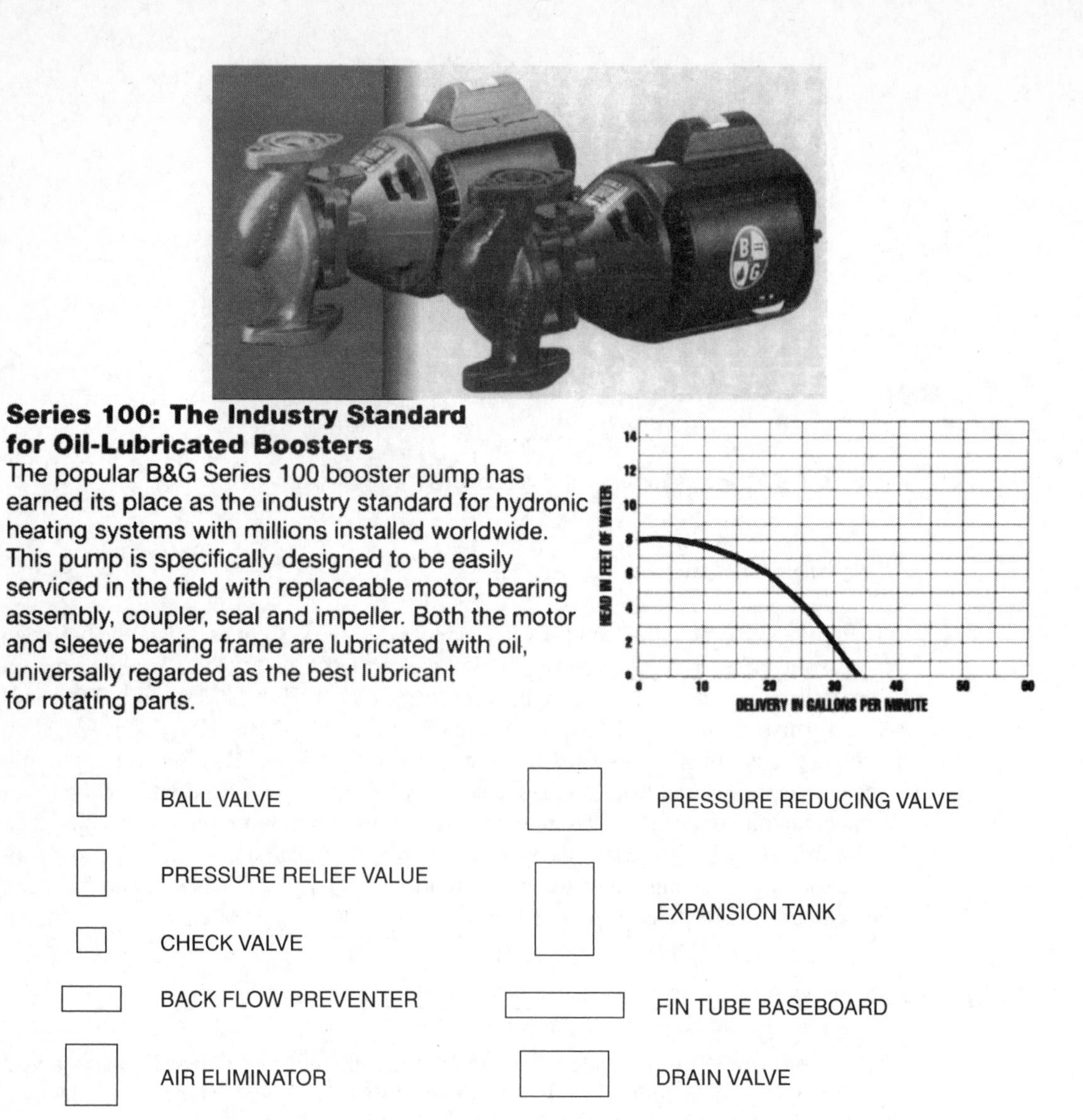

CIRCULATOR PUMP

ZONE VALVE

BALANCING VALVE

DIVERTED TEE

THREE WAY VALVE

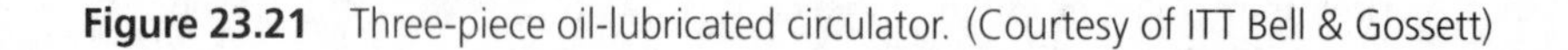

Figure 23.21 Three-piece oil-lubricated circulator. (Courtesy of ITT Bell & Gossett)

Red Fox: Easy to Install, Sure Start-ups

Engineered with features for installer and owner alike, the Red Fox system-lubricated booster pump is designed for quick, easy installation and long life. It features:

- The highest wet rotor starting torque in the industry for dependable start-ups, season after season.
- Pre-stripped leads to save installation time.
- Quick electrical connection, with full slotted ground screws on both sides of conduit box.

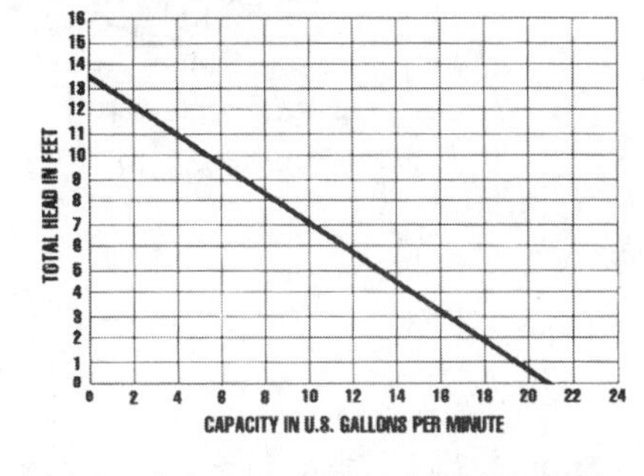

Figure 23.22 Wet rotor series NRF circulator. (Courtesy of ITT Bell & Gossett)

collected. Depending on the selection of accessories, entrained air can be vented at the compression tank. Even if the air is completely exhausted when the system is filled, air will continue to be generated when the water is heated. Oxygen and other gases are present in the form of ions in water at room temperature and atmospheric pressure. When heated, they form molecules of gases that must be removed from the system, or they will collect in the piping system and create noise and possibly inhibit the circulator's ability to move the water. Air and other gases always separate from the water in the boiler where temperature is highest, and the velocity pressure is lowest.

Water Volume Control

The compression tank allows for the change in volume of the water as the temperature changes. The tank holds water and air. The space occupied by the air will vary as the water volume changes. There are two types of tanks, the air cushion and the pressurized diaphragm.

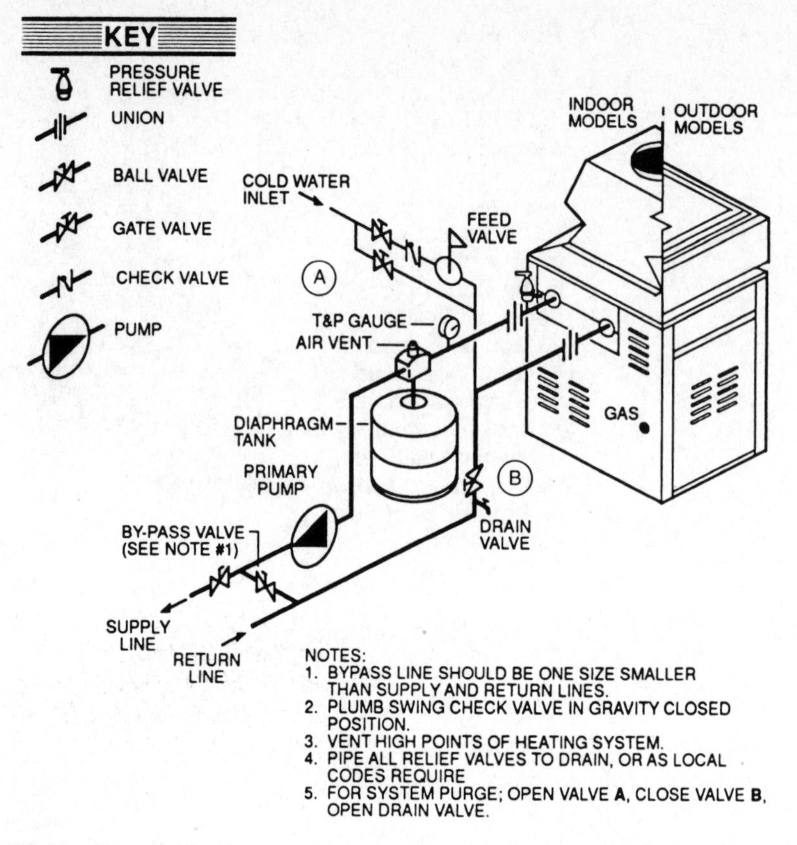

Figure 23.23 Piping hydronics system, single zone with diaphragm tank. (Courtesy of Raypak, Inc.)

Diaphragm Compression Tank

The compression tank has a flexible rubber diaphragm separating the air from the water. This tank is precharged with 12 psig of air, and therefore is much smaller than the steel compression tank. On startup the air side of the diaphragm is fully expanded against the inside of the tank. As the feed water enters the system the pressure pushes on the diaphragm and squeezes the air like a balloon. The tank's charge of 12 psig matches the head-pressure needs of a two-story house. The air pressure should always be checked before the tank is installed.

The Enhanced Heating Module shown in Figure 23.24 illustrates the ideal location of valves, circulator pumps, compression tanks, and the low water cutoff. Hydro-Flo systems can be installed on new or retrofit residential or light commercial heating installations.

The pressure-regulating valve or "feed valve" is adjustable to maintain a cold-water fill pressure greater than 4 psig at the top of the system while making sure that

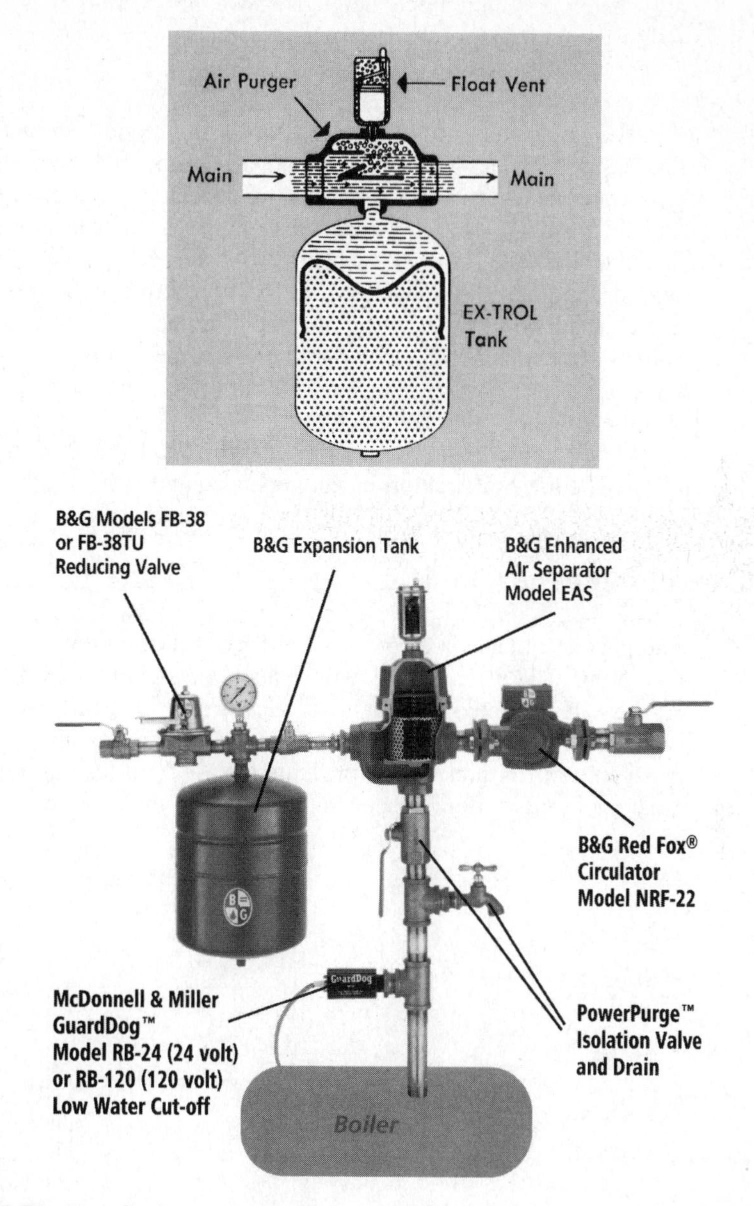

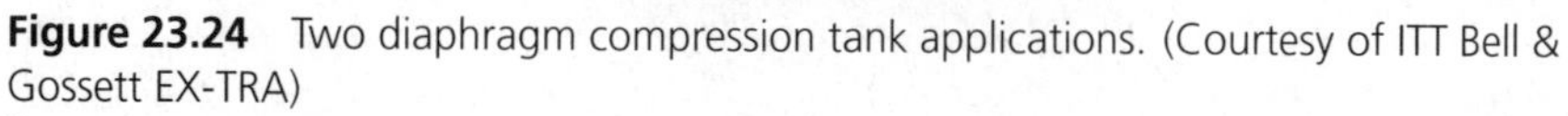

Figure 23.24 Two diaphragm compression tank applications. (Courtesy of ITT Bell & Gossett EX-TRA)

the pressure is not too high for the boiler pressure relief valve. The pressure-regulating valve is normally factory set at 12 psig. This setting is sufficient for a two-story building where the system piping height is above the location of the pressure regulator. The actual setting may be calculated as:

$$(\text{elevation} \times 0.43) + 4 \text{ psig} = \text{set pressure}$$

The low-water cutoff switch shown in the module stops the burner in the event that the system piping no longer contains water in the boiler. The probe, a conductive-type water detection control, senses the presence of water in the system. The probe must be installed at or above the minimum safe water level established by the boiler manufacturer. The low-water sensing probe may be installed directly in the boiler, if a suitable tapping is available, or other location above the waterline level in the boiler.

In steam boilers the control must be installed in a suitable tapping provided in the boiler or in another equalizing line; however, the control should be located below the level of the primary low-water cutoff but above the lowest permissible waterline as specified by the boiler manufacturer.

Should there be a power failure, the probe, located in the boiler, automatically resets. The low-water cutoff is recommended and even required by some local codes.

Plain-Steel Compression Tank

The plain-steel tanks are compression tanks with no moving parts. In a system full of cold water at 12 psig, the tank will be approximately two-thirds water and one-third air. As the water is heated it expands and squeezes the air cushion, as shown in Figure 23.25.

The air eventually leaves the tank and work its way up into the radiators. The Bell & Gossett airtrol tank fitting prevents the air from leaving the compression tank by creating a gravity-floor "check valve" between the tank and the system. This style of

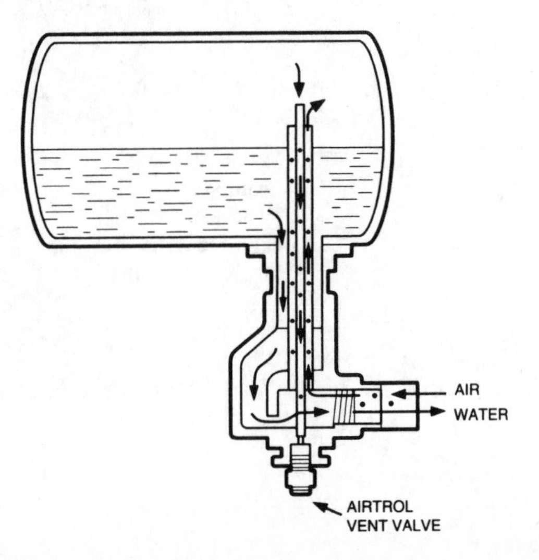

Figure 23.25 Plain-steel tank with airtrol fitting. (Courtesy of ITT Bell & Gossett)

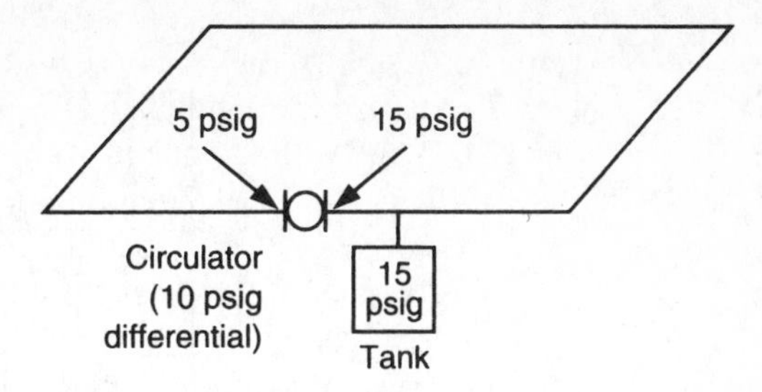

Figure 23.26 Circulator pumping toward the compression tank.

tank was used on earlier hydronic heating systems, but today it has been replaced with a diaphragm-type tank.

The location of the compression tank on a hydronic system is the point of no pressure change. The tank maintains the static pressure determined by the setting of the feed valve. In Figure 23.26 the circulator is pumping toward the compression tank. In this example, the tank pressure will remain a constant 15 psig. For the circulator to create a 10 psig pressure differential, its suction side will have to decrease to 5 psig. Pumping toward the compression tank will cause the overall system pressure to drop whenever the circulator starts. This sudden drop in pressure will release dissolved air from the water, creating noise and air-binding problems.

Pumping away from the compression tank will add the circulator pressure of 10 psig to the system's static fill pressure of 15 psig. This will result in a net rise in the system pressure, as shown in Figure 23.27. To eliminate potential circulating and air problems, it is always best to pump away from the compression tank, which is the point of no pressure change. The best location for the compression tank is as close as possible to the boiler discharge.

Safety Controls

When the boiler flue passages become blocked, the flame will roll out from the front of the combustion chamber. The flame rollout switch then interrupts power to the gas valve, preventing unsafe operation. This is either a single-use device that must be replaced if it is tripped, or a manual reset type device that can be reset by depressing the reset button. The rollout switch is located in the front of the combustion cabinet (see Figure 23.28).

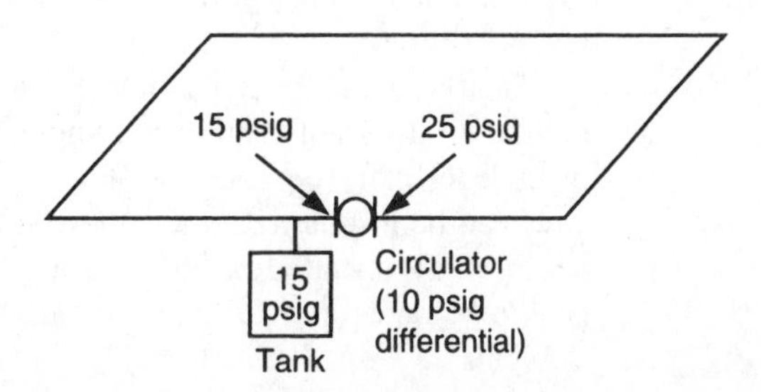

Figure 23.27 Circulator pumping away from the compression tank.

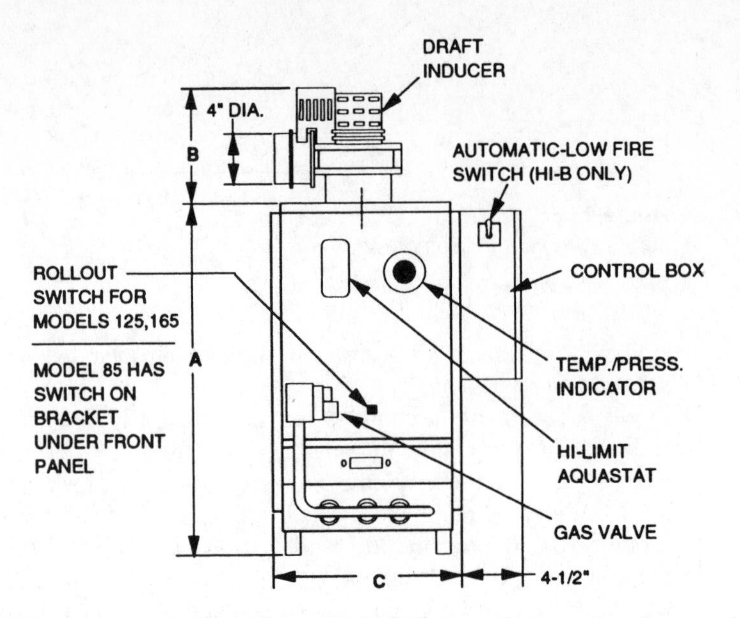

Figure 23.28 Location of safety rollout switch, hi-limit, and temperature/pressure indicator. (Courtesy of Hydro Therm Corp.)

Caution: If the flame rollout switch trips repeatedly during startup or operation, it indicates a hazardous condition that must be corrected immediately.

Special Provisions for Zoning

When circulators are used in place of zone valves to provide zoning in a system, Flo-control valves must also be installed in each zone to prevent gravity circulation. These valves require pump operating pressure in order to open up and allow flow. In a zone, when the pump is off, gravity flow of the water will not create sufficient pressure to open the Flo-control valve. Flo-control valves may be manually opened to allow for gravity flow during a power outage. There are situations in which these valves are used on both the supply side and the return side, as shown in Figure 23.29. This arrangement prevents gravity circulation through the return piping from a boiler into the "off" zone.

The problem of gravity flow can also occur in a system with zone valves. Gravity circulation does not require a complete piping loop. As long as the hot water is at a lower level than the cooler water, it will rise. To avoid this problem, a Flo-control valve can be installed on the supply side if the zone valve is on the return, or a check valve can be installed on the return if the zone valve is on the supply. This may not have to be done on every system, but if gravity heating occurs, this will eliminate the problem.

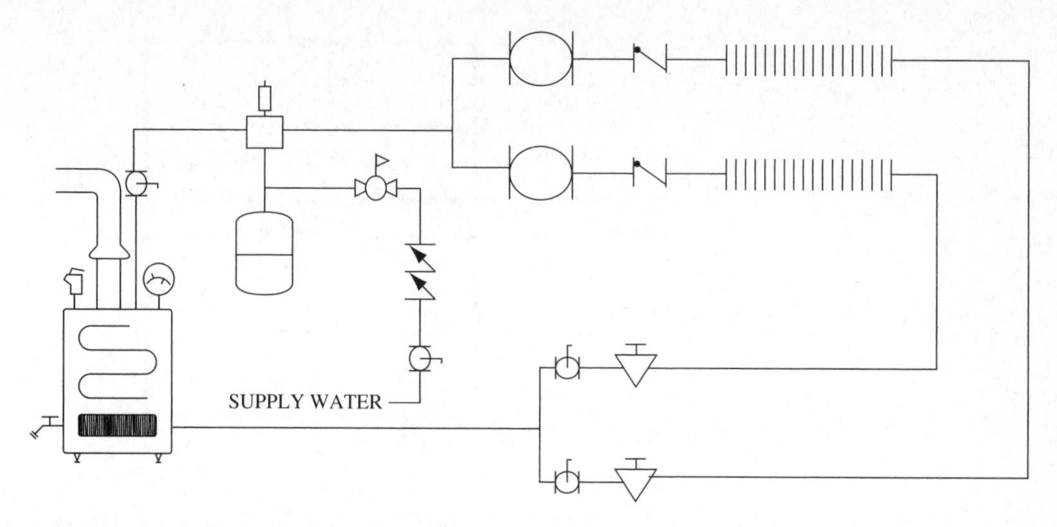

Figure 23.29 Zoning with circulators.

Selecting the Control System

There are two types of controls, elecromechanical and solid-state. Most boilers are shipped from the manufacturers prewired. Field wiring includes thermostats, pumps, zone valves, outdoor reset, and so on. A general sequence of operation is as follows:

1. The thermostat can control operation of the circulator. The boiler water temperature can be maintained by an operating aquastat. This is particularly suitable in an installation where domestic hot water is furnished using boiler heat.
2. The thermostat can turn on the boiler and the circulator at the same time. This arrangement has a distinct advantage in mild weather, since lower water temperatures can be used to heat the building. Also, energy is saved by not fully heating the boiler when the higher-temperature water is not required. Energy-code requirements mandate this as the preferred method.
3. Where zone valves are used, opening any zone valve should start the circulator. Boiler water temperature is controlled as described in arrangement 1 or 2 above.

Aquastats

Aquastats can be either the immersion type or the surface type. The immersion type is preferred wherever it can be used. A safety temperature-limiting aquastat is **always** required. An operating aquastat can be used to maintain boiler water temperature.

Reverse-Acting Aquastats

The reverse-acting aquastat is a control alternative for existing systems experiencing condensation in the combustion chamber on startup. It works best in single-zone sys-

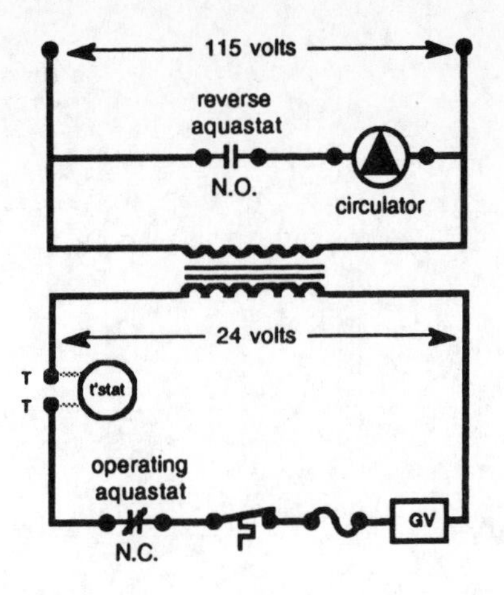

Figure 23.30 Wiring of the reverse-acting aquastat. (Courtesy of Hydro Therm Corp.)

tems. Wired in series with the circulator, this control starts the circulator after the boiler reaches an acceptable temperature, as shown in Figure 23.30.

Relays

Because the circulator operates on line voltage (120 V), and the thermostat is usually low voltage (24 V), a relay is required. Relays are often combined with an aquastat and transformer to form an aquastat relay.

Wiring Diagrams

The following figures show typical wiring diagrams for a hot-water heating system. The controls included in these diagrams depend on the components used in the system. All electrical wiring must be in accordance with the latest edition of the National Electrical Code® NFPA-70. Pictorial and schematic wiring diagrams for a Hydro Therm boiler, Model HI-B, are shown in Figures 23.31 and 23.32. A separate 115-V (60-HZ) power supply is recommended for the boiler. The inducer fan is field-wired to terminal strip TS1 as follows: dual white leads to terminal 3, red lead (low fire) to terminal 5, and black lead (high fire) to terminal 6. Connect the return-water aquastat leads as follows: blue lead to the Lo-fire Automatic Switch SW1, red lead to terminal 5 on terminal strip TSI, and black lead to terminal 6. The return-water aquastat contacts R-B open on temperature rise, closing contact R-W. Connect two-conductor low-voltage thermostat wiring from the thermostat to terminals 1 and 2 on the terminal strip TS2.

Pulse Boilers

The pulse boiler is a high-efficiency automatic gas-fired boiler that utilizes the pulse combustion principles. The wiring code that applies to the HI-B also applies to the pulse boiler. The wiring diagrams shown in Figures 23.33 and 23.34 are for the Hydro Therm Model

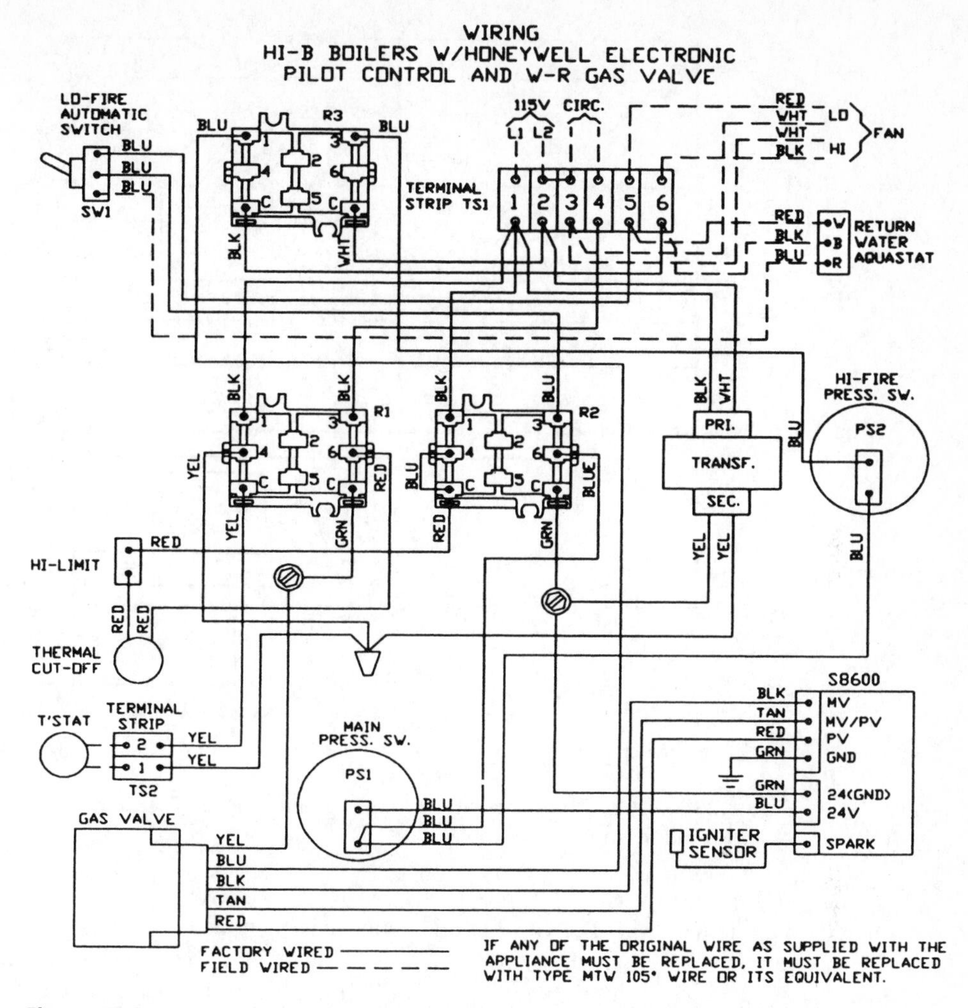

Figure 23.31 Pictorial wiring diagram for Hydro Therm Model H1-B (Courtesy of Hydro Therm Corp.)

AMI00. These systems require highly trained technicians for wiring and for service. The manufacturer's installation and troubleshooting guide should always be consulted.

Pulse Boiler Sequence of Operation

Line switch SW1 is closed, powering the 115/24-V transformer and the amber ON/OFF light. When the thermostat or operating control calls for heat the T-T contacts close. For residential applications, circulator relay R-1 (AM-100 and 150) is energized, providing power to the circulator pump. If the hi-limit is open, the GC-4 ignition control control waits. When or if the hi-limit closes, the red indicator light comes on, and the green

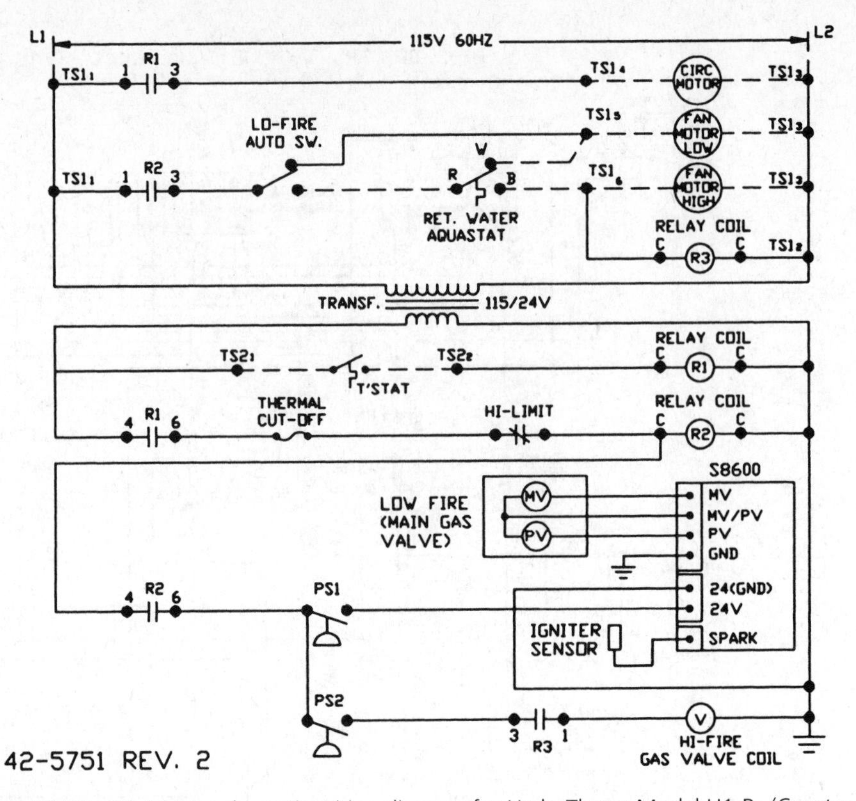

Figure 23.32 Matching schematic wiring diagram for HydroTherm Model H1-B. (Courtesy of Hydro Therm Corp.)

light remains off. The GC-4 ignition control is energized and checks the position of the combustion proving switch PS1 (and lockout switch PS3 on AM-150). If the proving switch PS1 is closed (and/or switch PS3 on AM-150 is open), the GC-4 control is on standby. When or if switch PS1 opens (and switch PS3 on AM-150 closes), the GC-4 control begins to attempt the ignition sequence, and the combustion fan starts.

After approximately 35 seconds, the GC-4 control will check the position of the fan proving switch PS2. If switch PS2 is open, the fan will continue to run, but there will be no attempt at ignition. If switch PS2 remains open, the fan will run for 3½ minutes, after which the GC-4 control will go into a 15-minute wait mode and then will again attempt ignition. If switch PS2 remains open, the attempt-at-ignition cycle will occur a total of 13 times at 15-minute intervals, after which the GC-4 control will go into a lockout mode requiring a line-voltage or thermostat reset.

When or if switch PS2 is closed, the GC-4 control provides a high-voltage spark and opens the gas valve for an 8-second trial for ignition **(red indicator light on; green light comes on).**

Trial for ignition: If ignition occurs during the 8-second trial, the combustion chamber pressure is sensed by pressure switch PS1, closing contacts R-W and completing the

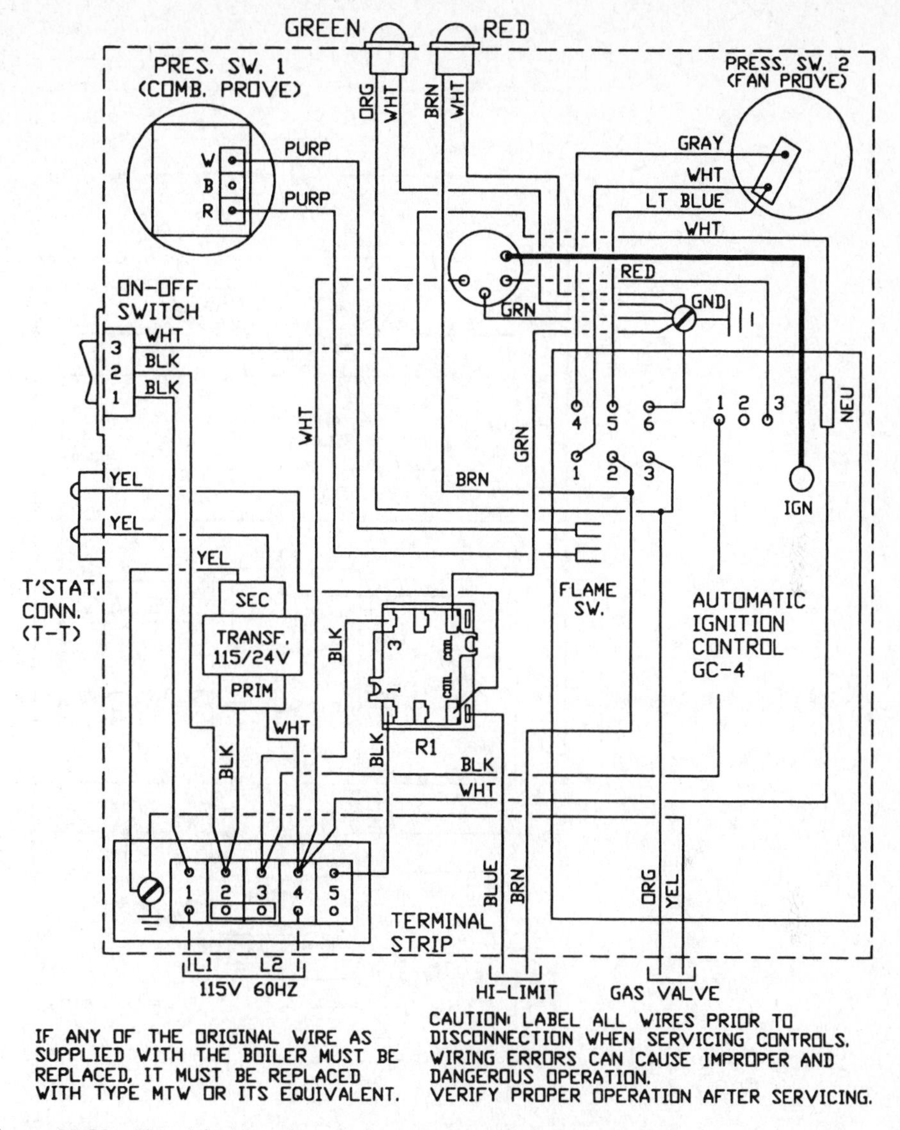

Figure 23.33 Pictorial wiring diagram for AM-100 Pulse boiler (Courtesy of Hydro Therm Corp.)

sensing circuit. Fan proving switch PS2 now opens; however, the gas valve remains energized through internal circuitry of the GC-4 control. The fan and igniter circuit are de-energized when contacts R-W close. All timer circuits are reset at the end of the heating call. **(Red and green indicator lights remain on).**

If ignition does not occur during the 8-second trial, the spark and the gas valve will shut off **(red indicator light on; green light shuts off),** but the fan will continue

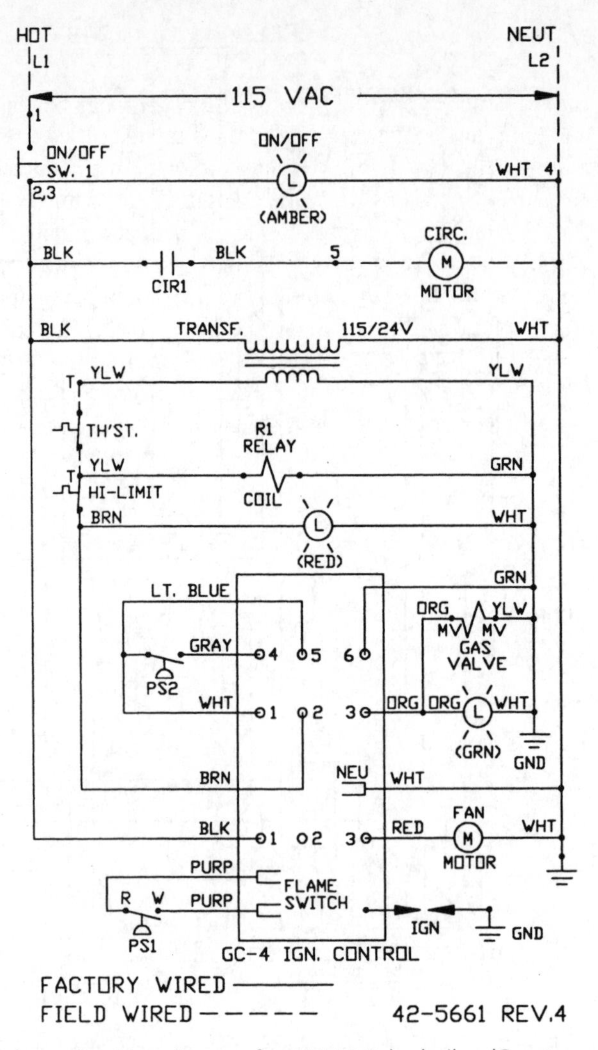

Figure 23.34 Schematic wiring diagram for AM-10 Pulse boiler. (Courtesy of Hydro Therm Corp.)

to run. After 26 seconds, the GC-4 control will check the fan proving switch PS2 position. If it is closed, the GC-4 control will initiate another 5-second trial for ignition. This sequence of 26 seconds off, 8 seconds on will occur three more times, after which the gas valve, the high-tension spark, and the fan will be de-energized (leaving only the circulator running on residential applications). The unit is now in a 15-minute wait mode. The attempt at ignition cycle will occur a total of 13 times at 15-minute intervals, after which GC-4 control will go into a lockout mode requiring a line voltage or thermostat reset.

Integrated Hot-Water System

The integrated hot-water system utilizes a water heater to supply hot water to the air handler or heat exchanger. Hot water is tapped off near the top of the water heater and is circulated through a fin-tube heat exchanger in an air handler or duct coil that circulates warm air to the conditioned space. Figure 23.35 shows a typical heating/cooling system. The horizontal air handler includes a fan, heat exchanger coil, and a circulating pump.

If a hot-water boiler is installed in connection with a water chiller, the chilled water must be piped in parallel with the boiler, using appropriate valves to prevent the chilled medium from entering the boiler (see Figure 23.36.) When boilers are connected to heating coils located in air-handling units where they may be exposed to refrigerated air, the boiler piping system must be equipped with flow-control valves or other automatic means to prevent gravity circulation of the boiler water during the cooling cycle.

Lennox "CompleteHeat" Integrated System

The Lennox "CompleteHeat" space heating/hot-water heating system, shown in Figure 23.37, combines high-efficiency space heating with high-efficiency water heating in two modules. This system stores 30 gallons of 110°F (43°C) to 170°F (76°C) hot

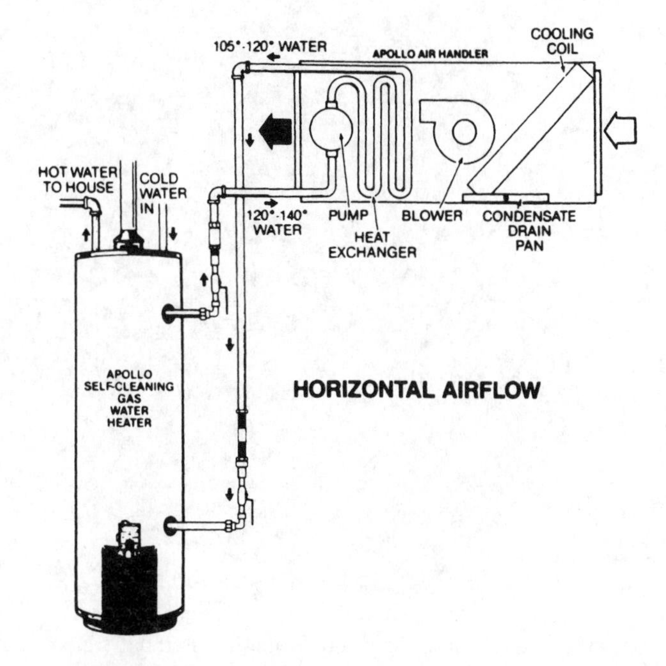

Figure 23.35 Integrated hot-water system and air handlers. (Courtesy of Apollo Hydro Heat Division of State Industries)

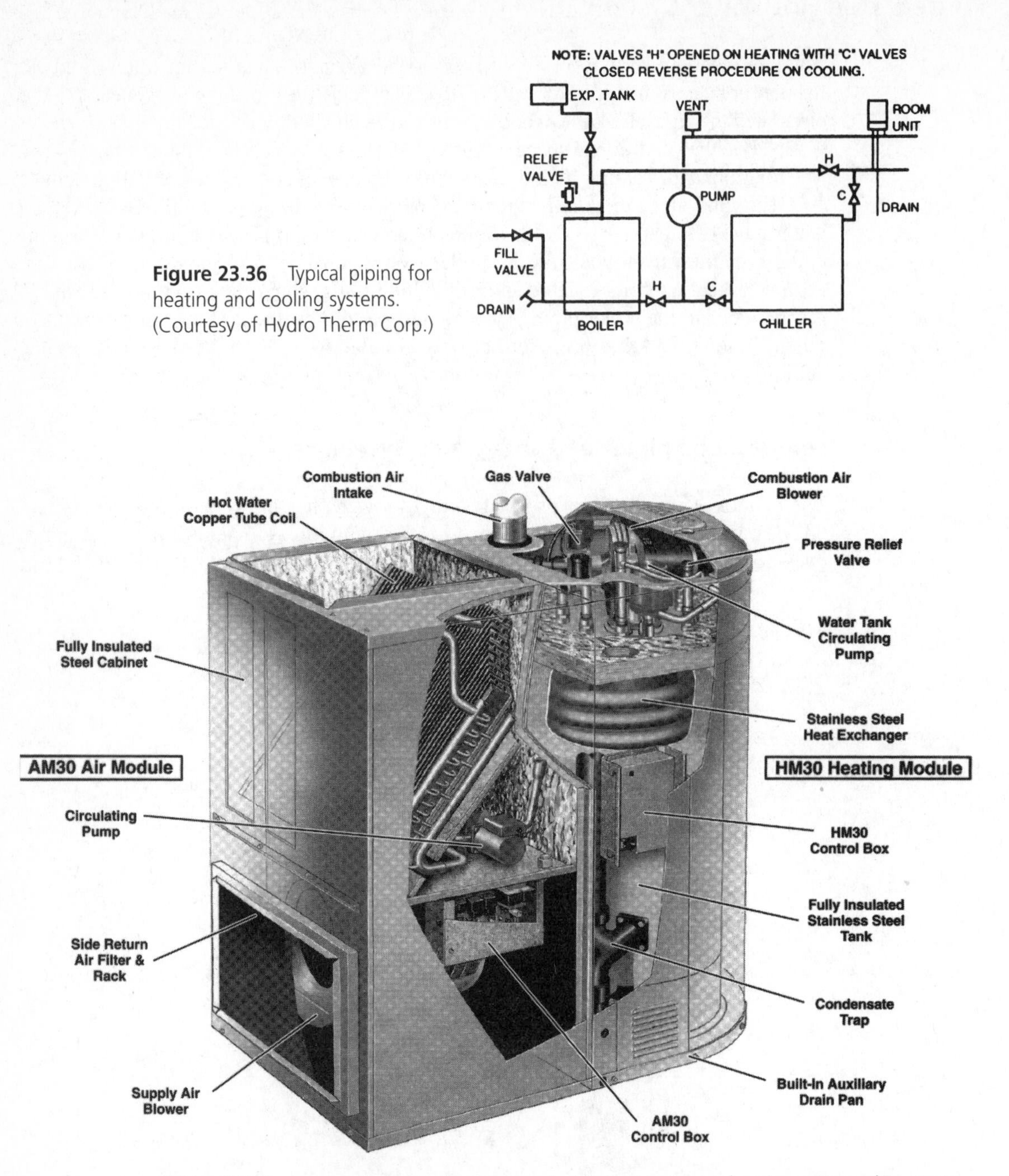

Figure 23.36 Typical piping for heating and cooling systems. (Courtesy of Hydro Therm Corp.)

Figure 23.37 Lennox CompleteHeat integrated hot-water/space-heating system. (Courtesy of Lennox Industries, Inc.)

water for domestic water use. On demand for space heating it circulates hot water through the integral hot-water coil. The supply-air blower extracts heat from the coil and distributes heated air throughout the conditioned space.

An add-on evaporator coil with remote condensing unit, electronic air cleaner, and automatic humidifier can be added for a complete all-season system (Figure 23.38.) The HM30 water-heating module may also be used with a radiant heating system, and the AM30 air-handling module may also be used with other makes of water heaters. The air-handling module may be installed in upflow, downflow, or horizontal position, either close-coupled or remote (Figure 23.39.)

Figure 23.40 shows the installation of the water piping and various components, such as the required thermal expansion tank. Water enters the systems from the main domestic cold-water supply through a pressure-reducing valve, past the thermal expansion tank, to the heat exchanger. A 6-in.-offset heat pipe trap must be installed to prevent reverse-gravity heat flow. The optional antiscald water-mixing valve and anti-thermal-siphon kit should be installed whenever tank temperatures above 140°F (60°C) are desired or when local codes require them. An auxiliary base pan drain line is required whenever the water heating module is installed where water damage could occur.

During operation, the HM30 water heating module heats potable water with a gas power burner and a helical-tube heat exchanger. To maintain the preselected temperature, a thermistor-type sensor controls the burner. The heated water is stored in an integral stainless steel tank until there is a demand for either domestic hot water or space heating.

The AM30 air module receives a call for heating from the room thermostat. This call activates the water-circulating pump and the gas burner at high or low gas input

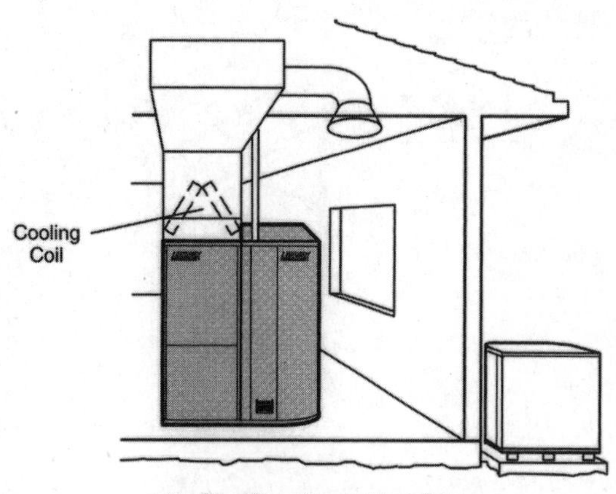

Figure 23.38 Upflow close-coupled installation with cooling coil. (Copyright, Lennox Industries, Inc. 1997, used with permission)

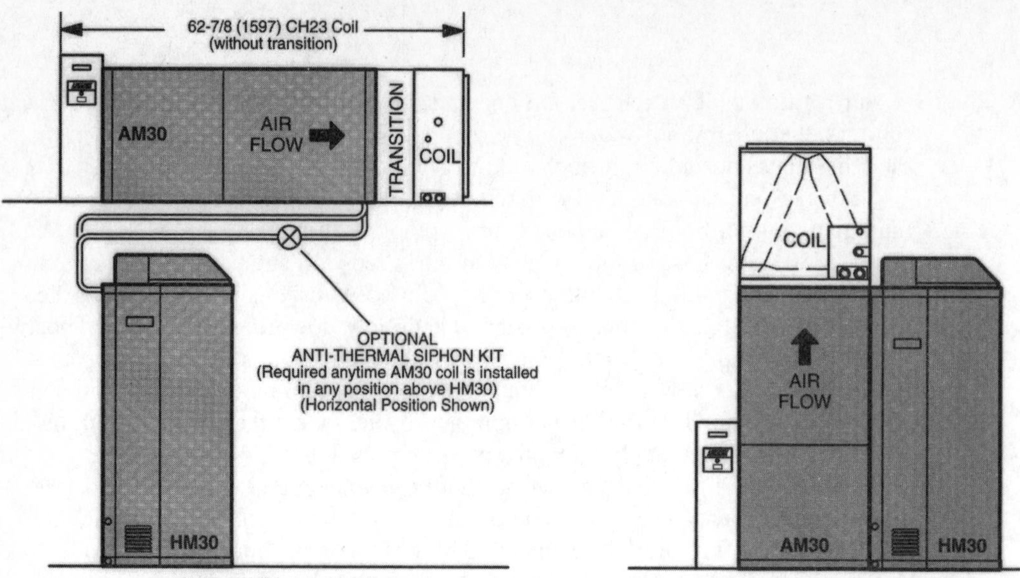

Figure 23.39 Lennox CompleteHeat systems in upflow and horizontal applications. (Copyright, Lennox Industries, Inc. 1997, used with permission)

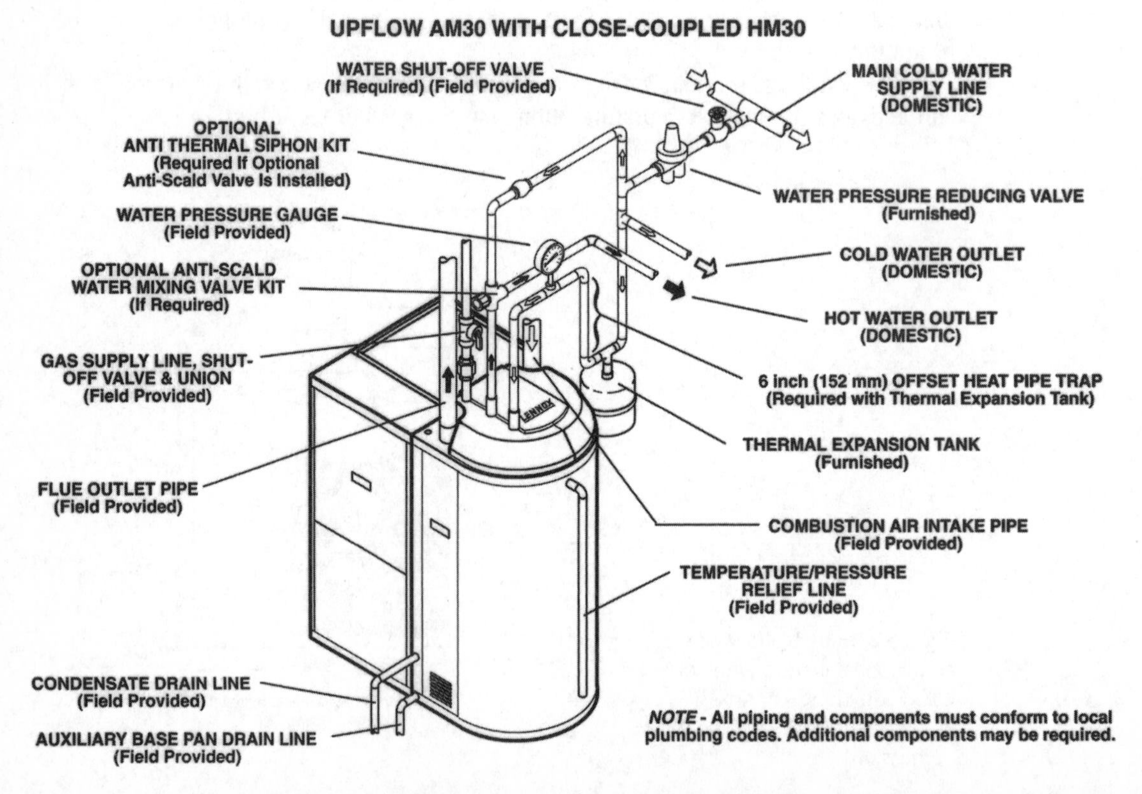

Figure 23.40 Application schematic of the AM30 with close-coupled HM30 in the upflow position. (Copyright, Lennox Industries, Inc. 1997, used with permission)

depending on the water temperature. After an adjustable time-on delay, the blower energizes at heating speed. When the heating demand is satisfied, the circulating pump shuts off. After a fixed time-off delay, the blower stops.

If the tank temperature falls to 20°F (11°C) below the tank temperature setpoint, the air module shuts off, giving priority to domestic hot water.

STUDY QUESTIONS

Answers to the study questions may be found in the section noted in brackets.

23-1 Name the five common piping arrangements employed in residential systems. *[Types of Systems]*

23-2 Which type of system is not normally a residential system? *[Types of Systems]*

23-3 What type of material is used to pipe a panel hot-water system? *[Types of Systems]*

23-4 Describe a primary/secondary hydronic system. *[Types of Systems]*

23-5 Where in a room should the heating elements be located and why. *[Sketching the Piping Layout]*

23-6 What formula is used for determining the gallons per minute circulated? *[Sizing the Circulator and Trunk Lines]*

23-7 What codes govern the installation of a boiler? *[Boiler Installation Codes]*

23-8 Which type of boiler requires constant water circulation? *[Selecting the Boiler]*

23-9 Which type of boiler has the highest efficiency and why? *[Selecting the Boiler]*

23-10 What are the combustion-air requirements for a boiler? *[Boiler Combustion-Air Requirements]*

23-11 Describe where the circulator should be located in a hydronic system and why. *[Location of the Circulator Pump]*

23-12 What is a wet rotor pump? *[Circulator Pump]*

23-13 Why is air purging of a hydronic system important? *[Air-Purging Arrangement]*

23-14 How is change in water volume allowed for? *[Diaphragm Compression Tank]*

23-15 Describe the physical and operational differences between the diaphragm tank and the steel tank. *[Diaphragm Compression Tank and Plain-Steel Compression Tank]*

23-16 What safety controls are required in a typical hydronic system? *[Safety Controls]*

23-17 Explain the use of a reverse-acting aquastat. *[Reverse-Acting Aquastats]*

23-18 Explain the use of a Flo-control valve. *[Special Provisions for Zoning]*

23-19 Explain the integrated hot-water system. *[Integrated Hot-Water System]*

23-20 At what temperature difference does the domestic hot water take preference over the heating system? *[Integrated Hot-Water System]*

CHAPTER 24

Heat Pumps and Integrated Systems

OBJECTIVES

After studying this chapter, the student will be able to:

- Describe the refrigeration cycle
- Describe the various types of heat pumps
- Explain the operation of a heat pump
- Describe the installation of add-on air conditioning

Refrigeration Cycle

A thorough understanding of the refrigeration cycle is required to troubleshoot heat pumps and add-on air conditioners. The cycle is based on two laws of science: (1) heat always travels from a warmer substance to a cooler substance, and (2) an increase in pressure of a gas will always produce an increase in temperature. In the refrigeration cycle, heat travels from the warm room air into the cool refrigerant. After an increase in pressure of the refrigerant, this heat is transferred from the hot refrigerant to the warm outdoors.

In Figure 24.1, the refrigerant enters the evaporator coil from the capillary tube as a cool, 40°F, low-pressure liquid. The evaporator has warm room air blowing over it. The heat from the air causes the cool liquid to boil (evaporate) into a 40°F vapor (gas). Additional heat brings this vapor above its 40°F boiling point to around 50°F. The added 10°F is called *superheat,* which is a very useful tool in service. The compressor then receives the low-pressure, low-temperature refrigerant vapor from the evaporator and compresses it into high-pressure, high-temperature vapor with a boiling point of 123°F. Since the 95°F outdoor air in the figure is below the 123°F temperature of the vapor, the heat will transfer from the refrigerant to the air. As heat leaves the vapor it condenses into a liquid at 123°F. More heat is removed from the liquid and it ends up

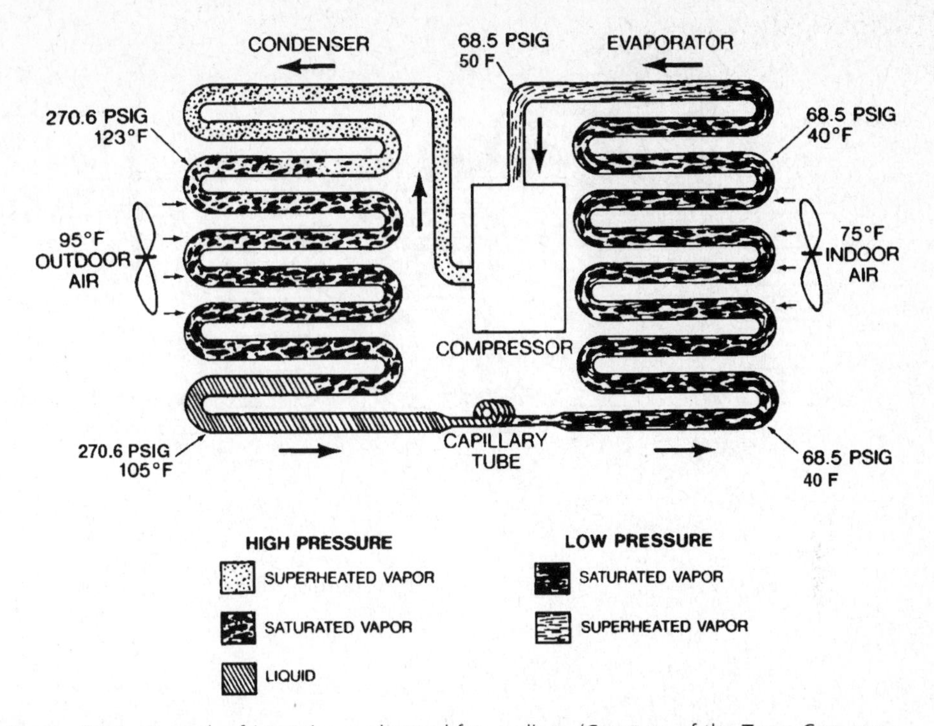

Figure 24.1 Typical refrigeration cycle used for cooling. (Courtesy of the Trane Company Unitary Products Group)

at 105°F. The additonal drop of 18°F is called *subcooling*. This subcooled liquid is then passed through an orifice (metering device), such as the capillary tube, attached to the inlet side of the evaporator. The evaporator is a large open space compared with the pipe and orifice, so the pressure and temperature are again reduced, and the cycle is repeated.

Heat is exchanged in three places: in the condenser, where the heat from the refrigerant is dissipated; in the evaporator, where heat is absorbed into the refrigerant; and in the compressor, where electrical energy from the motor and compression cycle are converted into heat and added to the refrigerant. The heat produced by the compressor is added to the heat absorbed by the evaporator and is exhausted to the outside air. The heat absorbed by the evaporator causes the indoor air to be cooled.

In the refrigeration cycle in a heat pump, the positions of the two coils are reversed. The condenser is placed inside the building to heat the air, and the evaporator is placed in the outdoor air to absorb heat. Needless to say, it is not practical to physically alter the position of these coils in changing from cooling to heating. To accomplish the same thing, a reversing (switchover) valve is used in the heat pump, as shown in Figure 24.2.

With a reversing valve in the position shown in Figure 24.2A, the cycle cools the indoor air. Note the arrangement at the bottom of the drawing for reversing the

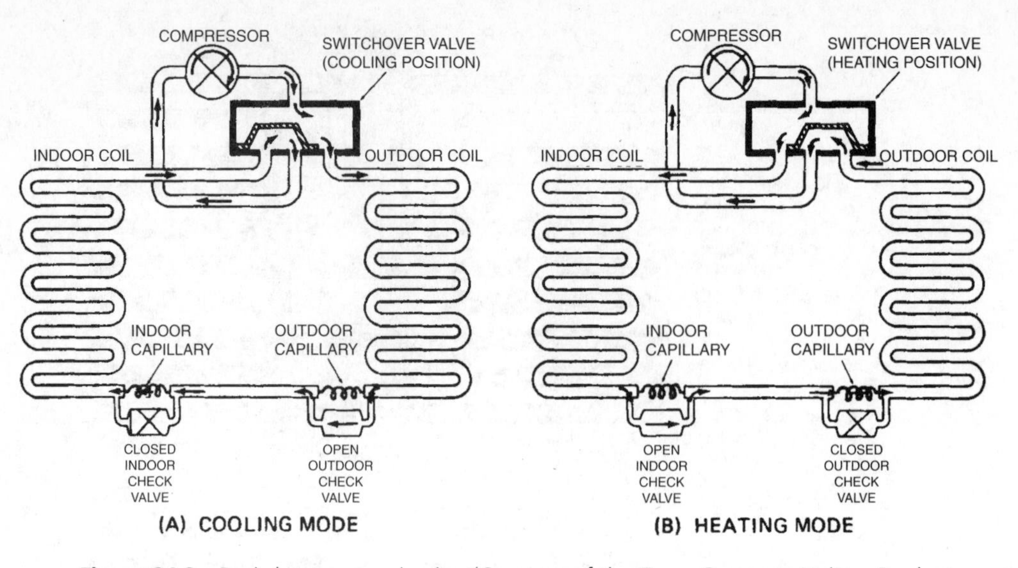

Figure 24.2 Basic heat-pump circuits. (Courtesy of the Trane Company Unitary Products Group)

direction of refrigerant flow through the metering device when changing from cooling to heating. A check valve in parallel with each metering device functions to direct the flow through the metering device being used, bypassing the unused one.

Compressor

Three types of compressors are commonly used in residential air conditioning. The *rotary compressor* is best for smaller units. It is very efficient, but it requires an exact refrigerant charge. Factory-charged and -sealed window units work well with this type of compressor. The *reciprocating compressor* uses one or more pistons to compress and move the refrigerant vapor. This type has been the most widely applied compressor in air-conditioning and heat-pump systems. Proper charging procedures must be followed, since an overcharge of refrigerant in units with orifice and capillary–type metering devices can damage the valves sealed within the compressor casing. The newest type is the *scroll compressor* (Figure 24.3). It is smaller in physical size and operates more quietly. It can handle an incorrect charge of refrigerant better than the other two compressors. This compressor consists of two involute spiral scrolls. During compression, one scroll remains stationary while the other scroll orbits around it. Gas is pulled in at the outer edge of the scrolls. As the orbiting movement pushes pockets of gas toward the center of the scrolls the volume is reduced and the gas is compressed. Gas at high pressure is forced out of the port located in the center of the scrolls. During compression, several pockets are compressed simultaneously, resulting in a smooth, continuous cycle.

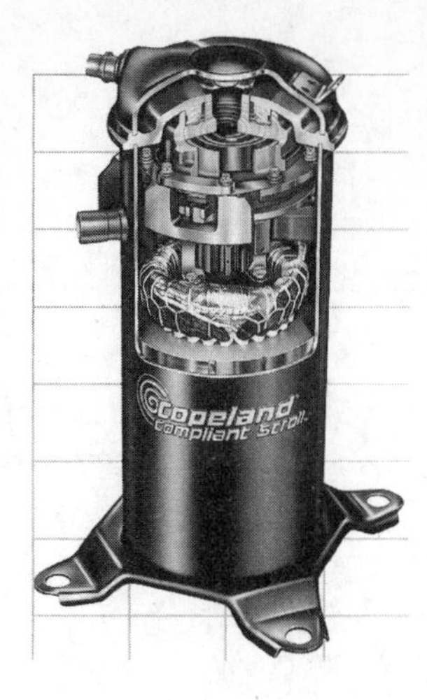

Figure 24.3 Compliant scroll compressor. (Copyright, Lennox Industries Inc. 1999 used with permission)

Condenser

The condenser in the cycle is the component from which heat is removed from the refrigerant. For air conditioning it is usually placed outside the building so that the heat can be dissipated in an area where it is not objectionable. In a heat-pump application, the condenser is the name given to the outdoor coil during cooling operation and to the indoor coil during heating operation. To provide heating the refrigerant must give up its heat on the inside of the building. The condenser is on the high-pressure side of the system.

Evaporator

The evaporator is the component from which heat is absorbed into the refrigerant. For air conditioning it is always located in the indoor air stream such as on top of the furnace, in an air handler, or in a combination heating/cooling rooftop unit. In a heat pump the evaporator is the name given to the outdoor coil during heating operation. Add-on evaporators are often called the A-coil because of the popular shape (Figure 24.4). Slant coils and horizontal flat coils are also used on top of furnaces. In horizontal duct applications, the vertical flat coil is the most common (Figure 24.5). The evaporator is on the low-pressure side of the system, often called the *suction side.*

Metering Devices

The metering device is used to provide for a separation of the high and low pressures in the system and to control the required refrigerant flow. Three types of metering

Figure 24.4 Evaporator A-coil. (Courtesy of Armstrong Air Conditioning Inc.)

devices are used by the various air-conditioning manufacturers. The *orifice* and the *capillary tube* work essentially in the same manner. They both restrict the flow much like a straw does. The smaller the size, the less the flow. A set of capillary tubes can be seen in Figure 24.4. The correctly sized orifice is installed in the smaller liquid line entering the evaporator.

The *thermostatic expansion valve* (TXV) has a superheat-sensing bulb on the outlet of the evaporator to control the flow of refrigerant (Figure 24.6.) If the superheat

Figure 24.5 Vertical flat evaporator coil for horizontal applications. (Courtesy of Armstrong Air Conditioning Inc.)

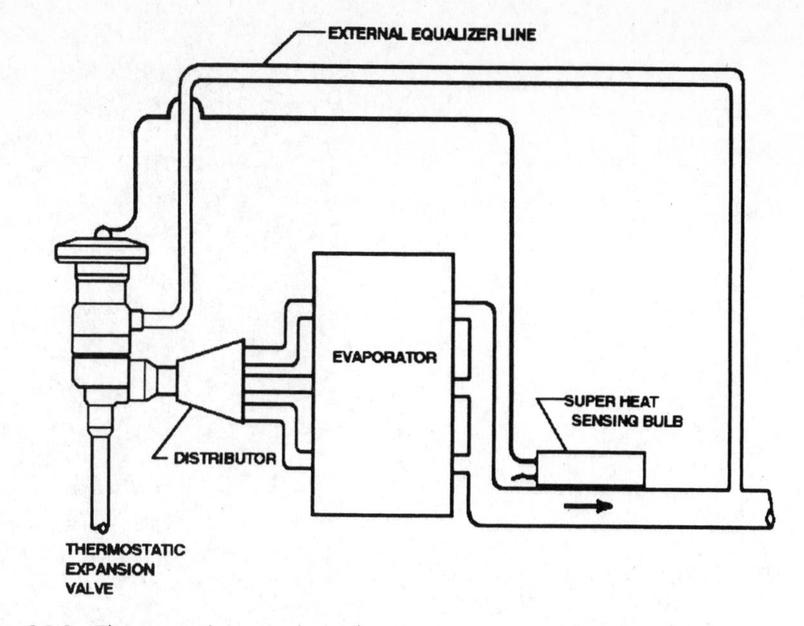

Figure 24.6 Thermostatic expansion valve.

becomes higher than its approximate setpoint of 10°F, the valve allows for more refrigerant flow. The valve also assures that the superheat does not fall below the setpoint by constricting the flow.

Accessories

Several devices are installed to make the system operate more safely. These devices protect not just the user or service technician but also the more expensive main components of the system. They also protect against any accidental release of refrigerant, which would then hurt the stratospheric ozone layer.

The *high-pressure control* shown in Figure 24.7 shuts the system down before it reaches pressures high enough to release refrigerant to the atmosphere. Usually, a defective condenser fan motor or dirty condenser fins activate this control. In most cases, this control must be manually reset, to provide an opportunity to repair the condition that caused the high pressure.

The low-pressure control, which looks like the high-pressure control but has no reset button, shuts down the system if the suction pressure drops below the setpoint. This may be because of a loss of refrigerant or because insufficient warm air is flowing across the evaporator. The control resets automatically and may cause the compressor to cycle on and off.

To prevent moisture and small particles from icing or plugging the metering device, a *filter–drier* (Figure 24.8) is installed in the line from the condenser to the

Figure 24.7 High-pressure control. (Courtesy of Johnson Controls, Inc.)

metering device. If the refrigeration system is ever opened, a new filter must be installed, which is preferably one size larger than the original installed at the factory. Larger units include a sight glass to indicate, with a change in color, if any moisture is present.

Some air-conditioning units with varying loads, and most heat pumps, utilize a *suction accumulator* (Figure 24.9) in the line between the evaporator and the compressor. The accummulator allows any liquid that may leave the evaporator to turn into a vapor before reaching the compressor. Because liquids cannot be compressed, severe damage can occur in the compressor should liquid refrigerant enter the cylinders.

Figure 24.8 Liquid line filter–drier. (Courtesy of Sporlan Valve Company)

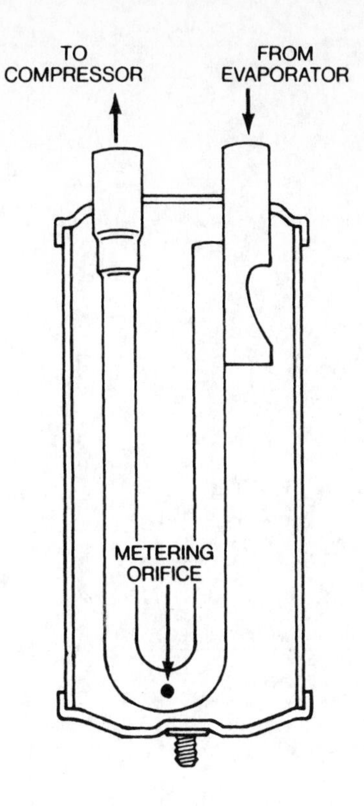

Figure 24.9 Suction line accumulator. (Courtesy of the Trane Company Unitary Products Group)

Heat-Pump Characteristics

A *heat pump* is a reversible air conditioner. During summer operation, the evaporator absorbs heat from the indoors and the condenser rejects that heat outdoors. For winter operation, a reversing valve changes the direction of refrigerant flow. The outdoor coil becomes the evaporator and absorbs sufficient heat from the winter air or from a coil in the ground. The building is conditioned by the heat rejected from the indoor coil.

Figure 24.10 shows the heat-pump heating cycle with an indoor air handler and an outdoor compressor unit. At point 1 cool outdoor air is circulated over the outdoor coil. At point 2 colder outdoor air is discharged. Within the coil, low-temperature liquid refrigerant boils into a gas between points 11 and 3. This gas travels at point 4 to the compressor (5), where it is squeezed to a higher pressure and temperature. At point 6, warm compressed gas travels to the indoor air-handler coil. When the cool house air (8) is blown across the gas in the coil (7), the refrigerant gives off heat to the air being pushed into the duct (9). The refrigerant at point 7 condenses, and the liquid travels past point 10 to the metering device at the entrance of the outdoor coil. At point 11 the liquid again absorbs the heat (12) from the outside air, and the cycle repeats.

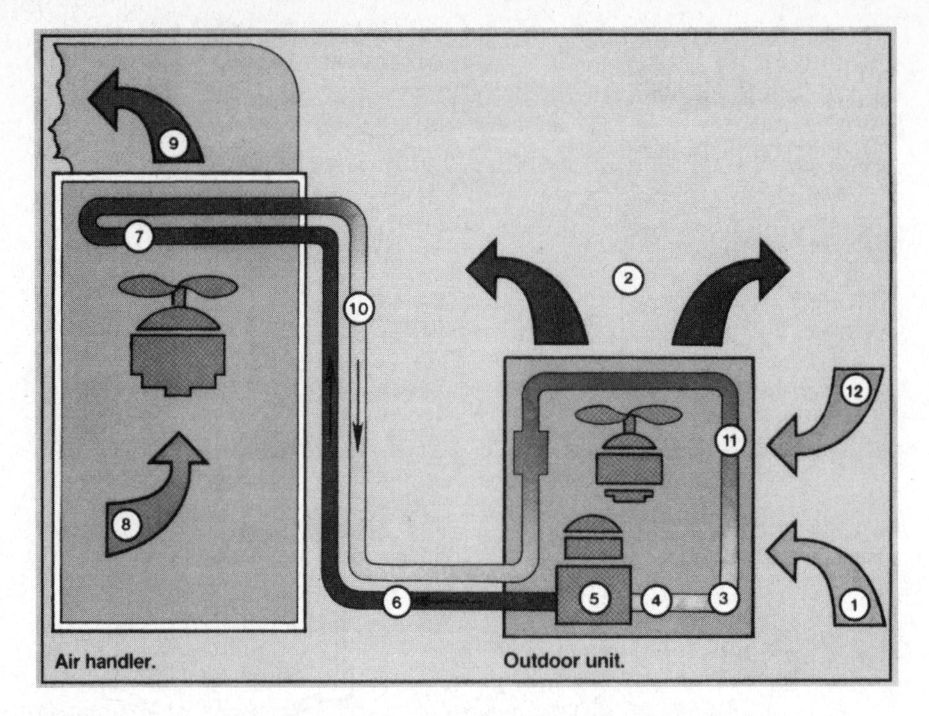

Figure 24.10 How a heat pump works. (Courtesy of The Trane Company)

Heat pumps are usually sized for the cooling load of the building. In warmer climates, such as Florida, the large cooling loads require heat pumps with greater capacities. Because the equipment size is large, the heating loads are sufficiently taken care of. In colder climates, such as Wisconsin, sizing the heat pump to the lesser cooling load will provide insufficient heating. For this reason supplementary electric heat is installed in the air handler. Once the outdoor temperature reaches a preset balance point, the first bank of strip heat is energized in addition to the compressor (Figure 24.11.) Any further drop in temperature will activate more balance points. While the electric heaters are on the compressor continues to operate. No matter how small the amount of heat provided by the compressor it is still less expensive than strip heat. Also, stopping and restarting the compressor at very low temperatures increases chances of valve damage.

In the *emergency heating mode,* supplementary heat is supplied in stages to meet the requirements of the thermostat. The heat-pump portion of the unit, consisting of the compressor, outdoor fan, and reversing valve, is inoperative. The heat is supplied by electric strip heaters.

An alternative method of providing supplementary heat is to install the heat-pump coil on top of a fossil-fuel furnace. The control system, which includes an outdoor thermostat, operates only the heat pump when the outdoor temperature is above that at which the operating costs are less than that of the furnace. Below the calculated temperature setpoint, the heat pump shuts off, and only the furnace operates.

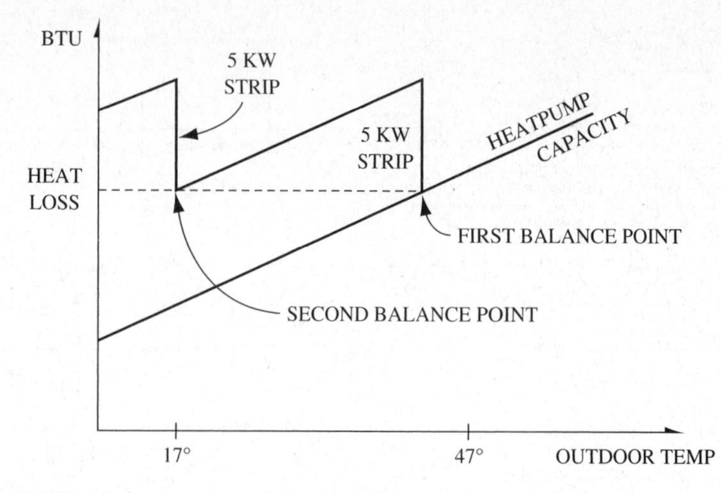

Figure 24.11 Balance points plotted on a heat-pump-capacity graph.

Defrost Cycle

When air is the medium used to absorb heat into the outdoor coil of a heat pump, sometimes during heating the refrigerant temperature in the outdoor coil falls below freezing, and ice forms on the coil, restricting the airflow. To prevent inefficient heating as a result of this condition, the outdoor coil is automatically defrosted when necessary.

One way to control the defrost cycle is to use a Dwyer defrost arrangement, as shown in Figure 24.12. Two conditions are necessary to initiate the defrost cycle:

1. The temperature of the refrigerant must be below 32°F.
2. The coil must be sufficiently frosted to restrict the airflow 70 to 90%.

Position B in Figure 24.12 shows the defrost termination switch open, indicating that the refrigerant temperature is above 32°F. Position A shows the vacuum-operated switch contacts open, indicating that the coil is not iced up sufficiently to require defrost. Position C shows the defrost termination switch closed, indicating that the refrigerant temperature is below 32°F. Position D shows the vacuum-operated switch contacts closed, indicating the need for defrosting the coil.

Based on the two switches (positions C and D) in defrost, the reversing valve automatically places the cycle in the cooling mode (Figure 24.2A), supplying hot gas from the compressor to the outside coil and defrosting it. The defrost cycle continues until the refrigerant temperature reaches 55°F, opening the defrost termination switch (position B) and automatically placing the system back in the heating mode (Figure 24.2B).

In the defrost mode, the compressor, the outdoor fan, and the indoor fan operate in the same manner as in the cooling mode in order to supply heat to the outdoor coil to defrost it. The air-pressure drop, the refrigerant temperature, and time are all factors in controlling the defrost cycle on a demand basis.

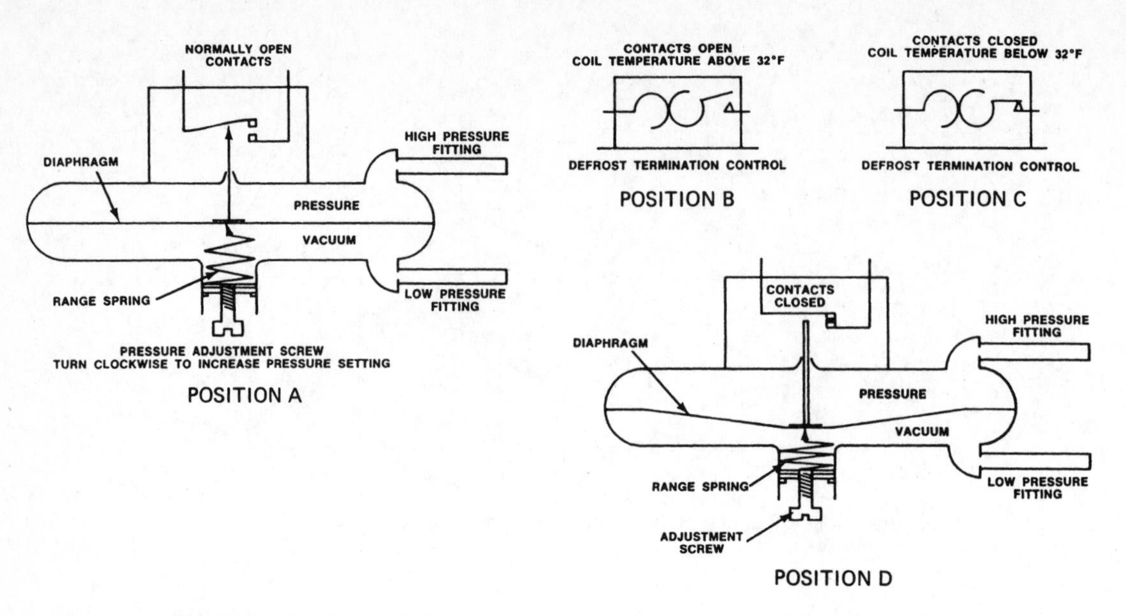

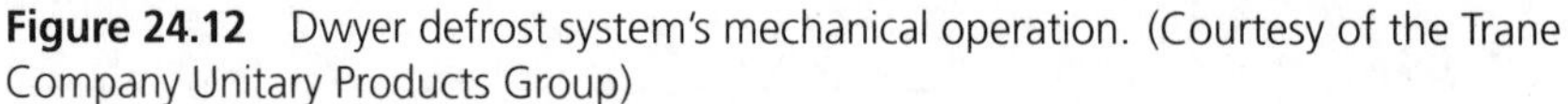

Figure 24.12 Dwyer defrost system's mechanical operation. (Courtesy of the Trane Company Unitary Products Group)

Types of Heat Pumps

Three types of heat pumps are currently in use. The *air-to-air* system uses outdoor air as the heat-transfer medium. The *water-to-air* system uses lake, river, or well water as the heat-transfer medium. The *GeoExchange* (previously known as *geothermal*) heat pump uses ground temperatures as the heat-transfer medium.

Air-to-Air Systems

Many of the items included in this section, such as certain rating definitions, the calculations for supply-air temperature rise, the type of thermostat used, the refrigeration piping, and details of condensate drains also apply to the water-to-air heat pump. The study of water-to-air systems is included in these sections and in GeoExchange systems.

Two principal arrangements of equipment are available:

1. Split system
2. Single package

The *split system* normally has two components, as shown in Figure 24.13. The compressor section, sometimes called the condensing unit, is located outside the building. The coil section and furnace or air handler are located inside the building.

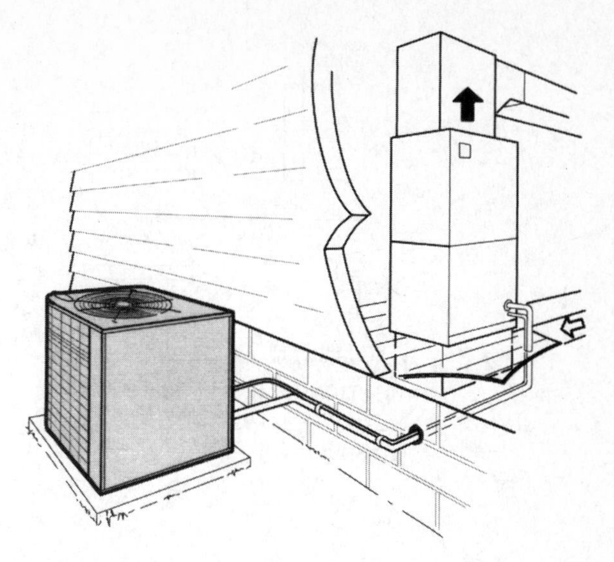

Figure 24.13 Components of a split-system heat pump. (Copyright, Lennox Industries Inc. 1999 used with permission)

A typical *single-package system* is shown in Figure 24.14. All components are housed in a single enclosure, which is installed outside the building. Wherever this arrangement can be applied, field labor is reduced.

Ratings Figure 24.15 shows typical heat-pump performance data. Ratings are based on Air Conditioning and Refrigeration Institute (ARI) standards. Standards and terms for rating heat pumps are as follows:

1. *Cooling standard* is based on an indoor temperature of 80°F db, 67°F wb and an outdoor temperature of 95°F db.
2. *High-temperature heating standard* is based on an indoor temperature of 70°F db and an outdoor temperature of 47°F db, 43°F wb.
3. *Low-temperature heating standard* is based on an indoor temperature of 70°F db and an outdoor temperature of 17°F db, 15°F wb.
4. *Coefficient of performance (COP)* is a rating given to heat pumps that is equivalent to the ratio of Btuh output to Btuh input. The input is converted to Btuh by multiplying watt input by 3.413 Btu per watt.
5. *Heating seasonal performance factor (HSPF)* is the ratio of heat output in Btu to heating input in watts.
6. *Seasonal energy efficient ratio (SEER)* is applied to the cooling performance and is the ratio of cooling output in Btus to cooling input in watts.
7. *Sound rating number (SRN)* is the noise-level rating in decibels.
8. *Total capacity (T.C.)* is the cooling or heating capacity in thousands of Btuh.
9. *Cubic feet per minute (cfm)* is the air-volume rating of the indoor fan.

All ratings are based on ARI Standard 240 and DOE test procedures at 450 cfm indoor air volume per ton (12,000 Btuh) cooling capacity with 25 ft of connecting refrigerant lines.

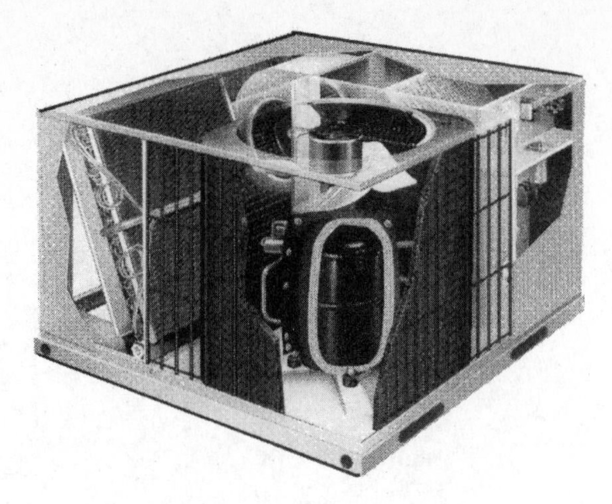

Figure 24.14 Single-package heat pump. (Copyright, Lennox Industries Inc. 1999 used with permission)

Indoor-Air-Temperature Rise Referring to Figure 24.15, note that at top performance the largest unit shown, with 47°F outside air, delivers 34,000 Btu using 1175 cfm. To determine the temperature rise (TR) on heating, the following formula is applied:

$$\text{Btu} = \text{cfm} \times 1.08 \times \text{TR}$$

$$\text{TR} = \frac{34{,}000\ \text{Btu}}{1175\ \text{cfm} \times 1.08} = 27°\text{F}$$

Based on a return-air temperature of 70°F, the delivered air is 97°F (70°F + 27°F). Thus, a heat pump delivers relatively low temperature air compared with a conventional furnace. This fact is important to keep in mind when designing the distribution system. Otherwise, discomfort can be the result.

				AR1 Standing Ratings (1–Ph)							
					Cooling		Hi-Temp Heat		Lo-Temp Heat		
Nominal Tons	Indoor Compressor Section	Outdoor Section	Indoor Fan Section	CFM	TC	SEER w/lls	TC	C.O.P.	TC	C.O.P.	HSPF
2.0	IC-024	OS-940	IF-024	850	23.8	9.80	25.2	2.90	15.3	2.05	7.65
2.5	IC-030	OS-940	IF-030	102.5	27.6	9.55	29.6	2.90	18.0	2.05	7.55
3.0	IC-036	OS-940	IF-036	1175	32.4	9.85	34.0	2.95	20.0	2.0	7.65

Figure 24.15 Typical heat-pump performance data.

Locating the Outdoor Coil Sections The outdoor coil section must be installed free of obstructions to permit full airflow to and from the heat-exchange surface, as illustrated in Figure 24.16. It should not be closer than 30 in. to a wall or 48 in. to a roof overhang, and it should be mounted on a base separate from the structure of the building to isolate any vibration.

Heat-Pump Thermostat The heat-pump thermostat differs from the normal furnace thermostat in that it has an emergency position that is used to operate the supplementary heat manually when the heat pump requires service.

Condensate Drain The inside coil section must be provided with a condensate drain. The drain pan must be level or sloping slightly toward the drain connection. The flow is actuated by gravity and must be trapped, as shown in Figure 24.17. If the indoor coil is installed in the attic or above construction that may be damaged, an auxilliary pan is installed to catch any overflow. The water draining from this pan must be visible to quickly warn of a service condition.

Refrigerant Piping Connections It is good practice to use preinsulated tubing for the refrigerant lines with brazed connections at each end. If the condensing unit is below the inside coil, the line should slope toward the outside unit. If the condensing unit is above the inside coil, the suction line should be trapped, as shown in Figure 24.18.

GeoExchange Systems

The geothermal or GeoExchange heat pump is also known as the *ground-source* heat pump because of the types of heat sources utilized for this system. Figure 24.19 illus-

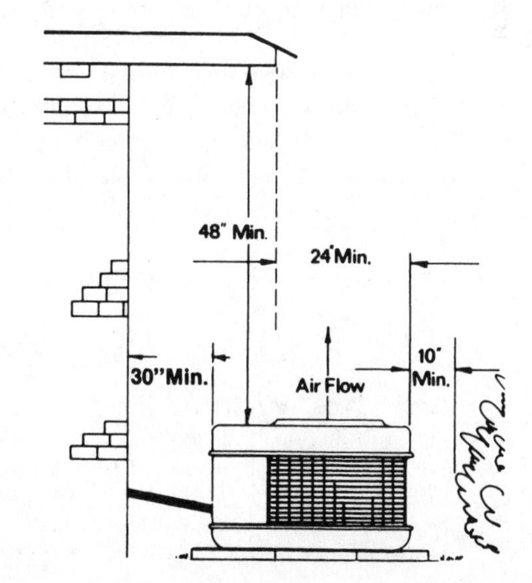

Figure 24.16 Typical outdoor coil section location. (Courtesy of Addison Products Company and Knight Energy Institute)

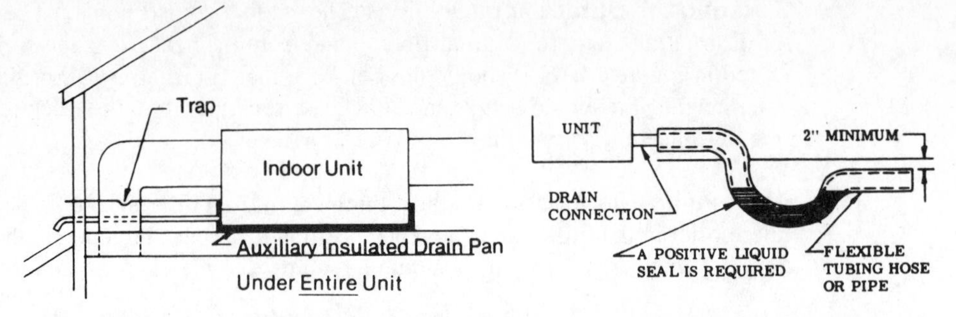

Figure 24.17 Typical condensate drain trap installation. (Courtesy of Addison Products Company and Knight Energy Institute)

trates the various configurations of the ground-coupled heat pump, the groundwater heat pump, and the surface-water heat pump. Ground-coupled systems have been widely used since the mid-1980s. Currently, horizontal systems constitute about 50% of all installations, vertical 35%, and pond and others approximately 15%. Groundwater systems have been popular since the early 1970s. The ground or water loop normally contains an intermediate fluid for transferring heat. Refrigerant-containing loops have been the least common of the GeoExchange systems.

The U.S. Environmental Protection Agency (EPA) has found that GeoExchange heat pumps can reduce energy consumption and corresponding emissions by 40% compared with that of air-source heat pumps. To meet Energy Star® qualifications, ratings of at least 2.8 COP and 13 SEER are required. The Geothermal Heat Pump Consortium has a goal of 400,000 annual installations, which will reduce emmisions equivalent to removing over half a million cars off the road. A GeoExchange heat pump can be installed in virtually any area of the country.

Equipment The most commonly used unit is the single-package water-to-air heat pump. All components, including the compressor, are contained in an enclosure about the size of a small gas furnace (Figure 24.20.) Because there is no outdoor unit, less refrigerant R-22 is required, and factory brazing reduces the potential for leaks. The

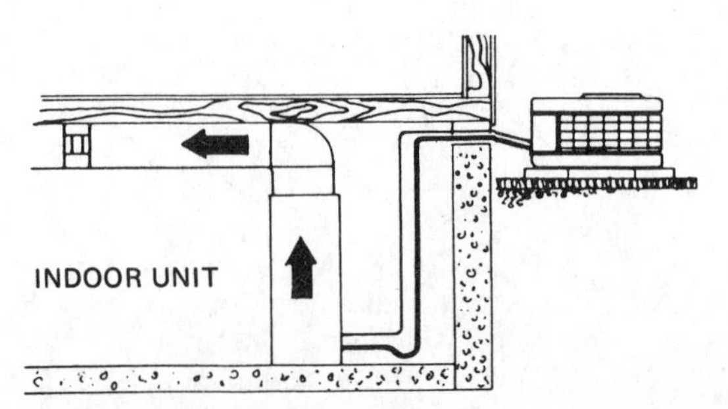

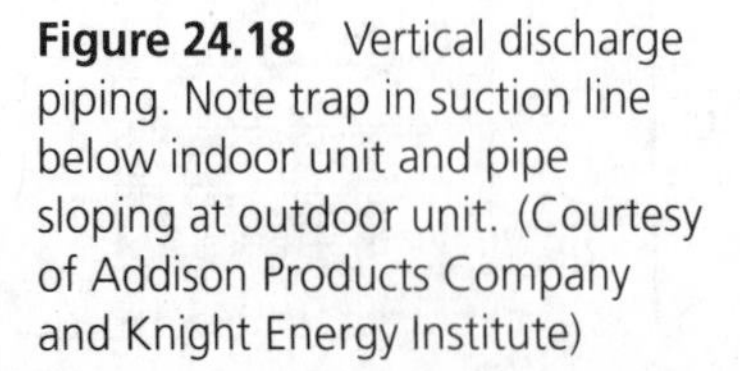

Figure 24.18 Vertical discharge piping. Note trap in suction line below indoor unit and pipe sloping at outdoor unit. (Courtesy of Addison Products Company and Knight Energy Institute)

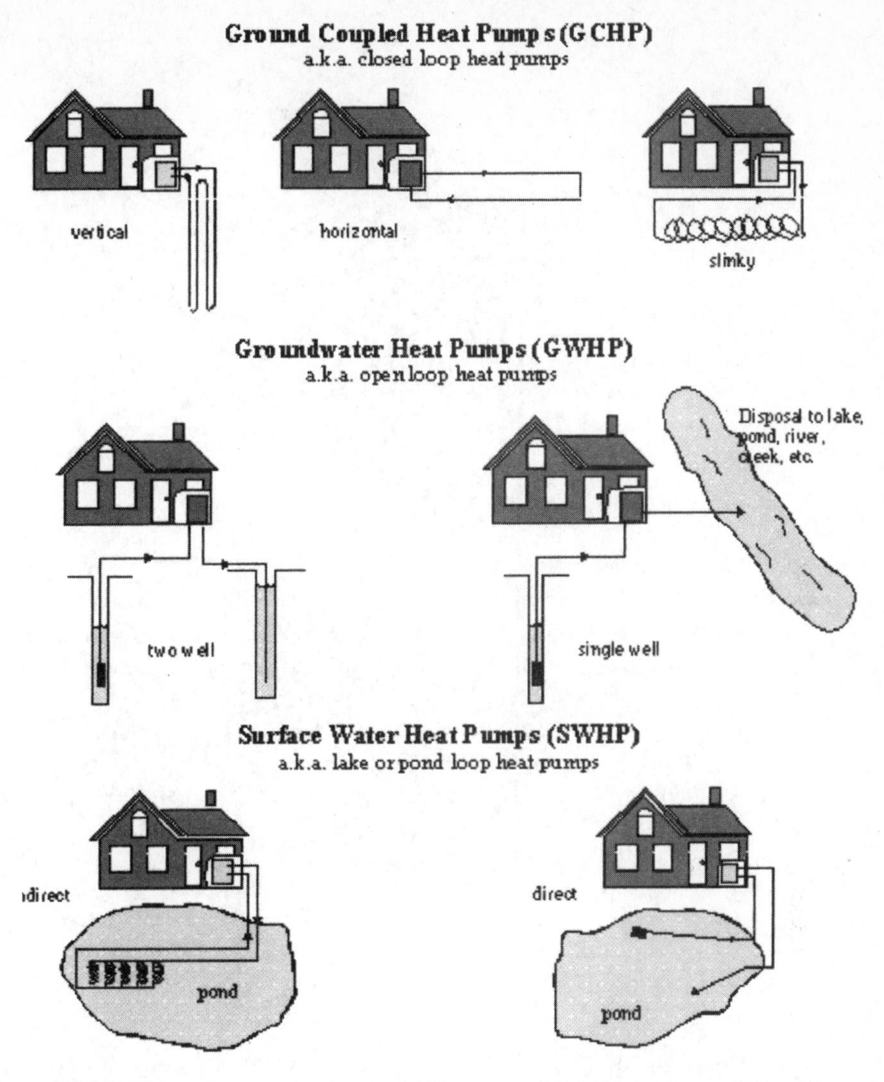

Figure 24.19 Ground-source heat pumps. (Courtesy of U.S. Department of Energy)

water-to-refrigerant coil replaces the outdoor air coil in the refrigeration cycle. The water-heating coil, located next to the compressor, may be used with a small pump for domestic hot-water heating. Field-installed piping connects this unit to the water heater. High-efficiency equipment generally contains a high-efficiency compressor, larger air coil, higher-efficiency fan motor, and larger refrigerant-to-water heat exchanger.

Water Loops The *open-loop* system picks up water in one location and returns it to another. Water quality and hardness are issues with open-loop systems. If the water

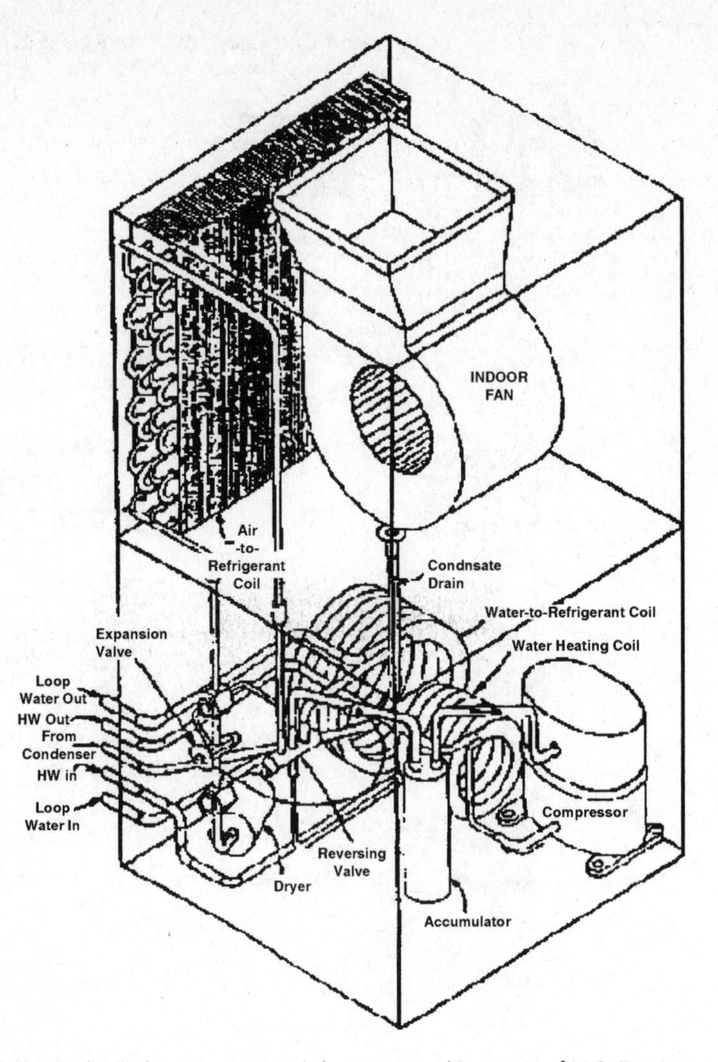

Figure 24.20 Single-package water-to-air heat pump. (Courtesy of U.S. Department of Energy)

quality is good and the building domestic water is to be supplied by a well, open loop is a good consideration. Used water is normally returned to a pond, lake, river, or injection well. If the water is supplied from a pond or lake that might freeze, the water intake must be a minimum of 20 ft below the surface of the lake to prevent icing in the return line.

The *closed-loop* system can be horizontal or vertical piping. The advantage of the horizontal loop is lower initial cost and contractor capability. The high-density polyethylene piping is buried 4 to 6 ft deep. The advantage of the vertical loop is the stability of the temperature in the loop. At a 100- to 400-ft depth, there is no temperature fluctuation. In the more extreme northern climates, this is a major consideration. For the vertical system, 4-in.-diameter holes are drilled about 20 ft apart. Into these holes go two pipes that are connected at the bottom with a U-bend to form a loop.

The pipes are filled with an environmentally friendly antifreeze/water solution that acts as a heat exchanger. In the winter, the fluid in the pipes extracts heat from the earth and carries it to the water-to-refrigerant coil. In the summer, the process reverses and deposits heat into the cooler ground.

Air System Conventional ductwork delivers the conditioned air throughout the building; however, the correct airflow is critical with all heat pumps. In the winter, the indoor air-handling coil is used as the condenser. A dirty condenser, insufficient airflow, or too warm an air temperature will increase operating costs and reduce capacity. Changing the air filter on a regular basis is critical—monthly at a minimum. All registers and grilles must remain open and unobstructed. If balancing dampers are partially closed, an airflow reading must taken to assure at least 450 cfm per ton of system capacity. For heating operation return-air grilles should be located near the floor to provide cooler temperatures.

Direct GeoExchange System The direct geothermal system uses refrigerant, rather than another fluid, in the ground loops. To maintain stable refrigerant conditions, special controls are necessary to manage the refrigerant in the long evaporator–condenser that is buried in direct contact with the heat source. Because the greatest heat transfer occurs during boiling and condensing, the subcooling and superheating should be kept to a minimum for highest efficiency. The EarthLinked™ system, shown in Figure 24.21 and Figure 24.22, manages the refrigerant flow with the use of the Active Charge Control (ACC) and the Liquid Flow Control (LFC).

The ACC (Figure 24.23) consists of a thermally insulated reservoir that replaces the standard suction accumulator. Its purpose is to constantly deliver refrigerant vapor and oil to the compressor at the correct quantity and with the least amount of superheat. Due to the design of the ACC, if any of the incoming vapor is superheated, its contact with the liquid evaporates some of the stored liquid, thus reducing the superheat to near zero. If any of the incoming vapor contains liquid, that liquid will be trapped in the reservoir. The exiting vapor will contain neither liquid nor superheat. The ACC also provides a reserve of refrigerant, so that the amount of charge in the system is not critical. The sight glasses mounted on the side of the ACC provide a means of quickly determining if the system is properly charged without the need for gauges and thermometers.

The LFC (Figure 24.24) replaces the conventional metering device. It senses the amount of liquid condensed and passes it on to the evaporator. Because no excess liquid is kept in the condenser, subcooling is kept at a minimum. Thus more of the area of the condenser actively condenses vapor into a liquid. As a result, the compressor discharge pressure is reduced, and refrigerant flow is increased.

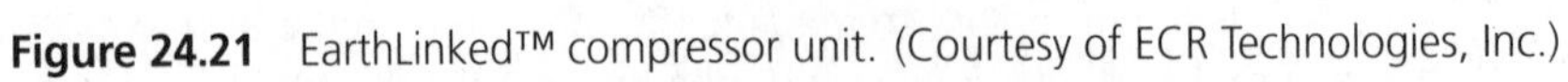

Figure 24.21 EarthLinked™ compressor unit. (Courtesy of ECR Technologies, Inc.)

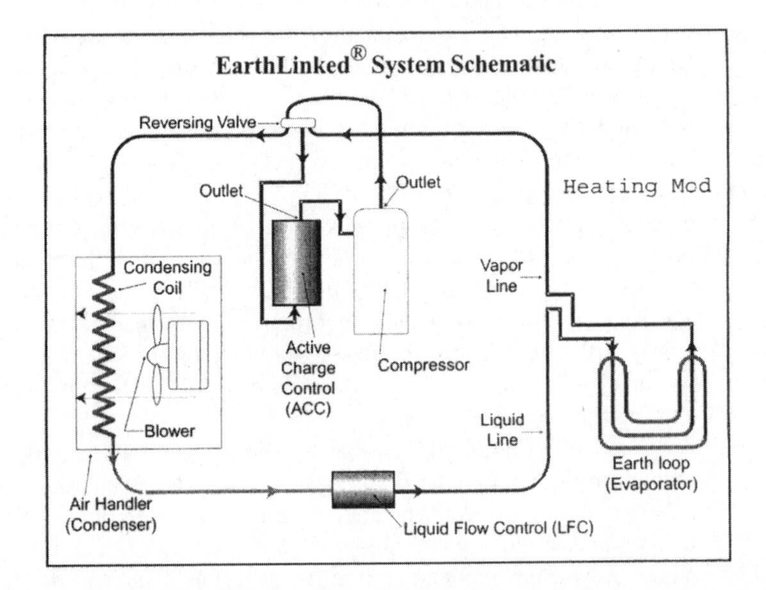

Figure 24.22 EarthLinked® system schematic. (Courtesy of ECR Technologies, Inc.)

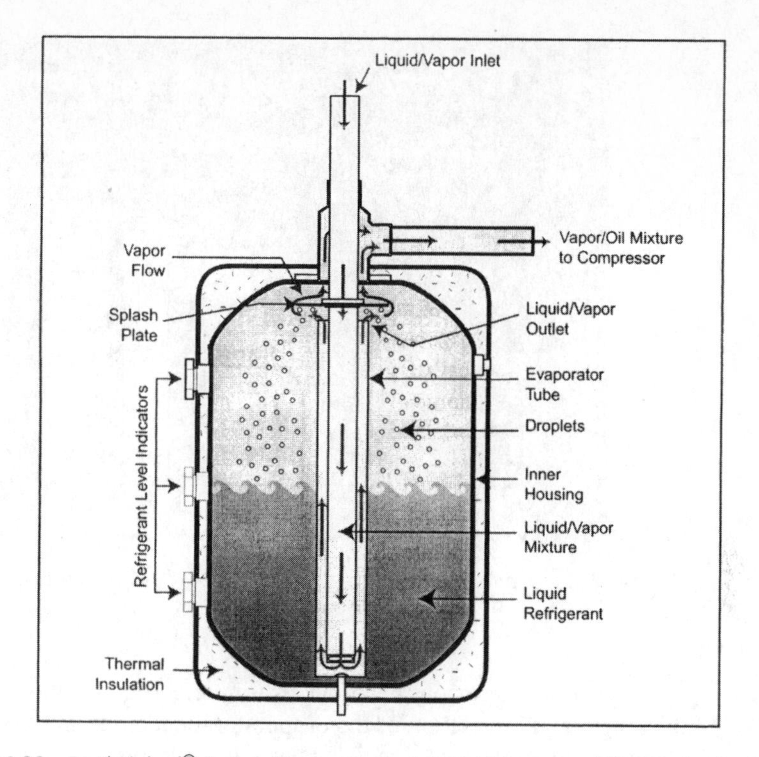

Figure 24.23 EarthLinked® Active Charge Control ACC. (Courtesy of ECR Technologies, Inc.)

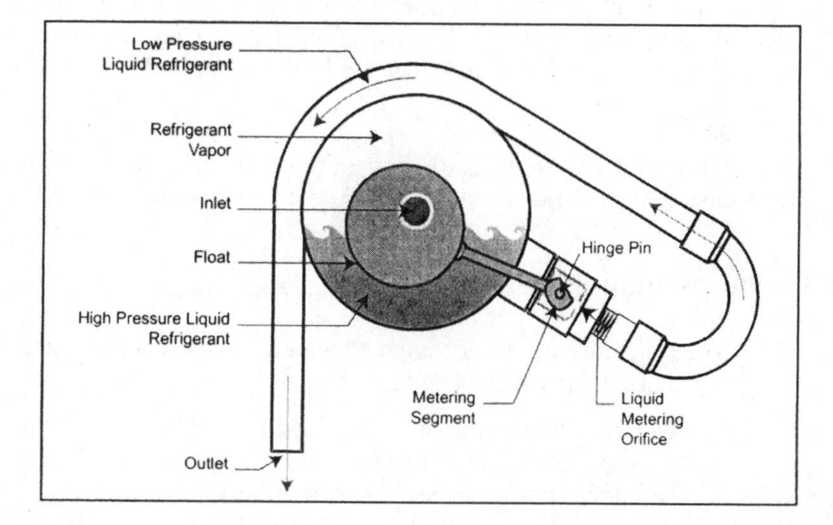

Figure 24.24 EarthLinked® Liquid Flow Control (LFC). (Courtesy of ECR Technologies, Inc.)

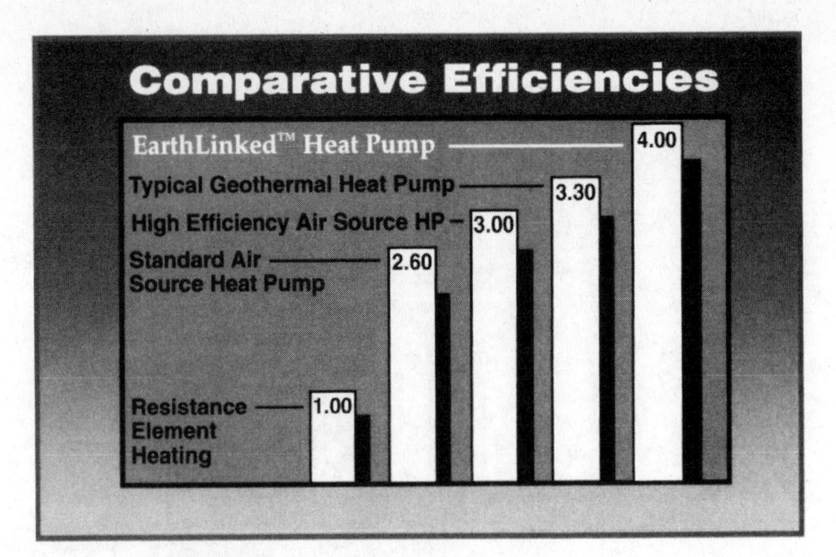

Figure 24.25 Comparative efficiencies. (Courtesy of ECR Technologies, Inc.)

The use of these controls provides for increased operational efficiency and makes the length and orientation of the evaporator–condenser and the amount of refrigerant charge in circulation not crucial. Thus the earth loops can be installed in horizontal, vertical, or diagonal configurations. Installation and service of the EarthLinked™ system are simplified by the ease and certainty of determining a correct charge without the need for gauges or weighing in of refrigerant.

Because no fuel is consumed, up to 75% of the heat delivered by EarthLinked™ comes from the renewable energy in the earth. This is 400% more efficient than electric strip heat and up to 33% more efficient than a high-efficiency air-to-air heat pump (Figure 24.25). The average heating COP for this system is 3.9, with a cooling-mode SEER of 17.

Defrost Because the earth temperature remains well above freezing, even during operation, defrosting is not required with properly installed GeoExchange systems.

Add-On Heat-Pump System

The major advantages of having a high-efficiency heat pump and a gas furnace combined in one system are the following:

1. Heating costs are reduced.
2. A comfort level is maintained.
3. The source of heat can be either gas or electricity.

The add-on heat-pump system is shown in Figure 24.26. High-efficiency heat pumps produce a more “even” heat at milder outdoor temperatures, above 30°F.

Figure 24.26 Cutaway view of a typical add-on heat pump system. (Courtesy of the Trane Company Unitary Products Group)

With the lower discharge-air temperature of approximately 100°F, the heat pump allows the indoor room temperature to remain more constant. Even though the heat pump may run for extended periods, its efficiency remains high. The efficiency begins to decrease at outdoor temperatures lower than 30°F.

The advantage of a gas furnace is that its efficiency does not change. As the outdoor temperature drops, the furnace runs longer to maintain the desired indoor temperature, but the operating efficiency (Btu of heat produced per cubic feet of gas consumed) does not change.

Figure 24.27 illustrates a comparison of heat-pump and gas-furnace heating. These curves depict operating costs compared with outdoor temperature. Actual operating costs depend on the efficiency of the heat pump and the gas furnace, the hours of operation, and the cost of gas and electricity. In general, the heat pump operates most efficiently in the temperature range from 65°F down to approximately 30°F. Below this temperature, the gas furnace takes over, operating more efficiently at the lower temperatures when compared with a heat pump plus electric heat.

Wiring Diagram

The wiring diagram in Figure 24.28 is for a typical split-system heat pump. The components are identified, so that the thermostat is in the lower right corner, the air handler is to the left of the thermostat, and the outdoor unit occupies the remainder of the diagram.

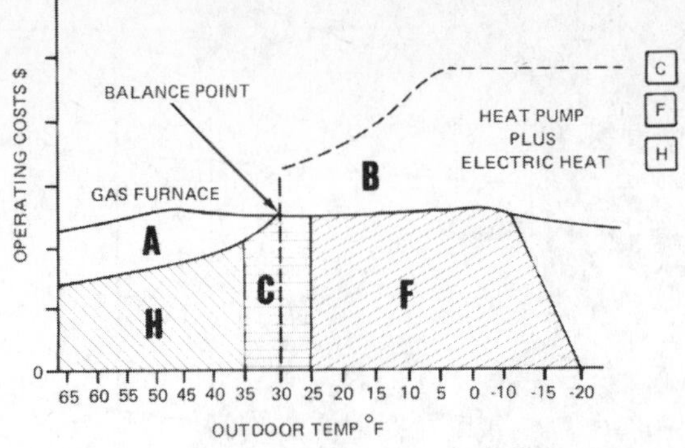

Figure 24.27 Performance curve for an add-on heat-pump system.

In studying a heat-pump wiring diagram, the technician must verify the mode of powering the reversing valve. Some manufacturers energize the valve during cooling operation, whereas others energize the valve during heating. The first method assures that heating is available even with a reversing-valve coil failure. The second method avoids the noise created in switchover when the customer may be present near the equipment in the summer.

The Trane Company heat pump also uses a thermistor in series with the heating heat anticipator. The thermistor is located in the outdoor unit and increases in resistance as the exterior temperature drops. This reduces the current flow through the heat anticipator, provides for longer cycles, and prevents a drop in room temperature during constant compressor operation.

Heat-Pump Servicing

A useful method of diagnosing refrigeration service problems is to refer to Figure 24.29. This chart lists the more frequent problems and suggested solutions. The same methods used for evacuating, leak testing, and refrigerant charging air-conditioning systems are also used to service heat pumps. Remember, a heat pump must have proper air flow and proper refrigerant charge.

Solid-State Control Certain manufacturers replace some of the standard electrical or mechanical controls with a solid-state control module. A typical control system of this type is shown in Figure 24.30. This logic module responds to the demand signal of the thermostat, examines the input from four sensors ("outdoor," "discharge," "defrost," and "liquid"), and determines when the heat pump or the supplementary heaters will operate.

Balance-Point Adjustment The location of the balance-point adjustment is shown in Figure 24.30. The balance point is the lowest ambient temperature at which the heat

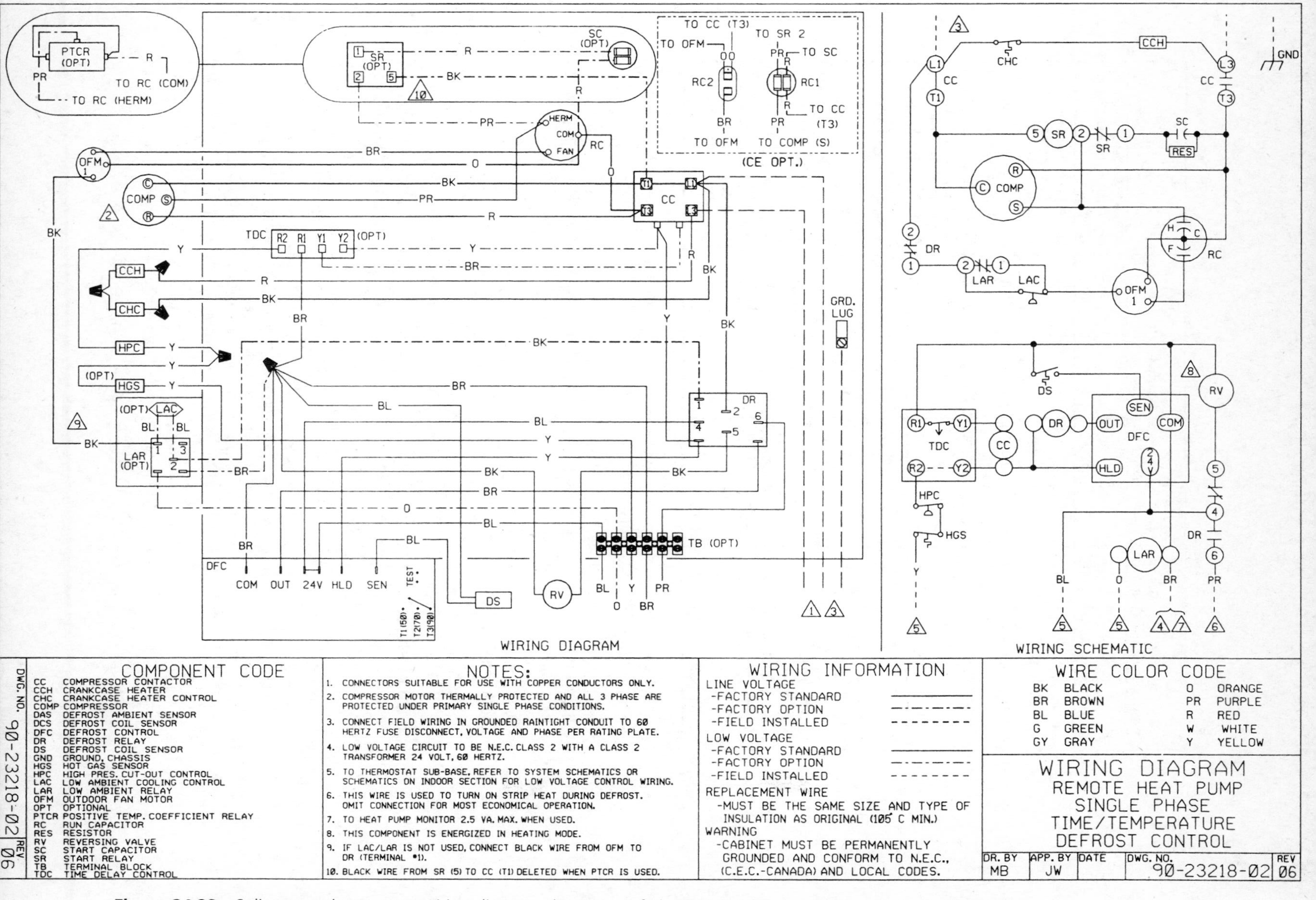

Figure 24.28 Split-system heat-pump wiring diagram. (Courtesy of The Trane Company, Unitary Products Group)

Problem		Suction Pressure	Head Pressure	Superheat	Compressor Amperage	Temp. Diff. Over Indoor Coil	Possible Solutions
Low Refrigerant Charge		LOW	LOW	HIGH	LOW	LOW	Possible icing of evaporator. Repair leak, evacuate and Charge by Weight.
RESTRICTION		LOW	LOW	HIGH	LOW	LOW	Most Restrictions occur in strainer. Possible icing of evaporator.
Low Air Flow Over Indoor Coil	Heating	HIGH	HIGH	LOW	HIGH	HIGH	Cause: Dirty Filter; Dirty Blower Wheel; Duct too small. Supply outlets closed.
	Cooling	LOW	LOW	LOW	LOW	HIGH	
Low Air Flow Over Outdoor Coil	Heating	LOW	LOW	LOW	LOW	LOW	Cause: Dirty Coil; Bad Fan Motor; Fan Blade; Ice on coil; air blocked by building, etc.
	Cooling	HIGH	HIGH	LOW	HIGH	LOW	
Leaking Check Valve Heating or Cooling Mode		HIGH	HIGH	LOW	HIGH	LOW	Switch System to opposite Mode. If conditions correct themselves this proves check valve was leaking.
Compressor running but not pumping		HIGH	LOW	HIGH OR NONE	LOW	LOW	If not pumping at all, suction and head pressure will be equal and no refrigeration.
Reversing Valve Leaking or Stuck Mid Position		HIGH	LOW	HIGH OR NONE	LOW	LOW	All four lines of reversing valve will be hot.
Unit Not Defrosting		LOW	LOW	LOW	LOW	LOW	Outdoor Coil will be iced over. Check Refrigerant Charge, defrost controls.

Figure 24.29 Heat-pump service problems and possible solutions. (Courtesy of Addison Products Company and Knight Energy Institute)

pump can operate without the use of supplementary heat. The balance point is set by the local contractor based on:

1. Outdoor design temperature
2. Building heat loss
3. Unit capacity

Service Analyzer The service analyzer, shown in Figure 24.31, is used by the service person to determine any malfunction in the operation of the heat pump. The analyzer indicates the service problem by a fault code:

Fault Code	*Failure Mode*
2	Discharge pressure reaches approximately 400 psig
3	Discharge temperature reaches approximately 275 °F
4	Discharge temperature did not reach approximately 90 °F within 1 h of compressor operation
5	Defrost failure
6	Outdoor temperature sensor failure
7	Liquid line temperature sensor failure
8	Bonnet sensor shorted

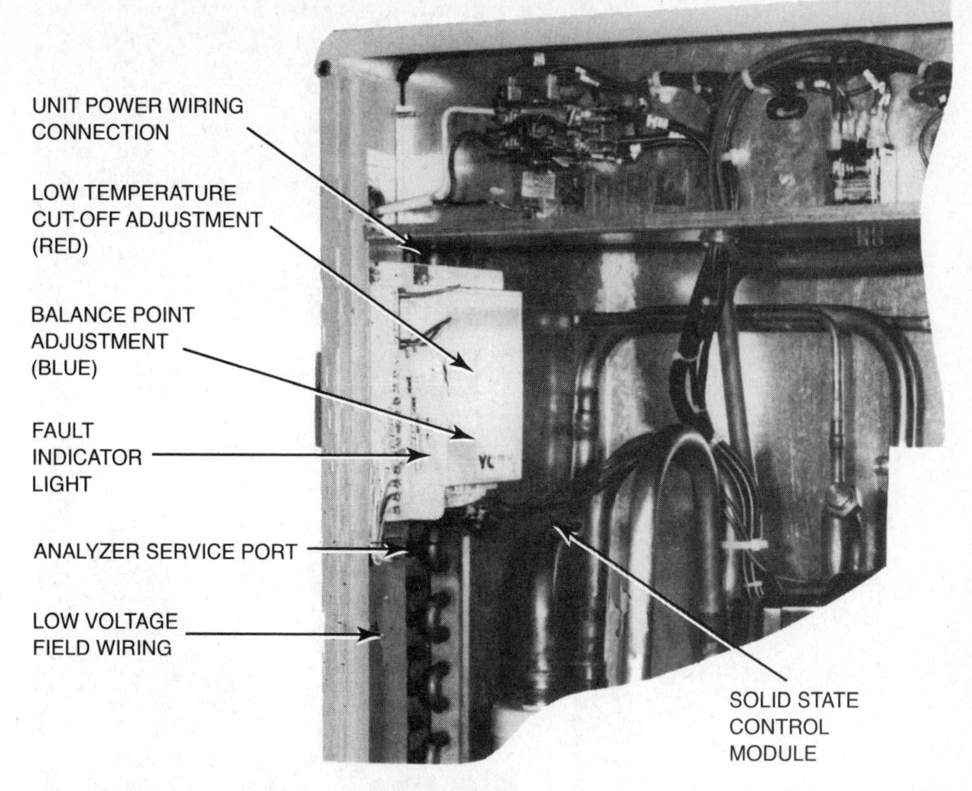

Figure 24.30 Location of the solid-state control module in the outdoor compressor unit of a heat-pump system. (Courtesy of York Corporation)

Add-On Air Conditioning

The furnace is just one part of a total comfort system that includes heating, cooling, humidification, and air cleaning. Even in the cooler northern climates, air conditioning is now considered a necessity. The heating and cooling portions can be contained in one housing such as a rooftop unit, but for residential applications individual components are more likely. As the name implies, add-on air conditioning can be installed on the furnace at the initial heating installation or at any time thereafter. Most furnaces are considered "air-prep" and require no modifications.

Sizing Prior to installation of a cooling system the required size must be determined. The best way is to perform an accurate heat gain calculation using ACCA *Manual J* or a similar method. Many rule-of-thumb methods exist, but they should not be relied on.

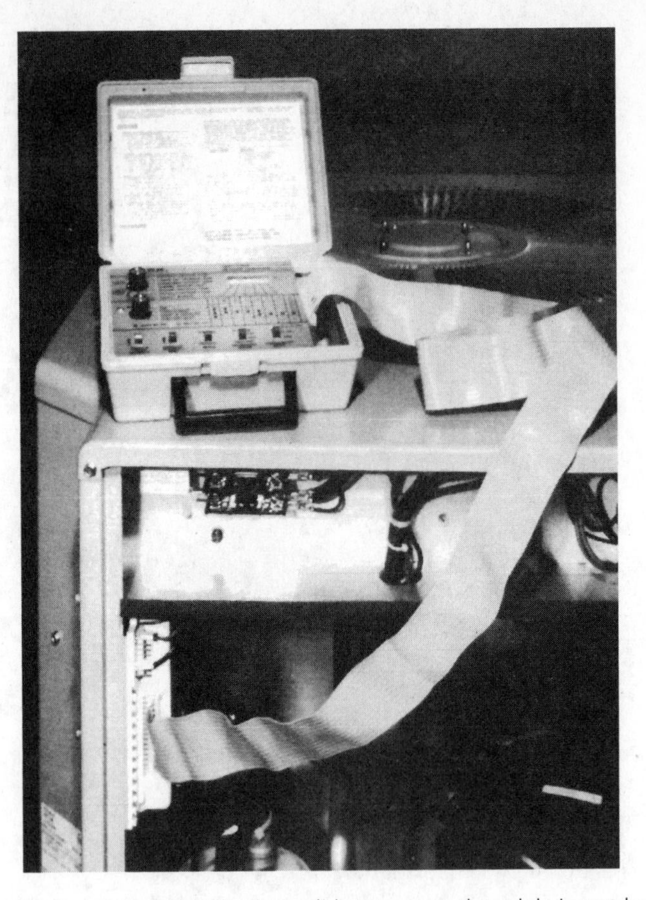

Figure 24.31 Service analyzer attached to solid-state control module in outdoor compressor unit. (Courtesy of York Corporation)

The size of the existing ductwork must be checked. Figure 24.32 lists the minimum duct requirements for various cooling loads. If the duct is insufficient, there are two alternatives: (1) replace all or portions of the duct, or (2) reduce the cooling load. A reduction in load may be achieved with sun shading or an attic power venter.

Condensing Unit The typical add-on air-conditioning system (Figure 24.33) looks just like an add-on heat pump. The condensing unit, containing the compressor, is located outdoors on a level slab. The slab may be poured concrete or premanufactured and should not touch the building, to prevent transmission of vibration. When the slab is placed, care must be taken that it is not located under a roof overhang. The condenser fan must have free airflow and should not be subject to falling icicles. Most condensing units require access on the side facing the building and therefore must have sufficient clearance from the wall. Personnel traffic, noise, piping, wiring, and zoning ordinances must also be considered.

Air Conditioner Size (tons)	Minimum Trunk Size	Minimum Number of 6″ Branches
1.5	14 × 8	6
2	18 × 8	8
2.5	21 × 8	10
3	24 × 8	12
3.5	28 × 8	14
4	24 × 10	16
5	30 × 10	20

Figure 24.32 Minimum duct sizes for cooling loads.

Piping Refrigerant piping should be limited to 25 ft but may be installed up to 50 ft for precharged systems. Any lengths over 50 ft must be correctly engineered. All piping must be secured with noncorrosive hangers. Bare copper tubing should not touch any concrete or cinder materials. Power wiring requires the services of a licensed electrician.

Evaporator Coil The evaporator coil must always be installed on the outlet side of the furnace to prevent heat exchanger corrosion. If the existing furnace is a downflow type, it must be raised and the coil installed under the furnace. Most evaporator coils have a

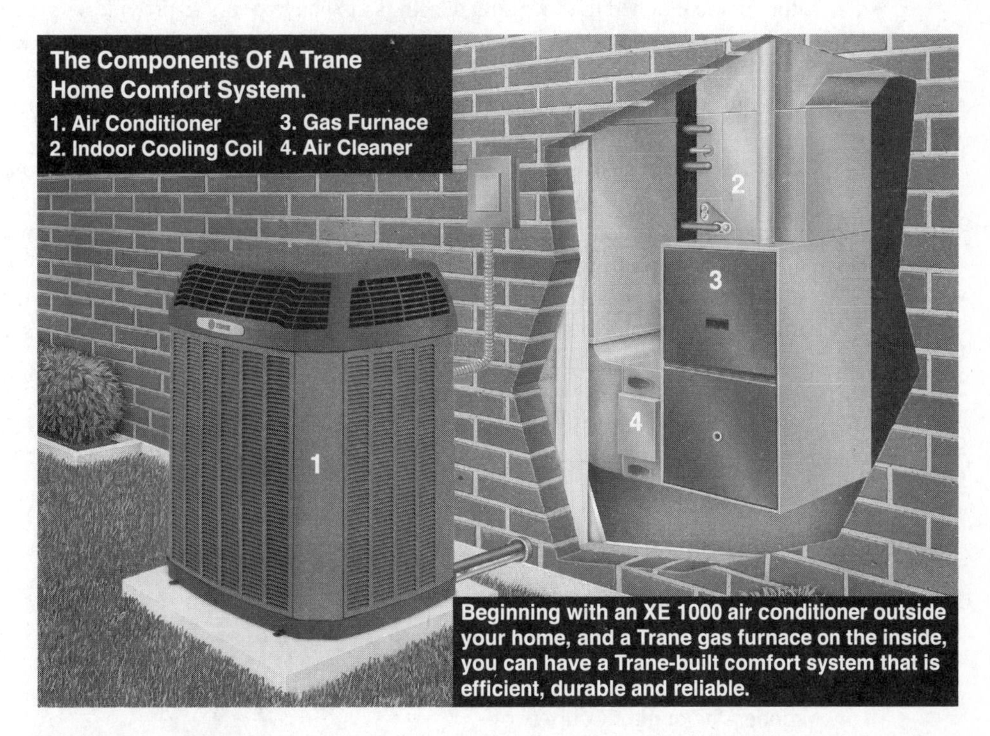

Figure 24.33 Total-comfort add-on air-conditioning system. (Courtesy of The Trane Company, Unitary Products Group)

Figure 24.34 Cased evaporator coil. (Courtesy of Armstrong Air Conditioning Inc.)

factory casing available (Figure 24.34). This allows for a neat installation with the existing duct reattached to the casing. Casings also have doors to provide service access to the coil; however, if the coil is installed in the existing sheet-metal plenum, it must be set on a pan 4 in. above the top edge of the furnace to provide proper airflow around the outside sections of the heat exchanger. A removable panel for service must also be provided.

Condensate During air conditioning, water is condensed out of the air. The evaporator coil has a built-in pan to collect and dispose of this condensate. Piping from the drain connection on the coil must not be reduced in size. It is usually constructed with CPVC plastic piping and includes a trap, as described in Figure 24.17. Condensate piping must never be directly connected to the building's plumbing system. The pipe must end above a floor drain, service sink, or sump pit. If gravity flow is not possible, a condensate pump can be used (Figure 24.35.)

Figure 24.35 Self-contained condensate pump. (Courtesy of Little Giant Pump Company)

Thermostat If the heating thermostat was wired with only a two-conductor wire, a new low-voltage cable must be installed. If the original wire is not stapled in the wall, the new cable can be secured to the old and pulled from the furnace to the existing thermostat location. Any wire obstruction will require the use of a chain in the wall to "fish" the wire through (Figure 24.36.)

Service

Air-conditioning service requires a good knowledge of the refrigerant cycle and proper airflow. The following discussion is a quick overview of the service steps to be followed.

The measured air-temperature drop across the evaporator coil should be between 17°F and 22°F. Any amount greater than 22°F indicates insufficient indoor airflow. Less than 17°F shows a problem with the refrigeration circuit.

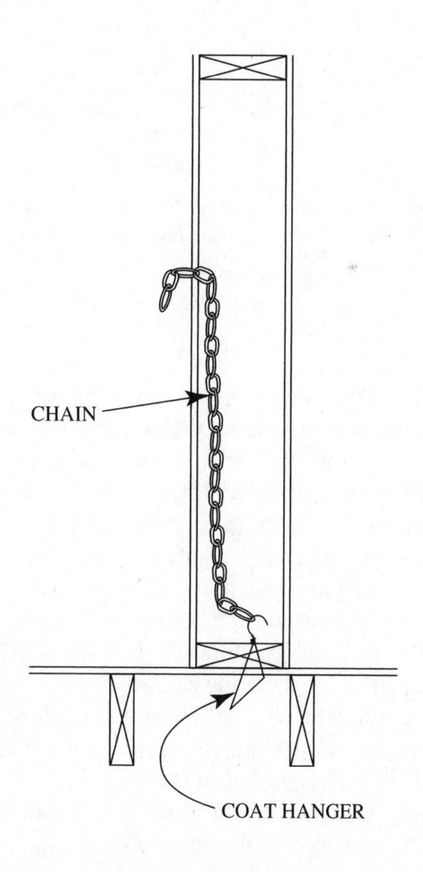

Figure 24.36 Fishing wire through a wall.

TEMPERATURE	REFRIGERANT				TEMPERATURE	REFRIGERANT			
°F	12	22	134a	502	°F	12	22	134a	502
12	15.8	34.7	13.2	43.2	42	38.8	71.4	37.0	83.8
13	16.4	35.7	13.8	44.3	43	39.8	73.0	38.0	85.4
14	17.1	36.7	14.4	45.4	44	40.7	74.5	39.0	87.0
15	17.7	37.7	15.1	46.5	45	41.7	76.0	40.1	88.7
16	18.4	38.7	15.7	47.7	46	42.6	77.6	41.1	90.4
17	19.0	39.8	16.4	48.8	47	43.6	79.2	42.2	92.1
18	19.7	40.8	17.1	50.0	48	44.6	80.8	43.3	93.9
19	20.4	41.9	17.7	51.2	49	45.7	82.4	44.4	95.6
20	21.0	43.0	18.4	52.4	50	46.7	84.0	45.5	97.4
21	21.7	44.1	19.2	53.7	55	52.0	92.6	51.3	106.6
22	22.4	45.3	19.9	54.9	60	57.7	101.6	57.3	116.4
23	23.2	46.4	20.6	56.2	65	63.8	111.2	64.1	126.7
24	23.9	47.6	21.4	57.5	70	70.2	121.4	71.2	137.6
25	24.6	48.8	22.0	58.8	75	77.0	132.2	78.7	149.1
26	25.4	49.9	22.9	60.1	80	84.2	143.6	86.8	161.2
27	26.1	51.2	23.7	61.5	85	91.8	155.7	95.3	174.0
28	26.9	52.4	24.5	62.8	90	99.8	168.4	104.4	187.4
29	27.7	53.6	25.3	64.2	95	108.2	181.8	114.0	201.4
30	28.4	54.9	26.1	65.6	100	117.2	195.9	124.2	216.2
31	29.2	56.2	26.9	67.0	105	126.6	210.8	135.0	231.7
32	30.1	57.5	27.8	68.4	110	136.4	226.4	146.4	247.9
33	30.9	58.8	28.7	69.9	115	146.8	242.7	158.5	264.9
34	31.7	60.1	29.5	71.3	120	157.6	259.9	171.2	282.7
35	32.6	61.5	30.4	72.8	125	169.1	277.9	184.6	301.4
36	33.4	62.8	31.3	74.3	130	181.0	296.8	198.7	320.8
37	34.3	64.2	32.2	75.8	135	193.5	316.6	213.5	341.2
38	35.2	65.6	33.2	77.4	140	206.6	337.2	229.1	362.6
39	36.1	67.1	34.1	79.0	145	220.3	358.9	245.5	385.0
40	37.0	68.5	35.1	80.5	150	234.6	381.5	262.7	408.4
41	37.9	70.0	36.0	82.1	155	249.5	405.1	280.7	432.9

Figure 24.37 Pressure–temperature chart. For an outdoor temperature of 85°F plus 25°F, the pressure of R-22 is 226.4 psig.

The high-side pressure corresponds to the outdoor temperature plus 25°F. That relationship is found on the pressure–temperature chart (Figure 24.37). High-efficiency units, over 12 SEER, may operate below that pressure, and older units will operate above that pressure. If the the high-side pressure is abnormally high, it is frequently due to a dirty condenser. A low pressure is often assumed to be due to a low refrigerant charge, but it may also be due to insufficient indoor airflow.

The low-side pressure corresponds to return-duct temperature minus 35°F. Again, a low pressure is often assumed to be due to a low refrigerant charge, but it may also be due to insufficient indoor airflow. In both cases, an accurate superheat reading must be taken. With a thermostatic expansion valve, the superheat should be about 10°F. With a capillary tube or orifice as the metering device, the superheat should correspond to the values in the chart shown in Figure 24.38. The required wet bulb temperature is taken with a sling psychrometer.

If the superheat reading is higher than normal, the problem could be a low refrigerant charge. If the superheat reading is normal or below, the problem could be insuf-

Superheat Charging Table
For Superheat at the Suction Service Valve

Outdoor Temp (°F)	INDOOR COIL ENTERING AIR °F wb													
	50	52	54	56	58	60	62	64	66	68	70	72	74	76
55	9	12	14	17	20	23	26	29	32	35	37	40	42	45
60	7	10	12	15	18	21	24	27	30	33	35	38	40	43
65	-	6	10	13	16	19	21	24	27	30	33	36	38	41
70	-	-	7	10	13	16	19	21	24	27	30	33	36	39
75	-	-	-	6	9	12	15	18	21	24	28	31	34	37
80	-	-	-	-	5	8	12	15	18	21	25	28	31	35
85	-	-	-	-	-	-	8	11	15	19	22	26	30	33
90	-	-	-	-	-	-	5	9	13	16	20	24	27	31
95	-	-	-	-	-	-	-	6	10	14	18	22	25	29

Figure 24.38 Superheat chart. For an outdoor temperature of 85°F and a return-duct wet bulb temperature of 66°F, the required superheat is 15°F.

ficient indoor airflow. Any caplillary tube or orifice system can be critically charged with refrigerant by using the superheat method.

Safety

Most residential air conditioners installed prior to publication of this text use refrigerant R-22. This refrigerant is safe in normal service exposure; however, it does displace oxygen and should not be inhaled. It also decomposes into dangerous vapors when in contact with a flame. These vapors can permanently damage the lungs or even cause death. When any brazing operation is performed, all refrigerant must be removed from the equipment and the area ventilated.

Refrigerant R-22 is slowly being phased out as of January 2004, and many new refrigerants are being marketed. Refrigerant R-410A is a popular product for use in residential units; however, the operating pressures are higher than many service technicians have been used to with R-22 (Figure 24.39.) The refrigeration gauges and hoses used with R-410A must be designed for this refrigerant and be in good condition to avoid injury.

Most refrigerants are considered harmful fo the environment and must be recovered with the use of EPA-approved equipment. Chapter 8 illustrates such equipment.

Genetron® AZ-20® R-410A
Genetron® R22
Pressure – Temperature Chart

(°F)	PSIG AZ-20	PSIG R22	(°F)	PSIG AZ-20	PSIG R22	(°F)	PSIG AZ-20	PSIG R22
-40	10.8	0.6	0	48.3	24.0	40	118.1	68.6
-39	11.5	1.0	1	49.6	24.9	41	120.3	70.0
-38	12.1	1.4	2	50.9	25.7	42	122.7	71.5
-37	12.8	1.8	3	52.2	26.5	43	125.0	73.0
-36	13.5	2.2	4	53.6	27.4	44	127.4	74.5
-35	14.2	2.6	5	55.0	28.3	45	129.8	76.1
-34	14.9	3.1	6	56.3	29.2	46	132.2	77.6
-33	15.6	3.5	7	57.8	30.1	47	134.7	79.2
-32	16.3	4.0	8	59.2	31.0	48	137.2	80.8
-31	17.1	4.5	9	60.7	31.9	49	139.7	82.4
-30	17.8	4.9	10	62.2	32.8	50	142.2	84.1
-29	18.6	5.4	11	63.7	33.8	51	144.8	85.7
-28	19.4	5.9	12	65.2	34.8	52	147.4	87.4
-27	20.2	6.4	13	66.8	35.8	53	150.1	89.1
-26	21.1	6.9	14	68.3	36.8	54	152.8	90.8
-25	21.9	7.4	15	69.9	37.8	55	155.5	92.6
-24	22.7	8.0	16	71.5	38.8	56	158.2	94.4
-23	23.6	8.5	17	73.2	39.9	57	161.0	96.1
-22	24.5	9.1	18	74.9	40.9	58	163.8	98.0
-21	25.4	9.6	19	76.6	42.0	59	166.7	99.8
-20	26.3	10.2	20	78.3	43.1	60	169.6	101.6
-19	27.2	10.8	21	80.0	44.2	61	172.5	103.5
-18	28.2	11.4	22	81.8	45.3	62	175.4	105.4
-17	29.2	12.0	23	83.6	46.5	63	178.4	107.3
-16	30.1	12.6	24	85.4	47.6	64	181.5	109.3
-15	31.1	13.2	25	87.2	48.8	65	184.5	111.2
-14	32.2	13.9	26	89.1	50.0	66	187.6	113.2
-13	33.2	14.5	27	91.0	51.2	67	190.7	115.3
-12	34.2	15.2	28	92.9	52.4	68	193.9	117.3
-11	35.3	15.9	29	94.9	53.7	69	197.1	119.4
-10	36.4	16.5	30	96.8	55.0	70	200.4	121.4
-9	37.5	17.2	31	98.8	56.2	71	203.6	123.5
-8	38.6	17.9	32	100.9	57.5	72	207.0	125.7
-7	39.8	18.7	33	102.9	58.8	73	210.3	127.8
-6	40.9	19.4	34	105.0	60.2	74	213.7	130.0
-5	42.1	20.1	35	107.1	61.5	75	217.1	132.2
-4	43.3	20.9	36	109.2	62.9	76	220.6	134.5
-3	44.5	21.7	37	111.4	64.3	77	224.1	136.7
-2	45.7	22.4	38	113.6	65.7	78	227.7	139.0
-1	47.0	23.2	39	115.8	67.1	79	231.3	141.3

(°F)	PSIG AZ-20	PSIG R22	(°F)	PSIG AZ-20	PSIG R22	(°F)	PSIG AZ-20	PSIG R22
80	234.9	143.6	104	334.9	207.7	128	463.2	289.2
81	238.6	146.0	105	339.6	210.8	129	469.3	293.0
82	242.3	148.4	106	344.4	213.8	130	475.4	296.9
83	246.0	150.8	107	349.3	216.9	131	481.6	300.8
84	249.8	153.2	108	354.2	220.0	132	487.8	304.7
85	253.7	155.7	109	359.1	223.2	133	494.1	308.7
86	257.5	158.2	110	364.1	226.4	134	500.5	312.6
87	261.4	160.7	111	369.1	229.6	135	506.9	316.7
88	265.4	163.2	112	374.2	232.8	136	513.4	320.7
89	269.4	165.8	113	379.4	236.1	137	520.0	324.8
90	273.5	168.4	114	384.6	239.4	138	526.6	329.0
91	277.6	171.0	115	389.9	242.8	139	533.3	333.2
92	281.7	173.7	116	395.2	246.1	140	540.1	337.4
93	285.9	176.4	117	400.5	249.5	141	547.0	341.6
94	290.1	179.1	118	405.9	253.0	142	553.9	345.9
95	294.4	181.8	119	411.4	256.5	143	560.9	350.3
96	298.7	184.6	120	416.9	260.0	144	567.9	354.6
97	303.0	187.4	121	422.5	263.5	145	575.1	359.0
98	307.5	190.2	122	428.2	267.1	146	582.3	363.5
99	311.9	193.0	123	433.9	270.7	147	589.6	368.0
100	316.4	195.9	124	439.6	274.3	148	596.9	372.5
101	321.0	198.8	125	445.4	278.0	149	604.4	377.1
102	325.6	201.8	126	451.3	281.7	150	611.9	381.7
103	330.2	204.7	127	457.3	285.4			

AZ-20 (R-410A) pressure is about 60% (1.6 times) greater than R-22.
Only use servicing equipment and components designed for AZ-20 (R-410A).
- Recovery cylinder service pressure rating 400 psig, e.g. DOT 4BA400 and DOT BW400.
- Manifold sets must be 800 psig high-side and 250 psig low-side with 550 psig low-side retard.
- Use hoses with 800 psig service pressure rating.
- Use filter driers with service pressure rating of at least 600 psig.
- Do not install a suction-line drier in the liquid line.

ARI color code is rose (PMS color code is 507).
- Read the label to confirm cylinder contents since lighting conditions may hinder color identification.

Liquid charge AZ-20.
- Use a commercial metering device in the manifold hose.

Use an R-410A TXV on indoor sections originally equipped with a TXV; do not use an R-22 TXV.

Existing capillary tube indoor coils will not operate properly.

A liquid-line filter drier is recommended.

Use a polyol ester (POE) lubricant approved by the equipment manufacturer.
- POE lubricants are not always interchangeable.
- Minimize exposure of POE lubricant to the atmosphere; use a pump to transfer lubricant.
- Use a filter drier to remove moisture from POE lubricant in the system – a vacuum pump will not work.

Leak detectors should detect HFC (hydrofluorocarbon) refrigerants.
- Halide torches will not work effectively.
- Instruments designed for CFCs and HCFCs will not have enough sensitivity for AZ-20.

Do not vent AZ-20 to the atmosphere.

Never pressurize a mixture of air and refrigerant since it may become flammable.

Always read the Genetron AZ-20 material safety data sheet (MSDS) before using the material.

Honeywell Genetron Refrigerants
P.O. Box 1053
Morristown, New Jersey 07962-1053
For more information call us at 1-800-631-8138

P.O. Box 1053
Morristown, New Jersey 07962-1053
1-800-631-8138

Printed in USA ©2000 Honeywell

G525-093 1-00-5K

Figure 24.39 Refrigerant AZ-20 (410A) presure–temperature chart. (Courtesy of Honeywell, Genetron Refrigerants)

STUDY QUESTIONS

Answers to the study questions may be found in the section noted in brackets.

24-1 In which direction does heat travel? *[Refrigeration Cycle]*

24-2 For a gas, what is the relationship between pressure and temperature? *[Refrigeration Cycle]*

24-3 Describe the operation of the refrigeration cycle. *[Refrigeration Cycle]*

24-4 Describe the operation of the heating cycle of a heat pump. *[Heat-Pump Characteristics]*

24-5 List the three types of heat pumps and their heat sources. *[Types of Heat Pumps]*

24-6 For the air-to-air system, what is the difference between a single-package and a split system? *[Air-to-Air System]*

24-7 What are the conditions for a high-temperature and a low-temperature rating? *[Ratings]*

24-8 How is the air temperature rise determined for a heat pump? *[Indoor-Air-Temperature Rise]*

24-9 In a GeoExchange system, what is used as the condenser in the cooling mode? *[GeoExchange Systems, Equipment]*

24-10 Describe the difference between open loop and closed loop. *[Water Loops]*

24-11 What does COP mean? *[Ratings]*

24-12 What is the major difference between a *direct* GeoExchange system and others GeoExchange Systems. *[Direct GeoExchange System]*

24-13 How is the proper refrigerant charge determined in the EarthLinked® GeoExchange system? *[Direct GeoExchange System]*

24-14 How is the service analyzer used? *(Service Analyzer]*

24-15 Name the three major advantages of an add-on heat-pump system. *[Add-on Heat-Pump System]*

24-16 Above what temperature does the heat pump operate most efficiently? *[Add-on Heat-Pump System]*

24-17 Who sets the temperature of the balance point? *[Add-on Heat-Pump System]*

24-18 What must be considered when placing the condensing unit? *[Add-on Air Conditioning]*

24-19 How should condensate piping be connected to the building's sewer system? *[Add-on Air Conditioning]*

24-20 What does low operating pressure indicate? *[Add-on Air Conditioning]*

Index for Appendices Service Information

Gas Furnace

Oil Furnace

Hydronic Systems

Miscellaneous

SERVICE GUIDE FOR GAS-FIRED FURNACES

CONDITIONS	POSSIBLE CAUSES	POSSIBLE CURES
FLAME TOO LARGE	1. PRESSURE REG. SET TOO HIGH. 2. DEFECTIVE REGULATOR 3. BURNER ORIFICE TOO LARGE.	1. RESET, USING MANOMETER 2. REPLACE 3. REPLACE WITH CORRECT SIZE.
NOISY FLAME	1. TOO MUCH PRIMARY AIR. 2. NOISY PILOT 3. BURR IN ORIFICE	1. ADJUST AIR SHUTTERS 2. REDUCE PILOT GAS· 3. REMOVE BURR OR REPLACE ORIFICE
YELLOW TIP FLAME	1. TOO LITTLE PRIMARY AIR. 2. CLOGGED BURNER PORTS 3. MISALIGNED ORIFICES 4. CLOGGED DRAFT HOOD	1. ADJUST AIR SHUTTERS 2. CLEAN PORTS 3. REALIGN 4. CLEAN
FLOATING FLAME	1. BLOCKED VENTING 2. INSUFFICIENT PRIMARY AIR	1. CLEAN 2. INCREASE PRIMARY AIR SUPPLY
DELAYED IGNITION	1. IMPROPER PILOT LOCATION. 2. PILOT FLAME TOO SMALL. 3. BURNER PORTS CLOGGED NEAR PILOT 4. LOW PRESSURE	1. REPOSITION PILOT. 2. CHECK ORIFICE, CLEAN, INCREASE PILOT GAS. 3. CLEAN PORTS 4. ADJUST PRESSURE REGULATOR.
FAILURE TO IGNITE	1. MAIN GAS OFF 2. BURNED OUT FUSE 3. LIMIT SWITCH DEFECTIVE 4. POOR ELECTRICAL CONNECTIONS. 5. DEFECT GAS VALVE. 6. DEFECTIVE THERMOSTAT.	1. OPEN MANUAL VALVE. 2. REPLACE 3. REPLACE 4. CHECK, CLEAN AND TIGHTEN. 5. REPLACE 6. REPLACE
BURNER WON'T TURNOFF	1. POOR THERMOSTAT LOCATION. 2. DEFECTIVE THERMOSTAT. 3. LIMIT SWITCH MALADJUSTED. 4. SHORT CIRCUIT 5. DEFECTIVE OR STICKING AUTOMATIC VALVE	1. RELOCATE 2. CHECK CALIBRATION. CHECK SWITCH AND CONTACTS. REPLACE. 3. REPLACE 4. CHECK OPERATION AT VALVE. LOOK FOR SHORT AND CORRECT. 5. CLEAN OR REPLACE.
RAPID BURNER CYCLING	1. CLOGGED FILTERS. 2. EXCESSIVE ANTICIPATION. 3. LIMIT SETTING TOO LOW. 4. POOR THERMOSTAT LOCATION.	1. CLEAN OR REPLACE. 2. ADJUST THERMOSTAT ANTICIPATOR FOR LONGER CYCLES. 3. READJUST OR REPLACE LIMIT. 4. RELOCATE.
RAPID FAN CYCLING	1. FAN SWITCH DIFF. TOO LOW. 2. BLOWER SPEED TOO HIGH.	1. READJUST OR REPLACE. 2. READJUST TO LOWER SPEED.
BLOWER WON'T STOP	1. MANUAL FAN "ON". 2. FAN SWITCH DEFECTIVE. 3. SHORTS	1. SWITCH TO AUTOMATIC. 2. REPLACE 3. CHECK WIRING AND CORRECT.

II MOTOR AND BLOWER

CONDITION·	POSSIBLE CAUSES	POSSIBLE CURES
NOISY	1. FAN BLADES LOOSE. 2. BELT TENSION IMPROPER. 3. PULLEYS OUT OF ALIGNMENT. 4. BEARINGS DRY. 5. DEFECTIVE BELT. 6. BELT RUBBING.	1. REPLACE OR TIGHTEN. 2. READJUST (USUALLY 1 INCH SLACK) 3. REALIGN 4. LUBRICATE 5. REPLACE 6. REPOSITION

Figure A.1 (Courtesy of Robertshaw Controls Company.)

S87 DIRECT-SPARK IGNITION SYSTEM TROUBLESHOOTING

Start the system by setting the temperature controller to call for heat. Observe the system response and establish the type of malfunction or deviation from normal operation using the appropriate table.

Use the table by following the instructions in the boxes. If the condition is true or okay (answer: yes), go down to the next box. If the condition is not true or not okay (answer: no), go to the box to the right. Continue checking and answering conditions in each box until a problem and/or repair is explained. After any maintenance or repair, the troubleshooting sequence should be repeated until normal system operation is obtained.

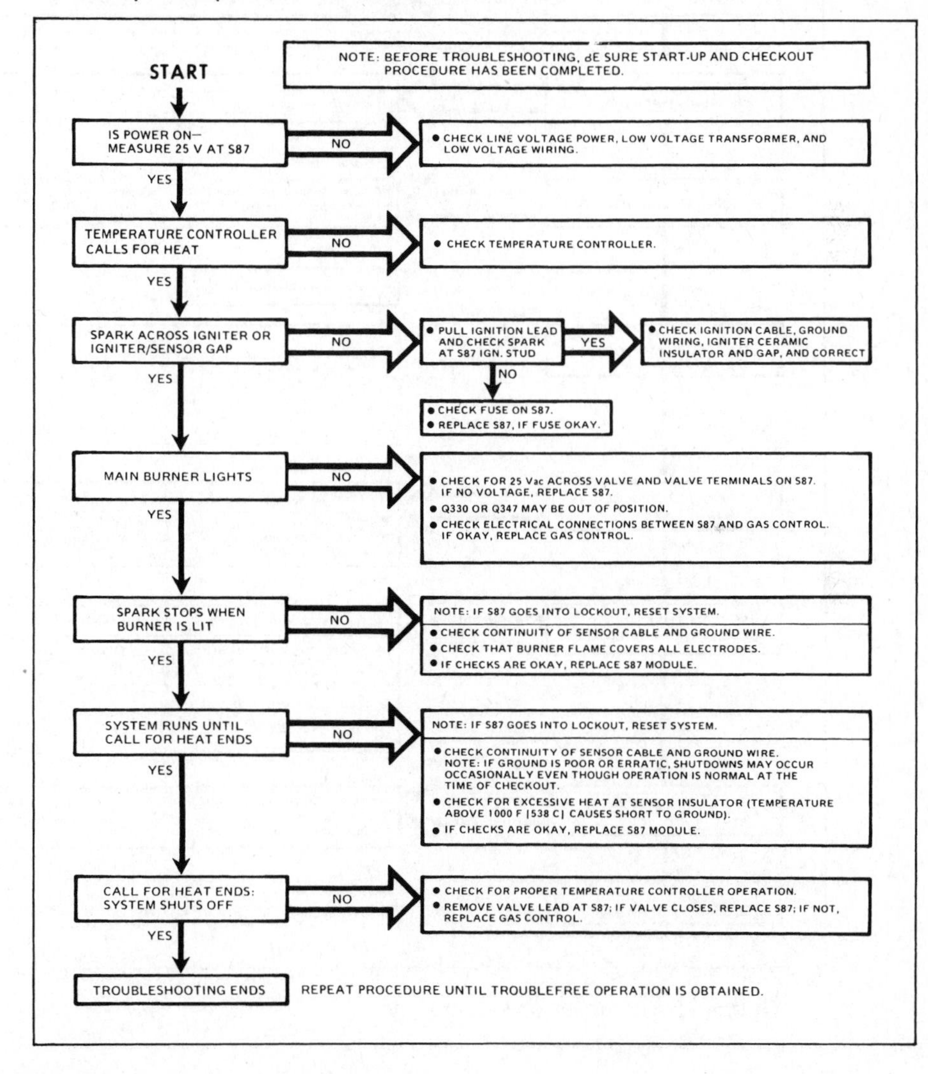

Figure A.2 (Courtesy of Honeywell Inc.)

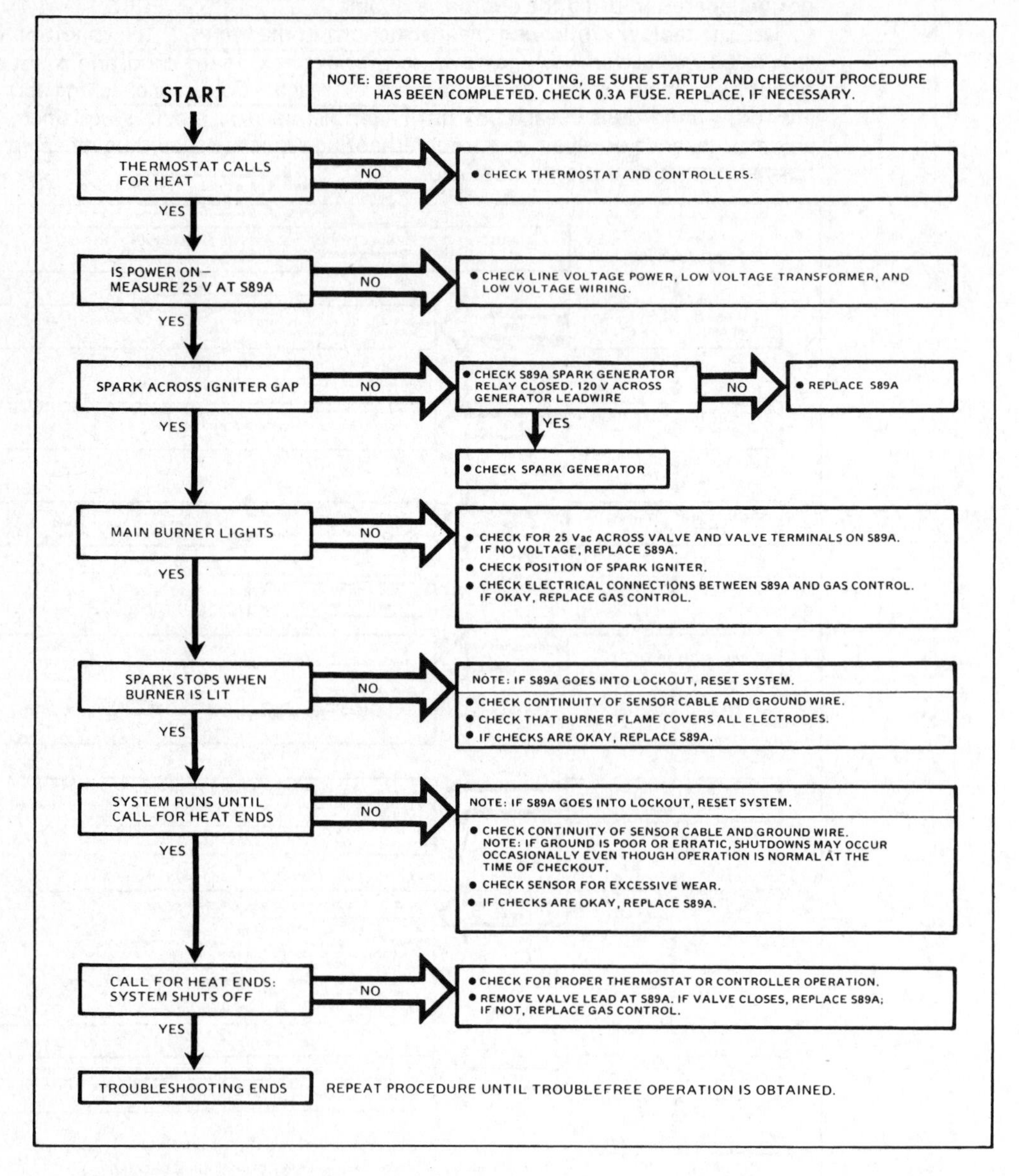

Figure A.3 (Courtesy of Honeywell Inc.)

S86 INTERMITTENT PILOT SYSTEM TROUBLESHOOTING

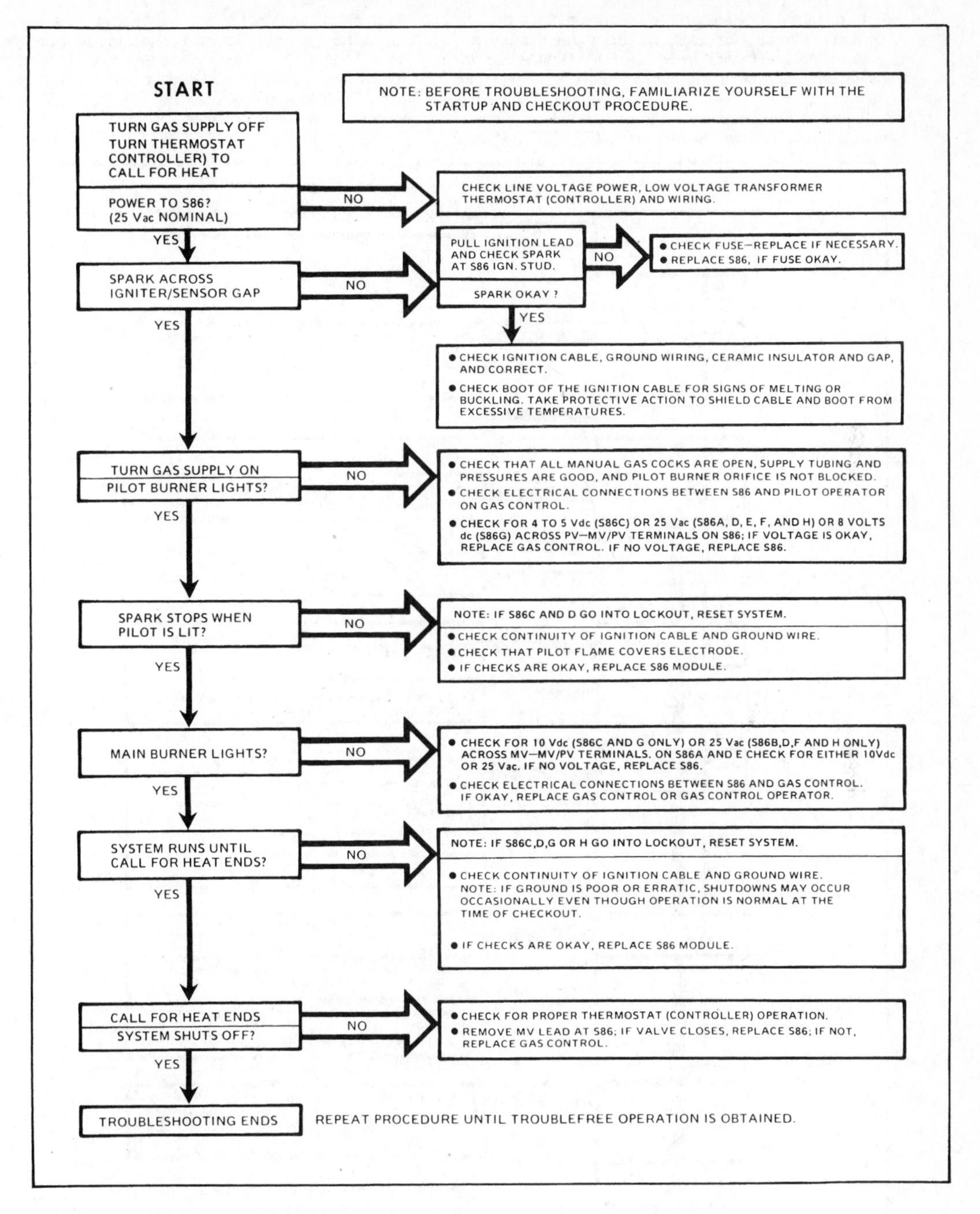

Figure A.4 (Courtesy of Honeywell Inc.)

S89C HOT-SURFACE IGNITION-CONTROL TROUBLESHOOTING

Start the system by setting the thermostat (temperature controller) to call for heat. Observe the system response and establish the type of malfunction or deviation from normal operation by using the flow chart below.

Follow the instructions in the boxes. If the condition is true or okay (answer is yes), go down to the next box. If the condition is not true or not okay (answer is no), go to the box at right. Continue checking and answering conditions in each box until a problem and/or repair is explained. After any maintenance or repair, the troubleshooting sequence should be repeated until normal system operation is obtained.

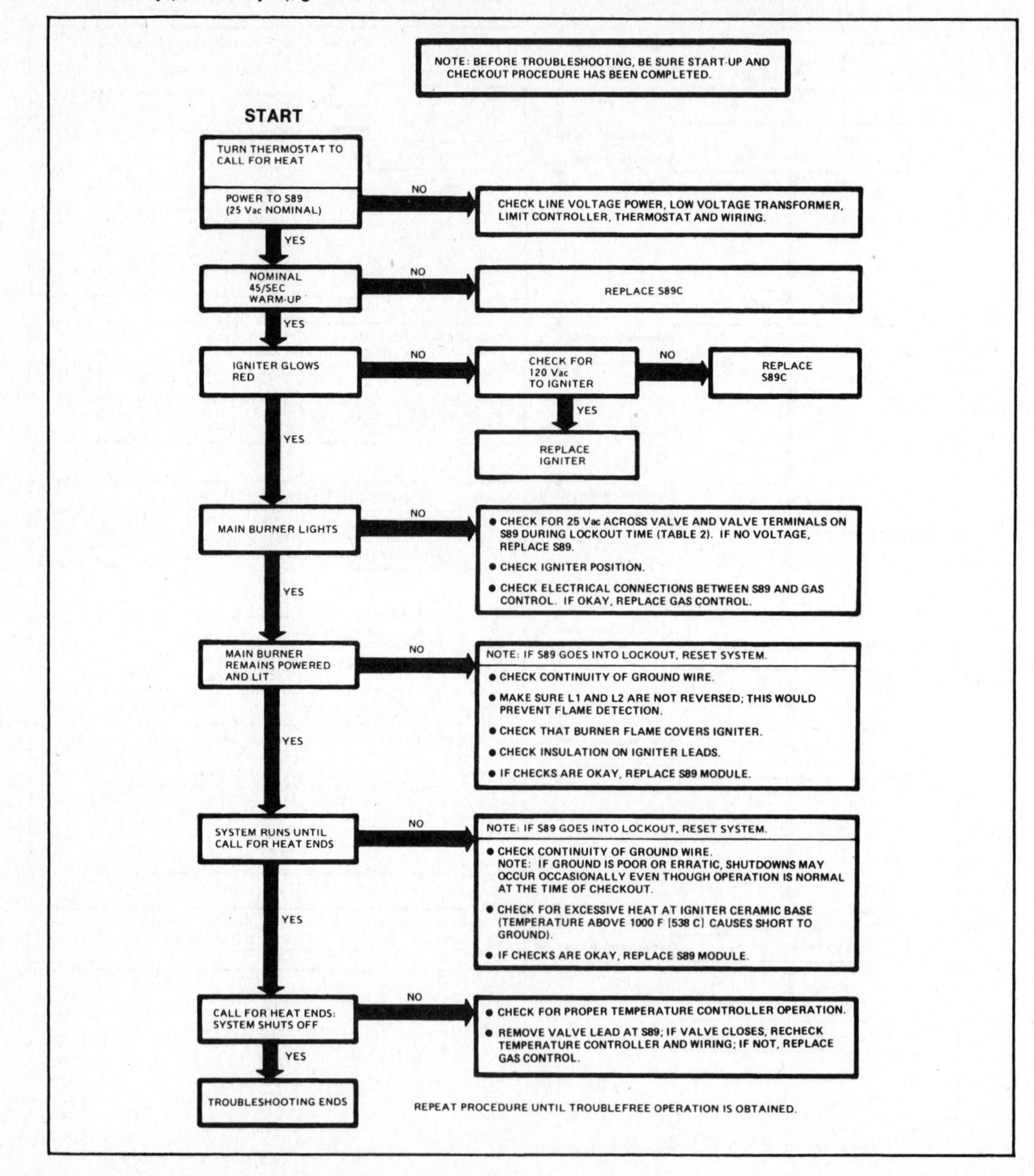

Figure A.5 (Courtesy of Honeywell Inc.)

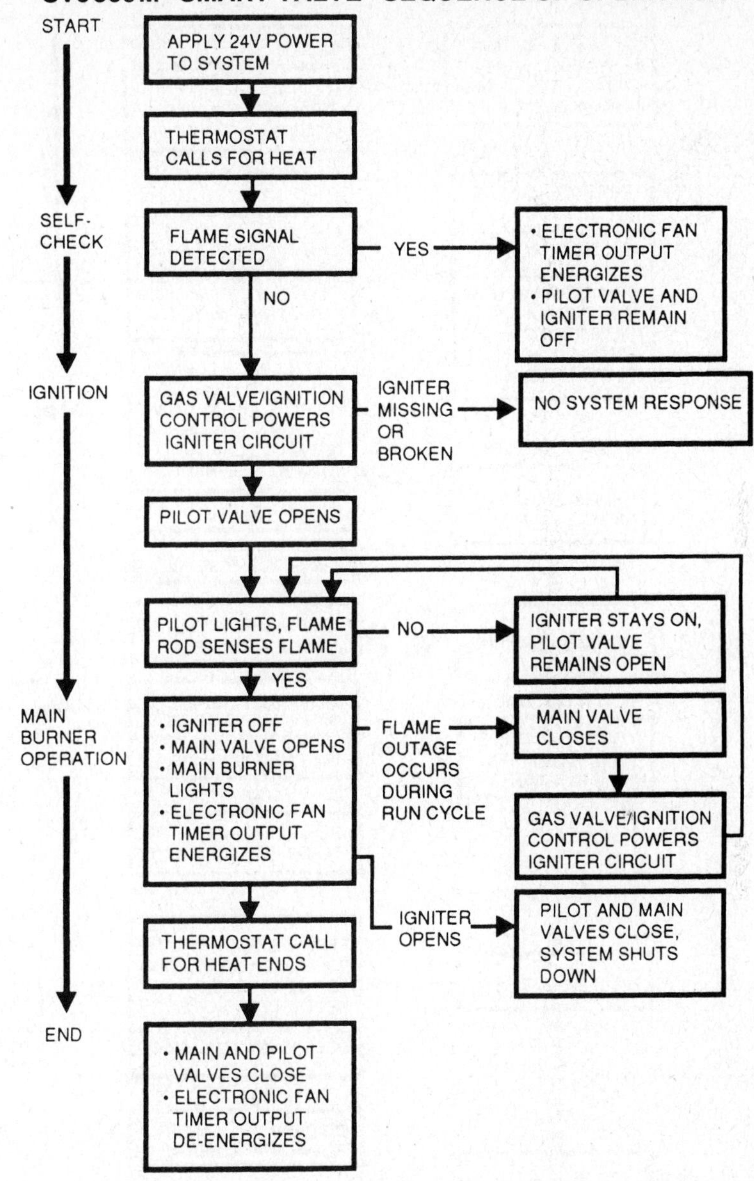

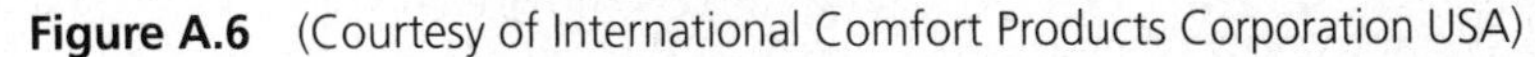
Figure A.6 (Courtesy of International Comfort Products Corporation USA)

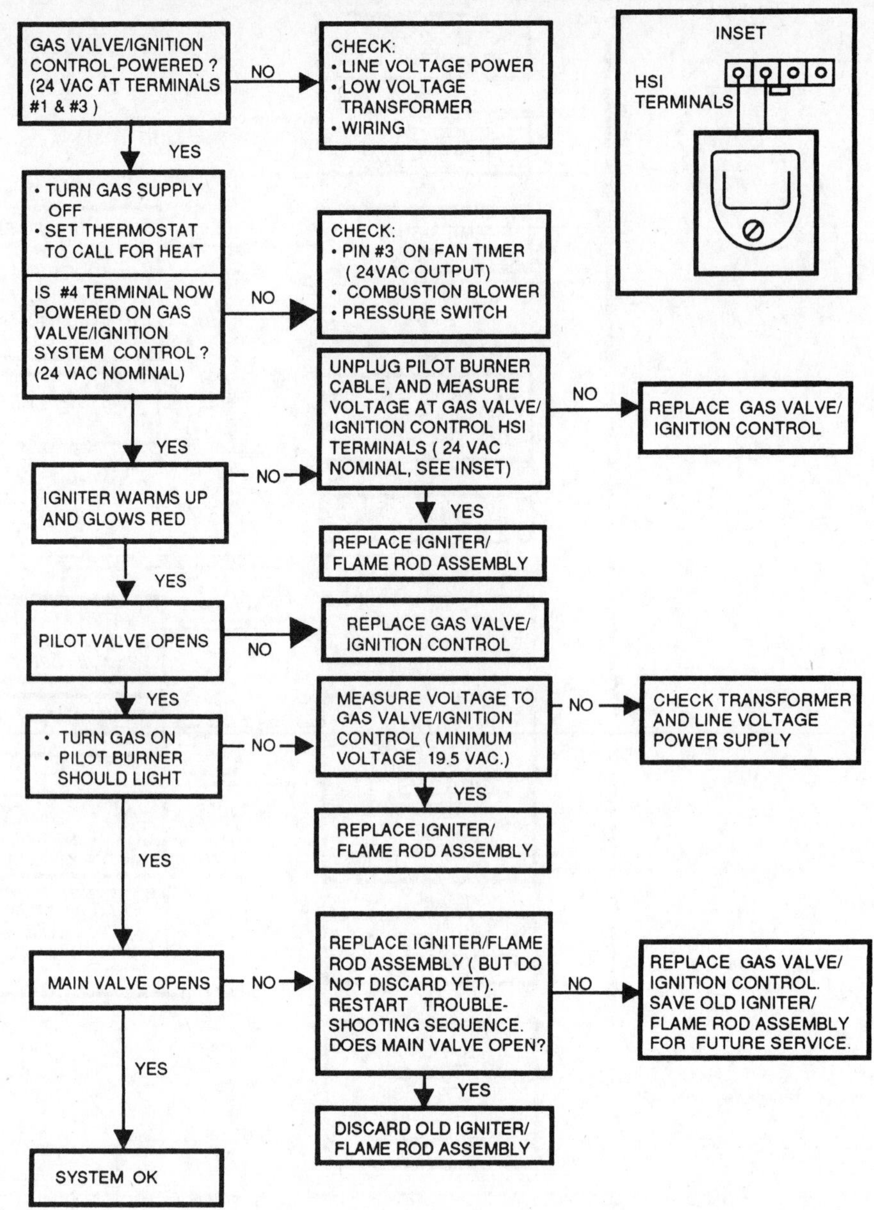

Figure A.7 (Courtesy of International Comfort Products Corporation USA)

"GENERAL 90" TOTAL FURNACE CONTROL TROUBLESHOOTING CHART

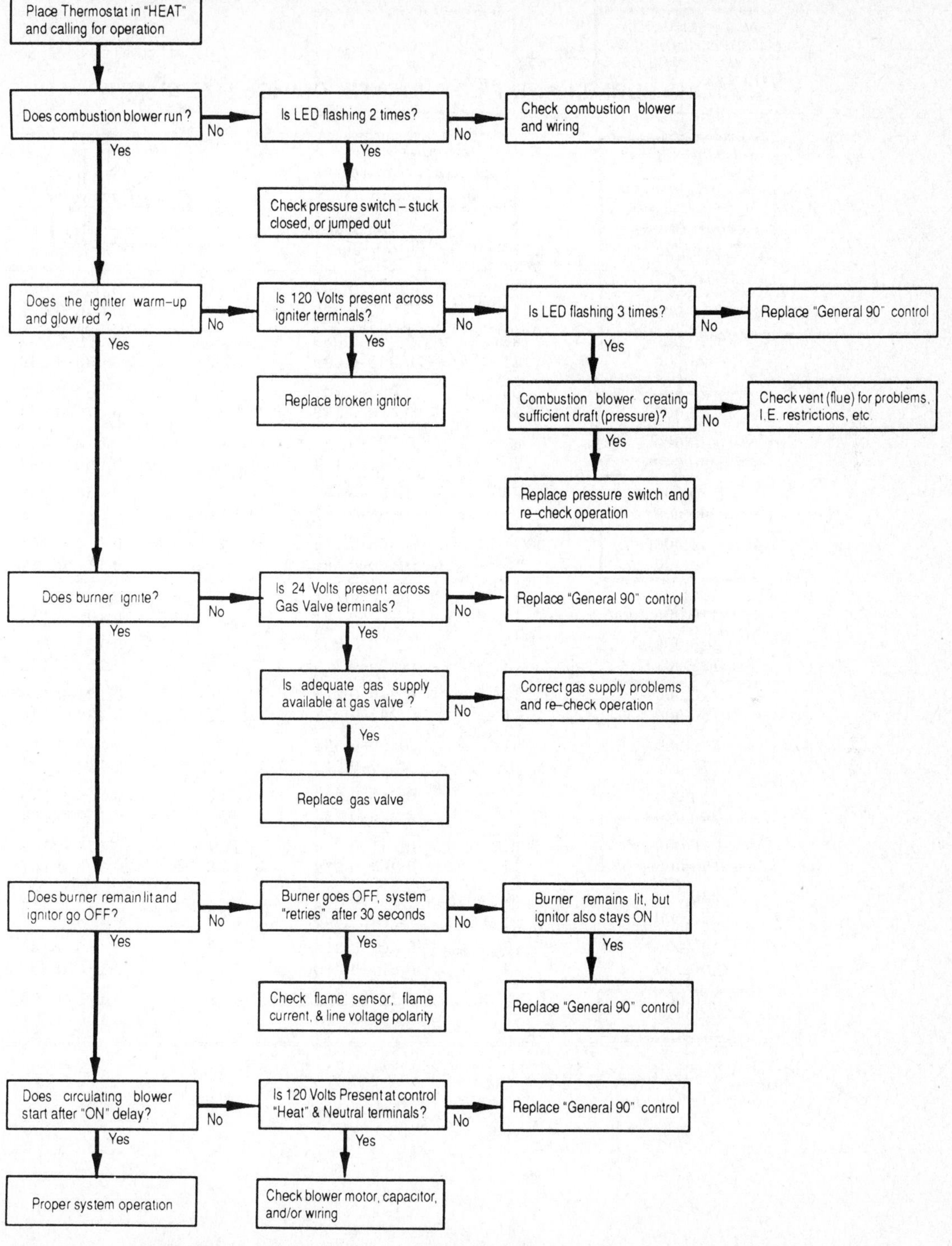

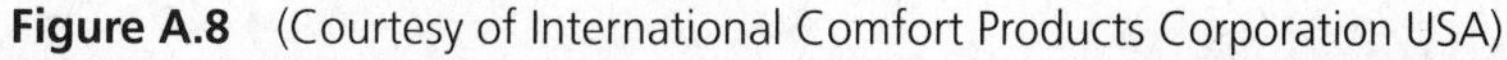
Figure A.8 (Courtesy of International Comfort Products Corporation USA)

MAIN BURNER AND PILOT BURNER, ORIFICE SIZE CHART

	Natural gas: 1020 Btu — 0.65 SG 3 ½ in. WC manifold		*Propane: 2500 Btu — 1.5 SG 11 in. WC manifold*	
Input (Btu/h) per spud	*Drill size*	*Decimal tolerance*	*Drill size*	*Decimal tolerance*
12,000	51	0.064-0.067	60	0.038-0.040
15,000	48	0.073-0.076	58	0.040-0.042
20,000	43	0.086-0.089	55	0.050-0.052
25,000	41	0.093-0.096	53	0.056-0.059
27,500	39	0.097-0.100	53	0.060-0.063
40,000	32	0.113-0.116	49	0.070-0.073
50,000	30	0.124-0.128	46	0.078-0.081
60,000	27	0.140-0.144	43	0.086-0.089
70,000	22	0.153-0.157	42	0.090-0.093
80,000	20	0.156-0.161	40	0.095-0.098
90,000	17	0.168-0.173	38	0.098-0.101
100,000	13	0.180-0.185	35	0.107-0.110
105,000	11	0.186-0.191	34	0.108-0.111
110,000	10	0.188-0.193	33	0.109-0.113
125,000	5	0.200-0.205	1/8	0.121-0.125
135,000	3	0.208-0.213	30	0.124-0.128
140,000	7/32	0.214-0.219	30	0.124-0.128
150,000	1	0.223-0.228	29	0.132-0.136
160,000	A	0.229-0.234	28	0.136-0.140
175,000	C	0.237-0.242	27	0.140-0.144
190,000	E	0.245-0.250	25	0.145-0.149
200,000	F	0.252-0.257	23	0.150-0.154
210,000	H	0.261-0.266	21	0.154-0.159
220,000	I	0.267-0.272	20	0.156-0.161
240,000	K	0.276-0.281	18	0.164-0.169
260,000	M	0.290-0.295	16	0.172-0.177
280,000	5/16	0.307-0.312	13	0.180-0.185
300,000	0	0.311-0.316	11	0.186-0.191
310,000	P	0.318-0.323	9	0.191-0.196
320,000	21/64	0.323-0.328	7	0.196-0.201

Figure A.9 (Courtesy of Luxaire, Inc.)

OIL FURNACE ROUTINE PERFORMANCE CHECKS AND TROUBLESHOOTING

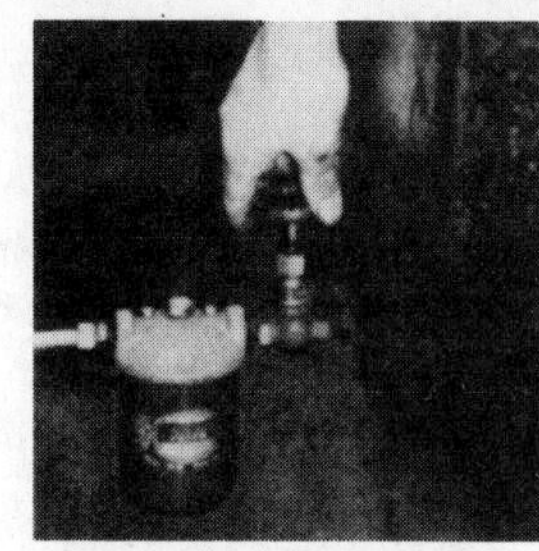

1. Check Shut-Off Valve and Line Filter. Replace or clean cartridge in line filter if dirty. Be sure to open shut-off valve.

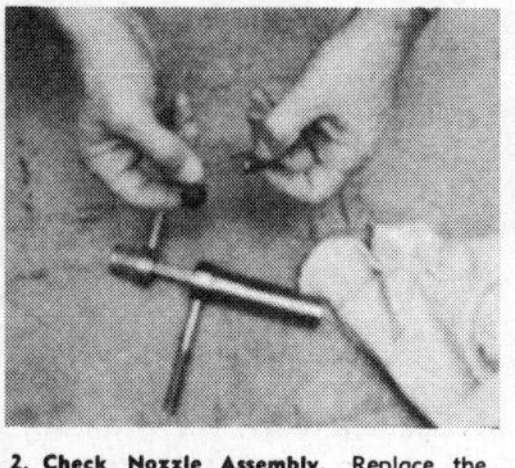

2. Check Nozzle Assembly. Replace the nozzle according to manufacturer's recommendations when needed.

Important:
Use proper designed tools for removal of nozzle from firing head.

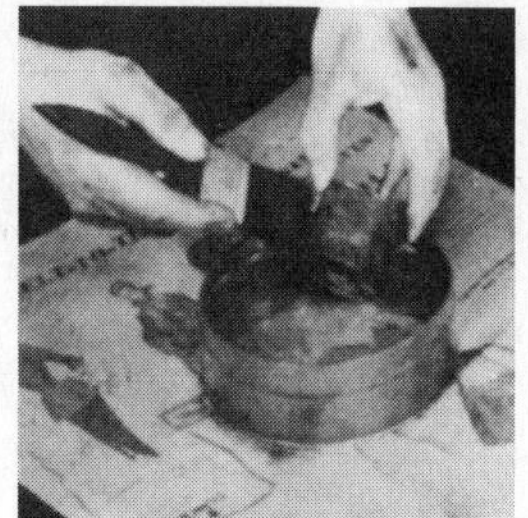

3. Check Strainer. Clean strainer using clean fuel oil or kerosene. Install new cover gasket. Replace strainer if necessary.

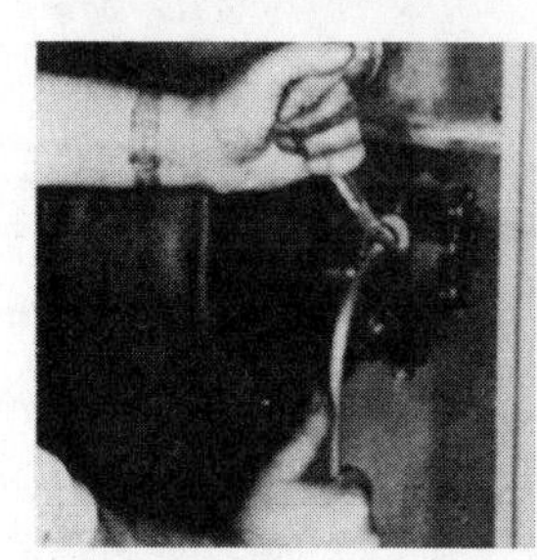

4. Check Connections. Tighten all connections and fittings in the intake line and unused intake port plugs.

5. Pressure Setting. Insert pressure gage in gage port. Normal pressure setting should be at 100 PSI. Check manufacturer's pressure setting recommendation on each installation being serviced.

6. Insert Vacuum gage in unused intake port. Check for abnormally high intake vacuum.

TROUBLESHOOTING

	cause	remedy
NO OIL FLOW AT NOZZLE	Oil level below intake line in supply tank	*Fill tank with oil.*
	Clogged strainer or filter	*Remove and clean strainer. Repack filter element.*
	Clogged nozzle	*Replace nozzle.*
	Air leak in intake line	*Tighten all fittings in intake line. Tighten unused intake port plug. Tighten in-line valve stem packing gland. Check filter cover and gasket.*
	Restricted intake line (High vacuum reading)	*Replace any kinked tubing and check any valves in intake line.*
	A two pipe system that becomes airbound	*Check and insert by-pass plug.*
	A single-pipe system that becomes airbound (Model J unit only)	*Loosen gage port plug or easy flow valve and drain oil until foam is gone in bleed hose.*
	Slipping or broken coupling	*Tighten or replace coupling.*
	Rotation of motor and fuel unit is not the same as indicated by arrow on pad at top of unit	*Install fuel unit with correct rotation.*
	Frozen pump shaft	*Return unit to approved service station or Sundstrand factory for repair. Check for water and dirt in tank.*

Figure A.10 (Courtesy of Sundstrand Hydraulics, Inc.)

OIL FURNACE ROUTINE PERFORMANCE CHECKS AND TROUBLESHOOTING continued

	cause	remedy
OIL LEAK	Loose plugs or fittings	*Dope with good quality thread sealer.*
	Leak at pressure adjusting end cap nut	*Fibre washer may have been left out after adjustment of valve spring. Replace the washer.*
	Blown seal (single pipe system)	*Check to see if by-pass plug has been left in unit. Replace fuel unit.*
	Blown seal (two pipe system)	*Check for kinked tubing or other obstructions in return line. Replace fuel unit.*
	Seal leaking	*Replace fuel unit.*
NOISY OPERATION	Bad coupling alignment	*Loosen fuel unit mounting screws slightly and shift fuel unit in different positions until noise is eliminated. Retighten mounting screws.*
	Air in inlet line	*Check all connections.*
	Tank hum on two-pipe system and inside tank	*Install return line hum eliminator.*
PULSATING PRESSURE	Partially clogged strainer or filter	*Remove and clean strainer. Replace filter element.*
	Air leak in intake line	*Tighten all fittings and valve packing in intake line.*
	Air leaking around cover	*Be sure strainer cover screws are tightened securely.*
LOW OIL PRESSURE	Defective gage	*Check gage against master gage, or other gage.*
	Nozzle capacity is greater than fuel unit capacity	*Replace fuel unit with unit of correct capacity.*

IMPROPER NOZZLE CUT-OFF

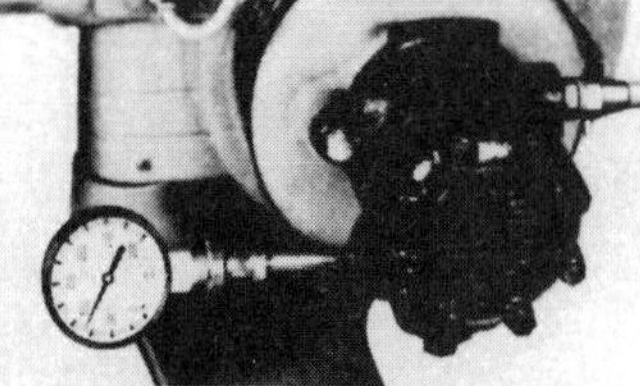

To determine the cause of improper cut-off, insert a pressure gage in the nozzle port of the fuel unit. After a minute of operation shut burner down. If the pressure drops and stabilizes above 0 P.S.I., the fuel unit is operating properly and air is the cause of improper cut-off. If, however, the pressure drops to 0 P.S.I., fuel unit should be replaced.

cause	remedy
Filter leaks	*Check face of cover and gasket for damage.*
Strainer cover loose	*Tighten 8 screws on cover.*
Air pocket between cut-off valve and nozzle	*Run burner, stopping and starting unit, until smoke and after-fire disappears.*
Air leak in intake line	*Tighten intake fittings and packing nut on shut-off valve. Tighten unused intake port plug.*
Partially clogged nozzle strainer	*Clean strainer or change nozzle.*

Figure A.11 (Courtesy of Sundstrand Hydraulics, Inc.)

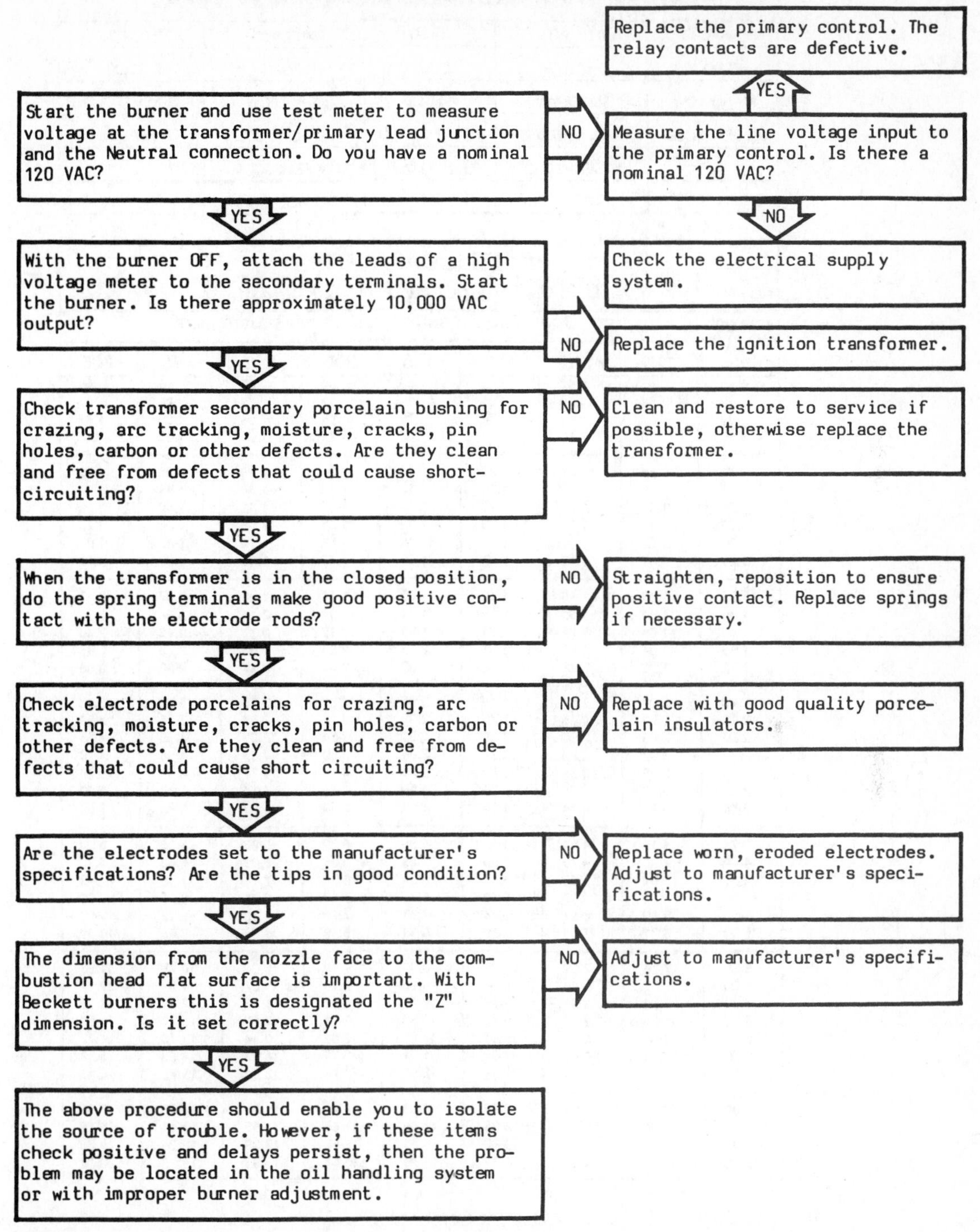

Figure A.12 (Courtesy of R.W. Beckett Corporation.)

NOZZLE MANUFACTURERS AND SPRAY PATTERNS

DANFOSS	DELAVAN	HAGO	MONARCH	STEINEN
AS-SOLID	A-HOLLOW	ES-SOLID	R-SOLID	S-SOLID
AH-HOLLOW	B-SOLID	P-SOLID	NS-HOLLOW	SS-SEMI-SOLID
AB-SEMI-SOLID	W-ALL PURPOSE	SS-SEMI-SOLID	AR-SPECIAL SOLID	H-HOLLOW
	SS-SEMI-SOLID	H-HOLLOW	PLP-SEMI-SOLID	
			PL-HOLLOW	

NOZZLE CAPACITIES

U.S. Gallons per Hour No. 2 Fuel Oil

rate gph @ 100 psi	Operating Pressure: pounds per square inch							
	125	140	150	175	200	250	275	300
.40	.45	.47	.49	.53	.56	.63	.66	.69
.50	.56	.59	.61	.66	.71	.79	.83	.87
.60	.67	.71	.74	.79	.85	.95	1.00	1.04
.65	.73	.77	.80	.86	.92	1.03	1.08	1.13
.75	.84	.89	.92	.99	1.06	1.19	1.24	1.30
.85	.95	1.01	1.04	1.13	1.20	1.34	1.41	1.47
.90	1.01	1.07	1.10	1.19	1.27	1.42	1.49	1.56
1.00	1.12	1.18	1.23	1.32	1.41	1.58	1.66	1.73
1.10	1.23	1.30	1.35	1.46	1.56	1.74	1.82	1.91
1.20	1.34	1.42	1.47	1.59	1.70	1.90	1.99	2.08
1.25	1.39	1.48	1.53	1.65	1.77	1.98	2.07	2.17
1.35	1.51	1.60	1.65	1.79	1.91	2.14	2.24	2.34
1.50	1.68	1.77	1.84	1.98	2.12	2.37	2.49	2.60
1.65	1.84	1.95	2.02	2.18	2.33	2.61	2.73	2.86
1.75	1.96	2.07	2.14	2.32	2.48	2.77	2.90	3.03
2.00	2.24	2.37	2.45	2.65	2.83	3.16	3.32	3.46
2.25	2.52	2.66	2.76	2.98	3.18	3.56	3.73	3.90
2.50	2.80	2.96	3.06	3.31	3.54	3.95	4.15	4.33
2.75	3.07	3.25	3.37	3.64	3.90	4.35	4.56	4.76
3.00	3.35	3.55	3.67	3.97	4.24	4.74	4.97	5.20
3.25	3.63	3.85	3.98	4.30	4.60	5.14	5.39	5.63
3.50	3.91	4.14	4.29	4.63	4.95	5.53	5.80	6.06
3.75	4.19	4.44	4.59	4.96	5.30	5.93	6.22	6.50
4.00	4.47	4.73	4.90	5.29	5.66	6.32	6.63	6.93
4.50	5.04	5.32	5.51	5.95	6.36	7.11	7.46	7.79
5.00	5.59	5.92	6.12	6.61	7.07	7.91	8.29	8.66
5.50	6.15	6.51	6.74	7.27	7.78	8.70	9.12	9.53
6.00	6.71	7.10	7.35	7.94	8.49	9.49	9.95	10.39
6.50	7.26	7.69	7.96	8.60	9.19	10.28	10.78	11.26
7.00	7.82	8.28	8.57	9.25	9.90	11.07	11.61	12.12
7.50	8.38	8.87	9.19	9.91	10.61	11.86	12.44	12.99
8.00	8.94	9.47	9.80	10.58	11.31	12.65	13.27	13.86
8.50	9.50	10.06	10.41	11.27	12.02	13.44	14.10	14.72
9.00	10.06	10.65	11.02	11.91	12.73	14.23	14.93	15.59
9.50	10.60	11.24	11.64	12.60	13.44	15.02	15.75	16.45
10.00	11.18	11.83	12.25	13.23	14.14	15.81	16.58	17.32
10.50	11.74	12.42	12.86	13.89	14.85	16.60	17.41	18.19
11.00	12.30	13.02	13.47	14.55	15.56	17.39	18.24	19.05
12.00	13.42	14.20	14.70	15.88	16.97	18.97	19.90	20.79

Figure A.13 (Courtesy of R.W. Beckett Corporation)

Helpful Conversion Factors for Heating Systems

Multiply...	by...	to obtain:
Atmosphere (atm)	14.70	lb. per sq. in. (psi)
Boiler Horsepower	34.50	pounds steam/hr @ 212° F.
Boiler Horsepower	33,475	BTU per hour
Cubic feet (cu. ft.)	7.481	gallons (gal.)
Degrees - Centigrade (°C)	(°Cx1.8) + 32	degrees - Fahrenheit (°F)
Degrees - Fahrenheit (°F)	(°F -32) x 5/9	degrees - Centigrade (°C)
Gallon, # 2 Oil	7.1	pounds (lbs.) approx.
Gallon, Water	8.337	pounds (lbs.)
Horsepower	.3	GPH
Horsepower - U.S. & British	33,000	foot pounds per minute
Horsepower - U.S. & British	42.42	BTU per minute
Inches Mercury (in. Hg)	13.60	inches water (in. w.c.)
Inches Mercury (in. Hg)	0.4912	pounds per square in. (psi)
Inches Water (in. w.c.)	0.0735	inches mercury (in. Hg)
Inches Water (in. w.c.)	0.0361	pounds per square in. (psi)
Kilocalorie (kcal)	.0284	GPH
Kilocalorie (kcal)	3.968	BTU
Kilogram/hr (kgh)	.3125	GPH
Kilowatt (kw)	.0244	GPH
Kilowatt hours (kwh)	3,413	BTU
Kilowatt (kw)	1.341	horsepower - U.S. & British
Liter	.2642	GPH
Pound - Steam @ 212°F	970.3	BTU (heat of evaporation)
Sq. Ft. Steam	240	BTU
Sq. Ft. Water	150	BTU
Watt Hour (whr)	3.413	BTU

Divide...	by...	to obtain:
BTU input	140,000	GPH #2 Oil
Gross Sq. Ft. Steam	360	GPH #2 Oil
Net BTUH (Output)	112,000	GPH #2 Oil
Net MBH (Output)	112	GPH #2 Oil
Net Sq. Ft. Steam	466	GPH #2 Oil
Sq. Ft. Heating Surface	27	GPH #2 Oil

Boiler Tube Sizing Pg. 7-205 Hoffman data book

$$\frac{\text{Sq.ft. heating surface} \times 1.15}{10} = \text{Hp} \times .3 = \text{Gph}$$

Water Capacity Per Foot of Pipe

Pipe Size	1/2"	3/4"	1"	1-1/4"	1-1/2"	2"	2-1/2"	3"	3-1/2"	4"	5"	6"
Gallons Per Foot	.016	.023	.040	.063	.102	.17	.275	.39	.53	.69	1.1	1.5

Most Common pipe sizes		Sq. Ft. of surface (ext) per 1 foot pipe
3" pipe	—	0.916
3-1/2" pipe	—	1.047
4" pipe	—	1.178

Oil Pressure vs. Nozzle Flow Rate Conversion

1. Divide new pressure by 100 psi (Standard nozzle pressure)
   EXAMPLE: 140 ÷ 100 = 1.4
2. Obtain square root. (Use calculator)
   EXAMPLE: 1.4 [√] = 1.1832159
3. Multiply result by nozzle size at 100 psi.
   EXAMPLE: 1.1832159 x 1.00 = 1.18gph
   This is your new gph flow rate @ 140 psi

How To Figure Domestic Water

1 gal. #2 oil per hr. = 140,000 BTUH
140,000 x 80% = 112,000 gross BTUH

1 gal. water = 8.3 lb.
To raise 1 lb Water 1°F = 1 BTU
To raise 8.3 lb Water 100°F in
1 minute = 8.3 x 100°F = 830 BTU/min.
To raise 8.3 lb Water 80°F in
1 minute = 8.3 x 80 = 664

$$\frac{112{,}000 \text{ BTU Gross hr.}}{60 \text{ min}} = 1866.66 \text{ BTU Gross per min}$$

$$\frac{1866.66}{830} = 2.2489 \text{ gal. per min } 100°\text{F rise.}$$

$$\frac{1866.66}{664} = 2.81 \text{ gal. per min. } 80°\text{F rise.}$$

40°F inlet water + 100°F rise = 140°F.
40°F inlet water + 80°F rise = 120°F.

Figure A.14

Pump Electrical Troubleshooting

WHEN WORKING WITH ELECTRICAL CIRCUITS, USE CAUTION TO AVOID ELECTRICAL SHOCK. It's recommended that care be taken to have metal terminal boxes and motors properly grounded.

WARNING: Failure to ground the pump may result in serious electrical shock.

Preliminary Tests

SUPPLY VOLTAGE

How to Measure

By means of a volt meter, which has been set to the proper scale, measure the voltage at the pump terminal box or starter.

On three-phase units, measure between the legs (phases).

What it Means

When the motor is under load, the voltage should be within ± 5% of the nameplate voltage. Larger voltage variation may cause winding damage.

Large variations in the voltage indicate a poor electrical supply and the pump should not be operated until these variations have been corrected.

If the voltage constantly remains high or low, the motor should be changed to the correct supply voltage.

CURRENT MEASUREMENT

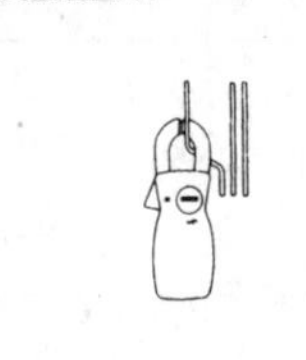

How to Measure

By use of an ammeter, set on the proper scale, measure the current on each power lead at the terminal box or starter. See the Electrical Data Table for motor amp draw information.

Current should be measured when the pump is operating at constant discharge pressure when the motor is fully loaded.

What it Means

If the amp draw exceeds the amps listed on the nameplate or if the current imbalance is greater than 5% between each leg on three-phase units, check the following:

1. Loose wires in terminal box or possible wire defect. Check winding and insulation resistances.
2. Too high or low supply voltage.
3. Motor windings are shorted.
4. Pump is damaged causing a motor overload.

INSULATION RESISTANCE

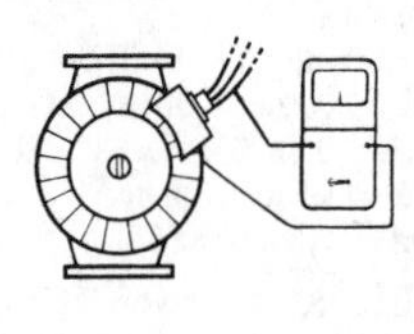

How to Measure

Turn off power and disconnect the supply power leads in the pump terminal box. Using an ohm or mega ohm meter, set the scale selector to Rx 100K and zero adjust the meter.

Measure the resistance between each of the terminals and ground.

What it Means

Motors of all HP, voltage, phase and frequencies have the same value of insulation resistance. Resistance value for a new motor must exceed 1,000,000 ohms. If it does not, the motor should be repaired or replaced.

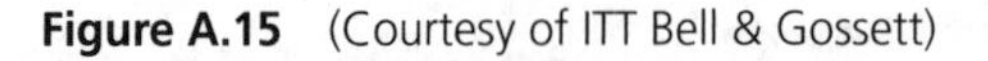

Figure A.15 (Courtesy of ITT Bell & Gossett)

PUMP TROUBLESHOOTING CHART

Fault	*Possible Causes*	*How to Check*	*How to Correct*
A. Pump Does Not Run	**1.** No power at pump panel.	Check for voltage at panel.	If no voltage at panel, check feeder panel for tripped circuits.
	2. Fuses are blown or circuit breakers are tripped.	Remove fuses and check for continuity with ohm meter.	Replace blown fuses or reset circuit breaker. If new fuses blow or circuit breaker trips, the electrical installation and motor must be checked.
	3. Defective controls.	Check all safety and pressure switches for operation. Inspect contact in control devices.	Replace worn or defective parts.
	4. Motor is defective.	Turn off power. Disconnect wiring. Measure the lead to lead resistances with the ohm meter (RX-1). Measure lead to ground values with ohm meter (RX-100K). Record measured values.	If the motor windings are open or grounded, replace motor.
	5. Defective capacitor (single phase pumps)	Turn off the power, then discharge capacitor. Disconnect leads and check with an ohm meter (RX-100K).	When the meter is connected, the needle should jump towards "0" ohms and slowly drift back to infinity. Replace capacitor if defective.
B. Pump Runs But at Reduced Capacity	**1.** Wrong rotation.	Check for proper electrical connection in terminal box.	Correct wiring and change leads as required.
	2. Discharge piping or valve leaking.	Examine system for leaks.	Repair leaks.
	3. Stainer is clogged.	Remove screen and inspect.	Clean, repair, rinse out screen and re-install.
	4. Pump worn.	Install pressure gauge, start pump, gradually close the discharge valve and read pressure at shut-off.	Refer to the specific pump curve for shut-off head for that pump model. If head is close to curve, pump is probably OK. If not, remove pump and inspect.
	5. Foreign material lodged in impeller.	Shut isolation valves. Drain pump. Remove stator housing allen bolts and remove stator/rotor assembly.	Inspect impeller for stuck foreign material. Remove and re-assemble pump. Check to insure o-ring between stator housing and volute is not damaged during re-assembly.
C. Fuses Blow or Circuit Breakers Trip	**1.** High or low voltage.	Check voltage at starter panel or terminal box.	If not within ±5%, check wire size and length of run to pump panel.
	2. Three phase current unbalance.	Check current draw on each lead.	Must be within ±5%. If not, contact power company.
	3. Terminal box wiring.	Check that actual wiring matches wiring diagram. Check for loose or broken wires or terminals.	Correct as required.
	4. Defective capacitor (single phase pumps).	Turn off the power, then discharge capacitor. Disconnect leads and check with an ohm meter (RX-100K).	When the meter is connected, the needle should jump towards "0" ohms and slowly drift back to infinity. Replace capacitor, if defective.

Figure A.16 (Courtesy of ITT Bell & Gossett)

ZONING MADE EASY Rules of Thumb

FLOW RATE

$$\frac{\text{Net Btuh Load}}{10{,}000} = \text{Flow Rate}$$

MAXIMUM FLOW RATE

Pipe Size (Copper)	Maximum Flow Rate
½″	1½ gpm
¾″	4 gpm
1″	8 gpm
1¼″	14 gpm

MAXIMUM FLOW RATE & HEAT CARRYING CAPACITY

Pipe Size (Copper)	Maximum Flow Rate	Heat Carrying Capacity
½″	1½ gpm	15,000 Btuh
¾″	4 gpm	40,000 Btuh
1″	8 gpm	80,000 Btuh
1¼″	14 gpm	140,000 Btuh

(Based on a 20-degree temperature drop across the system)

MAXIMUM LENGTH OF FIN-TUBE BASEBOARD LOOP

Baseboard Size Size (Copper)	Typical Btuh Per Linear Foot	Maximum Length of Baseboard Loop
½″	600	25 feet
¾″	600	67 feet
1″	770	104 feet
1¼″	790	177 feet

(Based on 180-degree average water temperature and a 20-degree temperature drop across the system)

TOTAL CONVECTORS A PIPE CAN SERVE

Pipe Size (Copper)	Maximum Btuh Capacity of Pipe	Total Convectors (6″ x 36″ x 24″ 5,100 Btuh each)
½″	15,000	3
¾″	40,000	8
1″	80,000	16
1¼″	140,000	27

(Based on 180-degree average water temperature and a 20-degree temperature drop across the system)

SHARED PIPING SIZE

Pipe Size	Maximum Flow Rate
½″ copper	1½ gpm
¾″ copper	4 gpm
1″ copper	8 gpm
1¼″ copper	14 gpm
1½″ copper	22 gpm
2″ copper	45 gpm
1¼″ iron pipe	17 gpm
1½″ iron pipe	25 gpm
2″ iron pipe	50 gpm

ZONE-CIRCULATOR SIZING FOR HEATING ZONES

Zone Supply Pipe Size (Copper)	Bell & Gossett Circulator to Use
½"	NRF-22, NRF-33 or Series 100
¾"	NRF-22, NRF-33 or Series 100
1"	NRF-22, NRF-33 or Series 100
1¼"	Series HV or NRF-33

"PUMP HEAD"

1. Measure the longest run in feet.
2. Add 50% to this.
3. Multiply that by .04, and
4. That's the pump head!

CIRCULATOR SIZING FOR SYSTEMS WITH ZONE VALVES

1. An NRF-22 or Series 100 can be used with:
   a. Up to three ¾" heating zones, or
   b. Two ¾" heating zones and one 1" zoned domestic water storage tank.
2. An NRF-33 or Series 100 can be used with:
   a. Up to five ¾" heating zones, or
   b. Three ¾" heating zones and one 1" zoned domestic water storage tank.

Figure A.17 (Courtesy of ITT Bell & Gossett)

TROUBLESHOOTING "DEAD SPOTS"

"Dead spot" is a common term for a certain orientation of the rotor at which the motor (PSC *or* split phase) will not start. Dead spots can be caused by two things. First, there could be a break in one of the aluminum bars inside the rotor, due to a fault in the molding process. If this occurs and the rotor happens to be in that particular spot when the motor is turned on, the motor may not have enough torque to start the burner. This fault in the rotor is quite rare, and cannot be repaired. Secondly, if the start switch of a split phase motor is unevenly worn, the contacts may become slightly separated if the rotor is in a particular location. No current will be able to flow through the start winding, inhibiting the motor from being able to start.

PSC MOTOR TROUBLESHOOTING		
CONDITION	**CAUSE**	**RECOMMENDED ACTION**
Motor does not start.	No power to motor.	Check wiring and power from primary control lead. If necessary, replace control, limit controller, or fuses (time-delay type).
	Insufficient voltage supply.	Check power from primary control.
	Thermal protector has tripped.	Determine and repair cause of thermal overload and reset (if manually resettable).
	Pump shaft will not turn.	Disconnect motor from pump. Turn coupling to ensure free rotation of pump shaft.
	Capacitor or windings have failed.	Check capacitor and windings (see page 3).
	Motor bearings have failed.	Turn the motor shaft, which should turn easily.
Motor starts but does not reach full speed.	Motor is overloaded.	Disconnect pump from motor. Turn pump shaft to ensure free rotation.
	Insufficient voltage supply.	Check power from primary control. Voltage should be 110 V – 120 V.
	Capacitor or windings have failed.	Check capacitor and windings (see page 3).
Motor vibrates or is noisy.	Bearings are worn, damaged, or fouled with dirt or rust.	Replace motor.
	Motor and pump are misaligned with each other or housing.	Check pump to motor, motor to housing, and pump to housing alignment.
	Blower wheel or wheel balancing weight (if applicable) is loose.	Check blower wheel and balancing weight (if applicable) for location and tightness.
Motor draws excessive current (>10% over rated current).	Motor and pump misaligned with each other or housing.	Check pump to motor, motor to housing, and pump to housing alignment.
	Motor is undersized for the application.	See Table 1, note 1. Increase motor size if necessary.
	Motor windings are damaged.	Check windings. If damaged, replace motor.

Figure A.18 (Courtesy of R.W. Beckett Corporation)

MEASUREMENT OF AIR VELOCITY WITH A MANOMETER

To measure air velocity, connect a Dwyer Durablock inclined manometer to a Pitot tube in the air stream as shown. This method requires only a static tap plus a simple tube in center of duct to pick up total pressure. The differential pressure reading on the manometer is velocity pressure, which may be converted to air velocity by calculation or reference to conversion charts.

Dwyer stainless steel Pitot tubes are made in numerous lengths and configurations to serve in the smallest to the largest duct sizes.

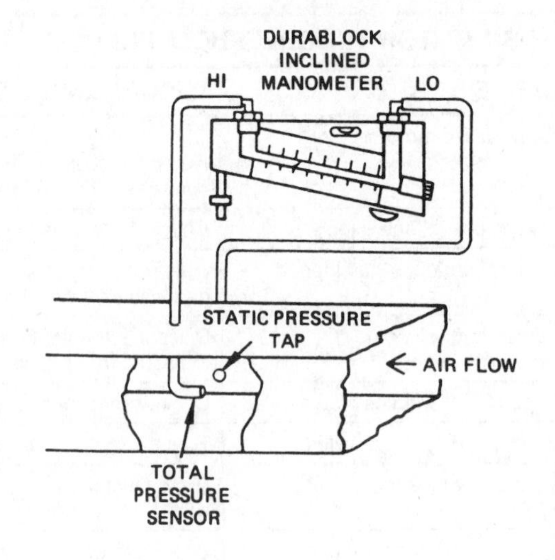

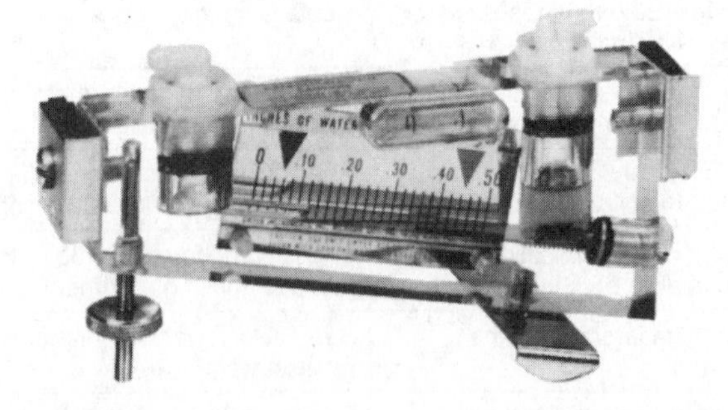

Figure A.19 (Courtesy of Dwyer Instruments, Inc.)

Duct and Round Pipe System Design

For Systems up to 60 Ft. from Unit to Register*
(Not Recommended Where any Line In System is Over 60 Foot. For Other Systems Use Duct Calculator)

C.F.M.	Heating BTU	Cooling BTU	Round Pipe	Square Duct	Floor Diffuser	Sidewall Diffuser⌟	Baseboard Diffuser	Return Grille•
32	2,400	970	4					
60	4,400	1,820	5			10x6	24"	
100	7,400	3,030	6	3 1/4x10	2 1/4x10		48"	10x6
120	8,900	3,640		3 1/4x12	2 1/4x12	12x6		12x6
145	10,700	4,390	7	3 1/4x14	4x14	14x6		14x6
180	13,300	5,450		6x8				
210	15,600	6,360	8					
270	20,000	8,200		8x8				24x6
290	21,500	8,800	9					
300	22,200	9,100						30x6
370	27,400	11,200		10x8				
390	28,900	11,800	10					
460	34,000	13,900		12x8				
560	41,500	17,000		14x8				
620	45,900	18,800	12					
660	48,900	20,000		16x8				
800	59,300	24,200		18x8				
900	66,700	27,300		20x8				
930	68,900	28,200	14					
1000	74,100	30,300		22x8				
1100	81,500	33,300		24x8				
1200	88,900	36,400		26x8				
1300	96,300	39,400	16	28x8				
1400	103,700	42,400		30x8				
1500	111,100	45,500	17	24x10				
1700	125,900	51,500		26x10				
1800	133,300	54,500	18					
1900	140,700	57,600		28x10				
2000	148,100	60,600		30x10				
2200	163,000	66,700		30x12				
2300	170,400	69,700		32x12				
2400	177,800	72,700	20					
2500	185,200	75,800		34x12				

* Based on .1 supply and .1 return and average number of fittings. For one or two long branch lines suggest one size larger branch pipe

▪ Use proper size Floor Diffuser based on 6 to 7 1/2 ft. throw based on actual CFM required (see register spec.)

⌟ Use proper size Wall Diffuser based on Vertical throw of 6 to 7 1/2 ft. above floor based on actual CFM required (see register spec.)

• One full stud space must be provided for 10, 12 and 14" grilles and two full stud spaces for 24 and 30" grilles.

System Design Instructions: With the room by room heat loss figured by the method in Manual J and the room by room heat gain figured proceed as follows. Where the system is to be used for heating only disregard the references to cooling.

1. Select equipment by comparing BTU capacity of units with requirements of building.
2. Adjust BTU for each room if building requirement varies more than 10% from equipment capacity.
3. Select register location, size and number (see chart) based on adjusted BTU for each room. Where system is to be used for both heating and cooling use the greater CFM equal to adjusted BTU for each room on heating and cooling.
4. Size branch pipe from chart based on step 3 CFM or BTU and locate trunk line.
5. If trunk line is less than 30 ft., suggest the use of extended plenum system.
   a. Add together CFM equal to BTU adjusted (heating or cooling whichever greater) for all branch pipe coming off of trunk line. Size trunk line from chart to carry total CFM.
6. For step down trunk, start at end of trunk furthest from unit and add together CFM of branch pipes until step up of size is desired. Select trunk size from chart.
7. For Return Air system repeat steps 3, 4, 5, and 6 above.

Figure A.20 (Courtesy of ACME Manufacturing Co.)

REFRIGERANT RECOVERY

GS1 Recovery Diagram

DIAGRAM 1:
Liquid
Recovery

NOTE: ***Do Not*** connect liquid line to recovery unit. Compressor could be damaged.

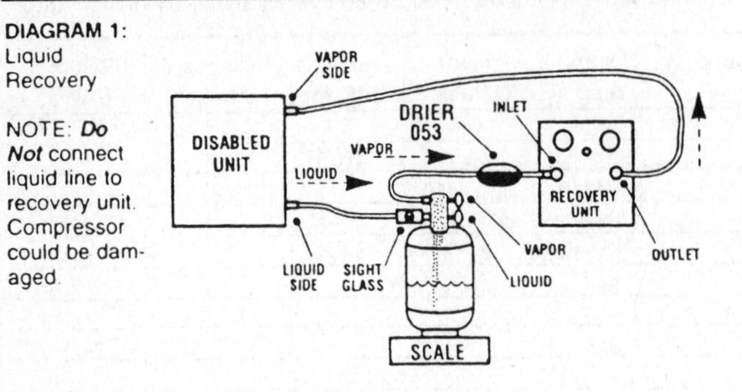

DIAGRAM 2:
Vapor Recovery

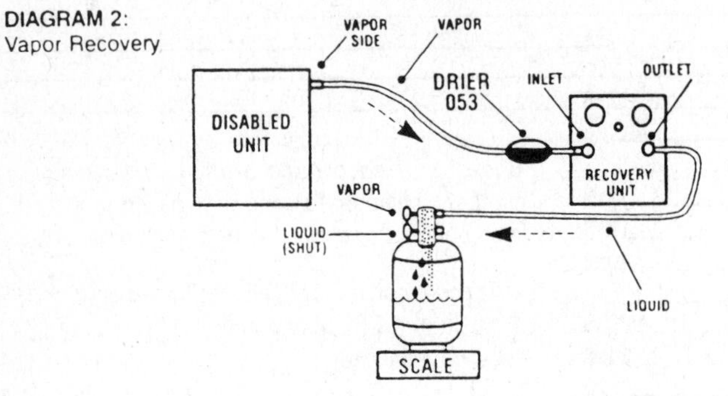

GS1 UL Recovery Diagram

DIAGRAM 1:
Liquid
Recovery

NOTE: ***Do Not*** connect liquid line to recovery unit. Compressor could be damaged.

DIAGRAM 2:
Vapor Recovery

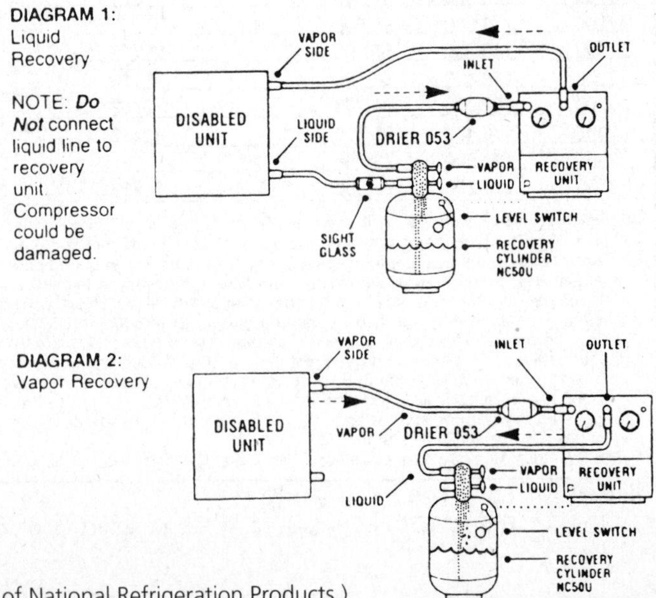

Figure A.21 (Courtesy of National Refrigeration Products.)

DuPont™ Suva® refrigerants

R-410A SPECIAL HANDLING TIPS:

Did You Know?

More and more residential and light commercial A/C equipment is being built with the refrigerant R-410A. There are several brands being marketed as R-410A, most common are Suva® 410A and Puron®. These and others have the same ASHRAE number R-410A. Many of the major U.S. a/c original equipment manufacturer's (OEM) have air conditioning units available with R-410A.

IMPORTANT TO NOTE: R-410A can only be used in equipment designed for R-410A. R-410A CANNOT be used to retrofit existing R-22 A/C equipment due to significantly higher operating pressures as outlined below.

Saturation Pressure (psig): R-22 vs. R-410A

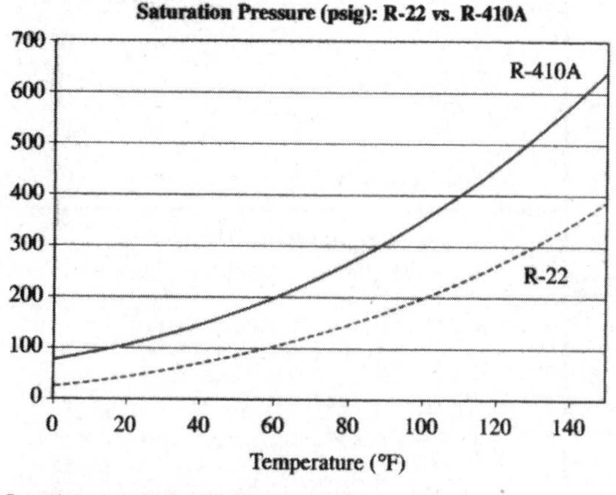

Puron® is a registered trademark of the Carrier Corp.
Suva® is a registered trademark of the DuPont Comapny.

For more information contact DuPont at 800-235-SUVA or www.suva.dupont.com

Safe handling of R-410A also known as Suva® 410A and Puron®:

- head pressure is significantly higher than R-22
- special equipment is required as follows:

Service Equipment
High pressure manifold gauge set
High pressure recovery machine
High pressure recovery tank (DOT 4BA400 or 4BW 400)

A/C System Considerations:
Because of its enhanced properties, R-410A systems will operate differently than R-22 systems:

Properties	R-410A vs. R-22
Discharge pressure:	+ 50 to 70%
Cooling capacity:	+40%
Discharge temperature:	-10°F
Energy efficiency:	Units can be designed to meet the proposed DOE guidelines of 12 to 14 SEER
Lubricant:	POE

Because of higher pressure, most system components have been redesigned with increased wall thickness. When servicing this equipment, make sure you use reversing valves, expansion valves, filter-driers, and other components specifically designed for R-410A.

In addition to increased wall thickness, the expansion valve flow area required to provide the same tonnage as R-22 will be about 15% smaller.

R-410A is a blend of two refrigerants. To achieve optimum performance, remove as a liquid from the cylinder.

Consult your DuPont authorized distributor or the OEM for special equipment recommendations.

Figure A.22 (Courtesy of Dupont Company)

ELECTRICAL FORMULAS AND WIRE RATINGS

Formulas - Electrical

VOLTS =

$$\text{Amps x Ohms} \qquad \frac{\text{Watts}}{\text{Amps}} \qquad \sqrt{\text{Watts x Ohms}}$$

AMPS =

$$\frac{\text{Volts}}{\text{Ohms}} \qquad \frac{\text{Watts}}{\text{Volts}} \qquad \sqrt{\frac{\text{Watts}}{\text{Ohms}}}$$

WATTS =

$$\text{Volts x Amps} \qquad \text{Amps}^2 \text{ x Ohms} \qquad \frac{\text{Volts}^2}{\text{Ohms}}$$

OHMS =

$$\frac{\text{Volts}}{\text{Amps}} \qquad \frac{\text{Volts}^2}{\text{Watts}} \qquad \frac{\text{Watts}}{\text{Amps}^2}$$

$$\text{Power Factor} = \frac{KW}{KVA} = \cos \theta$$

	Single Phase	Three Phase
KW =	$\frac{\sqrt{\ x\ A\ x\ PF}}{1000}$	$\frac{\sqrt{3}\ x\ V\ x\ A\ x\ PF}{1000}$
KVA =	$\frac{V\ x\ A}{1000}$	$\frac{\sqrt{3}\ x\ V\ x\ A}{1000}$
AMPS =	$\frac{KVA\ x\ 1000}{V}$	$\frac{KVA\ x\ 1000}{\sqrt{3}\ x\ V}$

$$\sqrt{3} = 1.73$$

Approx. Motor KVA = Motor Horsepower (At Full Load)

Capacitors Connected In Parallel $C_1 + C_2 + C_3 = C \text{ Total}$

Capacitors Connected In Series

For Two

$$\frac{C_1 \times C_2}{C_1 + C_2} = C \text{ Total}$$

More Than Two

$$\frac{1}{\frac{1}{C_1} + \frac{1}{C_2} + \frac{1}{C_3}} = C \text{ Total}$$

VOLTAGE UNBALANCE

% Voltage Unbalance =

$$\frac{100 \text{ x Max. Voltage Deviation From Average Voltage}}{\text{Average Voltage}}$$

BOOST TRANS.:

Rating Plate F.L.A. x Rating Plate VOLTS = KVA

$$\frac{\text{Rating Plate VOLTS}}{\text{Rating Plate VOLTS — Norm. Line VOLTS}} = \text{FACTOR}$$

$$\frac{KVA}{FACTOR} = \text{Trans. KVA Rating}$$

$$\left(\frac{V_2}{V_1}\right)^2 \text{ x Heater Rating} = \text{Rating @ New Voltage}$$

V_1 Rated Volts V_2 = Measured Volts

Typical Ampere Wire Ratings*

COPPER CONDUCTORS, IN CONDUIT*			
AWG MCM	TEMP. RATING OF CONDUCTOR*		
	60°C*	75°C*	90°C*
14	15	15	25*
12	20	20	30°
10	30	30	40°
8	40	45	50
6	55	65	70
4	70	85	90
3	80	100	105
2	95	115	120
1	110	130	140
1/0	125	150	155
2/0	145	175	185
3/0	165	200	210
4/0	195	230	235
250	215	255	270
300	240	285	300
350	260	310	325
400	280	335	360
500	320	380	405
600	355	420	455
700	385	460	490
750	400	475	500
800	410	490	515
900	435	520	555
1000	455	545	585
1250	495	590	645
1500	520	625	700
1750	545	650	735
2000	560	665	775

Summary only, refer to NEC 310-16, -17, -18, -19 (and others) fo limitations.

Typical Electric Wire Size

MOTOR HP	SINGLE PH.		THREE PH.	
	115 VOLT	230 VOLT	230 VOLT	460 VOLT
1-1/3	14	14		
1/2	14	14	14	14
3/4	12	14	14	14
1	12	14	14	14
1-1/2	10	14	14	14
2		12	14	14
3		10	14	14
5			12	14
7-1/2			10	14
10			8	12

From Standards of the National Board of Fire Underwriters.

Correction Table For Watts - Amperes - Volts

WATTS	VOLTAGE (C - Single Phase)			
	120	208	240	277
	AMPERES			
500	4.2	2.4	2.1	1.8
1000	8.3	4.8	4.2	3.6
1500	12.5	7.2	6.3	5.4
2000	16.7	9.6	8.3	7.2
2500	20.9	12.0	10.4	9.0
3000	25.0	14.4	12.5	10.8
3500	29.2	16.8	14.6	12.6

Figure A.23 (Courtesy of Military Products, Group 2000)

COMPRESSOR ELECTRICAL TROUBLESHOOTING

Figure 37 Checking Locked Rotor Voltage

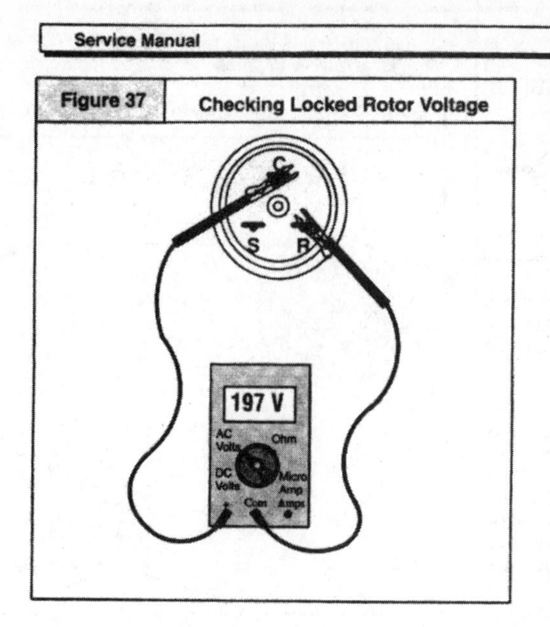

Locked Rotor Voltage readings of less than 197 volts will not allow the compressor to start. If your reading is less than 197 volts, problems may be indicated in the unit's electrical supply circuit. Some examples of these problems are undersized supply wiring, excessive length of supply run, and loose and/or dirty (high resistance) connections in the supply circuit. These conditions MUST be corrected before the compressor can be expected to start reliably.

The compressor contactor itself can also be a cause of low locked rotor voltage readings. To check and rule out this possibility, check the locked rotor voltage across the "L1" & "L2" terminals of the compressor contactor. If the voltage reading is the same as that obtained across the "T1 & "T2" terminals, then supply circuit problems are indicated. If, however, the locked rotor voltage across "L1" & "L2" is Higher than it is across "T1" & "T2", there is high resistance through the points of the compressor contactor causing the voltage drop, and the contactor should be replaced.

COMPRESSOR WINDING CHECKS

If the compressor fails to start, the compressor windings should be checked for open circuits and/or short circuits in order to determine their condition. Winding checks are made using a standard Ohmmeter (See Figure 38), with the power to the unit OFF.

WARNING

Electrical shock hazard.

Disconnect power at fuse box or service panel before checking compressor windings.

Failure to follow this warning can result in property damage, personal injury, and/or death.

Checking for open windings

With power to the unit OFF, disconnect wiring to the compressor. Resistance should be checked between terminals C & R, C & S, and S & R. The reading between C & R should indicate the LEAST resistance. The reading between C & S should indicate a HIGHER resistance (than between C & R). The Reading between S & R should indicate the TOTAL of the readings obtained between C & R and C & S. This check will indicate if any of the windings are open. A reading of infinity (∞) between any two terminals MAY indicate an open winding. If, however, a reading of infinity (∞) is obtained between C & R and C & S, accompanied by a resistance reading between S & R, an open internal overload is indicated. Should obtain this indication, allow the compressor to cool (may take up to 24 hours) then re–check before condemning the compressor.

Figure 38 Checking Compressor Windings

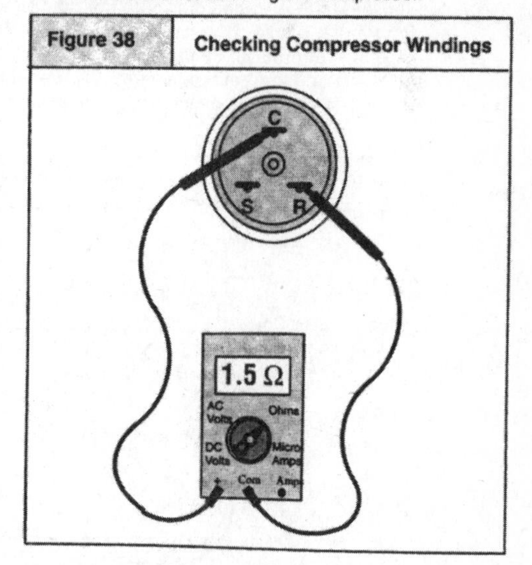

If an open internal overload is indicated, the source of its opening must be determined and corrected. Failure to do so will cause repeat problems with an open overload

Figure A.24 (Courtesy of AIRCOAIRE Air Conditioning & Heating)

and/or premature compressor failure. Some possible causes of an open internal overload include insufficient refrigerant charge, restriction in the refrigerant circuit, and excessive current draw.

Checking for shorted (grounded) windings

The Compressor should also be checked for shorted (grounded to case) windings anytime the fuse or circuit breaker to the unit is tripping. You should also check the compressor for shorted windings whenever there is a starting problem, since their may be enough resistance in the shorted winding to prevent the fuse or circuit breaker from tripping.

With power to the unit OFF, disconnect wiring to the compressor. Resistance should be checked (one terminal at a time) between terminals C, S, R, and the compressor case (the suction line may be used for this purpose). Be certain to insure that (when using the compressor case) the point of contact of the Ohmmeter probe is clean and free from paint. The reading between each terminal (C, S, & R) and the compressor case should indicate infinity (∞). Any reading obtained less than infinity (∞) is indicative of a shorted (grounded) winding, and the compressor should be replaced.

Figure 39	Checking For Shorted Windings

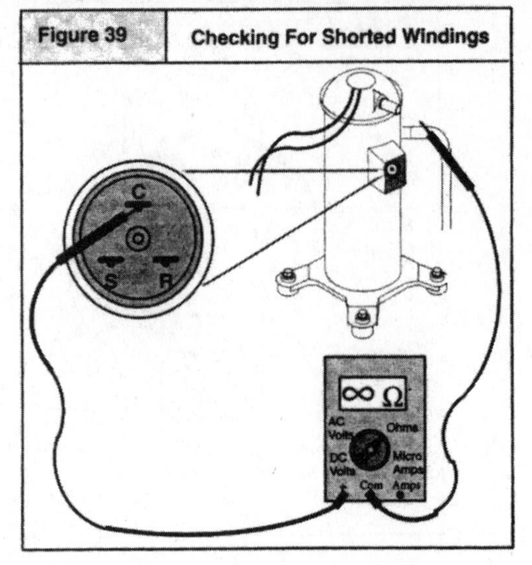

CONDENSER FAN CONTROL CIRCUIT

The condenser fan motor is controlled by the compressor contactor. Anytime the compressor is operating, the condenser fan motor should also be operating. When servicing a unit whose condenser fan motor will not run, both its capacitor (may be a dual capacitor shared with compressor) and low ambient control (if so equipped) should be suspect since they are part of the circuit.

Figure 40	Low Ambient Control

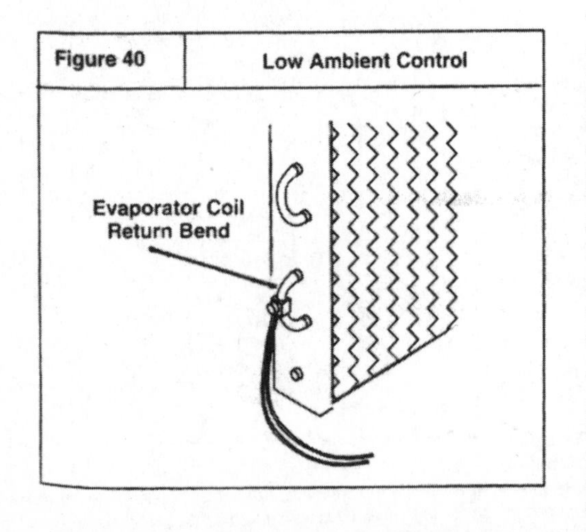

LOW AMBIENT CONTROL

Some units may be equipped with an optional low ambient control. The purpose of this control is to cycle the condenser fan motor "OFF" and "ON" to maintain head (discharge) pressure in the refrigeration system during low ambient (Outdoor Temperature) operation.

The control is a "Normally Closed" temperature operated switch (attached to one of the return bends of the evaporator coil) wired in series with the condenser fan motor. When the evaporator temperature drops below 30°F ±5°F the switch will open, breaking the circuit to the condenser fan motor. This will cause the discharge pressure to rise (due to lack of airflow across the condenser coil), which in turn will increase the suction pressure.

The increased suction pressure causes the evaporator temperature to rise. Then, when the evaporator temperature rises above 50°F ±6°F the switch will close, completing the circuit to the condenser fan motor.

The component part used for the low ambient control may be identical to the part used for the freeze thermostat described on page 31. Or, it may have different wire colors and/or lengths. It can be used interchangeably, however, as long as it is properly attached, and wired into the circuit.

Figure A.25 (Courtesy of AIRCOAIRE Air Conditioning & Heating)

CONVERSION FACTORS

MULTIPLY	BY	TO OBTAIN
Atmospheres (Std.) 760 MM of Mercury at 32°F.	14.696	Lbs./sq. inch
Atmospheres	76.0	Cms. of mercury
Atmospheres	29.92	In. of mercury
Atmospheres	33.90	Feet of water
Atmospheres	1.0333	Kgs./sq.cm.
Atmospheres	14.70	Lbs./sq. inch
Atmospheres	1.058	Tons/sq. ft.
Brit. Therm. Units	0.2520	Kilogram-calories
Brit. Therm. Units	777.5	Foot-lbs.
Brit. Therm. Units	0.000393	Horse-power-hrs.
Brit. Therm. Units	0.293	Watt-hrs.
BTU/min.	12.96	Foot-lbs./sec.
BTU/min.	0.02356	Horse-power
BTU/min.	0.01757	Kilowatts
BTU/min.	17.57	Watts
Calorie	0.003968	BTU
Centimeters	0.3937	Inches
Centimeters	0.03280	Feet
Centimeters	0.01	Meters
Centimeters	10	Millimeters
Centmtrs. of Merc.	0.01316	Atmospheres
Centimtrs. of merc.	0.4461	Feet of water
Centimtrs. of merc.	136.0	Kgs./sq. meter
Centimtrs. of merc.	27.85	Lbs./sq. ft.
Centimtrs. of merc.	0.1934	Lbs./sq. inch
Cubic feet	2.832×10^4	Cubic cms.
Cubic feet	1728	Cubic inches
Cubic feet	0.02832	Cubic meters
Cubic feet	0.03704	Cubic yards
Cubic feet	7.48052	Gallons U.S.
Cubic feet/minute	472.0	Cubic cms./sec.
Cubic feet/minute	0.1247	Gallons/sec.
Cubic foot water	62.4	Pounds @ 60°F.
Feet	30.48	Centimeters
Feet	12	Inches
Feet	0.3048	Meters
Feet	1/3	Yards
Feet of water	0.02950	Atmospheres
Feet of water	0.8826	Inches of mercury
Feet of water	0.03048	Kgs./sq. cm.
Feet of water	62.43	Lbs./sq. ft.
Feet of water	0.4335	Lbs./sq. inch
Feet/min.	0.5080	Centimeters/sec.
Feet/min.	0.01667	Feet/sec.
Feet/min.	0.01829	Kilometers/hr.
Feet/min.	0.3048	Meters/min.
Feet/min.	0.01136	Miles/hr.
Foot-pounds	0.001286	BTU
Gallons	3785	Cu. centimeters
Gallons	0.1337	Cubic feet
Gallons	231	Cubic inches
Gallons	128	Fluid ounces
Gallons	3.785	Liters
Gallons water	8.35	Lbs. water @ 60°F.
Horse-power	42.44	BTU/min.
Horse-power	33,000	Foot-lbs./min.
Horse-power	550	Foot-lbs./sec.
Horse-power	0.7457	Kilowatts
Horse-power	745.7	Watts
Horse-power (boiler)	33,479	BTU/hr.
Horse-power (boiler)	9.803	Kilowatts
Horse-power-hours	2547	BTU
Horse-power-hours	0.7457	Kilowatt-hours
Inches	2,540	Centimeters
Inches	25.4	Millimeters
Inches	0.0254	Meters
Inches	0.0833	Foot
Inches of mercury	0.03342	Atmospheres
Inches of mercury	1.133	Feet of water
Inches of mercury	13.57	Inches of water
Inches of mercury	70.73	Lbs./sq. ft.
Inches of mercury	0.4912	Lbs./sq. inch
Inches of water	0.002458	Atmospheres
Inches of water	0.07355	In. of mercury
Inches of water	0.5781	Ounces/sq. inch
Inches of water	5.202	Lbs./sq. foot
Inches of water	0.03613	Lbs./sq. inch
Kilowatts	56.92	BTU/min.
Kilowatts	1.341	Horse-power
Kilowatts	1000	Watts
Kilowatt-hours	3415	BTU
Liters	0.2642	Gallons
Liters	2.113	Pints (liq.)
Liters	1.057	Quarts (liq.)
Meters	100	Centimeters
Meters	3.281	Feet
Meters	39.37	Inches
Meters	1000	Millimeters
Meters	1.094	Yards
Ounces (fluid)	1.805	Cubic inches
Ounces (fluid)	0.02957	Liters
Ounces/sq. inch	0.0625	Lbs./sq. inch
Ounces/sq. inch	1.73	Inches of water
Pints	0.4732	Liter
Pounds (avoir.)	16	Ounces
Pounds of water	0.01602	Cubic feet
Pounds of water	27.68	Cubic inches
Pounds of water	0.1198	Gallons
Pounds/sq. foot	0.01602	Feet of water
Pounds/sq. foot	0.006945	Pounds/sq. inch
Pounds/sq. inch	0.06804	Atmospheres
Pounds/sq. inch	2.307	Feet of water
Pounds/sq. inch	2.036	In. of mercury
Pounds/sq. inch	27.68	Inches of water
Temp. (°C.) + 273	1	Abs. temp. (°C.)
Temp. (°C.) + 17.78	1.8	Temp. (°F.)
Temp. (°F.) + 460	1	Abs. temp. (°F.)
Temp. (°F.) − 32	5/9	Temp. (°C.)
Therm	100,000	BTU
Tons (long)	2240	Pounds
Ton, Refrigeration	12,000	BTU/hr.
Tons (short)	2000	Pounds
Watts	3.415	BTU
Watts	0.05692	BTU/min.
Watts	44.26	Foot-pounds/min.
Watts	0.7376	Foot-pounds/sec.
Watts	0.001341	Horse-power
Watts	0.001	Kilowatts
Watt-hours	3.415	BTU/hr.
Watt-hours	2655	Foot-pounds
Watt-hours	0.001341	Horse-power hrs.
Watt-hours	0.001	Kilowatt-hours

Figure A.26 (Courtesy of Robertshaw Controls Company)

Abbreviations for Text and Drawings

A	Amperes
AC	Auxiliary contacts
ACCA	Air Conditioning Contractors of America
ACS	Air Conditioning Service, National Response Center
AFUE	Annual fuel utilization efficiency
AGA	American Gas Association
ALS	Auxiliary limit switch
AMP	Amperes
API	American Petroleum Institute
ARI	Air Conditioning and Refrigeration Institute
ASHRAE	American Society of Heating, Refrigerating, and Air Conditioning Engineers, Inc.
AWG	American wire gauge
BK	Black-wire color
BL	Blue-wire color
BLWM	Blower motor
BOCA	Building Officials & Code Administrators International, Inc.
Btu	British thermal unit
Btuh	British thermal units per hour
C	Degrees Celsius
C	24V Common connection
C.A.	California seasonal efficiency percent
CAD	Cad cell
CAP	Capacitor
CB	Circuit breaker
CBR	Cooling blower relay
CC	Compressor contactor

CCH	Compressor crankcase heater
CFC	Chlorofluorocarbon
cfh	Cubic feet per hour
cfm	Cubic feet per minute
CGV	Combination gas valve
CH	Crankcase heater
CO	Carbon monoxide
CO_2	Carbon dioxide
COMP	Compressor
COP	Coefficient of Performance
CPVC	Chlorinated polyvinyl chloride
CR	Control relay
cu	Cubic
dB	Decibel
db	Dry bulb
DEG	Degree
DOE	Department of Energy
DOT	Department of Transportation
DPDT	Double-pole double-throw switch
DPST	Double-pole single-throw switch
DSI	Direct spark ignition
DSS	Draft safeguard switch
E	Electromotive force (volts)
EAC	Electrostatic air cleaner
EMF	Electromotive force
EP	Electric pilot
EPA	Environmental Protection Agency
ESP	External static pressure
ERV	Energy recovery ventilator
EWT	Entering water temperature
F	Degrees Fahrenheit
F	Farad
F	Fuse
FC	Fan control
FD	Fused disconnect
FL	Fuse link
FLA	Full load amperes

FM	Fan motor
FR	Fan relay
ft.	Feet
$ft.^3$	Cubic feet
$ft.^3/h$	Cubic feet per hour
$ft.^3/min.$	Cubic feet per min.
FU	Fuse
g	Grams
G	Green-wire color
G	Neutral-ground
GFCI	Ground fault circuit interrupter
gph	Gallons per hour
gpm	Gallons per minute
GV	Gas valve
H	Humidistat
H_2O	Water
HAM	Home access module
HCFC	Hydrochlorofluorocarbon
HCHO	Formaldehyde
HCT-2S	Two-stage heating-cooling thermostat
HE	High efficiency
HEF	High efficiency furnace
HEPA	High efficiency particulate arresting (filter)
HFR	Heating blower relay
HI	High speed
HP	High pressure (control)
HP	Horsepower
HR	Heat relay
HS	Humidification system
HSPF	Heating season performance factor
HTM	Heat transfer multiplier
HTM™	Heat Transfer Module™
HU	Humidifier
HVAC	Heating, ventilating, and air-conditioning
Hz	Cycles/second
I	Amperes
IAQ	Indoor air quality

I=B=R	Institute of Boiler and Radiator Manufacturers
ID	Inside diameter
IDM	Induced draft motor
IDR	Induced draft relay
IFR	Inside fan relay
in.Hg	Inches of mercury
IPS	Iron pipe size
J	Joule
kcal	Kilocalorie
kg	Kilogram
kW	Kilowatt
kWh	Kilowatt-hours
L	Change in latent heat per pound, Btu
L	Limit
L1 L2	Power supply
LA	Limit auxiliary
lb.	Pound
LED	Indicator lights
LO	Low speed
LP	Liquid petroleum
LR	Lockout relay
LRA	Locked rotor amperes
LS	Limit switch
m	Meter
M	One thousand
m/s	Meters per second
mA	Milliampere
MABS	Mastertrol automatic balancing system
MAZ	Mastertrol add-a-zone control panels
MBh	Thousand of BTU/hour
MCHP	Mastertrol changeover heat pump thermostat
MCTS	Mastertrol changeover two-stage thermostat
MFD	Microfarads
mm	Millimeter
mu m	Micrometer or micron
MuF	Microfarads
mV	Millivolt

N	Neutral
N	Nitrogen
NC	Normally closed
NFPA	National Fire Protection Association
NO	Normally open
O	Oxygen
OBM	Oil burner motor
OD	Outside diameter
OL	Overload
OSHA	Occupational Safety and Health Administration
OT	Outdoor thermostat
PC	Printed circuit
PCB1	Inducer motor controller
PCB2	Blower motor controller
PCB3	Microprocessor board
PF	Power factor
Ph	Phase
PPC	Pilot power control
ppm	Parts per million
PRS	Pressure switch
PSC	Permanent split phase capacitor
psi	Pressure per square inch
psia	Pressure per square inch, absolute
psig	Pounds per square inch gauge
PVC	Polyvinyl chloride
Q	Quantity of heat, Btu
R	Red-wire color (24-V power terminal)
R	Refrigerant
R	Resistance, Ohms
rev/min	Revolutions per minute
rh	Relative humidity
RH	Resistance heater
RV	Reversing valve
S	Cad cell relay terminal
s	Seconds
SAE	Society of Automotive Engineers
SEER	Seasonal energy efficient ratio

SH	Specific heat
SI	International System Units
Sol	Solenoid valve
SP	Static pressure
SPDT	Single-pole double-throw switch
SPST	Single-pole single-throw switch
SQ	Sequencer
SRN	Sound rating number
T	Thermostat terminal
T.C.	Total capacity
TD	Temperature difference, degrees F
TEL	Total equivalent length
TEV	Thermal expansion valve
TFS	Time fan start
THR	Thermocouple
TR	Temperature rise
TR	Transformer
TRAN	24-volt transformer
TRT	Temperature-reversing thermostat
U	Overall heat transfer coefficient
UBC	Uniform Building Code
UL	Underwriters' Laboratories
V	Volt (electromotive force)
VA	Volt-amp
VAC	Volts alternating current
VOCs	Volatile compounds
W	Watt
w	Weight, lb
W	White-wire color
W.C.	Water column
W/lls	(Units rated) with (accessory) liquid line solenoid
W1	First stage heat-thermostat
W2	Second state heat-thermostat
wb	Wet bulb
wc	Water column in inches
Y	Yellow-wire color
Y, R, G, W	Thermostat terminals

INDEX